# Progress in Biological Control

Volume 23

**Series Editors**

Heikki M. T. Hokkanen, Department of Environmental and Biological Sciences, University of Eastern Finland, Kuopio, Finland

Yulin Gao, Institute of Plant Protection, Chinese Academy of Agricultural Sciences, Beijing, China

Biological control of pests, weeds, and plant and animal diseases utilising their natural antagonists is a well-established but rapidly evolving field of applied ecology. Despite its documented applications and systematic development efforts for longer than a century, biological control still remains a grossly underexploited method of pest management. Its untapped potential represents the best hope to providing lasting, environmentally sound, and socially acceptable control of most problem pests in agriculture, and of invasive alien organisms threatening global biodiversity. Based on the overwhelmingly positive features of biological control, it is the prime candidate in the search for reducing dependency on chemical pesticides. Public demand for finding solutions based on biological control is the main driving force in the rapid developments in the various strategies of utilising natural enemies for controlling noxious organisms. This book series is intended to accelerate these developments through exploring the progress made within the various aspects of biological control, and via documenting these advances to the benefit of fellow scientists, students, public officials, and the public at large. Each of the books in this series is expected to provide a comprehensive, authoritative synthesis of the topic, likely to stand the test of time.

Nabil El-Wakeil • Ahmed Abdelkhalek
Editors

# Microbiomes and Plant Secondary Metabolites for Insect Pest Management

Springer

*Editors*
Nabil El-Wakeil
Plant Protection Science Program
Department of Arid Land Agriculture
College of Agricultural and Food Sciences
King Faisal University
Al-Ahsa, Saudi Arabia

Ahmed Abdelkhalek
Plant Protection and Biomolecular Diagnosis Department, Arid Lands Cultivation Research Institute
City of Scientific Research & Technological Applications
Alexandria, Egypt

ISSN 1573-5915 ISSN 2543-0076 (electronic)
Progress in Biological Control
ISBN 978-3-032-24329-4 ISBN 978-3-032-24330-0 (eBook)
https://doi.org/10.1007/978-3-032-24330-0



This Springer imprint is published by the registered company Springer Nature Switzerland AG
The registered company address is: Gewerbestrasse 11, 6330 Cham, Switzerland

# Preface

This book, entitled *Microbiomes and Plant Secondary Metabolites for Agricultural Pest Management*, is divided into two main parts comprising 19 chapters written by more than 20 professors, researchers, and scientists from different Egyptian and Arab universities and research institutions.

For the purpose of providing an introduction to this book, the editors have chosen to begin with the following title: “The Roadmap of the book: In general, it discusses the role that microbiomes and plant secondary metabolites play as sustainable tools for agricultural pest management. The purpose of this book is to apply these new field applications in agriculture, which are poised to play a key role in improving the agricultural productivity. The numerous applications of microbiomes and plant secondary metabolites as natural control products for the management of a variety of agricultural pests are included. A general discussion is included in the final chapter; it is concerned with the discussion of all the theme chapters in relation to each other and with the outcomes of international research that has been conducted up to this point, which have an impact on plant production globally. The purpose of this book is to provide a concise overview of low-cost methods for the management of a wide variety of agricultural pests.” Therefore, the editors confirmed that this book’s co-authors have made an effort, and every specialist has done their best in their respective fields to offer a synthesis of their ideas, facts, and research articles in order to produce this useful book for our scientists, colleagues, and students.

Al-Ahsa, Saudi Arabia
Alexandria, Egypt
01 January 2026

Nabil El-Wakeil
Ahmed Abdelkhalk

# General Introduction: The Roadmap of the Book

Agriculture is the support of human civilization, providing food, fiber, and fuel to sustain increasing populations. However, the growing demand for agricultural production is challenged by the persistent threat of many agricultural insect pests, which cause major crop losses and economic damage worldwide. Traditional pest management policies, heavily reliant on synthetic pesticides, have raised concerns about environmental pollution, biodiversity loss, and increased pesticide resistance. In this context, there is an urgent need for sustainable, eco-friendly alternative tactics to manage agricultural insect pests successfully. This book, *Microbiomes and Plant Secondary Metabolites for Agricultural Pest Management*, explores the emerging frontiers of biological pest control by harnessing the power of plant-associated microbiomes and plant secondary metabolites. These natural systems offer inventive solutions to improve crop resilience, lessen pest populations, and endorse sustainable agricultural practices.

The book is divided into two key thematic areas.

## Phytomicrobiome and Insect Microbiome for Pest Management

In the face of intensifying global challenges such as climate change, biodiversity loss, and the growing demand for sustainable agricultural practices, the need for innovative and eco-friendly pest management approaches has never been more urgent. Traditional methods of pest control, often reliant on chemical pesticides, have proven to be detrimental to ecosystems, human health, and nontarget organisms. This has spurred a paradigm shift toward harnessing the power of nature's own systems to fight pest-related issues in agriculture and beyond.

The first part of this book, *Phytomicrobiome and Insect Microbiome for Pest Management*, researches the fascinating and rapidly evolving fields of plant-associated microbiomes (phytomicrobiomes) and insect-associated microbiomes (insect microbiomes). These microbial societies play essential roles in shaping the health, behavior, and interactions of plants and insects, offering a treasure trove of opportunities for maintainable pest management tools.

The phytomicrobiome, encompassing bacteria, fungi, viruses, and other microorganisms associated with plants, impacts plant growth, nutrient uptake, and resilience to biotic and abiotic stresses. By understanding and manipulating these microbial communities, the authors believe they are capable of boosting plant defenses against insect pests and reducing reliance on chemical interventions. Similarly, the insect microbiome, which includes symbiotic bacteria, fungi, and other microbes, plays a serious role in insect physiology, behavior, and susceptibility to pathogens. By targeting these microbial partners, the authors may develop novel strategies to manage the pest populations, interrupt their life cycles, and mitigate their impact on crops and their production.

This book part provides a comprehensive overview of the latest research, technologies, and practical applications in the fields of phytomicrobiome and insect microbiome studies. It explores the intricate interactions among plants, insects, and their associated microbes, highlighting how these relationships can be leveraged for integrated pest management (IPM). Seven topics covered include:

1. The Role of Tree-Insect-Microbiome Interactions in Herbivory and Applications in Pest Management
2. Tri-trophic Interactions: The Role of the Plant and Insect Microbiome in Shaping Pest-Parasitoid Dynamics
3. Use of Plant Beneficial Microbes to Control Bayoud Disease in Date Palms
4. Effect of Plant Microbiome on Plant Physiology to Overcome Climate Changes
5. The Interconnected Microbiomes of Plants and Insects: Implications for Yield Quality and Pest Management
6. The Plant microbiome: An Engine for Sustainable Crop Growth and Yield
7. Effectiveness of Microbiota of Honeybees on Their Activities and Productivity

The editors as well as authors of different topics have tried to write this useful book for researchers, students, agricultural professionals, and policymakers; this book part aims to bridge the gap between fundamental science and practical applications. It stresses the importance of interdisciplinary approaches, combining microbiology, entomology, plant science, and biotechnology to develop supportable solutions for pest management. *Phytomicrobiome and Insect Microbiome for Pest Management* provides a forward-looking perspective on how the authors could harness the power of microbial ecosystems to protect different crops' kinds, preserve biodiversity, and ensure food security for future generations. The editors believe that this book arrives at a key juncture where science meets sustainability.

Summarily, the plant microbiome, containing bacteria, fungi, viruses, and other microorganisms, plays a critical role in plant health and defense. Recent advances in microbiome research have revealed how beneficial microbes could manage pest populations, induce systemic resistance in plants, and increase nutrient uptake.

This section investigates the mechanisms by which microbiomes interact with insect pests and pathogens, the potential for microbiome engineering, and the application of microbial consortia as biological control agents.

## Secondary Metabolism of Plants for Managing Agricultural Pests

Agriculture has always been at the head of human civilization, providing sustenance and economic steadiness. However, the persistent challenge of agricultural insect pests and pathogens has threatened crop yields and food security for many centuries. In the search for sustainable pest management, the focus has progressively shifted toward natural resolutions, predominantly the role of plant secondary metabolites. These compounds, though not directly involved in plant growth and development, play a serious role in plant defense mechanisms against insect herbivores, pathogens, and environmental stressors. This book part, *Secondary Metabolism of Plants for Managing Agricultural Pests*, investigates the fascinating world of plant secondary metabolites and their potential as eco-friendly tools for sustainable pest management.

Secondary metabolites, such as alkaloids, terpenoids, phenolics, and glucosinolates, are bioactive compounds that plants produce to deter many insect pests, attract beneficial organisms, and enhance their survival and role in pest management. Over the years, scientific research has uncovered the immense potential of these compounds in developing natural pesticides, repellents, and antifeedants. Unlike synthetic chemicals, which often have detrimental effects on the environment and nontarget organisms, plant-derived metabolites offer sustainable and biodegradable alternative tools.

This book part provides a comprehensive exploration of the role of secondary metabolites in plant-pest interactions, emphasizing their application in modern agriculture. It covers the biochemical pathways involved in the synthesis of these compounds, their ecological significance, and their mechanisms of action against pests. Additionally, the book highlights case studies and practical applications, showcasing how secondary metabolites can be harnessed to develop pioneering pest management strategies. Ten topics are presented in this section of the valuable book as follows:

1. Using plant metabolic products to develop relationship among plants, insect pests and their biocontrol agents
2. Efficiency of plant metabolic products on insect pests and their parasitoids
3. Role of secondary metabolites in plant defense and their effects on insect pests and predatory arthropods
4. Efficacy of plant metabolic products on insect pests and the associated entomopathogenic organisms

5. Efficacy of plant metabolic products to suppress the plant pathogenic nematodes
6. Efficacy of plant metabolic products to suppress the plant pathogenic fungi
7. Efficacy of plant metabolic products to suppress plant pathogens organisms
8. Antifungals produced by bacteria and their role in biological control
9. Biological control of viruses using *Actinomycetes* species
10. Effect of mineral nutrition and metabolic products on barley yield under drought conditions

In an age where sustainable agriculture is dominant, *Secondary Metabolism of Plants for Managing Agricultural Pests* serves as a valuable resource for researchers, entomologists, agronomists, and policymakers. By bridging the gap between plant biochemistry and agricultural practices, this book aims to motivate the adoption of nature-based solutions to manage insect pest challenges, ensuring a healthier and more resilient agricultural system for future generations.

## Synthesis and Future Directions

There is a discussion chapter, which summarizes the topics covered in this book with an update on the recently published research, highlights the extracted conclusions from the chapters, and presents a set of recommendations for future research on the role of IPM in the sustainable agricultural environment. The suppression level of any insect may show the role of microbiomes and plant secondary metabolites, and it is highly likely that any contribution they make to insect suppression is useful to the integrated management of that insect pest.

In summary, plants produce an assorted array of secondary metabolites, such as alkaloids, terpenoids, phenolics, and glucosinolates, which serve as chemical defenses against insect herbivores and pathogens. This section explores the biosynthesis, regulation, and ecological roles of these compounds, as well as their potential for use in pest management strategies. It also discusses various tactics for optimizing the production of secondary metabolites through breeding, genetic engineering, and elicitation.

Drawing on cutting-edge research in microbiology, plant biochemistry, ecology, and agronomy, this book provides a comprehensive overview of the potential of microbiomes and plant secondary metabolites to revolutionize agricultural pest management. It is designed for researchers, students, and practitioners in the fields of agriculture, plant science, and environmental biology, offering both foundational knowledge and practical insights into the development of sustainable pest control approaches.

As the authors face the dual challenges of feeding a growing global population and preserving the health of our planet, the integration of microbiomes and plant secondary metabolites into agricultural systems represents a promising pathway toward a more resilient and sustainable future. This book aims to encourage and guide the next generation of innovations in agricultural science, paving the way for a greener and more productive world. Welcome to a journey into the hidden world of microbes as well as secondary metabolites, and additionally, their transformative potential in the fight against insect pests and pathogens.

Nabil El-Wakeil

Department of Arid Land Agriculture,
College of Agricultural and Food
Sciences, King Faisal University,
Al-Ahsa, Saudi Arabia

Pests & Plant Protection Department,
Agricultural & Biological Sciences
Institute, NRC, Cairo, Egypt

# Acknowledgments

Many colleagues within National Research Centre and other research institutions, as well as colleagues in Egyptian and Arabian universities—entomologists, acarologists, bacteriologists, virologists, mycologists, nematologists, plant pathologists, and botanists in agricultural research institutions and universities—and in agricultural research institutions and universities have provided valuable information and comments, often providing more accurate and up-to-date accounts of the various projects than are available in publications and from their experience in case studies. Although it is not possible to acknowledge all of the many individuals involved, sincere thanks are due to all who contributed and helped us, and we regret that we have omitted their names here.

The editors would like to express their special thanks to all those who contributed in one way or another to making this high-quality book a real source of information on agricultural pest management, using the role of microbiomes and plant secondary metabolites for that purpose, supported by the latest findings in the field. The book is designed to support graduate students, researchers, scientists, and decision/policy makers in all Arab countries and everywhere in the world who are interested in this topic.

Much appreciation and great thanks are due to all the authors who contributed to this book. Without the efforts and patience of all the contributors in writing, reviewing, and revising the different versions of the chapters, it would not have been possible to produce this unique book titled *Microbiomes and Plant Secondary Metabolites for Agricultural Pest Management*. Also, special thanks are due to the Springer team and editors, who largely supported the authors and editors during the production of this book. The book editor would be happy to receive any comments to improve future editions. Comments, feedback, suggestions for improvement, or new chapters for next editions are welcome and should be sent directly to the book editor.

## AI Use Disclosure

The authors declare the limited, assistive use of DeepSeek-V3 AI to improve the language and clarity of specific manuscript sections. They emphasize that these AI-generated points were thoroughly reviewed and verified by them, and that they accept full responsibility for the final work, which has also passed through a standard peer-review process.

## Fund Acknowledgment

The authors would like to express their gratitude to the Deanship of Scientific Research at King Faisal University in Al-Ahsa, Saudi Arabia, for providing all the required facilities under project number (KFU251139), which allowed them to provide all of the necessary literature, including books, manuscripts, and other literary materials.

Prof. Dr. Nabil El-Wakeil, KFU, Al Ahsaa, Saudi Arabia
Dr. Ahmed Abdelkhalk, Burj Al Arab, Alexandria, Egypt
01 January 2026

# Contents

# Part I
# Phytomicrobiome and Insect Microbiome for Pest's Management

# Chapter 1
# The Role of Tree-Insect-Microbiome Interactions in Herbivory and Applications in Pest Management

**Nabil El-Wakeil and Nawal Gaafar**

## 1.1 Introduction

Insect pests cause significant ecological and economic disturbances to forest ecosystems, which are exacerbated by climate change and the introduction of foreign species (Aukema et al. 2011; Boyd et al. 2013). These insects can attack various tree organs, causing damage that ranges from a decrease in tree aesthetic value to slower tree growth or even tree death. Although not all insect herbivores are agricultural pests, background insect herbivory can cause significant loss of growth in urban and forest yards (Kozlov and Zvereva 2017).

There is a difference in the duration of the interactions that take place between microbes and insects. Some of these interactions are long-lasting, in which microorganisms are found in individual insects that belong to a population and are passed down from parents to children. These long-term microbial partners are considered to be symbionts since they are considered to be the core of the microbiota. Through gametes, egg smearing, coprophagy, or adult-nymph trophallaxis, they can be taken regularly from the environment or transmitted vertically. Additionally, they can be transmitted vertically. The opposite is true for temporary microbiota, which can be obtained through the feeding of insects or by the consumption of plants (Mason et al. 2019a) or soil (Hannula et al. 2019). In the process of opportunistically traversing

N. El-Wakeil (✉)
Department of Arid Land Agriculture, College of Agricultural & Food Sciences, King Faisal University, Al-Ahsa, Saudi Arabia
e-mail: nelwakeil@kfu.edu.sa

N. Gaafar
Pests & Plant Protection Department, Agricultural & Biological Sciences Institute, NRC, Cairo, Egypt

 
N. El-Wakeil, A. Abdelkhalek (eds.), *Microbiomes and Plant Secondary Metabolites for Insect Pest Management*, Progress in Biological Control 23,
https://doi.org/10.1007/978-3-032-24330-0_1

with the insect, it may reside in the insect's gut or mouthparts, or it may adhere to the insect's exoskeleton, potentially overlapping with the microbiota of the tree.

The efficiency of insect herbivores and the injury they cause to trees and shrubs are subsequently controlled by a number of direct and indirect exchanges with bacteria from the insect and microbiomes in trees and shrubs. These interactions can be either in direct or indirect ways. Frequent studies, similar the ones conducted by Gange et al. (2019), Fernandez-Conradi et al. (2018), and Koricheva et al. (2009), have provided indication and evaluated the impact that bacteria have on the relationships between insects and their hos plants (trees and shrubs). Forest entomologists and ecologists have only lately come to a thorough considerate of the intricate nature of microbial societies and the holobiont structures of both trees and insect herbivores.

New detections concerning the microbial assemblies that pointedly contribute to tree resistance against insect pests and change insect feeding behavior and performance are something that the authors need to conduct a deep investigation. Additionally, they want to highlight the influence that microbial exchanges among trees, shrubs and insect holobionts have on the health of these trees. In the following paragraphs, the authors try to present the possibility of incorporating tree and insect microbiomes into forest health management systems. In the end, the authors enlarge their research to comprise the forest level and deal with the importance of tree-insect-microbiome relations for the regeneration of forests and the biodiversity of such forest trees and shrubs.

While experiential research on microbiome-mediated herbivore resistance is certainly more advanced in well-defined agricultural systems, these prototypes often lack the complex ecological networks found in natural environments. Forest ecosystems, with their vast phylogenetic assortment, long-lived hosts, and stable co-evolutionary histories, serve as idyllic model systems to uncover the essential principles of multi-trophic plant-microbe-insect interactions as stated by Busby et al. (2017a). The complex relationships perceived in forests, where tree chemistry, insect herbivores, and a suite of microbial symbionts form a complex web of defense and counter-defense, represent a rich, untapped source of ecological knowledge (Bender et al. 2016). Therefore, this chapter purposely examines these complex forest systems to extract core mechanisms and evolutionary insights. The subsequent discussion will explicitly translate these philosophies into a blueprint for modernizing pest management in agricultural landscapes, arguing that the complexity of forests provides the necessary depth to inform sustainable approaches for simplified cropping systems (Stam et al. 2014).

### *1.1.1 Tree Microbiota and Its Effects*

A holobiont is formed by the tree and its microorganisms (Theis et al. 2016). Trees have a long history of co-evolution with microorganisms, which has been essential for the terrestrial colonization of plants (Strullu-Derrien et al. 2018). The microorganisms that inhabit a variety of tree organs are exceedingly diverse, encompassing bacteria, archaea, fungi, oomycetes, and nematodes. These bacteria are acquired through horizontal transmission from the environment (soil, atmosphere) and vertical transmission from the maternal tree to its progeny via seeds (Lemanceau et al. 2017a; Shade et al. 2017). The plant microbiome is frequently defined as the microbial population that resides on the surfaces of aerial tissues (phyllosphere; Vacher et al. 2016a), subterranean tissues (rhizosphere; Berendsen et al. 2012), and seeds (spermosphere; Shade et al. 2017), or in the internal tissues of the plant (endosphere; Brader et al. 2017). Additionally, the microbiota assemblages of generalist predators within the landscape are included (Saqib et al. 2023).

A broader concept is the plant microbiome. It includes the microbiota, which is the population of living microorganisms, the complete array of chemicals that are either synthesized by or constitutive of the microbes, as well as the mobile genetic elements (e.g., transposons, viruses) and residual DNA that are incorporated into the microbial environment (Berg et al. 2020). Certain components of the plant microbiota have neutral or detrimental effects on their host plant, while others are crucial for the plant's fitness due to their ecological roles, which improve the plant's ability to withstand environmental challenges, particularly insect herbivores (Hacquard et al. 2014; Vandenkoornhuyse et al. 2015).

### *1.1.2 Interactions Between the Microbiome of Trees and Insects*

Multiple studies indicate that the tripartite interactions among trees, insects, and their microbiomes significantly influence herbivore damage in forest ecosystems. Recent studies on the plant microbiome suggest that it may be feasible to manipulate the microbiomes of planted trees across different stages of their life cycles, from seed orchards to mature forests, to enhance the resilience of forest plantations to insect pests. Thus, the combination of the tree microbiome into forest health management policies may be feasible (Vacher et al. 2021). They mentioned that forest researchers must combine reductionist procedures with holistic objectives, as has been globally practiced, to realize these important goals. Experimental systems should be established to manipulate microbial populations, facilitating a deeper understanding of how microbial fluxes between trees and insects impact forest ecosystem health and explaining the underlying mechanisms. To get from small-scale mechanisms to forecasts at the forest ecosystem level, it is essential to assimilate information on the interactions among trees, insects, and microbiomes into spatial models of forest dynamics (Vacher et al. 2021).

### 1.1.3 Tree Health Influenced by Interactions Among Trees, Insects, and Microbiomes

The result of tree-herbivore interactions is shaped by the interrelated microbial societies established by the microbiomes of insects, shrubs and trees. The insect microbiome enriches feeding abilities, while key components of the tree microbiota converse resistance to insect pests. Ultimately, direct exchanges of microorganisms between insect and tree holobionts affect both insect infestation and tree health, the latter characterized as the particular tree's capacity to withstand ecological stressors, including insect pests, as shown in Fig. 1.1, as well as further clarified by Döring et al. (2011).

### 1.1.4 The Influence of Entomopathogenic Endophytes on Herbivorous Insects

Both the aerial and subterranean portions of plants are linked to microorganisms that provide herbivory resistance (Pineda et al. 2017; Gange et al. 2019). By functioning as pathogens or producing insecticidal poisons, several of these bacteria have a direct effect on insect herbivores. Others affect plant development, strengthen the immune system, or make plant hosts more resistant, all of which have an indirect effect (Pineda et al. 2017). Induced systemic resistance, which involves several phytohormones like ethylene, salicylic acid, and jasmonic acid, can be caused by both bacteria and fungi (Pieterse et al. 2014; Pineda et al. 2017). Microorganisms also

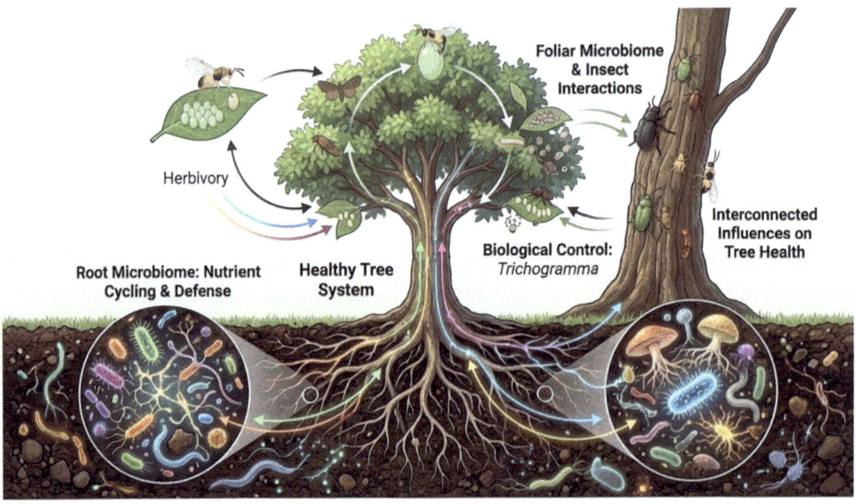


**Fig. 1.1** Interactions among trees, insects, and microbiomes and their influence on tree health

affect the release of volatile organic compounds from host plants in response to insect pests (herbivore-induced plant volatiles) to attract insect parasitoids or predators (War et al. 2012; Pangesti et al. 2013).

### 1.1.5 Soil Microorganisms' Function in Integrated Crop Management Techniques

Subterranean microorganisms significantly influence insect herbivory (Koricheva et al. 2009; Pineda et al. 2010, 2017). Root mutualists, including endophytic fungi, mycorrhizal fungi, rhizobacteria, and rhizobia, influence plant development, tolerance, and defense, hence regulating the activities of airborne insect herbivores (Bennett et al. 2006). Plant growth-promoting rhizobacteria and mycorrhizal fungi enhance plant tissue development and nutritional quality, benefiting insect herbivores. Conversely, they can augment plant resistance to insect pests through the induction of systemic resistance and improve plant tolerance to herbivory by promoting tissue regrowth (Pineda et al. 2010; Kempel et al. 2010; Tao et al. 2016). Indeed, the feedback effects of insects on root colonization by microorganisms render the interactions between aboveground insect herbivores and belowground microbes even more intricate. Initial studies on pinyon pine (Pinus edulis) revealed that trees vulnerable to infestation by the bug species *Matsucoccus acalyptus* and *Dioryctria albovittella* had markedly less ectomycorrhizal colonization. Furthermore, mycorrhizal colonization rebounded after the elimination of insects from vulnerable trees (Gehring and Whitham 1991, 1994; Del Vecchio et al. 1993).

### 1.1.6 The Function of the Insect Microbiome in Herbivory

Composting plant tissue, addressing nutrient deficiencies, and mitigating the effects of plant-specific metabolites are all tasks that herbivorous insects are required to perform during consumption. According to Watanabe and Tokuda (2010), the majority of insects lack the enzymes that are necessary to break down lignins or cellulose. This is especially true for insect borers, which are responsible for feeding on or excavating woody tree structures. The microbiome of insects is necessary for the organism's sustenance.

Some microorganisms create obligate associations with herbivores, rendering them essential. Adults of *Anoplophora glabripennis* have recognized a fungal association in their gut to facilitate the feeding of alive woody debris (Geib et al. 2008). Wood wasps (Hajek et al. 2013), bark beetles, and ambrosia beetles (Six 2013; Skelton et al. 2019) also depend on fungus for nourishment. They circulate fungi into host trees by bringing them into specific organs on their exoskeleton. The facility of mycelial material, which supplies sterols and nitrogen, improves the dietary intake of insects by fungal symbionts, possibly serving as their exclusive food source in specific circumstances (Six 2012). It is increasingly recognized that these fungal mutualists fulfill roles beyond mere nutritional assistance. The microorganisms like yeasts associated with *Ips typographus* (European spruce bark beetle), can degrade the stilbene defense chemicals of the tree, *Picea* sp. (Hulcr and Stelinski 2017; Hammerbacher et al. 2013). Furthermore, they can inhibit the proliferation of insect diseases (Davis et al. 2010). Insect herbivores can mitigate the nutritional deficiencies of their food sources through interactions with various microorganisms, as well as through prevalent fungal symbioses seen in forest ecosystems. All Hemiptera members depend on bacteria, usually contained within specialized insect cells, to produce a few of the essential amino acids and vitamins in their diet, as insects only consume tree sap (xylem or phloem) (Douglas 2015). During evolution, herbivorous insects have assimilated these essential nutritional partners as loyal associates, possibly derived from the plant microbiome. Phylogenetic reconstructions indicate that bacterial symbionts offering nutrition are derived from phytopathogens (Husník et al. 2011; Manzano-Marín et al. 2019), whereas fungal mutualists likely evolved from fungi ingested by insects (Vanderpool et al. 2017; Biedermann and Vega 2020).

### 1.1.7 Microbiota of Forest Trees and Insect Activity

The interpretation of research is that most of the gut bacteria found in forest insect pests are probably acquired through opportunistic plant tissue consumption, making them transient and environment-dependent. The microbial communities in the guts of pine weevils and *Dendroctonus* species have a number of traits and a composition that overlaps with the host tree's microbiome (Berasategui et al. 2016; Gonzalez-Escobedo et al. 2018; Jones et al. 2019). The eastern spruce budworm *Choristoneura fumiferana* (Landry et al. 2015) and the moth of *Thaumetopoea pityocampa* (Strano et al. 2017) have been found to exhibit differences in the stomach microbiota of larvae that consume different host trees. Both culture-dependent and culture-independent methods have shown that the plant species that the larvae eat has an impact on the gut bacterial community in the gypsy moth (Broderick et al. 2004). Longhorned beetles that cover their eggs with fungal symbionts establish that some gut symbionts can be vertically communicated (Mason et al. 2019b).

### 1.1.8 Herbivorous Insects Modify the Tree Microbiome and Affect Tree Health

The tree microbiome is accountable for the regulation of insect pests. It is responsible for the attendance of entomopathogenic bacteria, the construction of insecticidal toxins, the stimulation of tree defenses, and the improvement of the insect gut microbiome, which makes it easier for insects to feed (Li et al. 2018). The indication suggesting the reverse association, in which insect pests change the shrub and tree microbiome and may have an influence on tree health, was analyzed in this section.

Throughout the spread of microbial plant diseases' infections, which can happen actively (in mouthparts or ovarian excretions) or passively (on the organism's surface or in frass), herbivory can have a direct effect on the microbiome and health of shrubs and trees. Insects transmit a number of infamous tree disease organisms, such as bacteria (Chatterjee et al. 2008; Marcone 2015; Bendix and Lewis 2017), viruses, and fungi (Ghelardini et al. 2016; Linnakoski and Forbes 2019). An ascomycete fungus belonging to the Ophiostoma (Ophiostomatales) family is the source of Dutch elm disease, a fungal disease spread by insect pests. Elm tree populations in North America, Asia, and Europe have been significantly reduced as a result of this disease. According to Santini and Battisti (2019), the spores of the fungus are carried by the exoskeleton of bark beetles and enter the tree through the tunnels they make. The microbial pathogens carried by insects, the microbial symbionts of these insects, and the insects themselves combine synergistically to cause a diversity of complicated tree diseases (Lamichhane and Venturi 2015; Ghelardini et al. 2016; Wingfield et al. 2016). Multiple factors working either sequentially or concurrently have cumulative effects on the health of the host tree in Acute Oak Decline Syndrome, which is characterized by necrosis of the inner bark and the larval galleries of the bark-boring beetle *Agrilus biguttatus* (Denman et al. 2018). It is uncertain how *A. biguttatus* and the microbiological origins causing acute oak decline are related (Brown et al. 2014). Complex relationships among a nematode, its insect vector, the Monochamus beetle, and related bacteria and fungi are the cause of pine wilt disease (Vicente et al. 2013; Alves et al. 2018).

By manipulating the chemistry of injured tissues and/or the physiology and defense mechanisms of the entire shrub or tree, herbivory has the potential to alter the microbiome of various tree species, either locally or systemically. Microbiome dysbiosis is a mechanism by which changes in the tree microbiome can affect the health of the tree (Griffiths et al. 2019; Smets and Koskella 2020). Müller et al. (2003) found that the consumption of leaves by lepidopteran larvae resulted in local changes in the assortment of culturable phyllosphere bacteria. These changes were attributed to the increased availability of nitrate and ammonia in the injured leaves (Stadler and Michalzik 2000; Müller et al. 2003). The presence of insect endophagous herbivores has been the cause of numerous localized changes in tree microbiota, as documented in abundant studies (Treu et al. 2014; Menkis et al. 2015; Pec et al. 2016; Koivusaari et al. 2018; Fernandez-Conradi et al. 2019). Beule et al. (2017) have shown that the composition of fungal populations in both the phyllosphere and the rhizosphere is changed as a consequence of the significant defoliation of Scots pine and the pine lappet *Dendrolimus pini* by the nun moth *Lymantria monacha*. Tree defense mechanisms help mitigate the remote impacts of herbivory on tree microbiota. Herbivores and microorganisms both stimulate and respond to plant defenses that rely on similar hormonal pathways. Consequently, herbivory has the potential to modify the plant microbiome, as demonstrated by recent experimental induction of anti-herbivore barriers (Humphrey and Whiteman 2020) and studies (Biere et al. 2016; Moreira et al. 2018).

### 1.1.9 *Advancing the Integration of Tree-Insect-Microbiome in Forest Pest Management*

A highly promising technique for strengthening plant defense against biotic and abiotic stressors, including herbivores, is the manipulation of the plant microbiome in crops. This strategy is recognized as having exceptional potential. According to Bakker et al. (2012), the recently developed notion of the "phytobiome" places a significant emphasis on the essential role that manipulation plays. It takes into account all of the biotic and abiotic factors that have an effect on the growth and health of plants working together. Bakker et al. (2012) mentioned that the phytobiome should be organized as an integrated system because it includes the plant microbiome as well as the microbiomes of herbivores. Plant breeding and agroecosystem management (Bakker et al. 2012) are the two basic tactics that could be utilized to change plant microbiomes and phytobiomes. Furthermore, the direct inoculation of microbial strains or consortia (Mulner et al. 2019) is another method that could be employed. The research carried out on the management of the phytobiome in planted trees is somewhat less advanced compared to the research conducted on agricultural crop plants. However, a number of studies have projected tree microbiome management as a viable tactic for enhancing the health of forest plantations (Witzell and Martín 2018; Vivas et al. 2015, 2020).

Currently, the idea of altering the insect microbiome in order to implement IPM strategies has come to light (Douglas 2018 and others). Changing the microbiome of insects may reduce their ability to devour plant tissues and/or increase the effectiveness of parasitoids in controlling insect pests. Both of these outcomes are possible under certain circumstances. To regulate the dynamics of pest populations, it is possible to make use of symbionts that have the ability to influence the reproduction of insects. Research by Hurwitz et al. (2011) and Carrington et al. (2017) demonstrates that modifying the bacterial endosymbionts of insect vectors is a promising strategy for reducing the transmission of insect-borne diseases in human populations.

Comparable manipulations in insect herbivores have been proposed as a method for controlling insect-vectored plant pathogens; however, their application in crops and orchards remains infrequent. Paratransgenic strategies that aim to obstruct the transmission of *Xylella fastidiosa* by sharpshooters were reviewed by Arora et al. (2018); however, these same strategies have not yet been explored for implementation in forest pest management strategies. Instead of focusing on the microbiome of insect pests, the scientists chose to focus on the microbiome of trees that had been planted.

### 1.1.10 Harnessing Microbial Consortia for Sustainable Pest Control

Microbial consortia could be utilized for the purpose of controlling a wide range of insect pests. Beneficial microorganisms, which may include pest antagonists, can be directly inoculated onto or into plant organs or soil, or they can be used as a seed coating for preventative or therapeutic purposes (Rivas-Franco et al. 2019; Lacey et al. 2015). These approaches have been established time and again. *Rhizobacteria* that excite plant growth, for instance, can be used to inoculate soil or coat seeds to increase the plant's resistance to stress and to enhance growth (Tailor and Joshi 2014; Dessaux et al. 2016). Modern research has utilized omic practices, synthetic communities, and microbial network inference to develop advantageous microbial groups. These consortia have the potential to afford better and more sustainable plant protection in contrast to individual microbial strains (Vannier et al. 2019; Toju et al. 2018). Similar to what was obtained by Mitter et al. (2017), who stated that one potential field of research for the management of many important forest insect pests is the introduction of beneficial bacteria into plant reproductive organs.

One of the goals of this book chapter is to improve the transfer of these bacteria to children through their mothers. In order to achieve integrated pest management, it is essential to use biocontrol of insect pests in planted forests. Microbial entomopathogens are promising biocontrol agents, according to Kenis et al.'s 2019 research. Both Augustyniuk-Kram and Kram (2012), Jaber and Ownely (2018) and O'Brien (2017) have demonstrated that fiber bands that have been impregnated with fungal entomopathogens have demonstrated good results in the control of a variety of insect pests. Agreeing with Frasz et al. (2014) and Nguyen et al. 2016, who stated that the utilization of the fungal endophyte *Phialocephala scopiformis* as a means of inoculating white spruce seedlings against the spruce budworm *Choristoneura fumiferana* results in a toxic effect that persists for almost 11 years. It is possible that the survival and reproduction rates of major insect pests, such as Ambrosia beetles, could be reduced through the inoculation of trees with fungal entomopathogens or bacterial endophytes (Xie et al. 2020; Castrillo et al. 2016). Omics and systems biology tactics are important in defining microbial consortia antagonists of insects and vectored microorganisms (Poudel et al. 2016; Vacher et al. 2016a; Rohart et al. 2017; Vannier et al. 2019). These approaches also play a role in understanding the complex interactions that occur between trees, insect pests, and pathogens. It is recommended that the effectiveness of these microbial antagonists be evaluated in situ on a variety of tree genotypes. This is because there is a possibility that there is a relationship between the genotype of the tree and the effectiveness of microbial biocontrol agents (Broderick et al. 2010).

### 1.1.11 Cultivating Trees That Foster Protective Microbial Communities

Bakker et al. (2012), Gopal and Gupta (2016), Nelson et al. (2017), and Busby et al. (2017b) all point to the artificial assortment of cultivars that engage with a beneficial microbiome as a possible research direction for improving natural defenses against disease organisms and insect pests. According to Pérez-Jaramillo et al. (2015), Weese et al. (2015), and Gopal and Gupta (2016), the processes of domestication, selection, and cultivation initiated at the dawn of agriculture may have diminished the capacity of modern crops to form associations with beneficial soil microbes. Restoring microbial diversity in agroecosystems through inoculations or cultural practices and developing crop breeding methods that take the plant microbiome into account are both major challenges (Wubs et al. 2016; Pineda et al. 2020; Gupta and Nair 2020).

So far, tree breeders have ignored the microbiota as a potential selection component. Multiple studies have shown that genetic determinants influence tree microbiomes. For example, in olive trees (Mina et al. 2020), Norway spruce (Elfstrand et al. 2019), and eucalypts (Messal et al. 2019), among others. Therefore, this area of inquiry deserves additional investigation, because it does not work for every type of tree (Mason et al. 2015). Additionally, seed orchards and tree nurseries may foster the growth of trees that contribute to the maintenance of beneficial microbial communities because of the conditions in which genetic material is created.

The nurturing environment and early-arriving microbial types could influence the tree microbiome's assembly trajectory from seeds to mature shrubs or trees (Vivas et al. 2015; Fukami 2015). Research on *Eucalyptus grandis* shows that maternal trees' environment affects seedlings' endophytic fungal populations and their responses to kinds of pathogens and insect pests (Vivas et al. 2017, 2020). Accepting these maternal impacts is critical for incorporating tree microbiomes into breeding programs. Other features affecting seed and seedling microbiomes comprise seed storage conditions (Xing et al. 2018), plantation founding methods, soil type, and potting mix treatment, such as heat treatment, as mentioned by van der Voort et al. (2016). Early soil microbiome consideration is vital for plant health (Wei et al. 2019).

### 1.1.12 Enhancing Microbiome-Driven Resistance to Herbivory Via Silvicultural Techniques

The management strategies of agroecosystems have the potential to foster the growth of a microbiome that is valuable in both aboveground and belowground crop habitation. A greater amount of consideration has been directed towards the management of the belowground microbiome (Dessaux et al. 2016). This is due to the fact that the belowground microbiome possesses a greater variety in comparison to the aboveground microbiome, which only contains a subset of the organisms that are found belowground (Amend et al. 2019). Moreover, the relations between the belowground microbiome and plant health are more thoroughly understood (Berendsen et al. 2012).

The microbiome of the soil could be altered through a diversity of methods, such as the introduction of new microbiomes, the incorporation of organic modifications, or the growing of particular plants in close proximity to one another (Pineda et al. 2017). The term "plant-soil feedback" refers to the impact that plants have on the microbiome of the soil, as well as the following influences that this association has on the performance of impending plants, particularly their resistance to insect pests. According to Pineda et al. (2017), this idea serves as the foundation for predictable agricultural practices such as crop rotation, intercropping, and cover cropping. Research conducted by Hu et al. (2018) and Erb et al. (2013) confirmed that root exudates have the ability to change the composition of microbial inhabitants in the rhizosphere, which in turn has an effect on the growth and defense mechanisms of the following plant types.

There is an indication to propose that the conformation of the soil microbiome in woodlands is regulated by the genotype and species of shrubs and trees (Ayres et al. 2009; Gao et al. 2013; Urbanová et al. 2015). Agreeing with Pineda et al. (2017), who stated that plant-soil feedback can be exploited to provide information about forming the soil microbiome prior to the growth of forest plantations. This could help to improve tree resistance to a diversity of stresses and reduce insect pests within the aboveground environment.

Heterospecific plant-soil feedback occurs when different plant species lead crop plantations toward favorable outcomes and chances for further manipulation (Pineda et al. 2017). The composition of the aboveground tree microbiome should be influenced by plant diversity and soil microbiome management. Tree species' richness at the forest plot level enhances leaf epiphytic bacteria diversity (Laforest-Lapointe et al. 2017).

However, effects on foliar fungal communities have not been observed in Norway spruce *Picea abies* or Scots pine *Pinus sylvestris* (Helander et al. 2006; Nguyen et al. 2017; Lynikienė et al. 2020). Plant diversity also impacts the aboveground microbiota in orchards, altering the quantity of certain phyllosphere bacteria on banana leaves (Kãberl et al. 2015).

### *1.1.13 Forest Management Shapes Tree Microbiomes for Pest Resistance*

The aboveground microbiome might be affected by water and dead wood. In olive orchards, phyllosphere bacterial groups exhibit compositional differences in *Olea europaea*, yet show no variation in richness and diversity between conventional, organic, and sustainable systems characterized by distinct management practices regarding soil, water, and pruning residues (Pascazio et al. 2015; Fausto et al. 2018). Wemheuer et al. (2020) found that agroforestry management practices also affect the diversity, abundance, and function of endophyte communities in *Theobroma cacao* leaves, including fungi and bacteria. When comparing cocoa tree farms grown using organic and conventional methods, Oliveira et al. (2014) mentioned that a variety of foliar fungal endophytes, the quantity of which varied. According to Müller et al. (2015), the composition of the leaf epiphytic bacterial communities in apple orchards is influenced by the cropping system, namely organic vs conventional.

These findings suggest that there may be above- and belowground effects of forest management practices on tree plantation microbiomes, which in turn alter the resistance of mature trees to forest pests. The best course of action for scientists is to identify which tree species benefit most from the presence of certain bacteria. Gange et al. (2019) and Vega (2018), among many other studies on herbaceous plants, found that fungal endophytes can protect plants from insect herbivores through their entomopathogenic properties and/or the production or stimulation of secondary metabolites (Nisa et al. 2015).

Forest management should prioritize the preservation of fungal species that protect trees from diseases. Microbiome management is a strategy aimed at enhancing crop plant growth and protection against infections and pests. This research, informed by the "phytobiome" idea, can be applied to forest plantings. Regulating the microbiome of cultivated trees at various production phases can enhance their resistance against insect pests and other biotic and abiotic challenges.

### *1.1.14 Tree-Insect Microbiome Interactions to the Ecosystem Level*

Much of the research on the relationships between plant-insect herbivore interactions and their microbiomes follows a reductionist tenet, concentrating on tripartite interactions involving a single species of plant and an even smaller number of microbial taxa associated with the insect host, or one of the two, but rarely both. Research on the interplay between tritrophs, microbiomes, and different herbivore and plant species, as well as on ecosystems as a whole, is lacking. The fact that tripartite interactions can be both geographically and temporally unpredictable is one reason why scaling them is difficult, as has been shown by studies by Aukema et al. (2005), Adams et al. (2010), and Aukema et al. (2010). The researchers set out to provide proof that interactions between trees, insects, and microbiomes affect three parts of forest ecosystem health: the control of insect groups, the maintenance of the soil microbiome, and the process of tree regeneration by natural resources.

### 1.1.15 *Microbiomes of Trees and Insects Regulate Forest Insect Communities*

The foraging behavior and performance of insect herbivores are influenced by their microbiome, which has community-wide implications. Although there are no case studies on the cascade effects of the insect microbiome on trophic networks within forest ecosystems, research on herbaceous and agricultural plants indicates that it directly influences the recruitment of parasitoids and alters plant interactions with natural enemies and competitors (Madden 1968; Sullivan and Berisford 2004; Boone et al. 2008; Frago et al. 2012; Kandasamy et al. 2016). Sanders et al. (2016) have demonstrated that protective symbionts can reduce the number of competing aphid species and result in the extinction of associated parasitoids through experimental investigations. The release of herbivore-induced tree/plant volatiles can be reduced by aphid endosymbionts (Frago et al. 2017). Additionally, herbivory is facilitated by bacteria and viruses that are associated with insects at the tritrophic level. Herbivory can be reduced by viruses that are associated with parasitoids, which can affect the molecular interactions between plants and their infested herbivores (Cusumano et al. 2018).

### 1.1.16 *Tree Microbiomes Shape Forest Ecosystems and Insect Dynamics*

The tree microbiome influences not only insect pest species but also the entire forest insect population. Powdery mildew on oak foliage affects insect groups at numerous trophic levels (Tack et al. 2012). Foliar fungal endophytes in Neotropical forests control insect herbivores by predatory ants (Hammer and Van Bael 2015), change leaf chemistry, and deter leaf-cutting ants. These ants act as ecosystem causes, modifying forest structure, microclimate, and regeneration, impacting the entire trophic web as mentioned by Estrada et al. (2013), Coblentz and Van Bael (2013), Griffin and Carson (2018). The belowground microbiome, including plant mutualists, pathogens, and saprotrophs, meaningfully modifies the plant community's composition, affecting the insect community (Schuldt et al. 2017).

These results indicate that multipartite relations among macro-organisms, microbes, and viruses likely contribute significantly to the conservation of forest health and biodiversity. Comprehensive analyses of these interaction networks would significantly improve our comprehension of the connections between biodiversity and ecosystem health (Hudson et al. 2006); although such analyses are occasional in highly biodiverse ecosystems like forests, the advancements in ecological network discovery are facilitated by high-throughput sequencing technologies (Vacher et al. 2016b; Bohan et al. 2017). The solution likely resides in recognizing essential species groups and elucidating their functional relationships by employing experimental means on simplified ecosystems.

### 1.1.17 Aboveground Insect Pests Adjust the Belowground Forest Microbiome

According to Kristensen et al. (2019), the belowground forest microbiome has an effect on the insect pest population that is positioned above ground, whereas insect herbivores have an effect on the belowground microbial societies. Insect outbreaks hinder the allocation of carbon to roots and root exudation, which in turn has an effect on the microbial communities that reside in shrub or tree roots. Moth outbreaks are responsible for the change from ectomycorrhizal fungi to arbuscular mycorrhiza in mountain birch forests that are located in subarctic regions (Saravesi et al. 2015). Herbivory alters belowground biogeochemical processes, adversely affecting ectomycorrhizal fungi and permeating the entire soil food web (Kristensen et al. 2019). Experimental methodologies are crucial for understanding the linkages between ecosystems and forecasting insect outbreaks' cumulative impacts.

### 1.1.18 Microbial Fluctuations Regulate Forest Regeneration Amid Herbivory Density

Understanding microbial fluxes is vital for ecosystem health, as reported by van Bruggen et al. (2019). Microorganisms are communicated through airborne or insect vectors and seeds, with the horizontal transmission administered by the 'Janzen-Connell hypothesis' (Connell 1971; Shade et al. 2017; Lemanceau et al. 2017b; Vannier et al. 2018). This hypothesis proposes that microbial infections and insect herbivores have a predilection for destroying seedlings that are positioned in high density close to the parent tree (Benítez et al. 2013; Comita et al. 2014), hence affecting the process of forest regeneration. Microbiological infections and insect herbivores are not the only factors that contribute to seedling death. The impact that these factors have on seedling mortality is dependent on mycorrhizae (Jia et al. 2020), and seed-borne bacteria may also play a role in this spectacle.

The vertical transmission of microbiota in trees may impact seedling recruitment under herbivory density, and the primary (Ganley and Newcombe 2006; Qi et al. 2012; de la Bastide et al. 2019), maternally inherited microbiome may influence the assembly trajectory and composition of the microbiome during the seedling phase via priority effects, as confirmed by Fukami (2015). Research on crop seeds has revealed that vertically transmitted microorganisms improve seedling health (Bergna et al. 2018; Morella et al. 2019) and generate volatile organic composites (Tyc et al. 2019), which contribute to both herbivores and their predators in recognizing the proper plant hosts. It is reported by Shugart et al. (2020) that future research needs to improve the understanding of the implications of horizontal and vertical microbial fluxes on seedling recruitment. Additionally, these fluxes need to be incorporated into modeling frameworks, such as forest gap models, in order to scale their effects to the level of the forest ecosystem.

### 1.1.19 Insect Herbivory on the Quantity and Diversity of Leaf-Associated Bacterial Species

Research by Humphrey and Whiteman (2020) shows that herbivory alters leaf microbiome and increases bacterial infection sensitivity, underlining the trade-off between plant defenses and illnesses. Insect herbivory is common in plant communities, but its consequences on microbial plant colonists are poorly studied. By normalizing sequencing-based bacterial recognition to absolute bacterial load, they found that insect herbivory increases the within-host abundance of most phyllosphere species colonizing native forbs. The amplification of herbivore-associated bacteria indicates a community-wide trend toward lower ecological variety, but absolute abundance is needed to quantify these compositional changes. Inducing anti-herbivore defenses in Pseudomonas species changed fitness hierarchies. Field isolates and improved a subset of putatively phytopathogenic *P. syringae* match field-scale trends (Humphrey and Whiteman 2020). Herbivore injury reduces plant reproduction and is linked to phytopathogens. Indigenous plant microbiomes and plant-infecting bacteria epidemiology are affected by insect herbivory. Non-focal biotic interactions between hosts and other organisms can profoundly affect host-associated microbiome population and community dynamics.

### 1.1.20 The Defensive Methods Employed by Plants, Insects, and Various Other Species

Defining the trade-offs that exist between different defense strategies is something that continues to be of interest. The majority of organisms are subjected to simultaneous selection from a wide variety of natural enemies, and the act of reacting to one selection pressure frequently results in a negative impact on that of reacting to another selection pressure. According to Thaler et al. (1999), the antagonistic relationship that exists between jasmonic acid and salicylic acid signaling in plants demonstrates quite clearly that the prevailing response determines whether or not plant defenses target herbivorous insects and necrotrophic diseases or biotrophic pathogens.

In certain situations, therapy with a bacterial symbiont could boost defenses that are enabled by both jasmonic acid and salicylic acid, according to research that was conducted by Haney et al. (2018) on the interactions between beneficial rhizosphere Pseudomonas species and the herbivorous cabbage looper. The growing role of the microbiome in modifiable host fitness and defenses against antagonists, such as parasitoids, predators, and herbivores (Wilkinson et al. 2019), highlights the necessity of considering the microbiome's engagement in plant-defense research. Research carried out by Berendsen et al. (2012) obtained that plants that are attacked by insect pests have the capability to recruit defense bacteria into their rhizosphere microbiome.

In modern years, there has been a lack of transparency on the impact that insect herbivores have on the microbiome of the leaves, which is separate from the microbiome of the rhizosphere. Recent results gained by Humphrey and Whiteman (2020) provide unique insights into the relations that exist among insect herbivores, plant microbiomes, and host plants. This research convincingly establishes that insect herbivory in bittercress dramatically modifies the leaf microbiome.

### 1.1.21 Interactions Between the Microbiomes of Plants and Insects (Defense and Offense)

The plant and gut microbiomes are influenced by host defense mechanisms and themselves, resulting in a continuous tug of war between the host and its bacteria (Kembel et al. 2014). A healthy plant builds a diverse microbiome, which is considered "healthy." Despite the difficulties associated with defining the health of the microbiome (which were addressed by Bäckhed et al. (2012)), this idea has become increasingly prominent in the context of the human microbiome. A healthy microbiome should be associated with a healthy host and should demonstrate community stability. Species richness and evenness are particularly important characteristics of a healthy microbiome. It controls host defense and opposes microbial intruders, defending the host from infections. Dysbiosis, a shift in the microbiome linked to host disease, is vital for understanding and reversing microbial communities to limit injury to the host plant (Bäckhed et al. 2012).

Humphrey and Whiteman (2020) examine the effect that herbivory by the common leaf-mining fly, *Scaptomyza nigrita*, has on the microbiome of bittercress leaves through their research. Bittercress was revealed to be affected by herbivory from *S. Nigrita, which* exhibited significantly greater bacterial densities compared to undamaged leaves, mostly attributable to the heightened prevalence of common leaf bacteria (Humphrey and Whiteman 2020). A remarkable increase in numbers was noted for a specific species of Pseudomonas, with 3 to 6 doublings as noted by Humphrey and Whiteman (2020). According to Humphrey and Whiteman (2020), the rise in microbial populations within the wounded leaf tissue is probably attributable to both the loss of nutrients and the plant's response. This is additionally corroborated by the prevalence of pathogenic pseudomonads and the notion that compromised leaf tissue may supply nutrients that are especially profitable to plant diseases.

### *1.1.22 Plants React to Insect Pests by Elevating Jasmonic Acid Levels*

One of the most common responses that plants have to insect herbivory is an increase in the amount of jasmonic acid that they produce. Moreover, in a supplementary experiment that utilized previously characterized strains of *Pseudomonas syringae*, it was observed that an increase in jasmonic acid had a useful effect on the commonness of certain strains within the host plant. However, it should be noted that not all strains were affected by this effect, as Humphrey and Whiteman (2020) mentioned. As observed in *Leptinotarsa decemlineata* larvae, which exude *Pseudomonas* sp. to decrease the jasmonic acid response and disrupt the plant's antiherbivore defense (Chung et al. 2013), an unproven idea proposes that the insects may have transported Pseudomonas to the leaves through oral emission. This hypothesis has not been validated. Therefore, the observed increase in total bacterial counts that was reported by Humphrey and Whiteman (2020) can be attributed to three possible factors: nutritional enhancement as a result of plant cell death, an increase in jasmonic acid, and bacterial inoculation by insect secretions. What appears to be an indication of dysbiosis is the changing of a large number of community characteristics. The remarkable proliferation of Pseudomonas and various other taxa is indicative of a decline in community evenness, a transformation previously suggested as an unfavorable deviation from the original healthy microbiome (Bäckhed et al. 2012).

The microbiomes of injured leaves show greater dissimilarity compared to undamaged leaves, indicating that a healthy host plant maintains robust selection on its microbiome (Kembel et al. 2014). In a scratched host, the microbiome becomes unstable and may undergo alterations due to random impacts. Herbivores' influence on the leaf microbiome varies from their influence on the rhizosphere microbiome, where defensive microbes are attracted (Berendsen et al. 2012). Humphrey and Whiteman (2020) proposed that herbivory positively contributed to the rise in possible pathogens; yet, the hypothesis that insect pests are drawn to microbial-infested areas of the leaf tissue remains reasonable. Although the causal relationship remains unclear, the simultaneous occurrence of possible plant diseases and insect herbivores relates to a reduced fruit set in the damaged plant, signifying diminished fitness (Humphrey and Whiteman 2020). They determined that the concurrent presence of insect pests and elevated levels of potential plant diseases is likely to be widespread, in addition to being detrimental to the plant. As Hökeberg et al. (1997) have described, the significance of their results is contingent upon the plant pathogenic potential of the *Pseudomonas* strains recognized in their study. The Pseudomonas genus encompasses a diverse spectrum of organisms, including biocontrol agents.

### 1.1.23 Agricultural Insect Pests Management Through the Potential of Various Organisms

Microbes belonging to the same genus, such as *Pseudomonas*, often have similar appearances, such as a propensity to cause a number of plant diseases, as mentioned by Humphrey and Whiteman (2020). These findings, they argue, might have broader inferences for interactions among plants, microbes, and herbivore insects. As a result, this discovery could have a major impact on farming practices and the study of the plant microbiome. Insect pests are already a major threat to crops, and the increased chance of plant diseases makes it much worse. This encourages farmers to take a more proactive approach to treat insect infestations. In addition, Humphrey and Whiteman (2020) suggest that the combined existence of insect herbivory and potential plant diseases could be a problem in agricultural settings. Hence, preventive actions could be taken to ward off these diseases before symptoms appear. Additional research is needed to regulate the most common combinations of plant-insect-pathogen in order to apply this strategy effectively. Secondly, the phyllosphere, or leaf microbiome, is under intense examination because of its potential for ecological study and practical uses. By studying the leaf microbiome for signs of dysbiosis, the authors may be able to develop treatments that help plants regain health, reduce damage, or both. Researchers in the phyllosphere need to discriminate between microbiomes that are healthy and those that are dysbiotic, since the two may react differently to treatments.

### 1.1.24 Role of Plant-Associated Microorganisms in Attracting Natural Predators of Insects

In investigations on biological control agents in the phyllosphere, it is vital to determine whether the introduction of these agents may attract insect herbivores before their widespread. Moreover, if correcting dysbiosis benefits the host plant, the enlargement of biocontrol medications that restore a healthy microbiome may aid a stressed host plant (Humphrey and Whiteman 2014). Although Humphrey and Whiteman (2020) had tried to inspect the communities at the compositional level in most microbiome surveys, they could not have recognized whether the absolute abundance of bacterial taxa decreased or increased. The observed increase in Pseudomonads owing to herbivory was associated with a decrease in the relative abundances of all other species in the same trial.

### 1.1.25 Numerous Adversaries Assault Organisms, Modifying Host Phenotypes

For numerous creatures, assaults by various adversaries are unavoidable and frequently transpire consecutively throughout the lifespan of individual hosts. Previous attacks can modify host phenotypes and influence the progression of subsequent attacks, frequently resulting in cascade consequences over broader spatial and temporal dimensions (Laine 2011; Tollenaere et al. 2016; Karvonen et al. 2018). Due to the significant impact of co-infection on host health and the population dynamics of parasites, the explicit research of co-infection is becoming more prevalent (Susi et al. 2015; Halliday et al. 2017; Karvonen et al. 2018). However, this viewpoint has seldom been used in research on varied host-associated microbial populations (microbiomes).

Primarily, microbiome studies investigate the impact of host genotype or abiotic variation on the diversity of microbiomes (Ottesen et al. 2009; Bulgarelli et al. 2012; Horton et al. 2014; Bodenhausen et al. 2014; Edwards et al. 2015; Wagner et al. 2016). This has led to a significant lacuna in the comprehension of the impact of host establishment by non-microbial antagonists on the population dynamics of host-associated microbial types. There is a significant interest in elucidating the structure and function of the plant microbiome for practical applications, such as growth improvement and disease resistance (Finkel et al. 2017; Orozco-Mosqueda et al. 2018), as well as for the study of host–microbial symbioses in a broader context. In both natural and farmed locations, insect herbivory poses a significant threat to plants (Maron and Crone 2006). The vulnerability of plants to insect and microbial attacks can be influenced by herbivory, which alters plant phenotypes through tissue injury and the activation of plant defenses (Agrawal 1998; Bressan et al. 2009; Thaler et al. 2012). Therefore, the colonization and growth of plant-associated microorganisms are likely to be influenced by factors that alter the effect of herbivores on hosts. Plant phenotypes and fitness are substantially influenced by insect herbivores (Maron and Crone 2006) and plant-associated microorganisms (Wagner et al. 2014); however, they are regularly considered in isolation.

### 1.1.26 The Correlation Among Herbivory, Bacterial Populations, and Plant Fitness

*Cardamine cordifolia*, known as bittercress rarely, is a native forb that belongs to the Brassicaceae family. Several studies have been conducted to study the influence that insect herbivory has on the abundance and variety of endophytic bacterial taxa in this plant, as reported by Humphrey and Whiteman (2020). Confirmation of sequence-based assessments of bacterial burden in host tissue was achieved by researchers through the utilization of marker gene sequencing and paired leaf culturing. By using this method, they were able to evaluate the differences in the distributions of bacterial abundances between hosts that had previously been herbivorous to the leaf-mining bug *Scaptomyza nigrita* (Drosophilidae) and those that had not been eaten by the insect. Another thing that they did was examine the degree of co-clustering that existed between the quantity of microorganisms and the intensity of insect herbivory at the plant patch size within their natural study populations. Using twelve genetically distinct isolates of Pseudomonas species taken from bittercress, Humphrey and Whiteman (2020) explored the variety in sensitivity to inducible plant defenses against chewing herbivores. They found that the sensitivity varied from plant to plant. They achieved this by experimentally infecting accessions of native bittercress, where earlier herbivory was replicated through exogenous pre-treatment with the plant defense hormone jasmonic acid (JA), which triggers classical responses in plants against chewing herbivores (El-Wakeil et al. 2010; El-Wakeil and Volkmar 2012; Thaler et al. 2012), including bittercress (Humphrey and Whiteman 2016).

### *1.1.27 Impact of Insects and Microbiomes on Host Plant and Microbial Symbionts*

Insects maintain a symbiotic association with their gut microbiota, which is essential for survival in severe settings. The uptake of bacteria by insects indicates the initiation of microbial symbiosis, resulting in bacterial species adaptation to the gastrointestinal environment. Modern research by Gupta and Nair (2020) designated that evolutionary mechanisms within an insect's physiology regulate the transformation of a free-living microorganism into an obligatory or facultative symbiont, hence forming and diversifying these symbiotic associations. The symbiotic partners experience genetic, biochemical, and physiological modifications that profoundly affect their life cycle and biology. Comprehending the dynamics of this relationship may facilitate sustainable and ecologically sound methods for controlling economically significant pests of agricultural crops (Gupta and Nair 2020).

Insects, a primordial and diverse living form, have been on Earth since the Paleozoic epoch. They have tolerated severe climatic conditions and hold significant ecological and commercial value (Misof et al. 2014). Insect pests, frequently exacerbated by climate change, result in over 20% yearly crop loss. Despite these limitations, insect pests have increased their host range, changed their behavior, and populated several agro-climatic zones (Deutsch et al. 2018). Their substantial population size, brief reproductive cycles, and elevated reproduction rates have enabled them to prosper in diverse habitats (Shrestha 2019).

### *1.1.28 Insect-Microbe Symbiosis: Survival, Immunity, and Pest Control*

Insects preserve a multifaceted connection with advantageous microorganisms, encircling bacteria and fungi. Certain bacteria inhabit specialized cells within insects, whilst others are located on the external body surface (Thompson and Simpson 2009). These bacteria are primarily situated in the digestive tract, serving as important regulators of the insect host's behaviors. The gut microbiota supports feeding, boosts immunity, compensates for nutrient-deficient diets, mediates communication, regulates mating and reproductive success, aids digestion, and provides essential amino acids, metabolic compounds, and nutrients (Douglas 2015; Wielkopolan and Obrêpalska-Stêplowska 2016). Insects depend seriously on their gut microbiota for survival and life cycle functions (Ferrari and Vavre 2011). Endosymbionts and their insect hosts have a multifaceted relationship, with some specialists claiming that they act as parasitoids. The insect may have created the association to ensure life, and the bacterial species in the stomach could show mutualism, commensalism, or destructive behavior (Dillon and Dillon 2004).

Insects encourage an immune response against dangerous bacteria while protecting beneficial microbes (Mikonranta et al. 2014). Symbiotic anti-microbial peptides (AMPs) effect symbiotic interactions by limiting the multiplication of symbiotic bacteria (Mergaert 2018), occasionally converting them into differentiated forms, and eradicating unwanted, vulnerable microorganisms. Insects include several structural families of antimicrobial peptides (AMPs), including apidaecin and hymenoptaecin (Kwong et al. 2017; Wu et al. 2018). Nonetheless, endosymbiotic bacteria do not cause antibacterial responses in insects (Bourtzis et al. 2000). Numerous symbionts engage distinctively with the insect immune system, with certain ones evading the cellular immunological response while others impact the melanization response (Thomas et al. 2011). Recent data shows that symbionts can circumvent insect immunity; however, the mechanisms behind this capability remain ambiguous.

Prior research has inspected the evolution of endosymbionts from free-living entities to obligate symbionts in insect pests. However, knowledge concerning the mechanisms facilitating this transformation is scarce. The bacterial genome alters to acclimate to the gut environment, and certain germs can exploit the host for survival (Yuval 2017). The emphasis on the microbiome of insects has improved comprehension of insect-microbe interactions, with the host influencing the microbiome (Engel and Moran 2013). Comprehending these dynamics may facilitate the improvement of sustainable pest management solutions.

### *1.1.29 Insect-Microbe Relationships*

Insect-microbe symbiosis depends on the gaining of microorganisms from the surrounding atmosphere. These bacteria change from free-living entities to intracellular parasites, depending on the host's life cycle (Engel and Moran 2013). Insects establish niches and gut compartments that facilitate microbial survival, manipulating colonization through gut pH. Various bacterial endosymbionts occupy the insect body (Engel and Moran 2013).

Obligatory mutualists infiltrate insect bodies and locate within bacteriocytes, offering benefits and enhancing fitness. They form a lasting relationship with their host, fulfilling their nutritional requirements by generating critical amino acids and uncommon vitamins. Insects such as *Buchnera aphidicola* and *Megacopta cribraria* depend on these obligatory symbionts for survival and reproductive success (Couret et al. 2019). Obligate symbionts, such as Buchnera, have developed a significant dependency on their host over time, resulting in the loss of genes crucial for amino acid synthesis, which are vital in vitamin biosynthesis paths (Hansen and Moran 2011; Russell et al. 2013).

The microorganisms identified as facultative bacteria are relocated vertically and provide their hosts with support in digestion and the detoxification of xenobiotics. For example, Burkholderia and Serratia are both able to be acquired from new sources with each new generation. Definite facultative symbionts are advantageous to the host, whilst others are commensal or harmful (Ojha et al. 2017). They reside in different sections of an insect's anatomy and create the active element of the microbiome. Certain facultative microorganisms are situated in the hemocoel, whereas others are confined to bacteriocytes (Marubayashi et al. 2014). They have developed approaches to facilitate their transfer and multiplication within an insect's body, including transmission via haemolymph or milk gland secretions (De Vooght et al. 2015). A multitude of experiments have demonstrated the characteristics of diverse bacterial symbionts across various insect orders.

Insects serve as a substantial reservoir of phytopathogenic microorganisms, which exist in their bodies as a result of their feeding habits. Nevertheless, the immobility of plants presents a considerable barrier to the dissemination of these microbes. Therefore, these microorganisms depend on insects as vectors for their dissemination and reproduction. The intricate link among these bacteria and their insect carrier is influenced by their dynamic interactions and modifications of the host, as confirmed by Nadarasah and Stavrinides (2011). Certain microorganisms can proliferate within the insect vector, but others cannot. This indicates that the insect functions as both a vector and an alternative host for these bacteria. Microorganisms can be easily communicated or latently propagated within an insect's body (Perilla-Henao and Casteel 2016). Upon entering the body, they are released into the hemolymph through the midgut or hindgut epithelium. During feeding, they are delivered to the plant by the salivary gland (Kwon et al. 1999). Insects gain an advantage from this connection and have changed the way they fight these diseases.

### *1.1.30 Pathway of Microorganisms Within the Insect Body*

Phytopathogenic microorganisms, residing in insect bodies, are mostly disseminated by plants due to their stationary characteristics. These microorganisms depend on insect vectors for dissemination and reproduction, establishing a complicated interrelationship. They engage with their host and alter it to their advantage. Certain microorganisms can proliferate within the vector, but others cannot. The insect functions as both a vector and an alternative host for these bacteria (Schubert and Vu 2016). Certain germs are semipersistent and non-circulative, whilst others circulate within the body of the insect and have been transmitted after a period of time that has been concealed. From the moment they enter the body of the insect, they move to the epithelium of the midgut or the hindgut, where they infiltrate the salivary gland, and then they are transported to the plant while it is feeding (Robertson 2005). Insects have developed methods to combat these infections and gain benefits from this relationship.

Insect guts harbor endosymbiotic microorganisms that have been adapted for their advantage. These bacteria possess distinctive genomes that elucidate their phylogenetic placement and lifestyle (Kikuchi 2009). Endosymbiotic bacteria possess a significantly diminished genome resulting from successive gene loss, predominantly influenced by the proliferation of mobile elements and the pseudogenization of non-essential genes (Dutta and Paul 2012). Serratia symbiotica in aphids possesses 550 pseudogenes, in contrast to its free-living counterpart, *S. proteamaculans*. According to Nicks and Rahn-Lee (2017), the ancient endosymbiont *Buchnera aphidicola* has suffered from pseudogenization of a number of genes that are not crucial to its survival. The fact that *Sodalis glossinidius*, a facultative bacterial symbiont of tsetse flies, has a remarkable and significant expression of pseudogenes suggests that it is a relatively new acquisition. Carsonella, a bacterial symbiont that is obligately associated with psyllids, has experienced significant gene loss, which has rendered it completely dependent on its host for its continued existence (Moran et al. 2009). A negative association was found between host dependency and the size of the symbiont genome, according to research conducted by Fisher et al. (2017) on 58 obligatory bacterial symbionts that were found across 89 different varieties of hosts. It appears from this that the reduction of the genome that occurs as a result of gene losses leads to an increase in complementarity and functional redundancy, which in turn strengthens the dependence of microorganisms on their host.

### *1.1.31 Instruments Facilitating the Transitional Evolution of Insect Endosymbionts*

Parasitoids and symbionts experience simplification instead of intricate metabolic pathways within their insect hosts, resulting in genome reduction as a prevalent evolutionary strategy. This suggests that these bacterial species are subject to selection at the community level within their host (the organism they are living in). According to the "black-queen hypothesis" (BQH), which was put up by Morris et al. (2012), the reductive genome development of endosymbiotic bacterial species is the source of the dependencies and partnerships that exist between insects and their microbiomes. In experimental populations of *Methylobacterium extorquens*, Lee and Marx (2012) discovered selection-induced genome reduction, which revealed parallel deletions that resulted in a 10% reduction in the size of the genome.

The deletion of genes in symbionts provides a selection gain by preserving energy and resources when gene function is unnecessary. Functional synchronization within an insect host alleviates the necessity for specific bacterial species to sustain their entire metabolic network. The adaptive genome reorganization of bacterial endosymbionts may be significantly advantageous, particularly for the insect host (Sabater-Muñoz et al. 2017). The alterations in the bacterial genome presumably confer structural and functional stability, with non-random and adaptive modifications likely influenced by selection pressures.

### *1.1.32 Insect Symbionts: Genomic Adaptation and Host Evolution*

Symbionts frequently forfeit DNA replication and repair processes early in their relationship with the host, resulting in heightened detrimental mutations and gene inactivation. This loss is preordained and adaptive, facilitating the incorporation of genomic variation when the organism encounters difficulties in adapting to a particular existence in an insect host. The findings of Giraud et al. (2001) indicated that mutant phenotypes in both wild and laboratory-based bacterial populations closely mimic those seen in the gut of an insect, which promotes symbiotic adaptation to the environment of the gut. This loss is also connected with the deletion of genes that encode proteins that are defended by an insect's protective system. This deletion is a reflection of the adaptation of bacteria to the insect physiology. There is a connection between the insect lineage and the flagellar genes found in endosymbiotic bacteria, as mentioned by Toft and Fares (2008).

Despite the fact that the bacterial population that lives within the gut of the insect is subject to relaxed selection, the selection pressures that the host is subjected to are rather significant. Most of the time, changes in the bacterial genome are beneficial to the host, but environmental stressors, dietary adjustments, and shifts in ecological niches can have a significant impact on the population of bacteria (Ng et al. 2018). It is possible for insects to adapt to changes in their environment by modifying their microbiome. One example of this is the gut microbiota of *Drosophila melanogaster*, which has a significant impact on the insect's ability to survive in harsh conditions. As a result of the tight connection that exists between the insect and its bacterial symbiont, selective pressures have the potential to influence the evolution of the endosymbiotic bacteria. The findings of a study conducted by Rennison et al. (2019) suggest that the microbial communities found in the gut undergo changes in tandem with the colonization, adaptation, and speciation of the host. This suggests that interactions between insects and microbes play a role in the development of microbial endosymbionts.

Fluctuations in insect microbiomes may arise from nutrient-rich environments that facilitate microbial cohabitation, whereas nutrient-poor situations lead to competitive exclusion. This indicates that the "resource ratio" competition concept, initially described by Tilman in 1977, may be applicable to insect stomach endosymbionts. Nonetheless, additional experimental validation is required to substantiate this idea. This may clarify the reported microbiome differences in insects when consuming resistant plants or refractory food sources.

### *1.1.33 The Microbiome: A Promising Strategy for Sustainable Pest Management*

Progress in science and technology has reformed the agricultural and horticultural sectors, ensuring global food production self-sufficiency. The growing population and escalating need for more food, feed, and fodder have introduced new worries that jeopardize global food, nutritional, and livelihood security. According to the World Resources Institute (WRI 2019), hunger persists in poor and developing nations, with food demand projected to rise by 60–90% by 2050 due to the rapidly growing human population. Agriculture experiences an annual production loss of 20–40% solely because of insect infestations (FAO 2019).

Anticipating agricultural lethal outbreaks helps avert such incidents; nevertheless, inadequate warning systems and the overuse of pesticides and nitrogenous fertilizers intensify the challenges encountered by farmers. Low-dosage molecules provided by firms are neither economically viable nor readily available, and traditional pest management approaches are demonstrating ineffectiveness (Kikuchi et al. 2016). Invasive pests pose a significant challenge for farmers, as biological treatment is frequently unfeasible due to the absence of natural enemies.

Researchers have confirmed that significant genome degradation makes obligatory symbionts more susceptible to ecological variations than the host organism. Insects are evolving at a significantly enhanced pace compared to their hosts, with modifications in climatic and environmental factors prompting these conversions in insect pests. Gut bacteria undergo significant evolutionary pressures within an insect host, enabling rapid alterations to the constantly changing environment. Comprehending the evolutionary history of insects would facilitate the assessment of their population structure and the prediction of their potential for invasion in a certain region (Gupta and Nair 2020).

Phytopathogenic bacteria, notably those belonging to the family Enterobacteriaceae, have developed into diseases that impact plants as a result of their introduction into the phloem by insect vectors while they were carrying out their feeding activities. According to Gonella et al. (2015) and (2019) research, having an understanding of the connections that exist between these bacteria and insect symbionts may make it easier to treat phloem-limited plant diseases in a way that is both ecologically friendly and sustainable. Numerous coleopterans, such as the Colorado potato beetle *Leptinotarsa decemlineata,* cereal leaf beetle *Oulema melanopus,* western corn rootworm *Diabrotica virgifera virgifera,* red flour beetle *Tribolium castaneum*, and rice weevil *Sitophilus oryzae* (Wielkopolan and Obrêpalska-Stêplowska 2016), are able to achieve greater levels of success as a result of tripartite interactions between microbes, plants, and insects. Comprehending the function of insect-associated microorganisms in crop protection may be essential for formulating successful pest control tactics in agriculture. Notwithstanding attempts to cultivate resistant crop varieties, such resistance frequently proves ephemeral, allowing insects to circumvent host defenses after a few generations. Focusing on the gut microbiota may serve as a useful method for creating a comprehensive, eco-friendly, and sustainable pest management strategy.

### *1.1.34 Investigating the Impact of Microorganisms on Plant Tissues and Host Organisms*

The research assessed the bacterial populations of two olive cultivars in twigs and leaves using genetic identification of cultivable isolates. The predominant phyla were Proteobacteria, Actinobacteria, and Firmicutes, with epiphytes exhibiting more diversity and abundance than endophytes. The host genotype influenced the makeup of bacterial communities, although the impact exerted by plant organs was less pronounced than that of plant genotype (Mina et al. 2020). Each olive genotype and plant organ exhibited selectivity towards particular bacterial operational taxonomic units (OTUs) (Mina et al. 2020), indicating that the advantages gained from the plant-bacteria interaction should be evaluated at the genetic levels beneath the species. The research that was carried out by Mina et al. (2020) indicated that the bacterial communities that are linked with olive trees are primarily determined by the host cultivar and the particular plant organ, with the latter having a more significant influence on the creation of epiphytic bacterial communities. The alterations that occur in the fungal communities of olive tree twigs as a result of bacterial disease are dependent on the host genotype, according to the findings of Gomes et al. (2019). The endophytic bacterial isolate Curtobacterium was able to reduce the salt stress response in soybean plants, according to Khan et al.'s 2019 research.

The phyllosphere, the aerial components of plants, serves as an essential habitat for diverse microorganisms (Bringel and Couée 2015), including bacteria (Lindow and Brandl 2003; Leach et al. 2017). Bacteria can inhabit the surface (epiphytes) or interior (endophytes) of plant tissues, establishing intricate microbial interactions that profoundly influence plant growth and productivity (Newton et al. 2010). This varied and plentiful assemblage of microorganisms is essential for plant vitality and productivity (Bulgarelli et al. 2013).

### 1.1.35 Traits of Tree Species and Microbes Influence Particular Natural Enemies

Natural enemies, including parasitoids, predators, and entomopathogenic organisms, have been established as vital determinants of insect community diversification. Mulberry, a ligneous tree utilized in sericulture and medicine, confronts various dangers from phytopathogens and insect pests. *Botrytis cinerea* is a prominent phytopathogenic fungus that is responsible for mulberry gray mold disease (Xie et al. 2020). The Gram-positive, endospore-forming bacteria ZJU1 was extracted from healthy mulberry trees and utilized alongside a combination of antimicrobials to combat *Botrytis cinerea*. Field trials with endophyte-free plants inoculated with ZJU1 showed a marked decrease in leaf necrosis and death. A higher level of expression of genes that confer disease resistance in mulberry was also observed after inoculation with endophytes, which enhanced the plant's systemic defense.

The generalist herbivore *Spodoptera litura* has a survival rate of less than 50% after being exposed to ZJU1, according to Xie et al. (2020). An improvement in the insecticidal efficiency against common lepidopteran pests may be possible once crystal toxin (Cry10Aa) was identified. Xie et al. (2020) highlight the unique traits of *Bacillus amyloliquefaciens* ZJU1 as evidence of its biocontrol capabilities. Both Barra et al. (2018) and Cordovez et al. (2019) stated that the microbes found on plant surfaces and in plant tissues are essential to the host plant's well-being and productivity. Bacteria are ubiquitous in plants and perform numerous useful roles, according to new high-throughput techniques like whole metagenome sequencing (Leach et al. 2017). Common phyllosphere bacteria shield Arabidopsis from phytopathogens by halting their multiplication, according to Innerebner et al. (2011). Research by Kwak et al. (2018) shows that vulnerable tomato cultivars could have fewer disease symptoms after receiving rhizosphere bacterial communities from resistant tomato varieties. This proves that microbiome engineering can potentially boost agricultural yields (Foo et al. 2017).

There have been a limited number of studies that have specifically explored foliar endophytic bacteria, as reported by Muller et al. (2015). In spite of the fact that they are found in low population densities within plants, foliar bacterial endophytes are capable of forming a mutualistic relationship with their hosts. According to Lopes et al. (2018), certain endophytic Bacillus species produce a wide variety of antimicrobials that target a variety of phytopathogens, insects, and nematodes. This observation suggests that these antimicrobials could be utilized in biocontrol. It is possible that a thorough investigation of endophytes in a variety of plants could result in the discovery of multiple microbial resources that could be utilized in biotechnology to improve agricultural production (He et al. 2013).

Perennial woody trees of the mulberry (*Morus* spp.) species are widely grown in temperate and subtropical regions (He et al. 2013). Mulberry plants face several threats from herbivores and phytopathogens. One such threat is *Botrytis cinerea*-induced gray mold, which is difficult to control or prevent because of the wide variety of hosts it can infect (Chen et al. 2018, 2019). Though biocontrol offers a real alternative to dealing with herbivores and these illnesses, it is an opportunistic human infection that preys on young plants (Ji et al. 2010).

One important way in which tree species can live together is by limiting the recruitment of seedlings at higher densities of conspecifics, a phenomenon known as negative conspecific density dependency. Natural enemies mediate conspecific density dependency, which impacts various tree populations; however, the severity of this dependency varies greatly among species (Jia et al. 2020). Community dynamics are complicated, and it's not easy to understand how species-specific traits influence the variability of conspecific density dependency. This was a community-scale, three-year manipulation study in a temperate forest. At higher adult conspecific densities, seedling recruitment and survival are both decreased by plant-associated fungi and, to a lesser degree, by insect herbivores (Jia et al. 2020). As previously suspected, ectomycorrhizal and shade-tolerant species are more likely to have fungal attachments that hinder seedling recruitment in close proximity to conspecific adults.

## 1.2 Conclusion

This book chapter aimed to investigate the impact that interactions between trees, insects, and microbiomes have on the destructive effects of herbivory in agricultural contexts. Both aboveground and belowground microbiomes have the ability to influence herbivory in a variety of ways, including direct effects on insects and indirect effects on plant or tree chemistry and defense mechanisms. The microbiome of trees or plants can either reduce or increase herbivory across a wide range of plant or tree species. When it comes to the ability of insect pests to feed, the insect microbiome makes a significant contribution. Increasing the resilience of farmed trees to insect pests can be accomplished by manipulating the microbiome of the trees during various stages of production, according to the authors' contention. Since the extent of the impact that flora, arboreal species, and insect microbiomes have on environmental processes, it is imperative that additional research and examination be conducted. Future studies should aim to integrate reductionist and holistic methods, study the interaction among microbiomes, insects, trees, and other organisms, and look into how microbiomes are passed down through populations and generations.

**Acknowledgments** In order to express their gratitude, the authors would like to thank the Deanship of Scientific Research at King Faisal University in Al-Ahsa, Saudi Arabia, for providing all of the necessary facilities under project number (KFU251140). These facilities enabled the authors to provide all of the necessary literature, which included books, manuscripts, and other literary materials.

**Declaration of AI Assistance** The authors used the artificial intelligence tool DeepSeek-V3 in a limited and assistive capacity to improve the linguistic phrasing and clarity of the following sections of the manuscript: the introduction, the key elements of the chapter, and the adjustment and formatting of the References. All outputs were reviewed, edited, and their accuracy was carefully verified by the authors themselves, who bear full responsibility for the final content of the work. The chapter was subsequently peer-reviewed by specialist reviewers.

## References

Adams AS, Adams SM, Currie CR, Gillette NE, Raffa KF (2010) Geographic variation in bacterial communities associated with the red turpentine beetle (Coleoptera: Curculionidae). Environ Entomol 39:406–414

Agrawal AA (1998) Induced responses to herbivory and increased plant performance. Science 279: 1201–1202

Alves M, Pereira A, Vicente C, Matos P, Henriques J, Lopes H et al (2018) The role of bacteria in pine wilt disease: insights from microbiome analysis. FEMS Microbiol Ecol 94(fiy077):1–13

Amend AS, Cobian GM, Laruson AJ, Remple K, Tucker SJ, Poff KE et al (2019) Phytobiomes are compositionally nested from the ground up. Peer J 7:e6609

Arora AK, Pesko KN, Quintero-Hernández V, Possani LD, Miller TA, Durvasula RV (2018) A paratransgenic strategy to block transmission of *Xylella fastidiosa* from the glassy-winged sharpshooter *Homalodisca vitripennis*. BMC Biotechnol 18:50

Augustyniuk-Kram A, Kram KJ (2012) Entomopathogenic fungi as an important natural regulator of insect outbreaks in forests (review). In: Blanco JA (ed) Forest ecosystems - more than just trees. In Tech, pp 265–290

Aukema BH, Werner RA, Haberkern KE, Illman BL, Clayton MK, Raffa KF (2005) Quantifying sources of variation in the frequency of fungi associated with spruce beetles: implications for hypothesis testing and sampling methodology in bark beetle–symbiont relationships. For Ecol Manag 217:187–202

Aukema BH, Zhu J, Moeller J, Rasmussen J, Raffa KF (2010) Interactions between below-and above-ground herbivores drive a forest decline and gap-forming syndrome. Forest Ecol Manage 259:374–382

Aukema JE, Leung B, Kovacs K, Chivers C, Britton KO, Englin J et al (2011) Economic impacts of non-native forest insects in the continental United States. PLoS One 6:e24587

Ayres E, Steltzer H, Berg S, Wallenstein MD, Simmons BL, Wall DH (2009) Tree species traits influence soil physical chemical, and biological properties in high elevation forests. PLoS One 4:e5964

Bäckhed F, Fraser CM, Ringel Y, Sanders ME, Sartor RB, Sherman PM, Versalovic J, Young V, Finlay BB (2012) Defining a healthy human gut microbiome: current concepts, future directions, and clinical applications. Cell Host Microbe 12:611–622

Bakker MG, Manter DK, Sheflin AM, Weir TL, Vivanco JM (2012) Harnessing the rhizosphere microbiome through plant breeding and agricultural management. Plant Soil 360:1–13

Barra PJ, Viscardi S, Jorquera MA, Duran PA, Valentine AJ, de la Luz MM (2018) Understanding the strategies to overcome phosphorus deficiency and aluminum-toxicity by ryegrass endophytic and rhizosphere phosphobacteria. Front Microbiol 9:1155. https://doi.org/10.3389/fmicb.2018.01155

de la Bastide PY, LeBlanc J, Kong L, Finston T, May EM, Reich R et al (2019) Fungal colonizers and seed loss in lodgepole pine orchards of British Columbia. Botany 97:23–33

Bender SF, Wagg C, van der Heijden MG (2016) An underground revolution: biodiversity and soil ecological engineering for agricultural sustainability. Trends Ecol Evol 31(6):440–452

Bendix C, Lewis JD (2017) The enemy within: phloem-limited pathogens. Mol Plant Pathol 19: 238–254

Benítez M-S, Hersh MH, Vilgalys R, Clark JS (2013) Pathogen regulation of plant diversity via effective specialization. Trends Ecol Evol 28:705–711

Bennett AE, Alers-Garcia J, Bever JD (2006) Three-way interactions among mutualistic mycorrhizal fungi plants, and plant enemies: hypotheses and synthesis. Am Nat 167:141–152

Berasategui A, Axelsson K, Nordlander G, Schmidt A, Borg-Karlson A-K, Gershenzon J et al (2016) The gut microbiota of the pine weevil is similar across Europe and resembles that of other conifer-feeding beetles. Mol Ecol 25:4014–4031

Berendsen RL, Pieterse CM, Bakker PA (2012) The rhizosphere microbiome and plant health. Trends Plant Sci 17:478–486

Berg G, Rybakova D, Fischer D, Cernava T, Vergès M-CC, Charles T et al (2020) Microbiome definition re-visited: old concepts and new challenges. Microbiome 8(103):1–22

Bergna A, Cernava T, Rändler M, Grosch R, Zachow C, Berg G (2018) Tomato seeds preferably transmit plant beneficial endophytes. Phytobiomes J 2:183–193

Beule L, Grüning M, Karlovsky P, l-M-Arnold A (2017) Changes of Scots pine phyllosphere and soil fungal communities during outbreaks of defoliating insects. Forests 8:316

Biedermann PHW, Vega FE (2020) Ecology and evolution of insect fungus mutualisms. Ann Rev Entomol 65:431–455

Biere A, Goverse A (2016) Plant-mediated systemic interactions between pathogens parasitic nematodes, and herbivores above- and belowground. Annu Rev Phytopathol 54:499–527

Bodenhausen N, Bortfeld-Miller M, Ackermann M, Vorholt JAA (2014) Synthetic community approach reveals plant genotypes affecting the phyllosphere microbiota. PLoS Genet 10: e1004283

Bohan DA, Vacher C, Tamaddoni-Nezhad A, Raybould A, Dumbrell AJ, Woodward G (2017) Next-generation global biomonitoring: large-scale automated reconstruction of ecological networks. Trends Ecol Evol 32:477–487

Boone CK, Six DL, Zheng Y, Raffa KF (2008) Parasitoids and dipteran predators exploit volatiles frommicrobial symbionts to locate bark beetles. Environ Entomol 37:150–161

Bourtzis K, Pettigrew MM, O'Neill SL (2000) Wolbachia neither induces nor suppresses transcripts encoding antimicrobial peptides. Insect Mol Biol 9:635–639. https://doi.org/10.1046/j.1365-2583.2000.00224.x

Boyd IL, Freer-Smith PH, Gilligan CA, Godfray HCJ (2013) The consequence of tree pests and diseases for ecosystem services. Science 342:1235773

Brader G, Compant S, Vescio K, Mitter B, Trognitz F, Ma L-J et al (2017) Ecology and genomic insights into plant-pathogenic and plant-nonpathogenic endophytes. Ann Rev Phytopathol 55: 61–83

Bressan M, Roncato MA, Bellvert F, Comte G, Haichar FZ, Achouak W, Berge O (2009) Exogenous glucosinolate produced by *Arabidopsis thaliana* has an impact on microbes in the rhizosphere and plant roots. ISME J 3:1243–1257

Bringel F, Couée I (2015) Pivotal roles of phyllosphere microorganisms at the interface between plant functioning and atmospheric trace gas dynamics. Front Microbiol 6:486. https://doi.org/10.3389/fmicb.2015.00486

Broderick NA, Raffa KF, Goodman RM, Handelsman J (2004) Census of the bacterial community of the gypsy moth larval midgut by using culturing and culture-independent methods. Appl Environ Microbiol 70:293–300

Broderick NA, Vasquez E, Handelsman J, Raffa KF (2010) Effect of clonal variation among hybrid poplars on susceptibility of gypsy moth (Lepidoptera: Lymantriidae) to *Bacillus thuringiensis* subsp. *kurstaki*. Forest Entomol 103:718–725

Brown N, Inward DJG, Jeger M, Denman S (2014) A review of *Agrilus biguttatus* in UK forests and its relationship with acute oak decline. Forestry 88:53–63

van Bruggen AHC, Goss EM, Havelaar A, van Diepeningen AD, Finckh MR, Morris JG (2019) One health - cycling of diverse microbial communities as a connecting force for soil plant, animal, human and ecosystem health. Sci Total Environ 664:927–937
Bulgarelli D, Rott M, Schlaeppi K, Themaat EVL, Ahmadinejad N et al (2012) Revealing structure and assembly cues for *Arabidopsis* root-inhabiting bacterial microbiota. Nature 488:91–95
Bulgarelli D, Schlaeppi K, Spaepen S, van Themaat EVL, Schulze-Lefert P (2013) Structure and functions of the bacterial microbiota of plants. Annu Rev Plant Biol 64:807–838. https://doi.org/10.1146/annurev-arplant-050312-120106
Busby PE, Sokol NW, Friesen ML et al (2017a) Facilitating reforestation through the plant microbiome: perspectives from the phyllosphere. Annu Rev Phytopathol 55:561–582
Busby PE, Soman C, Wagner MR, Friesen ML, Kremer J, Bennett A et al (2017b) Research priorities for harnessing plant microbiomes in sustainable agriculture. PLoS Biol 15:e2001793
Carrington LB, Tran BCN, Le NTH, Luong TTH, Nguyen TT, Nguyen PT et al (2017) Field- and clinically derived estimates of Wolbachia-mediated blocking of dengue virus transmission potential in *Aedes aegypti* mosquitoes. Proc Natl Acad Sci USA 115:361–366
Castrillo LA, Griggs MH, Vandenberg JD (2016) competition between biological control fungi and fungal symbionts of ambrosia beetles *Xylosandrus crassiusculus* and *X. germanus* (Coleoptera: Curculionidae): mycelial interactions and impact on beetle brood production. Biol Cont 103: 138–146
Chatterjee S, Almeida RPP, Lindow S (2008) Living in two worlds: the plant and insect lifestyles of Xylella fastidiosa. Ann Rev Phytopathol 46:243–271
Chen B, Du K, Sun C, Vimalanathan A, Liang X, Li Y et al (2018) Gut bacterial and fungal communities of the domesticated silkworm (*Bombyx mori*) and wild mulberry-feeding relatives. ISME J 12:2252–2262. https://doi.org/10.1038/s41396-018-0174-1
Chen B, Xie S, Zhang X, Zhang N, Feng H, Sun C et al (2019) Gut microbiota metabolic potential correlates with body size between mulberry feeding lepidopteran pest species. Pest Manag Sci 76:1313–1323. https://doi.org/10.1002/ps.5642
Chung SH, Rosa C, Scully ED, Peiffer M, Tooker JF, Hoover K, Luthe DS, Felton GW (2013) Herbivore exploits orally secreted bacteria to suppress plant defenses. Proc Natl Acad Sci USA 110:15728–15733
Coblentz KE, Van Bael SA (2013) Field colonies of leaf-cutting ants select plant materials containing low abundances of endophytic fungi. Ecosphere 4(66):1–10. https://doi.org/10.1890/ES13-00012.1
Comita LS, Queenborough SA, Murphy SJ, Eck JL, Xu K, Krishnadas M et al (2014) Testing predictions of the Janzen-Connell hypothesis: a meta-analysis of experimental evidence for distance- and density-dependent seed and seedling survival. J Ecol 102:845–856
Connell JH (1971) On the role of natural enemies in preventing competitive exclusion in some marine animals and in rain forest trees. Dynam Popul:298–312
Cordovez V, Dini-Andreote F, Carrión VJ, Raaijmakers JM (2019) Ecology and evolution of plant microbiomes. Ann Rev Microbiol 73:69–88. https://doi.org/10.1146/annurev-micro-090817-062524
Couret J, Huynh-Griffin L, Antolic-Soban I, Acevedo-Gonzalez TS, Gerardo NM (2019) Even obligate symbioses show signs of ecological contingency: impacts of symbiosis for an invasive stinkbug are mediated by host plant context. Ecol Evol 9:9087–9099. https://doi.org/10.1002/ece3.5454
Cusumano A, Zhu F, Volkoff A-N, Verbaarschot P, Bloem J, Vogel H et al (2018) Parasitic wasp-associated symbiont affects plantmediated species interactions between herbivores. Ecol Lett 21:957–967
Davis TS, Hofstetter RW, Foster JT, Foote NE, Keim P (2010) Interactions between the yeast Ogataea pini and filamentous fungi associated with the western pine beetle. Microb Ecol 61: 626–634

De Vooght L, Caljon G, Van Hees J, Van Den Abbeele J, Mcdonald JH (2015) Paternal transmission of a secondary symbiont during mating in the viviparous tsetse fly. Mol Biol Evol 32:1977–1980. https://doi.org/10.1093/molbev/msv077
Del Vecchio TA, Gehring CA, Cobb NS, Whitham TG (1993) Negative effects of scale insect herbivory on the ectomycorrhizae of juvenile pinyon pine. Ecology 74:2297–2302
Denman S, Doonan J, Ransom-Jones E, Broberg M, Plummer S, Kirk S et al (2018) Microbiome and infectivity studies reveal complex polyspecies tree disease in acute oak decline. ISME J 12: 386–399
Dessaux Y, Grandclément C, Faure D (2016) Engineering the rhizosphere. Trends Plant Sci 21: 266–278
Deutsch CA, Tewksbury JJ, Tigchelaar M, Battisti DS, Merrill SC, Huey RB et al (2018) Increase in crop losses to insect pests in a warming climate. Science 361:916–919. https://doi.org/10.1126/science.aat3466
Dillon RJ, Dillon VM (2004) The gut bacteria of insects: nonpathogenic interactions. Annu Rev Entomol 49:71–92. https://doi.org/10.1146/annurev.ento.49.061802.123416
Döring TF, Pautasso M, Finckh MR, Wolfe MS (2011) Concepts of plant health - reviewing and challenging the foundations of plant protection. Plant Pathol 61:1–15
Douglas AE (2015) Multiorganismal insects: diversity and function of resident microorganisms. Annu Rev Entomol 60:17–34. https://doi.org/10.1146/annurev-ento-010814-020822
Douglas AE (2018) Strategies for enhanced crop resistance to insect pests. Ann Rev Plant Biol 69: 637–660
Dutta C, Paul S (2012) Microbial lifestyle and genome signatures. Curr Genomics 13:153–162. https://doi.org/10.2174/138920212799860698
Edwards J, Johnson C, Santos-Medellín C et al (2015) Structure, variation, and assembly of the root-associated microbiomes of rice. Proc Natl Acad Sci USA 112:E911–E920
Elfstrand M, Zhou L, Baison J, Olson Å, Lundén K, Karlsson B et al (2019) Genotypic variation in Norway spruce correlates to fungal communities in vegetative buds. Mol Ecol 29:199–213
El-Wakeil NE, Volkmar C (2012) Effect of Jasmonic acid application on economically insect pests and yield in spring wheat. Gesunde Pflanzen 64:107–116. https://doi.org/10.1007/s10343-012-0278-x
El-Wakeil NE, Volkmar C, Sallam AA (2010) Jasmonic acid induces resistance to economically important insect pests in winter wheat. Pest Manag Sci 66:549–554. https://doi.org/10.1002/ps.1906
Engel P, Moran NA (2013) The gut microbiota of insects - diversity in structure and function. FEMS Microbiol Rev 37:699–735
Erb M, Huber M, Robert CAM, Ferrieri AP, Machado RAR, Arce CCM (2013) The role of plant primary and secondary metabolites in root-herbivore behaviour nutrition and physiology. Adv Insect Physiol 45:53–95
Estrada C, Wcislo WT, Van Bael SA (2013) Symbiotic fungi alter plant chemistry that discourages leaf-cutting ants. New Phytol 198:241–251
FAO (2019) The state of food & agriculture. Moving forward on food loss & waste reduction. FAO, Rome. https://openknowledge.fao.org/server/api/core/bitstreams/11f9288f-dc78-4171-8d02-92235b8d7dc7/content. 03.02.2025
Fausto C, Mininni AN, Sofo A, Crecchio C, Scagliola M, Dichio B et al (2018) Olive orchard microbiome: characterisation of bacterial communities in soil-plant compartments and their comparison between sustainable and conventional soil management systems. Plant Ecol Divers 11:597–610
Fernandez-Conradi P, Jactel H, Robin C, Tack AJM, Castagneyrol B (2018) Fungi reduce preference and performance of insect herbivores on challenged plants. Ecology 99:300–311
Fernandez-Conradi P, Fort T, Castagneyrol B, Jactel H, Robin C (2019) Fungal endophyte communities differ between chestnut galls and surrounding foliar tissues. Fungal Ecol 42: 100876

Ferrari J, Vavre F (2011) Bacterial symbionts in insects or the story of communities affecting communities. Philos Trans R Soc B Biol Sci 366:1389–1400. https://doi.org/10.1098/rstb.2010.0226

Finkel OM, Castrillo G, Herrera Paredes S, Salas Gonzlez I, Dangl JL (2017) Understanding and exploiting plant beneficial microbes. Curr Opin Plant Biol 38:155–163

Fisher RM, Henry LM, Cornwallis CK, Kiers ET, West SA (2017) The evolution of host-symbiont dependence. Nat Commun 8:15973

Foo JL, Ling H, Lee YS, Chang MW (2017) Microbiome engineering: current applications and its future. Biotechnol J 12:1600099. https://doi.org/10.1002/biot.201600099

Frago E, Dicke M, Godfray HCJ (2012) Insect symbionts as hidden players in insect-plant interactions. Trends Ecol Evol 27:705–711

Frago E, Mala M, Weldegergis BT, Yang C, McLean A, Godfray HCJ et al (2017) Symbionts protect aphids from parasitic wasps by attenuating herbivore-induced plant volatiles. Nat Commun 8:1860

Frasz SL, Walker AK, Nsiama TK, Adams GW, Miller JD (2014) Distribution of the foliar fungal endophyte *Phialocephala scopiformis* and its toxin in the crown of a mature white spruce tree as revealed by chemical and qPCR analyses. Can J For Res 44:1138–1143

Fukami T (2015) Historical contingency in community assembly: integrating niches species pools, and priority effects. Annu Rev Ecol Evol Syst 46:1–23

Gange AC, Koricheva J, Currie AF, Jaber LR, Vidal S (2019) Metaanalysis of the role of entomopathogenic and unspecialized fungal endophytes as plant bodyguards. New Phytol 223:2002–2010

Ganley RJ, Newcombe G (2006) Fungal endophytes in seeds and needles of *Pinus monticola*. Mycol Res 110:318–327

Gao C, Shi N-N, Liu Y-X, Peay KG, Zheng Y, Ding Q et al (2013) Host plant genus-level diversity is the best predictor of ectomycorrhizal fungal diversity in a Chinese subtropical forest. Mol Ecol 22:3403–3414

Gehring CA, Whitham TG (1991) Herbivore-driven mycorrhizal mutualism in insect-susceptible pinyon pine. Nature 353:556–357

Gehring CA, Whitham TG (1994) Interactions between aboveground herbivores and the mycorrhizal mutualists of plants. Trends Ecol Evol (Amst) 9:251–255

Geib SM, Filley TR, Hatcher PG, Hoover K, Carlson JE, Jimenez-GascoMdel M et al (2008) Lignin degradation in wood-feeding insects. Proc Natl Acad Sci USA 105:12932–12937

Ghelardini L, Pepori AL, Luchi N, Capretti P, Santini A (2016) Drivers of emerging fungal diseases of forest trees. For Ecol Manag 381:235–246

Giraud A, Matic I, Tenaillon O, Clara A, Radman M, Fons M et al (2001) Costs and benefits of high mutation rates: adaptive evolution of bacteria in the mouse gut. Science 291:2606–2608. https://doi.org/10.1126/science.1056421

Gomes T, Pereira JA, Lino-Neto T, Bennett AE, Baptista P (2019) Bacterial disease induced changes in fungal communities of olive tree twigs depend on host genotype. Sci Rep 9:5882. https://doi.org/10.1038/s41598-019-42391-8

Gonella E, Pajoro M, Marzorati M, Crotti E, Mandrioli M, Pontini M et al (2015) Plant-mediated interspecific horizontal transmission of an intracellular symbiont in insects. Sci Rep 5:15811

Gonella E, Tedeschi R, Crotti E, Alma A (2019) Multiple guests in a single host: interactions across symbiotic and phytopathogenic bacteria in phloem feeding vectors—a review. Entomol Exp Appl 167:171–185. https://doi.org/10.1111/eea.12766

Gonzalez-Escobedo R, Briones-Roblero CI, Pineda-Mendoza RM, Rivera-Orduña FN, Zúñiga G (2018) Bacteriome from *Pinus arizonica* and *P. durangensis*: diversity comparison of assemblages, and overlapping degree with the gut bacterial community of a bark beetle that kills pines. Front Microbiol 9:77. https://doi.org/10.3389/fmicb.2018.00077

Gopal M, Gupta A (2016) Microbiome selection could spur next generation plant breeding strategies. Front Microbiol 7:1971. https://doi.org/10.3389/fmicb.2016.01971

Griffin EA, Carson WP (2018) Tree endophytes: cryptic drivers of tropical forest diversity. In: Endophytes of forest trees, pp 63–103. https://doi.org/10.1007/978-3-319-89833-9_4
Griffiths SM, Galambao M, Rowntree J, Goodhead I, Hall J, O'Brien D et al (2019) Complex associations between cross-kingdom microbial endophytes and host genotype in ash dieback disease dynamics. J Ecol 108:291–309
Gupta A, Nair S (2020) Dynamics of insect–microbiome interaction influence host and microbial symbiont. Front Microbiol 11:1357. https://doi.org/10.3389/fmicb.2020.01357
Hacquard S, Schadt CW (2014) Towards a holistic understanding of the beneficial interactions across the Populus microbiome. New Phytol 205:1424–1430
Hajek AE, Nielsen C, Kepler RM, Long SJ, Castrillo L (2013) Fidelity among Sirex woodwasps and their fungal symbionts. Microb Ecol 65:753–762
Halliday FW, Umbanhowar J, Mitchell CE (2017) Interactions among symbionts operate across scales to influence parasite epidemics. Ecol Lett 20:1285–1294
Hammer TJ, Van Bael SA (2015) An endophyte-rich diet increases ant predation on a specialist herbivorous insect. Ecol Entomol 40:316–321
Hammerbacher A, Schmidt A, Wadke N, Wright LP, Schneider B, Bohlmann J et al (2013) A common fungal associate of the spruce bark beetle metabolizes the stilbene defenses of Norway spruce. Plant Physiol 162:1324–1336
Haney CH, Wiesmann CL, Shapiro LR, Melnyk RA, O'Sullivan LR, Khorasani S, Xiao L, Han J, Bush J, Carrillo J et al (2018) Rhizosphere-associated *Pseudomonas* induce systemic resistance to herbivores at the cost of susceptibility to bacterial pathogens. Mol Ecol 27:1833–1847
Hannula SE, Zhu F, Heinen R, Bezemer TM (2019) Foliar-feeding insects acquire microbiomes from the soil rather than the host plant. Nat Commun 10:1–9
Hansen AK, Moran NA (2011) Aphid genome expression reveals hostsymbiont cooperation in the production of amino acids. Proc Natl Acad Sci USA 108:2849–2854. https://doi.org/10.1073/pnas.1013465108
He N, Zhang C, Qi X, Zhao S, Tao Y, Yang G et al (2013) Draft genome sequence of the mulberry tree *Morus notabilis*. Nat Commun 4:2445. https://doi.org/10.1038/ncomms3445
Helander M, Wäli P, Kuuluvainen T, Saikkonen K (2006) Birch leaf endophytes in managed and natural boreal forests. Can J For Res 36:3239–3245
Hökeberg M, Gerhardson B, Johnsson L (1997) Biological control of cereal seedborne diseases by seed bacterization with greenhouse-selected bacteria. Eur J Plant Pathol 103:25–33
Horton MW, Bodenhausen N, Beilsmith K, Meng D, Muegge BD et al (2014) Genome-wide association study of *Arabidopsis thaliana* leaf microbial community. Nat Commun 5:5320
Hu L, Robert CAM, Cadot S, Zhang X, Ye M, Li B et al (2018) Root exudate metabolites drive plant-soil feedbacks on growth and defense by shaping the rhizosphere microbiota. Nat Commun 9:2738
Hudson PJ, Dobson AP, Lafferty KD (2006) Is a healthy ecosystem one that is rich in parasites? Trends Ecol Evol 21:381–385
Hulcr J, Stelinski LL (2017) The ambrosia symbiosis: from evolutionary ecology to practicalmanagement. Annu Rev Entomol 62:285–303
Humphrey PT, Whiteman NK (2020) Insect herbivory reshapes a native leaf microbiome. Nat Ecol Evol 4:221–229. https://doi.org/10.1038/s41559-019-1085-x
Humphrey PT, Nguyen TT, Villalobos MM, Whiteman NK (2014) Diversity and abundance of phyllosphere bacteria are linked to insect herbivory. Mol Ecol 23:1497–1515
Humphrey PT, Gloss AD, Alexandre NM, Villalobos MM, Fremgen MR et al (2016) Aversion and attraction to harmful plant secondary compounds jointly shape the foraging ecology of a specialist herbivore. Ecol Evol 6:3256–3268
Hurwitz I, Fieck A, Read A, Hillesland H, Klein N, Kang A et al (2011) Paratransgenic control of vector borne diseases. Int J Bio Sci 7:1334–1344
Husník F, Chrudimský T, Hypša V (2011) Multiple origins of endosymbiosis within the Enterobacteriaceae (γ-Proteobacteria): convergence of complex phylogenetic approaches. BMC Biol 9:1–18

Innerebner G, Knief C, Vorholt JA (2011) Protection of *Arabidopsis thaliana* against leaf-pathogenic *pseudomonas syringae* by *Sphingomonas* strains in a controlled model system. Appl Environ Microbiol 77:3202–3210. https://doi.org/10.1128/AEM.00133-11

Jaber LR, Ownley BH (2018) Can we use entomopathogenic fungi as endophytes for dual biological control of insect pests and plant pathogens? Biol Control 116:36–45

Ji X, Lu G, Gai Y, Gao H, Lu B, Kong L et al (2010) Colonization of *Morus alba* L. by the plant-growth-promoting and antagonistic bacterium *Burkholderia cepacia* strain Lu10-1. BMC Microbiol 10:243. https://doi.org/10.1186/1471-2180-10-243

Jia S, Wang X, Yuan Z, Lin F, Ye J, Lin G et al (2020) Tree species traits affect which natural enemies drive the Janzen-Connell effect in a temperate forest. Nat Commun 11:286. https://doi.org/10.1038/s41467-019-14140-y

Jones AG, Mason CJ, Felton GW, Hoover K (2019) Host plant and population source drive diversity of microbial gut communities in two polyphagous insects. Sci Rep 9:2792

Kãberl M, Dita M, Martinuz A, Staver C, Berg G (2015) Agroforestry leads to shifts within the gammaproteobacterial microbiome of banana plants cultivated in Central America. Front Microbiol 6:91

Kandasamy D, Gershenzon J, Hammerbacher A (2016) Volatile organic compounds emitted by fungal associates of conifer bark beetles and their potential in bark beetle control. J Chem Ecol 42:952–969

Karvonen A, Jokela J, Laine AL (2018) Importance of sequence and timing in parasite coinfections. Trends Parasitol 35:109–118

Kembel SW, O'Connor TK, Arnold HK, Hubbell SP, Wright SJ, Green JL (2014) Relationships between phyllosphere bacterial communities and plant functional traits in a neotropical forest. Proc Natl Acad Sci USA 111:13715–13720

Kempel A, Schmidt AK, Brandl R, Schädler M (2010) Support from the underground: induced plant resistance depends on arbuscular mycorrhizal fungi. Funct Ecol 24:293–300

Kenis M, Hurley BP, Colombari F, Lawson S, Jiang Hua S, Wilcken C, et al (2019) Guide to the classical biological control of insect pests in planted and natural forests. FAO Forestry Papers. p 182

Khan MA, Asaf S, Khan AL, Ullah I, Ali S, Kang S, Lee I (2019) Alleviation of salt stress response in soybean plants with the endophytic bacterial isolate *Curtobacterium* sp. SAK1. Ann Microbiol 1:12–808. https://doi.org/10.1007/s13213-019-01470-x

Kikuchi Y (2009) Endosymbiotic bacteria in insects: their diversity and culturability. Microbes Environ 24:195–204. https://doi.org/10.1264/jsme2.me09140s

Kikuchi Y, Tada A, Musolin DL, Hari N, Hosokawa T, Fujisaki K et al (2016) Collapse of insect gut symbiosis under simulated climate change. mBio 7:e01578–e01516. https://doi.org/10.1128/mBio.01578-16

Koivusaari P, Pohjanen J, Wäli PR, Ahonen SHK, Saravesi K, Markkola AM et al (2018) Different endophyte communities colonize buds of sprouts compared with mature trees of mountain birch recovered from moth herbivory. Tree Physiol 38:1437–1444

Koricheva J, Gange AC, Jones T (2009) Effects of mycorrhizal fungi on insect herbivores: a meta-analysis. Ecology 90:2088–2097

Kozlov MV, Zvereva EL (2017) Background insect herbivory: impacts patterns and methodology. In: Progress in botany, vol 79, pp 313–355

Kristensen JÅ, Rousk J, Metcalfe DB (2019) Below-ground responses to insect herbivory in ecosystems with woody plant canopies: a meta- analysis. J Ecol 108:1–14. https://doi.org/10.1111/1365-2745.13319

Kwak MJ, Kong HG, Choi K, Kwon SK, Song JY, Lee J et al (2018) Rhizosphere microbiome structure alters to enable wilt resistance in tomato. Nat Biotechnol 36:1100–1109. https://doi.org/10.1038/nbt.4232

Kwon MO, Wayadande AC, Fletcher J (1999) *Spiroplasma citri* movement into the intestines and salivary glands of its leafhopper vector, *Circulifer tenellus*. Phytopathology 89:1144–1151. https://doi.org/10.1094/phyto.1999.89.12.1144

Kwong WK, Mancenido AL, Moran NA (2017) Immune system stimulation by the native gut microbiota of honey bees. R Soc Open Sci 4:170003. https://doi.org/10.1098/rsos.170003
Lacey LA, Grzywacz D, Shapiro-Ilan DI, Frutos R, Brownbridge M, Goettel MS (2015) Insect pathogens as biological control agents: back to the future. J Invertebr Pathol 132:1–41
Laforest-Lapointe I, Paquette A, Messier C, Kembel SW (2017) Leaf bacterial diversity mediates plant diversity and ecosystem function relationships. Nature 546:145–147
Laine AL (2011) Context-dependent effects of induced resistance under co-infection in a plant-pathogen interaction. Evol Appl 4:696–707
Lamichhane JR, Venturi V (2015) Synergisms between microbial pathogens in plant disease complexes: a growing trend. Front Plant Sci 6:385
Landry M, Comeau AM, Derome N, Cusson M, Levesque RC (2015) Composition of the spruce budworm (Choristoneura fumiferana) midgut microbiota as affected by rearing conditions. PLoS One 10:e0144077
Leach JE, Triplett LR, Argueso CT, Trivedi P (2017) Communication in the phytobiome. Cell 169: 587–596. https://doi.org/10.1016/j.cell.2017.04.025
Lee MC, Marx CJ (2012) Repeated, selection-driven genome reduction of accessory genes in experimental populations. PLoS Genet 8:e1002651. https://doi.org/10.1371/journal.pgen.1002651
Lemanceau P, Barret M, Mazurier S, Mondy S, Pivato B, Fort T et al (2017a) Plant communication with associated microbiota in the spermosphere rhizosphere and phyllosphere. Adv Botan Res:101–133
Lemanceau P, Blouin M, Muller D, Moënne-Loccoz Y (2017b) Let the core microbiota be functional. Trends Plant Sci 22:583–595
Li CY, Zhou X, Zheng H (2018) Diversity and function of insect gut microbiome. Acta Microbiol Sin 58:1016–1024
Lindow SE, Brandl MT (2003) Microbiology of the phyllosphere. Appl Environ Microbiol 69: 1875–1883. https://doi.org/10.1128/AEM.69.4.1875-1883.2003
Linnakoski R, Forbes KM (2019) Pathogens the hidden face of forest invasions by wood-boring insect pests. Front Plant Sci 10:90. https://doi.org/10.3389/fpls.2019.00090
Lopes KBA, Carpentieri-Pipolo V, Fira D, Balatti PA, López SMY, Oro TH, Pagliosa ES, Degrassi G (2018) Screening of bacterial endophytes as potential biocontrol agents against soybean diseases. J App Microbiol 125 1466–1481. https://doi.org/10.1111/jam.14041
Lynikienė J, Marčiulynienė D, Marčiulynas A, Gedminas A, Vaičiukynė M, Menkis A (2020) Managed and unmanaged *Pinus sylvestris* forest stands harbour similar diversity and composition of the phyllosphere and soil fungi. Microorganisms 8:259
Madden JL (1968) Behavioural responses of parasites to the symbiotic fungus associated with Sirex noctilio. Nature 218:189–190
Manzano-Marín A, Coeur d'acier A, Clamens A-L, Orvain C, Cruaud C, Barbe V et al (2019) Serial horizontal transfer of vitaminbiosynthetic genes enables the establishment of new nutritional symbionts in aphids' di-symbiotic systems. The ISME J 14:259–273
Marcone C (2015) Current status of phytoplasma diseases of forest and landscape trees and shrubs. J Plant Pathol 1:9–36
Maron JL, Crone E (2006) Herbivory: effects on plant abundance, distribution and population growth. Proc Biol Sci 273:2575–2584
Marubayashi JM, Kliot A, Yuki VA, Rezende JAM, Krause-Sakate R, Pavan MA et al (2014) Diversity and localization of bacterial endosymbionts from whitefly species collected in Brazil. PLoS One 9:e108363. https://doi.org/10.1371/journal.pone.0108363
Mason CJ, Pfammatter JA, Holeski LM, Raffa KF (2015) Foliar bacterial communities of trembling aspen in a common garden. Can J Microbiol 61:143–149
Mason CJ, Jones AG, Felton GW (2019a) Co-option of microbial associates by insects and their impact on plant-folivore interactions. Plant Cell Environ 42:1078–1086
Mason CJ, Campbell AM, Scully ED, Hoover K (2019b) Bacterial and fungal midgut community dynamics and transfer between mother and brood in the Asian longhorned beetle (*Anoplophora glabripennis*), an invasive Xylophage. Microb Ecol 77:230–242

Menkis A, Marčiulynas A, Gedminas A, Lynikienė J, Povilaitienė A (2015) High-throughput sequencing reveals drastic changes in fungal communities in the phyllosphere of Norway spruce (*Picea abies*) following invasion of the spruce bud scale (Physokermes piceae). Microb Ecol 70: 904–911

Mergaert P (2018) Role of antimicrobial peptides in controlling symbiotic bacterial populations. Nat Prod Rep 35:336–356. https://doi.org/10.1039/c7np00056a

Messal M, Slippers B, Naidoo S, Bezuidt O, Kemler M (2019) Active fungal communities in asymptomatic *Eucalyptus grandis* stems differ between a susceptible and resistant clone. Microorganisms 7:375

Mikonranta L, Mappes J, Kaukoniitty M, Freitak D (2014) Insect immunity: oral exposure to a bacterial pathogen elicits free radical response and protects from a recurring infection. Front Zool 11:23. https://doi.org/10.1186/1742-9994-11-23

Mina D, Pereira JA, Lino-Neto T et al (2020) Epiphytic and Endophytic bacteria on olive tree Phyllosphere: exploring tissue and cultivar effect. Microb Ecol 80:145–157. https://doi.org/10.1007/s00248-020-01488-8

Misof B, Liu S, Meusemann K, Peters RS, Donath A, Mayer C et al (2014) Phylogenomics resolves the timing and pattern of insect evolution. Science 346:763–767

Mitter B, Pfaffenbichler N, Flavell R, Compant S, Antonielli L, Petric A et al (2017) A new approach to modify plant microbiomes and traits by introducing beneficial bacteria at flowering into progeny seeds. Front Microbiol 8:1–11

Moran NA, McLaughlin HJ, Sorek R (2009) The dynamics and time scale of ongoing genomic erosion in symbiotic bacteria. Science 323:379–382. https://doi.org/10.1126/science.1167140

Moreira X, Abdala-Roberts L, Castagneyrol B (2018) Interactions between plant defence signalling pathways: evidence from bioassays with insect herbivores and plant pathogens. J Ecol 106: 2353–2364

Morella NM, Zhang X, Koskella B (2019) Tomato seed-associated bacteria confer protection of seedlings against foliar disease caused by *Pseudomonas syringae*. Phytobiomes J 3:177–190

Morris JJ, Lenski RE, Zinser ER (2012) The black queen hypothesis: evolution of dependencies through adaptive gene loss. mBio 3:e00036–e00012. https://doi.org/10.1128/mBio.00036-12

Müller T, Müller M, Behrendt U, Stadler B (2003) Diversity of culturable phyllosphere bacteria on beech and oak: the effects of lepidopterous larvae. Microbiol Res 158:291–297

Müller H, Berg C, Landa BB, Auerbach A, Moissl-Eichinger C, Berg G (2015) Plant genotype-specific archaeal and bacterial endophytes but similar *Bacillus* antagonists colonize Mediterranean olive trees. Front Microbiol 6:138. https://doi.org/10.3389/fmicb.2015.00138

Mülner P, Bergna A, Wagner P, Sarajlić D, Gstöttenmayr B, Dietel K et al (2019) Microbiota associated with sclerotia of soilborne fungal pathogens a novel source of biocontrol agents producing bioactive volatiles. Phytob J 3:125–136

Nadarasah G, Stavrinides J (2011) Insects as alternative hosts for phytopathogenic bacteria. FEMS Microbiol Rev 35:555–575. https://doi.org/10.1111/j.1574-6976.2011.00264.x

Nelson R, Wiesner-Hanks T, Wisser R, Balint-Kurti P (2017) Navigating complexity to breed disease-resistant crops. Nat Rev Gen 19:21–33

Newton AC, Gravouil C, Fountaine JM (2010) Managing the ecology of foliar pathogens: ecological tolerance in crops. Ann Appl Biol 157:343–359. https://doi.org/10.1111/j.1744-7348.2010.00437.x

Ng SH, Stat M, Bunce M, Simmons LW (2018) The influence of diet and environment on the gut microbial community of field crickets. Ecol Evol 8:4704–4720. https://doi.org/10.1002/ece3.3977

Nguyen D, Boberg J, Ihrmark K, Stenström E, Stenlid J (2016) Do foliar fungal communities of Norway spruce shift along a tree species diversity gradient in mature European forests? Fungal Ecol 23:97–108

Nguyen D, Boberg J, Cleary M, Bruelheide H, Hönig L, Koricheva J et al (2017) Foliar fungi of Betula pendula: impact of tree species mixtures and assessment methods. Sci Rep 7

Nicks T, Rahn-Lee L (2017) Inside out: archaeal ectosymbionts suggest a second model of reduced-genome evolution. Front Microbiol 8:384. https://doi.org/10.3389/fmicb.2017.00384
Nisa H, Kamili AN, Nawchoo IA, Shafi S, Shameem N, Bandh SA (2015) Fungal endophytes as prolific source of phytochemicals and other bioactive natural products: a review. Microb Pathog 82:50–59
O'Brien PA (2017) Biological control of plant diseases. Australas Plant Pathol 46:293–304. https://doi.org/10.1007/s13313-017-0481-4
Ojha A, Sinha DK, Padmakumari AP, Bentur JS, Nair S (2017) Bacterial community structure in the Asian rice gall midge reveals a varied microbiome rich in proteobacteria. Sci Rep 7:9424
Oliveira RJV (2014) Endophytic fungal diversity in coffee leaves (*Coffea arabica*) cultivated using organic and conventional crop management systems. Mycosphere 5:523–530
Orozco-Mosqueda MDC, Rocha-Granados MDC, Glick BR, Santoyo G (2018) Microbiome engineering to improve biocontrol and plant growth-promoting mechanisms. Microbiol Res 208:25–31
Ottesen AR, White JR, Skaltsas DN, Newell MJ, Walsh CS (2009) Impact of organic and conventional management on the phyllosphere microbial ecology of an apple crop. J Food Protect 72:2321–2325
Pangesti N, Pineda A, Pieterse CMJ, Dicke M, van Loon JJA (2013) Two-way plant mediated interactions between root-associated microbes and insects: from ecology to mechanisms. Front Plant Sci 4:414
Pascazio S, Crecchio C, Ricciuti P, Palese AM, Xiloyannis C, Sofo A (2015) Phyllosphere and carposphere bacterial communities in olive plants subjected to different cultural practices. Int J Plant Biol 6:6011
Pec GJ, Karst J, Taylor DL, Cigan PW, Erbilgin N, Cooke JEK et al (2016) Change in soil fungal community structure driven by a decline in ectomycorrhizal fungi following a mountain pine beetle (*Dendroctonus ponderosae*) outbreak. New Phytol 213:864–873
Pérez-Jaramillo JE, Mendes R, Raaijmakers JM (2015) Impact of plant domestication on rhizosphere microbiome assembly and functions. Plant Mol Biol 90:635–644
Perilla-Henao LM, Casteel CL (2016) Vector-borne bacterial plant pathogens: interactions with hemipteran insects and plants. Front Plant Sci 7:1163. https://doi.org/10.3389/fpls.2016.01163
Pieterse CMJ, Zamioudis C, Berendsen RL, Weller DM, Wees SCMV, Bakker PAHM (2014) Induced systemic resistance by beneficial microbes. Annu Rev Phytopathol 52:347–375
Pineda A, Zheng S-J, van Loon JJA, Pieterse CMJ, Dicke M (2010) Helping plants to deal with insects: the role of beneficial soilborne microbes. Trends Plant Sci 15:507–514
Pineda A, Kaplan I, Bezemer TM (2017) Steering soil microbiomes to suppress aboveground insect pests. Trends Plant Sci 22:770–778
Pineda A, Kaplan I, Hannula SE, Ghanem W, Bezemer TM (2020) Conditioning the soil microbiome through plant-soil feedbacks suppresses an aboveground insect pest. New Phytol 226:595–608
Poudel R, Jumpponen A, Schlatter DC, Paulitz TC, Gardener BBM, Kinkel LL et al (2016) Microbiome networks: a systems framework for identifying candidate microbial assemblages for disease management. Phytopathol 106:1083–1096
Qi F, Jing T, Zhan Y (2012) Characterization of endophytic fungi from Acer ginnala Maxim. in an artificial plantation: media effect and tissue-dependent variation. PLoS One 7:e46785
Rennison DJ, Rudman SM, Schluter D (2019) Parallel changes in gut microbiome composition and function in parallel local adaptation and speciation. bioRxiv. https://doi.org/10.1101/736843
Rivas-Franco F, Hampton JG, Morán-Diez ME, Narciso J, Rostás M, Wessman P et al (2019) Effect of coating maize seed with Entomopathogenic fungi on plant growth and resistance against *Fusarium graminearum* and *Costelytra giveni*. Biocontrol Sci Tech 29:877–900
Robertson HM (2005) Insect genomes. Am Entomol 51:166–173
Rohart F, Gautier B, Singh A, Cao K-AL (2017) mixOmics: an R package for -omics feature selection and multiple data integration. PLoS Comput Biol 13:e1005752

Russell CW, Bouvaine S, Newell PD, Douglas AE (2013) Shared metabolic pathways in a coevolved insect-bacterial symbiosis. Appl Environ Microbiol 79:6117–6123. https://doi.org/10.1128/aem.01543-13

Sabater-Muñoz B, Toft C, Alvarez-Ponce D, Fares MA (2017) Chance and necessity in the genome evolution of endosymbiotic bacteria of insects. ISME J 11:1291–1304. https://doi.org/10.1038/ismej.2017.18

Sanders D, Kehoe R, van Veen FJF, McLean A, Godfray HCJ, Dicke M et al (2016) Defensive insect symbiont leads to cascading extinctions and community collapse. Ecol Lett 19:789–799

Santini A, Battisti A (2019) Complex insect pathogen interactions in tree pandemics. Front Physiol 10:3389. https://doi.org/10.3389/fphys.2019.00550

Saqib HSA, Sun L, Pozsgai G, Liang P, Goraya MU, Akutse KS, You M, Gurr GM, You S (2023) Gut microbiota assemblages of generalist predators are driven by local- and landscape-scale factors. Front Microbiol 14:1172184. https://doi.org/10.3389/fmicb.2023.1172184

Saravesi K, Aikio S, Wäli PR, Ruotsalainen AL, Kaukonen M, Huusko K et al (2015) Moth outbreaks alter root-associated fungal communities in subarctic mountain birch forests. Microb Ecol 69:788–797

Schubert I, Vu GTH (2016) Genome stability and evolution: attempting a holistic view. Trends Plant Sci 21:749–757. https://doi.org/10.1016/j.tplants.2016.06.003

Schuldt A, Bruelheide H, Buscot F, Assmann T, Erfmeier A, Klein AM et al (2017) Belowground top-down and aboveground bottom-up effects structure multitrophic community relationships in a biodiverse forest. Sci Rep 7:4222

Shade A, Jacques M-A, Barret M (2017) Ecological patterns of seed microbiome diversity transmission, and assembly. Curr Opin Microbiol 37:15–22

Shrestha S (2019) Effects of climate change in agricultural insect pest. Acta Sci Agric 3:74–80. https://doi.org/10.31080/asag.2019.03.0727

Shugart HH, Foster A, Wang B, Druckenbrod D, Ma J, Lerdau M et al (2020) Gap models across micro- to mega-scales of time and space: examples of Tansley's ecosystem concept. Forest Ecosyst 7. https://doi.org/10.1186/s40663-020-00225-4

Six DL (2012) Ecological and evolutionary determinants of bark beetle fungus symbioses. Insects 3:339–366

Six DL (2013) The bark beetle holobiont: why microbes matter. J Chem Ecol 39:989–1002

Skelton J, Johnson AJ, Jusino MA, Bateman CC, Li Y, Hulcr J (2019) A selective fungal transport organ (mycangium) maintains coarse phylogenetic congruence between fungus-farming ambrosia beetles and their symbionts. Proc Royal Soc B: Biol Sci 286:20182127

Smets W, Koskella B (2020) Microbiome: insect herbivory drives plant phyllosphere dysbiosis. Curr Biol 30:412–414

Stadler B, Michalzik B (2000) Effects of phytophagous insects onmicroorganisms and throughfall chemistry in forested ecosystems: herbivores as switches for the nutrient dynamics in the canopy. Basic Appl Ecol 1:109–116

Stam JM, Kroes A, Li Y et al (2014) Plant interactions with multiple insect herbivores: from community to genes. Annu Rev Plant Biol 65:689–713

Strano CP, Malacrinò A, Campolo O, Palmeri V (2017) Influence of host plant on Thaumetopoea pityocampa gut bacterial community. Microb Ecol 75:487–494

Strullu-Derrien C, Selosse M-A, Kenrick P, Martin FM (2018) The origin and evolution of mycorrhizal symbioses: from palaeomycology to phylogenomics. New Phytol 220:1012–1030

Sullivan BT, Berisford CW (2004) Semiochemicals from fungal associates of bark beetles may mediate host location behavior of parasitoids. J Chem Ecol 30:703–717

Susi H, Barrès B, Vale PF, Laine AL (2015) Co-infection alters population dynamics of infectious disease. Nat Commun 6:5975

Tack AJM, Gripenberg S, Roslin T (2012) Cross-kingdom interactions matter: fungal-mediated interactions structure an insect community on oak. Ecol Lett 15:177–185

Tailor AJ, Joshi BH (2014) Harnessing plant growth promoting rhizobacteria beyond nature: a review. J Plant Nutr 37:1534–1571

Tao L, Ahmad A, de Roode JC, Hunter MD (2016) Arbuscular mycorrhizal fungi affect plant tolerance and chemical defences to herbivory through different mechanisms. J Ecol 104:561–571

Thaler JS, Fidantsef AL, Duffey SS, Bostock RM (1999) Trade-offs in plant defense against pathogens and herbivores: a field demonstration of chemical elicitors of induced resistance. J Chem Ecol 25:1597–1609

Thaler JS, Humphrey PT, Whiteman NK (2012) Evolution of jasmonate and salicylate signal crosstalk. Trends Plant Sci 17:260–270

Theis KR, Dheilly NM, Klassen JL, Brucker RM, Baines JF, Bosch TCG et al (2016) Getting the hologenome concept right: an eco-evolutionary framework for hosts and their microbiomes. mSystems 1:e00028–e00016

Thomas P, Kenny N, Eyles D, Moreira LA, O'Neill SL, Asgari S (2011) Infection with the wMel and wMelPop strains of Wolbachia leads to higher levels of melanization in the hemolymph of *Drosophila melanogaster*, *Drosophila simulans* and *Aedes aegypti*. Dev Comp Immunol 35: 360–365. https://doi.org/10.1016/j.dci.2010.11.007

Thompson SNN, Simpson SJJ (2009) Nutrition. In: Resh VH, Cardé RT (eds) Encyclopedia of insects. Academic Press, Cambridge, MA, pp 715–720

Tilman D (1977) Resource competition between plankton algae: an experimental and theoretical approach. Ecology 58:338–348. https://doi.org/10.2307/1935608

Toft C, Fares MA (2008) The evolution of the flagellar assembly pathway in endosymbiotic bacterial genomes. Mol Biol Evol 25:2069–2076. https://doi.org/10.1093/molbev/msn153

Toju H, Peay KG, Yamamichi M, Narisawa K, Hiruma K, Naito K et al (2018) Core microbiomes for sustainable agroecosystems. Nat Plant 4:247–257

Tollenaere C, Susi H, Laine AL (2016) Evolutionary and epidemiological implications of multiple infection in plants. Trends Plant Sci 21:80–90

Treu R, Karst J, Randall M, Pec GJ, Cigan PW, Simard SW et al (2014) Decline of ectomycorrhizal fungi following amountain pine beetle epidemic. Ecology 95:1096–1103

Tyc O, Putra R, Gols R, Harvey JA, Garbeva P (2019) The ecological role of bacterial seed endophytes associated with wild cabbage in UK. Microbiol Open 9:e00954

Urbanová M, Šnajdr J, Baldrian P (2015) Composition of fungal and bacterial communities in forest litter and soil is largely determined by dominant trees. Soil Biol Biochem 84:53–64

Vacher C, Hampe A, Porté AJ, Sauer U, Compant S, Morris CE (2016a) The phyllosphere: microbial jungle at the plant-climate interface. Ann Rev Ecol Evol Syst 47:1–24

Vacher C, Tamaddoni-Nezhad A, Kamenova S, Peyrard N, Moalic Y, Sabbadin R et al (2016b) Learning ecological networks from next-generation sequencing data. In: Ecosystem services: from biodiversity to society, Part 2. Elsevier, pp 1–39

Vacher C, Castagneyrol B, Jousselin E, Schimann H (2021) Trees and insects have microbiomes: consequences for forest. Curr For Rep 7:81–96

Vandenkoornhuyse P, Quaiser A, Duhamel M, Le VA, Dufresne A (2015) The importance of the microbiome of the plant holobiont. New Phytol 206:1196–1206

Vanderpool D, Bracewell RR, McCutcheon JP (2017) Know your farmer: ancient origins and multiple independent domestications of ambrosia beetle fungal cultivars. Mol Ecol 27:2077–2094

Vannier N, Mony C, Bittebiere A-K, Michon-Coudouel S, Biget M, Vandenkoornhuyse P (2018) A microorganisms' journey between plant generations. Microbiome 6:79

Vannier N, Agler M, Hacquard S (2019) Microbiota-mediated disease resistance in plants. In: Zipfel C (ed) PLOS pathogens, vol 15, p e1007740

Vega FE (2018) The use of fungal entomopathogens as endophytes in biological control: a review. Mycologia 110:4–30

Vicente CSL, Nascimento FX, Espada M, Barbosa P, Hasegawa K, Mota M et al (2013) Characterization of bacterial communities associated with the pine sawyer beetle *Monochamus galloprovincialis* the insect vector of the pinewood nematode Bursaphelenchus xylophilus. FEMS Microbiol Lett 347:130–139

Vivas M, Kemler M, Slippers B (2015) Maternal effects on tree phenotypes: considering the microbiome. Trends Plant Sci 20:541–544
Vivas M, Kemler M, Mphahlele MM, Wingfield MJ, Slippers B (2017) Maternal effects on phenotype resistance and the structuring of fungal communities in *Eucalyptus grandis*. Environ Exp Bot 140:120–127
Vivas M, Wingfield MJ, Slippers B (2020) Maternal effects should be considered in the establishment of forestry plantations. Forest Ecol Manag 460:117909
van der Voort M, Kempenaar M, van Driel M, Raaijmakers JM, Mendes R (2016) Impact of soil heat on reassembly of bacterial communities in the rhizosphere microbiome and plant disease suppression. Ecol Lett 19:375–382
Wagner MR, Lundberg DS, Coleman-Derr D, Tringe SG, Dangl JL, Mitchell-Olds T (2014) Natural soil microbes alter flowering phenology and the intensity of selection on flowering time in a wild *Arabidopsis* relative. Ecol Lett 17:717–726
Wagner MR, Lundberg DS, del Rio TG, Tringe SG, Dangl JL, Mitchell-Olds T (2016) Host genotype and age shape the leaf and root microbiomes of a wild perennial plant. Nat Commun 7:12151
War AR, Paulraj MG, Ahmad T, Buhroo AA, Hussain B, Ignacimuthu S et al (2012) Mechanisms of plant defense against insect herbivores. Plant Signal Behav 7:1306–1320
Watanabe H, Tokuda G (2010) Cellulolytic systems in insects. Annu Rev Entomol 55:609–632
Weese DJ, Heath KD, Dentinger BTM, Lau JA (2015) Long-term nitrogen addition causes the evolution of less-cooperative mutualists. Evolution 69:631–642
Wei Z, Gu Y, Friman V-P, Kowalchuk GA, Xu Y, Shen Q et al (2019) Initial soil microbiome composition and functioning predetermine future plant health. Sci Adv 5:eaaw0759
Wemheuer F, Berkelmann D, Wemheuer B, Daniel R, Vidal S, Daghela HBB (2020) Agroforestry management systems drive the composition diversity, and function of fungal and bacterial endophyte communities in *Theobroma cacao* leaves. Microorganisms 8:405
Wielkopolan B, Obrêpalska-Stêplowska A (2016) Three-way interaction among plants, bacteria, and coleopteran insects. Planta 244:313–332. https://doi.org/10.1007/s00425-016-2543-1
Wilkinson SW, Magerøy MH, Sánchez AL, Smith LM, Furci L, Cotton TEA, Krokene P, Ton J (2019) Surviving in a hostile world: plant strategies to resist pests and diseases. Annu Rev Phytopathol 57:505–529
Wingfield MJ, Garnas JR, Hajek A, Hurley BP, de Beer ZW, Taerum SJ (2016) Novel and co-evolved associations between insects and microorganisms as drivers of forest pestilence. Biol Invasions 18:1045–1056
Witzell J, Martín JA (2018) Endophytes and forest health. In: Pirttilä AM, Frank AC (eds) Endophytes of forest trees. Springer International Publishing, Cham, pp 261–282
World Resources Institute [WRI] (2019) Creating a sustainable food future—a menu of solutions to feed nearly 10 billion people by 2050. Available online at: http://wrr-food.wri.org/sites/default/files/2019-07/WRR_Food_Full_Report_0.pdf
Wu Q, Patočka J, Kuča K (2018) Insect antimicrobial peptides, a mini review. Toxins 10:461. https://doi.org/10.3390/toxins10110461
Wubs ERJ, van der Putten WH, Bosch M, Bezemer TM (2016) Soil inoculation steers restoration of terrestrial ecosystems. Nat Plant 2016:2
Xie S, Vallet M, Sun C, Kunert M, David A, Zhang X, Chen B, Lu X, Boland W, Shao Y (2020) Biocontrol potential of a novel endophytic bacterium from Mulberry (Morus) tree. Front Bioeng Biotechnol 7:488. https://doi.org/10.3389/fbioe.2019.00488
Xing H-Q, Ma J-C, Xu B-L, Zhang S-W, Wang J, Cao L et al (2018) Mycobiota of maize seeds revealed by rDNA -ITS sequence analysis of samples with varying storage times. Microbiol Open 7:e00609
Yuval B (2017) Symbiosis: gut bacteria manipulate host behaviour. Curr Biol 27:R746–R747

# Chapter 2
# Tri-trophic Interactions: The Role of the Plant and Insect Microbiome in Shaping Pest-Parasitoid Dynamics

**Mahmoud Saleh, Nabil El-Wakeil, and Nawal Gaafar**

## 2.1 Introduction

The mounting demand for sustainable food production, coupled with the severe environmental and resistance issues posed by chemical insecticides, necessitates a paradigm shift towards harnessing the power of the microbiome as a novel, ecological result for pest management (Thakur et al. 2023; Nebo et al. 2025). One of the important microbial societies that resides either in the endosphere) or in the episphere, is called the plant microbiome or microbiota (Bosmans et al. 2018; Shao et al. 2024). According to Whipps et al. (1988) and Dastogeer et al. (2020), the plant microbiome has an influence on the physiology and ecology of plants, which in turn affects plant health and their production.

The plant microbiome is supposed to have evolved through natural selection processes, and critical genes impact the plant-microbiome association, also known as the plant holobiont (Compant et al. 2019). Factors affecting the plant, such as its species, organ, and health state, as well as ecological factors, such as climate conditions and different uses of land, influence the microbiome structures (Berendsen et al. 2012; Martínez-Solís et al. 2020). Biological pest management methods can benefit from plant microorganisms for two reasons: first, they boost plant health and productivity. Second, they increase plant defensive responses

M. Saleh (✉) · N. Gaafar
Pests & Plant Protection Department, Agricultural & Biological Sciences Institute, NRC, Cairo, Egypt

N. El-Wakeil
Department of Arid Land Agriculture, College of Agricultural & Food Sciences, King Faisal University, Al-Ahsa, Saudi Arabia
e-mail: nelwakeil@kfu.edu.sa

 
N. El-Wakeil, A. Abdelkhalek (eds.), *Microbiomes and Plant Secondary Metabolites for Insect Pest Management*, Progress in Biological Control 23,
https://doi.org/10.1007/978-3-032-24330-0_2

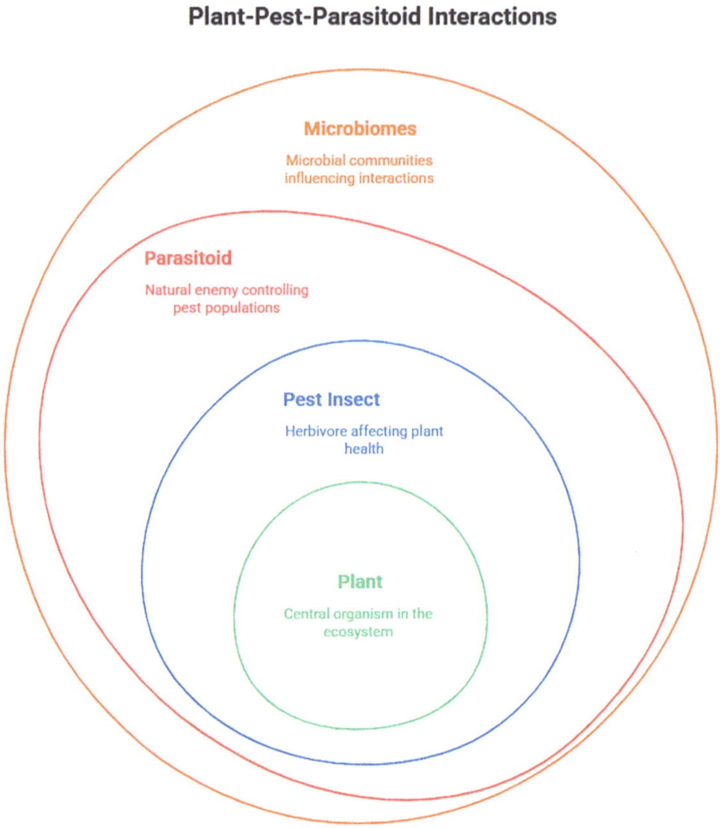


**Fig. 2.1** A diagram clarifying the multi-channel interactions among tritrophic levels (plant, insect, parasitoid) and their particular microbiomes

against insect pests' infestation. These microorganisms have the potential to yield antibiotics and chemicals that suppress the insect pests.

Censoriously, these defensive responses outspread beyond direct toxicity to include the cultured recruitment of the plant's tritrophic (third trophic) level: biocontrol agents of insect herbivores. This positions the plant microbiome as a central regulator of tri-trophic interactions, the dynamic interplay between the first trophic level (plants), the second trophic level (insect pests), and the third trophic level (parasitoids or predators). By facilitating communication and nutrition across these different levels, microbes primarily shape pest-parasitoid dynamics and eventually, the success of biocontrol tools (Rashid and Chung 2017; Pang et al. 2020).

The plant microbiome influences the production of plant volatiles (semiochemicals) and other secondary metabolites. These microbial-induced compounds are crucial for attracting the natural enemies of herbivores, thereby directly supporting the claim that the microbiome is a central regulator of tri-trophic interactions (Harvey et al. 2013; Pang et al. 2020) (Fig. 2.1). These exchanges are mediated by specific signaling molecules. For example, plants distinguish conserved Pathogen-Associated Molecular Patterns (PAMPs) and Herbivore-Associated Molecular Patterns (HAMPs) found in insect saliva, released from wounded tissues, all of which activate defense cascades (Turlings and Erb 2018). DAMPs (Damage-Associated Molecular Patterns): Plant-derived molecules released from damaged tissues that signal "self-danger" and stimulate protective responses (Turlings and Erb 2018).

### 2.1.1 *Role of Microorganisms in Enhancing Plant Defense and Biological Pest Control*

Microorganisms that inhabit bloomy nectar and insect honeydews can harvest volatile composites that attract some kinds of natural enemies, presenting an indirect biocontrol of insect pests. Using microbes for indirect pest management can lead to long-term biological control, especially when harmonizing with other biocontrol agents like predator insects and parasitoid wasps, according to Jaber and Araj (2018), Jaber and Alananbeh (2018), Cusumano et al. (2020). Therefore, it is vital to utilize every chance that microorganisms present to increase the plant defense. Pieterse et al. (2012) and El-Wakeil 2003, El-Wakeil et al. (2010) found that plant microorganisms can contribute to the plant's defense mechanism against insect pest infestation, namely induced systemic resistance (ISR), through different pathways such as jasmonic acid, salicylic acid, and ethylene. It was confirmed by Pineda et al. (2010) that root-associated bacteria are the ones responsible for jasmonic acid production, which is a principal elicitor of ISR. The insect's reaction to plant microorganisms could be determined by its specialization degree; for example, *Spodoptera exigua*, a generalist insect, was harmfully impacted by poisonous and deterrent metabolites, while *Pieris rapae*, a specialist insect, remained unaffected (Schaeffer et al. 2014; Van Oosten et al. 2008).

The rhizosphere of one gram of soil might theoretically comprise more than one million bacterial genomes (Gans et al. 2005; Talia et al. 2012) and a high assortment of operational taxonomic entities (Curtis et al. 2002; Curtis and Sloan 2005). Mycorrhizal fungi, associated with 80% of all plants, play a dynamic function in earthly ecology by governing the cycles of carbon and nutrients, allowing plants to accomplish a significant portion of their nitrogen and phosphate necessities. Soil microbial mutualists, including mycorrhizal and non-mycorrhizal fungi, and plant growth-promoting bacteria like *Pseudomonas fluorescens* (Battaglia et al. 2013), play a crucial role in plant defense against many insect pests both above and below the soil surface (Rasmann et al. 2017).

The most well-documented mechanism is the role of plant-associated microbes in altering the smellscape of an infested plant, making it more apparent to foraging parasitoids. For example, infection of maize plants with the rhizobacterium *Rhizobium etli* enhances the emission of induced volatile organic compounds (VOCs) following herbivory, which significantly increases the attraction of the generalist parasitoid *Cotesia marginiventris* (D'alessandro et al. 2014). Similarly, root colonization by beneficial fungi like *Trichoderma* can systemically prime the plant's volatile emissions, leading to improved recruitment of parasitoids like *Aphidius ervi* to aphid-infested beans (Battaglia et al. 2013).

While the ecological role of microbial volatiles is well-established, the specific genetic and biochemical pathways responsible for their production are an area of active discovery. In bacteria, many key volatiles are byproducts of primary metabolic pathways. For instance, ammonia is generated through amino acid deamination and urea hydrolysis, while many sulfur-containing compounds are derived from the metabolism of methionine and cysteine. In the case of the attractant dimethyl disulfide, a common parasitoid attractant, specific methanethiol oxidases and dimethyl sulfide-forming enzymes in bacteria like *Pseudomonas* are responsible (Stone et al. 2010; Rering et al. 2018). In yeasts, the signature fermentation products ethanol and acetaldehyde are generated via the pyruvate decarboxylase and alcohol dehydrogenase pathways, while fruity esters like ethyl acetate are synthesized by alcohol acetyltransferases that catalyze the condensation of alcohols and acyl-CoA derivatives. The specific manifestation and regulation of these pathway genes are highly dependent on the ecological context, such as nutrient accessibility and pH, which finally shapes the volatile blend perceived by insects (Schmidt et al. 2015; Rering et al. 2018).

Beyond indirect chemical prompts, microbes can directly impact parasitoid fecundity and longevity by modifying the nutritional superiority of the host insect. For instance, the larval ectoparasitoid *Habrobracon hebetor* exhibits increased longevity and fecundity when rearing on *Plodia interpunctella* larvae that possess a gut microbiome, compared to those nurtured on antibiotic-treated, microbiome-depleted hosts (Jones et al. 2020). Moreover, some parasitoids have forged intimate alliances with symbiotic viruses, known as Polydnaviruses (PDVs), which are combined into the parasitoid's genome and expressed in the ovary. When co-injected with the parasitoid egg into the host, these viral elements destroy the host's immune system, defending the hatching and emerging larva (Beckage and Gelman 2004; Strand 2014).

### *2.1.2 Plant-Microbe Interactions in Pest Control: Mechanisms of Induced Resistance and Parasitoid Attraction*

The association among plants and rhizosphere bacteria can be characterized as follows: (1) Microbes enhance plant growth and vitality, which may increase the nutritional content of plants and support the biological performance of insects. (2) Soil microbes have the potential to generate constituents that are venomous and repellent against the insect pests, leading to the development of plant resistance or an enhancement in the plant's defense mechanisms. The outcome of such positive and negative effects ultimately regulates insect performance (Francis et al. 2020). (3) Soil bacteria can modify volatile organic chemicals released from plants and inhibit plant signaling (Rassmann and Schmidt 2017). For example, treatment of *Arabidopsis* with suppressed the emission of aromatic compounds such as salicylate and terpenes, but enhanced plant defense by attracting a large number of parasitoid wasps (Kessler and Baldwin 2010).

Microbes populating the phyllosphere reside on the airborne surfaces of plants. This bacterial community is highly susceptible to ecological fluctuations and is vital for preserving plant health, function, and evolution (Friesen et al. 2011). While principally composed of bacteria, the phyllosphere also hosts a range of other microorganisms, such as fungi, viruses, algae, and archaea (Vorholt 2012). These microbes play important roles in encouraging plant growth and providing defensive mechanisms against insect pests as well as plant diseases (Saleem et al. 2017). Applying the leaf-colonizing yeast *Sporonolomyces roseus* to maize leaves was shown to prevent oviposition by *Ostrinia nubilalis*. Furthermore, inoculating leaves with soil-derived microorganisms can change the leaf microbiome and increase plant resistance against many pests (Badri 2013). Some of entomopathogenic fungi, similar to *Beauveria bassiana* and *Metarhizium anisopliae,* can also exhibit an endophytic phase that helps plants defend against insect pests.

Endosphere microorganisms populate plant tissues and directly impact plant health and its production, primarily consisting of fungi and bacteria (Vokou et al. 2012). Entomopathogenic fungi can colonize plant tissues and enhance plant defense against insect pests (Saikkonen et al. 2006). Secondary metabolites created by endosphere microorganisms, which involve deterrent as well as poisonous constituents for insects, are largely responsible for this heightened resistance (Vega 2008). Endophytic entomopathogenic fungi can also guard plants from serious diseases, encourage plant growth, elevate the transport of soil nutrients, and increase tolerance to abiotic stress (Bamisile et al. 2018). However, data on the impact of endophytes on parasitoid wasps is limited (Urrutia et al. 2007; Muvea et al. 2014; Fuchs and Krauss 2019). Additionally, endophytes can effect higher trophic levels; for example, hyperparasitoids growing in aphids that were nurtured on endophyte-infected grass (*Lolium perenne*) had a shortened lifetime (Harri et al. 2008; Gao et al. 2021).

### 2.1.3 *Elicitors and Effectors in Plant-Microbe and Plant-Insect Interactions*

In plant-microbe and plant-insect relations, numerous elicitors induce plant signaling cascades that result in gene expression and plant responses (Zhang et al. 2022, 2023). Effectors contain proteins and minor compounds produced in response to pest attacks, which alter the structure and function of plant cells, causing the release of some volatiles. The molecules linked to pathogens are referred to as pathogen-associated molecular patterns (PAMPs), while those linked to insect herbivores are herbivore-associated molecular patterns (HAMPs). HAMPs are chemicals derived from insect pests that interact with plants during feeding, including mouth excretions such as regurgitant and saliva (Pang et al. 2023). The caterpillar saliva of *Heliothis zea* contains glucose oxidase (GOX) synthesized by the insect's symbiotic bacteria *Enterobacter ludwigii*. Injection of GOX into tomato plants improved their cautious responses to insect pests' attacks, indicating that symbiotic bacteria indirectly simplified the plant-insect relationship by stimulating salivary elicitors (Wang et al. 2017, 2020, 2023). Pectinase and polyphenol oxidase found in the saliva of *Sitobion avenae* elicited volatile emissions in wheat that attracted the parasitoid wasps of aphid (Tariq et al. 2013). Volatiles formed as a product of pest attacks are termed herbivore-induced plant volatiles (HIPVs), which are very selective and operative in attracting biocontrol agents (Dicke and Baldwin 2010). Injury-associated molecular patterns (DAMPs), which are plant-derived chemicals signaling tissue damage, were also recognized as elicitors of plant guard (Turlings and Erb 2018).

### 2.1.4 Nectar and Honeydew Microbes

Nectar from flowers and honeydew produced by sap-feeding insects are exceptional sugar sources for parasitoid wasps and provide a helpful atmosphere for microbial evolution. These nutrient solutions host a varied microorganism communities, primarily yeasts and bacteria, which are modified to high osmotic pressure and an imbalanced carbon-to-nitrogen ratio (van Borm et al. 2002; Pozo et al. 2015; Lievens et al. 2015). These microbes are transmitted among flowers primarily via insect vectors, as mentioned by Vannette and Fukami (2017). Microbial metabolism can meaningfully change nectar substances, including their sugar composition, concentration, acidity, and ethanol content. This motion can even raise nectar temperature (Herrera and Pozo 2010) and produce volatiles or fermentation byproducts, thereby modifying the nectar's flavor and aroma profile (Rering et al. 2018). These microbial-mediated changes directly impact plant-insect interactions. For instance, nectar-inhabiting yeasts can adjust the volatile blends, enhancing attraction to parasitoid wasps (Herrera et al. 2013). Similarly, certain nectar bacteria have been revealed to increase the survival percentages of aphid parasitoids such as *Aphidius* spp. (Lenaerts et al. 2017).

### 2.1.5 *Gut Microbial Diversity and Functional Potential in Parasitoid Wasps*

The gut microbiota plays a significant role in insects' digestion, development, and resistance to infections (Jing et al. 2020). Studies on parasitoid wasps like *Diadromus collaris* and *Cotesia vestalis* have helped characterize the diversity and function of their gut microbial communities (Schmidt and Engel 2021). The phyla Proteobacteria and Firmicutes constitute the bulk of the bacteria discovered in the majority of parasitoid wasps (Dicke et al. 2020; Gwokyalya et al. 2024).

The genus *Enterococcus* is prevalent in many insects' digestive tracts and is a dominant gut bacterium in both *C. vestalis* and *D. collaris*. *Enterococcus* may help parasitoids metabolize carbs and amino acids, suggesting it could act as a probiotic, improving their health and growth (Ueda et al. 2009). For instance, *Enterococcus mundtii*, isolated from *Plutella xylostella's* midgut, utilized several carbon compounds (Zhao et al. 2014; Lin et al. 2015; Li et al. 2021; Yang et al. 2021; Unban et al. 2022), and *Enterococcus casseliflavus* from *Bombyx mori* improved L-tryptophan synthesis (Liang et al. 2022).

In *C. vestalis*, the genus *Rhizobium* is also found in relatively high proportions. While known for promoting plant growth, *Rhizobium* in the insect gut appears to play a crucial part in the biodegradation and metabolism of xenobiotics, combating poisonous phytochemicals derived from the diet (Hu et al. 2024). Conversely, the larval intestine of *D. collaris* regularly contains species of *Serratia*, such as *Serratia marcescens*. *Serratia* was found to have considerable positive associations with the immune system in *D. collaris* and may be involved in lipid metabolism, possibly linked to its adaptation to the lipid-rich pupae of its host (Hu et al. 2024).

The gut microbiota performs various functions in food digestion, including the digestion of lipids, proteins, and carbohydrates. The gut microbiota of *C. vestalis* showed an enrichment in functions associated with the metabolism of carbohydrates, amino acids, and energy, reflecting its nutrition from host hemolymph. In contrast, *D. collaris*, which grows in the host pupa and preys on the caterpillar, presented greater activity in functions like cell formation and death, biodegradation of xenobiotics, and metabolic processes (Hu et al. 2024). These differences likely reflect the distinct nutritional conformation and microbial atmosphere of their particular host phases (larval vs. pupal).

### 2.1.6 *Gut Microbiota and Chitin Metabolism in Parasitoid Wasps*

Insect growth depends on chitin in midgut cells and peritrophic atmospheres (Moussian 2010). The gut microbiota is crucial for creating structurally integrated peritrophic matrices, as confirmed by Zha et al. (2021). Hu et al. (2024) found that parasitoid wasp gut microbiota contribute a majority of CAZymes family genes involved in extinguishing and synthesizing chitin. For instance, the most abundant CAZyme family in *C. vestalis*, GT2, was related to bacterial chitin synthases (Fernández et al. 2019). In *D. collaris*, the most abundant CAZyme genes were linked to GT47, which has chitin synthase homology. The gut commensal microbiota is also involved in the immune defense response, helping to attack foreign pathogens (Zeng et al. 2022).

### *2.1.7 Host-Parasitoid Microbiome Dynamics*

The bacterial microbiome of caterpillars such as *Pieris brassicae* and *P. rapae* is primarily made up of Pseudomonadota (Proteobacteria), Bacillota (Firmicutes), and Actinomycetota (Actinobacteria) (Flury et al. 2016; Teoh et al. 2021; Gloder et al. 2021, 2024). The microbiomes of laboratory-reared caterpillars were largely dominated by a specific species of *Enterococcus*. The results also reveal that the parasitoid larvae extracted from the host worms possess their own associated microbiota (Sedláček et al. 2013; Ramirez-Ahuja et al. 2019), with specific parallels to the internal microbiome of the caterpillar host, indicating a potential connectivity between both microbiomes.

### *2.1.8 Host-Specific Effects of Cotesia Parasitism on Caterpillar Microbial Communities*

Parasitism by *Cotesia* parasitoids significantly alters the internal microbial community of the host caterpillar, and these effects are highly host-dependent (Rattan et al. 2011; Nedoluzhko et al. 2017; Park et al. 2019). For example, the occurrence of the dominant *Enterococcus* zOTU was highly reliant on the host; its relative abundance was much lower in parasitized *P. rapae* compared to nonparasitized caterpillars, but remained high in parasitized *P. brassicae* (Chen et al. 2022; Gloder et al. 2024). Similarly, species of *Pseudomonas* were more abundant on the exterior of parasitized *P. rapae*, while a *Serratia* species decreased markedly (Raymann et al. 2018; Bai et al. 2019; Duan et al. 2020; Du et al. 2020; Zhou et al. 2021; Pons et al. 2022). The physiological differences between *P. brassicae* and *P. rapae* may contribute to these host-specific outcomes.

### 2.1.9 *Parasitoid-Specific Microbiome Alterations in Host Caterpillars*

Beyond host-dependent variations, parasitism-induced alterations in the host's microbiome are also influenced by the specific species of the parasitoid. Caterpillars of both *P. brassicae* and *P. rapae* parasitized by *C. glomerata* contained a substantial quantity of *Wolbachia* (zOTU2), which was absent in non-parasitized caterpillars and those parasitized by *C. rubecula* (Gloder et al. 2024). This *Wolbachia* strain was also detected in high concentrations in the developing *C. glomerata* larvae within the hosts. Since *Wolbachia* was detected in adult *C. glomerata* rearing females but not in non-parasitized caterpillars, it is likely that it was transmitted into the caterpillars during oviposition (Ahmed et al. 2015a, b). The presence of *Wolbachia* in parasitized caterpillars has been shown to alter their body odors, which could provide cues for hyperparasitoids (Hedges et al. 2008; Werren et al. 2008; van Nouhuys et al. 2016; Bourne et al. 2023).

A *Nosema* microsporidian (zOTU9) was also abundantly present within caterpillars parasitized by either *Cotesia* species, and in the developing parasitoid larvae, but not in non-parasitized caterpillars (Choi et al. 2002; Yaman et al. 2014). This suggests it was also transferred from the parasitoids, potentially during the rearing process. The mechanisms causing these parasitoid-dependent changes are likely linked to the transfer of parasitoid-associated microorganisms during oviposition or from developing larvae, as well as the activity of parasitoid effectors that alter host immune function and physiology.

### 2.1.10 *Microbial Potential for Biological Control of Insect Pests*

Many species of microbes have the capability for indirect biological control of insect pests (see Table 2.1). Fungi such as *Metarhizium*, *Beauveria*, and *Aspergillus*, and bacteria like *Bacillus* and *Pseudomonas*, can be produced in large quantities. Insect symbiotic microorganisms that affect host physiology and ecology should be recognized as significant participants in tritrophic interactions and as sources for pest control. Formulation of microorganisms or their secondary metabolites is crucial for their effective application.

### 2.1.11 *The Role of Insect Gut Bacteria in Mediating Natural Enemy Behavior*

Biological control agents of herbivorous insects use volatile chemical cues to detect their hosts. Volatile constituents produced by the gut bacteria of insect pests can

**Table 2.1** Examples of interactions between plant microorganisms, insect pests and their natural enemies in indirect biological control of insect pests

| Soilborne microbes | Insect pests | Beneficials | Effects | Ref |
|---|---|---|---|---|
| *Pseudomonas fluorescens* WCS417r | *Myzus persicae* Sultzer | *Diaeretiella rapae* M'Intosh | Increasing parasitoid attraction to plant | Pineda et al. (2010, 2013) |
| *Pseudomonas fluorescens* WCS417r | *Mamestra brassicae* L. | *Microplitis mediator* Haliday | Plant growth and parasitoid attraction | Pangesti et al. (2015) |
| *Enterobacter aerogenes* | *Spodoptera littoralis* Boisduval | *Cotesia marginiventris* Cresson | Increasing parasitoid attraction to plant | D'Alessandro et al. (2014) |
| *Pseudomonas syringae* | *Myzus persicae* | *Coccinella septempunctata* L. | Plant pest resistance and parasitoid attraction | Song and Ryu (2013), Song et al. (2018) |
| Endophytes and plant symbionts | | | | |
| *Beauveria bassiana, Metarhizium brunneum* | *Myzus persicae* | *Aphidius colemani* Vierick | Reduction of pest development and parasitoid biology | Jaber and Araj (2018) |
| *Aspergillus flavus, A. niger* | *S. littoralis* | *Bracon hebetor* Say | Pest resistance and beneficial attraction | Kaur et al. (2021) |
| *Neotyphodium lolii* | *Agrotis ipsilon* (Hufnagel) | *Copidosoma bakeri* (Howard) | Negative development of parasitoid | Bixby-Brosi and Potter (2012) |
| *Beauveria bassiana* | *Aphis gossypii* Glover | *Chrysoperla carnea, A. colemani* | Reduction of pest and beneficial performances | González-Mas et al. (2019) |
| Nectar and honeydew microbes | | | | |
| *Metschnikowia gruessii, M. reukaufii* | Not studied | *Aphidius ervi* | Attraction and dispersal of parasitoids | Sobhy et al. (2018) |
| *Bacillus* strains | Not studied | *Aphidius colemani* | Attraction of parasitoids | Goelen et al. (2020) |
| *Staphylococcus sciuri* | *Acyrthosiphon pisum* | *Episyrphus balteatus* Degeer | Attraction and longevity of parasitoids | Leroy et al. (2011) |

impact this behavior. For example, volatile complexes released by *Staphylococcus sciuri* in aphid honeydew could attract the hoverfly *Episyrphus balteatus*, as mentioned by Leroy et al. (2011). Conversely, other honeydew-associated bacteria can decrease the attraction of the aphid parasitoid *Aphidius colemani* (Goelen et al. 2020). Bacteria from the larvae frass also could produce volatiles used by other parasitoids such as *Diadromus pulchellus* to find its host (Thibout et al. 1995; Visôtto et al. 2009). These findings propose that various natural enemies may have grown to observe chemical signals from insect gut microbiota.

### 2.1.12 *The Expanding Role of Gut Bacteria in Modern Insect Pest Management*

Management of different insect pests is gradually leveraging the potential of gut bacteria as a sustainable biological control approach. Whereas entomopathogenic bacteria *Bacillus thuringiensis* (Bt) are widely used, emerging resistance in pests (Pickett et al. 2017: Jurat-Fuentes et al. 2021) has driven research on the road to alternative microbial solutions. Insect gut microbiomes, which affect host physiology, immunity, and behavior, are now predictable as promising goals for pest management (Douglas 2015; Mabbott 2018; Li et al. 2018; Cavichiolli de Oliveira and Cônsoli 2020).

The current studies highlight gut bacteria as sources of innovative insecticidal complexes and as intermediaries of RNA interference (RNAi)-based pest control. Additionally, paratransgenesis, genetically modifying symbiotic gut bacteria to distribute antipathogen molecules, shows potential for managing vector-borne diseases. Moreover, integrating gut bacteria with other management strategies such as Sterile Insect Technique (SIT) improves precision in pest population suppression. For example, employing *Wolbachia* symbionts has successfully compacted mosquito-borne disease transmission (Moreira et al. 2009; Furihata et al. 2015; van Nouhuys et al. 2016; Sanaei et al. 2021). As molecular tools advance, gut bacteria-based strategies offer eco-friendly substitutions to chemical insecticides.

## 2.2 Conclusion

This chapter produces the multifaceted roles of plant and insect microbiomes in mediating tri-trophic interactions. A central theme is that microbial populations, whether residing in the rhizosphere, phyllosphere, nectar, or insect gut, are not mere passengers but active regulators of ecological dynamics among tritrophic levels (plants, insect pests, and parasitoids). The results have revealed that plant-associated microbes improve indirect defense by amplifying the emission of HIPVs, thereby recruiting many of the biological control agents. Conversely, the gut microbiota of parasitoid wasps is essential for their nutrition, development, and immune function, with its composition often reflecting the host's internal habitats. Analytically, parasitoid wasps act as ecosystem agents, actively relocating symbiotic microbes (e.g., *Wolbachia*, *Nosema*) and deploying effectors that reconfigure the host caterpillar's immune system and microbiome, producing a favorable niche for their offspring. This interconnectedness emphasizes the concept of the holobiont, where the fitness of each organism is a product of its own genome and that of its microbial associates.

### 2.2.1 *Future Directions and Implications for Biological Control Strategy*

To translate these vital facts into robust, field-applicable approaches, future research must bridge several key gaps:

*Mechanistic Depth*: Moving beyond relationships, studies must identify the specific microbial genes and metabolites (i,e., volatiles, proteins) that drive observed behaviors in parasitoid wasps and physiological changes in their hosts.

*Synergistic Formulations*: The future of microbial biocontrol lies in consortia. Merging plant growth-promoting rhizobacteria (PGPR), endophytic entomopathogens, and probiotics for natural enemies could create resilient, self-sustaining defensive ecosystems that outperform single-strain applications.

Finally, viewing pest control tools through the lens of the microbiome offers a paradigm shift from direct eradication to ecological modulation. By harnessing and steering these invisible microbial alliances, the authors could develop precisely targeted, durable, and sustainable strategies that enhance the stability of landscape ecosystems.

**Acknowledgments** The authors would like to show their appreciation to the Deanship of Scientific Research at King Faisal University in Al-Ahsa, Saudi Arabia, for providing all of the important facilities for the project that was assigned the number (KFU251142) during the writing of this book chapter.

**AI Use Disclosure** The authors declare that during the preparation of this work, they used DeepSeek-V3 to improve linguistic phrasing and clarity in the introduction and specific chapter elements. The AI was also used to adjust reference formatting. All AI-generated content was severely edited and verified by the authors, who take full responsibility for the publication's content. The final book chapter was reviewed through peer review.

## References

Ahmed MZ, Araujo-Jnr EV, Welch JJ et al (2015a) *Wolbachia* in butterflies and moths: geographic structure in infection frequency. Front Zool 12:16

Ahmed MZ, Li S-J, Xue X et al (2015b) The intracellular bacterium *Wolbachia* uses parasitoid wasps as phoretic vectors for efficient horizontal transmission. PLoS Pathog 11:e1004672

Badri DV (2013) Inoculation of plant leaves with soil microorganisms can alter the leaf microbiome and improve plant resistance to pests. J Plant Interact 8(4):280–294

Bai L, Wang L, Vega-Rodríguez J, Wang G, Wang S (2019) A gut symbiotic bacterium *Serratia marcescens* renders mosquito resistance to *Plasmodium* infection through activation of mosquito immune responses. Front Microbiol 10:1580. https://doi.org/10.3389/fmicb.2019.01580

Bamisile BS, Dash CK, Akutse KS, Keppanan R, Wang L (2018) Fungal endophytes: beyond herbivore management. Front Microbiol 9:544. https://doi.org/10.3389/fmicb.2018.00544

Battaglia D, Bossi S, Cascone P, Digilio MC, Duran Prieto J, Fanti P, Guerrieri E, Iodice L, Lingua G, Lorito M, Maffei ME, Massa N, Ruocco M, Sasso R, Trotta V (2013) Tomato below ground–above ground interactions: *Trichoderma longibrachiatum* affects the performance of *Macrosiphum euphorbiae* and its natural antagonists. Mole Plant-Microbe Interact 26(10): 1249–1256

Beckage NE, Gelman DB (2004) Wasp parasitoid disruption of host development: implications for new biologically based strategies for insect control. Annu Rev Entomol 49:299–330. https://doi.org/10.1146/annurev.ento.49.061802.123324

Berendsen RL, Pieterse CM, Bakker PA (2012) The rhizosphere microbiome and plant health. Trends Plant Sci 17:478–486

Bixby-Brosi AJ, Potter DA (2012) Endophyte-mediated tritrophic interactions between a grass-feeding caterpillar and two parasitoid species with different life histories. Arthropod Plant Interact 6:27–34. https://doi.org/10.1007/s11829-011-9163-2

van Borm S, Buschinger A, Boomsma JJ, Billen J (2002) *Tetraponera* ants have gut symbionts related to nitrogen-fixingroot-nodule bacteria. Proc Biol Sci 269:2023–2027. https://doi.org/10.1098/rspb.2002.2101

Bosmans L, Pozo MI, Verreth C et al (2018) Habitat-specific variation in gut microbial communities and pathogen prevalence in bumblebee queens (*Bombus terrestris*). PLoS One 13:e0204612

Bourne ME, Gloder G, Weldegergis BT et al (2023) Parasitism causes changes in caterpillar odours and associated bacterial communities with consequences for host-location by a hyper parasitoid. PLoS Pathog 19:e1011262

Cavichiolli de Oliveira N, Cônsoli FL (2020) Beyond host regulation: changes in gut microbiome of permissive and non-permissive hosts following parasitization by the wasp *Cotesia flavipes*. FEMS Microbiol Ecol 96:fiz206

Chen B, Mason CJ, Peiffer M, Zhang D, Shao Y, Felton GW (2022) Enterococcal symbionts of caterpillars facilitate the utilization of a suboptimal diet. J Insect Physiol 138:104369. https://doi.org/10.1016/j.jinsphys.2022.104369

Choi J, Kim JG, Choi YC et al (2002) *Nosema* sp. isolated from cabbage white butterfly (*Pieris rapae)* collected in Korea. J Microbiol 40:199–204

Compant S, Samad A, Faist H, Sessitsch A (2019) A review on the plant microbiome: ecology, functions, and emerging trends in microbial application. J Adv Res 19:29–37

Curtis TP, Sloan WT (2005) Exploring microbial diversity—a vast below. Science 309(5739): 1331–1333

Curtis TP, Sloan WT, Scannell JW (2002) Estimating prokaryotic diversity and its limits. Proc Natl Acad Sci 99(16):10494–10499

Cusumano A, Harvey JA, Bourne ME et al (2020) Exploiting chemical ecology to manage hyperparasitoids in biological control of arthropod pests. Pest Manag Sci 76:432–443

D'alessandro M, Erb M, Ton J, Brandenburg A, Karlen D, Zopfi J, Turlings TCJ (2014) Volatiles produced by soil-borne endophytic bacteria increase plant pathogen resistance and affect tritrophic interactions. Plant Cell Environ 37:813–826. https://doi.org/10.1111/pce.12220

Dastogeer KMG, Tumpa FH, Sultana A, Akter MA, Chakraborty A (2020) Plant microbiome–an account of the factors that shape community composition and diversity. Curr Plant Biol 23: 100161

Dicke M, Baldwin IT (2010) The evolutionary context for herbivore-induced plant volatiles: beyond the 'cry for help'. Trends Plant Sci 15(3):167–175

Dicke M, Cusumano A, Poelman EH (2020) Microbial symbionts of parasitoids. Annu Rev Entomol 65:171–190. https://doi.org/10.1146/annurev-ento-011019-024939

Douglas AE (2015) Multiorganismal insects: diversity and function of resident microorganisms. Annu Rev Entomol 60:17–34

Du M, Yu J, Zhou Y, Wang X, Ma T, Tan X, Wan F, Zhou H (2020) Differentiation of symbiotic bacteria is a new evidence for two genetic clades of *Aphelinus Mali* in China. Orient Insects 54: 447–464. https://doi.org/10.1080/00305316.2019.1671252

Duan R, Xu H, Gao S, Gao Z, Wang N (2020) Effects of different hosts on bacterial communities of parasitic wasp *Nasonia vitripennis*. Front Microbiol 11:1435. https://doi.org/10.3389/fmicb.2020.01435

El-Wakeil N (2003) New aspects of biological control of *Helicoverpa armigera* in organic cotton production, PhD disseration Göttingen University, Germany (140pp); full text in http://webdoc.sub.gwdg.de/diss/2003/el-wakeil/index.html
El-Wakeil NE, Volkmar C, Sallam AA (2010) Jasmonic acid induces resistance to economically important insect pests in winter wheat. Pest Manag Sci 66:549–554. https://doi.org/10.1002/ps.1906
Fernández MDM, Meeus I, Billiet A, Van Nieuwerburgh F, Deforce D, Vandamme P, Viñuela E, Smagghe G (2019) Influence of microbiota in the susceptibility of parasitic wasps to abamectin insecticide: deep sequencing, esterase and toxicity tests. Pest Manag Sci 75:79–86. https://doi.org/10.1002/ps.5195
Flury P, Aellen N, Ruffner B et al (2016) Insect pathogenicity in plant- beneficial pseudomonads: phylogenetic distribution and comparative genomics. ISME J 10:2527–2542
Francis I, Holsters M, Vereecke D (2020) The gram-positive side of plant-microbe interactions. Environ Microb 22(10):4049–4062
Friesen ML, Porter SS, Stark SC, von Wettberg EJ, Sachs JL, Martinez-Romero E (2011) Microbially mediated plant functional traits. Ann Rev Ecol Evol Syst 42:23–46
Fuchs B, Krauss J (2019) Can Epichloë endophytes enhance direct and indirect plant defence? Fungal Ecol 38:98–103
Furihata S, Hirata M, Matsumoto H et al (2015) Bacteria endosymbiont, *Wolbachia,* promotes parasitism of parasitoid wasp *Asobara japonica.* PLoS One 10:e0140914
Gans J, Wolinsky M, Dunbar J (2005) Computational improvements reveal great bacterial diversity and high metal toxicity in soil. Science 309(5739):1387–1390
Gao X, Niu R, Zhu X, Wang L, Ji J, Niu L, Wu C, Zhang S, Luo J, Cui J (2021) Characterization and comparison of the bacterial microbiota of *Lysiphlebia japonica* parasitioid wasps and their aphid host *Aphis gosypii.* Pest Manag Sci 77:2710–2718. https://doi.org/10.1002/ps.6299
Gloder G, Bourne ME, Verreth C et al (2021) Parasitism by endoparasitoid wasps alters the internal but not the external microbiome in host caterpillars. Anim Microbiome 3:73
Gloder G, Bourne ME, Cuny MAC, Verreth C, Crauwels S, Dicke M, Poelman EH, Jacquemyn H, Lievens B (2024) Caterpillar–parasitoid interactions: species-specific influences on host microbiome composition. FEMS Microbiol Ecol 100:fiae115. https://doi.org/10.1093/femsec/fiae115
Goelen T et al (2020) Volatiles of bacteria associated with parasitoid habitats elicit distinct olfactory responses in an aphid parasitoid and its hyperparasitoid. Funct Ecol 34:507–520
González-Mas N, Cuenca-Medina M, Gutiérrez-Sánchez F et al (2019) Bottom-up effects of endophytic *Beauveria bassiana* on multitrophic interactions between the cotton aphid, *Aphis gossypii*, and its natural enemies in melon. J Pest Sci 92:1271–1281. https://doi.org/10.1007/s10340-019-01098-5
Gwokyalya R, Herren J, Weldon C et al (2024) Shaping the microbial landscape: parasitoid-driven modifications of *Bactrocera dorsalis* microbiota. Microb Ecol 87:81
Harri SA, Krauss J, Müller CB (2008) Fungal endosymbionts of plants reduce lifespan of an aphid secondary parasitoid. J Anim Ecol 77(3):565–571
Harvey J, Poelman E, Tanaka T (2013) Intrinsic inter- and intraspecific competition in parasitoid wasps. Annu Rev Entomol 58:333–351
Hedges LM, Brownlie JC, O'Neill SL et al (2008) *Wolbachia* and virus protection in insects. Science 322:702
Herrera CM, Pozo MI (2010) Nectar yeasts warm the flowers of a winter-blooming plant. Proc R Soc B: Biol Sci 277(1689):1827–1834
Herrera CM, Pozo MI, Bazaga P (2013) Yeasts in nectar of an early-blooming herb: sought by bumble bees, detrimental to plant fecundity. Ecology 94(2):273–279
Hu NN, Wang ZQ, Zhang SJ, Wang ZZ, Xue-xin Chen XX (2024) Characterization of larval gut microbiota of two endoparasitoid wasps associated with their common host, *Plutella xylostella.* Microbiol Spectrum 12:e01208–e01224. https://doi.org/10.1128/spectrum.01208-24

Jaber LR, Alananbeh KM (2018) Fungal entomopathogens as endophytes in plant protection: a systematic review. Egypt J Biol Pest Control 28(1):1–11

Jaber LR, Araj SE (2018) Interactions among endophytic fungal entomopathogens (Ascomycota: Hypocreales), the green peach aphid *Myzus persicae* Sulzer (Homoptera: Aphididae), and the aphid endoparasitoid *Aphidius colemani* Viereck (Hymenoptera: Braconidae). Biol Control 116: 53–61. https://doi.org/10.1016/j.biocontrol.2017.04.005

Jing TZ, Qi FH, Wang ZY (2020) Most dominant roles of insect gut bacteria: digestion, detoxification, or essential nutrient provision? Microbiome 8:38. https://doi.org/10.1186/s40168-020-00823-y

Jones EJ, Gorman JV, You M, Lietze VU (2020) The gut microbiome of the caterpillar, *Plodia interpunctella*, and its interaction with the parasitoid wasp, *Habrobracon hebetor*. J Insect Sci 20(6):29

Jurat-Fuentes JL, Heckel DG, Ferre J (2021) Mechanisms of resistance to insecticidal proteins from *Bacillus thuringiensis*. Annu Rev Entomol 66:121–140

Kaur M, Saraf I, Kumar R, Singh IP, Kaur S (2021) Biological effects of secondary metabolites of *Inula racemosa* on the parasitoid *Bracon hebetor*. Èntomol Exp Appl 169:743–749. https://doi.org/10.1111/eea.13070

Kessler A, Baldwin IT (2010) Soil-borne microbes alter plant volatile organic compound emission and predator attraction. Plant Cell Environ 33(10):1687–1697

Lenaerts M, Goelen T, Paulussen C et al (2017) Nectar bacteria affect life history of a generalist aphid parasitoid by altering nectar chemistry. Funct Ecol 31(11):2061–2069

Leroy P, Sabri A, Heuskin S et al (2011) Microorganisms from aphid honeydew attract and enhance the efficacy of natural enemies. Nat Commun 2:348. https://doi.org/10.1038/ncomms1347

Li CY, Zhou X, Zheng H (2018) Diversity and function of insect gut microbiome. Acta Microbiol Sin 58:1016–1024

Li W, Zhang J, Zhu Y, Li F (2021) Probiotic characterization of *Enterococcus mundtii* isolated from larval gut of the diamondback moth, *Plutella xylostella*. J Kansas Entomol Soc 93:196–210. https://doi.org/10.2317/0022-8567-93.3.196

Liang X, He J, Zhang N, Muhammad A, Lu X, Shao Y (2022) Probiotic potentials of the silkworm gut symbiont *Enterococcus casseliflavus* ECB140, a promising L-tryptophan producer living inside the host. J Appl Microbiol 133:1620–1635. https://doi.org/10.1111/jam.15675

Lievens B, Hallsworth JE, Pozo MI et al (2015) Microbiology of sugar-rich environments: diversity, ecology, and system constraints. Environ Microbiol 17(2):278–298

Lin XL, Pan QJ, Tian HG, Douglas AE, Liu TX (2015) Bacteria abundance and diversity of differentlife stages of *Plutella xylostella* (Lepidoptera: Plutellidae), revealed by bacteria culture-dependent and PCR-DGGE methods. Insect Sci 22:375–385. https://doi.org/10.1111/1744-7917.12079

Mabbott NA (2018) The influence of parasite infections on host immunity to co-infection with other pathogens. Front Immunol 9:2579

Martínez-Solís M, Collado MC, Herrero S (2020) Influence of diet, sex, and viral infections on the gut microbiota composition of *Spodoptera exigua* caterpillars. Front Microbiol 11:753

Moreira LA et al (2009) A Wolbachia symbiont in *Aedes aegypti* limits infection with dengue, chikungunya, and plasmodium. Cell 139:1268–1278

Moussian B (2010) Recent advances in understanding mechanisms of insect cuticle differentiation. Insect Biochem Mol 40:363–375. https://doi.org/10.1016/j.ibmb.2010.03.003

Muvea AM, Meyhöfer R, Subramanian S, Poehling HM, Eilenberg J, Maniania NK (2014) Colonization of onions by endophytic fungi and their impacts on the biology of *Thrips tabaci*. PLoS One 9(9):e108242

Nebo UC, Ugwuja AN, Lawal AM, Agbo SO (2025) Pesticides and microbiome shifts. In: Babaniyi BR, Babaniyi EE (eds) The interplay of pesticides and climate change. Springer, Cham. https://doi.org/10.1007/978-3-031-81669-7_13

Nedoluzhko AV, Sharko FS, Tsygankova SV, Boulygina ES, Sokolov AS, Rastorguev SM, Kadnikov VV, Mardanov AV et al (2017) Metagenomic analysis of microbial community of a parasitoid wasp *Megaphragma amalphitanum*. Genom Data 11:87–88. https://doi.org/10.1016/j.gdata.2016.12.007

van Nouhuys S, Kohonen M, Duplouy A (2016) *Wolbachia* increases the susceptibility of a parasitoid wasp to hyperparasitism. J Exp Biol 219:2984–2990

Pang Z, Chen J, Wang T, Gao C, Li Z, Guo L, Xu J, Cheng Y (2020) Linking plant secondary metabolites and plant microbiomes: a review. Front Plant Sci 11:621

Pang L, Dong Z, Liu Z et al (2023) Larval secretions of parasitoid wasps are new effectors that impair host immune defences. Crop Health 1:11. https://doi.org/10.1007/s44297-023-00011-y

Pangesti N, Weldegergis BT, Langendorf B, van Loon JJA, Dicke M, Pineda A (2015) Rhizobacterial colonization of roots modulates plant volatile emission and enhances the attraction of a parasitoid wasp to host-infested plants. Oecologia 178:1169–1180

Park R, Dzialo MC, Spaepen S et al (2019) Microbial communities of the house fly *Musca domestica* vary with geographical location and habitat. Microbiome 7:147

Pickett BR, Gulzar A, Ferre J, Wright DJ (2017) *Bacillus thuringiensis* Vip3Aa toxin resistance in *Heliothis virescens* (Lepidoptera: Noctuidae). Appl Environ Microbiol 83:e03506–e03516

Pieterse CMJ, Van der Does D, Zamioudis C, Leon-Reyes A, Van Wees SCM (2012) Hormonal modulation of plant immunity. Ann Rev Cell Dev Biol 28:489–521

Pineda A, Zheng SJ, van Loon JJA, Pieterse CMJ, Dicke M (2010) Helping plants to deal with insects: the role of beneficial soil-borne microbes. Trends Plant Sci 15:507–514

Pineda A, Soler R, Weldegergis BT, Shimwela MM, Van Loon JJA, Dicke M (2013) Non-pathogenic rhizobacteria interfere with the attraction of parasitoids to aphid-induced plant volatiles via jasmonic acid signalling. Plant Cell Environ 36:393–404

Pons I, Scieur N, Dhondt L et al (2022) Pervasiveness of the symbiont *Serratia symbiotica* in the aphid natural environment: distribution, diversity and evolution at amultitrophic level. FEMS Microbiol Ecol 98:fiac012

Pozo MI, Lievens B, Jacquemyn H (2015) Impacts of microbes on plant reproduction and pollen ecology. Crit Rev Plant Sci 34(1–3):125–137

Ramirez-Ahuja MD, Gomez-Govea MA, Lugo-Trampe A, Borrego-Soto G, Delgado-Enciso I, Ponce-Garcia G, Martinez-Fierro ML, Ramirez-Valles EG, Trevino V, Flores-Suarez AE, Rodriguez-Sanchez IP (2019) Microbiota of *Telenomus tridentatus* (Platygastroidea: Scelionidae): an unwanted parasitoid. J Appl Entomol 143:834–841. https://doi.org/10.1111/jen.12656

Rashid MH, Chung YR (2017) Induction of systemic resistance against insect herbivores in plants by beneficial soil microbes. Front Plant Sci 8:1816

Rasmann S, Bennett A, Biere A, Karle A, Guerrieri E (2017) Root symbionts: powerful drivers of plant above- and belowground indirect defenses. Insect Sci 24(6):947–960

Rassmann S, Schmidt R (2017) The influence of soil bacteria on plant volatile emissions. Trends Plant Sci 22(11):911–913. https://doi.org/10.1016/j.tplants.2017.09.010

Rattan RS, Hadapad AB, Reineke A et al (2011) Molecular evidence for the presence of the endosymbiontic bacteria *Wolbachia* in *Cotesia* populations (Hymenoptera: Braconidae). J Asia Pac Entomol 14:183–185

Raymann K, Coon KL, Shaffer Z, Salisbury S, Moran NA (2018) Pathogenicity of *Serratia marcescens* strains in honey bees. mBio 9:1–15. https://doi.org/10.1128/mBio.01649-18

Rering CC, Beck JJ, Hall GW et al (2018) Nectar-inhabiting microorganisms influence nectar volatile composition and attractiveness to a generalist pollinator. New Phytol 220(3):750–759

Saikkonen K, Lehtonen P, Helander M, Koricheva J, Faeth SH (2006) Model systems in ecology: dissecting the endophyte–grass literature. Trends Plant Sci 11(9):428–433

Saleem M, Meckes N, Pervaiz ZH, Traw MB (2017) Microbial interactions in the phyllosphere increase plant performance under herbivore biotic stress. Front Microbiol 8:137

Sanaei E, Charlat S, Engelstädter J (2021) *Wolbachia* host shifts: routes, mechanisms, constraints and evolutionary consequences. Biol Rev 96:433–453

Schaeffer RN, Mei YZ, Andico A et al (2014) Microbial metabolites elicit distinct olfactory and gustatory preferences in bumblebees. Biol Lett 10(9):20140562

Schmidt K, Engel P (2021) Mechanisms underlying gut microbiota-host interactions in insects. J Exp Biol 224:1–16. https://doi.org/10.1242/jeb.207696
Schmidt R, Etalo MW, de Jager V, Gerards S, Zweers H, de Boer W, Garbeva P (2015) Microbial small talk: volatiles in fungal–bacterial interactions. Front Microbiol 6:1495
Sedláček I, Holochová P, Mašlaňová I, Kosina M, Spröer C, Bryndová H, Vandamme P, Rudolf I, Hubálek Z, Švec P (2013) *Enterococcus ureilyticus* sp. nov. and *enterococcus rotai* sp. nov., two urease-producing enterococci from the environment. Int J Syst Evol Microbiol 63:502–510. https://doi.org/10.1099/ijs.0.041152-0
Shao Y, Mason CJ, Felton GW (2024) Toward an integrated understanding of the lepidoptera microbiome. Annu Rev Entomol 69:117–137
Sobhy IS, Baets D, Goelen T, Herrera-Malaver B, Bosmans L, Van den Ende W, Verstrepen KJ, Wäckers F, Jacquemyn H, Lievens B (2018) Sweet scents: nectar specialist yeasts enhance nectar attraction of a generalist aphid parasitoid without affecting survival. Front Plant Sci 9. https://doi.org/10.3389/fpls.2018.01009
Song GC, Ryu CM (2013) Two volatile organic compounds trigger plant self-defense against a bacterial pathogen and a sucking insect in cucumber under open field conditions. Int J Mol Sci 14:9803–9819. https://doi.org/10.3390/ijms14059803
Song X, Wang M, Dong L, Zhu H, Wang J (2018) PGRP-LD mediates *A. stephensi* vector competency by regulating homeostasis of microbiota-induced peritrophic matrix synthesis. PLoS Pathog 14:e1006899. https://doi.org/10.1371/journal.ppat.1006899
Stone BA, Jacobs AK, Hrmova M, Burton RA, Fincher GB (2010) Biosynthesis of plant cell wall and related polysaccharides by enzymes of the GT2 and GT48 families. In: Ulvskov P (ed) Plant polysaccharides, vol 41, pp 109–165. https://doi.org/10.1002/9781444391015.ch5
Strand MR (2014) Teratocytes and their functions in parasitoids. Curr Opin Insect Sci 6:68–73
Talia P, Sede SM, Campos E, Rorig M, Principi D, Tosto D, Hopp HE, Grasso D, Cataldi A (2012) Biodiversity characterization of cellulolytic bacteria present on native Chaco soil by comparison of ribosomal RNA genes. Res Microbiol 163:221–232. https://doi.org/10.1016/j.resmic.2011.12.001
Tariq M, Wright DJ, Bruce TJ, Staley JT (2013) Drought and root herbivory interact to alter the response of above-ground parasitoids to aphid infested plants and associated plant volatile signals. PLoS One 8:e69013
Teoh M-C, Furusawa G, Veera Singham G (2021) Multifaceted interactions between the pseudomonads and insects: mechanisms and prospects. Arch Microbiol 203:1891–1915
Thakur N, Nigam M, Mann NA et al (2023) Host-mediated gene engineering and microbiome-based technology optimization for sustainable agriculture and environment. Funct Integr Genomics 23:57. https://doi.org/10.1007/s10142-023-00982-9
Thibout E et al (1995) Origin and identification of bacteria which produce kairomones in the frass of *Acrolepiopsis assectella* (Lep., Hyponomeutoidea). Experientia 51:1073–1075
Turlings TCJ, Erb M (2018) Tritrophic interactions mediated by herbivore-induced plant volatiles: mechanisms, ecological relevance, and application potential. Annu Rev Entomol 63:433–452
Ueda M, Ohata K, Konishi T, Sutrisno A, Okada H, Nakazawa M, Miyatake K (2009) A novel goose-type lysozyme gene with chitinolytic activity from the moderately thermophilic bacterium *Ralstonia* sp. A-471: cloning, sequencing, and expression. Appl Microbiol Biotechnol 81:1077–1085. https://doi.org/10.1007/s00253-008-1676-y
Unban K, Klongklaew A, Kodchasee P, Pamueangmun P, Shetty K, Khanongnuch C (2022) Enterococci as dominant xylose utilizing lactic acid bacteria in Eri silkworm midgut and the potential use of *Enterococcus hirae* as probiotic for Eri culture. Insects 13:1–13. https://doi.org/10.3390/insects13020136
Urrutia MA, Wade RN, Keyes AA (2007) Do endophytic fungi alter the direct and indirect defense of plants against insect herbivores? A literature review. Ecol Entomol 32(6):573–582
Van Oosten VR, Bodenhausen N, Reymond P, Van Pelt JA, Van Loon LC, Dicke M, Pieterse CMJ (2008) Differential effectiveness of microbially induced resistance against herbivorous insects in Arabidopsis. Mole Plant-Microbe Interact 21(7):919–930

Vannette RL, Fukami T (2017) Dispersal enhances beta diversity in nectar microbes. Ecol Lett 20(7):901–910
Vega FE (2008) Insect pathology and fungal endophytes. J Inverteb Pathol 98(3):277–279
Visôtto LE, Oliveira MGA, Guedes RNC, Ribon AOB, Good-God PIV (2009) Contribution of gut bacteria to digestion and development of the velvetbean caterpillar, *Anticarsia gemmatalis*. J Insect Physiol 55:185–191. https://doi.org/10.1016/j.jinsphys.2008.10.017
Vokou D, Chalkos D, Karamanlidou G, Yiangou M (2012) Activation of soil respiration and shift of the microbial population balance in soil as a response to *Lavandula stoechas* essential oil. J Chem Ecol 38(6):658–666
Vorholt JA (2012) Microbial life in the phyllosphere. Nat Rev Microbiol 10(12):828–840
Wang J, Peiffer M, Hoover K, Rosa C, Zeng R, Felton GW (2017) *Helicoverpa zea* gut-associated bacteria indirectly induce defenses in tomato by triggering a salivary elicitor(s). New Phytol 214(3):1294–1306
Wang Y, Zhu J, Fang J et al (2020) Diversity, composition and functional inference of gut microbiota in indian cabbage white *Pieris canidia*. Life 10:254
Wang Y, Gao P, Zheng J et al (2023) Effects of parasitism by a parasitoid wasp on the gut microbiota in a predaceous lady beetle host. Pest Manag Sci 79:4501–4507
Werren JH, Baldo L, Clark ME (2008) *Wolbachia:* master manipulators of invertebrate biology. Nat Rev Microbiol 6:741–751
Whipps JM, Lewis K, Cooke RC (1988) Mycoparasitism and plant disease control. In: Burge N (ed) Fungi in biological control systems. Manchester University Press, pp 161–187
Yaman M, Bekircan Ç, Radek R et al (2014) *Nosema pieriae* sp. n. (Microsporida, Nosematidae): a new microsporidian pathogen of the cabbage butterfly *Pieris brassicae* L. (Lepidoptera: pieridae). Acta Protozool 53:223–332
Yang J et al (2021) Profiling of microRNAs in midguts of *Plutella xylostella* provides novel insights into the *Bacillus thuringiensis* resistance. Front Genet 12:739849
Zeng T, Jaffar S, Xu Y, Qi Y (2022) The intestinal immune defense system in insects. Int J Mol Sci 23:1–18. https://doi.org/10.3390/ijms232315132
Zha XL, Wang H, Sun W, Zhang HY, Wen J, Huang XZ, Lu C, Shen YH (2021) Characteristics of the peritrophic matrix of the silkworm, *Bombyx mori* and factors influencingits formation. Insects 12:1–14. https://doi.org/10.3390/insects12060516
Zhang S, Huang J, Wang Q et al (2022) Changes in the host gut microbiota during parasitization by parasitic wasp *Cotesia vestalis*. Insects 13:760
Zhang Y, Zhang S, Xu L (2023) The pivotal roles of gut microbiota in insect plant interactions for sustainable pest management. npj Biofilms Microbiome 9:66. https://doi.org/10.1038/s41522-023-00435-y
Zhao W, Dong S, Shi M, Chen XX (2014) Morphology and development of immature stage of *Diadromus collaris*, an important endoparasitoid of *Plutella xylostella*. Ann Entomol Soc Am 107:234–241. https://doi.org/10.1603/an12132
Zhou X, Ling X, Guo H, Zhu-Salzman K, Ge F, Sun Y (2021) *Serratia symbiotica* enhances fatty acid metabolism of pea aphid to promote host development. Int J Mol Sci 22:1–13. https://doi.org/10.3390/ijms22115951

# Chapter 3
# Use of Plant Beneficial Microbes to Control Bayoud Disease in Date Palms

**Rachid Sabbahi and Virginia Hock**

## 3.1 Introduction

The date palm, *Phoenix dactylifera* L., is an ancient and vital fruit crop in the dry and semi-dry regions of North Africa and the Middle East, where it offers food, income, and ecological benefits for millions of people (Al-Khayri et al. 2015). The date palm is thought to have originated in the Arabian Peninsula and was cultivated and dispersed by humans to other parts of the world, such as Africa, Asia, America, and Europe (Zehdi-Azouzi et al. 2015). The date palm is a very adaptable and resilient plant that can withstand harsh environmental conditions, such as drought, salinity, and high temperatures, and produces high-quality fruits with various health and medicinal properties (Al-Farsi et al. 2005). However, *P. dactylifera* production is facing several challenges and threats due to biotic and abiotic factors that affect its growth, yield, and quality.

Among the biotic factors, the most destructive and widespread disease of date palm is the bayoud disease or *Fusarium* wilt of date palm, caused by the soil-living fungus *Fusarium oxysporum* f. sp. *albedinis* (Foa) (Bouissil et al. 2022). The bayoud disease is a vascular wilt that affects the xylem vessels of the date palm, leading to the wilting, yellowing, necrosis, and death of the leaves, and eventually the whole plant (Dihazi et al. 2011). The disease was first reported in the Draa valley in the south of Morocco in 1870, where it has wiped out more than 12 million date palms, reducing the total areas from 150,000 to 44,000 ha (Meddich et al. 2018; Oulad

R. Sabbahi (✉)
Research Team in Science and Technology, Higher School of Technology of Laayoune, Ibn Zohr University, Laayoune, Morocco
e-mail: r.sabbahi@uiz.ac.ma

V. Hock
Department of Biology, Dawson College, Montreal, QC, Canada

 
N. El-Wakeil, A. Abdelkhalek (eds.), *Microbiomes and Plant Secondary Metabolites for Insect Pest Management*, Progress in Biological Control 23,
https://doi.org/10.1007/978-3-032-24330-0_3

Ziane et al. 2023; Samo and Ismailli 2020). The disease spread to Algeria and is also present in Mauritania, Tunisia, and Libya, and poses a serious threat to other date palm-growing countries, such as Egypt, Iran, and Saudi Arabia (Haq and Khan 2020; Saleh et al. 2017).

The current methods of controlling bayoud disease are mainly based on the use of fungicides, cultural practices, and resistant cultivars (Boulahouat et al. 2022; Sedra and Lazrek 2011). However, these methods have several limitations and drawbacks, such as high cost, environmental pollution, health hazards, and the development of pathogen resistance (Khan et al. 2018). Therefore, there is a need for alternative and sustainable strategies to manage bayoud disease, such as biological control using beneficial microbes. Biological control is the application of live organisms or their products to control the population or activity of a pathogen or pest, thereby reducing its damage to the host plant (Ortiz and Sansinenea 2022). Biological control agents (BCAs) can be bacteria, fungi, viruses, nematodes, insects, or plants that react with the host plant and the pathogen in various ways, such as parasitism, induction of plant defense mechanisms, antibiosis, or competition (Lahlali et al. 2022). There have been several studies pointing to the possibility of using beneficial microbes, especially fungi and bacteria, as novel BCAs for biocontrol of *F. oxysporum* in many crops, among them date palms (Ali et al. 2023; Boulahouat et al. 2022, 2023; Bubici et al. 2019; Slama et al. 2019). Some studies have investigated the possibility of using plant-derived bioactive compounds to control *F. oxysporum* (Bouhlali et al. 2020; Jangir et al. 2021).

The main objectives of this chapter are to provide an overview and discussion on the potential utilization of beneficial microbes as BCAs for the biocontrol of *F. oxysporum*, with a focus on bayoud disease of the date palm. The chapter will cover the following aspects: (1) the mechanisms and modes of action of beneficial microbes against *F. oxysporum*; (2) the factors affecting the efficacy and performance of beneficial microbes in the field; (3) the methods and techniques for isolating, screening, and selecting effective BCAs from different sources; (4) the application methods and formulations of beneficial microbes for biocontrol purposes; and (5) the challenges and opportunities for the development and commercialization of beneficial microbes as biocontrol products.

## 3.2 Foa Biology and Hosts

### *3.2.1 Biology*

*Fusarium oxysporum* is a soil-living fungus that causes vascular wilt diseases in many crop plants, such as banana, tomato, cotton, melon, potato, and date palm (Boutaj et al. 2022). This fungus infects *P. dactylifera* and makes them wilt and die. The fungus can survive as chlamydospores (Fig. 3.1) in any dead parts of the plant, mainly in decaying roots, and in the soil (Benzohra et al. 2015). When the dead parts break down, the chlamydospores go into the soil, where they can last for over 8 years

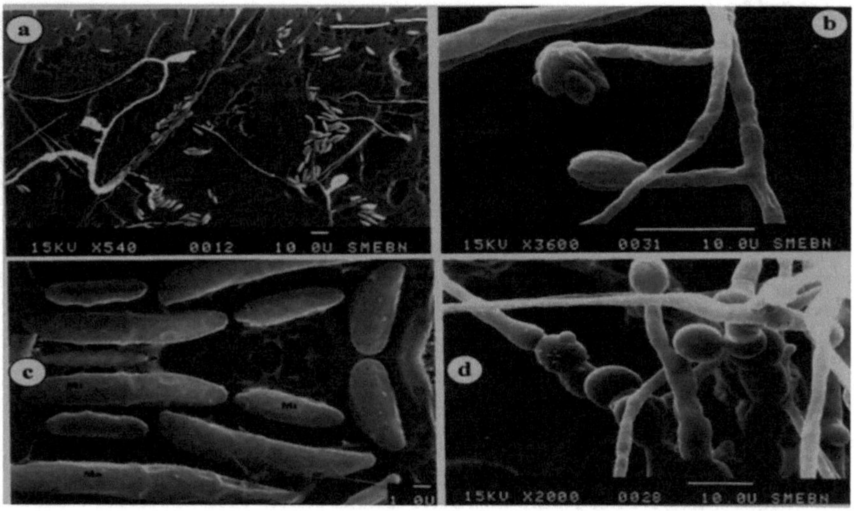


**Fig. 3.1** SEM asexual reproductive organs of *Fusarium oxysporum f.* sp. *albedinis*. (**a**): mycelium and microconidia, (**b**): microphialides, (**c**): micro and macroconidia, (**d**): chlamydospores formed from mycelial fragments. (Rahmania 2000)

(EFSA Panel on Plant Health et al. 2018). The fungus has been found in soil up to 30 cm deep, and in some cases over a meter (Tantaoui 1989). A few chlamydospores can initiate the disease and if only a few roots are infected, the plant can perish. When the fungus encounters the roots of the host, the chlamydospores germinate and invade the roots' vessels. Then the fungus rapidly grows inside the vessels and reaches the stem (Ghaemi et al. 2011; Laurence et al. 2012). Next, the fungus produces micro- and macroconidia (Fig. 3.1) in the vessels, which are carried up to the rest of the plant by the water flow (Benzohra et al. 2015). As the spores go up, they meet a wall that obstructs the vessel, the spores germinate, the germlings break the wall and then the fungus produces more spores on the opposite side of the wall. These new spores are transported to the subsequent wall. This continues until the fungus gets to the top of the plant and makes it wilt and die. After the plant is dead, the fungus keeps growing in the soft tissue and makes a lot of chlamydospores (Sutherland et al. 2013). Generally, Foa grows better when the host does too. The optimal temperatures for the fungus to grow are between 21 and 27 °C; the fungus grows slowly at 18 and 32 °C and ceases growth at 7 and 37 °C (Shabani et al. 2014; Shabani and Kumar 2013). The fungus can be spread through different means, such as soil, irrigation, infected plant material, and contaminated tools (Sedra 2018).

### *3.2.2 Hosts*

Bayoud disease has wiped out over 12 million date palm trees in Morocco over a 100 years. In 1981, more than 150,000 palm trees were removed in the Draa Valley, Morocco (Djerbi 1983). The fungus also destroyed nearly three million palm trees in Algeria (Benzohra et al. 2015). According to Toutain (1968), bayoud disease began to affect the Figuig oasis in eastern Morocco in the early twentieth century, destroying more than 50% of the high-quality and local date palm varieties. The mortality rate from this disease declined to 4.3% at the beginning of the twenty-first century, but some high-quality cultivars still remain vulnerable (Hakkou and Bouakka 2004). There is no effective way to prevent or cure this disease, so Morocco has developed new date palm cultivars that are resistant to bayoud disease. Morocco's strategy is to replace the infected palms with these new varieties and increase the date palm production (Moujaoui et al. 2021).

Foa impacts all premium date palm varieties in North Africa, including Mejhoul, Bou Feggous, and Deglet Nour. However, resistance has been observed in certain varieties, namely Black Bou Sthammi, Moussa, Sair Layalet, White Bou Sthammi, Tadment, Iklane, and Bou Feggous from Morocco, as well as Takerboucht from Algeria (Benzohra et al. 2015; Saaidi 1990, 1992). Despite this, only Takerboucht and Sair Layalet yield dates of acceptable quality which, while satisfactory, do not match the superior quality of Deglet Nour or Mejhoul. Research on various cultivars for the purpose of selecting resistant varieties since the 1980s has uncovered a seventh resistant variety, Bou Khanni, along with new strains that offer both resistance and high-quality dates (Sedra 1995, 2011a). Despite the availability of many varieties that can resist bayoud disease, most of the marketable ones, like Mejhool and Bouffegous, are very susceptible and need rigorous control measures to prevent the infection (Bouhlali et al. 2020). These varieties are mainly grown in new farms that are close to areas affected by bayoud, which poses a serious threat of spreading the disease to uninfected locations (Parween et al. 2019).

Some other plants that grow in date plantations are also hosts of Foa: *Trifolium* spp., *Medicago sativa* (lucerne), *and Lawsonia inermis* (henna) (EFSA Panel on Plant Health et al. 2018). These plants do not show symptoms of the disease, but they can spread the fungus. They are grown in Near East and North African countries (Bengyella et al. 2012; Benzohra et al. 2015). The fungus strains that come from these plants are not as harmful as those that come from diseased palm leaves (Sedra 2011b).

## 3.3 Bayoud Disease: Symptoms, Diagnosis, and Management

### 3.3.1 Symptoms

Bayoud disease originates from the soil-dwelling fungus Foa, which infects the vascular system of the palm and causes wilting, whitening and death of the fronds (Fig. 3.2), followed by the collapse of the whole palm (Essarioui et al. 2018). The symptoms of this disease is influenced by factors such as the palm's age and vulnerability, environmental conditions, as well as the pathogen's virulence (Moujaoui et al. 2021). The initial visible sign, discernible by those familiar with the disease, manifests as a dull or ash-grey hue on one or several middle crown leaves, which subsequently dry up distinctively (Benzohra et al. 2015). The deterioration of the frond begins at the bottom and moves upward on one side, then reverses from the top down on the opposite side. Along the back of the rachis, a longitudinal brown discoloration emerges, marking the trail of the mycelium through the vascular bundles. The affected fronds eventually dry up and hang down around the trunk, giving the palm a characteristic "skirt" appearance (Boulahouat et al. 2022). The disease may also affect the inflorescences, which become necrotic and abort. The palm may survive for several years in a state of decline, or may die rapidly within a few months (Ahmed et al. 2019).

**Fig. 3.2** First symptoms (**a**) and final stage of bayoud disease (**b**) caused by *Fusarium oxysporum* f. sp. *albedinis*. (Hakkou et al. 2012)

### 3.3.2 Diagnosis

The diagnosis of bayoud disease primarily involves observing external symptoms, isolating and identifying the pathogen from affected tissues, and employing molecular tools such as Polymerase Chain Reaction (PCR) and Restriction Fragment Length Polymorphism (RFLP) for confirmation (Fernandez et al. 1998; Samo and Ismailli 2020; Tantaoui et al. 1996). Development of tools capable of distinguishing isolates during the analysis of large sample volumes is essential. Equally important is the advancement of methods for pathogen detection in infected hosts. Furthermore, the creation of highly precise markers would be invaluable for securing patents on non-pathogenic beneficial strains with commercial potential (Sabbahi et al. 2009). Foa can be isolated from the rachis, the petiole, the roots, and the offshoots of the diseased palms, and cultured on potato dextrose agar or other suitable media (Meddich et al. 2018). The pathogen identification is achieved on the basis of morphological, physiological, and biochemical characteristics, such as the production of micro- and macro-conidia, the growth rate and colony color, the ability to assimilate different carbon and nitrogen sources, and the vegetative compatibility groups (Djerbi 1990). Molecular tools can provide a fast and accurate diagnosis of the pathogen, as well as its genetic diversity and population structure (El Bouhssini and Faleiro 2018; Raja et al. 2017).

### 3.3.3 Management

The management of bayoud disease is challenging and requires an integrated approach combining preventive, cultural, chemical, and biological methods (Bouhlali et al. 2020). The preventive methods involve the utilization of tolerant or resistant varieties, the certification of healthy planting material, the quarantine and eradication of infected palms, and the monitoring and surveillance of the disease (Khan et al. 2023; Sedra 2018). The cultural methods include the improvement of soil and water quality, soil solarization, the removal and burning of diseased fronds and offshoots, the disinfection of tools and equipment, and the avoidance of injury and stress to the palms (Essarioui et al. 2018; Essarioui and Sedra 2017). The chemical methods include the application of synthetic fungicides, such as copper oxychloride, carbamate, thiophanate-methyl, and fosetyl-aluminum (Bengyella et al. 2012; Loth et al. 2015; Qin et al. 2012). However, these methods have limited efficacy, high cost, and negative environmental and health impacts (de la Perrière et al. 1995). This makes it difficult to tackle Foa, particularly in the delicate ecosystems of the oases, which cannot withstand conventional pesticide control (Oulad Ziane et al. 2023). Biological methods include the utilization of beneficial microbes, such as fungi and bacteria, as well as bioactive plant products that can antagonize or suppress the pathogen, or induce systemic resistance in the plant (Bouhlali et al. 2020; Dihazi et al. 2012; El Hassni et al. 2007; Rahmouni et al.

2019). Table 3.1 presents a comparison of the advantages and disadvantages of these methods, illustrating the various tailored control strategies. However, the integrated disease management of Foa is still under research and development, and requires further evaluation and optimization to adopt and adapt the latest technologies and strategies that can improve date palm health and sustainability (Abumaali et al. 2023).

## 3.4 Beneficial Microbes as BCAs of *F. oxysporum*: Concepts and Mechanisms

Biological control involves employing living organisms or their products to inhibit the growth or activity of harmful organisms, such as plant pathogens. It is eco-friendly and sustainable, unlike synthetic pesticides that harm human health, soil quality, and biodiversity (Jaiswal et al. 2022; Sabbahi et al. 2022; Tariq et al. 2020). For the past 20 years, researchers have shown the efficacy of microorganisms such as *Bacillus* spp., *Pseudomonas* spp., *Trichoderma* spp., and non-pathogenic *F. oxysporum*, among other fungi and bacteria, to act as BCAs against *Fusarium* wilt. BCAs were tested throughout different ways of acting, such as hyperparasitism, antibiosis, induction of host resistance, and competition, to assess their biocontrol potential, which includes the ability to produce antibiotics, lytic and antioxidant enzymes, siderophores, and the stimulation of host resistance. These BCAs can be deployed singly or in synergy in combination, adopting various formulations and application systems like seed treatment, soil amendment, or a combination of both. Their efficacy is tested under a range of conditions—considering factors like host plant variety, pathogen type, inoculum level, BCA variant, ambient temperature, and planting time—in laboratory, greenhouse, and field environments to determine optimal performance (Purohit et al. 2023).

BCAs enhance disease management through microbial antagonism, which encompasses both direct interactions with the pathogen and indirect actions affecting the host (Ayaz et al. 2023; Köhl et al. 2019). Direct interactions often involve antagonistic behaviors like antibiosis, competition for nutrients, or parasitism by the biocontrol agent against the pathogen. Indirectly, BCAs can bolster plant defense systems to resist pathogen invasion. The combined impact of these interactions is evident in the use of certain BCAs (Pertot et al. 2017).

BCAs can engage in competition with pathogens over resources such as nutrients and space, or for potential infection sites on roots or within the rhizosphere, thereby diminishing the pathogen's numbers and harmfulness (Çakar and Tozlu 2022). For instance, certain variants of *B. subtilis* and *P. fluorescens* are known to foster plant growth by enhancing nutrient absorption via siderophores and organic acids that facilitate phosphorus solubilization. They also produce phytohormones and act as bacterial antagonists, either by depriving pathogens of vital elements or by triggering the plant's own resistance mechanisms. Additionally, they can break down fungal

**Table 3.1** Comparative analysis of strategies for managing *Fusarium oxysporum f.* sp. *albedinis* in date palm groves (Benzohra et al. 2015; Essarioui et al. 2018; Haq and Khan 2020; Khan et al. 2023; Sedra 2018)

| Method | Example | Pros | Cons |
|---|---|---|---|
| Chemical control | Applying fungicides with varied mechanisms of action in a rotational manner to reduce the likelihood of resistance emergence. | Effectiveness: Efficient in reducing the spread of diseases.<br>Rapid Action: Suppresses the pathogen swiftly.<br>Broad Spectrum: Certain chemicals can target a wide range of pathogens. | Environmental Impact: Potential harm to beneficial organisms.<br>Resistance Development: Pathogens may develop resistance over time.<br>Health Hazards: Could be harmful to humans and animals. |
| Biological control | Employing beneficial microbes and bioactive plant products | Environmentally Friendly: Employs natural enemies to suppress pathogens.<br>Sustainability: Does not harm the environment and reduces dependency on chemicals.<br>Long-Term Protection: Offers enduring protection against diseases. | Specificity: Some agents may only target specific pathogens.<br>Variable Efficacy: Effectiveness can depend on environmental conditions.<br>Research Requirement: Continuous study and research are needed. |
| Cultural practices | Implementation of practices like soil solarization, removing and incinerating diseased fronds and offshoots, and the avoidance of removing the spines by pulling | Preventive Measures: Helps in preventing the spread of diseases.<br>Cost-Effectiveness: These practices are generally low-cost and easily accessible.<br>Integrated Approach: Complements other disease control methods effectively. | Labor Intensity: Requires consistent and dedicated effort.<br>Non-curative: Cannot cure already infected trees.<br>Compliance Dependence: The success of these practices heavily relies on strict adherence. |
| Resistant varieties | Cultivation of naturally resistant palm varieties | Inherent Resistance: Certain palm varieties have natural resistance to bayoud disease.<br>Low Maintenance: Requires minimal care once established.<br>Sustainability: Does not harm the environment and reduces dependency on chemicals. | Limited Options: Not all palm varieties are resistant.<br>Slow Transition: Adoption of resistant varieties takes time.<br>Genetic Diversity Concerns: Over-reliance on certain varieties can reduce overall genetic diversity. |
| Prophylactic measures | Implementation of monitoring and surveillance of disease spread | Disease Introduction Prevention: Helps in delaying the introduction of Foa into healthy | Limited Relevance: Not very useful in regions already affected by the disease. |

(continued)

**Table 3.1** (continued)

| Method | Example | Pros | Cons |
|---|---|---|---|
| | | regions.<br>Awareness Enhancement: Increases grower awareness about disease prevention.<br>Phytosanitary Controls: Monitors the exchange of materials (soil, shoots, machinery) within and between palm groves. | Eradication Limitations: Can slow down but not completely stop the disease. |

structures by producing and releasing hydrolytic enzymes (Beneduzi et al. 2012; Khan et al. 2017). Furthermore, some BCAs are capable of synthesizing plant growth regulators like auxins, cytokinins, and gibberellins, which aid in root development and expand the root surface area, consequently lowering the likelihood of pathogen invasion (Lahlali et al. 2022). An extensive root system is beneficial for plant survival as it ensures continued access to essential nutrients and water, even when portions of the roots are compromised by infection (Nadeem et al. 2023).

BCAs can produce antimicrobial substances, such as lytic enzymes and antibiotics, which can inhibit the growth or activity of the pathogen (Sehrawat and Sindhu 2019). For example, certain isolates of *T. harzianum* and *Clonostachys rosea* can secrete chitinases, proteases, and glucanases that can dismantle the cell walls of *F. oxysporum* (Sahgal 2021; Sharma et al. 2016). These enzymes are also capable of regulating the plant's immune response and fostering resistance to pathogens. Moreover, certain BCAs can emit volatile organic compounds (VOCs) like alcohols, aldehydes, terpenes, and ketones, which can propagate through the atmosphere to suppress the growth and spore formation of pathogens (Ayaz et al. 2023). These compounds can also affect the plant metabolism and induce systemic resistance.

Certain non-pathogenic isolates of *F. oxysporum* are recognized for their ability to defend crops from *F. oxysporum* pathogens (Iida et al. 2022). They can colonize the xylem vessels and prevent the pathogen from reaching the upper parts of the plant (Narayanasamy 2013). These strains can also produce bioactive secondary metabolites, such as fusaric acid and cyclosporines, that can inhibit the pathogen's growth and toxin production (Al-Ani 2019). Furthermore, some BCAs can form endophytic associations with the plant and protect it from the pathogen by producing antibiotics or inducing resistance. For instance, the endophytic *F. oxysporum* strain Fo47 is one of the best-studied BCAs (Aimé et al. 2013). The strain Fo47 establishes itself on the root surface and in the adjacent soil, where it competes with pathogens for nutrients, as reported by Iida et al. (2022). Furthermore, Fo47 triggers the plant's defense mechanisms, but according to de Lamo et al. (2021), not through the commonly recognized pathways involving pattern-triggered immunity, salicylic acid, ethylene, or jasmonic acid. Additionally, non-pathogenic *F. oxysporum* strains can stimulate the host plant's defense against pathogens, enhance tolerance to environmental stressors, and promote the synthesis of hormones like gibberellins and auxins (Iida et al. 2022; Sajeena et al. 2020).

BCAs can initiate a variety of plant defense actions, including the generation of phytoalexins, pathogenesis-related proteins, and reactive oxygen species, thereby bolstering the plant's defense or tolerance against pathogens, as noted by Guo et al. (2019) and Lahlali et al. (2022). According to Kloepper et al. (1992), the introduction of beneficial microbes leads to a state of heightened resistance known as induced systemic resistance (ISR), which equips the host with defenses against a wide array of pathogens. When ISR is activated, plants engage in extensive systemic signaling to shield distal tissues, prompting swift and potent immune responses to pathogen attacks. ISR signal transmission has traditionally been viewed as dependent on jasmonic acid and ethylene, but not on salicylic acid, as per Yu et al. (2022). However, certain pathogenic microbes are also capable of triggering systemic acquired resistance, a durable and wide-ranging resistance that can be passed down to subsequent plant generations (Durrant and Dong 2004).

Research has demonstrated the potential use of some BCAs against *Fusarium* wilt for fruit, legumes and other crops (e.g. Banana, melon, onion, tomato, and potato) (Arie 2019; Çakar and Tozlu 2022; Shin et al. 2023). However, it is imperative to promptly comprehend the action mechanisms of these BCAs, as well as assessing them under varying physiological and environmental scenarios. Additionally, conducting extensive field trials is vital for the successful commercialization of efficient BCAs. Leveraging the progress in high-throughput DNA sequencing and diverse omics technologies, this accumulated knowledge can facilitate the commercial application of biological control strategies in combating *Fusarium* wilt in fruits, legumes, and other crops.

## 3.5 Biological Control of Bayoud Disease: Potential and Challenges

Efforts to identify biological strategies for managing bayoud disease in date palm groves are still in the early stages. Current knowledge about the microorganisms capable of shielding palm trees from Foa is sparse, and even less is understood about how these organisms can trigger the plant's defense mechanisms. Consequently, exploring the microflora associated with *P. dactylifera* to find native bacteria that could serve as antagonists is considered the most promising approach to combat this affliction, as suggested by Boulahouat et al. (2022). In their research, El Hassni et al. (2007) evaluated 21 bacterial strains from the rhizosphere of unaffected date palms for their capacity to counteract Foa both in the lab and in living plants. Among these, four strains—*B. cereus* X16, *B. pumilus* W1, *Rahnella aquatilis* W2, and an unidentified strain S1—demonstrated significant suppression of Foa's mycelial expansion (70–77%) and spore production (80–95% compared to the control). Additionally, the Foa mycelium that developed near the zone of inhibition exhibited noticeable cytological changes. Introducing these antagonistic bacteria to date palm seedlings initiated defense responses, including the build-up of non-constitutive hydroxycinnamic acid derivatives like the sinapic acid derivative I2, which is crucial for the date palm's resistance to Foa. This defensive response was more evident in the resistant cultivar "Bousthami Noir" than in the susceptible "Jihel." The potential of leveraging both the indirect and direct effects of these antagonistic bacteria as part of a biocontrol strategy against bayoud is seen with optimism.

The study by Dihazi et al. (2012) explored the efficacy of the antagonistic bacterial species *B. amyloliquefaciens* strain Ag1 and *Burkholderia cepacia* strain Cs5 in combating bayoud disease. Both strains demonstrated the ability to suppress the growth and sporulation of Foa. They secreted substances that altered the cytological structure of Foa's mycelium. A month post-inoculation, the defensive response of date palms to Foa varied based on the bacterial strain applied. *B. cepacia* was associated with an increased presence of caffeoylshikimic acid isomers, while *B. amyloliquefaciens* prompted the production of novel phenolic compounds, specifically hydroxycinnamic acid derivatives. Additionally, peroxidase activity was notably enhanced, differing depending on the type of bacterial strain and the presence of Foa. Further research by Belaidi et al. (2021) assessed the impact of other antagonistic fungal species, *Trichoderma harzianum* and *Aspergillus flavus*, on controlling Foa in the Deglet-Nour date palm variety, a commercially susceptible variety, under greenhouse conditions. These bacterial strains were sourced from the rhizosphere soil surrounding date palm trees. The trial involved treating 3- to 4-month-old date palm seedlings with a 5 ml inoculum suspension. The outcome revealed total resistance in seedlings treated with *T. harzianum*, whereas *A. flavus* decreased the mortality rate to 25%. Consequently, further investigations are necessary to refine the application of antagonistic fungi to safeguard date palm groves from bayoud disease.

The study of mycorrhization in date palms is a critical research field, encompassing the plant's growth and yield as well as its role in managing bayoud disease. To this end, researchers tested both arbuscular mycorrhizal fungi (AMF) and plant growth-promoting rhizobacteria (PGPR) consortia for their effectiveness in controlling *Fusarium* wilt. The individual application of either PGPR or AMF consortia led to a reduction in disease occurrence, severity, and death rate of date palm seedlings (Oulad Ziane et al. 2023). However, the combined use of these consortia yielded the highest control efficacy, indicating a substantial synergistic impact. This enhanced control was paralleled by a boost in plant immunity, marked by the activation of the phenylpropanoid pathway, as evidenced by increased phenylalanine ammonia-lyase (PAL) activity and higher levels of polyphenols and lignin in the roots. Antagonistic in vitro tests against Foa showed that PGPR strains, especially *B. cereus* strain 263AG5, inhibited Foa growth by 60.59%. Furthermore, 27 weeks post-inoculation with Foa, the dual application of consortia resulted in a threefold increase in mycorrhization intensity, arbuscules, and vesicles, and a 1.5-fold rise in mycorrhization frequency (Oulad Ziane et al. 2023). The notable synergy between PGPR and AMF consortia in shielding date palms from Foa may be due to their combined antagonistic effects and potent stimulation of the plant's innate defenses, suggesting that this dual approach could be a viable strategy for pathogen mitigation.

El Hassni et al. (2004) observed that treating date palm seedlings with a less aggressive strain of *F. oxysporum* could shield them from subsequent invasions by a more virulent strain of the fungus. The less aggressive strain didn't directly counteract the virulent strain; instead, it hastened the build-up of certain phenolic compounds and triggered an earlier response of peroxidase enzymes in both susceptible and resistant date palm varieties. Additionally, Jaiti et al. (2007) investigated the impact of AMF from the Aoufous date palm grove in southern Morocco on the defense responses of date palms to Foa. They found that AMF application decreased disease severity by 8–77%, depending on the AMF strain used. Additionally, all AMF strains tested significantly improved shoot growth, biomass, and leaf count. The plants associated with the Aoufous complex showed the most substantial growth and were highly effective in reducing bayoud disease incidence. The AMF also modified the activity of defense enzymes such as peroxidases and polyphenoloxidases, leading to a significant increase in non-constitutive hydroxycinnamic acid derivatives, especially the sinapic derivative I2, which is essential for the date palm's resistance to the pathogen (Jaiti et al. 2007).

Plants harbor a wealth of bioactive compounds that remain largely unexplored, providing a solid foundation for developing novel biological strategies to effectively combat bayoud disease. Bouhlali et al. (2020) assessed the antifungal properties of extracts from five plant species—*Acacia cyanophylla*, *Cupressus atlantica*, *Eucalyptus torquata*, *Nerium oleander*, and *Schinus molle*—against Foa. Notably, the extracts from *E. torquata* and *C. atlantica* exhibited potent antifungal effects, inhibiting mycelial growth, sporulation, and spore germination in a concentration-dependent manner. The antifungal efficacy of these extracts correlated with their content of polyphenols and flavonoids, as well as their antioxidant activities, suggesting that their inhibitory actions are partially attributed to their abundance of antioxidative secondary metabolites. Utilizing these plants as composted material could serve as a sustainable and eco-friendly alternative to traditional fungicides. Further research is necessary to refine and apply a compost-based biological method for managing bayoud disease in date palm oases. In a subsequent study, Bouderba et al. (2022) explored the in vitro antifungal effects of flavonoid extracts from the leaves of two medicinal plants, *Salvia officinalis* and *Ocimum basilicum*, against Foa. The flavonoid extract from *O. basilicum* inhibited mycelium growth by 73.04% at a concentration of 3.17 µg/ml, while the extract from *S. officinalis* achieved a 57.42% inhibition at 1.13 µg/ml. These encouraging findings support further investigation into identifying the most potent active compounds and examining their synergistic effects, which could yield even more effective results in field conditions.

Rahmouni et al. (2019) conducted a study on the influence of essential oils (EOs) and their major compounds from five medicinal and aromatic plants (MAPs) on Foa. The findings revealed that EOs from *Origanum compactum* and *Thymus satureioides* inhibited Foa growth at minimal fungicidal concentration (MFC) of 5 and 13 µl/ml, respectively, outperforming the synthetic fungicide mancozeb (MFC of 6 µl/ml) and significantly surpassing the EOs from *Rosmarinus officinalis*,

*Lavandula dentata*, and *Myrtus communis* (MFC of 40 µl/ml). The compounds thymol and carvacrol were notably potent, with MFCs of 0.94 and 2.08 µl/ml, respectively, followed by α-terpineol (MFC of 10 µl/ml), linalool (MFC of 13.00 µl/ml), and borneol (MFC of 26.67 µl/ml). Other components like α-pinene, eucalyptol, and myrtenyl acetate also demonstrated fungicidal properties but required higher concentrations (20–80 µl/ml). The efficacy of EOs from *Origanum* and *Thymus* species is likely tied to the high levels of carvacrol and thymol. Considering the detrimental effects of Foa on date palms and the environmental and health concerns associated with fungicides, employing EOs from *T. satureioides* and *O. compactum*, or their key elements carvacrol and thymol, could offer a viable alternative to conventional synthetic fungicides.

Utilizing compost for the biological control of soil-dwelling plant pathogens is a well-recognized practice in horticulture. In this vein, Mohamed et al. (2020) assessed the impact of compost extracts on controlling Foa and explored their mechanisms of action, as well as the factors indicating compost maturity. Compost derived from date palm residuals, processed via a forced aeration system, proved non-toxic to plants, as evidenced by a germination index of 83.78%. This was in stark contrast to two commercial composts that exhibited phytotoxicity. A negative correlation was observed between compost maturity and factors such as carbon/nitrogen ratio, salinity, and total soluble phenols. Pathogen indicator levels remained below the recommended thresholds across all compost samples. Efficacy tests revealed that compost made from date palm waste outperformed others; a 10% concentration of its unsterilized extract inhibited Foa completely (100% inhibition rate) on the sixth day of incubation. However, the inhibition rate for the microfiltered extract did not surpass 30%. All sterilized extracts lost their inhibitory capabilities against the target fungus, indicating that the suppression is largely attributable to the native microbial population. These results underscore the importance of assessing compost maturity to determine the quality and the effectiveness of date palm waste compost in suppressing Foa.

Recent advancements in biocontrol strategies include the enhancement of plant resistance through the activation of their innate defenses, a method that has demonstrated considerable promise in safeguarding crops within sustainable agricultural systems (Jamiołkowska 2020). Formulations that stimulate plant defenses, some derived from seaweed polysaccharides, have been developed and are now commercially available (Ben Salah et al. 2018; Krishna Perumal et al. 2023). Specifically, Bouissil et al. (2022) discovered that alginate from the brown seaweed *Bifurcaria bifurcata* boosts the natural defense mechanisms in date palm roots, notably by elevating PAL activity and the build-up of phenolic compounds. Following a 17-week period after Foa inoculation, alginate treatment decreased plant mortality by up to 80% compared to untreated controls, causing the activation of defense-related gene expression and the fortification of date palm's natural defenses within 24 h of alginate application. These findings suggest a bright future for alginate use in agriculture as a stimulant that activates date palm immunity against Foa. In a separate study, Dihazi et al. (2011) conducted histochemical and ultrastructural studies to evaluate the structural and biochemical alterations in date palm roots pre-treated with salicylic acid and then exposed to Foa. The presence of flavonoids, induced proteins, and heightened peroxidase activity in the roots of salicylic acid-prepared plants post-Foa exposure was indicative of increased resistance to the pathogen. The defensive response in these plants included the obstruction of intercellular spaces, accumulation of electron-dense materials at fungal entry points, and causing various damages to the fungal cells. Conversely, plants that were not pre-treated exhibited significant cell wall breakdown and complete cytoplasmic disarray upon Foa inoculation, highlighting the protective role of salicylic acid.

## 3.6 Conclusion and Future Perspectives

Bayoud disease is still devastating date palm trees and threatens the oasis ecosystem as well as the date industry in North Africa. Employing beneficial microbes offers a viable solution for the sustainable management of bayoud disease and, therefore, the preservation of *P. dactylifera* both as a valuable crop and as an important part of local cultural heritage. Research has identified and characterized various BCAs from diverse sources, demonstrating their efficacy against Foa both in vitro and in vivo. These BCAs can inhibit or reduce the pathogen by different mechanisms, such as competition, antagonism, parasitism, or induced resistance. They can also promote plant growth and yield by producing plant growth regulators, siderophores, or enzymes, or by improving the soil quality and fertility. Despite their potential, the application of beneficial microbes for bayoud disease faces some challenges and limitations, such as the variability of their efficacy under different environmental conditions, the compatibility and stability of their formulations, the regulation and registration of their products, and their acceptance and adoption for use by the farmers. Moreover, there are still major gaps in the research and development of beneficial microbes as BCAs for bayoud disease, such as the identification and characterization of novel and effective BCAs, the elucidation of their modes of action and interactions with plants and pathogens, the optimization of their production and application methods, the evaluation of their long-term effects and safety, and their integration with other management strategies.

Therefore, some recommendations and directions for future research on the use of beneficial microbes as BCAs for bayoud disease are suggested, as follows:

- The screening and selection of BCAs should be based on multiple criteria, such as their broad-spectrum and consistent antagonistic activity against Foa, their high colonization and survival ability on date palm roots or within its endosphere or rhizosphere, their high adaptability and stability under different environmental conditions, their beneficial impact on the plant growth and yield, their low risk of pathogenicity, toxicity, or resistance, and their compatibility with other BCAs or management practices.
- The characterization and identification of BCAs should be performed using molecular and biochemical tools, such as PCR, RFLP, sequencing,

MALDI-TOF, and HPLC, to determine their taxonomic and functional diversity, and to assess their genetic and metabolic potential.
- The mode of action and interaction of BCAs should be investigated using omics and imaging techniques, such as genomics, transcriptomics, proteomics, metabolomics, and microscopy, to reveal their molecular and cellular mechanisms of antagonism or resistance induction, and to visualize their spatial and temporal distribution and colonization on the plant and with regard to the pathogen.
- The production and application of BCAs should be optimized using biotechnological and engineering approaches, such as fermentation, encapsulation, microfluidics, and nanotechnology, in order to enhance their yield, viability, stability, delivery, and release.
- The evaluation and validation of BCAs should be conducted using in vivo and in situ experiments, such as greenhouse and field trials, to measure their efficacy, durability, and safety under realistic conditions, and to compare their performance with conventional methods.
- The integration and implementation of BCAs should be supported by multidisciplinary and participatory research, involving agronomists, microbiologists, biotechnologists, plant pathologists, extension agents, and farmers, in order to develop and disseminate appropriate and acceptable protocols and practices for the use of BCAs in date palm.

**AI Declaration** AI (DeepSeek-V3) was used only for language polishing and grammar refinement. All scientific content, structure, conceptual development, and critical analysis are the results of our work. All chapter paragraphs were reviewed, edited, and verified by the authors, who take full responsibility for the final content. The chapter also underwent formal peer review.

## References

Abumaali DA, Al-Hadidi SH, Daly Yahia MN, Erfanian MB, Ahmed TA, Alatalo JM (2023) The date palm microbiome: a review. Ecol Genet Genomics 29:100212. https://doi.org/10.1016/j.egg.2023.100212

Ahmed Y, Hussien A, El-badry N, Soliman MS (2019) Validation of a diagnostic protocol for the detection of fusarium oxysporum f.sp. albedinis, the causal agent of Bayoud disease of date palm. Egypt J Phytopathol 47:297–312. https://doi.org/10.21608/ejp.2019.126144

Aimé S, Alabouvette C, Steinberg C, Olivain C (2013) The endophytic strain Fusarium oxysporum Fo47: a good candidate for priming the defense responses in tomato roots. Mol Plant Microbe Interact 26:918–926. https://doi.org/10.1094/MPMI-12-12-0290-R

Al-Ani LKT (2019) Secondary metabolites of non-pathogenic fusarium: scope in agriculture. In: Singh HB, Keswani C, Reddy MS, Sansinenea E, García-Estrada C (eds) Secondary metabolites of plant growth promoting rhizomicroorganisms: discovery and applications. Springer, Singapore, pp 59–76. https://doi.org/10.1007/978-981-13-5862-3_3

Al-Farsi M, Alasalvar C, Morris A, Baron M, Shahidi F (2005) Comparison of antioxidant activity, anthocyanins, carotenoids, and phenolics of three native fresh and sun-dried date (Phoenix dactylifera L.) varieties grown in Oman. J Agric Food Chem 53:7592–7599. https://doi.org/10.1021/jf050579q

Ali M, Hussain S, Nadeem A, Ullah S, Yasin M (2023) Studies on biological management of Fusarium Wilt of tomato. Gesunde Pflanzen 75:1475–1483. https://doi.org/10.1007/s10343-022-00800-5

Al-Khayri JM, Jain SM, Johnson DV (eds) (2015) Date palm genetic resources and utilization: volume 2: Asia and Europe. Springer Netherlands, Dordrecht. https://doi.org/10.1007/978-94-017-9707-8

Arie T (2019) Fusarium diseases of cultivated plants, control, diagnosis, and molecular and genetic studies. J Pestic Sci 44:275–281. https://doi.org/10.1584/jpestics.J19-03

Ayaz M, Li C-H, Ali Q, Zhao W, Chi Y-K, Shafiq M, Ali F, Yu X-Y, Yu Q, Zhao J-T, Yu J-W, Qi R-D, Huang W-K (2023) Bacterial and fungal biocontrol agents for plant disease protection: journey from lab to field, current status, challenges, and global perspectives. Molecules 28: 6735. https://doi.org/10.3390/molecules28186735

Belaidi H, Toumi-Benali F, Benzohra IE (2021) Biocontrol of Bayoud disease (Fusarium oxysporum F. Sp. Albedinis) on Deglet-nour variety of date palm (Phoenix dactylifera L.) in South Western oases of Algeria. Agric Sci Dig 41(3):450–454

Ben Salah I, Aghrouss S, Douira A, Aissam S, El Alaoui-Talibi Z, Filali-Maltouf A, El Modafar C (2018) Seaweed polysaccharides as bio-elicitors of natural defenses in olive trees against verticillium wilt of olive. J Plant Interact 13:248–255. https://doi.org/10.1080/17429145.2018.1471528

Beneduzi A, Ambrosini A, Passaglia LMP (2012) Plant growth-promoting rhizobacteria (PGPR): their potential as antagonists and biocontrol agents. Genet Mol Biol 35:1044–1051. https://doi.org/10.1590/S1415-47572012000600020

Bengyella L, Pranab R, Elsie Y, Sayanika DW (2012) The farmers cry: impact of heat stress on Fusarium oxysporum f. sp. dianthi, interaction with fungicides. Asian J Plant Pathol 6:19–24. https://doi.org/10.3923/ajppaj.2012.19.24

Benzohra IE, Megateli M, Berdja R (2015) Bayoud disease of date palm in Algeria: history, epidemiology and integrated disease management. Afr J Biotechnol 14:542–550. https://doi.org/10.5897/AJBX2014.14292

Bouderba NN, Khalid KM, Kadi H (2022) In vitro inhibitory effect of some secondary metabolites extract from Ocimum basilicum and Salvia officinalis against Fusarium oxysporum f.sp albedinis of Phoenix dactylifera. L. Appl Biochem Biotechnol 194:5655–5665. https://doi.org/10.1007/s12010-022-04067-4

Bouhlali EDT, Derouich M, Ben-Amar H, Meziani R, Essarioui A (2020) Exploring the potential of using bioactive plant products in the management of Fusarium oxysporum f.sp. albedinis: the causal agent of Bayoud disease on date palm (Phoenix dactylifera L.). Beni-Suef Univ J Basic Appl Sci 9:46. https://doi.org/10.1186/s43088-020-00071-x

Bouissil S, Guérin C, Roche J, Dubessay P, El Alaoui-Talibi Z, Pierre G, Michaud P, Mouzeyar S, Delattre C, El Modafar C (2022) Induction of defense gene expression and the resistance of date palm to Fusarium oxysporum f. sp. Albedinis in response to alginate extracted from Bifurcaria bifurcata. Mar Drugs 20:88. https://doi.org/10.3390/md20020088

Boulahouat S, Cherif-Silini H, Silini A, Chenari Bouket A, Luptakova L, Saadaoui N, Alenezi FN, Belbahri L (2022) Critical evaluation of biocontrol ability of Bayoud infected date palm phyllospheric Bacillus spp suggests that in vitro selection does not guarantee success in planta. Agronomy 12:2403. https://doi.org/10.3390/agronomy12102403

Boulahouat S, Cherif-Silini H, Silini A, Bouket AC, Luptakova L, Alenezi FN, Belbahri L (2023) Biocontrol efficiency of rhizospheric Bacillus against the plant pathogen fusarium oxysporum: a promising approach for sustainable agriculture. Microbiol Res 14:892–908. https://doi.org/10.3390/microbiolres14030062

Boutaj H, Meddich A, Roche J, Mouzeyar S, El Modafar C (2022) The effects of mycorrhizal fungi on vascular wilt diseases. Crop Prot 155:105938. https://doi.org/10.1016/j.cropro.2022.105938

Bubici G, Kaushal M, Prigigallo MI, Gómez-Lama Cabanás C, Mercado-Blanco J (2019) Biological control agents against Fusarium wilt of banana. Front Microbiol 10:616. https://doi.org/10.3389/fmicb.2019.00616

Çakar G, Tozlu E (2022) The biological control of fusarium oxysporum, the causal agent of potato rot. Gesunde Pflanzen 74:305–315. https://doi.org/10.1007/s10343-021-00610-1

de la Perrière RAB, Amir H, Bounaga N (1995) Prospects for integrated control of 'bayoud' (Fusarium wilt of the date palm) in Algerian plantations. Crop Prot 14:227–235. https://doi.org/10.1016/0261-2194(95)00007-9

de Lamo FJ, Šimkovicová M, Fresno DH, de Groot T, Tintor N, Rep M, Takken FLW (2021) Pattern-triggered immunity restricts host colonization by endophytic fusaria, but does not affect endophyte-mediated resistance. Mol Plant Pathol 22:204–215. https://doi.org/10.1111/mpp.13018

Dihazi A, Serghini MA, Jaiti F, Daayf F, Driouich A, Dihazi H, El Hadrami I (2011) Structural and biochemical changes in salicylic-acid-treated date palm roots challenged with *Fusarium oxysporum* f sp *albedinis*. J Pathog 2011:e280481. https://doi.org/10.4061/2011/280481

Dihazi A, Jaiti F, WafaTaktak, kilani-Feki O, Jaoua S, Driouich A, Baaziz M, Daayf F, Serghini MA (2012) Use of two bacteria for biological control of bayoud disease caused by *Fusarium oxysporum* in date palm (*Phoenix dactylifera* L) seedlings. Plant Physiol Biochem 55:7–15. https://doi.org/10.1016/j.plaphy.2012.03.003

Djerbi M (1983) Report on consultancy mission on date palm pests and diseases AGRIS.FAO2880/043

Djerbi M (1990) Méthodes de diagnostic du bayoud du palmier dattier. EPPO Bull 20:607–613. https://doi.org/10.1111/j.1365-2338.1990.tb00184.x

Durrant WE, Dong X (2004) Systemic acquired resistance. Annu Rev Phytopathol 42:185–209. https://doi.org/10.1146/annurev.phyto.42.040803.140421

EFSA Panel on Plant Health, Jeger M, Bragard C, Caffier D, Candresse T, Chatzivassiliou E, Dehnen-Schmutz K, Gilioli G, Grégoire J-C, Jaques Miret JA, MacLeod A, Navajas Navarro M, Niere B, Parnell S, Potting R, Rafoss T, Urek G, Van Bruggen A, Van der Werf W, West J, Winter S, Armengol Forti J, Vloutoglou I, Bottex B, Rossi V (2018) Pest categorisation of Fusarium oxysporum f. sp. albedinis. EFSA J 16:e05183. https://doi.org/10.2903/j.efsa.2018.5183

El Bouhssini M, Faleiro JRH (2018) Date palm pests and diseases: integrated management guide. International Center for Agricultural Research in the Dry Areas (ICARDA)

El Hassni M, J'Aiti F, Dihazi A, Ait Barka E, Daayf F, El Hadrami I (2004) Enhancement of defence responses against Bayoud disease by treatment of date palm seedlings with an hypoaggressive Fusarium oxysporum isolate. J Phytopathol 152:182–189. https://doi.org/10.1111/j.1439-0434.2004.00824.x

El Hassni M, El Hadrami A, Daayf F, Chérif M, Barka EA, El Hadrami I (2007) Biological control of bayoud disease in date palm: selection of microorganisms inhibiting the causal agent and inducing defense reactions. Environ Exp Bot 59:224–234. https://doi.org/10.1016/j.envexpbot.2005.12.008

Essarioui A, Sedra MH (2017) Lutte contre la maladie du bayoud par solarisation et fumigation du sol. Une expérimentation dans les palmeraies du Maroc. Cah Agric 26:45010. https://doi.org/10.1051/cagri/2017043

Essarioui A, Ben-Amar H, Khoulassa S, Meziani R, Amamou A, Mokrini F (2018) Gestion du Bayoud du palmier dattier dans les oasis marocaines. Rev Marocaine Sci Agron Vét 6:537–543

Fernandez D, Ouinten M, Tantaoui A, Geiger JP, Daboussi MJ, Langin T (1998) Fot 1 insertions in the Fusarium oxysporum f. sp. albedinis genome provide diagnostic PCR targets for detection of the date palm pathogen. Appl Environ Microbiol 64:633–636. https://doi.org/10.1128/AEM.64.2.633-636.1998

Ghaemi A, Rahimi A, Banihashemi Z (2011) Effects of water stress and Fusarium oxysporum f. sp. lycoperseci on growth (leaf area, plant height, shoot dry matter) and shoot nitrogen content of tomatoes under greenhouse conditions. Iran Agric Res 29:51–62. https://doi.org/10.22099/iar.2011.136

Guo Q, Li Y, Lou Y, Shi M, Jiang Y, Zhou J, Sun Y, Xue Q, Lai H (2019) Bacillus amyloliquefaciens Ba13 induces plant systemic resistance and improves rhizosphere microecology against tomato yellow leaf curl virus disease. Appl Soil Ecol 137:154–166. https://doi.org/10.1016/j.apsoil.2019.01.015

Hakkou A, Bouakka M (2004) Oasis de Figuig: état actuel de la palmeraie et incidence de la fusariose vasculaire. Sci Chang Planétaires/Sécheresse 15:147–158

Hakkou A, Chakroune K, Souna F, Bouakka M (2012) La fusariose vasculaire du palmier dattier (Bayoud): méthodes de lutte. Communication présentée à la Conférence multidisciplinaire sur les bases écologiques et culturelles de l'oasis de Figuig (Maroc), Oujda, Maroc

Haq IU, Khan NA (2020) Fungal diseases of date palm (Phoenix dactylifera): etiology and management. In: Ul Haq I, Ijaz S (eds) Etiology and integrated management of economically important fungal diseases of ornamental palms, sustainability in plant and crop protection. Springer, Cham, pp 169–196. https://doi.org/10.1007/978-3-030-56330-1_10

Iida Y, Ogata A, Kanda H, Nishi O, Sushida H, Higashi Y, Tsuge T (2022) Biocontrol activity of nonpathogenic strains of fusarium oxysporum: colonization on the root surface to overcome nutritional competition. Front Microbiol 13:826677. https://doi.org/10.3389/fmicb.2022.826677

Jaiswal DK, Gawande SJ, Soumia PS, Krishna R, Vaishnav A, Ade AB (2022) Biocontrol strategies: an eco-smart tool for integrated pest and diseases management. BMC Microbiol 22:324. https://doi.org/10.1186/s12866-022-02744-2

Jaiti F, Meddich A, El Hadrami I (2007) Effectiveness of arbuscular mycorrhizal fungi in the protection of date palm (*Phoenix dactylifera* L.) against bayoud disease. Physiol Mol Plant Pathol 71:166–173. https://doi.org/10.1016/j.pmpp.2008.01.002

Jamiołkowska A (2020) Natural compounds as elicitors of plant resistance against diseases and new biocontrol strategies. Agronomy 10:173. https://doi.org/10.3390/agronomy10020173

Jangir M, Sharma S, Sharma S (2021) Development of next-generation formulation against Fusarium oxysporum and unraveling bioactive antifungal metabolites of biocontrol agents. Sci Rep 11:22895. https://doi.org/10.1038/s41598-021-02284-1

Khan N, Maymon M, Hirsch AM (2017) Combating fusarium infection using bacillus-based antimicrobials. Microorganisms 5:75. https://doi.org/10.3390/microorganisms5040075

Khan N, Martínez-Hidalgo P, Ice TA, Maymon M, Humm EA, Nejat N, Sanders ER, Kaplan D, Hirsch AM (2018) Antifungal activity of bacillus species against fusarium and analysis of the potential mechanisms used in biocontrol. Front Microbiol 9:2363. https://doi.org/10.3389/fmicb.2018.02363

Khan RR, Haq IU, Naqvi SA (2023) Pest and disease management in date palm. In: Date palm, crop production science in horticulture, pp 297–338. https://doi.org/10.1079/9781800620209.0010

Kloepper JW, Tuzun S, Kuć JA (1992) Proposed definitions related to induced disease resistance. Biocontrol Sci Tech 2:349–351. https://doi.org/10.1080/09583159209355251

Köhl J, Kolnaar R, Ravensberg WJ (2019) Mode of action of microbial biological control agents against plant diseases: relevance beyond efficacy. Front Plant Sci 10:845. https://doi.org/10.3389/fpls.2019.00845

Krishna Perumal P, Huang C-Y, Chen C-W, Anisha GS, Singhania RR, Dong C-D, Patel AK (2023) Advances in oligosaccharides production from brown seaweeds: extraction, characterization, antimetabolic syndrome, and other potential applications. Bioengineered 14:2252659. https://doi.org/10.1080/21655979.2023.2252659

Lahlali R, Ezrari S, Radouane N, Kenfaoui J, Esmaeel Q, El Hamss H, Belabess Z, Barka EA (2022) Biological control of plant pathogens: a global perspective. Microorganisms 10:596. https://doi.org/10.3390/microorganisms10030596

Laurence MH, Burgess LW, Summerell BA, Liew ECY (2012) High levels of diversity in Fusarium oxysporum from non-cultivated ecosystems in Australia. Fungal Biol 116:289–297. https://doi.org/10.1016/j.funbio.2011.11.011

Loth M, Meddah B, Touil AT, Len C, Verstraet W (2015) Antifungal activity of synthesized dithiocarbamate derivatives on Fusarium oxysporum f sp. albedinis in Algeria. J Chem Pharm Res 7(10):49–54

Meddich A, Ait El Mokhtar M, Bourzik W, Mitsui T, Baslam M, Hafidi M (2018) Optimizing growth and tolerance of date palm (Phoenix dactylifera L.) to drought, salinity, and vascular fusarium-induced wilt (Fusarium oxysporum) by application of arbuscular mycorrhizal fungi (AMF). In: Giri B, Prasad R, Varma A (eds) Root biology, soil biology. Springer, Cham, pp 239–258. https://doi.org/10.1007/978-3-319-75910-4_9

Mohamed O-Z, Yassine B, Hilali Rania E, El Hassan A, Abdellatif H, Rachid B (2020) Evaluation of compost quality and bioprotection potential against Fusarium wilt of date palm. Waste Manag 113:12–19. https://doi.org/10.1016/j.wasman.2020.05.035

Moujaoui N, Hariri E, Elhoumaizi MA (2021) Bayoud and Belaat diseases of date palm (Phoenix dactylifera L.) in Figuig oasis of Morocco. IOP Conf Ser Earth Environ Sci 782:042065. https://doi.org/10.1088/1755-1315/782/4/042065

Nadeem M, Yahya M, Tong J, Shah L, Khan SU, Ali A, Sher A, Ullah N, Waheed A (2023) Improving nitrogen acquisition and utilization through root architecture remodelling: insight from legumes. J Plant Growth Regul 42:5295–5310. https://doi.org/10.1007/s00344-023-10938-9

Narayanasamy P (2013) Mechanisms of action of fungal biological control agents. In: Narayanasamy P (ed) Biological management of diseases of crops: volume 1: characteristics of biological control agents, progress in biological control. Springer Netherlands, Dordrecht, pp 99–200. https://doi.org/10.1007/978-94-007-6380-7_3

Ortiz A, Sansinenea E (2022) The role of beneficial microorganisms in soil quality and plant health. Sustainability 14:5358. https://doi.org/10.3390/su14095358

Oulad Ziane S, El Alaoui Talibi Z, Ibnsouda koraichi S, Douira A, Amir S, Meddich A, El Modafar C (2023) Synergistic effects of arbuscular mycorrhizal fungi associated to plant growth-promoting rhizobacteria in suppression of soil-borne Fusarium wilt of date palm. Biocatal Agric Biotechnol 51:102753. https://doi.org/10.1016/j.bcab.2023.102753

Parween D, Sahu BB, Kumari M, Pudake RN (2019) Plant metabolites involved in plant–pathogen interactions. In: Varma A, Tripathi S, Prasad R (eds) Plant biotic interactions: state of the art. Springer, Cham, pp 61–84. https://doi.org/10.1007/978-3-030-26657-8_5

Pertot I, Giovannini O, Benanchi M, Caffi T, Rossi V, Mugnai L (2017) Combining biocontrol agents with different mechanisms of action in a strategy to control *Botrytis cinerea* on grapevine. Crop Prot 97:85–93. https://doi.org/10.1016/j.cropro.2017.01.010

Purohit A, Ghosh S, Chaudhuri RK, Chakraborti D (2023) Biological control of *Fusarium* wilt in legumes. In: Ghorbanpour M, Adnan Shahid M (eds) Plant stress mitigators. Academic, pp 435–454. https://doi.org/10.1016/B978-0-323-89871-3.00019-7

Qin Y, Liu S, Xing R, Yu H, Li K, Meng X, Li R, Li P (2012) Synthesis and characterization of dithiocarbamate chitosan derivatives with enhanced antifungal activity. Carbohydr Polym 89: 388–393. https://doi.org/10.1016/j.carbpol.2012.03.018

Rahmania F (2000) Contribution à la connaissance des relations histo-cyto-physiologiques des relations parasitaires entre le palmier dattier (Phoenix dactylifera L.) et Fusarium oxysporum f. sp. albedinis agent causal bayoud. Thèse de doctorat. Université des sciences et de la technologie Houari-Boumédiène, Alger

Rahmouni A, Saidi R, Khaddor M, Pinto E, Da Silva Joaquim Carlos Gomes E, Maouni A (2019) Chemical composition and antifungal activity of five essential oils and their major components against Fusarium oxysporum f. sp. albedinis of Moroccan palm tree. Euro-Mediterr J Environ Integr 4:27. https://doi.org/10.1007/s41207-019-0117-x

Raja HA, Miller AN, Pearce CJ, Oberlies NH (2017) Fungal identification using molecular tools: a primer for the natural products research community. J Nat Prod 80:756–770. https://doi.org/10.1021/acs.jnatprod.6b01085

Saaidi M (1990) Amélioration génétique du palmier dattier. Critères de sélection, techniques et résultats. In: Dollé V, Toutain G (eds) Les systèmes agricoles oasiens. CIHEAM, Montpellier,

pp 133–154. (Options Méditerranéennes : Série A. Séminaires Méditerranéens; n. 11). Les Systèmes Agricoles Oasiens, 1988/11/19-21, Tozeur (Tunisia)
Saaidi M (1992) Comportement au champ de 32 cultivars de palmier dattier vis-à-vis du bayoud: 25 années d'observations. Agronomie 12:359–370. https://doi.org/10.1051/agro:19920502
Sabbahi R, Lavallée R, Merzouki A, Guertin C (2009) Differentiation of entomopathogenic fungus Beauveria bassiana (Ascomycetes: Hypocreales) isolates by PCR-RFLP. Phytoprotection 90: 49–56. https://doi.org/10.7202/044022ar
Sabbahi R, Hock V, Azzaoui K, Saoiabi S, Hammouti B (2022) A global perspective of entomopathogens as microbial biocontrol agents of insect pests. J Agric Food Res 10:100376. https://doi.org/10.1016/j.jafr.2022.100376
Sahgal M (2021) Fungal enzymes in biocontrol of phytopathogens. In: Satyanarayana T, Deshmukh SK, Deshpande MV (eds) Progress in mycology: biology and biotechnological applications. Springer Nature, Singapore, pp 327–356. https://doi.org/10.1007/978-981-16-3307-2_12
Sajeena A, Nair DS, Sreepavan K (2020) Non-pathogenic Fusarium oxysporum as a biocontrol agent. Indian Phytopathol 73:177–183. https://doi.org/10.1007/s42360-020-00226-x
Saleh AA, Sharafaddin AH, El-Komy MH, Ibrahim YE, Hamad YK, Molan YY (2017) Fusarium species associated with date palm in Saudi Arabia. Eur J Plant Pathol 148:367–377. https://doi.org/10.1007/s10658-016-1095-3
Samo MH, Ismailli NJ (2020) Isolation and characterization of Fusarium oxysporum f. Sp. albedinis causing bayoud disease of date palm tree from Gambat, district Khairpur, Sindh, Pakistan. Plant Prot 4:43–47. https://doi.org/10.33804/pp.004.01.3234
Sedra M (1995) Triage d'une collection de génotype de palmier dattier pour la résistance au Bayoud causé par Fusarium oxysporum f. sp. albedinis. Al-Awamia 90:9–18
Sedra M (2011a) Development of new moroccan selected date palm varieties resistant to Bayoud and of good fruit quality. In: Jain SM, Al-Khayri JM, Johnson DV (eds) Date palm biotechnology. Springer Netherlands, Dordrecht, pp 513–531. https://doi.org/10.1007/978-94-007-1318-5_24
Sedra M (2011b) Le Bayoud (fusariose vasculaire) du palmier dattier en Afrique du Nord: situation, acquis de la recherche et applications. Proceedings of the First Symposium international sur le palmier dattier en Algérie, p 59–75
Sedra MH (2018) Management of diseases of date palm. In: El Bouhssini M, Faleiro JR (eds) Date palm pests and diseases integrated management guide, Beirut, Lebanon, p 139. International Center for Agricultural Research in the Dry Areas (ICARDA)
Sedra M, Lazrek BH (2011) Fusarium oxysporum f. sp. albedinis toxin characterization and use for selection of resistant date palm to Bayoud disease. In: Jain SM, Al-Khayri JM, Johnson DV (eds) Date palm biotechnology. Springer Netherlands, Dordrecht, pp 253–270. https://doi.org/10.1007/978-94-007-1318-5_13
Sehrawat A, Sindhu SS (2019) Potential of biocontrol agents in plant disease control for improving food safety. Def Life Sci J 4:220–225. https://doi.org/10.14429/dlsj.4.14966
Shabani F, Kumar L (2013) Risk levels of invasive Fusarium oxysporum f. sp. in areas suitable for date palm (Phoenix dactylifera) cultivation under various climate change projections. PLoS One 8:e83404. https://doi.org/10.1371/journal.pone.0083404
Shabani F, Kumar L, Esmaeili A (2014) Future distributions of *Fusarium oxysporum* f. spp. in European, middle eastern and north African agricultural regions under climate change. Agric Ecosyst Environ 197:96–105. https://doi.org/10.1016/j.agee.2014.08.005
Sharma V, Salwan R, Sharma PN (2016) Differential response of extracellular proteases of Trichoderma harzianum against fungal phytopathogens. Curr Microbiol 73:419–425. https://doi.org/10.1007/s00284-016-1072-2
Shin J-H, Lee H-K, Lee S-C, Han Y-K (2023) Biological control of Fusarium oxysporum, the causal agent of fusarium basal rot in onion by bacillus spp. Plant Pathol J 39:600–613. https://doi.org/10.5423/PPJ.OA.08.2023.0118
Slama HB, Cherif-Silini H, Chenari Bouket A, Qader M, Silini A, Yahiaoui B, Alenezi FN, Luptakova L, Triki MA, Vallat A, Oszako T, Rateb ME, Belbahri L (2019) Screening for

Fusarium antagonistic bacteria from contrasting niches designated the endophyte bacillus halotolerans as plant Warden against Fusarium. Front Microbiol 9:3236

Sutherland R, Viljoen A, Myburg AA, van den Berg N (2013) Pathogenicity associated genes in Fusarium oxysporum f. sp. cubense race 4. South Afr J Sci 109:10–10. https://doi.org/10.1590/sajs.2013/20120023

Tantaoui A (1989) Contribution à l'étude de l'écologie du Fusarium oxysporum f.sp. albedinis agent causal du bayoudh. Densité et répartition de l'inoculum au sein du peuplement fusarien. D.E.S., University Cadi Ayyad, Marrakech, Morocco

Tantaoui A, Ouinten M, Geiger J, Fernandez D (1996) Characterization of a single clonal lineage of *Fusarium oxysporum* f. sp. *albedinis* causing Bayoud disease of date palm in Morocco. Phytopathology 86:787. https://doi.org/10.1094/Phyto-86-787

Tariq M, Khan A, Asif M, Khan F, Ansari T, Shariq M, Siddiqui MA (2020) Biological control: a sustainable and practical approach for plant disease management. Acta Agric Scand Sect B Soil Plant Sci 70:507–524. https://doi.org/10.1080/09064710.2020.1784262

Toutain G (1968) Essais de comparaison de la resistance au bayoud des varietes de palmier-dattier. II Notes sur l'experimentation en cours concernant les varieties marocaines et tunisiennes Awamia

Yu Y, Gui Y, Li Z, Jiang C, Guo J, Niu D (2022) Induced systemic resistance for improving plant immunity by beneficial microbes. Plants 11:386. https://doi.org/10.3390/plants11030386

Zehdi-Azouzi S, Cherif E, Moussouni S, Gros-Balthazard M, Abbas Naqvi S, Ludeña B, Castillo K, Chabrillange N, Bouguedoura N, Bennaceur M, Si-Dehbi F, Abdoulkader S, Daher A, Terral J-F, Santoni S, Ballardini M, Mercuri A, Ben Salah M, Kadri K, Othmani A, Littardi C, Salhi-Hannachi A, Pintaud J-C, Aberlenc-Bertoss F (2015) Genetic structure of the date palm (Phoenix dactylifera) in the Old World reveals a strong differentiation between eastern and western populations. Ann Bot 116:847. https://doi.org/10.1093/aob/mcv132

# Chapter 4
# Effect of Plant-Microbiome on Plant Physiology to Overcome Climate Changes

**M. M. S. Abdallah, A. A. Abd El-Monem, T. N. M. El-Sebai, and H. M. S. El-Bassiouny**

## 4.1 Introduction

### *4.1.1 Diversity, Functions, and Stress Responses of Soil Microorganisms*

The plant microbiomes are referred to as the communal populations of plant-associated microorganisms. Examination of teeny microscopic creatures in the soil and their roles and approaches to soil properties is called soil microbiology. Microorganisms in the soil are involved in several biotic and abiotic stress circumstances in the rhizosphere (Newman et al. 2016). The soil microorganisms transmute the organic material from the soil, which acts as a large cistern of organic carbon (Yadav and Pandey 2015) and plant nutrients (Jacoby et al. 2017). In Plants, carbon rivets are attached to the soil via their secretions, such as amino acids, aliphatic acids, fatty acids, flavonoids and sugars, which allure valuable microorganisms while preventing and murdering damaging ones (Jimtha et al. 2017). Microorganisms in the soil are crucial to the earth's life cycle. They act as an essential function in the soil to underpin the ecosystem (Gerba et al. 2023) and fertility (Tahat et al. 2020). The microscopic creature's diameter is less than 0.1 mm in mutable batches, shapes, and sizes (Neufeld and Mohn 2005). These organisms have an impact on the biological processes of the soil, which are serious for the animals and plant life (Du and Liu 2012).

M. M. S. Abdallah (✉) · A. A. Abd El-Monem · H. M. S. El-Bassiouny
Botany Department, Agricultural and Biological Research Institute, National Research Centre El Behouth St., (Former El-Tahrir St.) 12622, Dokki, Giza, Egypt

T. N. M. El-Sebai
Agricultural Microbiology Department, Agricultural and Biological Research Institute, National Research Centre, El Behouth St., (Former El-Tahrir St.) 12622, Dokki, Giza, Egypt



N. El-Wakeil, A. Abdelkhalek (eds.), *Microbiomes and Plant Secondary Metabolites for Insect Pest Management*, Progress in Biological Control 23,
https://doi.org/10.1007/978-3-032-24330-0_4

The organic deposits from different sources are supplemented to the soil and snarfed by soil microorganisms to establish a black-colored, multifaceted organic complex called humus (Leckie et al. 2004). Humus remains are disintegrated by several types of microorganisms (Jan et al. 2020), producing carbon dioxide and water during the synthesis procedure. Moreover, the minerals that are carried on the organic remains become free through the mineralization process with obtainable forms for plant growth (Verma and Pandey 2016). The soil atmosphere has more complex biological populations and roles of biodiversity as compared with tropical forests (Tongway and Ludwig 2005). Originally, Beijerinck (1901) established an "enrichment culture technique" that was used to isolate several soil microbes containing aerobic symbiotically nitrogen-fixing bacterium *Azotobacterchrococcum*, *Acetobacter*, green algae, *Lactobacillus species,* and the yeast *Schizosaccharomyces octosporus*, and many others. In addition, Beijerinck detaches *Bacillus radicicola,* which causes nodulation in the legumes for the fixation of nitrogen. The first recognized sulfate-reducing bacterium *Thiobacillus denitrificans*, replaced oxygen with sulfate as an electron acceptor. Also, a sulfur-oxidizing bacterium, *Beggiatoa*, consumes carbon dioxide as a carbon source and inorganic $H_2S$ as an energy source (chemoautotrophs).

### *4.1.2 Environment and Soil Microorganisms*

The changes in the global environment made by Humans led to the sixth major disappearance occasion in life history and also triggered universal variations in the global spreading of creatures. These fluctuations in biodiversity modify several processes in the ecosystem (Weisser et al. 2017) and alter the flexibility of ecosystems to environmental change (Correia and Lopes 2023). Microorganisms' activities play a principal role in the global fluctuations of biogenic gases (such as carbon dioxide, methane, and nitrous oxide) (Kharitonov et al. 2021). Until now the procedures used by creatures to adjust terrestrial greenhouse gas fluctuation are not obvious due to the complex collaborations that ensue among microorganisms (Watt et al. 2006a, b) and other biotic and abiotic factors (Sharma et al. 2008). Pretended climate change disturbs creatures, nematodes, biodiversity and density in cold climate conditions soil. It is well known that the main source of organic carbon is agricultural waste. Therefore, the carbon stock in the soil decreases due to the decomposition of organic matter in the soil through the combined relations of soil animals and microorganisms and also physical spread through leaching and erosion (Lal 2004; Kuzyakov 2010; Panagos et al. 2015).

Water stress in the soil, especially in arid and semi-arid areas, reduces the presence of organic matter and is at a very low level compared to unstressed areas (Tongway and Ludwig 2005; IPCC 2006). Organic matter in the soil is an important measure to estimate soil fertility because it is a vital nutrient reservoir and a suitable habitat for nutrient rotation (Jacoby et al. 2017). The variations in the climatic circumstances directly encourage moisture content in soil organic matter, affect the physical and chemical properties of the soil, and consequently, the biological performance of the soil (Furtak and Wolińska 2023). Soil microbial populations intermediate critical cycles of the ecosystem (carbon and nutrients) (Malik et al. 2020). These rotations of nutrients (nitrogen, phosphorus, sulfur, etc.) manage several ecosystem processes, including carbon requisitioning in both soil and plant (Penman et al. 2010). Climate change directly and indirectly influences nutrient accessibility in soil (Hui et al. 2016). In addition, Poor soil organization, such as deforestation, conversion from perennial to annual plant types and overgrazing, affects soil carbon and causes changes in soil temperature, leading to changes in its combination of microbial diversity (Schmidt et al. 2011). For example, carbon-compound-oxidizing bacteria (methanotrophs) need trace metal copper for methane oxidation. So, they collect Cu in their body and ensure a vast potential for bioremediation in soil and bio-transformations of chemicals and bioenergy (Hakemian and Rosenzweig 2007; Ma et al. 2009). Nature's primary biological tool for destroying atmospheric intensities of methane, a potent greenhouse gas, is methane monooxygenases (Kapahi and Sachdeva 2019).

### *4.1.3 Stress Responses of Plant Microbiome*

Soil microorganisms' biodiversity is convoluted in different stress conditions in the rhizosphere (Qiang et al. 2012). It interacts with various abiotic and biotic stresses via their numerous genetic compositions (Newman et al. 2016). Soil microorganisms are exposed to several stress conditions, such as nutrients, temperature, water potential, etc. (Herron et al. 2010; Miransari 2013). The cell wall of *Escherichia coli* bacteria is convoluted in the adaptation, preservation, and defense alongside stress conditions (Ayalaja and Pedro 2012). The harsh environments also affect soil fertility via mineralization, microbial-activity-mediated decomposition, and nitrogen fixation processes (Kaiser and Kalbitz 2012). The moisture and temperature stress highly affect the activity of soil microorganisms (Siebielec et al. 2020). The component of the mold cell wall reacts to fluctuations in the soil environment to keep the cell from the host's immune response through the creation of a cell surface epitope (Latge 2017). Moreover, soil microorganisms may modify their metabolic responses according to stress conditions, via alterations in the genetic structure (Hartmann et al. 2017). Both fungi and bacteria induce antibiotic metabolites in contradiction to pathogenic creatures (Hoffmeister and Keller 2007). Microorganisms have numerous physiological adaptation techniques to persist (Fig. 4.1). This impacts the growth and biochemical reactions of plants via several accomplishments (Devi et al. 2022).

#### 4.1.3.1 Abiotic Stresses and Soil Microbiome

Several abiotic aspects adjust the physical environment of the soil and affect microbial groups and activities such as drying/rewetting, freezing/thawing, oxic/

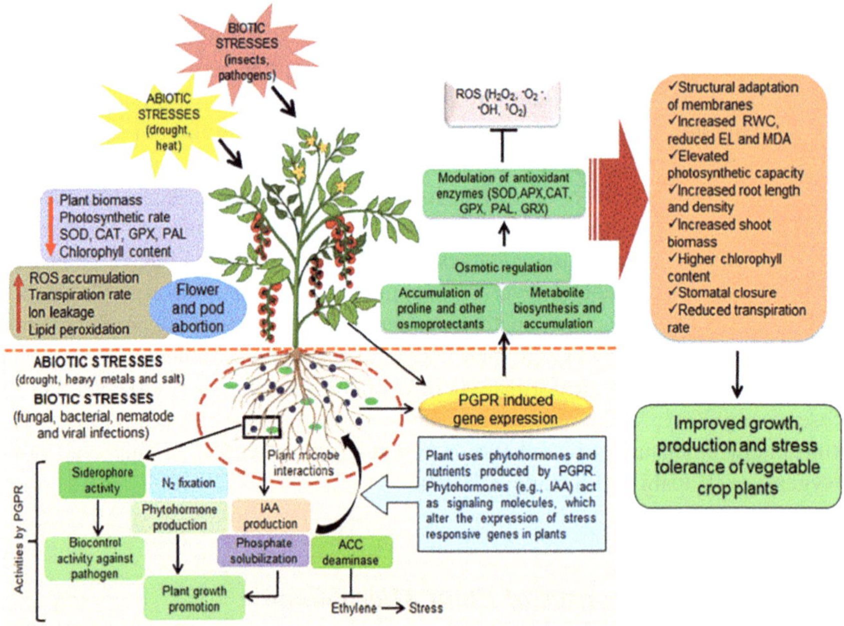


**Fig. 4.1** Schematic picture of plant-growth-promoting rhizobacteria (PGPR)-mediated growth advancement and stress tolerance in vegetable crops. The model illustrates stress-induced declines in plant biomass; photosynthetic rate; chlorophyll content; activities of CAT, GPX, PAL and SOD; and rises in reactive oxygen species (ROS), flower and pod abortion, transpiration rate, ion leakage, and lipid peroxidation. Plants injected with PGPR avoid pathogen infections via biocontrol activity and increase stress tolerance via ACC deaminase activity. It also encourages stress-responsive gene expression, causing the buildup of numerous osmoprotectant compounds and the detoxification of ROS in cells. Inflection of antioxidants avoids cell injury and preserves homeostasis, which recovers growth, yield, and stress tolerance in vegetable crops. (Adapted from Kumar et al. 2021)

anoxic and water percolation conditions (Borken and Matzner 2009). The microbiome of the rhizosphere exhibits an imperative role in plants surviving under unfavorable environmental conditions (Jorquera et al. 2012). Some bacterial strains in the soil maintain plant growth under halotolerant and osmotic stress (Berg et al. 2013). The bacillus bacteria have been isolated from saline-stress soils (Upadhyay et al. 2009). Soil microorganisms' action is inconstant at various stages of plant root growth with precise time and space (Barret et al. 2011). Microbial genes are elaborate in nutrient acquisition, stress reactions, and secretions in the rhizosphere (Rainey 1999). A greater variation in moisture and temperature highly affects microbial population and functions (Ibáñez et al. 2023). The configuration and variety of plant species, affected by grazing, consequently increase soil carbon loss, parallel to their effects on soil-borne microbial population (Klumpp et al. 2009). Effects of tillage on carbon dynamics and soil microbial activities through the breakdown process have been described by Srour et al. (2020).

Water stress in soil restricts organic matter involvement as a result of carbon reserve in dry land being about half of that soil in unstressed condition (IPCC 2006). The soil microbial biomass and their enzyme activities in the rhizosphere are affected by drought stress (Sanaullah et al. 2011). In addition, rainfall occasions prompt moisture instabilities, causing stress that fluctuates soil microbes' respirations and nutrient mineralization. Generally, drought stress reduces microfauna sooner than bacteria and fungi. Also, fungi are more tolerant than bacteria (Manzoni et al. 2012). In this regard, Kawasaki et al. (2012) mentioned that polycyclic fragrant hydrocarbons disturb microbial groups indirectly in leguminous plants, via declines in nitrogen fixation in the root. Moreover, pH and toxic metals as abiotic factors mainly affect the microbial inhabitants and consequently plant growth (Singh et al. 2015).

## 4.2 Harnessing the Plant Microbiome for Improved Abiotic Stress Tolerance

### *4.2.1 An Overview of Abiotic Stresses in Plants*

Crop plants suffer from various abiotic stresses, including salinity, drought, flooding, extreme temperatures and nutrient deficiency, which are ordinarily increased as a result of climatic changes (Villalobos-López et al. 2022; Sánchez-Bermúdez et al. 2022). These pressures cause a major risk to agricultural output (Capell et al. 2004). Global water scarcity as a result of impressive climatic variations is the main reason for abiotic stresses. Water shortage is the most important abiotic stress that harmfully affects crop plants' productivity and spreads worldwide (Marasco et al. 2013). Drought stress only affects 44% of the global agricultural soil, with an extraordinary human population (39%), parallel to improved food demands (Hussain et al. 2012). Plant development and progress suffer from harmful impacts of either short-term or long-term water inequity. Plants reveal several responses at physiological, metabolic, and molecular levels to persist or stand over the hostile conditions, such as amplified aquaporin and $H^+$-pyrophosphatase activity, stomatal closure and buildup of a range of compatible solutes (Marasco et al. 2013; Hussain et al. 2016).

The effects of water shortage on plant growth and development were detected; however, the injury triggered to plants at the cellular and molecular levels under drought circumstances is still unclear (Kapoor et al. 2020; Seleiman et al. 2021). In addition, salinity restricts agricultural assembly in dry and semiarid regions, especially those characterized by little annual rainfall (Agrawal et al. 2013). Amplified concentrations of salt ions such as $Ca^{2+}$, $Cl^-$, $K^+$, $CO_3^{2-}$, $Na^+$, $NO_3^-$, $SO_4^{2-}$ in soil diminish water uptake via roots that promote the addition of toxic salts in plant cells (Tester and Davenport 2003). Plants can tolerate this problem through keeping translocation by ATP-dependent ion flow. Irrigation water with extreme salinity levels leads to osmotic stress. So, any interruption in intracellular ion homeostasis results in negative properties, for example, cell signaling pathways containing high salinity levels induce the synthesis of osmoprotectants as, assimilation, nutritional disorders, specific proteins, membrane disorganization, decreased cell division and expansion, genotoxicity and definite free radical scavenging enzymes that govern ion and water instability (Zhang et al. 2009a, b) consequently diminished best growth and development. Various abiotic stresses are frequently consistent and encourage a similar set of plant responses via activating similar or cooperating pathways (Seki et al. 2002; Kreps et al. 2002).

Plants often respond to stress through the accumulation of osmotic substances, the creation of stress proteins and antioxidants at the cellular level in various diverse crops for all types of stresses (Bartels and Hussain 2008; Hussain et al. 2012). Little success has been achieved in enhancing plant tolerance under abiotic stress conditions as a result of inadequate awareness of the stress tolerance techniques in plants. Several studies have been conducted on the combined techniques convoluted in stress reaction and alteration such as accommodating metabolism, stress signaling and reprogramming gene expression to promote stress lenience (Ma et al. 2011; Marasco et al. 2016). The improvement of stress-tolerant plants is a possible method for agricultural and environmental sustainability, which appears obsessive to achieve the increasing requirements for food quality (Castiglioni et al. 2008). Yet present refinement systems do not have enough proper methodological resources to achieve crop production in unfavorable environmental conditions (Ashraf and Foolad 2007). Developing stress-tolerant plants by genetic engineering can play a key role in environmental sustainability. Moreover, incorporating transgenic methods with current breeding procedures can be used to evolve improved stress tolerance of plants (Anwar and Kim 2020).

Transfer of one or several dissimilar genes complicated in numerous pathways is the main objective of present transgenic methods to control transcription reasons, compatible solutes/osmoprotectants (proline, glycine betaine, polyamines) and proteins (aquaporin, heat shock) for producing stress-tolerant plants (Kathuria et al. 2007; Sreenivasulu et al. 2007; Marasco et al. 2016). The documentation of basic genes and their procedure in transgenic crops with enhanced stress tolerance without sacrificing yield is the main threat to transgenic approaches (Bartels and Hussain 2008). Numerous important and sudden findings connected to the plant-associated microbiome were documented by molecular tools combined with new microscopic techniques (Berg et al. 2016; Timmusk et al. 2017). A wide variety of agriculturally vital microorganisms has been utilized to improve growth and disease control in plants. Plant-correlated valuable microbiomes can considerably contribute to ameliorating abiotic pressures, consuming a diversity of techniques (Mapelli et al. 2013; Vejan et al. 2016). The use of microbiomes becomes more significant for crop sustainability, especially for challenged plants that need nonconventional solutions under environmental stresses (Schaeppi and Bulgarelli 2015).

Reinforcement of useful microbial features to plants, the environment, or both deals with an encouraging opportunity for the progress of sustainable agriculture. Utilization of microbial groups can be a respected tool and key factor in the treatment

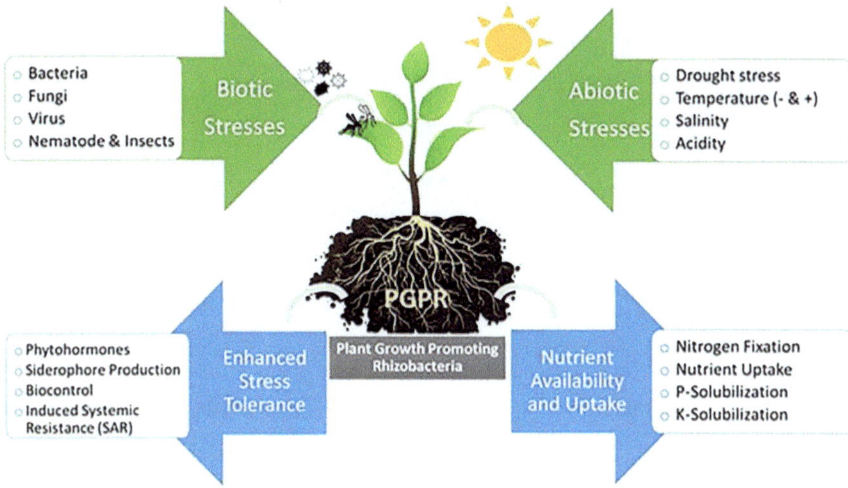


**Fig. 4.2** Interactions between plant growth-promoting rhizobacteria (PGPR) and plants in the rhizosphere. (Adapted from Khan 2023)

of plant strength and efficiency under circumstances of abiotic and biotic stresses (Marasco et al. 2016). Moreover, Khan (2023) exploits phytomicrobiome organisms as a virtual approach to alleviate definite biotic and abiotic stresses that affect plant growth and production (Fig. 4.2). The important information (identification, characterization, and utilization) on useful microbiomes that ameliorate abiotic stress tolerance in plants would assist in sustaining the following generations of agriculture in the world (Jorquera et al. 2012; Nadeem et al. 2014). So, these highlight the mastery of the plant-associated microbiome in escalating tolerance to diverse abiotic stresses, which cause a severe risk to global crop productivity.

## 4.2.2 *Investigating the Plant-Associated Microbiome for Successful Abiotic Stress Tolerance in Plants*

### 4.2.2.1 Drought Stress

The plant accompanied by microbiome can encourage numerous plant characters containing growth and abiotic and biotic hassle tolerance (Marasco et al. 2013; Sugiyama et al. 2013; Berg et al. 2014; Panke-Buisse et al. 2015). The impact of the plant accompanied by the microbiome on plant adaptation to overcome water stress is insufficient. Tiwari et al. (2024) proved that plant growth-promoting activity isolated from over 510 strains is stress-dependent in grapevine and *Arabidopsis* growth. Also, injected pepper plants with certain strains under water stress conditions showed a stress-dependent plant growth, encouraging configuration via

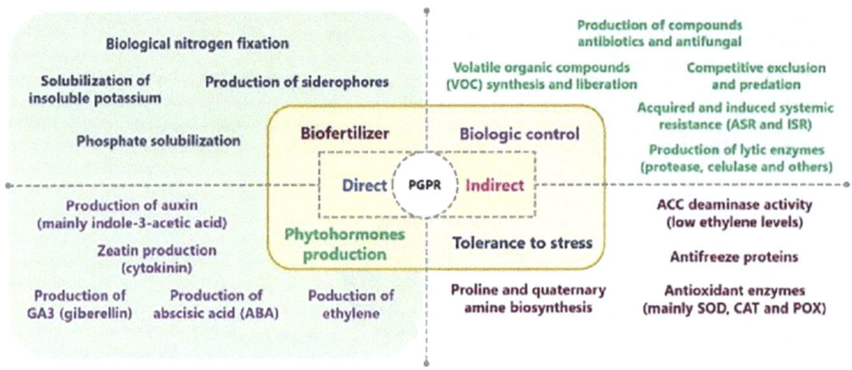


**Fig. 4.3** Mechanisms (direct and indirect) of plant growth-promoting bacteria. (Adapted from Timmusk et al. 2014)

improved leaf and shoot biomass, shoot length and promoted photosynthetic pigment contents in grapevine-stressed plants. The beneficial function of soil microbes in the growth of plants, controlling nutrients and biocontrol activity is competently recognized. These valuable creatures inhabit the rhizosphere/endorhizosphere of plants and encourage plant progress via several direct and indirect techniques.

The promising description for the rhizobacteria AQ: as a tool of plant drought tolerance contains (1) accumulation of osmolytes (2) making of phytohormones like indole-3-acetic acid (IAA); cytokinin, abscisic acid (ABA) and gibberellic acid (3) decrease the level of ethylene in the roots through 1-aminocyclopro pane-1-carboxylate ACC deaminase (4) bacterial exo-polysaccharides (5) encouraged systemic tolerance by bacterial compounds (6) antioxidant defense and inoculations (7) alteration of root morphology (Fig. 4.3) (Timmusk et al. 2014; Sánchez-Bermúdez et al. 2022). This current review covers a summary of the previous work describing PGPR's function in assisting plants to manage drought stress (Babou and Lisna 2019). Induced Systemic Tolerance (IST) has been created to investigate the physical and chemical microbial changes prompted in plants, concomitant with enhancing abiotic stress tolerance (Vurukonda et al. 2016). *Achromobacter piechaudii* ARV8, as PGPR augments drought stress tolerance in tomato and pepper plants via (ACC) deaminase. Conversely, the analysis of plants indicates decreasing in ethylene production that eventually moderates plant stress and allows normal development of the plant (Arshad et al. 2008; Zhang et al. 2020). The progressive role of bacterial instructing on crop seedlings under water stress conditions is cleared (Sánchez-Bermúdez et al. 2022). This treatment improved plant yield by 79% and enhanced photosynthetic pigment contents five and a half folds under extreme drought conditions. Additionally, some volatiles released from bacterially supplied to drought-stressed wheat seedlings are exploited to determine plant implementation under water shortage in the primary periods of stress progress (Timmusk et al. 2017).

Inoculation with *Burkholderia phytofirmans* PsJN improved chlorophyll content, photosynthesis, water use efficiency and wheat grain yield grown under drought conditions (Naveed et al. 2014). Moreover, inoculated maize with either *B. phytofirmans or Enterobacter* sp. FD17 accomplished recovery as related to controls (Naveed et al. 2014). Marulanda et al. (2009) proved the plant recovery after being treated with bacterial strains *Pseudomonas putida, Pseudomonas* sp. *and Bacillus megaterium*, isolated from extremely water-stressed soil. In this connection, Sandhya et al. (2009) stated that soil amendment with *Pseudomonas* sp. strain GAP-45 improved persistence and sunflower production under water-stress conditions. Amendment with bacteria causes professional colonization of the root-sticking soil, resulting in constant soil collections and improved stress tolerance. *Pseudomonas* strain GAP-45 inoculated in maize plants augmented antioxidants and osmoregulatory solutes under different levels of drought conditions (Sandhya et al. 2010). Isolated microbes from the plant's roots developing under particular drought conditions enhanced the growth of alternative hosts in tomato, grapevine, olive, and pepper plants (Marasco et al. 2013). Kohler et al. (2008) concluded the possible role of microbes in the mitigation of oxidative stress under a drought environment, through the injection of lettuce with *Pseudomonas mendocina and arbuscular mycorrhizal* fungi, which induced the activity of antioxidant catalase. Also, the buildup of specific proteins with ascorbate and glutathione preserves plant metabolic functions and conversing protection under drought conditions. Lavender plants injected with *Glomus intraradices and Glomus* sp. strain enhanced water contents, N and P contents, and root biomass, revealed high drought tolerance (Porcel et al. 2006; Marulanda et al. 2007). Generally, plant growth-promoting bacteria enhanced development in several plant species (common bean, pepper, pea, lettuce, rice, and sorghum) under water stress conditions (Arshad et al. 2008; Perez-Montano et al. 2014; Marasco et al. 2016).

#### 4.2.2.2 Salinity Stress

Over the past few decades, extreme climatic conditions and the misuse of agricultural land have led to extreme salinity, which is a main regulating aspect of crop productivity (Wicke et al. 2011). Numerous techniques belong to molecular tools, involved in controlling salinity such as soil management and reclamation performances. Nevertheless, these techniques are costly and not continually useful and maintainable for governing salinity. The utilization of accepted plant growth-promoting bacteria as inoculants for salt-affected plants is gaining strength (Shabala et al. 2013; Qin et al. 2014; Ruiz et al. 2015). Microbial populations raise crop productivity and increase plant health after hostile conditions (Zuppinger-Dingley et al. 2014; Sloan and Lebeis 2015). Microbes populating revealed numerous stress circumstances change adaptive tolerant behaviors and are promising as plant growth agents under unfavorable circumstances (Yang et al. 2016b). Halotolerant microbes flourish and have precise behaviors to support plants living under high salinity. Upadhyay et al. (2009) reported that several rhizobacterial strains isolated from wheat plants exhibited significant potential in alleviating salinity stress, with 24 isolates showing effective tolerance and stress-mitigating capacity. Arora et al. (2014) confirmed that most of the isolated halophyte bacteria and other salt-tolerant plant species were well developed at high levels of saline cultures. Growth of several plant species (canola, pepper, tomato, bean, wheat, and lettuce) has improved with plant-associated microbiomes (Zhao et al. 2016; Abdelaal et al. 2021).

#### 4.2.2.3 How Halotolerant PGPB-Transmit Salinity-Stress Tolerance

Halotolerant PGPB is used in preparing several systems for the performance of salinity amelioration (Shilev 2020). The halotolerant PGPB achieves frequent systems (Fig. 4.4) that directly or indirectly contribute to the improvement of salinity (Egamberdieva et al. 2015). Plant-growth-promoting Rhizobacteria produce an extensive collection of hormones that act as signal molecules, such as auxins and cytokinins, in the rhizospheric area that collaborate with plant roots and improve plant development. Halotolerant PGPB emanates phytohormones that play an essential role in controlling metabolic pathways and consequently physiological processes in plants grown under salinity stress. Plant Growth Promoting Rhizobacteria containing *Bacillus amyloliquefacians*, *Pseudomonas fluorescence, and Bradyrhizobium japonicum* considerably secrete plant growth hormones such as Auxins, Gibberellins, Cytokinins, Ethylene and Abscisic acid (Boiero et al. 2007) (Fig 4.5).

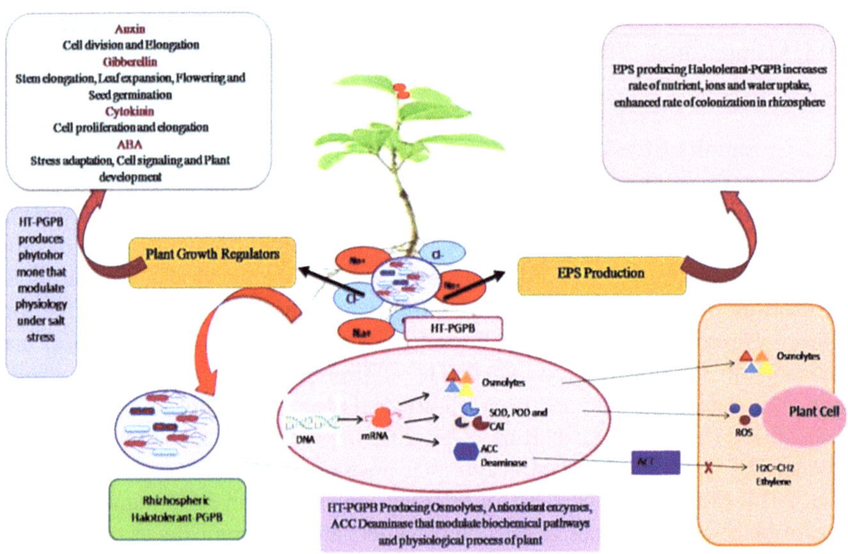


**Fig. 4.4** Halotolerant plant growth promoting bacteria in alleviating salt stress in plants. (Adapted from Sagar et al. 2022)

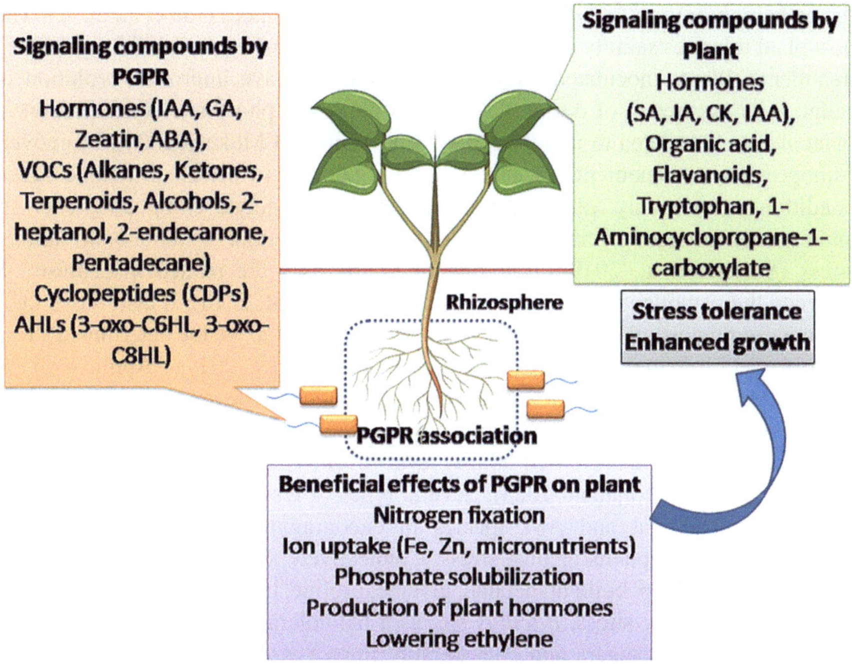


**Fig. 4.5** Plants and PGPR produce signaling compounds for the promotion of the valuable rhizospheric relationship. These compounds comprise hormones (IAA, ABA, GA, Zeatin, SA, JA, CK), ACC deaminase, acyl homoserine lactones (AHLs) like 3-oxo-C6HL, 3-oxo-C8HL, cyclopeptides (CDPs), volatile organic compounds (Alkanes, Alcohols, Ketones, 2-heptanol, 2-endecanone, Terpenoids, Pentadecane), which activates plant signaling, aiding in plant growth elevation and stress tolerance. The concomitant PGPR increases plant growth by providing necessary minerals via nitrogen fixation, and ion uptake (micronutrients, Fe, Zn). (Adapted from Jalmi and Sinha 2022)

Plant-growth-promoting bacteria produce ACC deaminase (Egamberdieva et al. 2015), exopolysaccharides (EPS) (Raghuvanshi et al. 2018), and siderophores (Shameer and Prasad 2018; Sayyed et al. 2019). They also mobilize nutrients, such as phosphorus solubilization (Etesami et al. 2017), nitrogen-fixing (Zaidi et al. 2017; Tang et al. 2020), and osmolytes (glycine betaines, proline, and trehalose), volatile organic compounds (Forni et al. 2017), and stimulate the antioxidative enzymes and adjust plant resistance systems under salinity conditions (Abbas et al. 2019) (Fig. 4.3). These combinations alleviate the harmful effects of abiotic stress on plants, permitting the plant to grow well.

This tolerance is recognized in diverse hormones, genes, and proteins such as indole-3-acetic acid (IAA), gibberellin and siderophores. Correspondingly, some halotolerant bacterial strains improved plant growth under saline conditions by decreasing ethylene assembly by ACC deaminase activity (Siddikee et al. 2010). The obtainability of other halotolerant diazotrophic bacteria, with ACC deaminase activity, phosphorus solubilization and IAA isolated from *Salicornia brachiate* (acme halophyte) roots characterizes other possible nominees (Jha et al. 2012). The host plant tolerates salinity stress as a result of *Arbuscular mycorrhizal* fungi (AMF) treatments. Plants inoculated with AMF *Glomus* sp. have improved evolution in saline soils as a result of decreased $Na^+$ and improved phosphate concentration in plant shoots, as related to non-treated controls (Giri and Mukerji 2004). Moreover, osmoprotectants benefit plants and/or microorganisms to persist in hostile saline conditions. Commonly, plants accumulate organic-osmolyte compounds such as amino acids, betaine, glycine, polyamines, proline, and others to ameliorate salinity stress (Sharma et al. 2019). Halotolerant PGPB uses the related mechanism to alleviate the osmotic stress of alkaline soils (Mishra et al. 2018). It also raises their $K^+$ ions, adding osmolytes, and their cytoplasmic content to aid the determined stress reaction and escape water loss (Bremer and Krämer 2019). Application of *Arbuscular mycorrhizal* fungi enhanced salt tolerance in clover, cucumber, maize, mung bean and tomato as a result of improved osmoregulation, proline accumulation, P acquisition and decreased salinity level (Evelin et al. 2009, 2019; Wahab et al. 2023; Sun and Shahrajabian 2023). Several types of research are needed to clarify the role of bacterial and AM species in encouraging defensive proteins and osmoprotectants in plants against stress. Connectively, Shilev (2020) documented that *Azospirillum* spp. build up betaine, glycine, proline, trehalose, and sustain plants to withstand salinity stress. *Bacillus* sp., as a halotolerant PGPB strain, builds up proline and soluble sugars and considerably improves the progress of *Zea maize* under drought and salinity conditions. Numerous studies explore the implication of osmoprotectant trehalose and generating halotolerant PGPB that contains the genes for trehalose biosynthetic pathways under stress conditions (Qin et al. 2018; Forni et al. 2017; Shim et al. 2019).

Commonly, approximately all rhizosphere bacteria build up surface polysaccharides or exopolysaccharides (EPSs). The structure and amount of EPS vary in various halotolerant PGPB cultures; sufficient EPS is produced in aggressive circumstances. This acts as a physical hitch around the root and keeps the plant progressing under extreme saline levels (Vaishnav et al. 2016). In addition, Singh and Jha (2016) proved the enriched uptake of ions in inoculated plants with EPS-producing halotolerant PGPB under saline conditions. In chickpeas, *Mesorhizobium ciceris*, the ACC-deaminase and EPS-producing halotolerant PGPB, improved plant growth, steadied the soil texture, and promoted colonization under salinity conditions. Regarding yield development, EPS-producing halotolerant PGPB is markedly employed as the seed priming agent and to assist in augmenting sprouting (Abbas et al. 2019). Moreover, Atouei et al. (2019) reported that wheat inoculated with *Marinobacter lipolyticus* SM19 and *B. subtilis* (EPS-producing and salt-tolerating) lowers the hostile effects of abiotic stresses. Chu et al. (2019) detected that *Pseudomonas PS01, as an* EPS-producing ST-PGPR, alleviated *Arabidopsis thaliana* development and regulated their genes under extreme saline conditions. They added that the LOX2 gene was upregulated and contributed to the jasmonic acid [JA] pathway and improved its buildup in plants under unfavorable stress conditions.

Halotolerant PGPB-inoculated plants conflict with the harmful effects of oxidative stress via manufacturing antioxidative enzymes (Vaishnav et al. 2020; Supriya et al. 2021). Numerous mechanisms of antioxidants revealed by the halotolerant PGPB, including enzymes, such as catalase, glutathione reductase, peroxidases, nitrate, and superoxide dismutase, as well as secretive molecules, keep tissues from oxidative injury by subduing and detoxifying ROS (Sharma et al. 2019; Gupta et al. 2021). *Rhizobium* sp., *Azotobacter* sp., *Pseudomonas* sp., *and Serratia* sp., which are PGPB strains, are used as inoculants for successful plant growth under salinity stress (Mishra et al. 2018; Khan et al. 2019; Khan et al. 2021). The delicate physiological indicators of salt stress are quantified by antioxidant production and metabolite synthesis that induce upregulation of the antioxidant gene response (Khanna et al. 2018; Rajput et al. 2019). In this regard, El-Esawi et al. (2018) established that chickpea plants inoculated with *Azospirillum lipoferum* FK1 showed the progressive expression of antioxidant genes responsible for enhancements in the antioxidant enzymes, non-enzymatic metabolite enlargement, flavonoid content, nutrient uptake, phenol, and the growth of the plant. Also, Habib et al. (2016) record the greater salt tolerance of okra plants when inoculated with *Enterobacter* sp. UPMR18 through augmented antioxidant enzyme activities and the ROS cascade gene expression under salinity stress conditions. Stress-tolerant PGPR, containing *Pseudomonas pseudoalcali* genes and *Enterobacter cloacae*, enhanced the intensities of CAT and APX in *Jatropha* leaves, increased the biomass, potassium, phosphorus, the uptake of nutrients, and the photosynthetic pigment content, stimulated the development of the root in response to salinity stress in different tissues of the plant (Egamberdieva et al. 2019). Moreover, in tomato plants injected with halotolerant PGPB under saline conditions, stimulated antioxidant enzymes involved in ROS scavenging activities and ascorbate peroxidase concomitantly improved tolerance were detected (Vaishnav et al. 2020).

#### 4.2.2.4 Temperature Stress

According to the Intergovernmental Panel on Climate Change (IPCC: 2007), climate alteration will give rise to global temperatures to increase by 1.8–3.6 °C by the end of this century. Both crop productivity and microbial colonization are significantly hampered by high temperatures, which also cause significant cellular damage such as protein aggregation and breakdown. Heat shock proteins are a certain class of polypeptides that are created in all organisms in restraint to high temperatures (HSPs). According to Srivastava et al. (2008), stress adaptation in microorganisms is a complicated multilevel regulatory process that may include many genes. As a result, germs evolve various coping mechanisms to resist stress. According to Hussain et al. (2012), certain microorganisms are more effective at high temperatures, and these microbes might be crucial for agricultural plants in high-temperature situations (Yang et al. 2016a). In the drought-stressed rhizosphere of chickpea, Srivastava et al. (2008) identified P. putida strain NBR10987, which showed thermotolerance and was linked to thick biofilm development and the overexpression of the stress sigma factor δs. To withstand stress, some bacterial strains produce exopolysaccharides (EPS), which have special cementing and water-holding properties and are essential for the development of biofilms and stress tolerance.

Ali et al. (2009), sorghum seedlings inoculated with the Pseudomonas AKM-P6 strain exhibited increased physiological and metabolic activity, showing a sole interplay of desirable proteins in heat tolerance utilizing microorganisms. These seedlings also showed improved resistance to heat stress. Because low temperatures negatively influence plant growth and development, they are a significant factor limiting agricultural yield. Likewise, endophyte inoculation produced a greater and quicker build-up of stress-associated proteins and metabolites, resulting in enhanced resistance to low temperatures and suggesting that plants had benefited from a beneficial priming effect (Theocharis et al. 2012). In a similar vein, Barka et al. (2006) reported that at low temperatures (4 °C), grapevine roots infected with B. phytofirmans PsJN showed improved root development, higher plant biomass, and enhanced physiological activity. Subsequent investigation showed that, in comparison to uninoculated control plantlets, bacterized plantlets dramatically raised proline, starch, and phenolic levels, which improved the ability of grapevine plantlets to withstand cold temperatures. The symbiotic activities of soybeans (nodule infection and nitrogen fixation) are often repressed by low temperatures. Nevertheless, inoculating soybeans with both *Bradyrhizobium japonicum and Serratia proteamaculans* led to quicker development at 15 °C (Zhang et al. 1996).

In addition, Mishra et al. (2011) found that inoculated wheat seedlings with psychrotolerant Pseudomonads exhibited decreased $Na^+/K^+$ ratio and electrolyte leakage, as well as increased physiologically attainable iron, anthocyanin, proline, protein, and relative water contents, all of which contributed to their improved cold tolerance. Numerous investigations have looked at various bacterial strains to help plants better withstand cold stress (Selvakumar et al. 2008a, b, 2009, 2010a, b). The aforementioned investigations demonstrate that *B. phytofirmans* PsJN has a broad host range, encompassing soybean, sorghum, wheat, and switchgrass, all of which have encouraging responses to various abiotic stressors.

#### 4.2.2.5 Heavy Metal Stress

The contamination of heavy metals brought on by increasing industrialization has received significant attention recently. Because heavy metals cannot decay (Rajkumar et al. 2012). Numerous biological and physicochemical methods that were created to eliminate pollutants have failed because they are costly, hazardous to the environment, and inadmissible to the general population (Vidali 2001; Doble and Kumar 2005). Utilizing plants to remove toxins from soil, a process known as phytoremediation, is widely accepted as a low-cost, ecologically benign method (Fester et al. 2014; Arslan et al. 2015). The use of plant-associated microbiomes is another workable and promising option. This involves changing the mobility and bioavailability of metals through microbial activity in the rhizosphere, which increases plant metal uptake (Ma et al. 2011; Yang et al. 2012). Plant-associated microbiomes generate a variety of growth-promoting substances that aid in plant development in heavy metal-contaminated soils, like siderophores, ACC deaminase, and plant growth hormones (IAA, cytokinins, and gibberellins) (Ijaz et al. 2016; Santoyo et al. 2016). High soil pollution may inhibit plant growth, particularly root growth, because of oxidative stress, which limits the proportion of phytoremediation (Hu et al. 2016). Phytoremediation is further restricted by low microbial density and a shortage of nutrients (Gerhardt et al. 2009).

Manganese, zinc, copper, cadmium, lead, chromium, and aluminum are examples of common heavy metals. Arsenic and antimony are two examples of metalloids that exhibit toxicity (Wuana and Okieimen 2011; Pandey 2012). Among the microorganisms engaged in phytoremediation, rhizosphere bacteria are particularly important (Arora et al. 2005), as they may directly increase the practicality of the process by modifying the pH of the soil and oxidation/reduction processes (Gondal et al. 2021). According to Sheng et al. (2008), lead (Pb) solubility in Brassica napus was greatly enhanced by *Microbacterium* sp. G16 and *Pseudomonas fluorescens* G10 when compared to uninoculated controls. This increase was mostly ascribed to IAA, siderophores, ACC deaminase, and phosphate solubilization. Similar to this, co-inoculation of Rhizobium *leguminosarum or Azotobacter chrococcum* with *Zea mays* increased biomass and plant development in soil polluted with lead (Hadi and Bano 2010; Hussain et al. 2013). In soil polluted with metals, numerous endophyte genera, such as Agrobacterium, *Enterobacter, Bacillus* sp., *Serratia, and Burkholderia* sp., have enhanced the amount of phytoremediation and biomass creation Suman et al. 2016; Singh et al. 2016; Feng et al. 2017). Furthermore, because mycorrhizal fungi are hyperaccumulators of heavy metals and have heavy metal tolerance, they are important players in phytoremediation (Zarei et al. 2010; Orłowska et al. 2011).

#### 4.2.2.6 Mechanisms of HM Treatment Via HMT-PGP Microbes

Mitigation of HM in soil mostly relies on its availability. Yet, because HM toxicity results in poor plant development, the bioavailability of HM might further impede the phytoremediation process. These two main problems are dealt with synchronously via HMT-PGP microorganisms in the rhizosphere, which modify plant development in addition to altering the physicochemical properties of the soil to raise metal bioavailability, which in turn causes fast detoxification or removal of heavy metals. A schematic illustration of the many pathways involved in the detoxification and remediation of heavy metals (HM) in polluted soils is shown in Fig. 4.6 (Mishra et al. 2017).

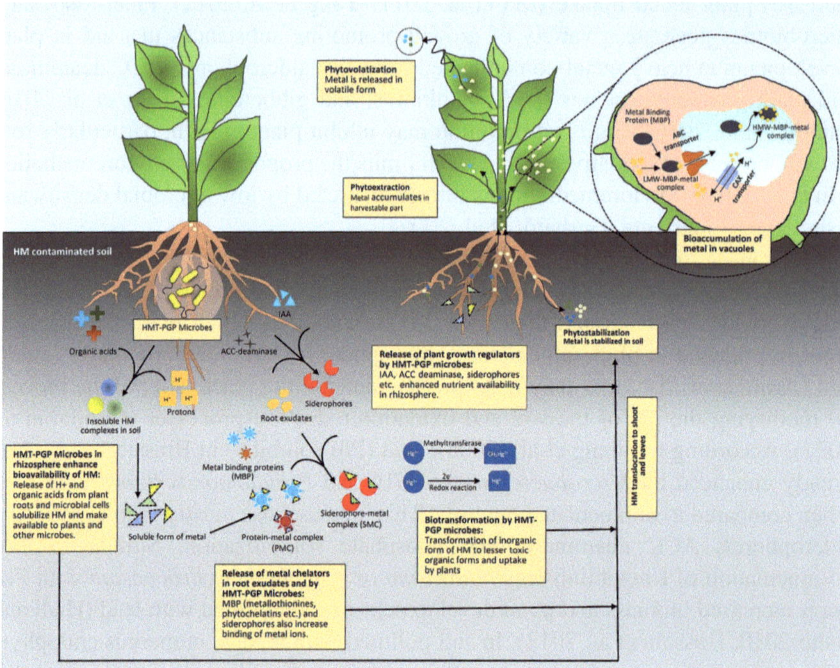


**Fig. 4.6** Mechanisms involving the interaction between HMT-PGP bacteria and plants to remediate HM-contaminated soil. (Adapted from Mishra et al. 2017)

#### 4.2.2.7 Nutrient Deficiency Stresses

Current agricultural systems may be made more sustainable by utilizing beneficial bacteria. Significant contributions to plant nutrient management are made by members of the rhizosphere microbiome (Adhya et al. 2015). Nitrogen-fixing rhizobia and mycorrhizal fungi that aid in phosphorus absorption are two well-known examples (Richardson et al. 2009; Miransari 2011). The rhizosphere and phyllosphere are often where plants obtain their nutrients (Turner et al. 2013). According to Miao et al. (2011), plant nutrient management calls for the best possible utilization of soil, water, atmospheric conditions, and NPK fertilizers. A useful microbiome can also aid in increasing nutrient usage efficiency. Numerous studies have been conducted about the use of symbionts, like mycorrhizal, in directing minerals, including phosphorus, and further crucial macro- and microelements from the soil to developing plants (Salvioli et al. 2016; Wahab et al. 2023) and for modeling and enhancing soil aggregates and structure (Omomowo and Babalola 2019; Wahab et al. 2023) to suit the nutritional needs of crops such as oilseeds, grains, pulses, and fruits (Johnson et al. 2012; Salvioli and Bonfante 2013). In addition to *Bradyrhizobium* and *Rhizobium*, several additional bacterial endophytes have been shown to form symbiotic relationships with plants to fix nitrogen in a bioavailable manner in unspecialized host tissue, even when nodules are absent (Guimaraes et al. 2012; Santi et al. 2013). For instance, blue-green algae generate hetero cysts that are appropriate for biological nitrogen fixation (BNF) using nitrogenase.

Blue-green algae have symbiotic associations with a variety of plants from diverse clades, including lichens, gunnera, and cycads (Santi et al. 2013). According to different research (Leite et al. 2014), of the 102 bacteria linked to sugarcane roots, 74 and 77 are successful in fixing nitrogen and solubilizing phosphorus, respectively. Similarly, 16S rRNA sequencing research of the cowpea rhizosphere showed that cowpea may be nodulated and supported by BNF by species such as *Burkholderia, Achromobacter, Rhizobium, and Bradyrhizobium* (Guimaraes et al. 2012). According to several findings (Shridhar 2012; Hasan 2013), algal taxa including Anabaena, Aphanocapsa, and Phormidium may fix atmospheric nitrogen in paddy fields. Finding bacterial species that influence macro- and micronutrient absorption in plant species under various hazardous and deficient situations makes sense, given the significance of vital plant nutrients (Mapelli et al. 2012; Singh et al. 2022). Additionally, microbiomes can help absorb several trace minerals, including zinc (Vaid et al. 2014) and iron (Prity et al. 2020). Plant microbiome members work together to break down organic chemicals that are necessary for plants to thrive in nutrient-poor soils as well as for their survival (Bhattacharyya et al. 2015; Hunter 2016).

#### 4.2.2.8 Improvement of Plant Nutritional Status by PGPR

Phosphorous

One of the key nutrients that stunts development is phosphorus, which is mostly found in soils as insoluble inorganic phosphate. This limits the amount of soluble phosphate that plants can absorb (Yang et al. 2009; Yu et al. 2022). Plant-associated bacteria from the Pseudomonas, Bacillus, or Rhizobium genera are effective phosphate solubilizers (Kumar 2016; Yu et al. 2022). This ability of some PGPR to solubilize phosphate is a crucial attribute for encouraging plant growth. Although PGPR can solubilize phosphate in a variety of ways, the main process is widely understood to be the synthesis of low molecular weight organic acids. These organic acids are produced and secreted, which either lowers the pH of the soil or chelates mineral ions, leading to the solubilization of phosphate (Glick 2012; Khan et al. 2014). Several of the most common organic acids implicated in phosphate solubilization by *Pseudomonas* sp. include glutamic and 2-ketogluconic acids (Spohn et al. 2020; Yu et al. 2022). Nevertheless, some phosphate-solubilizing bacteria also create additional organic acids, including malonic, isovaleric, isobutyric, acetic, lactic, glycolic, oxalic, and succinic (Khan et al. 2014). As an alternative, PGPR can secrete enzymes such as phosphatases, phytases, phosphonatases, and lyases to liberate organic and inorganic phosphorus from soil components. The primary mechanism for the enzymatic phosphate solubilization is the activity of acid phosphatases, which are among these extracellular enzymes and are found in large quantities in bacteria associated with plants (Richardson et al. 2011; Khan et al. 2014; Mei et al. 2021).

Nitrogen

Molybdenum nitrogenases, which are present in diazotrophic (nitrogen-fixing) bacteria, are responsible for the majority of biological N2 fixation activities (Bishop and Joerger 1990; Glick 2015). Of these, rhizobia are the one that causes the majority of bacterially mediated N2 fixation and produces root nodules in leguminous plants (Glick 2012; Maróti and Kondorosi 2014). Though their contribution to the overall amount of $N_2$ that is physiologically available for plants is debatable, non-symbiotic bacteria may likewise fix $N_2$ and make it attainable to plants in addition to rhizobia (Glick 2015; Lu et al. 2017). Free-living strains of the Azospirillum genus are the preferred distinction of these non-symbiotic $N_2$-fixing bacteria; however, their ability to promote plant growth has mostly been linked to the synthesis of bioregulators and an enhancement in the absorption of minerals through plant roots (Devi et al. 2022).

Potassium

Numerous microorganisms may solubilize potassium (K) in the soil and are intimately involved in reciprocal interactions with plants, particularly fungal and bacterial species (Gundala et al. 2013; Setiawati and Mutmainnah 2016). Research on the influences of plant growth-promoting microbiome on plant growth showed that the K-solubilizing by microorganisms' induced secretion of organic acids was related to increased potassium availability, which in turn led to growth promotion (Badr et al. 2006; Sheng and He 2006). Organic acids (acetate, oxalate, citrate, coumaric acid, and ferulic acid), produced by soil microbiome, play a vital role in enhancing mineral weathering through proton release. This process leads to acidification of the rhizosphere, thereby increasing mineral dissolution and facilitating the solubilization and availability of potassium (K) (Ahmad et al. 2016; Olaniyan et al. 2022). So, researchers concluded that using PGPM, like potassium-solubilizing bacteria, could be a dependable biofertilizer to achieve biological development for sustainable agriculture. This will increase plant nutrient availability and allow for a reduction in the usage of chemical fertilizers (Kumar et al. 2021).

Zinc

However, most agricultural soils lack zinc or have it in fixed forms, which prevents plants from using it (Sadeghzadeh 2013). More than 50% of soils worldwide are zinc-deficient, according to reports from the Food and Agriculture Organization

(FAO) (FAO 2002). This is primarily because zinc is associated with naturally existing mineral shapes like zincite, sphalerite, smithsonite, zinc silicates, willemite, and zinc sulfide (Saravanan et al. 2011), which are often not available for plant uptake. Applying inorganic fertilizers is one method of treating zinc shortage, although it has certain negative environmental effects and, as said, leaves a large portion of the metal inaccessible to plants. Utilizing plant growth-promoting rhizobacteria, which are recognized for their function in solubilizing naturally existing cation-bearing minerals, may be a more effective approach to solving this issue. According to Hussain et al. (2015), PGPR has been demonstrated to solubilize forms of zinc that are inaccessible throughout chelation, interchange response mechanisms, acidification, and dissolution procedures via excreting organic acids into the soil. Subramanian et al. (2009) stated that microorganisms improve Zn mineral obtainability in the rhizosphere by solubilizing the unavailable forms of Zn. Numerous studies show that zinc mobilization by PGPR inoculation greatly boosts cereal crop yields, such as wheat (Ullah et al. 2015; Kamran et al. 2017), and rice (Vaid et al. 2014).

Iron

The obtainability of this fundamental micro-nutrient for organisms in soils is confined (Ramos et al. 2020). Nevertheless, in reply to the iron shortage, bacteria could excrete low molecular cluster iron-chelating molecules named siderophores. Siderophores can create stable siderophore-$Fe^{+}3$ complexes by scavenging ferric ions, which are the most prevalent type of iron in nature. The decrease of $Fe_{+}3$ and the liberation of soluble $Fe^{+2}$ into the cytoplasm are the ultimate outcomes of this process, which is initiated when certain transporters at the bacterial surface identify these complexes (Llamas et al. 2014). Hider and Kong (2010) have found approximately 500 distinct siderophores; of these, pyoverdines are the primary siderophores generated by fluorescent pseudomonads that promote plant growth. Plant growth and development can benefit from the impacts of siderophores produced by PGPR through various direct and indirect methods (Khoso et al. 2024). The plant's improved iron nutritional status and subsequent enhancement of plant development are examples of direct positive consequences. Rather, it has been demonstrated that plants are capable of assimilating Fe+3-pyoverdine complexes, which ultimately increases the amount of iron in plant tissues (Trapet et al. 2021). Bacterial siderophores can indirectly promote plant growth, mostly due to their capability to decrease the availability of iron to phytopathogens. Notably, it has been shown that plants perceive bacterial siderophores to activate plant defenses and induce systemic resistance (ISR) (Aznar and Dellagi 2015; Zhu et al. 2022). Nevertheless, the induction of the plant-resistant replies via microorganisms has been demonstrated to rely on the plant's iron position (Timofeeva et al. 2022).

## 4.3 Conclusion

The earth's life cycle is fundamentally sustained by the soil's microflora and microfauna, which form a dynamic and complex interface with plant roots. Plant-microbe interactions range from symbiotic cooperation to intense competition for resources and space. Within this delicate balance, soil microorganisms are indispensable ecosystem engineers. They drive the decomposition of organic matter, releasing locked-up nutrients like nitrogen, phosphorus, and sulfur into inorganic forms for plant uptake, while simultaneously returning $CO_2$ to the atmosphere for photosynthesis. Crucially, the formation of humus—the stable organic backbone of fertile soil—is mediated by microbial activity, directly enhancing soil structure, water retention, and resilience.

Recent research into the plant microbiome has yielded key findings on its role as a biological buffer against environmental stress. Against abiotic stresses like drought and salinity, specific microbial consortia enhance plant tolerance by inducing osmotic adjustment (e.g., proline accumulation), upregulating antioxidant enzymes (e.g., SOD, CAT) to scavenge reactive oxygen species, and modulating stress hormone signaling (e.g., via ACC deaminase activity). Against biotic stresses from pathogens and pests, beneficial microbes act through direct antagonism (antibiotic production, niche competition) and by priming the plant's own systemic defense responses, creating a more robust, system-wide immunity.

Looking forward, harnessing this potential faces significant challenges and opportunities. The primary hurdle is translating lab-proven results into consistent field performance, which requires moving beyond single-strain bioinoculants toward carefully formulated, multi-functional microbial consortia that can thrive in diverse, non-sterile soil environments. Furthermore, regulatory hurdles for microbial products, a lack of standardized testing protocols, and the need for farmer-friendly formulations present substantial barriers to widespread adoption.

Finally, the central thesis remains clear: the plant microbiome is not merely a passive component but an active cornerstone for sustainable, climate-resilient agriculture. By managing and integrating these microbial communities, we can develop agricultural systems that rely less on chemical inputs, enhance nutrient use efficiency, and fortify crops against increasing climatic volatility. Investing in the understanding and application of soil microbiology is, therefore, an investment in the foundational health of our agroecosystems and long-term food security.

**Declaration of AI Assistance** The authors declare that some of the artificial intelligence tools were used in the preparation, writing, and editing of this book chapter. All contents, including the text, references, and formatting, were developed, drafted, and refined solely by the authors. The chapter has been peer-reviewed in accordance with standard academic practices, and the authors take full responsibility for the originality, integrity, and accuracy of the final work.

## References

Abdelaal K, AlKahtani M, Attia K, Hafez Y, Király L, Künstler A (2021) The role of plant growth-promoting bacteria in alleviating the adverse effects of drought on plants. Biology 10(6):520

Abbas R, Rasul S, Aslam K, Baber M, Shahid M, Mubeen F, Naqqash T (2019) Halotolerant PGPR: a hope for cultivation of saline soils. J King Saud Univ Sci 31:1195–1201

Adhya TK, Kumar N, Reddy G, Podile AR, Bee H, Bindiya S (2015) Microbial mobilization of soil phosphorus and sustainable P management in agricultural soils. Curr Sci 108:1280–1287

Agrawal AA, Johnson MTJ, Hastings AP, Maron JL (2013) A field experiment demonstrating plant life-history evolution and its co-evolutionary feedback to seed predator populations. Am Nat 181:S35–S45

Ahmad M, Nadeem SM, Naveed M, Zahir ZA (2016) Potassium-solubilizing bacteria and their application in agriculture. In: Meena VS, Maurya BR, Verma JP, Meena RS (eds) Potassium solubilizing microorganisms for sustainable agriculture. Springer India, pp 293–313

Ali SZ, Sandhya V, Grover M, Kishore N, Rao LV, Venkateswarlu B (2009) Pseudomonas sp. strain AKM-P6 enhances tolerance of sorghum seedlings to elevated temperatures. Biol Fertil Soils 46:45–55

Anwar A, Kim J-K (2020) Transgenic breeding approaches for improving abiotic stress tolerance: recent progress and future perspectives. Int J Mol Sci 21(8):2695

Arora DK, Saikia R, Dwievdi R, Smith D (2005) Current status, strategy and future prospects of microbial resource collections. Curr Sci 89:488–495

Arora S, Patel PN, Vanza MJ, Rao GG (2014) Isolation and characterization of endophytic bacteria colonizing halophyte and other salt tolerant plant species from coastal Gujarat. Afr J Microbiol Res 8:1779–1788

Arshad M, Sharoona B, Mahmood T (2008) Inoculation with *Pseudomonas* spp. containing ACC deaminase partially eliminate the effects of drought stress on growth, yield and ripening of pea (*Pisum sativum* L.). Pedosphere 18:611–620

Arslan M, Imran A, Khan QM, Afzal M (2015) Plant–bacteria partnerships for the remediation of persistent organic pollutants. Environ Sci Pollut Res:1–15

Ashraf M, Foolad MR (2007) Roles of glycine betaine and proline in improving plant abiotic stress resistance. Environ Exp Bot 59:206–216

Atouei MT, Pourbabaee AA, Shorafa M (2019) Alleviation of salinity stress on some growth parameters of wheat by exopolysaccharide-producing bacteria. Iran J Sci Technol Trans A Sci 43:2725–2733

Ayalaja CF, Pedro A (2012) Stress response in microbiology. In: Reguena JM (ed). Caister Academic Press, Madrid

Aznar A, Dellagi A (2015) New insights into the role of siderophores as triggers of plant immunity: what can we learn from animals? J Exp Bot 66(11):3001–3010

Babou C, Lisna T (2019) Liquid biofertilizer formulated from coconut and its effect on growth and root characteristics of robusta coffee seedlings under drought conditions. J Plant Dev Sci 11(1): 61–64

Badr M, Shafei A, Sharaf El-Deen S (2006) The dissolution of K and Pbearing minerals by silicate dissolving bacteria and their effect on sorghum growth. Res J Agric Biol Sci 2:5–11

Barka EA, Nowak J, Clement C (2006) Enhancement of chilling resistance of inoculated grapevine plantlets with a plant growth-promoting rhizobacterium, Burkholderia phytofirmans strain PsJN. Appl Environ Microbiol 72:7246–7252

Barret M, Morrissey JP, O'Gara F (2011) Functional genomics analysis of plant growth-promoting rhizobacterial traits involved in rhizosphere competence. Biol Fertil Soils 47:729–747

Bartels D, Hussain SS (2008) Current status and implications of engineering drought tolerance in plants using transgenic approaches. CAB Rev Persp Agr Vet Sci Nutri Natu Sci 3:020

Beijerinck MW (1901) UberoligonitrophileMikrobenzblBackt 7:561–582

Berg G, Alvi M, Schmidt C, Zachow C, Egamberdieva D, Kamilova F, Lugtenberg B (2013) Biocontrol and osmoprotection for plants under salinated conditions. In: de Bruijn FJ (ed) Molecular microbiology, ecology of the rhizosphere. Wiley, Blackwell

Berg G, Grube M, Scholter M, Smalla K (2014) Unraveling the plant microbiome: looking back and future perspectives. Front Microbiol 5:148

Berg G, Rybakova D, Grube M, Koberl M (2016) The plant microbiome explored: implications for experimental botany. J Exp Bot 67:995–1002

Bhattacharyya PN, Sarmah SR, Dutta P, Tanti AJ (2015) Emergence in mapping microbial diversity in tea (*Camellia sinensis* L.) soil of Assam, North-East India: a novel approach. Eur J Biotechnol Biosci 3:20–25

Bishop PE, Joerger RD (1990) Genetics and molecular biology of alternative nitrogen fixation systems. Annu Rev Plant Physiol Plant Mol Biol 41:109–125

Boiero L, Perrig D, Masciarelli O, Penna C, Cassan F, Luna V (2007) Phytohormone production by three strains of Bradyrhizobiumjaponicum and possible physiological and technologicalimplications. Appl Microbiol Biotechnol 74:874–880

Borken W, Matzner E (2009) Reappraisal of drying and wetting effects on C and N mineralization and fluxes in soils. Glob Chang Biol 15:808–824

Bremer E, Krämer R (2019) Responses of microorganisms to osmotic stress. Ann Rev Microbiol 73:313–334

Capell T, Bassie L, Christou P (2004) Modulation of the polyamine biosynthetic pathway in transgenic rice confers tolerance to drought stress. Proc Natl Acad Sci USA 101:9909–9914

Castiglioni P, Warner D, Bensen RJ, Anstrom DC, Harrison J, Stoecker M, Abad M, Kumar G, Salvador S, D'ordine R, Navarro S, Back S, Fernandes M, Targolli J, Dasgupta S, Bonin C (2008) Bacterial RNA chaperones confer abiotic stress tolerance in plants and improved grain yield in maize under water limited conditions. Plant Physiol 147:446–455

Chu TN, Tran BTH, Van Bui L, Hoang MTT (2019) Plant growth-promoting *rhizobacterium Pseudomonas* PS01 induces salt tolerance in Arabidopsis thaliana. BMC Res Notes 12:11

Correia AM, Lopes LF (2023) Revisiting biodiversity and ecosystem functioning through the lens of complex adaptive systems. Diversity 15(8):895

Devi S, Sharma M, Manhas RK (2022) Investigating the plant growth promoting and biocontrol potentiality of endophytic *Streptomyces* SP. SP5 against early blight in *Solanum lycopersicum* seedlings. BMC Microbiol 22:285

Doble M, Kumar A (2005) Biotreatment of industrial effluents. Elsevier, Butterworth-Heinemann, Oxford, pp 1–5

Du L, Liu W (2012) Occurrence, fate and ecotoxicity of antibiotics in agroecosystems. A review. Agron Sustain Dev 32:309–327

Egamberdieva D, Wirth SJ, Bellingrath-Kimura DSD, Mishra J, Arora NK (2019) Salt-tolerant plant growth promoting rhizobacteria for enhancing crop productivity of saline soils. Front Microbiol 18:2791

Egamberdieva D, Jabborova D, Hashem A (2015) *Pseudomonas* induces salinity tolerance in cotton (*Gossypium hirsutum*) and resistance to *Fusarium* root rot through the modulation of indole-3-acetic acid. Saudi J Biol Sci 22:773–779

El-Esawi MA, Alaraidh IA, Alsahli AA, Alamri SA, Ali HM, Alayafi AA (2018) *Bacillus firmus* (SW5) augments salt tolerance in soybean (*Glycine max* L.) by modulating root system architecture, antioxidant defense systems and stress-responsive genes expression. Plant Physiol Biochem 132:375–384

Etesami H, Emami S, Alikhani HA (2017) Potassium solubilizing bacteria (KSB): mechanisms, promotion of plant growth, and future prospects—a review. J Soil Sci Plant Nutr 17:897–911

Evelin H, Devi TS, Gupta S, Kapoor R (2019) Mitigation of salinity stress in plants by Arbuscular Mycorrhizal Symbiosis: current understanding and new challenges. Front Plant Sci 10:470

Evelin H, Kapoor R, Giri B (2009) *Arbuscular mycorrhizal* fungi in alleviation of salt stress: a review. Ann Bot 104(7):1263–1280

FAO (2002) World agriculture: towards 2015/2030. Summary Report. Food and Agriculture Organization of the United Nations (FAO), Rome. 97 p

Feng NX, Yu J, Zhao HM, Cheng YT, Mo CH, Cai QY, Li YW, Li H, Wng MH (2017) Efficient phytoremediation of organic contaminants in soils using plant-endophyte partnerships. Sci Total Environ 583:352–368

Fester T, Giebler J, Wick LY, Schlosser D, Kästner M (2014) Plant–microbe interactions as drivers of ecosystem functions relevant for the biodegradation of organic contaminants. Curr Opin Biotechnol 27:168–175

Forni C, Duca D, Glick BR (2017) Mechanisms of plant response to salt and drought stress and their alteration by rhizobacteria. Plant Soil 410:335–356

Furtak K, Wolińska A (2023) The impact of extreme weather events as a consequence of climate change on the soil moisture and on the quality of the soil environment and agriculture -a review. Catena 231:107378

Gerba CP, Goss MJ, Unc A (2023) Septic systems and municipal waste disposal on land. In: Goss MJ, Oliver M (eds) Encyclopedia of soils in the environment, 2nd edn. Academic Press, pp 567–577

Gerhardt KE, Huang XD, Glick BR, Greenberg BM (2009) Phytoremediation and rhizoremediation of organic soil contaminants: potential and challenges. Plant Sci 176:20–30

Giri B, Mukerji KG (2004) Mycorrhizal inoculant alleviate salt stress in *Sesbania aegyptiaca* and *Sesbania grandiflora* under field conditions: evidence for reduced sodium and improved magnesium uptake. Mycorrhiza 14:307–312

Glick BR (2012) Plant growth-promoting bacteria: mechanisms and applications. Scientica 15: 963401

Glick BR (2015) Phytoremediation. Beneficial plant-bacterial interactions. In: Springer, pp 191–221

Gondal AH, Farooq QSS, Kumar SS, Toor MD, Zafar A, Rehman B (2021) Adaptability of soil pH through innovative microbial approach. Curr Res Agric Sci 8(2):71–79

Guimaraes AA, Jaramillo PMD, Nóbrega RSA, Florentino LA, Silva KB, de Souza Moreira FM (2012) Genetic and symbiotic diversity of nitrogen-fixing bacteria isolated from agricultural soils in the western Amazon by using cowpea as the trap plant. Appl Environ Microbiol 78: 6726–6733

Gundala P, Chinthala P, Sreenivasulu B (2013) A new facultative alkaliphilic, potassium solubilizing, Bacillus Sp. SVUNM9 isolated from micacores of Nellore District, Andhra Pradesh, India. Research and reviews. J Microbiol Biotechnol 2:1–7

Gupta A, Rai S, Bano A, Khanam A, Sharma S, Pathak N (2021) Comparative evaluation of different salt-tolerant plant growth-promoting bacterial isolates in mitigating the induced adverse effect of salinity in Pisum sativum. Biointerface Res Appl Chem 11:13141–13154

Habib SH, Kausar H, Saud HM (2016) Plant growth-promoting rhizobacteria enhance salinity stress tolerance in okra through ROS-scavenging enzymes. BioMed Res Int:6284547

Hadi F, Bano A (2010) Effect of diazotrophs *(Rhizobium and Azotobacter*) on growth of maize (*Zea mays* L.) and accumulation of Lead (Pb) in different plant parts. Pak J Bot 42:4363–4370

Hakemian AS, Rosenzweig AC (2007) The biochemistry of methane oxidation. Annu Rev Biochem 76:223–241

Hartmann FE, Vallet AS, McDonald BA, Croll D (2017) A fungal wheat pathogen evolved host specialization by extensive chromosomal rearrangements. ISME J 11:1189–1204. https://doi.org/10.1038/15mej.2016.196

Hasan MA (2013) Investigation on the nitrogen fixing Cyanobacteria (BGA) in rice fields of North-West region of Bangladesh. J Environ Sci Nat Resour 6:253–259

Herron PM, Gage DJ, Cardon ZG (2010) Microscale water potential gradients visualized in soil around plant root tips using microbiosensors. Plant Cell Environ 33:199–210

Hider RC, Kong X (2010) chemistry and biology of siderophores. Nat Prod Rep 27:637–657

Hoffmeister D, Keller NP (2007) Natural products of filamentous fungi: enzymes, genes and their regulation. Nat Prod Rep 24:393–416

Hu S, Gu H, Cui C, Ji R (2016) Toxicity of combined chromium (VI) and phenanthrene pollution on the seed germination, stem lengths, and fresh weights of higher plants. Environ Sci Pollut Res 23:15227–15235
Hui D, Deng Q, Tian H, Luo Y (2016) Climate change and carbon sequestration in forest ecosystems. In: Suzuki T, Lackner M (eds) Handbook of climate change migration and application. Springer, Berlin, pp 555–594
Hunter P (2016) Plant microbiomes and sustainable agriculture: deciphering the plant microbiome and its role in nutrient supply and plant immunity has great potential to reduce the use of fertilizers and biocides in agriculture. EMBO Rep 17(12):1696
Hussain A, Abbas N, Arshad F, Akram M, Khan ZI, Ahmad K, Mansha M, Mirzaei F (2013) Effects of diverse doses of Lead (Pb) on different growth attributes *of Zea-Mays* L. Agric Sci 4: 262–265
Hussain A, Arshad M, Zahir ZA, Asghar M (2015) Prospects of zinc solubilizing bacteria for enhancing growth of maize. Pak J Agri Sci 52(4):915–922
Hussain SS, Asif MA, Sornaraj P, Ali M, Shi BJ (2016) Towards integration of system based approach for understanding drought stress in plants. In: Ahmad P, Rasool SP (eds) Water stress and crop plants: a sustainable approach. Elsevier, Atlanta, pp 227–247
Hussain SS, Raza H, Afzal I, Kayani MA (2012) Transgenic plants for abiotic stress tolerance: current status. Arch Agron Soil Sci 58:693–721
Ibáñez A, Garrido-Chamorro S, Barreiro C (2023) Microorganisms and climate change: a not so invisible effect. Microbiol Res 14(3):918–947
Ijaz A, Imran A, Ul Haq MA, Khan QM, Afzal M (2016) Phytoremediation: recent advances in plant-endophytic synergistic interactions. Plant Soil 405:179–195
IPCC (2006) In: Egghleston HS, Bunendia L, Miwa K, Ngara T, Tanabe K (eds) IPCC guidelines for national greenhouse gas inventories, Tokyo
IPCC (2007) Climate change 2007. The physical science basis. Contribution of working group I to the fourth assessment report of the intergovernmental panel on climate change. Cambridge University Press, Cambridge, NY
Jacoby R, Peukert M, Succurro A, Koprivova A, Kopriva S (2017) The role of soil microorganisms in plant mineral nutrition-current knowledge and future directions. Front Plant Sci 8:1617
Jalmi SK, Sinha AK (2022) Ambiguities of PGPR-induced plant signaling and stress management. Front Microbiol 13:899563
Jan U, Feiwen R, Masood J, Chun SC (2020) Characterization of soil microorganism from humus and indigenous microorganism amendments. Mycobiology 48:392
Jha B, Gontia I, Hartmann A (2012) The roots of the halophyte *Salicornia brachiata* are a source of new halotolerant diazotrophic bacteria with plant growth-promoting potential. Plant Soil 356: 265–277
Jimtha CJ, Jishma P, Shreelekha S, Chitra S, Radhakrishnan EK (2017) Antifungal properties prodigiosin producing *rhizospheric serratia* sp. Rhizosphere 3(1):105–108
Johnson D, Martin F, Cairney JWG, Anderson IC (2012) The importance of individuals: intraspecific diversity of mycorrhizal plants and fungi in ecosystems. New Phytol 194:614–628
Jorquera MA, Shaharoona B, Nadeem SM, de la Luz MM, Crowley DE (2012) Plant growth promoting rhizobacteria associated with ancient clones of creosote bush (*Larrea tridentata*). Microb Ecol 64:1008–1017
Kaiser K, Kalbitz K (2012) Cycling downwards- dissolved organic matter in soils. Soil Biol Biochem 52:29–32
Kamran S, Shahid I, Baig DN, Rizwan M, Malik KA, Mehnaz S (2017) Contribution of zinc solubilizing bacteria in growth promotion and zinc content of wheat. Front Microbiol 8:2593
Kapahi M, Sachdeva S (2019) Bioremediation options for heavy metal pollution. J Health Poll 9(24):191203
Kapoor D, Bhardwaj S, Landi M, Sharma A, Ramakrishnan M, Sharma A (2020) The impact of drought in plant metabolism: how to exploit tolerance mechanisms to increase crop production. Appl Sci 10(16):5692

Kathuria H, Giri J, Tyagi H, Tyagi AK (2007) Advances in transgenic rice biotechnology. Crit Rev Plant Sci 26:65–103

Kawasaki A, Watson ER, Kertesz MA (2012) Indirect effect of polycyclic aromatic hydrocarbon contamination on microbial communities in legume and grass rhizospheres. Plant Soil 358:169–182

Khan AL (2023) The phytomicrobiome: solving plant stress tolerance under climate change. Front Plant Sci 14:1219366

Khan MA, Asaf S, Khan AL, Adhikari A, Jan R, Ali S, Imran M, Kim K-M, Lee I-J (2019) Halotolerant rhizobacterial strains mitigate the adverse effects of NaCl stress in soybean seedlings. BioMed Res Int:9530963

Khan MA, Sahile AA, Jan R, Asaf S, Hamayun M, Imran MA, Adhikari A, Kang S, Kim KM, Lee I (2021) Halo-tolerant bacteria mitigate the effects of salinity stress on soybean growth by regulating secondary metabolites and molecular responses. BMC Plant Biol 21:176

Khan AL, Waqas M, Kang S-M, Al-Harrasi A, Hussain J, Al-Rawahi A, Al-Khiziri S, Ullah I, Ali L, Jung HY, Lee IJ (2014) Bacterial endophyte Sphingomonas sp. LK11 produces gibberellins and IAA and promotes tomato plant growth. J Microbiol 52(8):689–695

Khanna K, Jamwal VL, Kohli SK, Gandhi SG, Ohri P, Bhardwaj R, Abd-Allah E, Hashem A, Ahmad P (2018) Plant growth promoting rhizobacteria induced Cd tolerance in Lycopersicon esculentum through altered antioxidative defense expression. Chemosphere 217:463–474

Kharitonov S, Semenov M, Sabrekov A, Kotsyurbenko O, Zhelezova A, Schegolkova N (2021) Microbial communities in methane cycle: modern molecular methods gain insights into their global ecology. Environments 8(2):16

Khoso MA, Wagan S, Alam I, Hussain A, Ali Q, Saha S, Poudel TR, Manghwar H, Liu F (2024) Impact of plant growth-promoting rhizobacteria (PGPR) on plant nutrition and root characteristics: current perspective. Plant Stress 11:100341

Klumpp K, Fontaine S, Attard E, Le Roux X, Gleixner G, Soussana JF (2009) Grazing triggers soil carbon loss by altering plant roots and their control on soil microbial community. J Ecol 97:876–885

Kohler J, Hernandez JA, Caravaca F, Rolden A (2008) Plant growth promoting *Rhizobacteria and Arbuscular mycorrhizal* fungi modify alleviation biochemical mechanisms in water stressed plants. Funct Plant Biol 35:141–151

Kreps JA, Wu Y, Chang HS, Zhu T, Wang X, Harper JF (2002) Transcriptome changes for *Arabidopsis* in response to salt, osmotic and cold stress. Plant Physiol 130:2129–2141

Kumar A (2016) Phosphate solubilizing bacteria in agriculture biotechnology: diversity, mechanism and their role in plant growth and crop yield. Int J Adv Res 4:116–124

Kumar M, Giri VP, Pandey S, Gupta A, Patel MK, Bajpai AB, Jenkins S, Siddique KHM (2021) Plant-growth-promoting Rhizobacteria emerging as an effective bioinoculant to improve the growth, production and stress tolerance of vegetable crops. Int J Mol Sci 22:12245

Kuzyakov Y (2010) Primary effects: interactions between living and dead organic matter. Soil Biol Biochem 42(9):1363–1371

Lal R (2004) Soil carbon sequestration impacts on global climate change and global security. Science 403:1623–1627

Latge JP (2017) Immune evasion: face changing in the fungal opera. Nat Microbiol 2:16266–16269

Leckie SE, Prescott CE, Grayston SJ, Neufeld JD, Mohn WW (2004) Comparison of chloroform fumigation extraction, phospholipid fatty acid and DNA methods to determine microbial biomass in forest humus. Soil Biol Biochem 36:532–536

Leite MCBS, de Farias ARB, Freire FJ, Andreote FD, Sobral JK, Freire MBGS (2014) Isolation, bioprospecting and diversity of salt-tolerant bacteria associated with sugarcane in soils of Pernambuco, Brazil. Rev Bras Eng Agrıc Amb 18:S73–S79

Llamas MA, Imperi F, Visca P et al (2014) Cell-surface signaling in Pseudomonas: stress responses, iron transport, and pathogenicity. FEMS Microbiol Rev 38:569–597

Lu J, Yang F, Wang S, Ma H, Liang J, Chen Y (2017) Co-existence of rhizobia and diverse non-rhizobial bacteria in the rhizosphere and nodules of *Dalbergia odorifera* seedlings inoculated with *Bradyrhizobium elkanii, rhizobium multihospitium*–like and *Burkholderia pyrrocinia*–like strains. Front Microbiol 8

Ma Y, Prasad MNV, Rajkumar M, Freitas H (2011) Plant growth promoting rhizobacteria and endophytes accelerate phytoremediation of metalliferous soils. Biotechnol Adv 29:248–258

Ma Z, Jacobson FE, Geidroc DP (2009) Coordination chemistry of bacterial metal transport and sensing. Chem Rev 109:4644–4681

Malik AA, Martiny JBH, Brodie EL, Martiny AC, Treseder KK, Allison SD (2020) Defining trait-based microbial strategies with consequences for soil carbon cycling under climate change. ISME J 14(1):1–9

Manzoni S, Schimel JP, Porporato A (2012) Responses of soil microbial communities to water stress: results from a meta-analysis. Ecology 93(4):930–938

Mapelli F, Marasco R, Rolli E, Barbato M, Cherif H, Guesmi A, Ouzari I, Daffonchio D, Borin S (2013) Potential for plant growth promotion of rhizobacteria associated with *Salicornia* growing in Tunisian hypersaline soils. Biomed Res 2013:248078. 2 Harnessing the plant microbiome for improved abiotic stress tolerance 38

Mapelli F, Marasco R, Rolli E, Cappitelli F, Daffonchio D, Borin S (2012) Mineral-microbe interactions: biotechnological potential of bio-weathering. J Biotechnol 157:473–481

Marasco R, Mapelli F, Rolli E, Mosqueira MJ, Fusi M, Bariselli P, Reddy M, Cherif A, Tsiamis G, Borin S, Daffonchio D (2016) *Salicornia strobilacea* (synonym of *Halocnemum strobilaceum*) growth under different tidal regimes selects rhizosphere bacteria capable of promoting plant growth. Front Microbiol 7:1286

Marasco R, Rolli E, Vigani G, Borin S, Sorlini C, Ouzari H, Zocchi G, Daffonchio D (2013) Are drought-resistance promoting bacteria cross-compatible with different plant models? Plant Signal Behav 8:e26741

Maróti G, Kondorosi É (2014) Nitrogen-fixing *Rhizobium*-legume symbiosis: are polyploidy and host peptide-governed symbiont differentiation general principles of endosymbiosis? Front Microbiol 5:326

Marulanda A, Barea JM, Azcón R (2009) Stimulation of plant growth and drought tolerance by native microorganisms (AM fungi and bacteria) from dry environments: mechanisms related to bacterial effectiveness. J Plant Growth Regul 28:115–124

Marulanda A, Porcel R, Barea JM, Azcon R (2007) Drought tolerance and antioxidant activities in lavender plants colonized by native drought tolerant or drought sensitive Glomus species. Microb Ecol 54:543–552

Mei C, Chretien RL, Amaradasa BS, He Y, Turner A, Lowman S (2021) Characterization of phosphate solubilizing bacterial endophytes and plant growth promotion in vitro and in greenhouse. Microorganisms 9(9)

Miao B, Stewart BA, Zhang F (2011) Long-term experiments for sustainable nutrient management in China. A review. Agron Sustain Dev 31:397–414

Miransari M (2011) Hyperaccumulators, arbuscular mycorrhizal fungi and stress of heavy metals. Biotechnol Adv 29:645–653

Miransari M (2013) Soil microbes and the availability of soil nutrients. Acta Physiol Plant 35(11): 3075–3084

Mishra J, Fatima T, Arora NK (2018) Role of secondary metabolites from plant growth-promoting Rhizobacteria in combating salinity stress. In: Egamberdieva D, Ahmad P (eds) Plant microbiome: stress response. Microorganisms for sustainability, vol 5. Springer, Singapore

Mishra J, Singh R, Arora NK (2017) Alleviation of heavy metal stress in plants and remediation of soil by rhizosphere microorganisms. Front Microbiol 8

Mishra PK, Bisht SC, Ruwari P, Selvakumar G, Joshi GK, Bisht JK et al (2011) Alleviation of cold stress in inoculated wheat (*Triticum aestivum L.*) seedlings with psychrotolerant Pseudomonads from NW Himalayas. Arch Microbiol 193:497–513

Nadeem SM, Ahmad M, Zahir ZA, Javaid A, Ashraf M (2014) The role of mycorrhizae and plant growth promoting rhizobacteria (PGPR) in improving crop productivity under stressful environments. Biotechnol Adv 32:429–448

Naveed M, Hussain B, Zahir A, Mitter B, Sessitsch A (2014) Drought stress amelioration in wheat through inoculation with *Burkholderia phytofirmans* strain PsJN. Plant Growth Regul 73:121–131

Neufeld JD, Mohn WW (2005) Unexpectedly high bacterial diversity in arctic tundra relative to boreal forest soils, revealed by serial analysis of ribosomal sequence tags. Appl Environ Microbiol 71:5710–5718

Newman MM, Hoilett N, Lorenz N, Dick RP, Liles MR, Ramsier C, Kloepper JW (2016) Glyphosate effects on soil rhizosphere-associated bacterial communities. Sci Total Environ 543:155–160

Olaniyan FT, Alori ET, Adekiya AO, Ayorinde BB, Daramola FY, Osemwegie OO, Babalola OO (2022) The use of soil microbial potassium solubilizers in potassium nutrient availability in soil and its dynamics. Ann Microbiol 72:45

Omomowo OI, Babalola OO (2019) Bacterial and fungal endophytes: tiny giants with immense beneficial potential for plant growth and sustainable agricultural productivity. Microorganisms 7:481

Orłowska E, Przybyłowicz W, Orlowski D, Turnau K, Mesjasz-Przybyłowicz J (2011) The effect of mycorrhiza on the growth and elemental composition of Ni-hyperaccumulating plant Berkheya coddii Roessler. Environ Pollut 159:3730–3738

Panagos P, Borrell P, Meusburger K, Alewell C, Lugato E, Montanarella L (2015) Estimating the soil erosion cover-management factor at European scale. Land Use Policy 48:38–50

Pandey VC (2012) Phytoremediation of heavy metals from fly ash pond by Azolla caroliniana. Ecotoxicol Environ Saf 82:8–12

Panke-Buisse K, Poole A, Goodrich J, Ley R, Kao-Kniffin J (2015) Selection on soil microbiomes reveals reproducible impacts on plant function. ISME J 9:980–989

Penman TD, Pike DA, Webb JK, Shine R (2010) Predicting the impact of climate change on Australia's most endangered snake, *Hoplocephalus bungaroides*. Diver Distribut 16:109–118

Perez-Montano F, Alías-Villegas C, Bellogin RA, del Cerro P, Espuny MR, Jimenez-Guerrero I, Lopez-Baena FJ, Ollero FJ, Cubo T (2014) Plant growth promotion in cereal and leguminous agricultural important plants: from microorganism capacities to crop production. Microbiol Res 169:325–336

Porcel R, Aroca R, Cano C, Bago A, Ruiz-Lozano JM (2006) Identification of a gene from the arbuscular mycorrhizal fungus Glomus intraradices encoding for a 14-3-3 protein that is upregulated by drought stress during the AM symbiosis. Microb Ecol 52:575–582

Prity SA, Sajib SA, Das U, Rahman MM, Haider SA, Kabir AH (2020) Arbuscular mycorrhizal fungi mitigate Fe deficiency symptoms in sorghum through phytosiderophore-mediated Fe mobilization and restoration of redox status. Protoplasma 257(5):1373–1385

Qiang X, Weiss M, Kogel KH, Schafer P (2012) *Piriformospora indica* and mutualistic basidiomycete with an exceptionally large plant host range. Mol Plant Pathol 13:508–518

Qin S, Feng WW, Zhang YJ, Wang TT, Xiong YW, Xing K (2018) Diversity of bacterial microbiota of coastal halophyte *Limonium sinense* and amelioration of salinity stress damage by symbiotic plant growth-promoting *Actinobacterium Glutamicibacter* halophytocola KLBMP 5180. Appl Environ Microbiol 84:e01533–e01518

Qin S, Zhang YJ, Yuan B, Xu PY, Xing K, Wang J, Jiang JH (2014) Isolation of ACC deaminase producing habitat-adapted symbiotic bacteria associated with halophyte *Limonium sinense* (Girard) Kuntze and evaluating their plant growth promoting activity under salt stress. Plant Soil 374:753–766

Raghuvanshi R, Ansari MA, Amardeep VAP (2018) Adaptation to climate change by farmers in Himalayan region of Uttarakhand. Res J Agric Sci 9(2):399–403

Rainey PB (1999) Adaptation of Pseudomonas fluorescens to the plant rhizosphere. Environ Microbiol 1:243–257

Rajkumar M, Sandhya S, Prasad MNV, Freitas H (2012) Perspectives of plant-associated microbes in heavy metal phytoremediation. Biotechnol Adv 30:1562–1574

Rajput RS, Ram RM, Vaishnav A, Singh HB (2019) Microbe-based novel biostimulants for sustainable crop production. In: Satynarayan T (ed) Microbial diversity in ecosystem sustainability and biotechnological applications. Springer, Singapore, pp 109–144

Ramos M, Moscuzza H, Cirelli A (2020) Total content and availability of micronutrients in soils and livestock manure. Rev Int Contam Ambient 36(1):115–126

Richardson AE, Barea JM, McNeill AM, Prigent-Combaret C (2009) Acquisition of phosphorus and nitrogen in the rhizosphere and plant growth promotion by microorganisms. Plant Soil 321: 305–339

Richardson AE, Lynch JP, Ryan PR, Delhaize E, Smith FA, Smith SE, Harvey PR, Ryan MH, Veneklaas EJ, Lambers H, Oberson A, Culvenor RA, Simpson RJ (2011) Plant and microbial strategies to improve the phosphorus efficiency of agriculture. Plant Soil 349:121–156

Ruiz KB, Biondi S, Martínez EA, Orsini F, Antognoni F, Jacobsen SE (2015) Quinoa—a model crop for understanding salt-tolerance mechanisms in halophytes. Plant Biosyst 150(357–371):2

Sadeghzadeh B (2013) A review of zinc nutrition and plant breeding. J Soil Sci Plant Nutr 13(4): 905–927

Sagar A, Rai S, Ilyas N, Sayyed R, Al-Turki A, El Enshasy H, Simarmata T (2022) Halotolerant rhizobacteria for salinity-stress mitigation: diversity, mechanisms and molecular approaches. Sustainability 14:490. https://doi.org/10.3390/su140104

Salvioli A, Ghignone S, Novero M, Navazio L, Venice F, Bagnaresi P, Bonfante P (2016) Symbiosis with an endobacterium increases the fitness of a mycorrhizal fungus, raising its bioenergetic potential. ISME J 10:130–144

Salvioli A, Bonfante P (2013) Systems biology and "omics" tools: a cooperation for next-generation mycorrhizal studies. Plant Sci 203:107–114

Sanaullah M, Blagodatskaya E, Chabbi A, Rumpel C, Kuzyakov Y (2011) Drought effects on microbial biomass and enzyme activities in rhizosphere of grasses depend on plant community composition. Appl Soil Ecol 48:38–44

Sánchez-Bermúdez M, del Pozo JC, Pernas M (2022) Effects of combined abiotic stresses related to climate change on root growth in crops. Front Plant Sci 13:918537

Sandhya V, Ali SZ, Grover M, Kishore N, Venkateswarlu B (2009) Pseudomonas sp. strain P45 protects sunflowers seedlings from drought stress through improved soil structure. J Oilseed Res 26:600–601

Sandhya V, Ali SZ, Grover M, Reddy G, Venkateswaralu B (2010) Effect of plant growth promoting *Pseudomonas spp.* On compatible solutes, antioxidant status and plant growth of maize under drought stress. Plant Growth Regul 62:21–30

Santi C, Bogusz D, Franche C (2013) Biological nitrogen fixation in non-legume plants. Ann Bot 111:743–767

Santoyo G, Moreno-Hagelsieb G, Orozco-Mosqueda MC, Glick BR (2016) Plant growth promoting bacterial endophytes. Microbiol Res 183:92–99

Saravanan VS, Kumar MR, Sa TM (2011) Microbial zinc solubilization and their role on plants. In: Maheshwari D (ed) Bacteria in agrobiology: plant nutrient management. Springer, Berlin, Heidelberg

Sayyed RZ, Seifi S, Patel PR, Shaikh SS, Jadhav HP, El Enshasy H (2019) Siderophore production in groundnut rhizosphere isolate, *Achromobacter* sp. RZS2 influenced by physicochemical factors and metal ions. Environ Sustain 1(3):295–301

Schaeppi K, Bulgarelli D (2015) The plant microbiome at work. Mol Plant-Microbe Interact 28: 212–217

Schmidt MWI, Torn MS, Abivens S, Dittmar T, Guggenberger G, Janssens IA, Kleber M, KögelKnabner I, Lehmann J, Manning DA, Nannipieri P, Rasse DP, Weiner S, Trumbore SE (2011) Persistence of soil organic matter as an ecosystem property. Nature 478:49–56

Seki M, Narusaka M, Ishida J, Nanjo T, Fujita M, Oono Y, Kamiya A, Nakajima M, Enju A, Sakurai T, Satou M, Akiyama K, Taji T, Yamaguchi-Shinozaki K, Carninci P, Kawai J, Hayashizaki Y, Shinozaki K (2002) Monitoring the expression profiles of 7000 *Arabidopsis* genes under drought, cold, and high salinity stresses using full length cDNA microarray. Plant J 31:279–292

Seleiman MF, Al-Suhaibani N, Ali N, Akmal M, Alotaibi M, Refay Y, Dindaroglu T, Abdul-Wajid HH, Battaglia ML (2021) Drought stress impacts on plants and different approaches to alleviate its adverse effects. Plants 10(2):259

Selvakumar G, Joshi P, Nazim S, Mishra PK, Bisht JK, Gupta HS (2009) Phosphate solubilization and growth promotion by Pseudomonas fragi CS11RH1 (MTCC 8984) a psychrotolerant bacterium isolated from a high altitude Himalayan rhizosphere. Biologia 64:239245

Selvakumar G, Joshi P, Suyal P, Mishra PK, Joshi GK, Bisht JK, Bhatt JC, Gupta HS (2010a) Pseudomonas lurida M2RH3 (MTCC 9245), a psychrotolerant bacterium from the Uttarakhand Himalayas, solubilizes phosphate and promotes wheat seedling growth. World J Microbiol Biotechnol 5:1129–1135

Selvakumar G, Kundu S, Joshi P, Nazim S, Gupta AD, Gupta HS (2010b) Growth promotion of wheat seedlings by Exiguobacterium acetylicum 1P (MTCC 8707) a cold tolerant bacterial strain from the Uttarakhand Himalayas. Ind J Microbiol 50:50–56

Selvakumar G, Kundu S, Joshi P, Nazim S, Gupta AD, Mishra PK, Gupta HS (2008a) Characterization of a cold-tolerant plant growth-promoting bacterium Pantoea dispersa 1A isolated from a sub-alpine soil in the north western Indian Himalayas. World J Microbiol Biotechnol 24:955–960

Selvakumar G, Mohan M, Kundu S, Gupta AD, Joshi P, Nazim S, Gupta HS (2008b) Cold tolerance and plant growth promotion potential of Serratia marcescens strain SRM (MTCC 8708) isolated from flowers of summer squash (Cucurbita pepo). Lett Appl Microbiol 46:171–175

Setiawati TC, Mutmainnah L (2016) Solubilization of potassium containing mineral by microorganisms from sugarcane rhizosphere. Agri Agri Sci Proc 9:108–117. https://doi.org/10.1016/j.aaspro.2016.02.134

Shabala S, Hariadi Y, Jacobsen SE (2013) Genotypic difference in salinity tolerance in quinoa is determined by differential control of xylem Na+ loading and stomatal density. J Plant Physiol 170:906–914

Shameer S, Prasad TNVKV (2018) Plant growth promoting rhizobacteria for sustainable agricultural practices with special reference to biotic and abiotic stresses. Plant Growth Reg 84:603–615

Sharma A, Shahzad B, Kumar V, Kohli SK, Sidhu GPS, Bali AS, Zheng B (2019) Phytohormones regulate accumulation of osmolytes under abiotic stress. Biomolecules 9:285

Sharma N, Sudarshan Y, Sharma R, Singh G (2008) RAPD analysis of soil microbial diversity in Western Rajasthan. Curr Sci 94(8):1058–1061

Sheng X, He L, Wang Q, Ye H, Jiang C (2008) Effects of inoculation of biosurfactant-producing Bacillus sp. J119 on plant growth and cadmium uptake in a cadmium-amended soil. J Hazard Mater 155:17–22

Sheng XF, He LY (2006) Solubilization of potassium-bearing minerals by a wild-type strain of Bacillus edaphicus and its mutants and increased potassium uptake by wheat. Can J Microbiol 52:66–72

Shilev S (2020) Plant-growth-promoting bacteria mitigating soil salinity stress in plants. Appl Sci 10:7326

Shim JS, Seo JS, Seo JS, Kim Y, Koo Y, Choi YD, Jung C (2019) Heterologous expression of bacterial trehalose biosynthetic genes enhances trehalose accumulation in potato plants without adverse growth effects. Plant Biotechnol Rep 13:409–418

Shridhar BS (2012) Review: nitrogen fixing microorganisms. Int J Microbiol Res 3:46–52

Siddikee MA, Chauhan PS, Anandham R, Han GH, Sa TM (2010) Isolation, characterization, and use for plant growth promotion under salt stress, of ACC deaminase-producing halotolerant bacteria derived from coastal soil. J Microbiol Biotechnol 20(11):1577–1584

Siebielec S, Siebielec G, Klimkowicz-Pawlas A, Gałązka A, Grządziel J, Stuczyński T (2020) Impact of water stress on microbial community and activity in Sandy and Loamy soils. Agronomy 10(9):1429

Singh B, Kaur T, Kaur S, Manhas RK, Kaur A (2016) Insecticidal potential of an endophytic *Cladosporiumvelox* against *Spodoptera litura* mediated through inhibition of alpha glycosidases. Pestic Biochem Physiol 131:46–52

Singh K, Pandey SN, Mishra A (2015) Preference of heavy metals accumulation, tolerance limit and biochemical responses of *Eichhornia crassipes* (Mart.) exposed to industrial waste water. Int J Curr Res 7(1):11818–11822

Singh RP, Jha PNA (2016) Halotolerant bacterium bacillus licheniformis HSW-16 augments induced systemic tolerance to salt stress in wheat plant (*Triticum aestivum*). Front Plant Sci 7: 1890

Singh SK, Wu X, Shao C, Zhang H (2022) Microbial enhancement of plant nutrient acquisition. Stress Biol 2(3):1–14

Sloan SS, Lebeis S (2015) Exercising influence: distinct biotic interactions shape root microbiome. Curr Opin Plant Biol 26:32–36

Spohn M, Zeißig I, Brucker E, Widdig M, Lacher U, Aburto F (2020) Phosphorus solubilization in the rhizosphere in two saprolites with contrasting phosphorus fractions. Geoderma 366:114245

Sreenivasulu N, Sopory SK, Kavi Kishor PB (2007) Deciphering the regulatory mechanisms of abiotic stress tolerance in plants by genomic approaches. Gene 388:1–13

Srivastava S, Yadav A, Seem K, Mishra S, Chaudhary V, Srivastava CS (2008) Effect of high temperature on Pseudomonas putida NBRI0987 biofilm formation and expression of stress sigma factor RpoS. Curr Microbiol 56:453–457

Srour AY, Ammar HA, Subedi A, Pimentel M, Cook RL, Bond J, Fakhoury AM (2020) Microbial communities associated with long-term tillage and fertility treatments in a corn-soybean cropping system. Front Microbiol 11:1363

Subramanian KS, Tenshia V, Jayalakshmi K, Ramachandran V (2009) Biochemical changes and zinc fractions in arbuscular mycorrhizal fungus (*Glomus intraradices*) inoculated and uninoculated soils under differential zinc fertilization. Appl Soil Ecol 43(1):32–39

Sugiyama A, Bakker MG, Badri DV, Manter DK, Vivanco JM (2013) Relationships between Arabidopsis genotype-specific biomass accumulation and associated soil microbial communities. Botany 91:123126

Suman A, Yadav AN, Verma P (2016) Endophytic microbes in crops: diversity and beneficial impact for sustainable agriculture. In: Microbial inoculants in sustainable agricultural productivity. Springer, New Delhi, pp 117–143

Sun W, Shahrajabian MH (2023) The application of Arbuscular Mycorrhizal fungi as microbial biostimulant, sustainable approaches in modern agriculture. Plants 12(17):3101

Supriya P, Kusale S, Attar Y, Sayyed R, Malek R, Ilyas N, Suriani N, Khan N, El Enshasy H (2021) Production of plant beneficial and antioxidants metabolites by Klebsiellavariicola under salinity stress. Molecules 26:1894

Tahat MM, Alananbeh MK, Othman AY, Leskovar ID (2020) Soil health and sustainable agriculture. Sustainability 12(12):4859

Tang Y, Dai X, Dong B, Guo Y, Dai L (2020) Humification in extracellular polymeric substances (EPS) dominates methane release and EPS reconstruction during the sludge stabilization of high-solid anaerobic digestion. Water Res 175:115686

Tester M, Davenport R (2003) Na+ tolerance and Na+ transport in higher plants. Ann Bot 91:503–527

Theocharis A, Bordiec S, Fernandez O, Paquis S, Dhondt-Cordelier S, Baillieul F, Clement C, Barka EA (2012) *Burkholderia phytofirmans* PsJN primes *Vitis vinifera* L. and confers a better tolerance to low nonfreezing temperatures. Mol Plant-Microbe Interact 25:241–249

Timmusk S, Behers L, Muthoni J, Aronsson AC (2017) Perspectives and challenges of microbe application for crop improvement. Front Plant Sci 8:49

Timmusk S, El-Daim IA, Cpolovici L, Tanilas T, Kannaste A, Behers L, Nevo E, Seisenbaeva G, Stenstrom E, Niinemets U (2014) Drought-tolerance of wheat improved y rhizosphere bacteria from harsh environments: enhanced biomass production and reduced emissions of stress volatiles. PLoS One 9:e96086

Timofeeva AM, Galyamova MR, Sedykh SE (2022) Bacterial siderophores: classification, biosynthesis, perspectives of use in agriculture. *Plants* 11(22)

Tiwari P, Bose SK, Park K-I, Dufossé L, Fouillaud M (2024) Plant-microbe interactions under the extreme habitats and their potential applications. Microorganisms 12(3):448

Tongway DJ, Ludwig JA (2005) Heterogeneity in arid and semi-arid lands. In: Lovett GM, Turner MG, Jones GG, Weathers KC (eds) Ecosystem function in heterogeneous landscapes. Springer, New York, pp 189–205

Trapet PL, Verbon EH, Bosma RR, Voordendag K, Pelt JAV, Pieterse CMJ (2021) Mechanisms underlying iron deficiency-induced resistance against pathogens with different lifestyles. J Exp Bot 72(6):2231–2241

Turner TR, James EK, Poole PS (2013) The plant microbiome. Genome Biol 14:209

Ullah R, Ayub G, Ilyas M, Ahmad M, Umar M, Mukhtar S, Farooq S (2015) Growth and yield of tomato (*Lycopersicon esculentum* L.) as influenced by different levels of zinc and boron as foliar application. Am Eurasian J Agric Environ Sci 15:2495–2498

Upadhyay SK, Singh DP, Saikia R (2009) Genetic diversity of plant growth promoting rhizobacteria from rhizospheric soil of wheat under saline conditions. Curr Microbiol 59:489–496

Vaid SK, Kumar B, Sharma A, Shukla A, Srivastava P (2014) Effect of Zn solubilizing bacteria on growth promotion and Zn nutrition of rice. J Soil Sci Plant Nutr 14(4):889–910

Vaishnav A, Kumari S, Jain S, Varma A, Tuteja N, Choudhary DK (2016) PGPR-mediated expression of salt tolerance gene in soy-bean through volatiles under sodium nitroprusside. J Basic Microbiol 56:1274–1288

Vaishnav A, Singh J, Singh P, Rajput RS, Singh HB, Sarma BK (2020) Sphingo bacterium sp. BHU-AV3 induces salt tolerance in tomato by enhancing antioxidant activities and energy metabolism. Front Microbiol 11:443

Vejan P, Abdullah R, Khadiran T, Ismail S, Nusrulhaq Boyce A (2016) Role of plant growth promoting rhizobacteria in agricultural sustainability: a review. Molecules 21:573

Verma P, Pandey SN (2016) Effect of integrated nutrient management in alluvial soil on growth and biochemical responses of radish. J Biol Chem Res 33(1):34–39

Vidali M (2001) Bioremediation: an overview. Pure Appl Chem 73:1163–1172

Villalobos-López MA, Arroyo-Becerra A, Quintero-Jiménez A, Iturriaga G (2022) Biotechnological advances to improve abiotic stress tolerance in crops. Int J Mol Sci 23(19):12053

Vurukonda SSKP, Vardharajula S, Shrivastava M, SkZ A (2016) Enhancement of drought stress tolerance in crops by plant growth promoting rhizobacteria. Microbiol Res 184:13–24

Wahab A, Muhammad M, Munir A, Abdi G, Zaman W, Ayaz A, Khizar C, Reddy SPP (2023) Role of *Arbuscular Mycorrhizal* fungi in regulating growth, enhancing productivity, and potentially influencing ecosystems under abiotic and biotic stresses. Plants 12(17):3102

Watt M, Kirkegaard JA, Passioura J (2006a) Rhizosphere biology and crop productivity- a review. Aust J Soil Res 44:299–317

Watt M, Silk W, Passioura J (2006b) Rates of root and organism growth, soil conditions and temporal and spatial development of the rhizosphere. Ann Bot 97:839–855

Weisser WW, Roscher C, Meyer ST, Ebeling A, Luo GJ, Allan E, Besser H, Barnard RL, Buchmann N, Buscot F, Engels C, Fischer C, Fischer M, Gessler A, Gleixner G, Halle S, Hildebrandt A, Hillebrand H, de Kroon H et al (2017) Biodiversity effects on ecosystem functioning in a 15-year grassland experiment: patterns, mechanisms, and open questions. Basic Appl Ecol 23:1–73

Wicke B, Smeets E, Dornburg V, Vashev B, Gaiser T, Turkenburg W, Faij A (2011) The global technical and economic potential of bioenergy from salt-affected soils. Energy Environ Sci 4: 2669–2681

Wuana RA, Okieimen FE (2011) Heavy metals in contaminated soil: a review of sources, chemistry, risks and best available strategies for bioremediation. ISRN Ecol 2011:402647

Yadav A, Pandey SN (2015) Effect of integrated nutrient management on the growth, biochemical constituents and yield of tomato (*Lycopersicon esculentum* Mill.). J Biol Chem Res 32(2): 835–838

Yang AZ, Akhtar SS, Amjad M, Iqbal S, Jacobsen SE (2016a) Growth and physiological responses of quinoa to drought and temperature stress. J Agron Crop Sci 202:421–589

Yang AZ, Akhtar SS, Iqbal S, Amjad M, Naveed M, Zahir ZA, Jacobsen SE (2016b) Enhancing salt tolerance in quinoa by halotolerant bacterial inoculum. Funct Plant Biol 43:632–642

Yang J, Kloepper JW, Ryu CM (2009) Rhizosphere bacteria help plants tolerate abiotic stress. Trends Plant Sci 14:1–4

Yang Q, Tu S, Wang G, Liao X, Yan X (2012) Effectiveness of applying arsenate reducing bacteria to enhance arsenic removal from polluted soils by Pteris vittata L. Int J Phytoremediation 14:89–99

Yu H, Wu X, Zhang G, Zhou F, Harvey PR, Wang L, Fan S, Xie X, Li F, Zhou H, Zhao X, Zhang X (2022) Identification of the phosphorus-solubilizing bacteria strain JP233 and its effects on soil phosphorus leaching loss and crop growth. Front Microbiol 13

Zaidi A, Khan MS, Saif S, Rizvi A, Ahmed B, Shahid M (2017) Role of nitrogen-fixing plant growth-promoting rhizobacteria in sustainable production of vegetables. In: Zaidi A, Khan MS (eds) Microbial strategies for vegetable production. Springer, Netherland, pp 49–59

Zarei M, Wubet T, Schäfer SH, Savaghebi GR, Jouzani GS, Nekouei MK (2010) Molecular diversity of arbuscular mycorrhizal fungi in relation to soil chemical properties and heavy metal contamination. Environ Pollut 158:2757–2765

Zhang F, Dashti N, Hynes R, Smith DL (1996) Plant growth promoting rhizobacteria and soybean [*Glycine max* (L.) Merr.] nodulation and nitrogen fixation at suboptimal root zone temperatures. Ann Bot 77:453–460

Zhang H, Dong H, Li W, Sun Y, Chen S, Kong X (2009a) Increased glycine betaine synthesis and salinity tolerance in AhCMO transgenic cotton lines. Mol Breed 23:289–298

Zhang H, Sun Y, Xie X, Kim MS, Dowd SE, Pare PW (2009b) A soil bacterium regulates plant acquisition of iron via deficiency-inducible mechanisms. Plant J 58:568–577

Zhang H, Zhao Y, Zhu JK (2020) Thriving under stress: how plants balance growth and the stress response. Dev Cell 55(5):529–543

Zhao S, Zhou N, Zhao ZY, Zhang K, Wu GH, Tian CY (2016) Isolation of endophytic plant growth-promoting bacteria associated with the halophyte *Salicornia europea* and evaluating their promoting activity under salt stress. Curr Microbiol 73:574–581

Zhu L, Huang J, Lu X, Zhou C (2022) Development of plant systemic resistance by beneficial rhizobacteria: recognition, initiation, elicitation and regulation. Front in Plant Sci 13

Zuppinger-Dingley D, Schmid B, Petermann JS, Yadav V, De Deyn GB, Flynn DF (2014) Selection for niche differentiation in plant communities increases biodiversity effects. Nature 515:108–111

# Chapter 5
# The Interconnected Microbiomes of Plants and Insects: Implications for Yield Quality and Pest Management

**Wael Elmenofy**

## 5.1 Introduction

Insects represent one of the oldest and most varied groups of organisms on Earth. They have the potential to cause huge destruction if their numbers surpass a certain limit. They have existed since the Paleozoic period, when Hemipterans and Orthopterans first appeared on Earth, based on the geological time scale (Misof et al. 2014). Since then, they have developed and are able to survive under a wide range of unsuitable weather conditions. While insects play a significant role in both the environment and the economy, insect pests-which are mostly the result of human-manipulated habitats and, in many cases, the effects of climate change-cause more than 20% of agricultural losses each year (Deutsch et al. 2018). Insect population dynamics, life cycle timing, and migration patterns are all being impacted by shifting climate conditions. Overcoming these obstacles has also allowed them to increase the range of hosts they can occupy, which has an impact on their behavior and biology and aids in their invasion and colonization of various agroclimatic zones worldwide (Shrestha 2019). Their enormous population size with quick reproductive cycles has allowed them to effectively adapt to circumstances not suitable for life. Because of their light weight, which allows them to be carried by wind currents, they have spread over the earth, attacked different parts of the planet and are currently found in almost every type of ecosystem.

Furthermore, their evolution and diversity have been greatly supported by the complex relationship they have with the helpful microbes (Janson et al. 2008).

W. Elmenofy (✉)
Department of Arid Land Agriculture, College of Agricultural and Food Sciences, King Faisal University, Al-Ahsa, Saudi Arabia

Agricultural Genetic Engineering Research Institute, Agricultural Research Center, Giza, Egypt
e-mail: welmenofy@kfu.edu.sa

 
N. El-Wakeil, A. Abdelkhalek (eds.), *Microbiomes and Plant Secondary Metabolites for Insect Pest Management*, Progress in Biological Control 23,
https://doi.org/10.1007/978-3-032-24330-0_5

Throughout their evolutionary history, insects have been linked to microorganisms like bacteria and fungi. Some types of bacteria are called "endosymbionts" because they live in specialized cells inside insects called bacteriocytes, while other species are called "ectosymbionts" because they are found on the outside of the body (Thompson and Simpson 2009). Nonetheless, their main location is in the digestive system, where they play a crucial role in regulating many dietary and ecological niches of their insect host. According to different researchers, the gut microflora of an insect has been shown to affect the insect life cycle by several factors: (1) enhance insect immune system and protect insects against pathogens; (2) allow feeding even on unappropriated food (3) allow recovery of the nutrient-missing diet; (4) regulate reproduction and mating; (6) facilitate digestion for complex food contents and (7) provide nutrients and essential amino acids (Douglas 2015). As Jing et al. (2020) have demonstrated, the provision of necessary nutrients is actually the main function of symbionts, with digestion and detoxification coming in second. It means that insects mainly depend on their gut microbiota to survive and carry out regular life cycle activities. Hence, symbiont association can be categorized as facultative (optional) or obligate depending on the level of dependency (Moran et al. 2008). Differentiating between these two types is challenging since, under specific conditions, facultative bacteria can transform to an obligatory one (Moran et al. 2008).

The interaction between the bacterial community and its host is complex and fascinating, with many facets that need to be investigated. These symbionts are characterized by some researchers as "intracellular parasites" since they have taken over the body of the insect and have developed a variety of defense systems to secure their survival while benefiting their host.

Despite not having an adaptive immune system like vertebrates do, insects have a number of defense mechanisms against pathogen infection. In insects, body barriers and behavioral adaptations provide the first line of defense. Insect hygiene practices, fewer interactions, and selective feeding are examples of behavioral adaptations. In the same context, exoskeleton, trachea, intestinal epithelium, and intestinal peritrophic membrane are examples of body barriers (de Roode and Lefèvre 2012; Erlandson et al. 2019). Through the use of pattern recognition receptors (PRRs) to identify microbe-associated molecular patterns (MAMPs), insects are able to identify a variety of bacteria (Lemaitre and Hoffmann 2007). The innate immune system of insects consists of two distinct types of immunological responses: humoral responses, which consist of the production of antimicrobial effectors such as antimicrobial peptides (AMPs) and lysozyme, and cellular responses, which include nodulation, encapsulation, and phagocytosis (Lemaitre and Hoffmann 2007). Hence, according to Dillon and Dillon (2004), bacterial species found in insect guts can be beneficial, symbiotic or pathogenic to the host.

## 5.2 Insect Immune Response

Insects can preserve the beneficial germs while also triggering an immune response against the harmful bacteria (Mikonranta et al. 2014). Experimental research on the specific roles of symbiotic anti-microbial peptides (AMPs) has shown that these AMPs regulate symbiotic interactions by inhibiting symbiotic bacteria's ability to reproduce, occasionally causing them to change into a differentiated shape, and getting rid of sensitive, unwanted bacteria (Mergaert 2018). Acquisition, colonization, and transmission are all aspects of microbial symbiosis. While some germs are easily acquired by insects throughout their life cycle, others are inherited or transferred vertically. The physical and physiological characteristics of the insect stomach have a significant impact on the survival and persistence of these organisms, and effective colonization is essential for this. Transmission between generations then becomes crucial, particularly for obligatory endosymbionts that rely entirely on their host for existence (Gupta and Nair 2020).

The immunodeficiency (IMD) pathway controls the generation of antimicrobial peptides (AMPs) in the midgut. Two upstream receptors, the intracellular receptor PGRP-LE, which functions exclusively in the midgut, and the cell membrane surface-bound receptor PGRP-LC, with its function in the whole gut (foregut, midgut, and hindgut), activate the IMD pathway in the gut (Bosco-Drayon et al. 2012; Neyen et al. 2012). The majority of Gram-negative bacteria and certain Gram-positive bacteria have cell walls that include diaminopimelic acid (DAP) type peptidoglycans, which are recognized by PGRP-LC and PGRP-LE (Leulier et al. 2003; Kaneko et al. 2004). Upon the insect host being infected via pathogens, intestinal enterocytes (ECs) release humoral immune components, including AMPs (Lemaitre and Miguel-Aliaga 2013). Insects can preserve the beneficial bacteria while also triggering an immune response against the harmful bacteria (Mikonranta et al. 2014). Experimental research on the specific roles of symbiotic anti-microbial peptides (AMPs) has shown that these AMPs regulate symbiotic interactions by inhibiting symbiotic bacteria's ability to reproduce, occasionally causing them to change into a differentiated shape, and get rid of sensitive and unwanted bacteria (Mergaert 2018).

Different proteins such as defensins, diptericins, apidaecin and ponericins are among the structural families of AMPs that have been identified in insects (Kwong et al. 2017; Wu et al. 2018). For example, hymenoptaecin and apidaecin are AMPs whose expression in honeybee gut tissue is up-regulated with microbial uptake (Kwong et al. 2017).

Furthermore, it has been noted that various symbionts have distinct effects on the insect immunological system; some are able to prevent the insect's cellular immunological response, while others have an impact on the melanization reaction, which is an insect immune system defense mechanism (Thomas et al. 2011). In this context, new findings showed multiple ways in which the symbionts can avoid a negative effect on insect immunity. It is still unknown, nevertheless, how endosymbionts came to be able to evade the immune system of the insect, and how the insect host itself can differentiate between both harmful bacteria and the beneficial ones. However, it is evident that both the bacteria and the insect go through a number of changes when a symbiotic association begins. These changes are likely caused by different evolutionary forces and may give the symbionts the ability to avoid the invasion of the immune system of the insect host (Gupta and Nair 2020).

## 5.3 The Environmental Adaptations of Insects

Insects are filled with beneficial microbial symbionts that confer useful features on their hosts, enabling them to adapt to new ecological changes (Sudakaran et al. 2017). Most microbial symbionts that have been functionally identified in insects assist their host in overcoming obstacles in the presence of biotic interaction partners, such as food plants (Douglas 2009), predators, parasites, or microbial pathogens (Florez et al. 2015). Different evidences revealed that symbiotic microorganisms can alter the host's heat tolerance level (Corbin et al. 2017; Renoz et al. 2019), desiccation resistance (Engl et al. 2018), and sensitivity to inorganic toxicants (Senderovich and Halpern 2013). It is becoming more and more evident that bacteria can increase the viable range of their insect host by offering characteristics that improve thermal tolerance. But in other circumstances, it may restrict adaptation to temperature changes. Because of this, microbial symbionts have been shown to have the ability to both increase and decrease the value of abiotic stress.

## 5.4 Association Between Insects and Their Microbiome

Normally, the foundation of insect-microbe symbiosis depends on the natural uptake of microorganisms by the insect host during continuous feeding. Following uptake, these symbioses gradually change from being free-living creatures to intracellular parasites. Their longevity within the insect's body is mostly affected by the host's life cycle. Insects that acquire microbe species not only alter their eating preferences, but they also develop gut compartments and specific habitats for storing the microorganisms, which facilitate and support microbial persistence (Engel and Moran 2013). The physicochemical characteristics of the gut, especially its pH, have a significant impact on microbial colonization. By maintaining the ideal pH in their guts, insects with large microbial communities give their bacterial symbionts a good habitat (Engel and Moran 2013).

After entering the insect's body, the obligatory mutualists settle inside bacteriocytes, where they benefit and increase the host's fitness. They are passed on from mothers to their progeny. They have a mutualistic bond with their host that is highly stable. By producing uncommon vitamins and important amino acids, they augment the nutritional needs of their host, particularly for hemimetabolous sap-sucking insects that consume diets low in nutrients. For instance, the obligatory symbiont *Buchnera aphidicola's* bacteria, the primary endosymbiont of aphids, meet the nutritional requirements of the aphids to such an extent that eliminating the bacteria has a major negative effect on the survival and fecundity of aphids (Feng et al. 2019).

Similar to this, Candidatus Ishikawaella capsulate, the principal bacterial symbiont of the sap-feeding insect *Megacopta cribraria*, and its symbiotic relationship are necessary for the host to survive to adulthood (Couret et al. 2019).

The facultative microorganisms, on the other hand, show a completely different picture. While some, like Wolbachia, Spiroplasma, and Cardinium, are vertically transmitted, others, like Burkholderia and Serratia, are acquired fresh after each generation. Furthermore, it has been demonstrated that the food and developmental stage of the host have a significant impact on the community structure of various facultative microorganisms in a number of examples, such as the Asian rice gall midge, a Dipteran pest of rice (Ojha et al. 2017). Moreover, the facultative symbionts help their insect host with the digestion and detoxification of xenobiotics. As an example, a gram-negative Gamma proteobacteria like Pseudomonas, which is located in *Spodoptera frugiperda,* was shown to enhance pesticide resistance to its host (De Almeida et al. 2017).

Furthermore, a number of facultative endosymbionts, including Cardinium, Rickettsia, Wolbachia, and Spiroplasma, are known to cause sexual abnormalities in a variety of insect orders (Kageyama et al. 2012). As a result, although some facultative microorganisms are commensals and even harmful, others are beneficial to the host (at least in some situations). They are known to live in different areas of the body. Some are limited to the bacteriocytes, whilst others are dispersed throughout the hemocoel (Marubayashi et al. 2014).

In addition to facultative and obligate symbionts, the feeding process causes the presence of a number of phytopathogenic microorganisms in insect bodies. Nevertheless, because they are stationary, plants pose a significant barrier to the spread of these phytopathogenic bacteria. As a result, these microorganisms became dependent on their vectors, mostly insects, for growth and spread. As a result, these bacteria are shown to be associated with their insect carrier, which in turn shapes the complex interaction that is seen between them. In addition to actively interacting with their insect host, these bacteria also alter it to suit their own needs. Certain bacteria are propagative (able to proliferate within their insect host), while others are not (non-propagative). This suggests that the insect acts as both a vector and a different host for these bacteria (Nadarasah and Stavrinides 2011).

According to Perilla-Henao and Casteel (2016), certain microorganisms circulate throughout the insect's body and are only transferred after a latent time, whereas others circulate through the body of the vector and are semi-persistent, non-circulative. They enter the insect's body and go to the epithelium of the midgut or hindgut before being expelled into the haemolymph. During eating, they are transferred to the plant via the salivary gland after leaving the haemolymph (Kwon et al. 1999). In response, insects have developed defense systems against these harmful organisms and have gained some benefits from the interaction between insects and these microbes.

### 5.4.1 Plant Microbiota

Plants offer numerous niches for a variety of microorganisms, such as viruses, nematodes, protists, fungi, and bacteria (together referred to as the "plant microbiota"). These microbes have the ability to remain in plants and affect the health and production of the plant in its habitat. Although this theory has not yet been proven beyond a reasonable doubt, it was demonstrated that the microbiota and the host plants constitute a "holobiont," in which evolutionary selection between plants and microbes affects the general stability of the system. Taxa from several phyla are present in complex plant microbial communities, which also include deep-branching lineages at lower phylogenetic resolution. With the advent of culture-independent high-throughput sequencing in recent years, the range of microorganisms that are known to live in, on, and around plants has significantly increased (Lundberg et al. 2012; Vandenkoornhuyse et al. 2015; Hamonts et al. 2018).

In particular, plants are abundant microbe hosts because they harbor microbial symbionts inside their cells (known as endophytes), and even on the surface (called epiphytes) (Hardoim et al. 2015). Algae, oomycetes, bacteria, fungi, and protozoa in general make up the plant microbiomes (Leach et al. 2017). Virtually every plant tissue, including flowers, buds, seeds, stems, leaves, and roots, may support the microbial symbionts (Hardoim et al. 2015). In most close symbioses, the mutualistic bacteria of plants are transmitted vertically through seeds, although they are normally obtained horizontally from the surrounding environment (Müller et al. 2016).

### 5.4.2 Insect Microbiome

Microbial symbionts colonize insects, much like they do plants. These symbionts' homes include the stomach, exoskeleton, the hemocoel, and cells. The insect's microbiota consists of bacteria, archaea, and eukaryota, including fungus and protists. Bacteriophage symbioses are created by intracellular microorganisms (Douglas 2015; Giron et al. 2017). The microbiomes of insects may be classified into two groups: closed associations, where the invasion is inhibited, and open associations, where exogenous microorganisms infiltrate. The stomach and exoskeleton, for instance, are vulnerable to invasion. On the fruit fly's exoskeleton, there are up to 1000 culturable bacteria (Ren et al. 2007). Although the environment in the stomach of insects is abundant, the microorganisms that enter with food are chosen. Usually, the areas with the largest microbial populations are the ileum and the hindgut (Douglas 2015). Along with the growth of the insect, the microbial gut populations can also undergo compositional changes. Within the digestive tract of the highly destructive agricultural pest known as the African cotton leafworm (*Spodoptera littoralis* [Boisduval]), Enterococcus, Pantoea, and Citrobacter bacterial species are prevalent in the early stages of life, with Clostridium emerging as the most prevalent class in adulthood (Chen et al. 2016).

Cuticular structures, seen in some insect species, are responsible for creating a niche for their microbiota. For example, the fungus or bacteria that the insect's mycangia host needs in order to produce offspring (Douglas 2015). Intracellular bacteria are prevalent in several insect groups and are located in cells known as bacteriocytes or mycetocytes that are specifically designed for this function. This is a closed relationship where the major symbionts, the endosymbionts that dominate the bacteriocyte or mycetocyte, are isolated from the environment (Douglas 2015). Since phloem sap is a poor source of nitrogen, endosymbionts play a crucial role in supplying necessary amino acids for their hosts in plant sap-feeding insects (Douglas 2006). Nonetheless, it is common for this class of insects to additionally have secondary symbionts connected to the mycetocytes or bacteriocytes. The secondary symbionts live in the hemolymph or insect sheath cells and are also transported by vertical transmission; they can also horizontally spread. The host may acquire characteristics from the secondary symbionts, such as resistance to fungi or parasitoids or tolerance to heat (Douglas 2015).

### *5.4.3 Plant-Insect Interaction*

Plants and insects, the two categories of higher creatures, often interact with one another and are extremely dependent on one another. It is still not quite known how much the microbiomes of the two creatures differ and how much they overlap, despite the body of evidence on their respective microbiomes growing.

Plants have evolved defense systems to identify the kind and presence of intruders. Specific reactions are triggered by signaling pathways that are activated upon detection of the intruder (Pieterse et al. 2014; Coolen et al. 2022). The two primary methods for identifying the intruders are effector molecule-based detection and pattern recognition. Elicitors and pattern recognition receptors (PRRs), which are capable of detecting a variety of chemical patterns, are the foundation of pattern recognition. These patterns are known as herbivorous insect-associated molecular patterns (HAMP) in the context of herbivory (Pieterse et al. 2014). Specifically, endogenous elements that are produced from the cell wall are known as damage-associated molecular patterns (DAMP) (Chagas et al. 2018). Pathogen-associated molecular patterns (PAMPs), which originate from chemicals found in microbes located in the insect host, may potentially be used to identify the insect (Chaudhary et al. 2014). The PAMP-triggered immunity (PTI) is activated by pattern recognition (Pieterse et al. 2014). In addition to recognizing molecular patterns, plants may identify pathogen effector molecules through the products of their resistance genes, which subsequently results in effector-triggered immunity (ETI). For example, effectors can be made of the ingredients in bug saliva. Effector-based recognition takes place intracellularly, while pattern-based recognition happens at the cell membrane (Chagas et al. 2018). Upon detecting an intruder, a defensive mechanism is frequently triggered. Generally, biotrophic pathogens initiate the salicylic acid (SA) signaling route, while wounds, herbivore insects, and necrotrophic infections initiate the jasmonic acid and ethylene (JA/ET) signaling system. Furthermore, systemic acquired resistance (SAR) is frequently triggered by the activation of local defensive responses through SA. This involves long-distance transmission and primes the undamaged tissues (Pieterse et al. 2014). Induced systemic resistance (ISR), which also results in priming, can be brought on by the plant's interaction with mutualistic microorganisms (Pieterse et al. 2014; Pangesti et al. 2016). Priming is the capacity to react to pathogen assaults more powerfully and faster (Conrath et al. 2006). Priming essentially increases a plant's sensitivity to JA and ET, which regulate ISR (Pieterse et al. 2014).

The gut microbes' substances integrated in the honeydew of rice brown planthopper (Nilaparvata lugens) can cause systemic resistance in the plant host, which enhances the strength of the plant's defenses against herbivores. This further causes continuous building of phytoalexins and the production of volatile chemical substances on rice leaves, which draw in herbivore adversaries (Wari et al. 2019). Herbivore insects, of whom only 10% are generalists, primarily target certain plant species. They must choose appropriate and superior plants for eating and reproduction (Ali and Agrawal 2012). Herbivorous insects can be classified in a number of ways, such as; florivorous (leaves), xylophagous (wood), algivorous (algae), frugivorous (flowers), granivorous (seeds), and mucivorous (sap) species based on the type of tissue they consume or depending on how they feed, into external, internal or chewing or sucking insects (Novotny et al. 2010). Furthermore, herbivores can be categorized as monophagous, oligophagous, or polyphagous feeders based on the degree of host species specialization (i.e., feeding on a single plant species, several closely related species or genera, and a wide variety of different plant groups, respectively) (Jones and Coaker 1978). Plants are able to identify the type of insect and modify their defensive mechanisms accordingly. All in all, herbivory can set off a series of interactions involving the entire community (Soler et al. 2013).

Plant defense chemicals cover a wide range of actions, including damage to insect membranes, reduction of insect metabolism, as well as disruption of development through dysregulated hormone regulation (Gatehouse 2002; Shavit et al. 2013). The eating, settling, fertility, and oviposition of insect herbivores can all be impacted by plant defensive responses (Shavit et al. 2013; Adams et al. 2013). Using constitutive or induced mechanisms, herbivores try to get past plant defenses (Shavit et al. 2013). Many of these mechanisms are dependent on symbiotic microbes (Coolen et al. 2022; Adams et al. 2013), which are covered in more detail below.

### *5.4.4 Plant-Associated Microbes*

Members of the plant microbiome can be quite important in this connection even if they are not directly related to insects. The microbiome has the ability to affect a plant's nutritional status, disrupt relationships between pathogens and plants, and alter the plant's ability to withstand abiotic stressors (Müller et al. 2016). Conversely, plant-microbe interactions can be impacted by abiotic variables and plant diseases (Coolen et al. 2022; Pangesti et al. 2016).

A significant portion of microbial interactions take place at the roots of plants, where microorganisms are often found in the soil. Legumes generate a rhizobial symbiosis when the plant host produces root nodules that enable atmospheric nitrogen to be fixed by bacteria (Hirsch et al. 2001). By promoting JA production and the emergence of systemic resistance, rhizobia can contribute to legumes' ability to withstand herbivory (Dean et al. 2014). Numerous plant species, particularly forest trees, and soil bacteria from the genus Frankia create an additional nitrogen-fixing symbiosis known as actinorhiza (Põlme et al. 2014). The black slug (Arion rufus L.) enhanced its herbivory of young leaves in the red alder (Alnus rubra Bong) due to Frankia symbiosis (Ballhorn et al. 2017). Nonetheless, the growing body of research on actinorhizal symbiosis indicates that, like rhizobia, these symbionts may cause plant resistance (Vincent et al. 2023). Plant growth-promoting rhizobacteria (PGPR), a kind of soil bacteria, are also widely distributed throughout the plant rhizosphere. In addition to increasing nutrient availability, influencing root shape and development, and facilitating the emergence of additional symbioses, such as mycorrhizal or rhizobial symbioses, the PGPR may fix nitrogen (Goswami et al. 2016). In particular, PGPR can help repel herbivores by causing the host plant to develop systemic resistance. For instance, the treatment of PGPR in cotton results in increased synthesis of the secondary metabolite gossypol as well as systemic resistance. In cotton plants treated with PGPR, these reactions decreased the larval activity of beet armyworms (*Spodoptera exigua*:Hübner) (Zebelo et al. 2016). Additionally, the enhanced generation of plant volatiles following PGPR inoculation has the potential to draw predatory earwigs (Dermaptera) towards the herbivore assault (Bell et al. 2020). Nonetheless, the combination of the herbivore species with the PGPR is crucial for the host plant's protection (Friman et al. 2021).

## 5.5 The Interactions Between Plant and Herbivorous Microbiomes

Herbivore-toxic chemicals are produced by a variety of fungal endophytes (Bacon et al. 1977; Faeth 2002). Occasionally, the harmful consequences may be transmitted to the subsequent herbivore's generation, irrespective of whether the herbivore consumes the endophyte-carrying plant (Jaber and Vidal 2010) or to a parasitoid, hyperparasitoid, or herbivore predator (Härri et al. 2008). The relationship is complex in forest trees because fungal endophytes horizontally invade tree leaves, resulting in highly localized infections (Hardoim et al. 2015). Endophytic fungi in conifers are frequently linked to their ability to ward off insect herbivores (Pirttilä and Wäli 2009). Carroll and Carroll (1978) proposed as early as 1978 that conifer foliar endophytes may be mutualistic partners.

Recent research revealed that the spruce budworm's ability to feed on white spruce needles was greatly inhibited by the toxin rugulosin, in which it is generated by *Phialocephala scopiformis* as fungal endophyte (Miller et al. 2002; Sumarah and Miller 2009).

Saikkonen et al. (1996) found in another investigation on Scots pine (Pinus sylvestris L.) that endophytic fungi-hosting 5-year-old trees had decreased pine sawfly (Neodiprion sertifer Geoffroy) needle performance in 40% of the needles. They discovered that on trees with greater mean endophyte abundances, the larvae had a longer larval phase and a lower relative growth rate. They did not see any comparable impact on the activity of the aphid *Scizolachnus pineti* (Fabricius), as another herbivore (Saikkonen et al. 1996). However, no connection was observed between herbivory and the presence of endophyte by examination of the birch leaf rolling weevil (Deporaus betulae L.), life-history behaviors and birch arge (Arge clavicornis Fabricius) upon the deciduous tree mountain birch endophyte infection (Saikkonen et al. 1996). They proposed that the invasion of tree leaves by endophytic fungus varies so much seasonally and geographically, depending on the surrounding vegetation, tree density, weather, and ground topography, when a stable and efficient defense against insects is not developed. For several years, Conifers carry leaves, unlike deciduous trees, which generate new leaves annually. This might account for the more constant effectiveness of endophytic fungus against insect herbivores in conifers. Consequently, it's possible that greater mutualism evolved from the interactions between endophytic fungus and conifer needles to fight off insect herbivores (Pirttilä and Wäli 2009).

## 5.6 Microbial Transmission (Insects as Vectors)

Many bacteria associated with plants, particularly diseases, can spread by using insects as vectors and may rely entirely or largely on them for transmission. Protozoa, phytoplasmas, bacteria, and fungus are among the microorganisms that are spread (Agrios 2004). Vector-borne plant diseases can spread either internally or externally inside insects, and their interactions with vectors are frequently particular (76). For instance, bacterial or fungal spores may smear the insect (Six 2003).

Since some plant pathogens can only survive in the insect vector for a short time, nonpersistent plant pathogens, sometimes referred to as stylet-borne pathogens, are obtained and injected following brief feeding periods. Semipersistent infections are plant pathogens that live for 1–4 days in the insect vector. Persistent or "circulative" diseases, on the other hand, collect within and are transmitted to another host plant via the insect feeding. Certain propagative infections have the ability to grow within the vector and persist in the insect for the duration of its life. There may be a period of incubation in some situations between the plant pathogen's acquisition and transformation (Agrios 2004; Six 2003; Mann et al. 2011), and there has also been evidence of sexual transmission between the vectors. The most researched plant disease vectoring insects are those that feed on phloem, such as the American sap-feeding leafhopper *Scaphoideus titanus* Ball, which has been demonstrated to transmit whole endophytic bacterial communities from one grapevine (Vitis vinifera L.) to another (Lòpez-Fernàndez et al. 2017). Several data indicate that insects are the means by which plant root-colonizing bacteria, specifically PGPR, spread to new plants (Galambos et al. 2021; Pronk et al. 2022).

A case study of fungal symbionts with bark beetles (Coleoptera: Curculionidae: Scolytinae) in a forest environment involves insect-mediated microbial transmission. The beetles spread a large number of fungal symbionts across trees and use the bark of conifers as a place to eat and breed (Taerum et al. 2013). There is a subspecies of bark beetle that attacks and destroys healthy trees, but the majority of these species exclusively inhabit dead or dying coniferous trees (Paine et al. 1997). Most fungal taxa that transfer fungus from trees to bark beetles belong to the order Ophiostomatales (Kirisits 2004). Mycangia, bark beetle exoskeletons, or mites carried by the beetles are the carrier by which the fungal spores are dispersed across trees. These fungi may coexist with their hosts as commensals, parasites, or pathogens (Whitney 1982). They could nourish the bark beetle larvae (Six 2003), and aid in the insects' ability to get past tree defenses (Lieutier et al. 2009). For instance, by breaking down plant defensive molecules called stilbenes, the fungus Ceratocystis polonica, which is carried by the European spruce bark beetle (Ips typographus L.), aids in the beetle's invasion of the Norway spruce (Picea abies L.) (Hammerbacher et al. 2013). The Ophiostomatalean fungus can also create volatiles that influence beetle behavior by acting as attractants or repellents. However, the behavior of the beetles can differ depending on the kind of fungal species (Kandasamy et al. 2019), the stage at which the beetles are developing mentally, and the host tree's genotype (Kandasamy et al. 2016).

## 5.7 Overlap Between Plant and Insect Microbiomes

Numerous distinct microbial symbionts can coexist with an insect or plant host, each of which serves a unique purpose inside the host (Hardoim et al. 2015; Chuche et al. 2017). While one bacterium with good benefits in the plant host may be dangerous to the insect, other bacteria with pathogenic effects in the plant host may be helpful for the insect host (Monnerat et al. 2009). For example, *Beauveria bassiana* is an entomopathogenic fungus that grows and infects more than 700 insect species (Ortiz-Urquiza 2021). However, the microbial makeup of the insect's stomach is shaped by its plant-based diet, which comprises plant matter and endophytic microorganisms (Shikano et al. 2017; Strano et al. 2018). This is related to the insect's capacity to withstand plant defenses (Chung et al. 2017). The gut populations of the grapevines and the phloem-feeding grasshopper (S. titanus) studied by Làpez-Fernàndez et al. (Lòpez-Fernàndez et al. 2017) showed similarities between insect communities and the plant endophytic. However, the Proteobacteria dominated the plant communities while Firmicutes were abundant in insect communities (Lòpez-Fernàndez et al. 2017). In a recent study, the gut microbiomes of nine polyphagous caterpillar species that feed on five distinct types of trees differed significantly in terms of bacterial species when compared to the host leaves. Nonetheless, there were parallels between the host leaves' and insects' fungal microbiomes (Šigut et al. 2022). Thus, the specific species involved most likely have a significant influence on how much the plant and insect microbiomes overlap.

On the other hand, the populations of P. syringae, which were originally tiny on the leaves, were significantly impacted by the aphid feeding, which was linked to the presence of honeydew (Smee and Hendry 2022). Lepidopterous larvae feeding on the leaves of beech and oak generated alterations in the bacterial populations, which were linked to an increase in available nitrogen (Müller et al. 2003). However, rather than being predominantly influenced by Vanessa cardui L. caterpillars, the strawberry's floral microbiome was shaped by the genotype of the plant. More specifically, the composition of the bacterial and fungal communities in the strawberry blossoms was mostly determined by plant volatiles, including terpenoids and benzenoids (Wei et al. 2022). This implies that herbivory and other plant-insect interactions do not significantly alter the plant microbiome.

## 5.8 Advantages and Drawbacks of New Microbiome Integration

Microbes may generally have a significant impact on the spectrum of insects that they can host (Feldhaar 2011). For instance, fungi that have made it possible for them to invade China from North America have been spread by the red turpentine beetle (D. valens LeConte) (Taerum et al. 2013; Sun et al. 2013). The beetle poses little danger to pine trees in its natural habitat, but on the other hand, it has killed out significant numbers of pines (Sun et al. 2013). A significant change in the assemblage of the invasive red pennant beetle fungal symbionts was noted by Taerum et al. (2013). They came to the conclusion that the beetle's invasion of China was made possible by the change in the fungus community (Taerum et al. 2013).

Through horizontal gene transfer from microorganisms to insects, adaptation to a new host can occur quickly (Acuña et al. 2012). In addition, the microbiome of the insect can facilitate quick adaptation to environmental perturbations (Ferrater et al. 2013). Microbes have the ability to detoxify defensive chemicals in plants and supply the necessary enzymes for plant material digestion (Hansen and Moran 2014). For instance, the gut bacteria of the genera Rahnella and Pseudomonas, which are able to break down terpenes, enable the mountain pine bee (D. ponderosae Hopkins) to ingest terpenes (Adams et al. 2013). The gut microbes of wood-boring Siricidae, Curculionidae, and Cerambycidae, along with some lepidopteran species, can create cellulases that can enhance food digestion (Xia et al. 2017). Similarly, the symbiotic bacteria may be able to overcome proteinase inhibitors provided by the plant host by producing an excessive amount of proteases (Visôtto et al. 2009). The midgut microorganisms of the Cerambycidae family may also fix nitrogen (Morales-Jiménez et al. 2013) and provide vital nutrients for the host (Ayayee et al. 2016).

In particular, along the course of evolution, the sap-feeding Hemiptera have evolved to become reliant on the host plant's gut microbiome for the production of vitamins and amino acids (Douglas 2015). On the other hand, mutualists of herbivorous insects that cause plant diseases can also contribute to plant invasion by overcoming its defenses. A symbiotic interaction with the host, plant, or insect may have given rise to this relationship, which was then followed by adaptation to the new host (Gonella et al. 2019). Through a variety of methods for modifying plant signaling, a bacterium associated with insects can inhibit the plant's defense (Wang et al. 2016). A combination of different microbes can also lead to weakened plant defenses (Chung et al. 2013). To assault the plant, for instance, a variety of phytopathogenic bacteria and fungi influence cytokinin signaling. Microbes have the ability to either generate their own cytokinins or alter those that are generated by the plant (Giron et al. 2013). The antagonistic relationship between the Salicylic acid (SA) and jasmonic acid (JA) signaling networks can be used to further influence the plant defense system (Derksen et al. 2013). Additionally, microbes can modify a plant's volatile production and visual signals, among other indirect effects (Mauck et al. 2016). For a successful microbiome exchange, the specific timing of interactions between microbial symbionts of plants and insects may be critical. In Silver birch (Betula pendula L.), herbivore survival was impacted by the phenological synchronization of the host plant's budding with the overwintering herbivorous gypsy moth (Lymantria dispar L.).

This was due to the insect's gut microbiota. Asynchronous larvae were less susceptible to the entomopathogenic bacterium Bacillus thuringiensis because there was a definite decrease in the variety of bacteria in their intestines compared to those of synchronous larvae (Martemyanov et al. 2016). The transferred bacteria can change the immunity, fertility and survival of the insect host, which can change the herbivore's fitness in addition to overcoming plant defenses and facilitating exploitation by the insect herbivore (Mauck et al. 2016). Insect life spans can be shortened, latent periods lengthened, disease transmission blocked, reproduction altered, and insect susceptibilities to natural enemies increased by modification of the insect microbiome. The change of insect behavior, such as modified host choices or feeding patterns, might refer to the newly acquired microorganisms (Mauck et al. 2016). In the lodgepole pine (Pinus ponder osa Douglas ex C. Lawson), the symbiont of the mountain pine beetle (Dendroctonus ponderosae) showed antagonistic reflection to the beneficial phloem endophyte Bacillus pumilus (Adams et al. 2008).

The effects of microbiome interaction in plants have not been well researched, aside from the vectoring of plant diseases in insects. The most well-known example is the well-researched and utilized B. thuringiensis, which functions as a helpful endophyte inside plants, but realizing crystal protein (Cry toxins) which is harmful to the insect larvae (Monnerat et al. 2009). The recipient grapevine vines were in good condition after receiving a whole endophytic population from the phloem-feeding grasshopper S. titanus. Given that previous research had demonstrated that endophytes promote development and shield grapevines from pathogens (Lòpez-Fernàndez et al. 2017), hypothesized that endophytes had a beneficial role on the grapevine (Campisano et al. 2015).

## 5.9 Microbiome-Mediated Enhancement of Nutritional and Cosmetic Yield Traits

Beyond its role in pest management, the plant-insect-microbiome nexus directly influences critical determinants of agricultural yield quality. Symbiont-mediated enhancements in plant physiological efficiency directly translate to improved harvestable traits. For instance, specific root-associated microbes improve nutrient solubilization and uptake, directly increasing the protein and mineral content in grains and legumes (White and Broadley 2009; Rashid et al. 2016). Furthermore, foliar and endophytic microbes can act as biochemical modulators, influencing the synthesis of secondary metabolites and phytohormones that govern fruit development, sugar accumulation, and oil composition (Vandenkoornhuyse et al. 2015). This microbial "bio-fortification" is a key pathway to enhancing the nutritional density and marketability of produce. Concurrently, microbiome-primed systemic resistance reduces the incidence and severity of pathogen infections and insect feeding damage that cause cosmetic defects—such as blemishes, scars, and fungal lesions—which are primary causes of post-harvest loss and quality downgrades in fresh markets (Berg et al. 2017). Therefore, managing the shared microbiome is not merely a defensive strategy but a proactive tool for shaping the biochemical and physical attributes that define premium crop quality.

## 5.10 Microbial Contributions to Yield Stability, Uniformity, and Safety

The stability and consistency of yield quality across growing seasons are paramount for food security and economic viability, and the microbiome is central to this resilience. A robust, host-associated microbial community buffers plants against abiotic stressors like drought and salinity, which are known to compromise yield size and uniformity, and induce nutritional deficiencies in crops (Ngumbi and Kloepper 2016). By mitigating stress, microbiomes help maintain optimal development, leading to more consistent fruit size, shape, and ripening—key metrics for grading standards. Perhaps more critically, a microbiome-balanced agroecosystem can reduce dependency on synthetic agrochemicals. This directly addresses a major consumer concern: minimizing pesticide residues on and in food products, which is a significant factor in perceived quality and safety (Bakker et al. 2018). Thus, leveraging cross-kingdom microbiome interactions offers a dual promise: it enhances intrinsic quality traits (nutrition, appearance) while also promoting "clean" production methods that improve the extrinsic quality and safety of the final yield, aligning with both market demands and sustainable agricultural goals.

## 5.11 Future Prospects and Challenges

This chapter examines the multifaceted roles of the insect microbiome, with a specific focus on gut-associated bacteria and their intricate interactions with host plants. These microbial symbionts are instrumental in mediating insect adaptation to plant defenses, directly influencing host selection and feeding behavior. Crucially, they engage in a dynamic dialogue with the plant, capable of suppressing or modulating plant immune responses and growth pathways. Simultaneously, they can alter the virulence of plant pathogens vectored by the insect, creating complex, multi-trophic interactions. Advances in biotechnology, including metagenomics and microbial engineering, have significantly broadened the potential applications of these bacteria for targeted pest management. Such biological control strategies present a sustainable alternative to conventional chemical pesticides, offering reduced environmental toxicity and lower risks of resistance development. Therefore, the strategic manipulation of these cascading, multispecies relationships represents a novel and ecologically sound paradigm for integrated pest management, aligning agricultural productivity with environmental stewardship.

## 5.12 Use of AI Technology

The AI language model DeepSeek-V3 was used during the drafting process to assist with phrasing, clarity, and reference formatting. The authors critically reviewed and validated all such output. The final manuscript, its content, and its conclusions remain the sole responsibility of the authors.

## References

Acuña R, Padilla BE, Flórez-Ramos CP, Rubio JD, Herrera JC, Benavides P, Lee S-J, Yeats TH, Egan AN, Doyle JJ, Rose JKC (2012) Adaptive horizontal transfer of a bacterial gene to an invasive insect pest of coffee. Proc Natl Acad Sci USA 109:4197–4202. https://doi.org/10.1073/pnas.1121190109

Adams AS, Aylward FO, Adams SM, Erbilgin N, Aukema BH, Currie CR, Suen G, Raffa KF (2013) Mountain pine beetles colonizing historical and naive host trees are associated with a bacterial community highly enriched in genes contributing to terpene metabolism. Appl Environ 79:3468

Adams AS, Six DL, Adams SM, Holben WE (2008) In vitro interactions between yeasts and bacteria and the fungal symbionts of the mountain pine beetle (Dendroctonus ponderosae). Microb Ecol 56:460–466. https://doi.org/10.1007/s00248-008-9364-0

Agrios G (2004) Plant pathology, 5th edn. Academic Press, pp 1–922

Ali JG, Agrawal AA (2012) Specialist versus generalist insect herbivores and plant defense. Trends Plant Sci 17:293–302. https://doi.org/10.1016/j.tplants.2012.02.006

Ayayee PA, Larsen T, Rosa C, Felton GW, Ferry JG, Hoover K (2016) Essential amino acid supplementation by gut microbes of a wood-feeding cerambycid. Environ Entomol 45:66–73. https://doi.org/10.1093/ee/nvv153
Bacon CW, Porter JK, Robbins JD, Luttrell ES (1977) Epichloë typhina from toxic tall fescue grasses. Appl Environ Microbiol 34:576–581. https://doi.org/10.1128/aem.34.5.576-581.1977
Bakker PA et al (2018) The soil-borne identity and microbiome-assisted agriculture: looking back to the future. Mol Plant 11(7):899–910
Ballhorn DJ, Elias JD, Balkan MA, Fordyce RF, Kennedy PG (2017) Colonization by nitrogen-fixing Frankia bacteria causes short-term increases in herbivore susceptibility in red alder (Alnus rubra) seedlings. Oecologia 184:497–506. https://doi.org/10.1007/s00442-017-3888-2
Bell K, Naranjo-Guevara N, dos Santos RC, Meadow R, Bento JMS (2020) Predatory earwigs are attracted by herbivore-induced plant volatiles linked with plant growth-promoting rhizobacteria. Insects 11:271. https://doi.org/10.3390/insects11050271
Berg G et al (2017) The plant microbiome explored: implications for experimental botany. J Exper Bot 67(4):995–1002
Bosco-Drayon V, Poidevin M, Boneca IG, Narbonne-Reveau K, Royet J, Charroux B (2012) Peptidoglycan sensing by the receptor PGRP-LE in the drosophila gut induces immune responses to infectious bacteria and tolerance to microbiota. Cell Host Microbe 12:153–165. https://doi.org/10.1016/j.chom.2012.06.002
Campisano A, Pancher M, Puopolo G, Puddu A, Lòpez-Fernàndez S, Biagini B, Yousaf S, Pertot I (2015) Diversity in endophyte populations reveals functional and taxonomic diversity between wild and domesticated grapevines. Am J Enol Vitic 66:12–21. https://doi.org/10.5344/ajev.2014.14046
Carroll GC, Carroll FE (1978) Studies on the incidence of coniferous needle endophytes in the Pacific northwest. Can J Bot 56:3034–3043. https://doi.org/10.1139/b78-367
Chagas FO, Pessotti R d C, Caraballo-Rodríguez AM, Pupo MT (2018) Chemical signaling involved in plant–microbe interactions. Chem Soc Rev 47:1652–1704. https://doi.org/10.1039/c7cs00343a
Chaudhary R, Atamian HS, Shen Z, Briggs SP, Kaloshian I (2014) GroEL from the endosymbiont Buchnera aphidicola betrays the aphid by triggering plant defense. Proc Natl Acad Sci USA 111:8919–8924. https://doi.org/10.1073/pnas.1407687111
Chen B, Teh B-S, Sun C, Hu S, Lu X, Boland W, Shao Y (2016) Biodiversity and activity of the gut microbiota across the life history of the insect herbivore Spodoptera littoralis. Sci Rep 6:29505. https://doi.org/10.1038/srep29505
Chuche J, Auricau-Bouvery N, Danet J-L, Thiéry D (2017) Use the insiders: could insect facultative symbionts control vector-borne plant diseases? J Pest Sci 90:51–68. https://doi.org/10.1007/s10340-016-0782-3
Chung SH, Rosa C, Scully ED, Peiffer M, Tooker JF, Hoover K, Luthe DS, Felton GW (2013) Herbivore exploits orally secreted bacteria to suppress plant defenses. Proc Natl Acad Sci USA 110:15728–15733. https://doi.org/10.1073/pnas.1308867110
Chung SH, Scully ED, Peiffer M, Geib SM, Rosa C, Hoover K, Felton GW (2017) Host plant species determines symbiotic bacterial community mediating suppression of plant defenses. Sci Rep 7:39690. https://doi.org/10.1038/srep39690
Conrath U, Beckers GJM, Flors V, García-Agustín P, Jakab G, Mauch F, Newman M-A, Pieterse CMJ, Poinssot B, Pozo MJ, Pugin A, Schaffrath U, Ton J, Wendehenne D, Zimmerli L, Mauch-Mani B (2006) Priming: getting ready for battle. Mol Plant-Microbe Interact 19:1062–1071. https://doi.org/10.1094/MPMI-19-1062
Coolen S, Magda R-M, Welte CU (2022) The secret life of insect-associated microbes and how they shape insect-plant interactions. FEMS Microbiol Ecol 98:fiac083. https://doi.org/10.1093/femsec/fiac083
Corbin C, Heyworth ER, Ferrari J, Hurst GDD (2017) Heritable symbionts in a world of varying temperature. Heredity 118:10–20

Couret J, Huynh-Griffin L, Antolic-Soban I, Acevedo-Gonzalez TS, Gerardo NM (2019) Even obligate symbioses show signs of ecological contingency: impacts of symbiosis for an invasive stinkbug are mediated by host plant context. Ecol Evol 9:9087–9099

De Almeida LG, de Moraes LAB, Trigo JR, Omoto C, Cônsoli FL (2017) The gut microbiota of insecticide-resistant insects houses insecticidedegrading bacteria: a potential source for biotechnological exploitation. PLoS One 12:e0174754. https://doi.org/10.1371/journal.pone.0174754

de Roode JC, Lefèvre T (2012) Behavioral immunity in insects. Insects 3:789–820. https://doi.org/10.3390/insects3030789

Dean JM, Mescher MC, de Moraes CM (2014) Plant dependence on rhizobia for nitrogen influences induced plant defenses and herbivore performance. Int J Mol Sci 15:1466–1480. https://doi.org/10.3390/ijms15011466

Derksen H, Rampitsch C, Daayf F (2013) Signaling cross-talk in plant disease resistance. Plant Sci 207:79–87. https://doi.org/10.1016/j.plantsci.2013.03.004

Deutsch CA, Tewksbury JJ, Tigchelaar M, Battisti DS, Merrill SC, Huey RB et al (2018) Increase in crop losses to insect pests in a warming climate. Science 361:916–919. https://doi.org/10.1126/science.aat3466

Dillon RJ, Dillon VM (2004) The gut bacteria of insects: nonpathogenic interactions. Annu Rev Entomol 49:71–92. https://doi.org/10.1146/annurev.genet.41.110306.130119

Douglas AE (2006) Phloem-sap feeding by animals: problems and solutions. J Exp Bot 57:747–754. https://doi.org/10.1093/jxb/erj067

Douglas AE (2009) The microbial dimension in insect nutritional ecology. Funct Ecol 23:38–47

Douglas AE (2015) Multiorganismal insects: diversity and function of resident microorganisms. Annu Rev Entomol 60:17–34. https://doi.org/10.1146/annurev-ento-010814-020822ece3.5454

Engel P, Moran NA (2013) The gut microbiota of insects - diversity in structure and function. FEMS Microbiol Rev 37:699–735. https://doi.org/10.1111/1574-6976.12025

Engl T, Eberl N, Gorse C, Krueger T, Schmidt THP, Plarre R, Adler C, Kaltenpoth M (2018) Ancient symbiosis confers desiccation resistance to stored grain pest beetles. Mol Ecol 27: 2095–2108

Erlandson MA, Toprak U, Hegedus DD (2019) Role of the peritrophic matrix in insectpathogen interactions. J Insect Physiol 117:103894. https://doi.org/10.1016/j.jinsphys.2019.103894

Faeth SH (2002) Are endophytic fungi defensive plant mutualists? Oikos 98:25–36. https://doi.org/10.1034/j.1600-0706.2002.980103.x

Feldhaar H (2011) Bacterial symbionts as mediators of ecologically important traits of insect hosts. Ecol Entomol 36:533–543. https://doi.org/10.1111/j.1365-2311.2011.01318.x

Feng H, Edwards N, Anderson CMH, Althaus M, Duncan RP, Hsu YC et al (2019) Trading amino acids at the aphid–Buchnera symbiotic interface. Proc Natl Acad Sci USA 116:16003–16011. https://doi.org/10.1073/pnas.1906223116

Ferrater JB, de Jong PW, Dicke M, Chen YH, Horgan FG (2013) Symbiontmediated adaptation by planthoppers and leafhoppers to resistant rice varieties. Arthropod Plant Interact 7:591–605. https://doi.org/10.1007/s11829-013-9277-9

Florez LV, Biedermann PHW, Engl T, Kaltenpoth M (2015) Defensive symbioses of animals with prokaryotic and eukaryotic microorganisms. Nat Prod Rep 32:904–936

Friman J, Pineda A, Gershenzon J, Dicke M, Loon JJA (2021) Differential effects of the rhizobacterium Pseudomonas simiae on above- and belowground chewing insect herbivores. J Appl Entomol 145:250–260. https://doi.org/10.1111/jen.12842

Galambos N, Compant S, Wäckers F, Sessitsch A, Anfora G, Mazzoni V, Pertot I, Perazzolli M (2021) Beneficial insects deliver plant growth-promoting bacterial endophytes between tomato plants. Microorganisms 9:1294. https://doi.org/10.3390/microorganisms9061294

Gatehouse JA (2002) Plant resistance towards insect herbivores: a dynamic interaction. New Phytol 156:145–169. https://doi.org/10.1046/j.1469-8137.2002.00519.x

Giron D, Dedeine F, Dubreuil G, Huguet E, Mouton L, Outreman Y, Vavre F, Simon J-C (2017) Chapter seven - influence of microbial symbionts on plant–insect interactions. In: Sauvion N, Thiéry D, Calatayud P-A (eds) Advances in botanical research. Academic Press, pp 225–257

Giron D, Frago E, Glevarec G, Pieterse CMJ, Dicke M (2013) Cytokinins as key regulators in plant–microbe–insect interactions: connecting plant growth and defence. Funct Ecol 27:599–609. https://doi.org/10.1111/1365-2435.12042

Gonella E, Tedeschi R, Crotti E, Alma A (2019) Multiple guests in a single host: interactions across symbiotic and phytopathogenic bacteria in phloem-feeding vectors—a review. Entomol Exp Appl 167:171–185. https://doi.org/10.1111/eea.12766

Goswami D, Thakker JN, Dhandhukia PC (2016) Portraying mechanics of plant growth promoting rhizobacteria (PGPR) a review. Cogent Food Agric 2

Gupta A, Nair S (2020) Dynamics of insect-microbiome interaction influence host and microbial symbiont. Front Microbiol 11:1357. https://doi.org/10.3389/fmicb.2020.01357

Hammerbacher A, Schmidt A, Wadke N, Wright LP, Schneider B, Bohlmann J, Brand WA, Fenning TM, Gershenzon J, Paetz C (2013) A common fungal associate of the spruce bark beetle metabolizes the stilbene defenses of Norway spruce. Plant Physiol 162:1324–1336. https://doi.org/10.1104/pp.113.218610

Hamonts K et al (2018) Field study reveals core plant microbiota and relative importance of their drivers. Environ Microbiol 20:124–140

Hansen AK, Moran NA (2014) The impact of microbial symbionts on host plant utilization by herbivorous insects. Mol Ecol 23:1473–1496. https://doi.org/10.1111/mec.12421

Hardoim PR, van Overbeek LS, Berg G, Pirttilä AM, Compant S, Campisano A, Döring M, Sessitsch A (2015) The hidden world within plants: ecological and evolutionary considerations for defining functioning of microbial endophytes. Microbiol Mol Biol Rev 79:293–320. https://doi.org/10.1128/MMBR.00050-14

Härri SA, Krauss J, Müller CB (2008) Fungal endosymbionts of plants reduce lifespan of an aphid secondary parasitoid and influence host selection. Proc Biol Sci 275:2627–2632. https://doi.org/10.1098/rspb.2008.0594

Hirsch AM, Lum MR, Downie JA (2001) What makes the rhizobia-legume symbiosis so special? Plant Physiol 127:1484–1492. https://doi.org/10.1104/pp.010866

Jaber LR, Vidal S (2010) Fungal endophyte negative effects on herbivory are enhanced on intact plants and maintained in a subsequent generation. Ecol Entomol 35:25–36. https://doi.org/10.1111/j.1365-2311.2009.01152.x

Janson EM, Stireman JO, Singer MS, Abbot P (2008) Phytophagous insect-microbe mutualisms and adaptive evolutionary diversification. Evolution 62:997–1012. https://doi.org/10.1111/j.1558-5646.2008.00348.x

Jing TZ, Qi FH, Wang ZY (2020) Most dominant roles of insect gut bacteria: digestion, detoxification, or essential nutrient provision? Microbiome 8:38

Jones OT, Coaker TH (1978) A basis for host plant finding in phytophagous larvae. Entomol Exp Appl 24:472–484. https://doi.org/10.1111/j.1570-7458.1978.tb02807.x

Kageyama D, Narita S, Watanabe M (2012) Insect sex determination manipulated by their endosymbionts: incidences, mechanisms and implications. Insects 3:161–199. https://doi.org/10.3390/insects3010161

Kandasamy D, Gershenzon J, Andersson MN, Hammerbacher A (2019) Volatile organic compounds influence the interaction of the Eurasian spruce bark beetle (Ips typographus) with its fungal symbionts. ISME J 13:1788–1800. https://doi.org/10.1038/s41396-019-0390-3

Kandasamy D, Gershenzon J, Hammerbacher A (2016) Volatile organic compounds emitted by fungal associates of conifer bark beetles and their potential in bark beetle control. J Chem Ecol 42:952–969. https://doi.org/10.1007/s10886-016-0768-x

Kaneko T, Goldman WE, Mellroth P, Steiner H, Fukase K, Kusumoto S et al (2004) Monomeric and polymeric gram-negative peptidoglycan but not purified LPS stimulate the Drosophila IMD pathway. Immunity 20:637–649. https://doi.org/10.1016/s1074-7613(04)00104-9

Kirisits T (2004) Fungal associates of european bark beetles with special emphasis on the ophiostomatoid fungi. In: Bark and wood boring insects in living trees in Europe: a synthesis, pp 181–236

Kwon MO, Wayadande AC, Fletcher J (1999) Spiroplasma citri movement into the intestines and salivary glands of its leafhopper vector, Circulifer tenellus. Phytopathology 89:1144–1151. https://doi.org/10.1094/phyto.1999.89.12.1144

Kwong WK, Mancenido AL, Moran NA (2017) Immune system stimulation by the native gut microbiota of honey bees. R Soc Open Sci 4:170003. https://doi.org/10.1098/rsos.170003

Leach JE, Triplett LR, Argueso CT, Trivedi P (2017) Communication in the phytobiome. Cell 169: 587–596. https://doi.org/10.1016/j.cell.2017.04.025

Lemaitre B, Hoffmann J (2007) The host defense of Drosophila melanogaster. Annu Rev Immunol 25:697–743. https://doi.org/10.1146/annurev.immunol.25.022106.141615

Lemaitre B, Miguel-Aliaga I (2013) The digestive tract of Drosophila melanogaster. Annu Rev Genet 47:377–404. https://doi.org/10.1146/annurev-genet-111212-133343

Leulier F, Parquet C, Pili-Floury S, Ryu JH, Caroff M, Lee WJ et al (2003) The Drosophila immune system detects bacteria through specific peptidoglycan recognition. Nat Immunol 4:478–484. https://doi.org/10.1038/ni922

Lieutier F, Yart A, Salle A (2009) Stimulation of tree defenses by ophiostomatoid fungi can explain attack success of bark beetles on conifers. Ann For Sci 66:801–801. https://doi.org/10.1051/forest/2009066

Lòpez-Fernàndez S, Mazzoni V, Pedrazzoli F, Pertot I, Campisano A (2017) A phloem-feeding insect transfers bacterial endophytic communities between grapevine plants. Front Microbiol 8: 834. https://doi.org/10.3389/fmicb.2017.00834

Lundberg DS et al (2012) Defining the core Arabidopsis thaliana root microbiome. Nature 488:86–90

Mann RS, Pelz-Stelinski K, Hermann SL, Tiwari S, Stelinski LL (2011) Sexual transmission of a plant pathogenic bacterium, Candidatus Liberibacter asiaticus, between conspecific insect vectors during mating. PLoS One 6:e29197. https://doi.org/10.1371/journal.pone.0029197

Martemyanov VV, Belousova IA, Pavlushin S, Dubovskiy IM, Ershov NI, Alikina TY, Kabilov MR, Glupov VV (2016) Phenological asynchrony between host plant and gypsy moth reduces insect gut microbiota and susceptibility to Bacillus thuringiensis. Ecol Evol 6:7298–7310. https://doi.org/10.1002/ece3.2460

Marubayashi JM, Kliot A, Yuki VA, Rezende JAM, Krause-Sakate R, Pavan MA et al (2014) Diversity and localization of bacterial endosymbionts from whitefly species collected in Brazil. PLoS One 9:e108363. https://doi.org/10.1371/journal.pone.0108363

Mauck KE, de Moraes CM, Mescher MC (2016) Effects of pathogens on sensory-mediated interactions between plants and insect vectors. Curr Opin Plant Biol 32:53–61. https://doi.org/10.1016/j.pbi.2016.06.012

Mergaert P (2018) Role of antimicrobial peptides in controlling symbiotic bacterial populations. Nat Prod Rep 35:336–356. https://doi.org/10.1039/c7np00056a

Mikonranta L, Mappes J, Kaukoniitty M, Freitak D (2014) Insect immunity: oral exposure to a bacterial pathogen elicits free radical response and protects from a recurring infection. Front Zool 11:23. https://doi.org/10.1186/1742-9994-11-23

Miller JD, Mackenzie S, Foto M, Adams GW, Findlay JA (2002) Needles of white spruce inoculated with rugulosin-producing endophytes contain rugulosin reducing spruce budworm growth rate. Mycol Res 106:471–479. https://doi.org/10.1017/S0953756202005671

Misof B, Liu S, Meusemann K, Peters RS, Donath A, Mayer C et al (2014) Phylogenomics resolves the timing and pattern of insect evolution. Science 346:763–767

Monnerat RG, Soares CM, Capdeville G, Jones G, Martins ES, Praça L, Cordeiro BA, Braz SV, dos Santos RC, Berry C (2009) Translocation and insecticidal activity of Bacillus thuringiensis living inside of plants. Microb Biotechnol 2:512–520. https://doi.org/10.1111/j.1751-7915.2009.00116.x

Morales-Jiménez J, Vera-Ponce de León A, García-Domínguez A, Martínez-Romero E, Zúñiga G, Hernández-Rodríguez C (2013) Nitrogen-fixing and uricolytic bacteria associated with the gut of Dendroctonus rhizophagus and Dendroctonus valens (Curculionidae: Scolytinae). Microb Ecol 66:200–210. https://doi.org/10.1007/s00248-013-0206-3

Moran NA, McCutcheon JP, Nakabachi A (2008) Genomics and evolution of heritable bacterial symbionts. Annu Rev Genet 42:165–190

Müller DB, Vogel C, Bai Y, Vorholt JA (2016) The plant microbiota: systems- level insights and perspectives. Annu Rev Genet 50:211–234. https://doi.org/10.1146/annurev-genet-120215-034952

Müller T, Müller M, Behrendt U, Stadler B (2003) Diversity of culturable phyllosphere bacteria on beech and oak: the effects of lepidopterous larvae. Microbiol Res 158:291–297. https://doi.org/10.1078/0944-5013-00207

Nadarasah G, Stavrinides J (2011) Insects as alternative hosts for phytopathogenic bacteria. FEMS Microbiol Rev 35:555–575. https://doi.org/10.1111/j.1574-6976.2011.00264.x

Neyen C, Poidevin M, Roussel A, Lemaitre B (2012) Tissue- and ligand-specific sensing of gram-negative infection in Drosophila by PGRP-LC isoforms and PGRP-LE. J Immunol 189:1886–1897. https://doi.org/10.4049/jimmunol.1201022

Ngumbi E, Kloepper J (2016) Bacterial-mediated drought tolerance: current and future prospects. Appl Soil Ecol 105:109–125

Novotny V, Miller SE, Baje L, Balagawi S, Basset Y, Cizek L, Craft KJ, Dem F, Drew RAI, Hulcr J, Leps J, Lewis OT, Pokon R, Stewart AJA, Samuelson GA, Weiblen GD (2010) Guild-specific patterns of species richness and host specialization in plant–herbivore food webs from a tropical forest. J Anim Ecol 79:1193–1203. https://doi.org/10.1111/j.1365-2656.2010.01728.x

Ojha A, Sinha DK, Padmakumari AP, Bentur JS, Nair S (2017) Bacterial community structure in the Asian rice gall midge reveals a varied microbiome rich in proteobacteria. Sci Rep 7:9424

Ortiz-Urquiza A (2021) The split personality of Beauveria bassiana: understanding the molecular basis of fungal parasitism and mutualism. mSystems 6:e00766–21. https://doi.org/10.1128/mSystems.00766-21

Paine TD, Raffa KF, Harrington TC (1997) Interactions among scolytid bark beetles, their associated fungi, and live host conifers. Annu Rev Entomol 42:179–206. https://doi.org/10.1146/annurev.ento.42.1.179

Pangesti N, Reichelt M, van de Mortel JE, Kapsomenou E, Gershenzon J, van Loon JJA, Dicke M, Pineda A (2016) Jasmonic acid and ethylene signaling pathways regulate glucosinolate levels in plants during rhizobacteria- induced systemic resistance against a leaf-chewing herbivore. J Chem Ecol 42:1212–1225. https://doi.org/10.1007/s10886-016-0787-7

Perilla-Henao LM, Casteel CL (2016) Vector-borne bacterial plant pathogens: interactions with hemipteran insects and plants. Front Plant Sci 7:1163. https://doi.org/10.3389/fpls.2016.01163

Pieterse CMJ, Zamioudis C, Berendsen RL, Weller DM, van Wees SCM, Bakker PAHM (2014) Induced systemic resistance by beneficial microbes. Annu Rev Phytopathol 52:347–375. https://doi.org/10.1146/annurev-phyto-082712-102340

Pirttilä AM, Wäli P (2009) Conifer endophytes. In: White J, Torres MS (eds) Defensive mutualism in microbial symbiosis. CRC Press, Boca Raton, FL, pp 235–246

Põlme S, Bahram M, Kõljalg U, Tedersoo L (2014) Global biogeography of Alnus-associated Frankia actinobacteria. New Phytol 204:979–988. https://doi.org/10.1111/nph.12962

Pronk LJU, Bakker PAHM, Keel C, Maurhofer M, Flury P (2022) The secret life of plant-beneficial rhizosphere bacteria: insects as alternative hosts. Environ Microbiol 24:3273–3289. https://doi.org/10.1111/1462-2920.15968

Rashid MI et al (2016) Bacteria and fungi can contribute to nutrients bioavailability and aggregate formation in degraded soils. Microbiol Res 183:26–41

Ren C, Webster P, Finkel SE, Tower J (2007) Increased internal and external bacterial load during drosophila aging without life-span trade-off. Cell Metab 6:144–152. https://doi.org/10.1016/j.cmet.2007.06.006

Renoz F, Pons I, Hance T (2019) Evolutionary responses of mutualistic insect–bacterial symbioses in a world of fluctuating temperatures. Curr Opin Insect Sci 35:20–26

Saikkonen K, Helander M, Ranta H, Neuvonen S, Virtanen T, Suomela J, Vuorinen P (1996) Endophyte-mediated interactions between woody plants and insect herbivores? Entomol Exp 80:269–271. https://doi.org/10.1111/j.1570-7458.1996.tb00932.x

Senderovich Y, Halpern M (2013) The protective role of endogenous bacterial communities in chironomid egg masses and larvae. ISME J 7:2147–2158

Shavit R, Ofek-Lalzar M, Burdman S, Morin S (2013) Inoculation of tomato plants with rhizobacteria enhances the performance of the phloemfeeding insect Bemisia tabaci. Front Plant Sci 4:306. https://doi.org/10.3389/fpls.2013.00306

Shikano I, Rosa C, Tan C-W, Felton GW (2017) Tritrophic interactions: microbe-mediated plant effects on insect herbivores. Annu Rev Phytopathol 55:313–331. https://doi.org/10.1146/annurev-phyto-080516-035319

Shrestha S (2019) Effects of climate change in agricultural insect pest. Acta Sci Agric 3:74–80. https://doi.org/10.31080/asag.2019.03.0727

Šigut M, Pyszko P, Šigutová H, Višnovská D, Kostovcík M, Kotásková N, Dornák O, Kolarík M, Drozd P (2022) Fungi are more transient than bacteria in caterpillar gut microbiomes. Sci Rep 12:15552. https://doi.org/10.1038/s41598-022-19855-5

Six DL (2003) Bark beetle-fungus symbioses. In: Bourtzis K, Miller TA (eds) Insect symbiosis. CRC Press, New York, pp 97–114

Smee MR, Hendry TA (2022) Context-dependent benefits of aphids for bacteria in the phyllosphere. Am Nat 199:380–392. https://doi.org/10.1086/718264

Soler R, Erb M, Kaplan I (2013) Long distance root–shoot signalling in plant– insect community interactions. Trends Plant Sci 18:149–156. https://doi.org/10.1016/j.tplants.2012.08.010

Strano CP, Malacrinò A, Campolo O, Palmeri V (2018) Influence of host plant on Thaumetopoea pityocampa gut bacterial community. Microb Ecol 75:487–494. https://doi.org/10.1007/s00248-017-1019-6

Sudakaran S, Kost C, Kaltenpoth M (2017) Symbiont acquisition and replacement as a source of ecological innovation. Trends Microbiol 25:375–390

Sumarah MW, Miller JD (2009) Anti-insect secondary metabolites from fungal endophytes of conifer trees. Nat Prod Commun 4(11):1497–1504

Sun J, Lu M, Gillette NE, Wingfield MJ (2013) Red turpentine beetle: innocuous native becomes invasive tree killer in China. Annu Rev Entomol 58:293–311. https://doi.org/10.1146/annurev-ento-120811-153624

Taerum SJ, Duong TA, de Beer ZW, Gillette N, Sun J-H, Owen DR, Wingfield MJ (2013) Large shift in symbiont assemblage in the invasive red turpentine beetle. PLoS One 8:e78126. https://doi.org/10.1371/journal.pone.0078126

Thomas P, Kenny N, Eyles D, Moreira LA, O'Neill SL, Asgari S (2011) Infection with the wMel and wMelPop strains of Wolbachia leads to higher levels of melanization in the hemolymph of Drosophila melanogaster, Drosophila simulans and Aedes aegypti. Dev Comp Immunol 35: 360–365. https://doi.org/10.1016/j.dci.2010.11.007

Thompson SNN, Simpson SJJ (2009) Nutrition. In: Resh VH, Cardé RT (eds) Encyclopedia of insects. Academic Press, Cambridge, MA, pp 715–720

Vandenkoornhuyse P, Quaiser A, Duhamel M, Le Van A, Dufresne A (2015) The importance of the microbiome of the plant holobiont. New Phytol 206:1196–1206

Vincent M, Boubakri H, Gasser M, Hay A-E, Herrera-Belaroussi A (2023) What contribution of plant immune responses in Alnus glutinosa-Frankia symbiotic interactions? Symbiosis 89(1): 27–26

Visôtto LE, Oliveira MGA, Guedes RNC, Ribon AOB, Good-God PIV (2009) Contribution of gut bacteria to digestion and development of the velvetbean caterpillar, Anticarsia gemmatalis. J Insect Physiol 55:185–191. https://doi.org/10.1016/j.jinsphys.2008.10.017

Wang J, Chung SH, Peiffer M, Rosa C, Hoover K, Zeng R, Felton GW (2016) Herbivore oral secreted bacteria trigger distinct defense responses in preferred and non-preferred host plants. J Chem Ecol 42:463–474. https://doi.org/10.1007/s10886-016-0712-0

Wari D, Kabir MA, Mujiono K, Hojo Y, Shinya T, Tani A, Nakatani H, Galis I (2019) Honeydew-associated microbes elicit defense responses against brown planthopper in rice. J Exp Bot 70: 1683–1696. https://doi.org/10.1093/jxb/erz041

Wei N, Whyle RL, Ashman T, Jamieson MA (2022) Genotypic variation in floral volatiles influences floral microbiome more strongly than interactions with herbivores and mycorrhizae in strawberries. Hortic Res 9:uhab005. https://doi.org/10.1093/hr/uhab005

White PJ, Broadley MR (2009) Biofortification of crops with seven mineral elements often lacking in human diets–iron, zinc, copper, calcium, magnesium, selenium and iodine. New Phytol 182(1):49–84

Whitney HS (1982) Relationships between bark beetles and symbiotic organisms. In: Mitton JB, Sturgeon KB (eds) Bark beetles in North American conifers. University of Texas Press, Austin, pp 183–211

Wu Q, Patočka J, Kuča K (2018) Insect antimicrobial peptides, a mini review. Toxins 10:461. https://doi.org/10.3390/toxins10110461

Xia X, Gurr GM, Vasseur L, Zheng D, Zhong H, Qin B, Lin J, Wang Y, Song F, Li Y, Lin H, You M (2017) Metagenomic sequencing of diamondback moth gut microbiome unveils key holobiont adaptations for herbivory. Front Microbiol 8:663. https://doi.org/10.3389/fmicb.2017.00663

Zebelo S, Song Y, Kloepper JW, Fadamiro H (2016) Rhizobacteria activates (1)-d -cadinene synthase genes and induces systemic resistance in cotton against beet armyworm (*Spodoptera exigua*). Plant Cell Environ 39:935–943. https://doi.org/10.1111/pce.12704

# Chapter 6
# The Plant Microbiome: An Engine for Sustainable Crop Growth and Yield

**Tarek A. A. Moussa and Neveen M. Khalil**

## 6.1 Introduction

Now, after more than twenty years have passed in the twenty-first century, we observe that the increase in the world population, accompanied by the human use of the surrounding environmental resources in various fields, including the agriculture industry, has led to a critical ecological imbalance. There is a continuous release of perilous chemicals and xenobiotics. In consequence, human life faces a high risk because of this escalating environmental deterioration (Moussa and Khalil 2022).

Given the escalating cropping practices and the changing climate, providing for the growing world population would require the optimization of environmental sustainability and the reduction of the impacts on the ecosystem of food production (Afridi et al. 2022). The sustainable and eco-friendly manipulation of microorganisms in agricultural production is widely accepted as a crucial technology for the future. The use of soil microbiome for optimization of crop production dates back to ~300 BC (Vessey 2003). Interestingly, the microbiome is endorsed as a keystone for the upcoming green revolution (Parnell et al. 2016). Moreover, the continuous increase in the global population demands food availability to be of the utmost concern. Regarding the data by DESA (2019), if the increase continues, food demands will increase by values of 8.5, 9.7, and 11 billion in 2030, 2050, and 2100, respectively. Therefore, meeting these demands will require novel, green methodologies that utilize microorganisms.

The environmental stressors led to tremendous alterations in the plant biochemistry and physiology, causing notable reductions in crop yield. The unfavorable environmental conditions resulted in 30–50% agricultural losses. These losses along

T. A. A. Moussa (✉) · N. M. Khalil
Botany and Microbiology Department, Faculty of Science, Cairo University, Giza, Egypt
e-mail: tarekmoussa@cu.edu.eg



N. El-Wakeil, A. Abdelkhalek (eds.), *Microbiomes and Plant Secondary Metabolites for Insect Pest Management*, Progress in Biological Control 23,
https://doi.org/10.1007/978-3-032-24330-0_6

with the continuous population growth, need at least a 60% lift in agronomic production to meet the food demands on a large scale (Wild 2003). Mostly, agriculture is aided by pesticides. It was estimated that globally, two million tons of pesticides are administered for controlling pests to maximize crop production (Foong et al. 2020). Simultaneously, using agrochemicals could affect soil fertility and biochemistry, biodiversity, agricultural sustainability, nutritional security, and food safety. However, excessive pesticide utilization not only leads to pollution but also influences animal and human health (Sharma et al. 2010; Fu et al. 2022). In addition, they prevent the growth of useful microorganisms and decrease the availability of nutrients that are needed for the plant (Meena et al. 2020).

Plant microbiome assists the plant by various mechanisms including hormone regulation, production of certain metabolites that elicit plant resistance against pathogens, repression of soil-borne pathogens, competition for nutrients, and antibiosis (Choudhary et al. 2007; Penton et al. 2014; Reitz et al. 2015; Zhang et al. 2019; Rodríguez et al. 2020). Plants, along with their associated microbes, form a unique biological entity which is called "the holobiont" (Marco et al. 2022).

Most of the published studies are focused on applying soil bacteria (Babalola 2010) or fungi (Yadav et al. 2020) in agriculture. Despite many bacterial and fungal species having been identified as plant supporters, two groups of microorganisms are receiving greater attention due to their potential in stressed habitats. These are the actinomycetes associated with plants, and the arbuscular mycorrhizal fungi (AMF), which have a symbiotic interaction with almost all plants' roots, including crop plants. AMFs are considered a highly important plant ally for interacting with their surrounding environment. They supply agricultural plants with several ecosystem services (Gosling et al. 2006; Antoine et al. 2021). Since in a natural ecosystem, plant roots are commonly colonized with AMF, the concept of the rhizosphere has expanded to incorporate fungal components of mutual symbiosis, leading to the term mycorrhizosphere (Johansson et al. 2004). The rhizosphere comprises the environment where there are bacterial-fungal interactions, where fungi and bacteria affect one another, for example, by supplying nutrients and water (Nerva et al. 2021). However, much research is needed to investigate their role in improving plant production and restoring plant-microbiome equilibrium in agricultural systems. That is why excavating the mechanisms on which this cooperation is based, besides trying to boost it using a more integrated view of breeding programs, would be very propitious. Moreover, the process of domestication, where plants are not able to survive by themselves away from human care, has led to modifying plants' capability for interacting with essential soil microbes compared with respective wild plant varieties (Hetrick et al. 1992; Sangabriel-Conde et al. 2014; Chen et al. 2015; Zeder 2015). On the other hand, Actinomycetes are often included in plants' core microbiome. They are regarded as great supporters in the hidden world of interactions between plants and microorganisms. Utilizing approaches of the next-generation sequencing (NGS) for surveying microorganisms established that phylum Actinobacteria is one of the five most dominating soil bacterial phyla (Lauber et al. 2009; Zarraonaindia et al. 2015; Johnston-Monje et al. 2016). Actinomycetes can occupy several environmental niches, including a wide range of pH and temperatures. They can alter their morphology to adapt to extreme habitats. Thus, they are considered ecologically divergent groups (Ramijan et al. 2018; Millán-Aguiñaga et al. 2019). Their beneficial trait for plants is shown in their plant growth-promoting (PGP) actions, besides their potential as bio-controlling agents. Studies conducted on *Bacillus* spp. and *Pseudomonas* spp. are proof (Harman 2006; McSpadden Gardener 2007; Choudhary and Johri 2009). It was reported that these bacteria associate with AMF, which justifies analyzing the potential of both groups in ameliorating plant production (Kamal et al. 2014).

Accordingly, engineering the plant microbiome is considered a prospective strategy that could be applied for better plant growth, health, and yield when grown under extreme conditions. Many recent studies proposed accessible approaches for plant microbiome engineering (Arif et al. 2020; Kumar and Dubey 2020). The aim is, therefore, to attain viable agronomic production through innovative tools for enhancing microbial bioengineering to replace the life-threatening agrochemicals (Afridi et al. 2022).

## 6.2 Plant Microbiota

Plant microbiome can be classified into phyllospheric, endospheric, and rhizospheric microbiota (Berg et al. 2014). Phyllospheric microbiota describes the microorganisms inhabiting the epigeal plant surfaces, including stems, leaves, flowers, and fruits. Endospheric microorganisms (endophytes) are those living inside the plant. Meanwhile, the rhizospheric microbiota are the most studied. The rhizosphere (the narrow soil area close to plants' roots) is occupied by prokaryotes (viruses, archaea, bacteria), and eukaryotes (fungi, protozoa, nematodes). These organisms are crucial for the development and growth of plants. It is affirmed that the number of microbes comprising the plant microbiome far exceeds that of plant cells (Mendes et al. 2013). Due to the vast diversity and number of microorganisms of the plant microbiome, they are considered an extension of plants' genomes, since they possess higher protein-coding as well as metabolic potentialities. The key tasks done by these microorganisms include plant stress suppression, nutrient acquisition, and controlling plant diseases. Moreover, the coordination between plants and microbes can aid in the remediation of environmental contaminants (Zhang et al. 2021).

### 6.2.1 *Epigeal Plant Microbiota*

The epigeal (aboveground) plant parts including stem, leaves, and flowers, offered unique habitat for phyllospheric and endospheric microbiota. However, major microbial diversity occurs between both groups (Hardoim et al. 2015; Compant et al. 2021). Endospheric microorganisms are systemically distributed within the plant tissues through the xylem (Compant et al. 2010), while their entry to the plant is through the aerial parts of the plant (Compant et al. 2011). Distinct endospheric communities are noticed in different plant compartments according to the plant's source allocation. Phyllospheric microorganisms can be driven from the soil by the plant and other environmental criteria (Vorholt 2012; Wallace et al. 2018). This contributes to microbial distribution at the genus and species levels in the endospheric and phyllospheric areas. In this relation, after examining the grapevine's phyllosphere or carposphere's structure, it was found that *Sphingomonas*, *Pseudomonas*, *Bacillus*, *Frigoribacterium*, *Enterobacter*, *Citrobacter*, *Pantoea*, *Methylobacterium*, *Erwinia*, *Acinetobacter*, and *Curtobacterium* were the predominating genera (Zarraonaindia et al. 2015; Kecskeméti et al. 2016). Meanwhile, the most dominant genera that appeared, upon analyzing grape berries endophytes, were *Ralstonia*, *Pseudomonas*, *Burkholderia*, *Mesorhizobium*, *Staphylococcus*, *Propionibacterium*, *Bacillus*, and *Dyella* (Campisano et al. 2014). According to the study done by Wallace et al. (2018), *Methylobacteria* and *Sphingomonads* were the predominant genera on maize leaf microbiome. It was concluded that the environmental elements affect the phyllosphere microbiome composition. Steven et al. (2018) researched flowers of apples which elucidated that *Pseudomonas* and Enterobacteriaceae were of high dominance. Furthermore, *Pseudomonas* is mostly found in apples, grapefruit, pumpkin, almonds, and tobacco flowers (Aleklett et al. 2014). While recently studying seed microorganisms, Proteobacteria, Firmicutes, Actinobacteria, and Bacteroidetes were observed as the dominant taxa (Liu et al. 2012; Barret et al. 2015; Rodríguez et al. 2018). There is evidence of a relationship occurring between seed and soil microorganisms, as well as between them and flower and fruit microorganisms (Compant et al. 2010; Glassner et al. 2018). Epigeal diversity of microorganisms emanates from seeds, soil, and air, and then their dwelling whether on or inside plant tissues follows. The microorganisms to reside in the tissues of plants are then determined according to different criteria including agricultural, environmental, and soil management techniques. The degree of association between a plant's epigeal microbial composition and its host is particular to each compartment where diversity is found; further research is needed to understand this association. These endospheric and epigeal microorganisms have the evident potential to promote the growth of plants, alleviate stresses, and improve resistance to diseases (Hardoim et al. 2015; Vishwakarma et al. 2020a).

### 6.2.2 *Hypogeal Plant Microbiota*

The organic matter found in the soil can affect the physical, chemical, and biological characteristics of the soil, rendering it an active part of a thriving agricultural system. Decomposers like fungi, bacteria, and insects are responsible for delivering nutrients to the soil. This is done by producing enzymes and organic acids that disintegrate complex compounds such as cellulose and lignin into simpler utilizable forms for the nutrition of plants and other microorganisms (Ray et al. 2020).

The hypogeal root microbiota can be conveyed horizontally, where they develop in soil with abundant different bacterial taxa comprising Bacteroidetes, Acidobacteria, Proteobacteria, Actinobacteria, and Planctomycetes (Fierer 2017). On the other hand, vertical conveyance occurs through seeds, acting as an important source of microorganisms from plant roots, leading to its development (Hardoim et al. 2012). Plant roots create unique and fascinating soil microbial niches that permit their colonization in the rhizosphere, the root, and, up to a certain point, the epigeal regions (Hartmann et al. 2009). In a study conducted by Donn et al. (2015) in a wheat cropping system, they concluded that more abundance of microorganisms occurred in the rhizosphere compared with the bulk soil.

Exudation of roots is secreting compounds into the rhizosphere region, essential for roots. Such compounds include sugars, organic acids, amino acids, flavonoids, polyphenols, nutrients, and hormones, which act as sources of nutrition for microbes surrounding the root area (Mendes et al. 2013; Compant et al. 2019). This is called the rhizosphere effect. It is established that microorganisms growing in the rhizosphere niche, by utilizing root exudates, help engage other microorganisms through a cross-feeding procedure. Accordingly, they create a novel niche for other microorganisms (Jacoby and Kopriva 2019). The niche is very specific for the species of plants and the secreted compounds. For instance, secondary metabolites aiding in plant defense, such as benzoxazinoids, which are secreted by the roots of maize, alter the microbiome constituents, especially affecting Proteobacteria and Actinobacteria (Hu et al. 2018). Furthermore, the dynamics of the bacterial community lying in the vicinity of the roots of *Avena barbata* were studied by Zhalnina et al. (2018). They observed that substrate selectivity, along with root exudate amalgamation, changed the structure of bacterial taxa in the rhizosphere. Meanwhile, Fitzpatrick et al. (2018) detected different *Pseudoxanthomonas* rhizobacterial species with differential occurrence patterns among thirty angiosperms. Besides, the colossal diversity and niche stipulations of rhizospheric microorganisms are controlled by alterations in physical and chemical circumstances as well as the rhizospheric spatiotemporal consortium (Vetterlein et al. 2020).

Endospheric (endophytic) microorganisms colonize roots internally. Factors like the capacity of endospheric microbes to colonize plants or the distribution of plant resources control the distribution of endophytes. The fungus *Piriformospora indica* is a main symbiotic root endophyte that has been significantly utilized in the field of agriculture. It enhances the uptake of phosphorus and helps protect crops from different stresses (Lahrmann et al. 2013). A report by Trivedi et al. (2013) showed that cyclophilin A–like protein obtained from this fungus was overexpressed in order to protect tobacco plants from salt stress. Another study by Bhuyan et al. (2015) indicated that *Azotobacter chrococum* was capable of promoting the physiological modulation of *P. indica* besides helping in upgrading its ability of nutrient acquisition during their synergistic relationship. Many endospheric fungi have been shown to display chemotaxis for chemicals that exude from roots. As an example, the activity of a non-pathogenic isolate of *Fusarium oxysporum* was studied against *Meloidogyne incognita*, a root-knot nematode. It was observed that exudates from tomato plants aided in *F. oxysporum* colonization while diminishing the nematode incidence. This suggests that exudates from plant roots show a preference toward some microorganisms in their vicinity. However, chemotaxis mediated by root exudates can also be responsible for attracting pathogenic microorganisms (Sikora and Dababat 2007). Gu et al. (2017) used fine biochar to suppress tomato bacterial wilt disease. The mechanism that took place was that biochar absorbed the root exudates, which display a significant chemotactic signal directed to the bacterium *Ralstonia solanacearum*. In consequence, bacterial swarming motility and activity were halted. In another study, *R. solanacearum* again followed chemotaxis for the secondary metabolites hydroxylated aromatic acids that are secreted by the roots of tomato (Hasegawa et al. 2019). The endoparasite *Pratylenchus coffeae* is a nematode that mechanically injures root tissues to invade plants (Das and Das 1986). Gene expression of the cell wall-degrading enzyme from *P. coffeae* was studied in the presence and in the absence of the exudates from roots. It was found that the expression level changed according to the host-specific exudates given in the test (Bell et al. 2019). A protozoan parasitic organism, *Trypanosoma brucei*, normally directs its motion far from the existing microorganisms. Nevertheless, it was noticed that they chemotactically approached *Escherichia coli* colonies. This was attributable to the soluble diffusible attractant signal that relies on *E. coli* (DeMarco et al. 2020).

Mycorrhizae represent the mutual alliance occurring in soil between fungi and roots of plants, where both parts benefit from one another (Brundrett 2002). Such profits can comprise the interchange of carbon from photosynthetic plants, with nutrients becoming available due to the dwelling mycorrhizae. This applies to the true type of arbuscular mycorrhizae, which is the Arum-type. However, other mycorrhizal types can obtain carbon from soil organic compounds or can snitch it from the plant (Allen 1991). In this context, it was recorded that, contrary to deriving carbon solely from photosynthesis as in the case of *Arum maculatum*, the *Paris quadrifolia* green leaves were shown to strikingly hold 50% fungal carbon. It is thus established that the Paris-type mycorrhizae, which are about 100,000 species of plants, undergo partial mycoheterotrophy. This has far-reaching consequences for our knowledge of the dealings of carbon in the plant microbiome (Giesemann et al. 2019). The benefits that mycorrhizae obtain from these dealings are under discussion. However, the gains are assumed to be obtaining shelter from the surrounding environment, a more dependable source of carbon, and a more soil expanse (Sapp 2004).

Mycorrhizae carve the plant communities and have an impact on their cohabitant diversity as well. Conveying plant carbon, such as sugars and amino acids, is via the fungal mycelium of mycorrhizae. This is for the sustainability of the microbiome (Tarkka et al. 2018). The research concerned with soil microbial ecology elucidated that different interactions are moderated in the mycorrhizosphere soil by mycorrhizal fungi. The interactions comprise pathogenic organisms and mutualistic ones that are capable of fixing nitrogen from the atmosphere, producing vitamins, uptaking phosphorus, and offering protection from other antagonistic organisms (Buée et al. 2009; Tedersoo et al. 2020). The ectomycorrhizosphere constructs a particular association between plant trees and the soil, hosting a wide range of microbes that possibly participate in the solubilizing and weathering of minerals (Uroz et al. 2007). Besides being carbon-rich, this rhizosphere bears enormous root-connected microbial communities. The mycorrhizosphere enhances minerals' weathering with the aid of the excreted organic acids, phenolics, siderophores, and protons (Drever and Vance 1994; Illmer et al. 1995).

Mycelia of the mycorrhizae offer a way for transporting photosynthetic carbon into the soil, contributing to the presence of areas rich in nutrients that are available to other microorganisms, especially bacteria. This is shown through the higher abundance of bacteria, as tested by 16S rDNA, in the vicinity of zucchini root soil that was inoculated with the mycorrhizal fungi *Glomus mosseae* and *Acaulospora laevis* (Qin et al. 2014). Furthermore, the ability of mycorrhizae to enhance the presence of Streptomycetes, Firmicutes, Oxalobacteraceae, and Comamonadaceae, which inhabit the mycorrhizosphere, was reported (Offre et al. 2007; Nuccio et al. 2013). On the other hand, some research revealed that bacterial communities in the soil can exert inhibiting actions on mycorrhizae or vice versa. Svenningsen et al. (2018) recorded that soil containing a higher presence of Acidobacteria inhibited the performance of extra-radicle hyphae of mycorrhizae. Meanwhile, another study showed that the mycorrhizal fungi *Glomus mosseae* and *Acaulospora laevis* caused inhibition of the coexisting soil microbiota (Welc et al. 2010).

## 6.3 The Holobiont (Plant-Microbe Interactions)

### *6.3.1 From Basic Association to Holobiont Concept*

Plants are supported by microorganisms due to their ability to provide nutrients and protect them from stressful biotic and abiotic factors. Accordingly, plants are always correlated with particular microbial groups, associating to form what is known as the "holobiont" (Bordenstein and Theis 2015; Theis et al. 2016). The microorganisms constituting the plant microbiome greatly influence plant health, and plants, in turn, shape their microbiome through mechanisms like root exudates, creating variable plant traits that selectively harness beneficial microbes (Mohanram and Kumar 2019; Mendes et al. 2018; Kumar and Dubey 2020). This reciprocal dialogue means they do not merely coexist but form a single ecological and evolutionary unit (Vandenkoornhuyse et al. 2015), collectively influencing development, fitness, and resilience (Simon et al. 2019).

### *6.3.2 Defining the Holobiont and its Engineering Imperative*

Therefore, the holobiont concept fundamentally shifts the perspective from the plant as an autonomous organism to the plant as a dynamic, multi-species nexus (Zilber-Rosenberg and Rosenberg 2008). The plant governs its microbiome through immune signals and exudates, while the microbiome extends the plant's functional capacities—enhancing nutrition, modulating stress, and providing defense (Hassani et al. 2018). This integration shapes a collective "holobiont phenotype" that determines health and adaptability (Trivedi et al. 2020). Consequently, the goal of engineering for climate resilience is not merely to modify the plant *or* its microbes in isolation, but to engineer the *holobiont system* itself—optimizing the synergistic partnerships between host genetics and microbial functions (Toju et al. 2018). This perspective sets the stage for exploring how synthetic biology can be used to forge pre-adapted holobionts for a changing climate (Lawson et al. 2019).

## 6.4 Plants and Microbes Are Liaising Through Signals

Exudates of the plant roots, besides their role in securing the fitness and longevity of the plant, are also of great importance for microorganisms. Microorganisms utilize these exudates as sources of carbon as well as nutrients for their growth (Kawasaki et al. 2016; Bukhat et al. 2020; Korenblum et al. 2020).

The concentration of the molecules found inside or in the vicinity of the roots of plants is affected by how far they occur from the point of their release. These molecules could be tissues or cells that are disposed of from roots, $CO_2$ released from the respiration process, $H^+$ efflux, and proteins, besides mucilage (Xiong et al. 2020). Other low molecular weight molecules can be encountered, including sugars, phenolics, amides, amino acids, and aromatic compounds. These molecules promote associations through functioning as attractants or repellents to manage the root microbiome (Lei et al. 2023).

Plant root exudates can cause a preference for certain microbes to exist in the rhizosphere. In this regard, citric acid secreted from cucumber roots displayed attraction towards *Bacillus amyloliquefaciens*, while fumaric acid exudates from the roots of bananas drew *B. subtilis*, inducing the formation of biofilms (Zhang et al. 2014). Sometimes organic compounds can enhance the formation of root nodules. Such compounds include flavonoid derivatives of the compound 2-phenyl-1,4-benzopyrone, which induce the nod genes of bacteria that are responsible for lipochitooligosaccharides, which induce the formation of root nodules. These organic compounds are similar to quorum sensing occurring in bacteria, therefore affecting the metabolism of bacteria (Hassan and Mathesius 2012). On the other hand, other organic compounds aid in phytohormone synthesis, which are requisites for the metabolism of plant growth-promoting rhizobacteria; for instance, tryptophan is responsible for the biosynthesis of indole acetic acid (Haichar et al. 2014). Furthermore, aminocyclopropane-1-carboxylic acid (ACC) is another plant root exudate required for ethylene, a stress hormone compound, biosynthesis, and also required for carbon and nitrogen assimilation by bacteria. Evidence for this appears in expressing the acdS gene in the microorganisms' dwelling roots of plants, which is involved in assimilating root exudates (Haichar et al. 2012). Plant growth-promoting rhizobacteria (PGPRs) that produce ACC deaminase can aid in utilizing

ACC. This is to reduce ACC amounts outside plant roots in order to even the concentration with that inside plant roots (Glick et al. 1998).

Bacteria are crucial in the dynamics of the root microbiome (Wang et al. 2023). Bacterial cells communicate using signaling molecules recognizable by community constituents. When the signal is perceived, this could lead to enhancing or suppressing gene expression, causing changes in the metabolism of the organism subjected to the signal (Atkinson and Williams 2009; Phour et al. 2020). According to the type of relationships (pathogenic, promoting, symbiotic), they can lead to negative or positive effects on the agricultural system (Li et al. 2024). Meanwhile, fungi can extend their branching arm-like mycelia to connect with the roots of plants (Karimi-Jashni and Yazdanpanah 2023). In addition to water, nutrients, and sugars, fungal mycelia can also deliver chemical signals. Fungi can be a means of indirect association of a plant with another surrounding plant (Zhang et al. 2020). Flow of different chemical signals occurs through the fungal networks (Potgieter 2022). For instance, high concentrations of phosphorus in soil notify other plants that there is an association between plants and fungi. In turn, they may respond to this signal by two schemes: first, they may release sugars to tempt these fungi to gain part of the nutrients, or second, they produce chemical compounds to diminish fungal ability to form nutrients, thus making their plant competitors more vulnerable (Van der Putten 2010). On the other hand, archaea can communicate directly with plants. Communications could be through supplying nutrients and biosynthesizing auxins for the promotion of the growth of plants, besides securing plants against osmotic and oxidative abiotic stresses (Taffner et al. 2018; Odelade and Babalola 2019).

Microbial volatile organic molecules act as signaling molecules directed toward plants. They are characterized by the low molecular weight and low boiling point (Fincheira et al. 2021). They have the ability to modulate various physiological activities in the plant as a way to enhance the plant's growth and stress endurance. *Bacillus subtilis* GB03 released volatile compounds that caused an elevation in the total measured leaf area of *Arabidopsis thaliana*. Moreover, they supported the plant to tolerate salts through the downregulation of expressing HKT1 in the plant's roots, hence lowering the amount of sodium ions (Ryu et al. 2003). How the plant would respond to volatile compounds secreted by microorganisms was then studied. It was shown that the two compounds 2-heptanone and 2-ethyl-1-hexanol, and tetrahydrofuran-3-ol, 2-heptanone, secreted by a *Bacillus* species, had the ability to induce the growth of *A. thaliana* and tomato (Jiang et al. 2019). Furthermore, dimethyl disulfide, produced by *Pseudomonas stutzeri*, enhanced tomato growth (Rojas-Solis et al. 2018).

Phytohormones are of great importance to the plant since they regulate its growth and development as well as its ability to adapt to the surrounding stresses. Besides, they participate in hypogeal communications occurring among the plant roots, the microorganisms, and the soil. They can be classified according to whether their chemical structure or their mode of action. Groups of cytokinins, auxins, abscisic acid, salicylic acid, ethylene, gibberellins, jasmonic acid, brassinolide, and strigolactone can be encountered (Lu et al. 2021). Plant growth-promoting rhizobacteria were found to secrete indole acetic acid, which controls the growth of plant roots and their construction (Vacheron et al. 2013). Despite that, its influence is managed by the concentration of indole acetic acid found inside the plant (Backer et al. 2018). Signals of the microbial phytohormones can apply their roles in the hypogeal area before moving to the xylem through roots (Borghi et al. 2016). They can collaborate or oppose one another (Jing and Strader 2019).

## 6.5 Recent Microbiome Analysis Approaches

Various techniques are available for the characterization of a microbial community for depicting the whole microbiome or detailed interpretation of a single component (Vishwakarma et al. 2020b). Among the recent approaches are amplicon sequencing, metagenomics, metaproteomics, metabolomics, and metatranscriptomics (Fig. 6.1).

### *6.5.1 Amplicon Sequencing Approach*

In the amplicon sequencing approach, two universal primers (forward and reverse) specific for a definite major conserved region inside the genome of the microorganism of interest are utilized. Microbial ecologists apply amplicon sequencing to explore various communities of microorganisms. The product of the polymerase chain reaction (PCR) is further sequenced, which contains the hypervariable region specific for each taxon (D'Amore et al. 2016). In the case of bacteria, the 16S rRNA gene is the most utilized amplicon (Kittelmann et al. 2013), while for fungi, there are

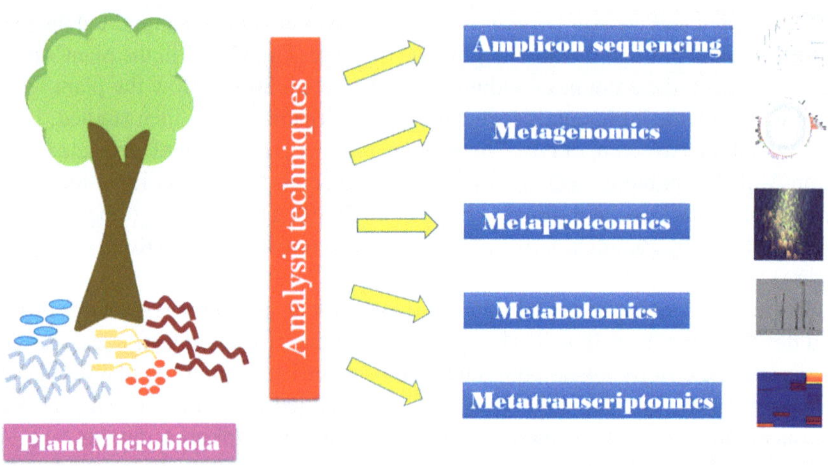


**Fig. 6.1** Recent approaches for analyzing plant microbiota. (Adapted from Arıkan and Muth (2023))

the 18S rRNA and the segments of the internal transcribed spacer (ITS). These loci constitute the data for microbial phylogeny. Despite that, it is worth mentioning that identification using this method is only accurate when reference databases are of high quality (Schoch et al. 2012).

### *6.5.2 Metagenomics Approach*

In the metagenomics approach, the shotgun method of the whole genome is performed to interpret the sequences of the complete DNA of a microorganism rather than targeted amplicon sequences. It is thus expected that the obtained data contain reads for different microorganisms as well as eukaryotes, including plants, animals, and humans. Accordingly, a variety of quality control methods are followed for trimming and filtering the collected data of metagenomics reads (Nayfach and Pollard 2016).

### *6.5.3 Metaproteomics Approach*

The metaproteomics approach studies all of the protein samples (resulting from translations of genes) obtained from an environmental source. The structure of the community of microorganisms can be explored by the quantification of the number of cells or biomass of each organism. Metaproteomics can be utilized to assess the structure of a microbial community by measuring the abundance of proteins to denote the amount of biomass of every single population (Kleiner et al. 2017). Furthermore, metaproteomics can boost our comprehension of the occurring interactions and the dynamic regulations of the studied microbial communities in response to environmental changes (Shrestha et al. 2021).

### *6.5.4 Metabolomics Approach*

Metabolomics is performing a comprehensive study by quantitative and qualitative measurement of all metabolites as well as molecules of low molecular weight in tested biological samples (Clish 2015). The advances in analytical chemistry, including mass spectrometric analyses using liquid and gas chromatography methods and nuclear magnetic resonance, paved the way for the metabolomics approach. Through this approach, the compounds in a plant or its rhizosphere are quantified, and there is elucidation of their chemical structure (Zhang et al. 2012).

### *6.5.5 Metatranscriptomics*

Metatranscriptomics gives information about the expressed genes of the entire microbial community. While utilizing the functional annotations of the genes that are expressed, there is a chance of obtaining deduced functional profiling of a microbiome under specific circumstances (Aguiar-Pulido et al. 2016). Here, the total RNA collected from an environmental sample is subjected to sequencing in order to show different occurring metabolic pathways in addition to the active microorganisms in the community (Urich et al. 2008). Both approaches of metabolomics and metatranscriptomics can be gathered for the interpretation of those genes in charge of producing signaling compounds in plant's rhizosphere (Yan et al. 2017).

## 6.6 Inoculation of Microorganisms to Partake in Boosting Crop Production

Employing microorganisms to interact with plants could be a means of sustainability. Plant probiotics, including fungi and bacteria, have been utilized to enhance microbial incidence in soil (Yadav et al. 2017). In agriculture, such recruited microbes are preferable to manufactured compost and chemical controls (Alori et al. 2017). Seeds amended with useful microorganisms were shown to have the capacity of colonizing roots, besides raising immunity against pathogenic organisms (Ahmad et al. 2018).

### *6.6.1 Diminishing of Plant Stress*

Stress factors, for example, salinity, drought, and heavy metals, often influence crops (Rho et al. 2018). Association of plants with microorganisms improves the plants' resilience to various stress conditions (Lugtenberg and Kamilov 2009). Utilization of such microorganisms could aid in higher crop production sustainability (Weyens et al. 2009). Some examples of utilizing microorganisms to aid plants against some stresses are explored in Table 6.1.

The physiological as well as the molecular aspects of halophytic plants, besides their associated microorganisms, have been points of research. In a study, the authors isolated forty-one bacterial isolates from the rhizospheric area of the halophytic plants *Allenrolfea occidentalis*, *Salicornia rubra,* and *Sarcocornia utahensi*. During their survey, the authors encountered two species belonging to the genera *Halomonas* and *Bacillus*. These species were inoculated into the seedlings of the alfalfa plant. They caused plant growth stimulation while 1% NaCl was present, suggesting their potential to enhance the growth of alfalfa if grown in saline soils

**Table 6.1** Some examples for role of microorganisms in diminishing plant stress

| Stress | Plant | Microorganism | Activity | Reference |
|---|---|---|---|---|
| Salinity | Alfalfa | *Halomonas* and *Bacillus* | Enhancement of plant growth | Kearl et al. (2019) |
| | *Vigna radiata* | *Aspergillus terreus* CR7 | Increase in plant biomass, oxidative stress parameters, chlorophyll, IAA phenolic compounds, relative water, and protein. | Chauhan et al. (2024) |
| Drought | Wheat (*Triticum aestivum* L.) | Species of *Bacillus, Enterobacter, Moraxella* and *Pseudomonas* | Production of auxin | Raheem et al. (2018) |
| | Wheat (*Triticum aestivum*) and maize (*Zea mays*) | *Bacillus* sp. (12D6) and *Enterobacter* sp. (16i) | Increase root proliferation. Production of phytohormones | Jochum et al. (2019) |
| | Wheat | *Bacillus subtilis* SIR1 | Enhanced plant biomass, antioxidant enzymes, and decrease in ROSs | Sood et al. (2020a) |
| | Maize | *Bacillus subtilis* DHK | Increase in plant biomass, antioxidant enzymes, nutrients, amino acids, and chlorophyll, and decrease in ROSs | Sood et al. (2020b) |
| | Wheat | *Azospirillum brasilense, Azospirillum lipoferum Bacillus* sp., and $SiO_2$ NPs | Improvements in plant growth, chlorophyll, gas exchange, relative water content, nutrients, osmolytes, and antioxidant enzymes | Akhtar et al. (2021) |
| | Tomato | *Trichoderma asperelloides*-NT33 | Higher shoot weight | Rawal et al. (2022) |
| | Soybean | *Rhizophagus irregularis* | Enhancement in growth, root proliferation, chlorophyll, photosynthesis, nitrogen, nitrate reductase, osmolytes, antioxidant systems, and MSI and decrease in ROSs | Begum et al. (2023) |
| Heavy metal | Wheat | *Pseudomonas chlororaphis* O6 | Better root length, less reaction against CuO nanoparticles | Wright et al. (2016) |
| | Okra | *Aspergillus violaceofuscus* Ch06 | Increase in plant biomass, chlorophyll, proteins, sugars, flavonoids, phenolics, proline, IAA, antioxidant enzymes, less translocation of Cd and Cr | Aziz et al. (2022) |

(Kearl et al. 2019). The endophytic fungal isolate *Aspergillus terreus* CR7 was obtained from *Catharanthus roseus*. The fungus exhibited plant growth promotion attributes as manifested in the production of IAA, solubilization of phosphate, and activity of ACC deaminase, besides its potential to develop at a 15% concentration of NaCl. When *Aspergillus terreus* CR7 was inoculated to salt-stressed *Vigna radiata*, the plant displayed improvements in biomass, lengths of both root and shoot, antiradical activity, oxidative stress parameters, and contents of chlorophyll, phenolic compounds, relative water, and protein. Besides, a notable increase in IAA level and protection against DNA damage occurred (Chauhan et al. 2024).

The root colonizing bacteria *Bacillus*, *Pseudomonas*, *Moraxella* and *Enterobacter* were recovered from *Acacia arabica*. They were utilized to alleviate the drought stress in wheat. These bacteria were producers of the auxins indole-3-acetic acid (IAA), indole-3-lactic acid (ILA), and indole-3-carboxylic acid (ICA). Inoculation of these bacteria to wheat plants improved its tillers and the number of spikelets. This indicated the significance of using bacteria that withstand drought to increase crop production in dry regions (Raheem et al. 2018). The plant growth-promoting rhizobacterial species *Enterobacter* sp. (16i) and *Bacillus* sp. (12D6) were inoculated into the rhizospheric areas of wheat and maize seedlings. The bacteria were capable of retarding the symptoms of drought. They enhanced root growth and proliferation. Furthermore, they produced the phytohormones salicylic acid (SA) and indole-3-acetic acid (IAA) (Jochum et al. 2019). Ten plant growth-promoting bacterial species were obtained from wheat roots and their vicinity. A *Bacillus subtilis* bacterial strain, namely SIR1, was inoculated into wheat at water stress levels. The results included induction in plant root and shoot yields, enhanced antioxidant enzyme activities, and a reduction in reactive oxygen species (ROSs), when compared to the uninoculated wheat (Sood et al. 2020a). The same previous authors conducted a similar survey on maize plants. They isolated a *Bacillus subtilis* strain DHK, which exhibited higher traits of plant growth promotion. The bacterium, when inoculated into maize plants under water stress levels, caused an elevation in maize root and shoot biomasses and antioxidant enzyme activities, accompanied by a decrease in ROSs. A notable increase in nutrients, amino acids, and chlorophyll levels was also achieved (Sood et al. 2020b). Applying a combination of plant growth-promoting rhizobacteria (*Azospirillum brasilense*, *Azospirillum lipoferum*, and *Bacillus* sp.) and $SiO_2$ nanoparticles (NPs) to wheat plants generated drought tolerance. This was through the improvement of growth and yield of plants, chlorophyll content, gas exchange, relative water content, uptake of nutrients, as well as production of osmolytes. Meanwhile, the antioxidant enzymes were up-regulated (Akhtar et al. 2021). The symbiotic fungus *Trichoderma* could enhance tolerance to abiotic stresses in plants. Rawal et al. (2022) found that some tested isolates of the genus *Trichoderma* were able to relieve the drought stress in tomato plants, especially as shown in the higher shoot weight for plants inoculated with *Trichoderma asperelloides* NT33. Furthermore, it was found that the inoculation of transgenic and wild types of soybean plants by the arbuscular mycorrhizal fungus (AMF) *Rhizophagus irregularis* mitigated the destructive effects initiated by drought on plants. For both soybean types, the results included enhancement in growth, root proliferation, chlorophyll content, photosynthesis, nitrogen, nitrate reductase activity and osmolytes, and antioxidant systems. Furthermore, the membrane stability index (MSI) increased along with a decrease in ROS accumulation (Begum et al. 2023).

While using fertilizers containing CuO nanoparticles, plants can suffer stress. It was found that wheat root colonization by *Pseudomonas chlororaphis* O6 exhibited a protective effect for wheat plants from the reduction in root length caused by CuO nanoparticles. Meanwhile, there was a reduction in expression levels of plant genes associated with the reaction of plants towards CuO nanoparticles (Wright et al. 2016). *Aspergillus violaceofuscus* Ch06 is an endophytic fungus recovered from the chili plant. It was able to produce higher amounts of IAA, proteins, sugar, flavonoids, and phenolic compounds in media containing cadmium (Cd) or chromium (Cr). When okra plants were colonized with this fungus and exposed to Cd or Cr, the plant biomass increased. Moreover, *A. violaceofuscus* Ch06 improved the plant's chlorophyll content, proteins, sugars, flavonoids, phenolic compounds, proline, IAA, and antioxidant enzymes. Meanwhile, the translocation of Cd and Cr was reduced to upper plant parts, thus diminishing their toxicity (Aziz et al. 2022).

### 6.6.2 Nutrient Uptake

The majority of the chief micro- and macronutrients required for the growth of plants are present as insoluble matter in soil. Plants can apply different techniques for obtaining soil nutrients. Microorganisms found in the root vicinity escalate acquiring micronutrients through acidification or producing enzymes, protons, or siderophores, thus they are solubilized or mineralized, rendering them available (Jog et al. 2014; Otieno et al. 2015; Li et al. 2016). Examples of microorganisms recruited to enhance nutrient uptake by plants are listed in Table 6.2.

The plant growth-promoting rhizobacterial (PGPR) *Pseudomonas* strains, *P. fluorescens* BHUJY29, *P. putida* BHUJY13, *P. putida* BHUJY23, *P. aeruginosa* BHUJY16, and *P. aeruginosa* BHUJY20, were termed a combination of *Pseudomonas* species culture (CPC). This CPC, *Azotobacter chroococcum,* and *Azospirillum brasilense,* along with phosphorus fertilizer doses, were used to inoculate the soil of the rice plant. The results were promising, as evidenced by the higher yield and growth of rice, as well as the nutrient content tested in rice straws and grains. It was attributed to the more available NPK (nitrogen, phosphorus, potassium) fertilizer in soil due to the action of the PGPRs (Lavakush et al. 2014).

The mycorrhizal fungus *Rhizophagus irregularis* was inoculated into the soil of *Medicago truncatula* plant. It was established that this fungus increased plant biomass and offered tolerance to the plant against deficiency of zinc (Zn) and phosphorus (P) in soil. This was ascribed to inducing the Zn transporter (ZIP) gene, MtZIP5, and the phosphate transporter PT gene, MtPT4, in the plant roots. Besides, it was found that protection against high concentrations of zinc also occurred (Nguyen et al. 2019).

**Table 6.2** Some examples of the role of microorganisms in enhancing nutrient uptake

| Plant | Microorganism | Activity | Reference |
|---|---|---|---|
| Rice | *Pseudomonas fluorescens* BHUJY29, *P. putida* BHUJY13, *P. putida* BHUJY23, *P. aeruginosa* BHUJY16, and *P. aeruginosa* BHUJY20, *Azotobacter chroococcum, Azospirillum brasilense* | Enhancing availability of N, P, and K | Lavakush et al. (2014) |
| *Medicago truncatula* | *Rhizophagus irregularis* | Increase in uptake of P and Zn | Nguyen et al. (2019) |
| Tomato (*Solanum lycopersicum* L.) | *P. gessardi* Tmt-16 *Pseudomonas koreensis* Ls-C21, *P. brassicacearum* Ltj-62 and *Acinetobacter calcoaceticus* LsC-58 | Tmt-16 fixed nitrogen. Ls-C21, Ltj-62 and LsC-58 solubilized potassium | Reyes-Castillo et al. (2019) |
| Tomato | *Bacillus amyloliquefaciens* and *Paenibacillus polymyxa* | Enhanced N transformation | Wang et al. (2021) |
| Pistachio (*Pistacia* spp.) | *Funneliformis mosseae* | Higher nutrients (N, P, and ions of Ca and Mg) uptake | Fattahi et al. (2021) |
| Wheat | *Arthrobacter, Bacillus, Burkholderia, Pseudomonas, Enterobacter, Mycobacterium, Rhizobia, Pantoea*, and other bacteria | Phosphate solubilization | Adnan et al. (2022) |
| Sugarcane (*Saccharum officinarum* L.) and cassava (*Manihot esculenta* Crantz) | *Aspergillus pseudotubingensis* SDBR-CMUO2, *A. chiangmaiensis* SDBR-CMUI4, and *A. pseudopiperis* SDBR-CMUI1 | Phosphate solubilization | Khuna et al. (2023) |
| Maize | *Azotoacter salinestris* and *Azospirillum oryzea* | Increase in N uptake | Shalaby et al. (2023) |
| Tomato | Encapsulated *Pseudomonas fluorescens* | More availability of N, P, and K | Isabel et al. (2024) |

*Pseudomonas koreensis, P. gessardi, P. marginalis, P. brassicacearum, Rahnella aquatica,* and *Acinetobacter calcoaceticus* are the plant growth-promoting rhizobacteria that were utilized to inoculate the soil of tomato (*Solanum lycopersicum* L.). The results showed that the nitrogen-fixing bacterium *P. gessardi* Tmt-16 enhanced root growth, kept the maximum content of nitrogen in tissues of plants, and increased the availability of nitrogen in soil. On the other hand, the potassium-solubilizing bacterial strains *Pseudomonas koreensis* Ls-C21,

*P. brassicacearum* Ltj-62, and *Acinetobacter calcoaceticus* LsC-58 were able to solubilize potassium (Reyes-Castillo et al. 2019).

A consortium of strains of PGPR, with *Bacillus amyloliquefaciens* and *Paenibacillus polymyxa* as the effectual organisms, were applied with biochar to the soil of the tomato plant. Positive results were obtained for soil microbial diversity, where the occurrence of *Bradyrhizobium* and *Nitrospira* increased in the soil. Meanwhile, soil enzyme activities and nitrogen transformation were enhanced, leading to an increase in tomato yield (Wang et al. 2021).

The nutritional uptake as well as the oxidative state of pistachio (*Pistacia* spp.) rootstocks were studied when the arbuscular mycorrhizal fungus (AMF) *Funneliformis mosseae* was used to inoculate the soil facing salinity and drought stresses. Higher nutrients (nitrogen, phosphorus, and ions of calcium and magnesium) uptake, chlorophyll concentrations and plant biomass were shown. There was also a decrease in oxidant contents. Accordingly, the tolerance to the studied stress increased (Fattahi et al. 2021).

Phosphorus (P) deficiency could be overcome by using microorganisms that can solubilize phosphate. P availability is enhanced through mineralization, desorption, and dissolution. A composite of the phosphate-solubilizing bacteria (PSB) was utilized as an inoculum for the calcareous soil of wheat plants. The inoculum was mainly composed of *Arthrobacter*, *Bacillus*, *Burkholderia*, *Pseudomonas*, *Enterobacter*, *Mycobacterium*, *Rhizobia*, *Pantoea*, and other unidentified bacteria. Phosphate solubilization occurred, along with the production of auxins, IAA, organic acids, and siderophores (Adnan et al. 2022). Phosphate-solubilizing fungi can contribute to the improvement of plant growth through enhancing the uptake of nutrients and the secretion of growth-promoting substances. Three fungal strains were studied for their ability to enhance plant growth. They were *Aspergillus pseudotubingensis* SDBR-CMUO2, *A. chiangmaiensis* SDBR-CMUI4, and *A. pseudopiperis* SDBR-CMUI1. They were capable of solubilizing the mineral salts of phosphate, $FePO_4$, and $Ca_3(PO_4)_2$. This happens through the production of organic acids (succinic, tartaric, and oxalic) as well as phosphatase enzymes. The tested fungi were inoculated to sugarcane (*Saccharum officinarum* L.) and cassava (*Manihot esculenta* Crantz) soils amended with $Ca_3(PO_4)_2$. The plants displayed escalated chlorophyll concentration, root and shoot weights, root length, and phosphate content (Khuna et al. 2023).

The outcome of utilizing the dual mixture of the nitrogen-fixing bacteria *Azotoacter salinestris* and *Azospirillum oryzea* was investigated along with mineral nitrogen on maize plants. The pronounced observations for the maize plant were enhancement of shoot and leaf growth parameters, chlorophyll and proline contents, antioxidant enzymes (polyphenol oxidase, PPO, and peroxidase, PO), and nitrogen and protein contents of grains (Shalaby et al. 2023).

The encapsulation technique, using a carrier material, offers protection to utilizable microbes that render nitrogen, phosphorus, and potassium available from soil. *Pseudomonas fluorescens* was encapsulated by chitosan, and then studies were performed in the soil of the tomato plant. Using this microbial fertilizer resulted in improvements in nutrient uptake, plant growth parameters, and resistance against diseases (Isabel et al. 2024).

### 6.6.3 Disease Subdue

Pathogens as well as insects attack different plants, causing a retard in their health, growth, and yield. Nevertheless, root-associated microorganisms were reported to secrete several bioactive substances capable of protecting plants and inducing their resistance against attackers (Singh et al. 2020). Some examples of microorganisms that were inoculated into plant soil for diminishing plant diseases are shown in Table 6.3.

**Table 6.3** Some examples of the role of microorganisms in disease subdue

| Microbial inoculant | Plant | Threatening pathogen | Activity | Reference |
|---|---|---|---|---|
| *Glomus mosseae* | Tomato (*Solanum lycopersicum* cv. Marmande) | *Meloidogyne incognita* | Suppression of *M. incognita* early phases | Vos et al. (2012) |
| *Rhizophagus intraradices* | Tomato | *Fusarium oxysporum* f. sp. *lycopersici* | Decrease in root rot index | Fierro-Coronado et al. (2013) |
| *Flavobacterium* | Tomato | *Ralstonia solanacearum* | Disease suppression | Kwak et al. (2018) |
| *Rhizophagus irregularis* and *Funneliformis mosseae* | Apple (*Malus pumila*) | *Neonectria ditissima* | Enhancement of plant's resistance | Berdeni et al. (2018) |
| *Funneliformis mosseae* and *Rhizophagus irregularis* | Rice cultivars | *Magnaporthe oryzae* | Induction of resistance in plant | Campo et al. (2020) |
| *Bacillus velezensis* JC-K3 and *Bacillus subtilis* HG-15 | Wheat | *Fusarium*, *Gibberella*, *Alternaria*, and *Cladosporium* | Reduction in diseased plants number, and disease incidence | Ji et al. (2022) |
| *Rhizophagus irregularis* and *Funneliformis mosseae* | Potato *Solanum tuberosum* L. | Virus Y (PVY) | Decrease in viral accumulation and oxidative damage caused by virus | Deja-Sikora et al. (2023) |
| *Rhizophagus irregularis* DAOM197198 | Rice | *Xanthomonas oryzae* pv *oryzae* | Mitigation of disease symptoms | Guigard et al. (2023) |

*Meloidogyne incognita* is a root-knot nematode that threatens agriculture. Colonization of the AMF *Glomus mosseae* was studied in roots of tomato (*Solanum lycopersicum* cv. Marmande). It was found that *G. mosseae* caused suppression of *M. incognita* during their early infecting stages of penetrating plant roots and the next developmental stage (Vos et al. 2012). *Rhizophagus intraradices* is an arbuscular mycorrhizal fungus. Its bio-protective action was investigated against *Fusarium* wilt disease of tomato caused by *Fusarium oxysporum* f. sp. *lycopersici*. It was found that the root rot index was lower in *F. oxysporum*-infested tomato plants colonized with *R. intraradices* in comparison with non-colonized ones. It was suggested that *R. intraradices* could be considered a protectant for tomato plants. The protecting mechanism may be in part arbitrated to enhance phosphate nutrition (Fierro-Coronado et al. 2013).

It was found that the variety Hawaii 7996 of tomato plant was resistant to the pathogen *Ralstonia solanacearum*, while the variety Moneymaker was a susceptible one. A *Flavobacterium*, named TRM1, was recovered from the rhizosphere of the resistant tomato variety. It was cultivated into a pot experiment and exhibited suppression of disease caused by *Ralstonia solanacearum* (Kwak et al. 2018).

The role of AMF was assessed in improving apple (*Malus pumila*) growth, nutrition, phenology, flowers, and fruit proliferation, in addition to disease resistance. The AMF *Rhizophagus irregularis* and *Funneliformis mosseae* were tested. They were able to cause significant enhancement of the plant's resistance towards the fungal pathogen *Neonectria ditissima*, which is a worldwide economically major apple pathogen. However, the AMF showed no effects on leaf growth, nutrient content, phenology, and flowers and fruits proliferation (Berdeni et al. 2018).

Inoculating the AMF *Funneliformis mosseae* and *Rhizophagus irregularis* was conducted on *japonica* rice cultivars. Symbiosis that occurred between the rice cultivars and the AMF granted protection against rice blast infection by the fungus *Magnaporthe oryzae*. According to rice genotype, the AMF induced resistance to different degrees in the tested rice cultivars. Meanwhile, there was an increase in grain yield in some of the tested cultivars (Campo et al. 2020).

A consortium of PGPR, comprising *Bacillus velezensis* JC-K3 and *Bacillus subtilis* HG-15, was introduced to the soil of wheat plants under low salt stress. The promising results included improvements in crop growth, photosynthesis, grain yield, and chemical properties of soil. Moreover, a reduction in the incidence of wheat disease was observed. This was shown by Illumina MiSeq sequencing, which proved the inhibition of the pathogenic organisms *Fusarium*, *Gibberella*, *Alternaria*, and *Cladosporium*. Concurrently, there was an enhancement in the arbuscular mycorrhizal community (Ji et al. 2022).

The effect of the AMF *Rhizophagus irregularis* and *Funneliformis mosseae* was studied on healthy potatoes as well as potatoes infected with virus Y (PVY). Colonization of potato roots varied, where *Rhizophagus irregularis* was 38%, while *Funneliformis mosseae* was 20%. *Rhizophagus irregularis* exhibited higher beneficial effects concerning the weight of the tuber and disease occurrence. The effects also comprised lowering of hydrogen peroxide concentration in infected plant leaves, besides modulating glutathione and ascorbate antioxidants. Both tested AMF lowered the peroxidation of lipids and oxidative damage caused by PVY. Moreover, they both decreased viral accumulation in the roots of potatoes (Deja-Sikora et al. 2023). Arbuscular mycorrhizal fungal (AMF) species can alleviate disease symptoms and load of the pathogenic microorganism via what is known as mycorrhizal-induced resistance (MIR). The symbiotic relationship was studied for the AMF, *Rhizophagus irregularis* DAOM197198 (RIR), *R. intraradices* FR121 (RIN), and *Funneliformis mosseae* FR140 (FM), with different rice cultivars. *R. irregularis* DAOM197198 proved to be the most efficient symbiont with rice. This was shown through higher colonization of rice roots and an increase in plant growth. In addition, there was mitigation of disease symptoms caused by the phytopathogen *Xanthomonas oryzae* pv *oryzae*, as elucidated by the RT-qPCR transcriptome analyses of rice leaf marker genes responsible for physiological responses to infection (Guigard et al. 2023).

## 6.7 Conclusion

The escalating global population presents a formidable challenge to food security, necessitating a paradigm shift towards sustainable agricultural intensification. In this context, plant growth-promoting microorganisms (PGPMs)—encompassing bacteria and fungi—emerge as a pivotal and achievable technology. Their application as biofertilizers and biopesticides offers a dual benefit: enhancing crop productivity through nutrient solubilization, phytohormone production, and stress tolerance, while simultaneously suppressing pathogens via competition and induced systemic resistance, thereby reducing reliance on synthetic agrochemicals.

However, the transition from promise to widespread, reliable practice requires deeper scientific understanding. The mechanisms governing the assembly, composition, and function of the plant microbiome (the phytomicrobiome) in different agricultural contexts remain a complex frontier. Plant growth and health outcomes are not guaranteed by simple inoculation; they are contingent upon a dynamic interplay of specific host genotypes, indigenous microbial communities, soil physicochemical properties, and environmental conditions. This context-dependency explains variable field results and underscores the need to move beyond single-strain solutions. A more robust and resilient approach lies in engineering synthetic microbial consortia. By strategically combining complementary PGPMs with synergistic functions, we can create multi-functional biologics that are more adaptable and effective across diverse agricultural settings than single-strain inoculants.

**Declaration of AI Assistance** The authors didn't use generative artificial intelligence (AI) and/or AI-assisted technologies in any of the writing processes.

## References

Adnan M, Fahad S, Saleem MH et al (2022) Comparative efficacy of phosphorous supplements with phosphate solubilizing bacteria for optimizing wheat yield in calcareous soils. Sci Rep 12(1):11997. https://doi.org/10.1038/s41598-022-16035-3

Afridi MS, Javed MA, Ali S et al (2022) New opportunities in plant microbiome engineering for increasing agricultural sustainability under stressful conditions. Front Plant Sci 13:899464. https://doi.org/10.3389/fpls.2022.899464

Aguiar-Pulido V, Huang W, Suarez-Ulloa V et al (2016) Metagenomics, metatranscriptomics, and metabolomics approaches for microbiome analysis. Evol Bioinformatics Online 12(Suppl 1): 5–16. https://doi.org/10.4137/EBO.S36436

Ahmad M, Pataczek L, Hilger TH et al (2018) Perspectives of microbial inoculation for sustainable development and environmental management. Front Microbiol 9:2992. https://doi.org/10.3389/fmicb.2018.02992

Akhtar N, Ilyas N, Mashwani ZU et al (2021) Synergistic effects of plant growth promoting rhizobacteria and silicon dioxide nano-particles for amelioration of drought stress in wheat. Plant Physiol Biochem 166:160–176

Aleklett K, Hart M, Shade A (2014) The microbial ecology of flowers: an emerging frontier in phyllosphere research. Botany 92:253–266. https://doi.org/10.1139/cjb-2013-0166

Allen MF (1991) The ecology of mycorrhizae. Cambridge University Press, Cambridge, p 184

Alori ET, Dare MO, Babalola OO (2017) Microbial inoculants for soil quality and plant fitness. In: Lichtfouse E (ed) Sustainable Agriculture Review. Springer, Berlin, pp 181–308. https://doi.org/10.1007/978-3-319-48006-0

Antoine S, Hériché M, Boussageon R et al (2021) A historical perspective on mycorrhizal mutualism emphasizing arbuscular mycorrhizas and their emerging challenges. Mycorrhiza 31:637–653. https://doi.org/10.1007/s00572-021-01053-2

Arif I, Batool M, Schenk PM (2020) Plant microbiome engineering: expected benefits for improved crop growth and resilience. Trends Biotechnol 38:1385–1396. https://doi.org/10.1016/j.tibtech.2020.04.015

Arıkan M, Muth T (2023) integrated multi-omics analyses of microbial communities: a review of the current state and future directions. Mol Omics 2023(19):607–623. https://doi.org/10.1039/D3MO00089C

Atkinson S, Williams P (2009) Quorum sensing and social networking in the microbial world. J R Soc Interface 6:959–978. https://doi.org/10.1098/rsif.2009.0203

Aziz L, Hamayun M, Rauf M et al (2022) *Aspergillus violaceofuscus* alleviates cadmium and chromium stress in Okra through biochemical modulation. PLoS One 17(10):e0273908. https://doi.org/10.1371/journal.pone.0273908

Babalola OO (2010) Beneficial bacteria of agricultural importance. Biotechnol Lett 32:1559–1570. https://doi.org/10.1007/s10529-010-0347-0

Backer R, Rokem JS, Ilangumaran G et al (2018) Plant growth-promoting rhizobacteria: context, mechanisms of action, and roadmap to commercialization of biostimulants for sustainable agriculture. Front Plant Sci 9:1473. https://doi.org/10.3389/fpls.2018.01473

Barret M, Briand M, Bonneau S et al (2015) Emergence shapes the structure of the seed microbiota. Appl Environ Microbiol 81:1257–1266. https://doi.org/10.1128/aem.03722-14

Begum N, Xiao Y, Wang L et al (2023) Arbuscular mycorrhizal fungus *Rhizophagus irregularis* alleviates drought stress in soybean with overexpressing the GmSPL9d gene by promoting photosynthetic apparatus and regulating the antioxidant system. Microbiol Res 273:127398. https://doi.org/10.1016/j.micres.2023.127398

Bell CA, Lilley CJ, McCarthy J et al (2019) Plant-parasitic nematodes respond to root exudate signals with host-specific gene expression patterns. PLoS Pathog 15:e1007503. https://doi.org/10.1371/journal.ppat.1007503

Berdeni D, Cotton TEA, Daniell TJ et al (2018) The effects of arbuscular mycorrhizal fungal colonisation on nutrient status, growth, productivity, and canker resistance of apple (*Malus pumila*). Front Microbiol 3(9):1461. https://doi.org/10.3389/fmicb.2018.01461
Berg G, Grube M, Schloter M et al (2014) Unraveling the plant microbiome: looking back and future perspectives. Front Microbiol 5:1–7. https://doi.org/10.3389/fmicb.2014.00148
Bhuyan SK, Bandyopadhyay P, Kumar P et al (2015) Interaction of *Piriformospora indica* with *Azotobacter chroococcum*. Sci Rep 5:13911. https://doi.org/10.1038/srep13911
Bordenstein SR, Theis KR (2015) Host biology in light of the microbiome: ten principles of holobionts and hologenomes. PLoS Biol 13. https://doi.org/10.1371/journal.pbio.1002226
Borghi L, Liu G-W, Emonet A et al (2016) The importance of strigolactone transport regulation for symbiotic signaling and shoot branching. Planta 243(6):1351–1360. https://doi.org/10.1007/s00425-016-2503-9
Brundrett MC (2002) Coevolution of roots and mycorrhizas of land plants. New Phytol 154:275–304. https://doi.org/10.1046/j.1469-8137.2002.00397.x
Buée M, De Boer W, Martin F et al (2009) The rhizosphere zoo: an overview of plant-associated communities of microorganisms, including phages, bacteria, archaea, and fungi, and of some of their structuring factors. Plant Soil 321:189–212. https://doi.org/10.1007/s11104-009-9991-3
Bukhat S, Imran A, Javaid S et al (2020) Communication of plants with microbial world: exploring the regulatory networks for PGPR mediated defense signaling. Microbiol Res 238:126486. https://doi.org/10.1016/j.micres.2020.126486
Campisano A, Antonielli L, Pancher M et al (2014) Bacterial endophytic communities in the grapevine depend on pest management. PLoS One 9:e112763. https://doi.org/10.1371/journal.pone.0112763
Campo S, Martín-Cardoso H, Olivé M et al (2020) Effect of root colonization by arbuscular mycorrhizal fungi on growth, productivity and blast resistance in Rice. Rice (N Y) 13(1):42. https://doi.org/10.1186/s12284-020-00402-7
Chauhan P, Singh M, Sharma A et al (2024) Halotolerant and plant growth-promoting endophytic fungus *Aspergillus terreus* CR7 alleviates salt stress and exhibits genoprotective effect in *Vigna radiata*. Front Microbiol 9(15):1336533. https://doi.org/10.3389/fmicb.2024.1336533
Chen YH, Gols R, Benrey B (2015) Crop domestication and its impact on naturally selected trophic interactions. Annu Rev Entomol 60:35–58. https://doi.org/10.1146/annurev-ento-010814-020601
Choudhary DK, Johri BN (2009) Interactions of *Bacillus* spp. and plants–with special reference to induced systemic resistance (ISR). Microbiol Res 164:493–513. https://doi.org/10.1016/j.micres.2008.08.007
Choudhary DK, Prakash A, Johri BN (2007) Induced systemic resistance (ISR) in plants: mechanism of action. Indian J Microbiol 47:289–297. https://doi.org/10.1007/s12088-007-0054-2
Clish CB (2015) Metabolomics: an emerging but powerful tool for precision medicine. Cold Spring Harb Mol Case Stud 1(1):a000588. https://doi.org/10.1101/mcs.a000588
Compant S, Clément C, Sessitsch A (2010) Plant growth-promoting bacteria in the rhizo- and endosphere of plants: their role, colonization, mechanisms involved and prospects for utilization. Soil Biol Biochem 42:669–678. https://doi.org/10.1016/j.soilbio.2009.11.024
Compant S, Mitter B, Colli-Mull JG et al (2011) Endophytes of grapevine flowers, berries, and seeds: identification of cultivable bacteria, comparison with other plant parts, and visualization of niches of colonization. Microb Ecol 62:188–197. https://doi.org/10.1007/s00248-011-9883-y
Compant S, Samad A, Faist H et al (2019) A review on the plant microbiome: ecology, functions and emerging trends in microbial application. J Adv Res 20:29–37. https://doi.org/10.1016/j.jare.2019.03.004
Compant S, Cambon MC, Vacher C et al (2021) The plant endosphere world - bacterial life within plants. Environ Microbiol 23(4):1812–1829. https://doi.org/10.1111/1462-2920.15240
D'Amore R, Ijaz UZ, Schirmer M et al (2016) A comprehensive benchmarking study of protocols and sequencing platforms for 16S rRNA community profiling. BMC Genomics 17:55. https://doi.org/10.1186/s12864-015-2194-9

Das S, Das SN (1986) Host-range of *Pratylenchus coffeae*. Indian J Nematol 16:180–184

Deja-Sikora E, Werner K, Hrynkiewicz K (2023) AMF species do matter: *Rhizophagus irregularis* and *Funneliformis mosseae* affect healthy and PVY-infected *Solanum tuberosum* L. in a different way. Front Microbiol 17(14):1127278. https://doi.org/10.3389/fmicb.2023.1127278

DeMarco SF, Saada EA, Lopez MA et al (2020) Identification of positive chemotaxis in the protozoan pathogen *Trypanosoma brucei*. mSphere 5(4):e00685-20. https://doi.org/10.1128/mSphere.00685-20

DESA (2019) World population prospects 2019: highlights | Multimedia Library - United Nations Department of Economic and Social Affairs. DESA, New York

Donn S, Kirkegaard JA, Perera G et al (2015) Evolution of bacterial communities in the wheat crop rhizosphere. Environ Microbiol 17:610–621. https://doi.org/10.1111/1462-2920.12452

Drever JI, Vance GF (1994) Role of soil organic acids in mineral weathering processes. In: Pittman ED, Lewan MD (eds) Organic acids in geological processes. Springer, Berlin, pp 138–161. https://doi.org/10.1007/978-3-642-78356-2_6

Fattahi M, Mohammadkhani A, Shiran B et al (2021) Beneficial effect of mycorrhiza on nutritional uptake and oxidative balance in pistachio (*Pistacia* spp.) rootstocks submitted to drought and salinity stress. Sci Hortic 281:109937. https://doi.org/10.1016/j.scienta.2021.109937

Fierer N (2017) Embracing the unknown: disentangling the complexities of the soil microbiome. Nat Rev Microbiol 15:579–590. https://doi.org/10.1038/nrmicro.2017.87

Fierro-Coronado RA, Castro-Moreno MG, Ruelas-Ayala RD et al (2013) Induced protection by *Rhizophagus intraradices* against *Fusarium* wilt of tomato. Interciencia 38(1):48–53. https://www.redalyc.org/articulo.oa?id=33926506008

Fincheira P, Quiroz A, Tortella G et al (2021) Current advances in plant-microbe communication via volatile organic compounds as an innovative strategy to improve plant growth. Microbiol Res 247:126726. https://doi.org/10.1016/j.micres.2021.126726

Fitzpatrick CR, Copeland J, Wang PW et al (2018) Assembly and ecological function of the root microbiome across angiosperm plant species. Proc Natl Acad Sci USA 115:E1157–E1165. https://doi.org/10.1073/pnas.1717617115

Foong SY, Ma NL, Lam SS et al (2020) A recent global review of hazardous chlorpyrifos pesticide in fruit and vegetables: prevalence, remediation and actions needed. J Hazard Mater 400: 123006. https://doi.org/10.1016/j.jhazmat.2020.123006

Fu H, Tan P, Wang R et al (2022) Advances in organophosphorus pesticides pollution: current status and challenges in ecotoxicological, sustainable agriculture, and degradation strategies. J Hazard Mater 424:127494. https://doi.org/10.1016/j.jhazmat.2021.127494

Giesemann P, Rasmussen HN, Liebel HT et al (2019) Discreet heterotrophs: green plants that receive fungal carbon through Paris-type arbuscular mycorrhiza. New Phytol 226:960–966. https://doi.org/10.1111/nph.16367

Glassner H, Zchori-Fein E, Yaron S et al (2018) Bacterial niches inside seeds of *Cucumis melo* L. Plant Soil 422:101–113. https://doi.org/10.1007/s11104-017-3175-3

Glick BR, Penrose DM, Li J (1998) A model for the lowering of plant ethylene concentrations by plant growth-promoting bacteria. J Theor Biol 190:63–68. https://doi.org/10.1006/jtbi.1997.0532

Gosling P, Hodge A, Goodlass G et al (2006) Arbuscular mycorrhizal fungi and organic farming. Agric Ecosyst Environ 113:17–35. https://doi.org/10.1016/j.agee.2005.09.009

Gu Y, Hou Y, Huang D et al (2017) Application of biochar reduces *Ralstonia solanacearum* infection via effects on pathogen chemotaxis, swarming motility, and root exudate adsorption. Plant Soil 415:269–281. https://doi.org/10.1007/s11104-016-3159-8

Guigard L, Jobert L, Busset N et al (2023) Symbiotic compatibility between rice cultivars and arbuscular mycorrhizal fungi genotypes affects rice growth and mycorrhiza-induced resistance. Front Plant Sci 24(14):1278990. https://doi.org/10.3389/fpls.2023.1278990

Haichar FEZ, Roncato M-A, Achouak W (2012) Stable isotope probing of bacterial community structure and gene expression in the rhizosphere of *Arabidopsis thaliana*. FEMS Microbiol Ecol 81:291–302. https://doi.org/10.1111/j.1574-6941.2012.01345.x

Haichar FEZ, Santaella C, Heulin T et al (2014) Root exudates mediated interactions belowground. Soil Biol Biochem 77:69–80. https://doi.org/10.1016/j.soilbio.2014.06.017

Hardoim PR, Hardoim CC, van Overbeek LS et al (2012) Dynamics of seedborne rice endophytes on early plant growth stages. PLoS One 7:e30438. https://doi.org/10.1371/journal.pone.0030438

Hardoim PR, van Overbeek LS, Berg G et al (2015) The hidden world within plants: ecological and evolutionary considerations for defining functioning of microbial endophytes. Microbiol Mol Biol Rev 79:293–320. https://doi.org/10.1128/mmbr.00050-14

Harman GE (2006) Overview of mechanisms and uses of *Trichoderma* spp. Phytopathology 96: 190–194. https://doi.org/10.1094/PHYTO-96-0190

Hartmann A, Schmid M, van Tuinen D et al (2009) Plant-driven selection of microbes. Plant Soil 321:235–275. https://doi.org/10.1007/s11104-008-9814-y

Hasegawa T, Kato Y, Okabe A et al (2019) Effect of secondary metabolites of tomato (*Solanum lycopersicum*) on chemotaxis of *Ralstonia solanacearum*, pathogen of bacterial wilt disease. J Agric Food Chem 67:1807–1813. https://doi.org/10.1021/acs.jafc.8b06245

Hassan S, Mathesius U (2012) The role of flavonoids in root rhizosphere signalling: opportunities and challenges for improving plant microbe interactions. J Exp Bot 63:3429–3444. https://doi.org/10.1093/jxb/err430

Hassani MA, Durán P, Hacquard S (2018) Microbial interactions within the plant holobiont. Microbiome 6(1):58. https://doi.org/10.1186/s40168-018-0445-0

Hetrick B, Wilson G, Cox T (1992) Mycorrhizal dependence of modern wheat varieties, landraces, and ancestors. Can J Bot 70:2032–2040. https://doi.org/10.1139/b92-253

Hu L, Robert CAM, Cadot S et al (2018) Root exudate metabolites drive plant-soil feedbacks on growth and defense by shaping the rhizosphere microbiota. Nat Commun 9:2738. https://doi.org/10.1038/s41467-018-05122-7

Illmer P, Barbato A, Schinner F (1995) Solubilization of hardly-soluble $AlPO_4$ with P-solubilizing microorganisms. Soil Biol Biochem 27:265–270. https://doi.org/10.1016/0038-0717(94)00205-f

Isabel JB, Balamurugan A, Renuka DP et al (2024) Chitosan-encapsulated microbial biofertilizer: a breakthrough for enhanced tomato crop productivity. Int J Biol Macromol 260(1):129462. https://doi.org/10.1016/j.ijbiomac.2024.129462

Jacoby RP, Kopriva S (2019) Metabolic niches in the rhizosphere microbiome: new tools and approaches to analyse metabolic mechanisms of plant–microbe nutrient exchange. J Exp Bot 70: 1087–1094. https://doi.org/10.1093/jxb/ery438

Ji C, Chen Z, Kong X et al (2022) Biocontrol and plant growth promotion by combined Bacillus spp. inoculation affecting pathogen and AMF communities in the wheat rhizosphere at low salt stress conditions. Front Plant Sci 8(13):1043171. https://doi.org/10.3389/fpls.2022.1043171

Jiang CH, Xie YS, Zhu K et al (2019) Volatile organic compounds emitted by Bacillus sp JC03 promote plant growth through the action of auxin and strigolactone. Plant Growth Regul 87(2): 317–328. https://doi.org/10.1007/s10725-018-00473-z

Jing H, Strader LC (2019) Interplay of auxin and cytokinin in lateral root development. Int J Mol Sci 20(3):486. https://doi.org/10.3390/ijms20030486

Jochum MD, McWilliams KL, Borrego EJ et al (2019) Bioprospecting plant growth-promoting rhizobacteria that mitigate drought stress in grasses. Front Microbiol 10:2106. https://doi.org/10.3389/fmicb.2019.02106

Jog R, Pandya M, Nareshkumar G et al (2014) Mechanism of phosphate solubilization and antifungal activity of *Streptomyces* spp. isolated from wheat roots and rhizosphere and their application in improving plant growth. Microbiol 160(4):778–788. https://doi.org/10.1099/mic.0.074146-0

Johansson JF, Paul LR, Finlay RD (2004) Microbial interactions in the mycorrhizosphere and their significance for sustainable agriculture. FEMS Microbiol Ecol 48:1–13. https://doi.org/10.1016/j.femsec.2003.11.012

Johnston-Monje D, Lundberg DS, Lazarovits G et al (2016) Bacterial populations in juvenile maize rhizospheres originate from both seed and soil. Plant Soil 405:337–355. https://doi.org/10.1007/s11104-016-2826-0

Kamal R, Gusain YS, Kumar V (2014) Interaction and symbiosis of AM fungi, actinomycetes and plant growth promoting rhizobacteria with plants: strategies for the improvement of plants health and defense system. Int J Curr Microb Appl Sci 3:564–585

Karimi-Jashni M, Yazdanpanah F (2023) Mycorrhizal networks: a secret interplant communication system. In: Rashad YM, Baka ZAM, Moussa TAA (eds) Plant mycobiome: diversity, interactions and uses. Springer, Cham, pp 447–467. https://doi.org/10.1007/978-3-031-28307-9_17

Kawasaki A, Donn S, Ryan PR et al (2016) Microbiome and exudates of the root and rhizosphere of *Brachypodium distachyon*, a model for wheat. PLoS One 11:e0164533. https://doi.org/10.1371/journal.pone.0164533

Kearl J, McNary C, Lowman JS et al (2019) Salt-tolerant halophyte rhizosphere bacteria stimulate growth of alfalfa in salty soil. Front Microbiol 10:1849. https://doi.org/10.3389/fmicb.2019.01849

Kecskeméti E, Berkelmann-Löhnertz B, Reineke A (2016) Are epiphytic microbial communities in the carposphere of ripening grape clusters (*Vitis vinifera* L.) different between conventional, organic, and biodynamic grapes? PLoS One 11:e0160852. https://doi.org/10.1371/journal.pone.0160852

Khuna S, Kumla J, Srinuanpan S et al (2023) Multifarious characterization and efficacy of three phosphate-solubilizing *aspergillus* species as biostimulants in improving root induction of cassava and sugarcane stem cuttings. Plants (Basel) 12(20):3630. https://doi.org/10.3390/plants12203630

Kittelmann S, Seedorf H, Walters W et al (2013) Simultaneous amplicon sequencing to explore co-occurrence patterns of bacterial, archaeal and eukaryotic microorganisms in rumen microbial communities. PLoS One 8:e47879. https://doi.org/10.1371/journal.pone.0047879

Kleiner M, Thorson E, Sharp CE et al (2017) Assessing species biomass contributions in microbial communities via metaproteomics. Nat Commun 16:8(1):1558. https://doi.org/10.1038/s41467-017-01544-x

Korenblum E, Dong Y, Szymanski J et al (2020) Rhizosphere microbiome mediates systemic root metabolite exudation by root-to-root signaling. Proc Natl Acad Sci USA 117:3874–3883. https://doi.org/10.1073/pnas.1912130117

Kumar A, Dubey A (2020) Rhizosphere microbiome: engineering bacterial competitiveness for enhancing crop production. J Adv Res 24:337–352. https://doi.org/10.1016/j.jare.2020.04.014

Kwak MJ, Kong HG, Choi K, et al (2018) Rhizosphere microbiome structure alters to enable wilt resistance in tomato. Nat Biotechnol. https://doi.org/10.1038/nbt.4232. Advance online publication. 10.1038/nbt.4232

Lahrmann U, Ding Y, Banhara A et al (2013) Host-related metabolic cues affect colonization strategies of a root endophyte. Proc Natl Acad Sci USA 110:13965–13970. https://doi.org/10.1073/pnas.1301653110

Lauber CL, Hamady M, Knight R et al (2009) Pyrosequencing-based assessment of soil pH as a predictor of soil bacterial community structure at the continental scale. Appl Environ Microbiol 75:5111–5120. https://doi.org/10.1128/AEM.00335-09

Lavakush, Yadav J, Verma JP et al (2014) Evaluation of PGPR and different concentration of phosphorus level on plant growth, yield and nutrient content of rice (*Oryza sativa*). Ecol Eng 62:123–128. https://doi.org/10.1016/j.ecoleng.2013.10.013

Lawson CE, Harcombe WR, Hatzenpichler R, Lindemann SR, Löffler FE, O'Malley MA et al (2019) Common principles and best practices for engineering microbiomes. Nat Rev Microbiol 17(12):725–741. https://doi.org/10.1038/s41579-019-0255-9

Lei X, Shen Y, Zhao J et al (2023) Root exudates mediate the processes of soil organic carbon input and efflux. Plants (Basel) 31:12(3):630. https://doi.org/10.3390/plants12030630

Li G, Kronzucker HJ, Shi W (2016) The response of the root apex in plant adaptation to iron heterogeneity in soil. Front Plant Sci 7:344. https://doi.org/10.3389/fpls.2016.00344

Li Y, Chen Y, Fu Y et al (2024) Signal communication during microbial modulation of root system architecture. J Exp Bot 75(2):526–537. https://doi.org/10.1093/jxb/erad263

Liu Y, Zuo S, Xu L et al (2012) Study on diversity of endophytic bacterial communities in seeds of hybrid maize and their parental lines. Arch Microbiol 194:1001–1012. https://doi.org/10.1007/s00203-012-0836-8

Lu Y, Wang E, Tang Z et al (2021) Roots and microbiome jointly drive the distributions of 17 phytohormones in the plant soil continuum in a phytohormone-specific manner. Plant Soil 470:153–165. https://doi.org/10.1007/s11104-021-04898-w

Lugtenberg B, Kamilov F (2009) Plant-growth-promoting rhizobacteria. Ann Rev Microbiol 63: 541–556. https://doi.org/10.1146/annurev.micro.62.081307.162918

Marco S, Loredana M, Riccardo V et al (2022) Microbe-assisted crop improvement: a sustainable weapon to restore holobiont functionality and resilience. Hortic Res 9:uhac160. https://doi.org/10.1093/hr/uhac160

McSpadden Gardener BB (2007) Diversity and ecology of biocontrol *Pseudomonas* spp. in agricultural systems. Phytopathol 97:221–226. https://doi.org/10.1094/PHYTO-97-2-0221

Meena R, Kumar S, Datta R et al (2020) Impact of agrochemicals on soil microbiota and management: a review. Land 9:34. https://doi.org/10.3390/land9020034

Mendes R, Garbeva P, Raaijmakers JM (2013) The rhizosphere microbiome: significance of plant beneficial, plant pathogenic, and human pathogenic microorganisms. FEMS Microbiol Rev 37: 634–663. https://doi.org/10.1111/1574-6976.12028

Mendes LW, Raaijmakers JM, De Hollander M et al (2018) Influence of resistance breeding in common bean on rhizosphere microbiome composition and function. ISME J 12(1):212–224. https://doi.org/10.1038/ismej.2017.158

Millán-Aguiñaga N, Soldatou S, Brozio S et al (2019) Awakening ancient polar Actinobacteria: diversity, evolution and specialized metabolite potential. Microbiology 165:1169–1180. https://doi.org/10.1099/mic.0.000845

Mohanram S, Kumar P (2019) Rhizosphere microbiome: revisiting the synergy of plant-microbe interactions. Ann Microbiol 69:307–320. https://doi.org/10.1007/s13213-019-01448-9

Moussa TAA, Khalil NM (2022) Extremozymes from extremophilic microorganisms as sources of bioremediation. In: Kuddus M (ed) Microbial extremozymes. Academic Press, pp 135–146. https://doi.org/10.1016/B978-0-12-822945-3.00005-1

Nayfach S, Pollard KS (2016) Toward accurate and quantitative comparative metagenomics. Cell 166:1103–1116. https://doi.org/10.1016/j.cell.2016.08.007

Nerva L, Giudice G, Quiroga G et al (2021) Mycorrhizal symbiosis balances rootstock-mediated growth-defence tradeoffs. Biol Fertil Soils 58:17–34. https://doi.org/10.1007/s00374-021-01607-8

Nguyen TD, Cavagnaro TR, Watts-Williams SJ (2019) The effects of soil phosphorus and zinc availability on plant responses to mycorrhizal fungi: a physiological and molecular assessment. Sci Rep 9(1):14880. https://doi.org/10.1038/s41598-019-51369-5

Nuccio EE, Hodge A, Pett-Ridge J et al (2013) An arbuscular mycorrhizal fungus significantly modifies the soil bacterial community and nitrogen cycling during litter decomposition. Environ Microbiol 15:1870–1881. https://doi.org/10.1111/1462-2920.12081

Odelade KA, Babalola OO (2019) Bacteria, fungi and archaea domains in rhizospheric soil and their effects in enhancing agricultural productivity. Int J Environ Res Public Health 16:3873. https://doi.org/10.3390/ijerph16203873

Offre P, Pivato B, Siblot S et al (2007) Identification of bacterial groups preferentially associated with mycorrhizal roots of *Medicago truncatula*. Appl Environ Microbiol 73:913–921. https://doi.org/10.1128/aem.02042-06

Otieno N, Lally RD, Kiwanuka S et al (2015) Plant growth promotion induced by phosphate solubilizing endophytic *Pseudomonas* isolates. Front Microbiol 6:745. https://doi.org/10.3389/fmicb.2015.00745

Parnell JJ, Berka R, Young HA et al (2016) From the lab to the farm: an industrial perspective of plant beneficial microorganisms. Front Plant Sci 7:1110. https://doi.org/10.3389/fpls.2016.01110

Penton CR, Gupta VVSR, Tiedje JM et al (2014) Fungal community structure in disease suppressive soils assessed by 28S LSU gene sequencing. PLoS One 9:e93893. https://doi.org/10.1371/journal.pone.0093893

Phour M, Sehrawat A, Sindhu SS et al (2020) Interkingdom signaling in plant-rhizomicrobiome interactions for sustainable agriculture. Microbiol Res 241:126589. https://doi.org/10.1016/j.micres.2020.126589

Potgieter L (2022) Words of a fungi, fungi's communicative connections. Berkeley Sci J 27:56–59

Qin H, Brookes PC, Xu J et al (2014) Bacterial degradation of Aroclor 1242 in the mycorrhizosphere soils of zucchini (*Cucurbita pepo* L.) inoculated with arbuscular mycorrhizal fungi. Environ Sci Pollut Res 21:12790–12799. https://doi.org/10.1007/s11356-014-3231-y

Raheem A, Shaposhnikov A, Belimov AA et al (2018) Auxin production by rhizobacteria was associated with improved yield of wheat (*Triticum aestivum* L.) under drought stress. Arch Agron Soil Sci 64(4):574–587. https://doi.org/10.1080/03650340.2017.1362105

Ramijan K, Ultee E, Willemse J et al (2018) Stress-induced formation of cell wall-deficient cells in filamentous actinomycetes. Nat Commun 9(1):5164. https://doi.org/10.1038/s41467-018-07560-9

Rawal R, Scheerens JC, Fenstemaker SM et al (2022) Novel *Trichoderma* isolates alleviate water deficit stress in susceptible tomato genotypes. Front Plant Sci 13:869090. https://doi.org/10.3389/fpls.2022.869090

Ray P, Lakshmanan V, Labbé JL et al (2020) Microbe to microbiome: a paradigm shift in the application of microorganisms for sustainable agriculture. Front Microbiol 11:622926. https://doi.org/10.3389/fmicb.2020.622926

Reitz MU, Gifford ML, Schäfer P (2015) Hormone activities and the cell cycle machinery in immunity-triggered growth inhibition. J Exp Bot 66:2187–2197. https://doi.org/10.1093/jxb/erv106

Reyes-Castillo A, Gerding M, Oyarzúa P et al (2019) Plant growth-promoting rhizobacteria able to improve NPK availability: selection, identification and effects on tomato growth. Chilean J Agric Res 79(3):473–485. https://doi.org/10.4067/S0718-58392019000300473

Rho H, Hsieh M, Kandel SL et al (2018) Do endophytes promote growth of host plants under stress? A meta-analysis on plant stress mitigation by endophytes. Microb Ecol 75:407–418. https://doi.org/10.1007/s00248-017-1054-3

Rodríguez CE, Mitter B, Barret M et al (2018) Commentary: seed bacterial inhabitants and their routes of colonization. Plant Soil 422:129–134. https://doi.org/10.1007/s11104-017-3368-9

Rodríguez M, Torres M, Blanco L et al (2020) Plant growth-promoting activity and quorum quenching-mediated biocontrol of bacterial phytopathogens by *Pseudomonas segetis* strain P6. Sci Rep 10:4121. https://doi.org/10.1038/s41598-020-61084-1

Rojas-Solis D, Zetter-Salmon E, Contreras-Perez M et al (2018) *Pseudomonas stutzeri* E25 and *Stenotrophomonas maltophilia* CR71 endophytes produce antifungal volatile organic compounds and exhibit additive plant growth promoting effects. Biocatal Agric Biotechnol 13:46–52. https://doi.org/10.1016/j.bcab.2017.11.007

Ryu C-M, Farag MA, Hu C-H et al (2003) Bacterial volatiles promote growth in *Arabidopsis*. Proc Natl Acad Sci USA 100(8):4927–4932. https://doi.org/10.1073/pnas0730845100

Sangabriel-Conde W, Negrete-Yankelevich S, Maldonado-Mendoza IE et al (2014) Native maize landraces from Los Tuxtlas, Mexico show varying mycorrhizal dependency for P uptake. Biol Fertil Soils 50:405–414. https://doi.org/10.1007/s00374-013-0847-x

Sapp J (2004) The dynamics of symbiosis: an historical overview. Can J Bot 82:1046–1056. https://doi.org/10.1139/b04-055

Schoch CL, Seifert KA, Huhndorf S et al (2012) Nuclear ribosomal internal transcribed spacer (ITS) region as a universal DNA barcode marker for fungi. Proc Natl Acad Sci USA 109:6241–6246. https://doi.org/10.1073/pnas.1117018109

Shalaby M, Elbagory M, El-Khateeb N et al (2023) Potential impacts of certain $N_2$-fixing bacterial strains and mineral N doses for enhancing the growth and productivity of maize plants. Plants (Basel) 12(22):3830. https://doi.org/10.3390/plants12223830

Sharma D, Nagpal A, Pakade YB et al (2010) Analytical methods for estimation of organophosphorus pesticide residues in fruits and vegetables: a review. Talanta 82:1077–1089. https://doi.org/10.1016/j.talanta.2010.06.043

Shrestha HK, Appidi MR, Villalobos Solis MI et al (2021) Metaproteomics reveals insights into microbial structure, interactions, and dynamic regulation in defined communities as they respond to environmental disturbance. BMC Microbiol 21:308. https://doi.org/10.1186/s12866-021-02370-4

Sikora R, Dababat AEF (2007) Influence of the mutualistic endophyte *Fusarium oxysporum* 162 on Meloidogyne incognita attraction and invasion. Nematol 9:771–776. https://doi.org/10.1163/156854107782331225

Simon JC, Marchesi JR, Mougel C, Selosse MA (2019) Host-microbiota interactions: from holobiont theory to analysis. Microbiome 7(1):5. https://doi.org/10.1186/s40168-019-0619-4

Singh A, Kumar M, Verma S et al (2020) Plant microbiome: trends and prospects for sustainable agriculture. In: Varma A, Tripathi S, Prasad R (eds) Plant microbe symbiosis. Springer, Cham, pp 129–151. https://doi.org/10.1007/978-3-030-36248-5_8

Sood G, Kaushal R, Sharma M (2020a) Significance of inoculation with *Bacillus subtilis* to alleviate drought stress in wheat (*Triticum aestivum* L.). Vegetos 33:782–792. https://doi.org/10.1007/s42535-020-00149-y

Sood G, Kaushal R, Sharma M (2020b) Alleviation of drought stress in maize (*Zea mays* L.) by using endogenous endophyte *Bacillus subtilis* in North West Himalayas. Acta Agric Scand, Sect B 70(5):361–370. https://doi.org/10.1080/09064710.2020.1743749

Steven B, Huntley RB, Zeng Q (2018) The influence of flower anatomy and apple cultivar on the apple flower phytobiome. Phytobiomes J 2:171–179. https://doi.org/10.1094/pbiomes-03-18-0015-r

Svenningsen NB, Watts-Williams SJ, Joner EJ et al (2018) Suppression of the activity of arbuscular mycorrhizal fungi by the soil microbiota. ISME J 12(5):1296–1307. https://doi.org/10.1038/s41396-018-0059-3

Taffner J, Erlacher A, Bragina A et al (2018) What is the role of archaea in plants? New insights from the vegetation of Alpine Bogs. mSphere 3:e00122–18. https://doi.org/10.1128/msphere.00122-18

Tarkka MT, Drigo B, Deveau A (2018) Mycorrhizal microbiomes. Mycorrhiza 28:403–409. https://doi.org/10.1007/s00572-018-0865-5

Tedersoo L, Bahram M, Zobel M (2020) How mycorrhizal associations drive plant population and community biology. Science 367(6480):eaba1223. https://doi.org/10.1126/science.aba1223

Theis KR, Dheilly NM, Klassen JL et al (2016) Getting the hologenome concept right: an eco-evolutionary framework for hosts and their microbiomes. MSystems 1:e00028–e116. https://doi.org/10.1128/mSystems.00028-16

Toju H, Peay KG, Yamamichi M, Narisawa K, Hiruma K, Naito K et al (2018) Core microbiomes for sustainable agroecosystems. Nat Plants 4(5):247–257. https://doi.org/10.1038/s41477-018-0139-4

Trivedi DK, Bhatt H, Pal RK et al (2013) Structure of RNA-interacting Cyclophilin A-like protein from *Piriformospora indica* that provides salinity-stress tolerance in plants. Sci Rep 3:3001. https://doi.org/10.1038/srep03001

Trivedi P, Leach JE, Tringe SG, Sa T, Singh BK (2020) Plant–microbiome interactions: from community assembly to plant health. Nat Rev Microbiol 18(11):607–621. https://doi.org/10.1038/s41579-020-0412-1

Urich T, Lanzén A, Qi J et al (2008) Simultaneous assessment of soil microbial community structure and function through analysis of the meta-transcriptome. PLoS One 3. https://doi.org/10.1371/journal.pone.0002527

Uroz S, Calvaruso C, Turpault M-P et al (2007) Effect of the mycorrhizosphere on the genotypic and metabolic diversity of the bacterial communities involved in mineral weathering in a forest soil. Appl Environ Microbiol 73:3019–3027. https://doi.org/10.1128/aem.00121-07

Vacheron J, Desbrosses G, Bouffaud M-L et al (2013) Plant growth-promoting rhizobacteria and root system functioning. Front Plant Sci 4:356. https://doi.org/10.3389/fpls.2013.00356

Van der Putten WH (2010) Impacts of soil microbial communities on exotic plant invasions. Trends Ecol Evol 25:512–519. https://doi.org/10.1016/j.tree.2010.06.006

Vandenkoornhuyse P, Quaiser A, Duhamel M, Le Van A, Dufresne A (2015) The importance of the microbiome of the plant holobiont. New Phytol 206(4):1196–1206. https://doi.org/10.1111/nph.13312

Vessey JK (2003) Plant growth promoting rhizobacteria as biofertilizers. Plant Soil 255:571–586. https://doi.org/10.1023/A:1026037216893

Vetterlein D, Carminati A, Kögel-Knabner I et al (2020) Rhizosphere spatiotemporal organization–a key to rhizosphere functions. Front Agron 2:8. https://doi.org/10.3389/fagro.2020.00008

Vishwakarma K, Kumar N, Shandilya C et al (2020a) Unravelling the role of endophytes in micronutrient uptake and enhanced crop productivity. In: Shrivastava N, Mahajan S, Varma A (eds) Symbiotic soil microorganisms. Springer, Cham, pp 63–85. https://doi.org/10.1007/978-3-030-51916-2_4

Vishwakarma K, Kumar N, Shandilya C et al (2020b) Revisiting plant–microbe interactions and microbial consortia application for enhancing sustainable agriculture: a review. Front Microbiol 11:560406. https://doi.org/10.3389/fmicb.2020.560406

Vorholt JA (2012) Microbial life in the phyllosphere. Nat Rev Microb 10:828–840. https://doi.org/10.1038/nrmicro2910

Vos C, Geerinckx K, Mkandawire R et al (2012) Arbuscular mycorrhizal fungi affect both penetration and further life stage development of root-knot nematodes in tomato. Mycorrhiza 22(2):157–163. https://doi.org/10.1007/s00572-011-0422-y

Wallace JG, Kremling KA, Kovar LL et al (2018) Quantitative genetics of the maize leaf microbiome. Phytobiomes J 2:208–224. https://doi.org/10.1094/pbiomes-02-18-0008-r

Wang Y, Li W, Du B et al (2021) Effect of biochar applied with plant growth-promoting rhizobacteria (PGPR) on soil microbial community composition and nitrogen utilization in tomato. Pedosphere 31(6):872–881. https://doi.org/10.1016/S1002-0160(21)60030-9

Wang D, Luo Q, Huang K et al (2023) Distinct mechanisms underlying assembly processes and interactions of microbial communities in two single-stage bioreactors coupling anammox with denitrification. Chem Eng J 452:139319. https://doi.org/10.1016/j.cej.2022.139319

Welc M, Ravnskov S, Kieliszewska-Rokicka B et al (2010) Suppression of other soil microorganisms by mycelium of arbuscular mycorrhizal fungi in root-free soil. Soil Biol Biochem 42:1534–1540. https://doi.org/10.1016/j.soilbio.2010.05.024

Weyens N, van der Lelie D, Taghavi S et al (2009) Exploiting plant-microbe partnerships to improve biomass production and remediation. Trends Biotechnol 27(10):591–598. https://doi.org/10.1016/j.tibtech.2009.07.006

Wild A (2003) Soils, land and food: managing the land during the twenty-first century. Cambridge University Press, Cambridge

Wright M, Adams J, Yang K et al (2016) A root-colonizing pseudomonad lessens stress responses in wheat imposed by CuO nanoparticles. PLoS One 11(10):e0164635. https://doi.org/10.1371/journal.pone.0164635

Xiong Y-W, Li X-W, Wang T-T et al (2020) Root exudates-driven rhizosphere recruitment of the plant growth-promoting rhizobacterium *Bacillus flexus* KLBMP 4941 and its growth-promoting effect on the coastal halophyte *Limonium sinense* under salt stress. Ecotoxicol Environ Saf 194:110374. https://doi.org/10.1016/j.ecoenv.2020.110374

Yadav G, Vishwakarma K, Sharma S et al (2017) Emerging significance of rhizospheric probiotics and its impact on plant health: current perspective towards sustainable agriculture. In: Kumar V, Kumar M, Sharma S et al (eds) Probiotics and plant health. Springer, Singapore, pp 233–251. https://doi.org/10.1007/978-981-10-3473-2_10

Yadav AN, Mishra S, Kour D et al (2020) Agriculturally important fungi for sustainable agriculture. Switzerland AG: Springer Nature, p 300. https://doi.org/10.1007/978-3-030-45971-0
Yan K, Wei Q, Feng R et al (2017) Transcriptome analysis of *Cinnamomum longepaniculatum* by high-throughput sequencing. Electron J Biotechnol 28:58–66. https://doi.org/10.1016/j.ejbt.2017.05.006
Zarraonaindia I, Owens SM, Weisenhorn P et al (2015) The soil microbiome influences grapevine-associated microbiota. MBio 6(2):e02527-14. https://doi.org/10.1128/mBio.02527-14
Zeder MA (2015) Core questions in domestication research. Proc Natl Acad Sci 112:3191–3198. https://doi.org/10.1073/pnas.1501711112
Zhalnina K, Louie K, Hao Z et al (2018) Dynamic root exudate chemistry and microbial substrate preferences drive patterns in rhizosphere microbial community assembly. Nat Microbiol 3:470–480. https://doi.org/10.1038/s41564-018-0129-3
Zhang A, Sun H, Wang P et al (2012) Modern analytical techniques in metabolomics analysis. Analyst 137:293–300. https://doi.org/10.1039/c1an15605e
Zhang N, Wang D, Liu Y (2014) Effects of different plant root exudates and their organic acid components on chemotaxis, biofilm formation and colonization by beneficial rhizosphere-associated bacterial strains. Plant Soil 374:689–700. https://doi.org/10.1007/s11104-013-1915-6
Zhang J, Liu Y-X, Zhang N et al (2019) NRT1.1B is associated with root microbiota composition and nitrogen use in field-grown rice. Nat Biotechnol 37:676–684. https://doi.org/10.1038/s41587-019-0104-4
Zhang W, Li X-G, Sun K et al (2020) Mycelial network-mediated rhizobial dispersal enhances legume nodulation. ISME J 14:1015–1029. https://doi.org/10.1038/s41396-020-0587-5
Zhang J, Cook J, Nearing JT et al (2021) Harnessing the plant microbiome to promote the growth of agricultural crops. Microbiol Res 245:126690. https://doi.org/10.1016/j.micres.2020.126690
Zilber-Rosenberg I, Rosenberg E (2008) Role of microorganisms in the evolution of animals and plants: the hologenome theory of evolution. FEMS Microbiol Rev 32(5):723–735. https://doi.org/10.1111/j.1574-6976.2008.00123.x

# Chapter 7
# Effectiveness of Microbiota of Honeybees on Their Activities and Productivity

**Tarek Essa Abd El-Wahab and Islam Mohammad Zidan**

## 7.1 Introduction

The honeybee, *Apis mellifera* is a social insect that has successfully colonized various ecosystems worldwide, playing a vital role in pollinating both wild and cultivated plants. This role has substantial implications for the global economy and natural ecosystems (Villalba et al. 2020). Honeybees are essential for food production, with their economic value in the United States alone estimated at USD 15 billion (Zheng et al. 2017). Beyond pollination, honeybees are valuable for producing commercial products like wax, pollen, propolis, royal jelly, and honey. They are also crucial for maintaining ecological balance.

Honeybees have complex interactions with their environment and a diverse range of microorganisms. Understanding these interactions is essential for creating a hospitable environment for both humans and bees. The honeybee microbiome is central to individual health, and disruptions can make bees susceptible to various problems. Research has focused on the intestinal microbiome of honeybees, its role in health, fitness, and metabolism, and its response to physical, biological, chemical, and environmental factors (Engel and Moran 2013a, b; Anderson and Ricigliano 2017; Anjum et al. 2018; Bonilla-Rosso and Engel 2018; Bommuraj et al. 2021). This broad perspective is necessary given the importance of honeybee health and its impact on environmental protection.

T. E. Abd El-Wahab (✉) · I. M. Zidan
Pests and Plant Protection Department, Agricultural and Biological Research Institute, National Research Centre (NRC), Giza, Egypt
e-mail: te.abdelwahab@nrc.sci.eg; im.zidan@nrc.sci.eg

 
N. El-Wakeil, A. Abdelkhalek (eds.), *Microbiomes and Plant Secondary Metabolites for Insect Pest Management*, Progress in Biological Control 23,
https://doi.org/10.1007/978-3-032-24330-0_7

## 7.2 What Is the Honeybee Microbiota?

A growing field of study that improves our comprehension of ecological and functional dynamics inside animal hosts is gut microbiome research. The gut microbes of honeybees, *Apis mellifera*, carry out a number of functions that are essential for many host organisms, including the breakdown of complex polysaccharides and immune system modulation (Hall et al. 2017). We can discover novel pathways and create preventative and therapeutic interventions for the health of humans and animals by comprehending the basic traits of symbiotic gut microbes. Research on *A. mellifera*'s gut microbiota has advanced quickly, identifying a stable and distinctive microbial community whose functional roles are still being investigated (Raymann and Moran 2018).

*Apis mellifera*'s gut environment is different from other insects for a number of reasons: first, like many social animals, the microbiota is passed on through direct contact during early social interactions with hive mates (Engel and Moran 2013a, b; Powell et al. 2014); second, the gut communities are ecologically resilient, retaining a unique group of organisms in spite of environmental changes (Moran et al. 2012); third, the microbiota's taxonomic composition, which primarily consists of species exclusive to social bees, suggests a highly specialised, coevolved relationship between the microbes and *A. mellifera* (Martinson et al. 2011); and finally, because *A. mellifera* is domesticated, they are an accessible system for microbiome studies.

These factors make the gut ecosystem of *A. mellifera* a source of new ecological data with a wide range of possible uses. The first review of techniques for culturing and characterizing *A. mellifera* gut microbial communities was given by Engel et al. (2013), and Zheng et al. (2018) compiled the traits that make honeybees an excellent experimental system for gut microbiome research, emphasizing similarities between the gut microbiota of honeybees and humans.

Numerous microorganisms live in the food sources (honey and pollen), honeycombs, floral nectar, and other anatomical areas of honeybee larvae and adults. According to Kaltenpoth (2011), eukaryotes' symbiotic relationships with microbes are a significant source of evolutionary innovation. The three microbes most frequently linked to honeybees are yeasts, molds, and bacteria. There are usually no internal bacteria in eggs, prepupae, pupae, or just emerged worker bees. Larvae excrete all bacteria they have acquired from a contaminated diet throughout development in single feces prior to pupation. Through food exchange with nest mates and pollen ingestion, newly emerging bees develop gut microbiota (Gilliam 1997).

## 7.3 Types of Honeybee Microbiota

Large colonies of honeybees are made up of one reproductive queen, hundreds of male drones, and thousands of nonreproductive female workers (Glenny et al. 2017). Core gut microbiota is either absent or diminished in newly emerged workers (Dong et al. 2020). Within a few days of emergence, microbial communities socialize with nurse bees and colonize their bodies (Martinson et al. 2012; Motta et al. 2018). Gut bacteria and the gut epithelium are expelled during pupae metamorphosis, and colonization starts over through trophallaxis, bee-to-bee contact, and exposure to the hive environment (Bleau et al. 2020). Three to five days following adult emergence, the gut's bacterial population peaks (Li et al. 2017).

After 3–8 days, taxonomic changes take place, indicating the existence of strains involved in pioneer or niche formation. The ileum, which is more changeable, completes its structure after 8 days, but the rectum community develops its structure in 3 days. This process is influenced by the host immunological response, the ileum's environmental alterations, and the prevalence of core species (Anderson and Ricigliano 2017). As possible routes of inoculation, newly emerged bees usually engage in hive maintenance and cleaning activities, interact with adult bees, use the comb, and eat bee bread (Powell et al. 2014; Engel et al. 2012).

The gut microbiota succession of *A. mellifera* workers from birth to senescence (0–40 days postemergence) was examined by Dong et al. (2020). At 1 dpe, the honeybee gut was colonised by the genera *Gilliamella*, *Frischella*, and *Snodgrassella*; at 3 dpe, *Lactobacillus*, *Bifidobacterium*, and *Commensalibacter* colonised, but *Gilliamella* decreased concurrently. Significant colonisation by L. kunkeei and Bartonella sp. took place at 12 dpe, whereas between 19 and 25 dpe, *Bacteroides* sp., *Escherichia* sp., *Shigella* sp., and *Porphyromonadaceae* declined. At 25 dpe, the abundance of *Commensalibacter* sp. and *Bifidobacterium* sp. declined (Martinson et al. 2012).

The microbiota of honeybees is located in different parts of the gut, including the crop, midgut, hindgut (ileum and rectum), and distal rectum (Engel et al. 2012; Vásquez et al. 2012). Only *Parasaccharibacter* sp. was found in relative abundance in worker hypopharyngeal glands (Corby-Harris et al. 2014a, b).

## 7.4 Characterizing Honeybee Gut's Microbiota

All healthy worker bees have five basic bacterial lineages that dominate the very simple microbiota of the honeybee gut. They belong to clusters within the genera *Bifidobacterium*, *Bombilactobacillus* (formerly known as *Lactobacillus* Firm-4 Zheng et al. 2020), *Gilliamella*, *Lactobacillus* (formerly known as *Lactobacillus* Firm-5 Zheng et al. 2020), and *Snodgrassella* (Kwong and Moran 2016). Bacterial contents are acquired orally after emerging from the pupal stage through social interaction and contact with hive compartments (Kwong and Moran 2016; Powell et al. 2014).

Although the absolute and relative abundances of these core bacteria vary by life stage, season, and geographic location, they consistently form a community of roughly 10^8–10^9 cells (D'Alvise et al. 2018; Jones et al. 2017). According to Kwong and Moran (2016), other non-core bacteria such *Frischella*, *Commensalibacter*, and *Bartonella* are also frequently found. Between one and five species have been formally described within each of these genera (Olofsson et al. 2014; Kwong and Moran 2013), while there may be more closely related species that have not been identified yet (Ellegaard and Engel 2019).

Some other studies conducted 16S rDNA community surveys and metagenomics of gut samples have revealed that nine bacterial species clusters are dominantly found in the guts of the honeybee workers, accounting for about 95–99.9% of the bacterial content in all individuals (Moran et al. 2012; Jeyaprakash et al. 2003; Babendreier et al. 2007; Martinson et al. 2011; Sabree et al. 2012; Engel et al. 2012).

Members of the Proteobacteria phylum, *Snodgrassella alvi* and *Gilliamella apicola*, are the two common Gram-negative species (Kwong and Moran 2013). Two species clusters within the Firmicutes phylum are also common and widespread among Gram-positive bacteria; these have been dubbed the *Lactobacillus* Firm-4 and *Lactobacillus* Firm-5 clades (Martinson et al. 2011; Sabree et al. 2012). The majority of adult worker bees also contain the Actinobacteria phylum's *Bifidobacterium asteroides* species cluster (Scardovi and Trovatelli 1969; Bottacini et al. 2012), albeit in lesser quantities.

The gut microbiota of honeybees is composed of these five clusters of bacterial species. Proteobacteria species *Frischella perrara* (Engel et al. 2013), *Bartonella apis* (Kešnerová et al. 2016), *Parasaccharibacter apium* (Corby-Harris et al. 2014a, b), and a genus of species related to *Gluconobacter* called Alpha2.1 (Sabree et al. 2012) are less common and less numerous.

These species may be less common in gut surveys because they occupy specific niches in the bee gut (like *F. perrara*) or are generalists found in the hive environment (like *P. apium* and the Alpha2.1 group). The main members of the bee gut microbiota are represented by these nine species groupings taken together.

These nine species clusters are specifically suited to coexist with their honeybee hosts; however, other bacteria may occasionally be present. As is typical in other animal microbiomes, these species' distinct localizations or metabolic niches in the gut are probably the cause of variations in abundance and prevalence (Donaldson et al. 2016). Worker honeybees' gut microbiome has been studied for this kind of organization. The crop, which is utilized to store and convey nectar for feeding larvae and producing honey, has very few microorganisms (Martinson et al. 2012). Species that live in nectar and hive materials that can be cultivated at atmospheric oxygen concentrations, including as members of the Enterobacteriaceae, *Lactobacillus kunkeei*, and *P. apium*, make up the majority of the crop's restricted bacterial population (Corby-Harris et al. 2014a, b; Anderson et al. 2013, 2014).

Additionally, according to Martinson et al. (2012) honeybees' midgut contains few microbiomes. Because of its constantly shedding chitinous peritrophic matrix, the insect midgut, which is in charge of food digestion and absorption, does not offer a stable substrate for bacterial colonization. On the other hand, according to 16S rDNA profiling, the hindgut, which is lined by a stable layer of cuticle, has a huge microbial community of between 10^8 and 10^9 bacterial cells, which accounts for more than 99% of the bacteria in adult worker bees (Martinson et al. 2012).

The rectum and ileum, the two sections of the hindgut, each have their own unique microbial populations. The primary Gram-negative bacterial species are found in the ileum, a thin tube with six longitudinal folds: *G. apicola* forms a layer on top of *S. alvi*, which produces a layer directly on the gut wall (Martinson et al. 2012; Engel et al. 2012).

According to Engel et al. (2015), this bacterial community spreads into the pylorus, a small area where the ileum, midgut, and Malpighian tubules converge. *F. perrara* is prevalent in the pylorus and settles next to the intestinal epithelium (Engel et al. 2015). The fermentative Gram-positive bacteria *Lactobacillus* Firm-4, *Lactobacillus* Firm-5, and the *B. asteroides* cluster predominate in the rectum, which is where fecal waste is deposited prior to feces. The rectum is where the two Lactobacillus species groups are most prevalent, while they are also present in the ileum lumen (Martinson et al. 2012; Engel et al. 2012). Although the molecular processes that allow colonization of the bee gut are yet unknown, the genomes of many gut's bacterial species contain genes involved in adhesion and biofilm formation (Engel et al. 2012; Kwong et al. 2014).

## 7.5 Functions of Honeybee Microbiota

### *7.5.1 Microbiota and Host Metabolism and Health*

The gut microbiota has a broad range of metabolic functions and plays a major role in the intestinal ecosystem's digestion of dietary carbohydrates in both insects and mammals (Flint et al. 2012). As the primary fermentation products of dietary fiber, short-chain fatty acids (SCFAs) like acetate, propionate, and butyrate are produced by the gut microbiota and can build up in the human colon at concentrations of 80–130 mM (Cummings et al. 1987). They are also a major source of energy for intestinal epithelial cells (Donohoe et al. 2011) and the host's primary respiratory substrate (Odelson and Breznak 1983). According to Zeitouni et al. (2016), microbial metabolites have a significant impact on gut physiology, affecting redox potential, pH, and oxygen concentration—all of which are critical for host health.

The gut microbiota produces SCFAs, which are neuroactive substances that have the ability to alter immunological and neurological pathways, hence impacting behavior and brain function (Cryan and Dinan 2012). According to genome-based studies, *G. apicola* strains in honeybees are capable of breaking down complex carbohydrates that the host would otherwise be unable to digest, such as pectin from pollen cell walls (Engel et al. 2012). According to a recent study, certain strains of *G. apicola* can also metabolize carbohydrates that are bad for bees (Zheng et al. 2016). Nevertheless, little is known about how the gut microbiota of bees affects host physiology and the gut microenvironment.

In order to address this, the effects of the gut microbiota on weight gain, the gene expression underlying hormone pathways, gut physicochemical conditions, and metabolite pools in the gut and hemolymph were examined by comparing germ-free (GF) honeybees to those with a typical gut community (CV). According to Zheng et al. (2017), altered host vitellogenin, insulin signaling, and gustatory response are probably the mechanisms by which the gut microbiome encourages weight gain in honeybees. Through the synthesis of SCFAs, microbial metabolism dramatically lowers gut pH and redox potential, and bacteria that are close to the gut wall create an oxygen gradient inside the intestine. In vitro, the main gut species' contribution to the SCFA profile was validated.

The metabolic profiles of the gut compartments and hemolymph are significantly influenced by the gut community, according to metabolomic investigations. This suggests that gut bacteria break down plant polymers from pollen, and the metabolites that are produced help the host survive. According to these results, the gut microbiota of bees plays roles that are comparable to those of other species, offering a model for researching host-microbe interactions.

### 7.5.2 *Pathogen Protection and Immune System*

Numerous studies revealed that the gut microbiota can shield honeybees from a variety of pathogens, such as opportunistic bacteria (Kwong et al. 2017; Horak et al. 2020; Steele et al. 2021) and fungi (Miller et al. 2021; Lang et al. 2023; Huang et al. 2023) and possibly RNA viruses (Dosch et al. 2021).

The effects of the entire community or particular core members on susceptibility to pathogen challenge were examined in these investigations using gnotobiotic honeybees with specified communities or bees whose natural microbiota had been disturbed by antibiotics or other agents.

Although the exact mechanisms of protection are frequently unknown, research indicates that increased resistance may be brought about by competition for nutrients and/or space (Steele et al. 2021), physical barriers that prevent pathogen invasion of the gut wall (Steele et al. 2021), stimulation of host immune pathways (Kwong et al. 2017; Horak et al. 2020; Steele et al. 2021; Wang et al.), or the generation of antimicrobial metabolites (Miller et al. 2021).

Bee immunity is greatly influenced by the gut microbiome. The expression of host immunity genes, including those encoding antimicrobial peptides (AMPs) (Kwong et al. 2017; Horak et al. 2020; Steele et al. 2021; Wang et al.) and the melanization cascade (Emery et al. 2017), is upregulated when the entire gut community or individual community members colonize. The innate immune system of honeybees, which can be roughly divided into humoral and cellular immunity, protects against opportunistic bacteria, fungi, and parasites (Evans et al. 2006). Pattern recognition receptors that identify substances like peptidoglycan and lipopolysaccharides from the bacterial outer membrane start both processes. AMPs that circulate in many bodily locations, including abaecin, apidaecin, defensin, and hymenoptaecin, are produced as part of humoral immunity.

Cellular immunity involves phagocytosis, nodulation, and encapsulation, often accompanied by melanization. The latter involves the activation of phenoloxidase, resulting in the formation of melanin, which encapsulates and kills invading microorganisms (Evans et al. 2006; Nappi and Christensen 2005). The details of how the bee gut microbiota influences the immune system remain unclear.

Proteobacteria, Firmicutes, and Actinobacteria are the three primary phyla that dominate the gut microbiota community of honeybees, *Apis* spp., which are social insects. Several species of these bacteria belong to the lactic acid bacteria (LAB) phylum, according to Niode et al. (2019). Because of their capacity to generate bioactive substances like organic acids (lactic, acetic, and formic acid), hydrogen peroxide ($H_2O_2$), ethanol, enzymes, benzoate, antimicrobial peptides (AMPs), free fatty acids, and volatile compounds that function as broad-spectrum antimicrobials against a variety of pathogens, certain LAB species may be developed as potential probiotics. Probiotic potential is demonstrated by a number of LAB strains that were isolated from the digestive system of honeybees. Numerous investigations have shown that honeybee LAB inhibits the growth of harmful microbes.

Formic acid and lactic acid produced by symbiotic LAB can lower the environmental pH of wounds, preventing the growth of pathogenic microbes. Additionally, volatile compounds produced by LAB are toxic, while small amounts of $H_2O_2$ are necessary for optimal wound recovery. LAB can produce broad-spectrum antibacterial compounds effective against *Serratia marcescens*, *Escherichia coli*, and methicillin-resistant *Staphylococcus aureus*.

### 7.5.3 Roles of Honeybee Gut Bacteria on Host Metabolism and Neurological Processes

According to recent research, honeybee gut microbiota significantly affects the brain's neurotransmitter levels, especially those of dopamine and serotonin, which are known to affect sensory sensitivity. These neurotransmitters are downregulated in bees infected by bacteria, which alters how they perceive their surroundings. The proboscis extension response (PER) test showed that honeybees' taste-related behaviors depend on a healthy gut microbiota, underscoring the significance of interactions between various gut bacterial species in influencing the physiological and behavioral responses of the host. Additionally, it has been demonstrated that certain gut bacteria can change the quantities of neurotransmitters, improving host perception and cognition (Zhang et al. 2022).

Because of their intricate and well-organized social systems, honeybees, *A. mellifera*, are frequently used as model organisms for research on perception, cognition, and social behavior (Menzel 2012). Honeybees are eusocial insects that display a variety of complex behaviors that are highly dependent on their senses. To find pollen and nectar sources, for example, foraging bees need to be able to precisely interpret optical, olfactory, gustatory, and mechanosensory information. Furthermore, honeybees rely on their sensory systems for actions like identifying nestmates and avoiding predators, which are facilitated by social information (Raine et al. 2006). Researchers have created standardized techniques to measure activities like memory, associative learning, sensory responsiveness, and hive dynamics in order to study the neurobiology and behavioral physiology of honeybees (Scheiner et al. 2013).

The way honeybees behave is greatly influenced by their nutrition. Normal behavioral development and function depend on essential nutrients such as lipids, amino acids, and carbs (Amdam et al. 2004). For instance, polyunsaturated fatty acid deficiencies may affect cognitive function (Arien et al. 2015). Certain biotic factors, including brain gene expression, especially those linked to foraging and division of labor, are also linked to behavioral changes in honeybees (Wheeler et al. 2015; Toma et al. 2000). Learning, foraging behavior, and regulating sensory sensitivity are all greatly impacted by neurotransmitters, including biogenic amines such as serotonin, dopamine, and octopamine (Scheiner et al. 2006).

The microbiota-gut-brain axis, which is the two-way communication between the gut microbiota and the brain, affects many facets of behavior in the animal kingdom. It has been demonstrated that the gut microbiome in honeybees uses chemical transmission to alter brain physiology and behavior. For instance, by modifications in cuticular hydrocarbon profiles, the gut microbiota can modify endocrine signaling, impacting nestmate detection and sucrose sensitivity (Zheng et al. 2017; Vernier et al. 2020). Lower levels of biogenic amines, which are essential for behavioral development and include serotonin, dopamine, and octopamine, are seen in newly emerging bees with an immature gut microbiome (Harris and Woodring 1992).

With eight to ten dominant bacterial phylotypes that make up more than 97% of the community, the honeybee gut microbiota is very basic but extremely specialized (Kwong and Moran 2016; Bonilla-Rosso and Engel 2018; Ellegaard and Engel 2019). *Gilliamella*, *Snodgrassella*, *Bifidobacterium*, *Lactobacillus* Firm4 and Firm5, and *Bartonella* are important bacterial families that can be cultivated in lab environments. Thanks to the availability of microbiota-free (MF) bees for experimental studies, recent studies have brought attention to the causative roles of these gut bacteria in host feeding, weight gain, and endocrine signaling (Zheng et al. 2018; Kešnerová et al. 2017). For example, *Bifidobacterium*, *Lactobacillus* Firm5, and *Gilliamella* may break down complex dietary polysaccharides, which helps pollen release its nutrient-rich content (Zheng et al. 2019; Brochet et al. 2021). Furthermore, the gut microbiota's interspecies interactions promote the synthesis of amino acids and the metabolism of carbohydrates, which promotes the host's nutrition (Brochet et al. 2021; Kwong et al. 2014).

According to metabolomic investigations, the presence of gut bacteria causes the honeybee's gut to accumulate different organic acids, which could have important functional implications for host physiology (Kešnerová et al. 2017; Zheng et al. 2019). Both the gut and hemolymph of the host are impacted by the gut microbiota; conventional (CV) microbiota-colonized bees show higher levels of insulin-related genes and heightened sensitivity to sweets (Zheng et al. 2017). Nevertheless, little is known about the precise mechanisms by which distinct gut bacterial species influence host behavior and neurological functions.

Recent research has filled this gap by using gnotobiotic honey bees that have been mono-colonized with particular gut bacteria to analyze the unique functions of these microorganisms in host metabolism and brain function. Gut bacteria have a major impact on brain transcriptional programs and circulation metabolic profiles, according to multi-omics investigations. These results highlight how crucial some gut bacteria are for modifying neurotransmitter levels and improving host perception and cognition (Zhang et al. 2022).

### *7.5.4 Pesticides and Microplastics*

The midgut has also been shown to experience some of these effects, such as cell destruction (Carvalho et al. 2013; Catae et al. 2018; Oliveira et al. 2014). Numerous pesticides, including coumaphos, imidacloprid, and chlorothalonil, have detrimental effects on bee health. They increase vulnerability to opportunistic infections by adversely affecting the microbiome's structure and function.

Honeybees are expected to be exposed to pesticide compounds, e.g., chlorothalonil, through contaminated water, nectar, and pollen, which significantly reduces the amount of *Lactobacillales* in the content of the microbiome (Motta et al. 2018).

Non-lethal concentrations of pesticides such as thiamethoxam, fipronil, imidacloprid, and coumaphos have been found to drastically diminish *Bifidobacterium* sp. and *Lactobacillus* sp. (Motta et al. 2018). Honeybees' gut microbiome is adversely affected by glyphosate exposure, which lowers the total number of bacteria in the gut, i.e., *Bifidobacterium* sp., *Lactobacillus* (Firm-4 and Firm-5), and *S. alvi* (Motta et al. 2018). The neonicotinoid insecticide nitenpyram has a considerable impact on the microbiome community, leading to significant metabolic changes in the host and a decrease in the effectiveness of the immune system (Zhu et al. 2020). Pesticides like imidacloprid and thiacloprid, which change several species like Bifidobacterium sp., *Frischella* sp., and *Lactobacillus* (Firm-4 and Firm-5), are the cause of intestinal dysbiosis (Diaz et al. 2019; Shilo et al. 2016).

The 5-enolpyruvylshikimate-3-phosphate synthase (EPSPS) enzyme in the shikimate pathway, which is present in plants and certain microbes, is the target of glyphosate, the main herbicide used worldwide to control weeds. Therefore, animals that live close to agricultural locations, including pollinators like bees, may have bacterial symbionts that are impacted by glyphosate (Motta et al. 2018). Major pollinators of blooming plants, including many crops, are honeybees and bumblebees. They may come into contact with glyphosate and other xenobiotics while foraging (Newman et al. 2016; Shehata et al. 2013). The growth of microorganisms is known to be impacted by this herbicide, and the unique gut microbial ecology of bees is inextricably linked to their health (Cariveau et al. 2014; Koch and Schmid-Hempel 2011). According to Motta et al. (2018), eight bacterial species that increase weight gain and lessen vulnerability to pathogens dominate the gut microbiota of honeybees.

Nearly all sequenced genomes of bee gut bacteria have the gene for EPSPS, suggesting that they may be glyphosate-susceptible. The study demonstrated that bees exposed to glyphosate at levels reported in the environment had lower relative and absolute abundances of dominating gut microbiota species. When young workers were treated with glyphosate, the death rate of bees later exposed to the opportunistic disease Serratia marcescens rose. Bees with class I (glyphosate-sensitive) or class II (glyphosate-insensitive) EPSPSs have different levels of glyphosate susceptibility in their gut microbiota.

A sensitive class I EPSPS is encoded by all strains of *S. alvi*, the gut species of the core bee, and a decrease in *S. alvi* levels was a consistent experimental finding. Nonetheless, certain strains of *S. alvi* seem to have a different glyphosate resistance mechanism. Accordingly, glyphosate exposure can disrupt the beneficial gut microbiome of bees, which may have an impact on their well-being and pollination efficiency (Motta et al. 2018).

The most widely used fungicide in China, carbendazim, has unknown effects on honeybees' nutritional intake levels and gut bacteria. Wang et al. evaluated the effects of carbendazim on honeybee mortality, pollen consumption, and gut microbiota composition in order to fill this study gap.

According to their findings, even at high concentrations (5000 mg/L), which are highly unlikely to occur in the field, honeybees are not acutely killed by carbendazim exposure. Neither in adult worker bees with established gut communities in the mid and hindgut nor in young worker bees during gut microbial colonization does carbendazim alter the composition of the microbiome in the gut. However, exposure to carbendazim dramatically reduces honeybees' consumption of pollen. Therefore, bee exposure to carbendazim can disrupt their positive nutrition balance, which may lower honeybee immunity and make them more vulnerable to infection by viruses. This could affect the bees' ability to pollinate and the general health of the colony (Wang et al.).

According to Wang et al. (2021) the effects of microplastics on honeybee fitness were studied by feeding microplastic particles to newly emerging bees for 2 weeks in a lab setting. Significant evidence was found of the buildup and breakdown of microplastics in the gut as well as their interactions with gut bacteria. The microbial population core changed, and the variety of the microbiota decreased. Determining the mechanism of action and appropriate measurement methods for environmental elements that impact the honeybee microbiome, such as air pollutants, heavy metals, and microplastics, is the true difficulty (Wang et al. 2021).

### 7.5.5 Antibiotic Effects

There have been reports of severe economic losses due to the sharp declines in bee colonies all around the world (Van Der Zee et al. 2012; Duan et al. 2020). According to Van Dooremalen et al. (2018), the primary causes include various agrochemicals, parasites, viruses, modifications to the distribution and structure of plants, or their interactions. The gut-dwelling microbial communities in animals, including humans and insects, are essential to the health of their hosts (Ley et al. 2008; Clemente et al.

2012). The host's physiology, behavior, and fitness characteristics are influenced by the gut symbionts. According to Nancy (2015), honeybee workers' intestines contain a unique ecosystem of bacterial species.

The primary gut microbiota of honeybees is made up of five core microbiota, i.e., *S. alvi*, *G. apicola*, *Lactobacillus* Firm-4, *Lactobacillus* Firm-5, and *Bifidobacterium*, and four additional bacterial strains: *Frischella perrara*, *Bartonella apis*, *Parasaccharibacter apium*, and *Alpha* 2.1 (Li et al. 2018). Numerous factors, including nutritional supplements, antibiotics, non-protein amino acids, and veterinary medications, as well as the honeybees' in-hive activities, influence the microbial profile (Cilia et al. 2020a, b). According to Kwong et al. (2017), this gut microbiome supports adult bees' body weight, nutrition provision, and food digestion. A colony's labor division is determined by the nutrient level of its adult bees: foragers with low nutrient states and nurse bees with high nutrient states (Ament et al. 2010). Precocious foraging results from low nutrition levels (Li et al. 2018; Antonio et al. 2008).

Low nutrition levels in adult bees whose gut flora have been eliminated or disturbed by antibiotics cause their gustatory responses to food requirements to increase (Zheng et al. 2017). Additionally, these bees shift from nursing bees to foragers earlier, suggesting that gut bacteria may have a role in controlling the behavioral development of honeybees (Ortiz-Alvarado et al. 2020). The findings of Duan et al. (2021) demonstrated the critical functions that gut bacteria play in bee larvae fitness by showing that antibiotics might impair food metabolism, disrupt the development process, and reduce the immunological competence of honeybee larvae. Therefore, in order to manage microbial illness in honeybee larvae, antibiotics should be avoided.

Attempts to cure colonies with chemical approaches have been made in response to the growing risk of colony collapse disorder (CCD). Antibiotics can affect the host by changing the gut microbiota species. Antibiotics have been shown to negatively impact honeybee production, immunology, and gut microbiota (Daisley et al. 2020). Several antibiotic and veterinary chemotherapeutic residues have been found in honey, indicating that honeybees are still exposed to these substances even though their use in beekeeping is prohibited in many countries (Reybroeck 2017; Raymann et al. 2018).

According to Li et al. (2017), these stressors cause the number of bacterial species in the honeybee's gut to decrease, which impairs immunity and makes them more vulnerable to illnesses. In one study, the microbiome of honeybees was eradicated after they were treated with antibiotics. Due to its detrimental effects on the immune system, as demonstrated by the reduction of the expression of genes encoding antimicrobial peptides, it was discovered that bees were more vulnerable to infections by Nosema ceranae, a common honeybee pathogen (Li et al. 2017). According to another study, tetracycline-induced disruption of the gut microbiota reduced honeybee longevity, which was linked to a higher vulnerability to the opportunistic pathogen Serratia sp. (Raymann et al. 2017).

Additionally, honeybee products may later include antibiotic residues examined how the physiology and behavior of bees were affected throughout their development by two marketed beekeeping antibiotics: Tylan (tylosin tartrate) and Terramycin (oxytetracycline). The findings demonstrated that both the quantity of lipids and the rate of behavioral development were elevated by antibiotic treatments. Cleaning came first, followed by breastfeeding and foraging, depending on when the antibiotic treatment began. While bees treated from larva to adulthood showed a delay in behavioral development and lipid loss, bees treated during the larva-pupa stages showed faster behavioral development and lipid loss. These antibiotic treatment side effects point to a possible function for microbiota in the brain-fat body relationship that is critical to honeybee behavioral development.

Zheng et al. (2018) provided a summary of current studies on antibiotic use. The richness of human gut populations may have been affected by long-term antibiotic use, which has led to high frequencies of resistance determinants (Blaser 2016). Antibiotic exposure has impacted honeybee gut populations in the United States and other nations where beekeepers have utilized antibiotics since the late 1940s to prevent or control larval bacterial illnesses (foulbrood) (Evans and Armstrong 2005; Reybroeck et al. 2012). Compared to the gut bacteria of honeybees from nations that forbid the use of antibiotics in beekeeping, this approach has led to high frequencies of antibiotic resistance determinants in core gut bacteria isolated from bees in the United States (Reybroeck 2017; Tian et al. 2012).

## 7.6 Conclusions and Future Perspectives

An important pollinator that has a big impact on plant genetic diversity and ecosystem stability is the honeybee, *A. mellifera*. Products from honeybees are employed extensively in integrative and traditional medicine in addition to their ecological importance. However, apiculture and biodiversity are seriously threatened by the worldwide collapse of honeybee colonies, which is caused by environmental pollution, pesticide exposure, and diseases. Using probiotic microorganisms to improve colony health is one of the synergistic measures needed to address this loss.

By supporting metabolic balance and pathogen resistance, probiotic supplementation can strengthen honeybees' natural gut microbiota. The environment and food that bees consume—nectar, pollen, and water—create the gut microbiota, which is essential to the health of the colony. The necessity for study into the molecular mechanisms of probiotics in pathogen defense is highlighted by the recommendation that beekeepers use probiotic-based techniques to reduce colony collapse. One of the main hurdles is finding microbial strains that have particular probiotic qualities for honeybees and creating formulations that have been scientifically proven.

Honeybee-derived lactic acid bacteria (LAB) have demonstrated potential for enhancing bee health and lowering the prevalence of pathogens. To fully comprehend their function in detoxifying neonicotinoid pesticides, more investigation is necessary. Cell culture models, such as the Caco-2 gut epithelial system, could assist in evaluating the bioavailability and detoxifying capability of pesticides and their metabolites in the presence of probiotics. The toxicity of pesticide metabolites generated by LAB strains is yet unknown, as is the relative toxicity of these chemicals to their parent molecules.

To further this research, a continuous honeybee cell line must be created. Although the AmE-711 fibroblast-type cell line, which is generated from embryonic tissues of A. mellifera, is accessible, its uses are restricted. The necessity for honeybee-specific models is highlighted by the fact that the majority of research uses insect cell lines.

Probiotic supplements taken over an extended period of time provide a viable substitute for antibiotics and chemical therapies. Natural probiotic formulations have the potential to function as biological control agents in agriculture, biocides for hive disinfection, and alternatives to traditional antibiotics in pathogen prevention. Optimizing probiotic applications, comprehending host-microbe interactions, and creating novel approaches to protect honeybee populations and their environments should be the main goals of future study.

**Declaration of AI Assistance** The authors used the artificial intelligence tools DeepSeek-V3, and Gemini-Google in a limited and assistive capacity to improve the linguistic phrasing and clarity of the following sections of the manuscript: the introduction and the key elements of the chapter. Hereby, we declare that AI assistance was made only for enhancing sentence structure and grammatical errors. All outputs were reviewed, edited, and their accuracy was carefully verified by the authors themselves, who bear full responsibility for the final content of the work. The chapter was subsequently peer-reviewed by specialist reviewers.

## References

Amdam GV, Norberg K, Fondrk MK, Page RE Jr (2004) Reproductive ground plan may mediate colony-level selection effects on individual foraging behavior in honey bees. Proc Natl Acad Sci USA 101:11350–11355. https://doi.org/10.1073/pnas.0403073101

Ament SA, Wang Y, Robinson GE (2010) Nutritional regulation of division of labor in honey bees: toward a systems biology perspective. WIREs Syst Biol Med 2:566–576

Anderson KE, Ricigliano VA (2017) Honey bee gut dysbiosis: A novel context of disease ecology. Curr Opin Insect Sci 22:125–132

Anderson KE et al (2013) Microbial ecology of the hive and pollination landscape: bacterial associates from floral nectar, the alimentary tract and stored food of honey bees (Apis mellifera). PLoS One 8:e83125

Anderson KE et al (2014) Hive-stored pollen of honey bees: many lines of evidence are consistent with pollen preservation, not nutrient conversion. Mol Ecol 23:5904–5917

Anjum SI, Shah AH, Aurongzeb M, Kori J, Azim MK, Ansari MJ, Bin L (2018) Characterization of gut bacterial flora of Apis mellifera from north-west Pakistan. Saudi J Biol Sci 25:388–392

Antonio DSM, Guidugli-Lazzarini KR, do Nascimento AM, Simoes ZLP, Hartfelder K (2008) RNAi-mediated silencing of vitellogenin gene function turns honeybee (*Apis mellifera*) workers into extremely precocious foragers. Naturwissenschaften 95:953–961

Arien Y, Dag A, Zarchin S, Masci T, Shafir S (2015) Omega-3 deficiency impairs honey bee learning. Proc Natl Acad Sci USA 112:15761–15766. https://doi.org/10.1073/pnas.1517375112

Babendreier D, Joller D, Romeis J, Bigler F, Widmer F (2007) Bacterial community structures in honeybee intestines and their response to two insecticidal proteins. FEMS Microbiol Ecol 59: 600–610

Blaser MJ (2016) Antibiotic use and its consequences for the normal microbiome. Science 352: 544–545
Bleau N, Bouslama S, Giovenazzo P, Derome N (2020) Dynamics of the Honeybee (*Apis mellifera*) gut microbiota throughout the overwintering period in Canada. Microorganisms 8(8):1146. https://doi.org/10.3390/microorganisms8081146
Bommuraj V, Chen Y, Birenboim M, Barel S, Shimshoni JA (2021) Concentration- and time-dependent toxicity of commonly encountered pesticides and pesticide mixtures to honeybees (*Apis mellifera* L.). Chemosphere 266:128974
Bonilla-Rosso G, Engel P (2018) Functional roles and metabolic niches in the honey bee gut microbiota. Curr Opin Microbiol 43:69–76. https://doi.org/10.1016/j.mib.2017.12.009
Bottacini F et al (2012) *Bifidobacterium asteroides* PRL2011 genome analysis reveals clues for colonization of the insect gut. PLoS One 7:e44229
Brochet S, Quinn A, Mars RA, Neuschwander N, Sauer U, Engel P (2021) Niche partitioning facilitates coexistence of closely related honey bee gut bacteria. elife 10. https://doi.org/10.7554/eLife.68583
Cariveau DP, Elijah Powell J, Koch H, Winfree R, Moran NA (2014) Variation in gutmicrobial communities and its association with pathogen infection in wild bumblebees (Bombus). ISME J8:2369–2379
Carvalho SM, Belzunces LP, Carvalho GA et al (2013) Enzymatic biomarkers as tools to assess environmental quality: A case study of exposure of the honeybee *Apis mellifera* to insecticides. Environ Toxicol Chem 32:2117–2124
Catae AF, Roat TC, Pratavieira M et al (2018) Exposure to a sublethal concentration of imidacloprid and the side effects on target and nontarget organs of *Apis mellifera* (Hymenoptera, Apidae). Ecotoxicology 27:109–121
Cilia G, Fratini F, Tafi E et al (2020a) Microbial profile of the ventriculum of honey bee *Apis mellifera ligustica* (Spinola, 1806) fed with veterinary drugs, dietary supplements and non-protein amino acids. Vet Sci 7:76
Cilia G, Fratini F, Tafi E et al (2020b) Changes of Western honey bee *Apis mellifera ligustica* (Spinola, 1806) ventriculus microbial profile related to their in-hive tasks. J Apic Res 60:198–202
Clemente JC, Ursell LK, Parfrey LW, Knight R (2012) The impact of the gut microbiota on human health: an integrative view. Cell 148:1258–1270
Corby-Harris V, Maes P, Anderson KE (2014a) The bacterial communities associated with honey bee (*Apis Mellifera*) foragers. PLoS One 9:e95056
Corby-Harris V, Snyder LA, Schwan MR, Maes P, McFrederick QS, Anderson KE (2014b) Origin and effect of Alpha 2.2 Acetobacteraceae in honey bee larvae and description of *Parasaccharibacter apium* gen. nov., sp. nov. Appl Environ Microbiol 80:7460–7472
Cryan JF, Dinan TG (2012) Mind-altering microorganisms: the impact of the gut microbiota on brain and behaviour. Nat Rev Neurosci 13:701–712
Cummings JH, Pomare EW, Branch WJ, Naylor CP, Macfarlane GT (1987) Short chain fatty acids in human large intestine, portal, hepatic and venous blood. Gut 28:1221–1227
Daisley BA, Pitek AP, Chmiel JA et al (2020) *Lactobacillus* spp. attenuate antibiotic-induced immune and microbiota dysregulation in honey bees. Commun Biol 3:534
D'Alvise P, Böhme F, Codrea MC et al (2018) The impact of winter feed type on intestinal microbiota and parasites in honey bees. Apidologie 49:252–264
Diaz T, Del-Val E, Ayala R, Larsen J (2019) Alterations in honey bee gut microorganisms caused by *Nosema* spp. and pest control methods. Pest Manag Sci 75(3):835–843
Donaldson GP, Lee SM, Mazmanian SK (2016) Gut biogeography of the bacterial microbiota. Nat Rev Microbiol 14:20–32
Dong ZX, Li HY, Chen YF et al (2020) Colonization of the gut microbiota of honey bee (*Apis mellifera*) workers at different developmental stages. Microbiol Res 231:126370
Donohoe DR et al (2011) The microbiome and butyrate regulate energy metabolism and autophagy in the mammalian colon. Cell Metab 13:517–526

Dosch C, Manigk A, Streicher T et al (2021) The gut microbiota can provide viral tolerance in the honey bee. Microorganisms 9:871. https://doi.org/10.3390/microorganisms9040871

Duan XL, Xiong MQ, Liu WB et al (2020) Effects of three fungicides on the activities of protective enzymes and detoxifying enzymes in *Apis mellifera*. Acta Pratac Sin 29:74–82

Duan XL, Zhao BA, Jin X et al (2021) Antibiotic treatment decrease the fitness of honeybee (*Apis mellifera*) larvae. Insects 12:301. https://doi.org/10.3390/insects12040301

Ellegaard KM, Engel P (2019) Genomic diversity landscape of the honey bee gut microbiota. Nat Commun 10:446. https://doi.org/10.1038/s41467-019-08303-0

Emery O, Schmidt K, Engel P (2017) Immune system stimulation by the gut symbiont *Frischella perrara* in the honey bee (*Apis mellifera*). Mol Ecol 26:2576–2590

Engel P, Moran NA (2013a) The gut microbiota of insects - diversity in structure and function. FEMS Microbiol Rev 37:699–735

Engel P, Moran NA (2013b) Functional and evolutionary insights into the simple yet specific gut microbiota of the honeybee from metagenomic analysis. Gut Microbes 4:60–65

Engel P, Martinson VG, Moran NA (2012) Functional diversity within the simple gut microbiota of the honey bee. Proc Natl Acad Sci USA 109:11002–11007. https://doi.org/10.1073/pnas.1202970109

Engel P, Kwong WK, Moran NA (2013) *Frischella perrara* gen. nov., sp. nov., a gamma proteobacterium isolated from the gut of the honeybee, *Apis mellifera*. Int J Syst Evol Microbiol 63:3646–3651

Engel P, Bartlett KD, Moran NA (2015) The bacterium *Frischella perrara* causes scab formation in the gut of its honeybee host. MBio 6:e00193–e00115

Evans J, Armstrong TN (2005) Inhibition of the American foulbrood bacterium, *Paenibacillus* larvae, by bacteria isolated from honey bees. J Apic Res 44:168–171

Evans JD et al (2006) Immune pathways and defence mechanisms in honey bees *Apis mellifera*. Insect Mol Biol 15:645–656

Flint HJ, Scott KP, Louis P, Duncan SH (2012) The role of the gut microbiota in nutrition and health. Nat Rev Gastroenterol Hepatol 9:577–589

Gilliam M (1997) Identification and roles of non-pathogenic microflora associated with honey bees. FEMS (Fed Eur Microbiol Soc) Microbiol Lett 155(1):1–10

Glenny W, Cavigli I, Daughenbaugh KF et al (2017) Honeybee (*Apis mellifera*) colony health and pathogen composition in migratory beekeeping operations involved in California almond pollination. PLoS One 12:e0182814

Hall AB, Tolonen AC, Xavier RJ (2017) Human genetic variation and the gut microbiome in disease. Nat Rev Genet 18:690–699

Harris JW, Woodring J (1992) Effects of stress, age, season, and source colony on levels of octopamine, dopamine and serotonin in the honey bee (*Apis mellifera* L.) brain. J Insect Physiol 38:29–35. https://doi.org/10.1016/0022-1910(92)90019-A

Horak RD, Leonard SP, Moran NA (2020) Symbionts shape host innate immunity in honey bees. Proc Biol Sci 287:20201184

Huang Q, Lariviere PJ, Powell JE, Moran NA (2023) Engineered gut symbiont inhibits microsporidian parasite and improves honey bee survival. Proc Natl Acad Sci USA 120: e2220922120

Jeyaprakash A, Hoy MA, Allsopp MH (2003) Bacterial diversity in worker adults of *Apis mellifera capensis* and *Apis mellifera scutellata* (Insecta: Hymenoptera) assessed using 16S rRNA sequences. J Invertebr Pathol 84:96–103

Jones JC, Fruciano C, Hildebrand F et al (2017) Gut microbiota composition is associated with environmental landscape in honey bees. Ecol Evol 8:441–451

Kaltenpoth M (2011) Honey bees and bumblebees share similar bacterial symbionts. Mol Ecol 20(3):439–440

Kešnerová L, Moritz R, Engel P (2016) Bartonella apis sp. nov., a honey bee gut symbiont of the class Alphaproteobacteria. Int J Syst Evol Microbiol 66:414–421

Kešnerová L, Mars RAT, Ellegaard KM et al (2017) Disentangling metabolic functions of bacteria in the honey bee gut. PLoS Biol 15:e2003467. https://doi.org/10.1371/journal.pbio.2003467

Koch H, Schmid-Hempel P (2011) Socially transmitted gut microbiota protect bumblebees against an intestinal parasite. Proc Natl Acad Sci USA 108:19288–19292

Kwong WK, Moran NA (2013) Cultivation and characterization of the gut symbionts of honey bees and bumble bees: description of *Snodgrassella alvi* gen. nov., sp. nov., a member of the family Neisseriaceae of the Betaproteobacteria, and *Gilliamella apicola* gen. nov., sp. nov., a member of Orbaceae fam. nov., Orbales ord. nov., a sister taxon to the order Enterobacteriales of the Gammaproteobacteria. Int J Syst Evol Microbiol 63:2008–2018. https://doi.org/10.1099/ijs.0.044875-0

Kwong WK, Moran NA (2016) Gut microbial communities of social bees. Nat Rev Microbiol 14:374–384. https://doi.org/10.1038/nrmicro.2016.43

Kwong WK, Engel P, Koch H, Moran NA (2014) Genomics and host specialization of honey bee and bumble bee gut symbionts. Proc Natl Acad Sci USA 111:11509–11514. https://doi.org/10.1073/pnas.1405838111

Kwong WK, Mancenido AL, Moran NA (2017) Immune system stimulation by the native gut microbiota of honey bees. R Soc Open Sci 4:170003

Lang H et al (2023) Engineered symbiotic bacteria interfering Nosema redox system inhibit microsporidia parasitism in honey bees. Nat Commun 14:2778

Ley RE, Hamady M, Lozupone C et al (2008) Evolution of mammals and their gut microbes. Science 320:1647–1651

Li JH, Evans JD, Li WF et al (2017) New evidence showing that the destruction of gut bacteria by antibiotic treatment could increase the honey bee's vulnerability to *Nosema* infection. PLoS One 12:e0187505

Li WF, Chen YP, Cook SC (2018) Chronic *Nosema ceranae* infection inflicts comprehensive and persistent immuno-suppression and accelerated lipid loss in host *Apis mellifera* honey bees. Int J Parasitol 48:433–444

Martinson VG, Danforth BN, Minckley RL et al (2011) A simple and distinctive microbiota associated with honeybees and bumble bees. Mol Ecol 20:619–628

Martinson VG, Moy J, Moran NA (2012) Establishment of characteristic gut bacteria during development of the honeybee worker. Appl Environ Microbiol 78(8):2830–2840

Menzel R (2012) The honeybee as a model for understanding the basis of cognition. Nat Rev Neurosci 13:758–768. https://doi.org/10.1038/nrn3357

Miller DL, Smith EA, Newton ILGA (2021) A bacterial symbiont protects honey bees from fungal disease. MBio 12:e0050321

Moran NA, Hansen AK, Powell JE, Sabree ZL (2012) Distinctive gut microbiota of honey bees assessed using deep sampling from individual worker bees. PLoS One 7:e36393

Motta E, Raymann K, Moran NA (2018) Glyphosate perturbs the gut microbiota of honey bees. Proc Natl Acad Sci USA 115(41):10305–10310

Nancy AM (2015) Genomics of the honey bee microbiome. Curr Opin Insect Sci 10:22–28

Nappi AJ, Christensen BM (2005) Melanogenesis and associated cytotoxic reactions: applications to insect innate immunity. Insect Biochem Mol Biol 35:443–459

Newman MM et al (2016) Glyphosate effects on soil rhizosphere-associated bacterial communities. Sci Total Environ 543:155–160

Niode NJ, Salaki CL, Rumokoy JM, Tallei TE (2019) Lactic acid bacteria from honey bees digestive tract and their potential as probiotics. Advances in biological sciences research, International Conference and the 10th Congress of the Entomological Society of Indonesia (ICCESI 2019)

Odelson DA, Breznak JA (1983) Volatile fatty acid production by the hindgut microbiota of xylophagous termites. Appl Environ Microbiol 45:1602–1613

Oliveira RA, Roat TC, Carvalho SM, Malaspina O (2014) Side-effects of thiamethoxam on the brain and midgut of the Africanized honeybee *Apis mellifera* (Hymenopptera: Apidae). Environ Toxicol 29:1122–1133

Olofsson TC, Alsterfjord M, Nilson B (2014) *Lactobacillus apinorum* sp. nov., *Lactobacillus mellifer* sp. nov., *Lactobacillus mellis* sp. nov., *Lactobacillus melliventris* sp. nov., *Lactobacillus kimbladii* sp. nov., *Lactobacillus helsingborgensis* sp. nov. and *Lactobacillus kullabbergensis* sp. nov., isolated from the honey stomach of the honeybee *Apis mellifera*. Int J Syst Evol Microbiol 64:3109–3119. https://doi.org/10.1099/ijs.0.059600-0

Ortiz-Alvarado Y, Clark DR, Vega-Melendez CJ et al (2020) Antibiotics in hives and their effects on honey bee physiology and behavioral development. Biol Open 9:bio053884

Powell JE, Martinson VG, Urban-Mead K, Moran NA (2014) Routes of acquisition of the gut microbiota of the honey bee *Apis mellifera*. Appl Environ Microbiol 80(23):7378–7387

Raine NE, Ings TC, Dornhaus A, Selah N, Chittka L (2006) Adaptation, genetic drift, pleiotropy, and history in the evolution of bee foraging behavior. Adv Stud Behav 36:305–354. https://doi.org/10.1016/S0065-3454(06)36007-X

Raymann K, Moran NA (2018) The role of the gut microbiome in health and disease of adult honey bee workers. Curr Opin Insect Sci 26:97–104

Raymann K, Shaffer Z, Moran NA (2017) Antibiotic exposure perturbs the gut microbiota and elevates mortality in honeybees. PLoS Biol 15:e2001861

Raymann K, Bobay LM, Moran NA (2018) Antibiotics reduce genetic diversity of core species in the honeybee gut microbiome. Mol Ecol 27:2057–2066

Reybroeck W (2017) Residues of antibiotics and chemotherapeutics in honey. J Apicult Res 57:97–112

Reybroeck W, Daeseleire E, De Brabander HF, Herman L (2012) Antimicrobials in beekeeping. Vet Microbiol 158:1–11

Sabree ZL, Hansen AK, Moran NA (2012) Independent studies using deep sequencing resolve the same set of core bacterial species dominating gut communities of honey bees. PLoS One 7: e41250

Scardovi V, Trovatelli D (1969) New species of bifidobacteria from *Apis mellifica* L. and *Apis indica* F. A. contribution to the taxonomy and biochemistry of the genus Bifidobacterium. Zentralbl Bakteriol Parasitenkd Infektionskr Hyg 123:64–88

Scheiner R, Baumann A, Blenau W (2006) Aminergic control and modulation of honeybee behaviour. Curr Neuropharmacol 4:259–276. https://doi.org/10.2174/157015906778520791

Scheiner R, Abramson CI, Brodschneider R et al (2013) Standard methods for behavioural studies of *Apis mellifera*. J Apicult Res 52:1–58. https://doi.org/10.3896/IBRA.1.52.4.04

Shehata AA, Schrödl W, Aldin AA, Hafez HM, Krüger M (2013) The effect of glyph-osate on potential pathogens and beneficial members of poultry microbiota in vitro. Curr Microbiol 66: 350–358

Shilo T, Zygier L, Rubin B, Wolf S, Eizenberg H (2016) Mechanism of glyphosate control of *Phelipanche Aegyptiaca*. Planta 244:1095–1107

Steele MI, Motta EVS, Gattu T, Martinez D, Moran NA (2021) The gut microbiota protects bees from invasion by a bacterial pathogen. Microbiol Spectrum 9:e0039421

Tian B, Fadhil NH, Powell JE et al (2012) Long-term exposure to antibiotics has caused accumulation of resistance determinants in the gut microbiota of honeybees. MBio 3:e00377–12

Toma DP, Bloch G, Moore D, Robinson GE (2000) Changes in period mRNA levels in the brain and division of labor in honey bee colonies. Proc Natl Acad Sci USA 97:6914–6919. https://doi.org/10.1073/pnas.97.12.6914

Van Der Zee R, Pisa L, Andonov S et al (2012) Managed honey bee colony losses in Canada, China, Europe, Israel and Turkey, for the winters of 2008-9 and 2009-10. J Apic Res 51:100–114

Van Dooremalen C, Cornelissen B, Poleij-Hok-Ahin C, Blacquière T (2018) Single and interactive effects of *Varroa destructor*, *Nosema* spp., and imidacloprid on honey bee colonies (*Apis mellifera*). Ecosphere 9:e02378

Vásquez A, Forsgren E, Fries I et al (2012) Symbionts as major modulators of insect health: lactic acid bacteria and honeybees. PLoS One 7:e33188

Vernier CL, Chin IM, Adu-Oppong B et al (2020) The gut microbiome defines social group membership in honey bee colonies. Sci Adv 6:eabd3431. https://doi.org/10.1126/sciadv.abd3431

Villalba A, Maggi M, Ondarza PM et al (2020) Influence of land use on chlorpyrifos and persistent organic pollutant levels in honeybees. Bee bread and honey: beehive exposure assessment. Sci Total Environ 713:136554

Wang K, Li J, Zhao L et al (2021) Gut microbiota protects honey bees (*Apis mellifera* L.) against polystyrene microplastics exposure risks. J Hazard Mater 402:123828

Wheeler MM, Ament SA, Rodriguez-Zas SL et al (2015) Diet and endocrine effects on behavioral maturation-related gene expression in the pars intercerebralis of the honey bee brain. J Exp Biol 218:4005–4014

Zeitouni NE, Chotikatum S, von Köckritz-Blickwede M, Naim HY (2016) The impact of hypoxia on intestinal epithelial cell functions: consequences for invasion by bacterial pathogens. Mol Cell Pediatr 3:14

Zhang Z, Mu X, Shi Y, Zhenga H (2022) Distinct roles of honeybee gut bacteria on host metabolism and neurological processes. Microbiol Spectrum 10(2). https://doi.org/10.1128/spectrum.02438-21

Zheng H, Nishida A, Kwong WK et al (2016) Metabolism of toxic sugars by strains of the bee gut symbiont *Gilliamella apicola*. MBio 7:e01326-e01316. https://doi.org/10.1128/mBio.01326-16

Zheng H, Powell JE, Steele MI et al (2017) Honeybee gut microbiota promotes host weight gain via bacterial metabolism and hormonal signaling. Proc Natl Acad Sci USA 114:4775–4780. https://doi.org/10.1073/pnas.1701819114

Zheng H, Steele MI, Leonard SP et al (2018) Honeybees as models for gut microbiota research. Nature Lab Anim 47:317–325. https://doi.org/10.1038/s41684-018-0173-x

Zheng H, Perreau J, Powell JE et al (2019) Division of labor in honey bee gut microbiota for plant polysaccharide digestion. Proc Natl Acad Sci USA 116:25909–25916. https://doi.org/10.1073/pnas.1916224116

Zheng J et al (2020) A taxonomic note on the genus *Lactobacillus*: description of 23 novel genera, emended description of the genus *Lactobacillus* Beijerinck 1901, and union of Lactobacillaceae and Leuconostocaceae. Int J Syst Evol Microbiol 70:2782–2858

Zhu L, Qi S, Xue X et al (2020) Nitenpyram disturbs gut microbiota and influences metabolic homeostasis and immunity in honey bee (*Apis mellifera* L.). Environ Pollut 258:113671

# Part II
# Secondary Metabolism of Plants for Managing the Agricultural Pests

# Chapter 8
# Using Plant Metabolic Products to Develop Relationship Among Plants, Insect Pests, and Their Natural Enemies

**Samar Sayed Ibrahim and Huda H. Elbehery**

## 8.1 Introduction

The world's population is rapidly rising, with an estimated increase of 1.940.235 billion people over the next 30 years. It could reach 9,735,034 billion people in 2050, up from 7,794,799 billion in 2020 (United Nations, Dep. of Economic and Social Affairs 2019). Food demand is increasing due to the world's rapidly growing population. At the same time, agricultural production suffers a one-third loss due to pest and disease damage in various crops and regions around the world (Oerke 2006). Insect pests are constantly adapting to crop hosts pose a threat at all stages of agricultural food production (Cingel et al. 2016). It has also been known that before and after harvest, insect pests lessen agricultural yields by approximately 30% (Bebber et al. 2013). Surprisingly, out of approximately 6 million recognized species only about 20–30 insect species were found important pests for major crops with notable commercial and agricultural consequences (García-Lara and Serna Saldivar 2016).

A pest is any animal, plant, or microorganism that is harmful or destructive to food or health (Mahmood et al. 2016). Pests' populations are dominant in agricultural ecosystems, and insect pests not only endanger homes, gardens, and water sources, but they also spread a variety of diseases since they are hosts of some parasites that may cause diseases. Pests also can develop resistance to chemicals, and have several generations per year, with female fecundity raising the risk of random genetic mutation and/or rapid accumulation of mutants that are resistant (Whalon et al. 2008; Bishop and Grafius 1996). Crop protection products including synthetic pesticides and herbicides have long been used to keep and boost crop output.

S. S. Ibrahim · H. H. Elbehery (✉)
Pests and Plant Protection Department, National Research Centre, Dokki, Egypt
e-mail: si.sayed@nrc.sci.eg



N. El-Wakeil, A. Abdelkhalek (eds.), *Microbiomes and Plant Secondary Metabolites for Insect Pest Management*, Progress in Biological Control 23,
https://doi.org/10.1007/978-3-032-24330-0_8

Pesticides are any compound or combination of components designed to stop, destroy, repel, or mitigate pests (FAO 2002). Pesticides are commonly classified based on their target, mechanism of action, and chemical composition (Laxmishree and Nandita 2017). Besides that, pesticides are classified into two broad categories: synthetic pesticides and biopesticides (Lyznicki et al. 1997). Biopesticides are further classified into four subgroups: (i) microbial biopesticides, (ii) plant incorporated protectants, (iii) botanical pesticides, and (iv) pheromones.

Over millions of years plants and insects have been living together and constitute the largest diversity. Both have improved techniques to avoid each other's defense systems through co-evolution (War et al. 2012). Numerous biotic and abiotic factors constantly threaten plants, with biotic stress from insect herbivory being a major factor limiting crop productivity (Sharma et al. 2009). Plants have, nevertheless, developed a number of highly dynamic defense mechanisms against insect herbivory. Herbivore injury causes morphological, biochemical, and molecular changes in plants (Sharma et al. 2009; War et al. 2011a). There are two types of plant resistance to herbivory: constitutive and induced (Sharma et al. 2009; Kessler and Baldwin 2001). Induced defenses are triggered in reaction to an herbivore attack, whereas constitutive resistance is expressed by the plant independently of external stimuli (Sharma et al. 2009; War et al. 2011a). Resistances in plants, both constitutive and induced, are crucial for protecting them from insect herbivory. Insect invasion has forced plants to protect themselves, since they couldn't escape from pest attack, consequently they have evolved a wide range of advanced defense systems to protect their tissues. Plants produce certain secondary metabolites and proteins that are toxic, repellent, and antifeedant to herbivores in order to defend against their attacks (Usha Rani and Jyothsna 2010; War et al. 2011b). In the following sections, plant defense systems, plant metabolites, and botanical pesticides as well as relation to natural enemies will be discussed.

## 8.2 Plant Defense Systems

Plants counter insect invasion by direct defense through affecting host plant preference or survival and reproductive success, and indirect defense by other species such as natural enemies of the insect pests. According to Aljbory and Chen (2018), plants use both direct and indirect defense against insects. Direct defenses are occurred by plant traits that impact the biology of herbivore such as mechanical defense on the surface's plant (e.g., hairs, spines, thicker leaves, trichomes, and thorns) or produce of toxic substances (such as phenols, alkaloids, terpenoids, and quinones). These morphological or chemical aspects influence insect growth, development, and survival (Howe and Jander 2008; War et al. 2018). To defend themselves from phytophagous herbivores, plants have developed a variety of secondary metabolites.

These secondary metabolites represent a significant barrier to herbivorous pests (Mithoefer and Boland 2012). These supply direct defense via repellent, toxic or anti-feeding effects on herbivorous pests (Bruce 2015). On the other hand, indirect defenses are provided through plant secondary metabolism, by attracting natural enemies of pests (van Loon et al. 2000; Heil 2008). In indirect defense, plants produce a special combination of volatile compounds, which are called herbivore-induced plant volatiles (HIPVs). HIPVs are a cue that various parasitoids and predators can utilize to locate their prey (Clavijo McCormick et al. 2012; Erb and Reymond 2019). Plant chemistry significantly influences the performance and behavior of natural enemies. Plant volatiles that are released specifically through pests' attack are used as host location cues by many different natural enemies. Whereas the release of a mixture of volatiles that especially draw the herbivores' natural enemies as well as the provision of shelter and food (floral nectar) can both act as indirect defenses against insects and boost efficiency of the natural enemies (Arimura et al. 2009). To prevent any resistance and/or adaptation to these chemical compounds, they work either in conjunction or synergistically against a specific stress through a variety of ways of action. It is crucial to research the synthesis and use of these compounds to include them in effective and reliable pest management programs because plant secondary metabolites are the primary plant defense chemicals against several stressors, including insect herbivory.

## 8.3 Plant Secondary Metabolites and Insect Pest Control

The majority of plants that are abundant in the environment have been used in medicines, ornamentals, spices, food, and feed (Srijita 2015). Plant products such as crude, extracts, and essential oils have been used as natural pesticides for centuries (El-Wakeil 2013). Plant products from various families can act as insect antifeedants, repellents, and growth regulators (Isman 2005), as well as interfere with the physiology of insects (Laxmishree and Nandita 2017). The conventional pesticide market is expected to shrink by 1.5% each year, while bio-pesticides are expected to account for 20% of the pesticide market by 2025 (Isman 2015). Botanical pesticides degrade quickly, are less toxic to humans, have multiple modes of action, do not pollute the environment, and are abundant in nature (Neeraj et al. 2017; Lengai et al. 2020). As a result, they alongside host resistance, the use of natural enemies (e.g., predators, parasitoids), good agricultural practices (GAP), microbial pesticides, and the confined usage of safe agrochemicals, are an essential element of integrated pest management (Lengai et al. 2020; Muthomi et al. 2017; Solomon et al. 2000). Botanical insecticides may occur as natural resources or derived constituents from plants. Crude botanical insecticides, such as neem in India, pyrethrum in Persia, rotenone in South America and East Asia, and sabadilla in South and Central America, have been utilized for centuries (Weinzierl 2000). Some botanical insecticides, such as rotenone and nicotine, may cause toxicity to humans (Regnault-Roger and Philogène 2008).

Plants from a variety of families, such as Zingiberaceae, Lauraceae, Myrtaceae, Rutaceae, Lamiaceae, and Apocynaceae, have been found to contain bioactive compounds (Lengai et al. 2020; Gakuubi et al. 2016). Secondary metabolites are

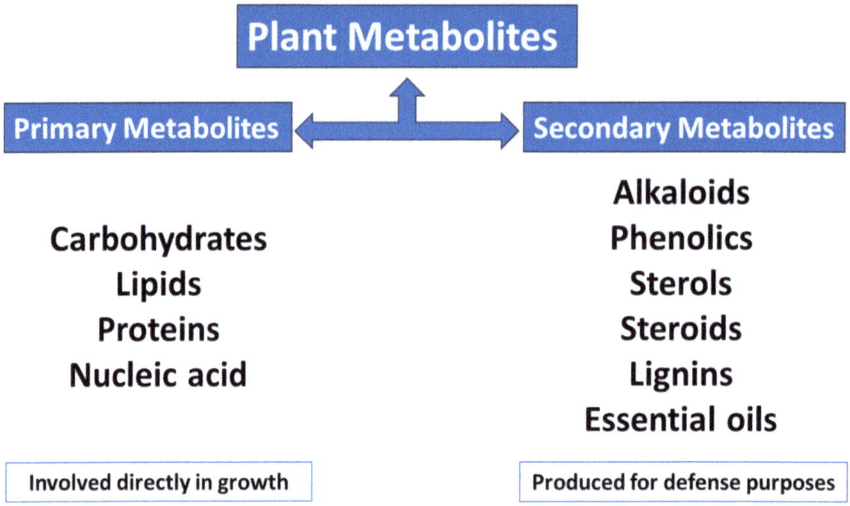


**Fig. 8.1** Plant metabolites

abundant in plants and contribute significantly to the distinctive odors, colors, and tastes of plants and serve as the plant's defensive system since it reacts to insect pests, pathogen infection, and other stresses (Bennett and Wallsgrove 1994). The three groups of secondary metabolites are (i) terpenes; sterols, plant volatiles, carotenoids, and glycosides, (ii) phenolics; phenolic acids, coumarins, lignans, stilbenes, flavonoids, tannins, and lignin, and (iii) nitrogen-containing compounds; alkaloids and glucosinolates (Agostini-Costa et al. 2012). Unlike primary metabolites, which represent main building blocks of living organisms (e.g., nucleic acids, carbohydrates, proteins, and lipids), secondary metabolites do not contribute to plant growth and development but instead protect the plant from herbivores by decreasing the palatability of its tissues (Fig. 8.1) (Agostini-Costa et al. 2012). Currently, four main groups of plant products are utilized to control insect pests; (i) pyrethrum, (ii) rotenone, (iii) neem, and (iv) essential oils, with sabadilla, nicotine, and Ryania being used in small portions (Isman 2006). As shown in Fig. 8.2, the mode of action of botanical insecticides differed according to the source of the product.

## 8.4 Mode of Action of Plant Metabolites on Insect Pests

The mode of action of a pesticide describes how it works, and it refers to how the pesticide affects the pest's particular systems. It depicts the specific biochemical interactivity that occurs when a pesticide generates its influence on pests. The affected enzyme, protein, or biological phase is usually included in the mode of action. The most other classifications relate to the controlled pests, physical features, and chemical properties, whereas the mechanism of action relates to which

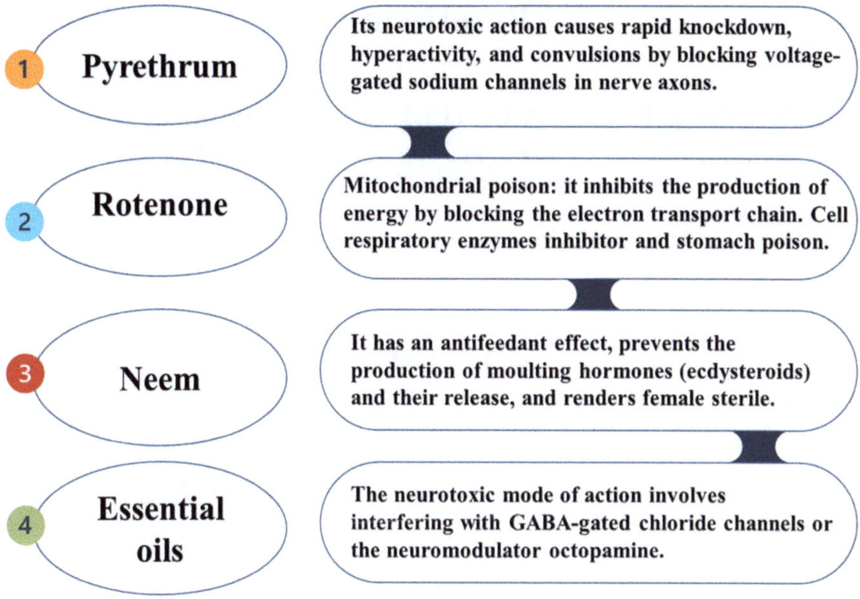


**Fig. 8.2** Major groups of botanical insecticides and their toxic effects on insects

biological function the pesticide disrupts (Bloomquist et al. 2008). Scientists must understand the mode of action to enhance a product's quality and sustainability. To realize how pesticides function (their mechanism of action), first recognize how the pests' specific target system works. Understanding how human systems work is also useful for identifying differences and similarities in both humans and the pests we are attempting to control. Understanding the modes of action of the pesticides is also critical to prevent acquired pest resistance for the used pesticide. Using pesticides with similar modes of action exacerbates that issue by causing death to susceptible pests and having left those with resistance to the whole group of pesticides with similar mechanisms (Brown 2005).

## *8.4.1 Botanical Metabolites and the Nervous System*

This category includes the vast majority of insecticides, both biological and synthetic, that act on a variety of targets in insects.

### 8.4.1.1 Voltage-Gated Sodium Channels

Pyrethrum is an oleoresin derived from the dried flowers of *Tanacetum cinerariifolium* (Asteraceae). It consists of two major compounds, Pyrethrins I and

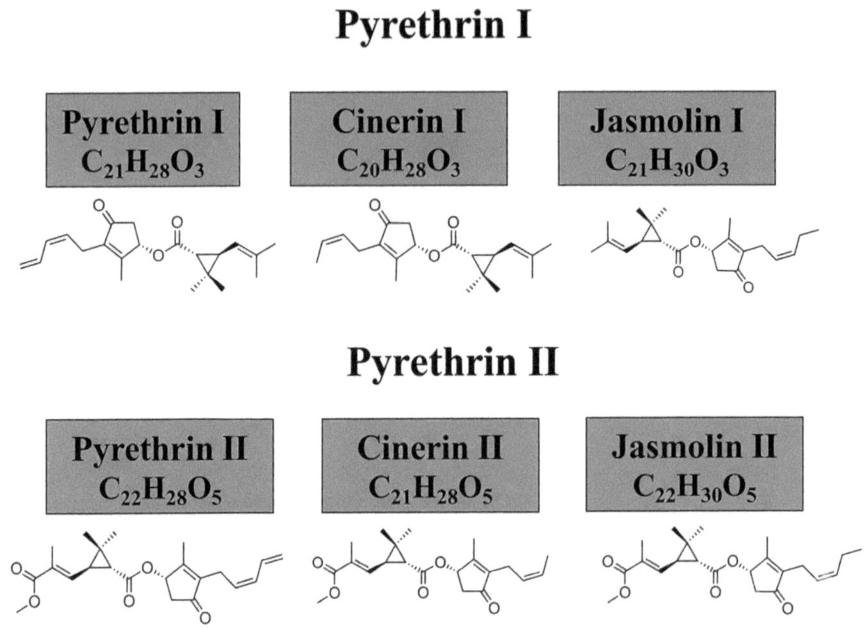


**Fig. 8.3** Chemical structure of some pyrethrins

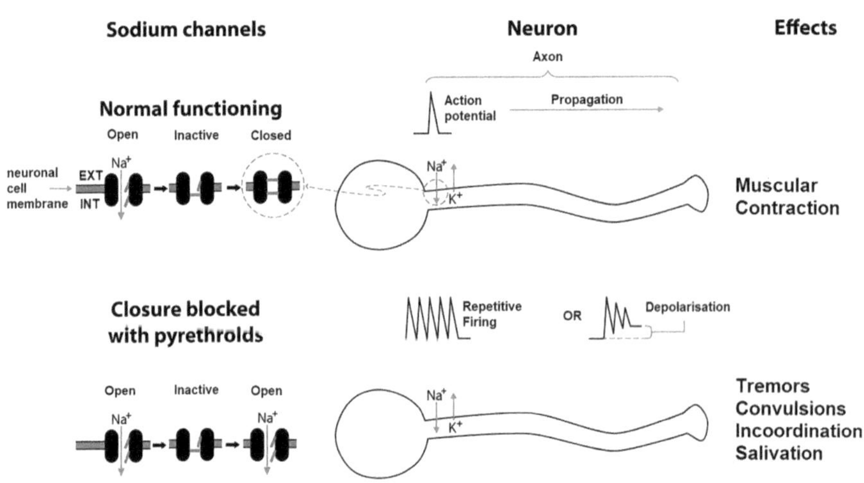


**Fig. 8.4** Pyrethroids' mode of action on neurons (Hénault-Ethier 2016)

II (Fig. 8.3) (Isman 2006), which act as a modulator on voltage-gated sodium channels. These channels are required for proper electrical signaling in the nervous system, resulting in sodium channel delay. This mode of action resembles that of other chemical insecticides (e.g., synthetic pyrethrins) (Fig. 8.4) (Hénault-Ethier

2016). On the other hand, the pyrethrins found naturally proved to be more target-specific than synthetic pyrethrin products (Du et al. 2013). Pyrethrins are contact and stomach poison with lower mammalian toxicity and a significant short residual activity as they degrade by sunlight, air, and moisture; thus, repeated application is essential (Casida and Quistad 1995). Pyrethrins are used to control a variety of insects and mites, which include flies, fleas, beetles, and spider mites (Casida and Quistad 1998). Because pyrethrins degrade rapidly in sunlight, they should be stored in the dark. Pyrethrins should not be blended with solutions of lime or soap because both highly alkaline and highly acidic conditions accelerate degradation. Their liquid formulations retain their effectiveness in storage, while powder preparations can lose approximately 20% in a year. Some pests require repeated applications to be effectively managed (Souto et al. 2021).

For many years, peoples of South and Central America have used sabadilla, an insecticide preparation derived from *Schoenocaulon officinale* (Melanthiaceae) pulverized seeds. Its alkaloid preparation is primarily composed of two main alkaloids, cevadine and veratridine, in a ratio of 2:1, and has a similar mechanism of action to that of pyrethrins, even though binding site would seem to be different (Copping and Duke 2007). One of the low toxic pesticide-active plant extracts is sabadilla; it is categorized as a contact and stomach poison with relatively low activity. When the main alkaloids of sabadilla are separated, they are highly toxic to humans and adversely influence stinks, squash bugs, thrips, leafhoppers, and caterpillars (Hare and Morse 1997).

#### 8.4.1.2 Voltage-Gated Calcium Channels

Ryanodine is an active component found in the roots and woody stems of the Trinidadian native *Ryania speciosa* (Salicaceae). It stimulates calcium channels in skeletal muscle cells' sarcoplasmic reticulum. Once calcium channels are stimulated, an excess of calcium ions is released into actin and myosin protein filaments, causing contraction of skeletal muscle and paralysis (Nauen 2006). Ryanodine is a fast-acting poison that boosts insecticidal action by contact or stomach, has low toxicity on mammals, and has a long residual activity, offering up to two weeks of control after the initial application. Piperonylbutoxide (PBO) synergizes the insecticidal activity of *Ryania* crude extracts, that are the most efficacious against caterpillars, worms, potato beetles, lace bugs, aphids, and squash bugs in hot water (Shivanandappa and Rajashekar 2014).

#### 8.4.1.3 Acetylcholinesterase Enzyme

Acetylcholinesterase (AChE) is an enzyme that hydrolyzes the neurotransmitter acetylcholine, which regulates cholinergic nervous impulse transmission and could be a biopesticide target. Coumaran (2,3-dihydrobenzofuran), an active component in *Lantana camara* L. (Verbenaceae), inhibits this enzyme, increasing acetylcholine

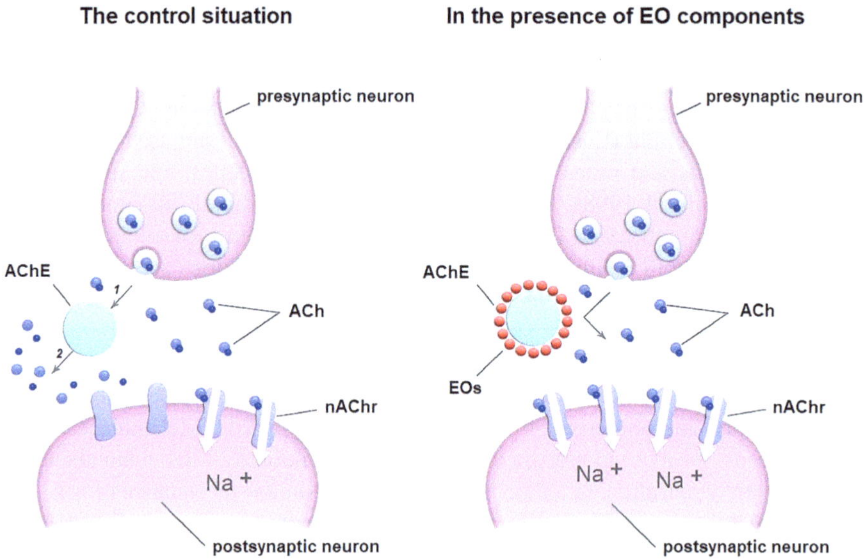


**Fig. 8.5** Components of essential oil (EO) inhibit the acetylcholinesterase (AChE) activity (Jankowska et al. 2018)

concentrations in the synapse cleft, resulting in extreme neuroexcitation because of the neurotransmitter's prolonged biding to its postsynaptic receptor, giving rise in restlessness, hyperexcitability, tremors, convulsion, paralysis and finally death (Fig. 8.5) (Jankowska et al. 2018). Despite its short residual activity, it is non-toxic to mammals and functions quickly on houseflies and stored product pests. Coumaran's mechanism of action is similar to that of the monoterpene 1,8-cineole, and other synthetic pesticides like organophosphates and carbamates (Hansen et al. 2021). It has been demonstrated the inhibitory activity of a methanolic extract of *Cassia fistula* L. (Fabaceae) roots and suggested indole alkaloids as new potential active agents (Khorshid et al. 2015). However, while many essential oils and terpenes have shown anti-AChE activity in vitro, their involvement to insect mortality is debatable (Isman and Tak 2017).

#### 8.4.1.4 Nicotinic Acetylcholine Receptors

Nicotinic acetylcholine receptors can be found on pre or postsynaptic nerve terminals, in addition to the cell bodies of inter neurons, motor neurons, and sensory neurons in the insect nervous system (Green et al. 2013). Nicotine, an alkaloid isolate obtained from *Nicotiana tabacum* L. (Solanaceae), can play a role as an acetylcholine receptor agonist to mimic acetylcholine, resulting in sodium ion influx and action potential generation. Acetylcholine's synaptic action is ended by AChE under normal conditions. Even so, because AChE cannot hydrolyze nicotine,

nicotine's persistent activation leads to overstimulation of cholinergic transmission, giving rise in convulsion, paralysis, as well as death (Green et al. 2013; Oguh et al. 2019). Nicotine is a highly effective nerve toxin against soft-bodied insects and mites. Moreover, it is among the most toxic of all botanicals and extremely dangerous to humans (Okwute 2012). Neonicotinoids are another category of insecticides based on the chemical formation of nicotine, and they include imidacloprid, acetamiprid, and thiamethoxam. Neonicotinoids, like nicotine, bind to acetylcholine receptors of nicotinic. They are more specific, however, being far more harmful to invertebrates such as insects than to mammals. Furthermore, they are highly water-soluble, which allows them to be applied easily on soils and, as a result, absorbed by plants, and boosting more efficacious defense (Elbert et al. 1996). They function as poison by ingestion or contact, causing feeding to cease within a few hours of contact and death soon after (Tomizawa and Casida 2005).

#### 8.4.1.5 GABA-Gated Chloride Channels

Insecticides could target γ-Aminobutyric acid GABA (gated chloride channels). Neuronal inhibition is reduced when its antagonists, for example, α-Thujone (*Artemisia absinthium*) or picrotoxine (*Anamirta cocculus*) are blocked, leading to central nervous system hyper-excitation, convulsion, and death. Thujone has been identified in a former study as a neurotoxin, it functions as an allosteric reversible modulator of GABA receptors (Höld et al. 2000). Furthermore, the monoterpenoids carvacrol, pulegone, and thymol may inhibit the GABA receptor (Tong and Coats 2010). Likewise, plant-derived natural compounds known as silphinene-type sesquiterpenes antagonize the action of aminobutyric acid via stabilizing non-conducting structures of chloride channel (Höld et al. 2000). GABA is known for feeding stimulation and invokes taste cell responses in most herbivorous insects. Chemicals that antagonize GABA receptors may be assumed as antifeedant or deterrent substances, particularly targeting beetles, aphids, and lepidopterans (González-Coloma et al. 2002).

#### 8.4.1.6 Octopamine Receptors

Octopamine is a multifunctional endogenous amine which functions as a neurotransmitter, neurohormone, and neuromodulator in invertebrates (Roeder 2005). Its receptors are widely distributed in the central and peripheral nervous systems of insects. It is crucial in mediating physiological and behavioral functions (Evans 1993). The octopamine1 receptor, for example, adjusts the myogenic rhythm of contraction in locust extensor-tibiae by changing intracellular calcium concentrations, while the octopamine2A and octopamine2B receptors do so by activating adenyl cyclase. Monoterpenes' quick action on some pests suggests a neurotoxic mechanism of action. This hypothesis has been confirmed by demonstrating the insecticidal and repellent action of eugenol against *Triatoma infestans*, the blood-sucking bug via activating the octopamine receptor (Reynoso et al. 2019). Former studies confirmed the presence of octopamine receptors in a wide range of insects, like lepidopterans, cockroaches, and flies (Kostyukovsky et al. 2002). Because the mentioned receptors do not fit into the receptor types found in vertebrates, agonists of octopamine receptors could be important candidate for a commercial pesticide once they are target-specific, less toxic to mammals, and have a different mechanism of actions than most of currently commercial pesticides in market (Reynoso et al. 2019).

### 8.4.2 *Botanical Secondary Metabolites and the Respiratory System*

In cellular respiration molecular process, nutrient compounds are converted into energy or adenosine triphosphate (ATP). The mitochondrial electron transport chain, which incorporates numerous key enzymes that could be insecticide targets, is in charge of this process (Souto et al. 2021). Rotenone is the most common natural product among rotenoids, derived from plants like *Derris* and *Lonchocarpus* (Fabaceae) and *Rhododendron* (Ericaceae), that are found in the East Indies, Malaya, and South America (Isman 2006). Rotenone is a complex I act as mitochondrial respiratory chain inhibitor that functions as contact and stomach poison. It prevents nicotinamide adenine dinucleotide (NADH) dehydrogenase, stopping electron flow from NADH to coenzyme Q and therefore preventing ATP formation from NADH, but allowing ATP formation via flavine adenine dinucleotide (FADH2) (Oguh et al. 2019). As a consequence, it is one of the slowest acting botanical insecticides, and yet it is easily degradable by air and sunlight, killing insects over several days. Furthermore, it has a broad spectrum of bioactivity against many pests such as leaf-feeding beetles, lice, caterpillars, mosquitoes, ticks, fire ants, and fleas [48]. However, rotenone is extremely dangerous to mammals and fish, and previous study has mentioned a possible link between its exposure and Parkinson's Disease (Sherer et al. 2007). Despite their severe toxicity, rotenoids have the potential to be a source of novel complex I inhibitors, providing safer and more efficient pesticide derivatives.

### 8.4.3 *Botanical Secondary Metabolites and Endocrine System*

Insect growth regulators (IGRs) are chemical substances that interfere with an insect's endocrine system. They may function as mimics or juvenile insect hormones inhibitors, and inhibitors of chitin synthesis. To maintain their immature state, insects produce juvenile hormones. Once growth is completed, the hormone production ceases, causing the molt to the adult stage (Singh et al. 2003). Triterpenes obtained from *Catharanthus roseus* (Apocynaceae) have shown IGR activity (Singh et al. 2003). Continuous IGR application to crops keeps the insects as larvae, preventing successful molting and resulting in efficient pest control (Kamboj and Saluja 2008). On the contrary the antijuvenile hormone activity of two chromenes found in *Ageratum conyzoides* L. (Asteraceae), precocene I and II has been reported to induce sterile, moribund, and dwarfish adults after exposure (Kamboj and Saluja 2008).

Azadirachtin is a complex tetranortriterpenoid limonoid particularly abundant in the seeds of *Azadirachta indica*, a plant species native to Burma, and now cultivated in more regions of Southeast Asia, Africa, the Americas, and Australia (Mordue and Nisbet 2000). Azadirachtin is a systemic contact poison and has two modes of action: (i) direct effects on cells and tissues and (ii) indirect effects via endocrine system interference. It primarily functions as feeding deterrent and IGR in a wide range of insect orders, for instance, lepidoptera, hemiptera, diptera, hymenoptera, and orthoptera (Mordue and Nisbet 2000). Azadirachtin prevents the release of morphogenetic peptide hormones (prothoracicotropic hormone, PTTH and allatostatins) via neurosecretory system of insect. These hormones are regulating the function of prothoracic glands and corpora allata. Both prothoracic glands and the corpora allata regulate the molting hormone that controls ecdysis and formation of new cuticle, as well as the juvenile hormone (JH), any interruption in this biochemical sequence can lead to molt disruption, molting malformations, or infertility. The direct effects on somatic and reproductive tissues severely affect feeding, development, and reproduction, whereas the molecule's affected indirectly through disruption of endocrine processes (Mordue and Nisbet 2000).

Chitin synthesis inhibitors prevent chitin production, which is a main constituent of nearly all fungi cell walls and a core part of the insect exoskeleton, which offers physical protection and osmoregulation. Because it is absent in plant and mammalian species but abundant in arthropods and most fungi, chitin biosynthesis has become an attractive objective for developing more specific insecticides and antifungal agents (Souto et al. 2021). Previously, 2-benzoyloxycinnamaldehyde (2-BCA), a natural product isolated from the roots of *Pleuropterus ciliinervis*, a plant species traditionally used in Chinese folk medicine to treat inflammation and various types of infection, was discovered to prevent chitin synthesis (Kang et al. 2007).

### 8.4.4 Botanical Secondary Metabolites and Water Balance

The insect body is covered in a thin layer of wax, which has an ecological purpose by stopping water loss from the cuticular surface. Saponins (natural soaps), as well as rice bran, cotton seed, and palm kernel crude oils, may act by interrupting this protective waxy covering, affecting the water balance of insects via quick water loss from the cuticle, and thus causing death by desiccation (Souto et al. 2021). Besides this, crude oils can control insects such as whiteflies, mites, caterpillars, leafhoppers, and beetles through interfering with insect respiration by plugging spiracles (Walia et al. 2014). Some selected botanical pesticides and their structures are illustrated in Fig. 8.6.

## 8.5 Essential Oils

Essential oils (EOs), which are produced as secondary metabolites in plants, are the primary group of botanical pesticides. These oils are volatile in nature and contain aromatic constituents. They have critical roles in life cycles of plants, function as defenses against pathogens and herbivores or attracting pollinators (Weinzierl 2000; Isman 2006). EOs are produced in plants via secondary metabolic processes, have a distinct odor, and are found as fluid droplets in the bark, flowers, fruits, leaves, stems, and roots of different plants (Ebadollahi et al. 2020). Several EOs have antimicrobial and antioxidant properties (Dorman et al. 2000). EOs are produced by glandular trichomes in Lamiaceae, secretory cavities in Myrtaceae and Rutaceae, and resin ducts in Asteraceae and Apiaceae (Fahn 1988). These structures burst open when herbivores feed or move on the surface of the plants, releasing vast quantities of compounds.

### *8.5.1 Essential Oil Constituents*

Essential oils contain two chemical categories according to their metabolic synthesis pathways: (i) phenylpropanoids with a small molecular weight and (ii) terpenoids (monoterpenes, sesquiterpenes). Methylerythritol phosphate pathway is responsible for monoterpenes production in the plastids. Whereas sesquiterpenes are produced in the cytosol by mevalonate pathway (Pavela and Benelli 2016). Benzene derivatives, hydrocarbons, terpenes, and other compounds are the four types of EO volatiles. Monoterpenoids account for roughly 90% of all EOs and have a diverse set of functions and structural features. Acids (chrysanthemic acid), acyclic alcohols (citronellol, geraniol), aldehydes (citronellal), cyclic alcohols (menthol), and phenols (thymol) are also related compounds (Koul et al. 2008). Chemical constituents in essential oils vary between species within the same genus, as well as between plant parts, geographical factors, harvest time, season, climate, and extraction method. For instance, 1,8-cineole concentration in *Eucalyptus citriodora* (18.9%), *E. globulus* (31%), *E. radiata* (63.3%), and *E. saligna* (45.2) varied (Mwamburi 2022).

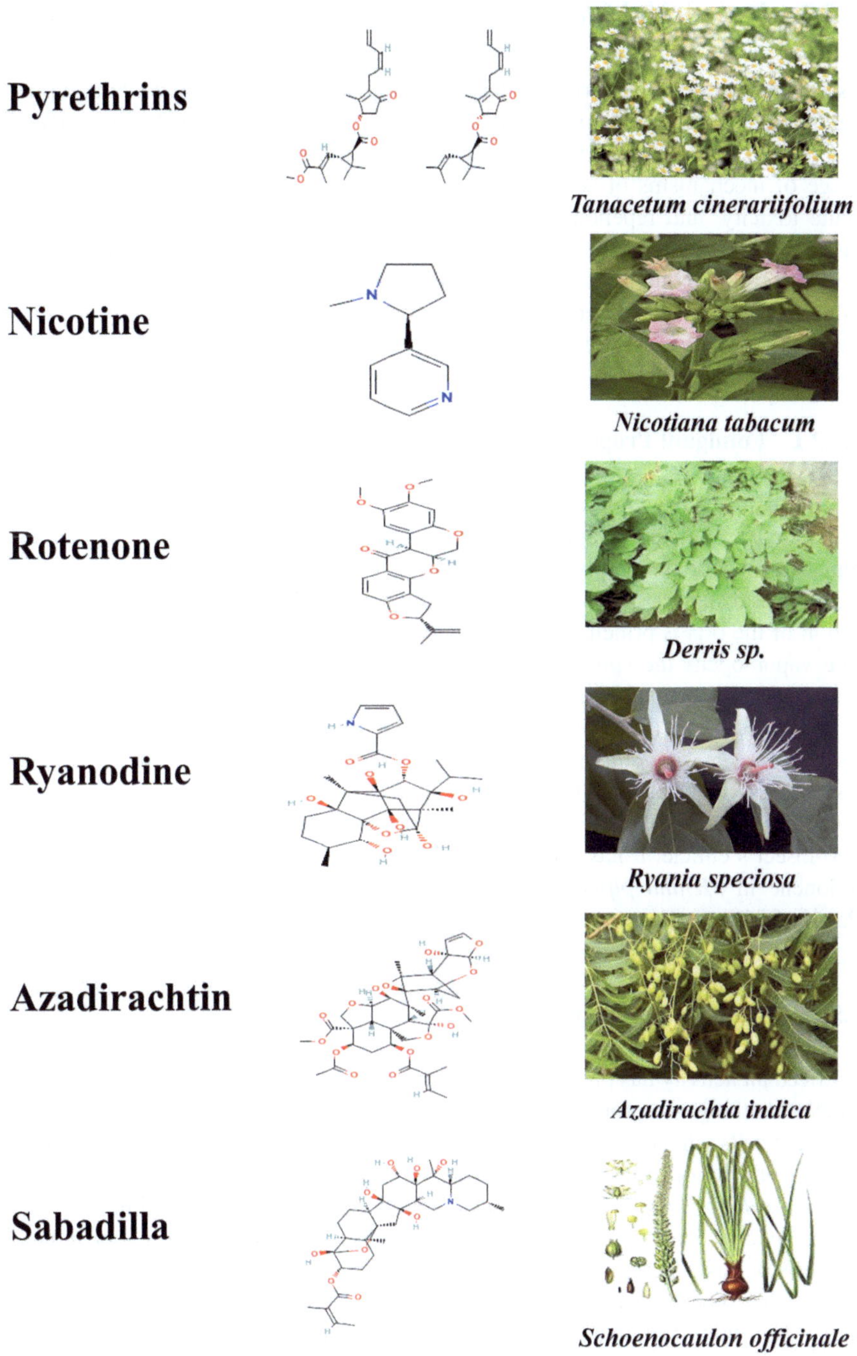


**Fig. 8.6** Names, structures, and plants of the major botanical pesticide groups

### 8.5.2 *Insecticidal Characteristics of Essential Oils*

Essential oils possess a variety of bio-pesticidal actions against coleoptera, diptera, hemiptera, isoptera, and lepidoptera, varying from lethal to sublethal effects (Pavela and Benelli 2016). The variations in components of diverse EOs account for a wide range of mechanisms of action, such as antinutritional, developmental inhibition, acute toxicity, and repellency (Isman 2006). The site where the toxin enters the insect is one of these factors. The typical modes of entry are inhalation, ingestion, or skin absorption by the insect (Ozols and Bicevskis 1979). As described in Table 8.1, aside from the different sequences of phytochemical action, EO toxicity has been correlated with a variety of other factors.

#### 8.5.2.1 Fumigant Properties

Due to their fast penetration, multiple insecticidal activities, and ease of conversion to vapor at room temperature, EO active compounds show promise as potential replacements for frequently used fumigants. Many research studies on the use of plant EO substances as insect fumigants have been performed. The mechanism of action of the oils is principally in the gaseous form and via the respiratory system. The vapor opens the spiracles because most insects breathe through the trachea. Suffocation happens as a result of tracheal respiration blockage, and the insect dies as a consequence (Schoonhoven 1978). The fumigation of EOs against stored product insects *Sitophilus* spp. and *T. castaneum* has piqued the interest of many researchers (Fang et al. 2010; Devi et al. 2020). Monoterpenoids are essentially volatile and cause toxic effects such as fumigants due to their ability of penetration into insect's cuticle. It has been stated that linalayl acetate, linalool, pulegone, and limonene in *Mentha piperita* caused fumigant toxicity to *S. oryzae* (Singh and Pandey 2018).

#### 8.5.2.2 Antifeedant Properties

Antifeedant activity has previously been linked to a chemoreception mechanism that entails blocking receptors which usually respond to phagostimulants or boosting deterrent cells which is considered primary antifeedancy. It has been mentioned that azadirachtin hinders the cholinergic transmission of neurons associated with *Drosophila melanogaster*'s suboesophageal ganglion, which is strongly linked to feeding behavior (Qiao et al. 2014). Besides this, consumption of food might be reduced because of toxic actions that promote astringency, bitter taste, or anti-digestive activity in some herbivores after the first food intake (Qiao et al. 2014). It was mentioned that schimperi, a gallotannin obtained from *Anogeissus schimperi* (Combretaceae), was responsible for the unpleasant taste in herbivores (Okwute and Nduji 1992). Antifeedant substances could deter feeding following contact, or

**Table 8.1** Botanical insecticides of different plant families and their insecticidal effects on different insect pests

| Plant Species | Plant Family | Major Bioactive Compounds | Target Insect | Insect Order | Mechanism of Action | Ref. |
|---|---|---|---|---|---|---|
| *Azadirachta indica* | Meliaceae | Azadirachtin | *Leptinotarsa decemlineata* | Coleoptera | Antifeedant effect | Zehnder and Warthen (1988) |
| *Allium sativum* | Amaryllidaceae | Diallyl sulfide, diallyl disulfide, and diallyl trisulfide | *Tribolium castaneum* | Coleoptera | Repellence, toxicity, and inhibition of progeny emergence | Ali et al. (2014) |
| *Curcuma longa* | Zingiberaceae | Curcumin | | | | |
| *Cinnamomum zeylanicum* | Lauraceae | Cinnamaldehyde, Eugenol | *Sitophilus oryzae* | Coleoptera | Toxicity | Vendan et al. (2017) |
| *Syzygium aromaticum* | Myrtaceae | Eugenol | | | | |
| *Citrus limon* | Rutaceae | D-Limonene and α-pinene | | | | |
| *Citrus aurantium* | Rutaceae | D-Limonene | | | | |
| *Agave americana* | Asparagaceae | Phenolics, flavonoids, and saponins | *Sitophilus oryzae* | Coleoptera | Repellence and enzyme inhibitory effects | Maazoun et al. (2019) |
| *Dittrichia viscosa* | Asteraceae | α- and γ- costic acid isomers | *Sitophilus granarius* | Coleoptera | Food deterrence | Rotundo et al. (2019) |
| *Artemisia annua* | Asteraceae | Monoterpenes and sesquiterpenes | *Tribolium castaneum* | Coleoptera | Repellence and enzyme inhibitory effects | Deb and Kumar (2020) |
| *Eucalyptus camaldulensis.* | Myrtaceae | 1,8-cineole | *Acanthoscelides obtectus* | Coleoptera | Fecundity and fertility inhibition | Hategekimana and Erler (2020) |
| *Mentha piperita* | Lamiaceae | L-menthol, neoisocarvomenthol, and eucalyptol | | | | |
| *Pimpinella anisum* | Apiaceae | Trans-anethole | | | | |

(continued)

**Table 8.1** (continued)

| Plant Species | Plant Family | Major Bioactive Compounds | Target Insect | Insect Order | Mechanism of Action | Ref. |
|---|---|---|---|---|---|---|
| *Tetrapleura tetraptera* | Mimosaceae | Phenols, flavonoids, saponins, and alkaloids | *Callosobruchus maculatus* | Coleoptera | Oviposition reduction and inhibition of adult emergence | Idoko and Ileke (2020) |
| *Annona muricata* | Annonaceae | Annonaceous acetogenins | | | | |
| *Aframomum melegueta* | Zingiberaceae | 3-arylalkanes, 6-paradol, 6-shogaol, 6-gingerol, pentacyclic triterpene, and oleanolic acid | | | | |
| *Eucalyptus globulus* | Myrtaceae | Gallic acid, ellagic acid, citramalic acid, and citric acid | | | | |
| *Ficus exasperata* | Moraceae | Terpenes, flavonoids, and phenols | | | | |
| *Cymbopogon citratus* | Poaceae | Citral, citronellol, geraniol, limonene | *Sesamia cretica* | Lepidoptera | Toxicity and induced biological and biochemical changes (Fig. 8.7) | Abdel-Hakim et al. (2021) |
| *Allium sativum* | Amaryllidaceae | Diallyl sulfide, diallyl disulfide, and diallyl trisulfide | | | | |
| *Citrus limon* | Rutaceae | D-Limonene, γ-Terpinene, and β-Pinene | *Agrotis ipsioln* | Lepidoptera | Toxicity and induce interruption in growth and adult emergence (Fig. 8.8) | Ahmed et al. (2023) |
| *Allium sativum* | Amaryllidaceae | Diallyl sulfide, diallyl disulfide, and diallyl trisulfide | *Spodoptera littoralis* | Lepidoptera | Toxicity and induce growth interruption | Ibrahim (2017) |
| *Pelargonium* sp. | Geraniaceae | Geraniol, eugenol, isomenthone, and linalool | *Agrotis ipsioln* | | inhibited hatchability and adult emergence | |
| *Spheranthus amaranthroids* | Asteracea | Alkaloids, phenolics, saponins, flavonoids, and tannins | *Spodoptera litura* | Lepidoptera | Larvicidal and enzyme inhibitory effects | Murfadunnisa et al. (2019) |

| | | | | | | |
|---|---|---|---|---|---|---|
| *Plumbago zeylanica* | Plumbaginaceae | Chitranone, plumbagin, sitosterol, and flavonoids | *Plodia interpunctella* | Lepidoptera | Inhibited hatchability and adult emergence | Adeyera and Akinneye (2020) |
| *Azadirachta indica* | Meliaceae | Azadirachtin | *Helicoverpa armigera* | Lepidoptera | Reduction in larval population | Fite et al. (2020) |
| *Millettia ferruginea* | Fabaceae | Alkaloids, anthraquinnes, and flavonoids | | | | |
| *Manihot esculenta* | Euphorbiaceae | 9, 12-Octadecadienoic acid and Lupeol | *Spodoptera litura* | Lepidoptera | Immune response and anti-oxidant enzyme activity | Manjula et al. (2020) |
| *Piper nigrum* | Piperaceae | β-caryophyllene, caryophyllene oxide, and α-copaene | *Megalurothrips sjostedti* | Thysanoptera | Repellence activity | Abtew et al. (2015) |
| *Cinnamomum zeylanicum* | Lauraceae | Cinnamaldehyde, trans-cadina-1 (6), and 4-diene | | | | |
| *Cinnamomum cassia* | Lauraceae | Cinnamaldehyde and (E)-ortho-methoxy cinnamaldehyde | | | | |

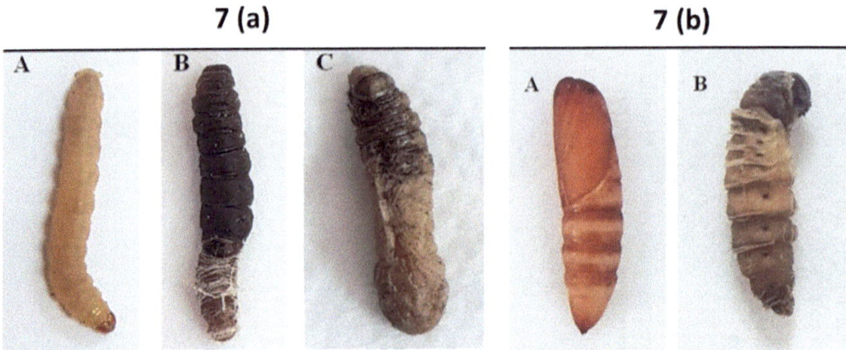


**Fig. 8.7** (**a**) Malformed *S. cretica* larvae after treatment of 6th instar diapausing-destined larvae with $LC_{50}$ of garlic oil. (A) Normal larva, (B, C) Malformed larvae. (**b**) Malformation resulted from treatment of the 6th instar diapausing-destined larvae with $LC_{50}$ of lemongrass oil. (A) Normal pupa, (B) larval–pupal intermediate (Abdel-Hakim et al. 2021)

**Fig. 8.8** Various stages malformations of *Agrotis ipsilon* treated with different concentrations of lemon peel essential oil (**a**) adult moth, (**b**) pupae, and (**c**) larvae (Ahmed et al. 2023)

they may act as repellents without coming into direct contact with the insects (Koul et al. 2008). The use of plant EOs in pest management is an important aspect of their antifeedant properties (Table 8.1). Deterrence of feeding can be measured using factors such as (i) feeding deterrence index (FDI), (ii) the efficiency of conversion of ingested food (ECI), (iii) the relative growth rate (RGR), and (iv) the relative consumption rate (RCR). For example, while the EOs of *Artemisia sieberi* and *A. scoparia* had antifeeding action on *T. castaneum*, the *A. sieberi* EO was more officious than *A. scoparia* EO and remarkably reduced RGR and RCR. In terms of FDI, *A. sieberi* oil performed better than *A. scoparia* oil (Negahban et al. 2007). In a previous study, it was found that *Carum copticum* and *Vitex pseudonegundo* EOs increased FDI when used against *C. maculatus* (Sahaf and Moharramipour 2008). Essential oils have the potential to influence nutritional parameters by interfering with the pre- and post-ingestive processes. As a result, an insect's feeding behavior may lead to less consumption and, consequently, a slower rate of growth. Despite the fact that many natural plant products act as antifeedants, no commercial products have been developed based on this characteristic (Huang and Renwick 1995). Insects' avoidance to feeding deterrents severely limits their utility in pest control.

#### 8.5.2.3 Repellent Properties

Many essential oils are primarily composed of monoterpenes and are highly efficient repellents. For example, *Eucalyptus globulus*, *Cymbopogon flexuosus*, *Thymus vulgaris*, *Rosmarinus officinalis*, and *Eugenia caryophyllus* (Tripathi et al. 2000). Repellents do not cause insect death but keep them away by emitting pungent vapors or exhibiting a mildly toxic effect. Previous research works on the activities of different essential oils as repellents have been published as described in Table 8.1. Repellents protect crops from insects while causing minimal environmental damage. The repellency of arthropods varies; for example, it was observed that the *Prangos acaulis* essential oil had a varied repellent action on *S. oryzae*, *C. maculatus*, and *T. castaneum* which was 83.6%, 71.6%, and 63.6%, respectively (Taghizadeh-Saroukolai et al. 2009).

#### 8.5.2.4 Attractants

Attractants are semiochemicals or communication substances that plants release to attract insects or natural predators of insects that feed on the plant. The release of (−) and (+) limonene from *Pinus strobus* (Pinaceae) has been linked to the attraction of the white pinecone beetle, *Conophthorus coniperda* (Curculionidae), as well as the predator beetle, *Enoclerus nigripes* (Cleridae), via the release of (−) α-pinene (Miller 2007; Degenhardt et al. 2009).

#### 8.5.2.5 Toxicants

The utilization of essential oils derived from plants as toxic substances is an appealing choice since they are efficient and have previously been used. Terpenes have been linked to the efficacy of essential oils against most insect pests [80]. Carvacrol, 1,8-cineol, thymol, eugenol, -pinene, and limonene, for example, showed toxicity to different insect pests. Insecticidal toxicity against *Aphis craccivora* was observed when *faba* beans were treated with a neem oil formulation (Sammour et al. 2011). Furthermore, in their study a cumulative mortality of adults up to 100% after seven days was observed. Terpenes such as carvone, linalool, terpeniol, phellandrine, and citronellol from garlic and mint oils exhibited insecticidal activity on *A. ipsilon* at different growth stages (Sharaby and El-Nujiban 2015).

#### 8.5.2.6 Growth Retardants and Developmental Inhibitors

Previous studies as described in Table 8.1 have found that plant EOs and their constituents interrupt insect growth and reduce body weight during different stages, thus extending the time for development. Also, extended effects on emergence rates of larvae, pupae, and adult can be observed (Koul et al. 2008). For instance, EOs obtained from azadirachtin and neem seed increased mortality of aphid nymphal by 80% and 77%, respectively, and caused longer maturation period to adult stage (Kraiss and Cullen 2008). It has also been demonstrated that basil oil prolonged the *Aphis craccivora* nymph stage, resulting in fewer adults (Sammour et al. 2011).

#### 8.5.2.7 Inhibitors of Fertility/Reproduction

Infertility can be caused by an induced insect sterility technique or by using a chemosterilant that inhibits reproduction. Chemosterilants can cause permanent or temporary sterility in male or female insects, as well as disrupt sexual development from larva to adult (Navarro-Llopis et al. 2011). Botanical insecticides, such as azadirachtin, interfere with the synthesis and release of molting hormones from the prothoracic gland, giving rise to imperfect ecdysis in immature insects and infertility in adult stages (Isman 2006). *Tenebrio molitor* pupae were found to be more susceptible to diallyl disulfide and diallyl sulfide than larvae and adult stages; this difference was linked to that how garlic constituents penetrate the insect body, and the insect's ability to dissolve these substances (Plata-Rueda et al. 2017).

#### 8.5.2.8 Essential Oils' Synergistic Action

Research results show that EO compound mixtures or combinations have additive, synergistic, and/or antagonist toxicity effects in many insect orders (Tak et al. 2016; Gaire et al. 2020a). The observed synergy in EOs might be due to their chemical constituents' different mechanisms of action. Using synergistically interacting groups of monoterpenoids found in products, for example, allows for higher insecticidal activity by using fewer active constituents (Tak et al. 2016). The reasoning behind combining EOs is to formulate a superior product with multiple modes of action, with the product providing a greater efficacy than the total effects of the individual Eos' known and unknown chemical components. Previous studies have indicated that components in EOs contain the highest camphor content function synergistically (Nerio et al. 2010). Camphor and 1,8-cineole combination, for example, increased cuticle penetration, resulting in a synergistic toxic effect on cabbage looper larvae. The modifications improved the two-oil mixture's ability to penetrate the cuticle, resulting in lower surface tension and increased solubility (Tak et al. 2016). The synergistic interaction of the mixture was most likely determined by factors relevant to the target site, such as the ability of monoterpenoids to function on different sites within an insect's nervous system (Gaire et al. 2020a).

#### 8.5.2.9 Phototoxins

Phototoxins, also known as light-activated compounds, are forms of phytoconstituents that, rather than being degraded by sunlight, are enhanced, or activated by two different mechanisms. The first uncommon mechanism in which phototoxin-derived molecular oxygen absorbs light energy, producing activated oxygen that eventually damages important biomolecules (Arnason et al. 1983). The second one is photo-genotoxic, in this mechanism phytochemicals damage DNA when activated by sunlight, whether oxygen present in the phototoxin or not. The phototoxin absorbs the photon in its ground state, allowing it to reach its excited state, where it interacts with ground state $O_2$ in the target's tissue, producing singlet oxygen and enabling insecticidal activity. Because phototoxins' mode of action is so dissimilar to that of conventional synthetic pesticides, cross resistance is unlikely (Towers 2011). Several classes of light-activated phototoxins include quinones, furanocoumarins, substituted acetylenes, and thiophenes. Several phototoxins have been reported, including 3-methyl-3-phenyl-1,4-pentadiyne, a phytoconstituent of *Artemisia monosperma* that, when exposed to sunlight, has mode of action similar to DDT against *Musca domestica* and *S. littoralis* larvae (Marchant and Cooper 1987).

## 8.6 Most Studied Monoterpenes with Insecticidal Activities

The number of studies on the insecticidal activities of monoterpenes has risen in recent decades. The insecticidal activities of some selected monoterpenes are discussed below.

### *8.6.1 Citronellal*

An acyclic citronellal monoterpene is found in citronella and eucalyptus oils. Citronellal and Citronellal-containing EOs exhibited fumigant activities. Citronellal is used as a fumigant against *Anopheles gambiae*, *Bemisia tabaci*, and *Drosophila suzukii* (Wu et al. 2020; Chae et al. 2014).

### *8.6.2 Linalool*

Linalool, acyclic monoterpene, with insecticidal and antimicrobial properties can be obtained from essential oils such as *Citrus sinensis* and *Pimpinella anisum*. It has fumigant and antifeedant toxicity to Adults of *B. tabaci*, larvae of *Aedes aegypti* and *A. albopictus*, as well as deltamethrin-resistant bed bugs *Cimex lectularius* (Oyedeji et al. 2020; Hashem et al. 2020; França et al. 2021).

### 8.6.3 Geraniol

Geraniol, an acyclic monoterpene, has been shown to exhibit toxicity against different insect species, phytopathogenic fungi, and nematode pests. For instance, it was toxic to *C. lectularius* that is resistant to deltamethrin, as well as showed fumigant toxicity to *B. tabaci* (Chae et al. 2014; Gaire et al. 2020b). Essential oil from *Melissa officinalis* containing geraniol has notable contact, fumigation, and repellent toxicity against *T. castaneum* (Upadhyay et al. 2019).

### 8.6.4 Citral

Citral, an acyclic monoterpene, is abundant in EOs derived from *Cymbopogon citratus* and *Rosmarinus officinalis* (Alves et al. 2019; Tak et al. 2017). Citral and citral-containing EOs showed toxic activities against insects, phytopathogenic fungi, and nematodes pests. Essential oil from *R. officinalis* was toxic to noctuid moth larvae, *Trichoplusia ni* [130]. It also has fumigation toxicity against *Plutella xylostella*, *C. maculatus*, *S. zeamais*, and *B. tabaci* (Chae et al. 2014; Oyedeji et al. 2020; Cai et al. 2020).

### 8.6.5 Eugenol

Eugenol is a monocyclic compound found in *Syzygium aromaticum* essential oil (Tabari et al. 2020). Laboratory studies revealed its high toxic effect against insects. It had a good contact toxicity against *Periplaneta americana*, *B. germanica*, and carpenter ants *Camponotus pennsylvanicus* (Tabari et al. 2020). *Bemisia tabaci*, *P. xylostella*, *Cimex lectularius*, and *S. oryzae*, all could be fumigated by it (Chae et al. 2014; Gaire et al. 2020b; Cai et al. 2020; Ibrahim et al. 2023).

## 8.7 Persistence of Botanical Pesticides

Because of their biological nature, botanical pesticides degrade quickly and do not remain in the environment, including water and soil, thus lowering the risk of pollution. Exposure to air, sunlight, moisture, and high temperatures degrades their constituents (Gunasekara 2005). Thymol, a common ingredient in *Satureja hortensis*, *Thymus vulgaris*, *Piper nigrum*, and *Zataria multiflora*, for instance, degrades in sunlight in about 28 h and in soils within about 8 days (Liu et al. 2017). Whereas azadirachtin has a half-life of 1–2 days in crops and soil (Yang et al. 2017). Furthermore, plant extracts used as pesticides, like neem, degrade when subjected to sunlight, implying that they degrade just as quickly after application (Gunasekara 2005). Also, pyrethrum-based insecticides last only a few hours under field conditions (Grdiša and Gršić 2013). It was stated that natural pyrethrins degrade from 100% to 1% in 5 hours when subjected to sunlight and oxygen (Gunasekara 2005). Microorganisms in ecological systems speed up biodegradation via oxidative metabolism due to an abundance of detoxification enzymes (Sehrawat et al. 2021). Ester bond hydrolysis is a familiar biodegradation pathway which is facilitated by carboxylesterases enzymes produced by microorganisms such as *Bacillus cereus* and *Aspergillus niger* (Cyco'n and Piotrowska-Seget 2016). Soil microorganisms produce enzymes which decompose botanical pesticides into less toxic degradable forms before converting those metabolites to a biologically inactive and non-toxic state (Velázquez-Fernández et al. 2012). The efficiency of such bioremediation is reliant on microbial population growth and survival factors (e.g., salinity, pH, water, oxygen supply). The nature of microorganisms, their interactions with pesticides, and environmental factors all have an impact on effective microbial pesticide degradation (Ortiz-Hernández et al. 2013). The rapid degradation of botanical pesticides benefits environmental conservation.

## 8.8 Safety of Botanical Pesticides

Botanical pesticides are attracting attention for pest management due to their low mammalian toxicity. However, some pure essential oil compounds (e.g., carvacrol, pulegone) showed slight toxicity with rat acute oral $LD_{50}$ values of 23 g kg, but an essential oil insecticide containing a mixture of essential oil compounds caused no mortality when fed to rats at 2 g kg (Enan et al. 1998). Since many essential oils and their constituents are commonly used as culinary herbs and spices, pesticide products containing specific essential oils and constituents are excluded from the EPA's toxicity data requirements. In static water toxicity tests using juvenile rainbow trout, eugenol is approximately 1500 times less toxic than the botanical insecticide pyrethrum and 15,000 times less toxic than the organophosphate insecticide azinphosmethyl based on 96 h-$LC_{50}$ values (Stroh et al. 1998). Moreover, laboratory tests demonstrate that eugenol and other essential oil components are not persistent in fresh water. Also, these components are non-persistent in soils under aerobic conditions at 23 °C, with the half-life of a-terpineol ranges from 30 to 40 h and degrade completely after 50 h (Misra and Pavlostathis 1997). In the soil, *Pseudomonas* bacteria completely degrade eugenol to common organic acids. Concerns about essential oil pesticide residues on food crops should be alleviated by taking into account the fact that some essential oil components obtained through diet are actually beneficial to human health (Huang et al. 1994).

## 8.9 Herbivores' Natural Enemies that Serve as an Indirect Defense

Predators and parasitoids are considered the two main groups of natural enemies of herbivorous insects that use plant volatiles as cues to locate their prey (Mumm and Dicke 2010). The five orders of insects, mites, and nematodes comprise the vast majority of the identified natural enemies (Clavijo McCormick et al. 2012).

### *8.9.1 Parasitoids*

Plant metabolites begin to have an impact when the adult parasitoids are looking for their herbivorous hosts. Parasitoids are attracted to plant odors in order to find the habitats of possible hosts. Once habitats are found, parasitoids may exploit the volatiles released by infected plants quite effectively. Only in response to damage caused by herbivores are these volatiles, which may be unique to a particular plant species, released in quite high quantities. The parasitoid employs these induced emissions as olfactory signals to pinpoint the precise location of plants or even plant portions that are afflicted with potential hosts. Additionally, plant toxins and digestibility inhibitors can delay the growth of herbivores and prolong the duration during which they are susceptible to parasite infection (Turlings and Benrey 1998). Even after a parasite has found a host, plant chemistry continues to affect it. Plant metabolites will also have an impact on the developing parasitoid on or in a herbivorous host. This impact can be either direct—if plant chemicals found in the host's tissues have an impact on the parasitoid as it develops—or indirect—when plant chemicals affect the health and viability of the herbivore, which serves as a host for developing parasitoids. Plants may benefit from the presence and activities of parasitoids and predators because they kill herbivores. Therefore, it's possible that plants have evolved to enhance the productivity of these herbivores' natural adversaries. Although it is doubtful that the primary function of plant metabolites is to interact with the third trophic level, since they do affect the beneficial creatures for the plant, natural selection may have resulted in some alterations enhancing these benefits. The plants have adapted their chemistry to promote the presence and effectiveness of these natural enemies of herbivores (Tumlinson et al. 1992). Chemicals utilized as host location cues are as diverse as the hosts that are attacked. A thorough analysis of the identities and origins of the compounds used in parasitoid host location is provided by Tumlinson et al. (1992). Understanding the biology of the host can give useful hints about the triggers that the host's natural enemies use.

The volatiles released by plants afflicted by insect herbivores have been found to attract 34 species of parasitic wasps across 10 families (Aljbory and Chen 2018). Braconidae, Mymaridae, Pteromalidae, Scelionidae, Eulophidae, Aphelinidae, Ichneumonidae, Figitidae, Encyrtidae, and Aphidiidae were among the families of parasitic wasps that drew attention. Among them, the parasites most frequently drawn to plants by volatiles released by lepidopterans, coleopterans, aphids, and some dipterans include those belonging to the family Braconidae (Hymenoptera), particularly *Cotesia* spp. (Turlings et al. 1990; Turlings and W¨ackers 2004; Bruinsma et al. 2009; Poelman et al. 2009; Girling et al. 2011). It has also been noted that plants harmed by insects attract parasitoids from different families of hymenopterans (Pettersson 2001; Lou and Cheng 2003; Snoeren et al. 2010). In addition to hymenopterans, it has been documented that parasitic dipterans are drawn to artificially harmed plants (Hanyu et al. 2009; Hanyu et al. 2011), or sticky traps that contain mixes of synthetic volatiles (James 2005).

### *8.9.2 Predators*

Plants attacked by different insect herbivores or artificial blends of volatiles have been shown to generate volatiles that attract 24 species of predators from 12 families of insects and mites (Aljbory and Chen 2018). Chrysopidae, Miridae, Geocoridae, Anthocoridae, Syrphidae, Empididae, and Coccinellidae are some of the families of predators that are drawn to them [(James 2005; Reddy 2002; James and Price 2004; James 2003; Hagenbucher et al. 2013)]. According to various studies (Krips et al. 2001; Horiuchi et al. 2003; van den Boom et al. 2004), the most frequently observed predators drawn to volatiles are lady beetles, such as *Stethorus punctum*, *S. gilvifrons*, *Coccinella septempunctata*, *Chilocorus kuwanae*, and *Aiolocaria hexaspilota*, as well as predatory mites, such as *Phytoseiulus persimilis*, *Amblyseius andersoni*, *A. finlandicus*, *A. californicus*, and *A. womerleyi*. The following predators have also been seen: 2-lined collops, 2-spotted stink bug, bigeyed bug, fruit fly, plant bug, green lacewings, predatory thrips, and minute pirate bug James 2003; Hagenbucher et al. 2013; Venzon et al. 1999). It is yet unknown how well-suited each predator species is to eradicate a particular type of nuisance insect.

### *8.9.3 Mites*

Certain predator mites may feed on many foods under various circumstances. Examples of these mites are those belonging to the families Tydeidae and Phytoseiidae. These predators can live by consuming other foods, such as foliar pathogenic fungi and occasionally pollen, when there is a low density of prey (Duso et al. 2005; Pozzebon et al. 2009). Plants harmed by herbivorous insects are susceptible to fungal secondary diseases. In tritrophic interactions, omnivores like predator mites drawn to the injured plants are advantageous (Tempfli et al. 2015).

### 8.9.4 Nematodes

Plant volatiles also draw invertebrates. For example, the western corn rootworm *Diabrotica virgifera virgifera* (Coleoptera: Chrysomelidae) population size is reduced in response to corn emissions by two species of parasitoid nematodes, *Heterorhabditis megidis* and *H. bacteriophora* (Rasmann and Turlings 2008). *H. bacteriophora* helps common milkweed (*Asclepias syriaca*) fend off attacks from the root-boring beetle *Tetraopes tetraophthalmus* below ground illnesses caused by fungus. In tritrophic interactions, omnivores like predator mites drawn to injured plants are advantageous (Rasmann et al. 2011).

## 8.10 Conclusions and Future Prospects

The adoption of biological control strategies is expanding, driven by increasing global demands for food, heightened public awareness of food quality, and pressing concerns for biodiversity and environmental protection. Plants have evolved a sophisticated arsenal of defense mechanisms against herbivory, including constitutive barriers and inducible responses. These encompass direct defenses—such as defensive proteins, metabolites, and morphological features that impair herbivore health—as well as indirect defenses, notably the emission of volatiles that recruit natural enemies. Harnessing these innate plant defenses, either by enhancing host plant resistance or deploying plant-derived compounds as biopesticides, offers a sustainable pathway to crop protection.

Plant secondary metabolites, in particular, demonstrate significant bioactivity as insecticides, growth regulators, and repellents. Their value is amplified by complex, multi-target modes of action, which can deter feeding and oviposition, cause toxicity, and disrupt physiological functions in pests. This complexity also reduces the likelihood of rapid pest resistance. However, despite the promising pesticidal properties of many botanical compounds, only a limited number have been successfully commercialized. Looking forward, translating this potential into reliable agricultural tools requires focused research and development in several key areas:

### 8.10.1 Standardization and Quality Control

A major hurdle for plant-derived products is the natural variability in their bioactive compound profiles, which can be influenced by genetics, geography, and extraction methods. Future efforts must prioritize the development of standardized protocols for cultivation, extraction, and compound quantification to ensure consistent efficacy, safety, and regulatory approval.

### 8.10.2 Advanced Formulation Technologies

The rapid biodegradability of many botanical pesticides, while environmentally desirable, often results in an unacceptably short field persistence and shelf-life. Innovations in formulation—such as microencapsulation, nanoemulsions, or carrier-based systems—are critical to protect active ingredients from degradation, enable controlled release, and enhance stability under real-world storage and application conditions.

### 8.10.3 Validation Through Field-Scale Trials

Promising laboratory results must be rigorously validated in practical agricultural settings. Comprehensive field trials are essential to evaluate the efficacy, phytocompatibility, and economic viability of botanical pesticides across diverse crops, climates, and pest pressures. This step is indispensable for bridging the gap between experimental proof-of-concept and adoption by farmers.

Ultimately, by addressing these challenges through interdisciplinary research, plant-based protection strategies can mature from promising alternatives into robust, mainstream components of integrated pest management, supporting a more resilient and sustainable agricultural system.

**Declaration of AI Assistance** The authors declare that no Artificial Intelligence (AI) tools were used to generate the intellectual content, data, analysis, or central ideas of this article. AI tools were utilized solely for the purpose of grammar and spelling correction support in the final editing stage. All information, data, and conclusions have been ethically reviewed, edited, and validated by the authors. No AI-generated text, data, or results are included in this work.

## References

Abdel-Hakim EA, Ibrahim SS, Salem NY (2021) Effect of garlic and lemongrass essential oils on some biological and biochemical aspects of corn stem borer *Sesamia cretica* larvae (Lepidoptera: Noctuidae) during diapausing phase. Proc Zool Soc 74:73–82

Abtew A, Subramanian S, Cheseto X et al (2015) Repellency of plant extracts against the legume flower thrips *Megalurothrips sjostedti* (Thysanoptera: Thripidae). Insects 6:608–625

Adeyera OJ, Akinneye JO (2020) Management of *Plodia interpunctella* (hübner) [Lepidoptera: pyralidae] using ethanolic oil extract of *Plumbago zeylanica* (LINN.). J Entom Nemat 12(1): 25–31

Agostini-Costa TS, Vieira RF, Bizzo HR et al (2012) Chromatography and its applications. In: Dhanarasu S (ed) plant secondary metabolites. Tech Publisher, Rijeka, pp 131–164

Ahmed HA, Nassrallah AA, Abdel-Raheem MA et al (2023) Lemon peel essential oil and its nano-formulation to control *Agrotis ipsilon* (Lepidoptera: Noctuidae). Sci Reports 13(1):17922

Ali S, Sagheer M, ul Hassan M et al (2014) Insecticidal activity of turmeric (*Curcuma longa*) and garlic (*Allium sativum*) extracts against red flour beetle, *Tribolium castaneum*: a safe alternative to insecticides in stored commodities. J Entom Zool Studies 2(3):201–205

Aljbory Z, Chen MS (2018) Indirect plant defense against insect herbivores: a review. Insect Sci 25: 2–23

Alves MDS, Campos IM, Brito DDMC et al (2019) Efficacy of lemongrass essential oil and citral in controlling *Callosobruchus maculatus* (Coleoptera: Chrysomelidae), a post-harvest cowpea insect pest. Crop Prot 119:191–196

Arimura GI, Matsui K, Takabayashi J (2009) Chemical and molecular ecology of herbivore-induced plant volatiles: proximate factors and their ultimate functions. Plant Cell Physiol 50: 911–923. https://doi.org/10.1093/pcp/pcp030

Arnason T, Towers GHN, PhilogÈNe BJR et al (1983) The role of natural photosensitizers in plant resistance to insects. In: American Chemical Society (ed) Plant resistance to insects, vol 208, Washington, DC, pp 139–151

Bebber DP, Ramotowski MAT, Gurr SJ (2013) Crop pests and pathogens move poleward in a warming world. Nat Clim Chang 3:985–988

Bennett RN, Wallsgrove RM (1994) Secondary metabolites in plant defence mechanisms. New Phytol 127:617–633

Bishop BA, Grafius EJ (1996) Insecticide resistance in the Colorado potato beetle. In: Jolive PH, Cox ML (eds) Chrysomelidae biology. SPB Academic, Amsterdam, pp 355–377

Bloomquist JR, Boina DR, Chow E et al (2008) Mode of action of the plant-derived silphinenes on insect and mammalian GABAA receptor/chloride channel complex. Pestic Biochem Physiol 91: 17–23

Brown AE (2005) Mode of action of insecticides and related pest control chemicals for production agriculture, ornamentals and turf. Pesticide Info Leaflet Nr 43:1–13

Bruce TJ (2015) Interplay between insects and plants: dynamic and complex interactions that have coevolved over millions of years but act in milliseconds. J Exp Bot 66(2):455–465

Bruinsma M, Posthumus MA, Mumm R et al (2009) Jasmonic acid-induced volatiles of *Brassica oleracea* attract parasitoids: effects of time and dose, and comparison with induction by herbivores. J Exp Bot 60:2575–2587

Cai YH, Hu XA, Wang P et al (2020) Biological activity and safety profile of monoterpenes against *Plutella xylostella* L. (Lepidoptera: Plutellidae). Environ Sci Pollut R 27:24889–24901

Casida JE, Quistad GB (1995) Pyrethrum flowers: production, chemistry, toxicology, and uses, 1st edn. Oxford University Press, Oxford

Casida J, Quistad G (1998) Golden age of insecticide research: past, present, or future? Annu Rev Entomol 43:1–16

Chae S, Kim S, Yeon SH et al (2014) Fumigant toxicity of summer savory and lemon balm oil constituents and efficacy of spray formulations containing the oils to B- and neonicotinoid-resistant Q-biotypes of *Bemisia tabaci* (Homoptera: Aleyrodidae). J Econ Entomol 107:286–292

Cingel A, Savic´ J, Lazarevic´ J et al (2016) Extraordinary adaptive plasticity of Colorado potato beetle: "Ten-Striped Spearman" in the era of biotechnological warfare. Int J Mol Sci 17:1538

Clavijo McCormick A, Unsicker SB, Gershenzon J (2012) The specificity of herbivore-induced plant volatiles in attracting herbivore enemies. Trends Plant Sci 17:303–310

Copping LG, Duke SO (2007) Natural products that have been used commercially as crop protection agents. Pest Manag Sci 63:524–554

Cyco'n M, Piotrowska-Seget Z (2016) Pyrethroid-degrading microorganisms and their potential for the bioremediation of contaminated soils: a review. Front Microbiol 7:1463

Deb M, Kumar D (2020) Bioactivity and efficacy of essential oils extracted from *Artemisia annua* against *Tribolium casteneum* (Herbst. 1797) (Coleoptera: Tenebrionidae): an eco-friendly approach. Ecotox environ safety 189:109988

Degenhardt J, Hiltpold I, Köllner TG et al (2009) Restoring a maize root signal that attracts insect-killing nematodes to control a major pest. Proc Natl Acad Sci USA 106:13213–13218

Devi MA, Sahoo D, Singh TB et al (2020) Toxicity, repellency and chemical composition of essential oils from *Cymbopogon* species against red flour beetle *Tribolium castaneum* Herbst (Coleoptera: Tenebrionidae). J Consum Prot Food Saf 15:181–191

Dorman HJD, Surai P, Dean SG (2000) In vitro antioxidant of a number of plant essential oils and phytoconstituents. J Essent Oil Res 12:241–248
Du Y, Nomura Y, Satar G et al (2013) Molecular evidence for dual pyrethroid-receptor sites on a mosquito sodium channel. Proc Natl Acad Sci USA 110:11785–11790
Duso C, Pozzebon A, Capuzzo C et al (2005) Grape downy mildew spread and mite seasonal abundance in vineyards: effects on *Tydeus caudatus* and its predators. Biol Control 32:143–154
Ebadollahi A, Ziaee M, Palla F (2020) Essential oils extracted from different species of the Lamiaceae plant family as prospective bioagents against several detrimental pests. Molecules 25:1556
Elbert A, Nauen R, Cahill M et al (1996) Resistance management for chloronicotinyl insecticides using imidacloprid as an example. Pflanzenschutz Nachr Bayer 49:5–54
El-Wakeil NE (2013) Botanical pesticides and their mode of action. Gesunde Pflanzen 65:125–149
Enan E, Beigler M, Kende A (1998) Insecticidal action of terpenes and phenols to cockroaches: effect on octopamine receptors. Paper presented at the International Symposium on Plant Protection, Gent
Erb M, Reymond P (2019) Molecular interactions between plants and insect herbivores. Annu Rev Plant Biol 70:527–557
Evans PD (1993) Molecular studies on insect octopamine receptors. EXS 63:286–296
Fahn A (1988) Secretory tissues in vascular plants. New Phytol 108:229–257
Fang R, Jiang CH, Wang XY et al (2010) Insecticidal activity of essential oil of *Carum carvi* fruits from China and its main components against two grain storage insects. Molecules 15:9391–9402
FAO, Food and Agriculture Organization of the United Nations (2002) International code of conduct on the distribution and use of pesticides. Retrieved on 2017–07–11
Fite T, Tefera T, Negeri M et al (2020) Effect of *Azadirachta indica* and *Milletia ferruginea* extracts against *Helicoverpa armigera* (Hubner) (Lepidoptera: Noctuidae) infestation management in chickpea. Cogent Food Agric 6(1):1712145
França LP, Amaral ACF, ADS R et al (2021) *Piper capitarianum* essential oil: a promising insecticidal agent for the management of *Aedes aegypti* and *Aedes albopictus*. Environ Sci Pollut R 28:9760–9776
Gaire S, Scharf ME, Gondhalekar AD (2020a) Synergistic toxicity interactions between plant essential oil components against the common bedbug (*Cimex lectularius* L.). Insects 11:133
Gaire S, Lewis CD, Booth W et al (2020b) Bed bugs, *Cimex lectularius* L., exhibiting metabolic and target site deltamethrin resistance are susceptible to plant essential oils. Pestic Biochem Phys 169:104667
Gakuubi MM, Wanzala WJ, Wagacha M et al (2016) Bioactive properties of *Tagetes minuta* L. (Asteraceae) essential oils: a review. Am J Essent Oil Nat Prod 2:27–36
García-Lara S, Serna Saldivar SO (2016) Insect pests. In: caballero B, Finglas PM, Toldrá F (eds) Encyclopedia of food and health. Academic Press, London, pp 432–436
Girling RD, Stewart-Jones A, Dherbecourt J et al (2011) Parasitoids select plants more heavily infested with their caterpillar hosts: a new approach to aid interpretation of plant headspace volatiles. Proc Biol Sci 278:2646–2653
González-Coloma A, Valencia F, Martín N et al (2002) Silphinene sesquiterpenes as model insect antifeedants. J Chem Ecol 28:117–129
Grdiša M, Gršić K (2013) Botanical insecticides in plant protection. Agric Conspect Sci 2:85–93
Green B, Welch K, Panter K et al (2013) Plant toxins that affect nicotinic acetylcholine receptors: a review. Chem Res Toxicol 26:1129–1138
Gunasekara AS (2005) Environmental fate of pyrethrins, environmental monitoring branch. California Dept. Pesticide Regulation
Hagenbucher S, Olson DM, Ruberson JR et al (2013) Resistance mechanisms against arthropod herbivores in cotton and their interactions with natural enemies. Crit Rev Plant Sci 32:458–482
Hansen MRH, Jørs E, Sandbæk A et al (2021) Organophosphate and carbamate insecticide exposure is related to lung function change among smallholder farmers: a prospective study. Thorax 76:780–789

Hanyu K, Ichiki RT, Nakamura S et al (2009) Duration and location of attraction to herbivore-damaged plants in the tachinid parasitoid *Exorista japonica*. Appl Entomol Zool 44:371–378

Hanyu K, Ichiki RT, Nakamura S, Kainoh Y (2011) Behavior of the tachinid parasitoid *Exorista japonica* (Diptera: Tachinidae) on herbivore-infested plants. Appl Entomol Zool 46:565–571

Hare J, Morse J (1997) Toxicity, persistence, and potency of Sabadilla alkaloid formulations to citrus thrips (Thysanoptera: Thripidae). J Econ Entomol 90:326–332

Hashem AS, Ramadan MM, Abdel-Hady AAA et al (2020) *Pimpinella anisum* essential oil nanoemulsion toxicity against *Tribolium castaneum*? Shedding light on its interactions with aspartate aminotransferase and alanine aminotransferase by molecular docking. Molecules 25: 4841

Hategekimana A, Erler F (2020) Fecundity and fertility inhibition effects of some plant essential oils and their major components against *Acanthoscelides obtectus* Say (Coleoptera: Bruchidae). J Plant Dis Prot 127:615–623

Heil M (2008) Indirect defence via tritrophic interactions. New Phytol 178(1):41–61

Hénault-Ethier L (2016) Backgrounder: Pyrethroids — just because we can use them at home doesn't mean that they're harmless. CAPE et Équiterre:1–13, 10.13140/RG.2.1.4822.2324

Höld K, Sirisoma N, Ikeda T et al (2000) a-Thujone (the active component of absinthe): γ-aminobutyric acid type A receptor modulation and metabolic detoxification. Proc Natl Acad Sci USA 97:3826–3831

Horiuchi J, Arimura G, Ozawa R et al (2003) A comparison of the responses of *Tetranychus urticae* (Acari: Tetranychidae) and *Phytoseiulus persimilis* (Acari: Phytoseiidae) to volatiles emitted from lima bean leaves with different levels of damage made by *T. urticae* or *Spodoptera exigua* (Lepidoptera: Noctuidae). Appl Entomol Zool 38:109–116

Howe GA, Jander G (2008) Plant immunity to insect herbivores. Ann Rev Plant Biol 59:41–66

Huang XP, Renwick JAA (1995) Cross habituation to feeding deterrents and acceptance of a marginal host plant by *Pieris rapae* larvae. Entomol Exp Appl 76:295–302

Huang MT, Ferraro T, Ho CT (1994) Cancer chemoprevention by phytochemicals in fruits and vegetables. Am Chem Soc Symp Ser 546:2–15

Ibrahim SS (2017) Applications of some biotechnological methods for controlling the most important pests of potatoes [thesis]. Fac. Sci, Al-Azhar University, Egypt

Ibrahim SS, Sahu U, Karthik P et al (2023) Eugenol nanoemulsion as bio-fumigant: enhanced insecticidal activity against the rice weevil, *Sitophilus oryzae* adults. J Food Sci Technol 60: 1435–1445. https://doi.org/10.1007/s13197-023-05690-7

Idoko JE, Ileke KD (2020) Comparative evaluation of insecticidal properties of essential oils of some selected botanicals as bio-pesticides against cowpea bruchid, *Callosobruchus maculatus* (Fabricius) [Coleoptera: Chrysomelidae]. Bull Nat Res Cent 44:119

Isman MB (2005) Problems and opportunities for the commercialization of botanical insecticides. In: Regnault-Roger C, Philogène BJR, Vincent C (eds) Biopesticides of plant origin. Lavoisier, Paris, pp 283–291

Isman MB (2006) Botanical insecticides, deterrents, and repellents in modern agriculture and an increasingly regulated world. Annu Rev Entomol 51:45–66

Isman MB (2015) A renaissance for botanical insecticides? Pest Manag Sci 71:1587–1590

Isman M, Tak JH (2017) Inhibition of acetylcholinesterase by essential oils and monoterpenoids: a relevant mode of action for insecticidal essential oils? Biopestic Int 13:71–78

James DG (2003) Synthetic herbivore-induced plant volatiles as field attractants for beneficial insects. Environ Entomol 32:977–982

James DG (2005) Further field evaluation of synthetic herbivore-induced plan volatiles as attractants for beneficial insects. J Chem Ecol 31:481–495

James DG, Price TS (2004) Field-testing of methyl salicylate for recruitment and retention of beneficial insects in grapes and hops. J Chem Ecol 30:1613–1628

Jankowska M, Rogalska J, Wyszkowska J et al (2018) Molecular targets for components of essential oils in the insect nervous system—a review. Molecules 23(1):34

Kamboj A, Saluja A (2008) *Ageratum conyzoides* L.: a review on its phytochemical and pharmacological profile. Int J Green Pharm 2:59–68

Kang TH, Hwang EI, Yun BS et al (2007) Inhibition of chitin synthases and antifungal activities by 2′-benzoyloxycinnamaldehyde from *Pleuropterus ciliinervis* and its derivatives. Biol Pharm Bull 30:598–602

Kessler A, Baldwin IT (2001) Defensive function of herbivore induced plant volatile emissions in nature. Science 291:2141–2144. https://doi.org/10.1126/science.291.5511.2141

Khorshid MA, Hassan AF, Abd El- Gawad M et al (2015) Effect of some plants and pesticides on acetylcholinesterase. Am J Food Technol 10:93–99

Kostyukovsky M, Rafaeli A, Gileadi C et al (2002) Activation of octopaminergic receptors by essential oil constituents isolated from aromatic plants: possible mode of action against insect pests. Pest Manag Sci 58:1101–1106

Koul O, Walia S, Dhaliwal GS (2008) Essential oils as green pesticides: potential and constraints. Biopestic Int 4:63–84

Kraiss H, Cullen EM (2008) Insect growth regulator effects of azadirachtin and neem oil on survivorship, development and fecundity of *Aphis glycines* (Homoptera: Aphididae) and its predator, *Harmonia axyridis* (Coleoptera: Coccinellidae). Pest Manag Sci 64:660–668

Krips OE, Willems PEL, Gols R et al (2001) Comparison of cultivars of ornamental crop *Gerbera jamesonii* on production of spider mite-induced volatiles, and their attractiveness to the predator, *Phytoseiulus persimilis*. J Chem Ecol 27:1355–1372

Laxmishree C, Nandita S (2017) Botanical pesticides– a major alternative to chemical pesticides: a review. Int J Life Sci 5(4):722–729

Lengai GMW, Muthomi JW, Mbeg ER (2020) Phytochemical activity and role of botanical pesticides in pest management for sustainable agricultural crop production. Sci Afr 7:e00239

Liu B, Chen B, Zhang J et al (2017) The environmental fate of thymol, a novel botanical pesticide, in tropical agricultural soil and water. Toxicol Environ Chem 99(2):223–232

Lou YG, Cheng JA (2003) Role of rice volatiles in the foraging behaviour of the predator *Cyrtorhinus lividipennis* for the rice brown planthopper *Nilaparvata lugens*. Biol Control 48: 73–86

Lyznicki JM, Kennedy WR, Young DC et al (1997) Educational and informational strategies to reduce pesticide risks. Prev Med 26(2):191–200. https://doi.org/10.1006/pmed.1996.0122

Maazoun AM, Hamdi SH, Belhadj F et al (2019) Phytochemical profile and insecticidal activity of *Agave Americana* leaf extract towards *Sitophilus oryzae* (L.) (Coleoptera: Curculionidae). Environ Sci Pollut R 26:19468–19480

Mahmood I, Imadi SR, Shazadi K et al (2016) Effects of pesticides on environment. In: Hakeem K, Akhtar M, Abdullah S (eds) Plant, soil and microbes. Springer, Cham

Manjula P, Lalitha K, Vengateswari G et al (2020) Effect of *Manihot esculenta* (Crantz) leaf extracts on antioxidant and immune system of *Spodoptera litura* (Lepidoptera: Noctuidae). Biocat Agric Biotech 23:101476

Marchant YY, Cooper GK (1987) Structure and function relationships in polyacetylene photoactivity. In: Light-activated pesticides, vol 339. American Chemical Society, Washington, DC, pp 241–254

Miller DR (2007) Limonene: attractant kairomone for white pinecone beetles (Coleoptera: Scolytidae) in an eastern white pine seed orchard in western North Carolina. J Econ Entomol 100:815–822

Misra G, Pavlostathis SG (1997) Biodegradation kinetics of monoterpenes in liquid and soil-slurry systems. Appl Microbiol Biotechnol 47:572–577

Mithoefer A, Boland W (2012) Plant defense against herbivores: chemical aspects. Annu Rev Plant Biol 63:431–450

Mordue AJ, Nisbet AJ (2000) Azadirachtin from the neem tree *Azadirachta indica*: its action against insects. An Soc Do Bras 29:615–632

Mumm R, Dicke M (2010) Variation in natural plant products and the attraction of bodyguards involved in indirect plant defense. Can J Zool 88:628–667

Murfadunnisa S, Vasantha-Srinivasan P, Ganesan R et al (2019) Larvicidal and enzyme inhibition of essential oil from *Spheranthus amaranthroids* (Burm.) against lepidopteran pest *Spodoptera litura* (Fab.) and their impact on non-target earthworms. Biocat Agri Biotec 21:101324

Muthomi J, Fulano AM, Wagacha JM et al (2017) Management of snap bean insect pests and diseases by use of antagonistic fungi and plant extracts. Sustain Agric Res 3:52

Mwamburi LA (2022) Role of plant essential oils in pest management. In: Mandal SD, Ramkumar G, Karthi S, Jin F (eds) New and future development in biopesticide research: biotechnological exploration. Springer, Singapore

Nauen R (2006) Insecticide mode of action: return of the ryanodine receptor. Pest Manag Sci 62: 690–692

Navarro-Llopis V, Vacas S, Sanchis J et al (2011) Chemosterilant bait stations coupled with sterile insect technique: an integrated strategy to control the Mediterranean fruit fly (Diptera: Tephritidae). J Econ Entomol 104:1647–1655

Neeraj GS, Kumar A, Ram S et al (2017) Evaluation of nematicidal activity of ethanolic extracts of medicinal plants to *Meloidogyne incognita* (kofoid and white) chitwood under lab conditions. Int J Pure Appl Biosci 1:827–831

Negahban M, Moharamipour S, Sefidkon F (2007) Fumigant toxicity of essential oil from *Artemisia sieberi* Besser against three stored product insects. J Stored Prod Res 43:123–128

Nerio LS, Oliver-Verbel J, Stashenko E (2010) Repellent activity of essential oils: a review. Bioresour Technol 101:372–378

Oerke EC (2006) Crop losses to pests. J Agric Sci 144:31–43

Oguh CE, Okpaka CO, Ubani CS et al (2019) Natural pesticides (Biopesticides) and uses in pest management—a critical review. Asian J Biotechnol Genet Eng 17:1–18

Okwute SK (2012) Plants as potential sources of pesticidal agents: a review. In: Soundararajan RP (ed) pesticides—advances in chemical and botanical pesticides; eBook, 1st edn. IntechOpen, London, pp 207–232

Okwute SK, Nduji AA (1992) Isolation of Schimperiin: a new gallotannin from the leaves of *Anogeissuss chimperi* (Combretaceae). Proc Natl Acad Sci USA 4:36–41

Ortiz-Hernández ML, Sánchez-Salinas E, Dantán-González E et al (2013) Pesticide biodegradation: mechanisms, genetics and strategies to enhance the process. In: Chamy R (ed) Biodegradation - life of science. InTech, Rijeka

Oyedeji AO, Okunowo WO, Osuntoki AA et al (2020) Insecticidal and biochemical activity of essential oil from *Citrus sinensis* peel and constituents on *Callosobrunchus maculatus* and *Sitophilus zeamais*. Pestic Biochem Phys 168:104643

Ozols G, Bicevskis M (1979) Respects for the use of Ips tyrographus attractant. In: Shumakov EM, Chekmenev SY, Ivanova TV (eds) Biologia Aktualis Veshchestva Zashchiva Rastenij. Izd. Kolos, Moscow, pp 49–51

Pavela R, Benelli G (2016) Essential oils as ecofriendly biopesticides? Challenges and constraints. Trends Plant Sci 21:1000–1007

Pettersson EM (2001) Volatile attractants for three pteromalid parasitoids attacking concealed spruce bark beetles. Chemoecology 11:89–95

Plata-Rueda A, Martinez LC, Santos MHD et al (2017) Insecticidal activity of garlic essential oil and their constituents against the mealworm beetle, *Tenebrio molitor* Linnaeus (Coleoptera: Tenebrionidae). Sci Reports 7:46406

Poelman EH, Oduor AMO, Broekgaarden C et al (2009) Field parasitism rates of caterpillars on *Brassica oleracea* plants are reliably predicted by differential attraction of *Cotesia* parasitoids. Funct Ecol 23:951–962

Pozzebon A, Loeb G, Duso C (2009) Grape powdery mildew as a food source for generalist predatory mites occurring in vineyards: effects on life-history traits. Ann Appl Biol 155:81–89

Qiao J, Zou X, Lai D et al (2014) Azadirachtin blocks the calcium channel and modulates the cholinergic miniature synaptic current in the central nervous system of *drosophila*. Pest Manag Sci 70:1041–1047

Rasmann S, Turlings TCJ (2008) First insights into specificity of belowground tritrophic interactions. Oikos 117:362–369

Rasmann S, Erwin AC, Halitschke R et al (2011) Direct and indirect root defences of milkweed (*Asclepias syriaca*): trophic cascades, trade-offs and novel methods for studying subterranean herbivory. J Ecol 99:16–25

Reddy G (2002) Plant volatiles mediate orientation and plant preference by the predator *Chrysoperla carnea* Stephens (Neuroptera: Chrysopidae). Biol Control 25:49–55

Regnault-Roger C, Philogène BJR (2008) Past and current prospects for the use of botanicals and plant allelochemicals in integrated pest management. Pharm Biol 46:41–52

Reynoso MMN, Lucia A, Zerba EN et al (2019) The octopamine receptor is a possible target for eugenol-induced hyperactivity in the blood-sucking bug *Triatoma infestans* (Hemiptera: Reduviidae). J Med Entomol 57:627–630

Roeder T (2005) Tyramine and octopamine: ruling behavior and metabolism. Annu Rev Entomol 50:447–477

Rotundo G, Paventi G, Barberio A et al (2019) Biological activity of *Dittrichia viscosa* (L.) Greuter extracts against adult *Sitophilus granarius* (L.) (Coleoptera, Curculionidae) and identification of active compounds. Sci Reports 9:6429

Sahaf BZ, Moharramipour S (2008) Fumigant toxicity of *Carum copticum* and *Vitex pseudonegundo* essential oils against eggs, larvae and adults of *Callosobruchus maculatus*. J Pestic Sci 81:213–220

Sammour EA, El Hawary FMA, Abdel-Aziz NF (2011) Comparative study on the efficacy of neemix and basil oil formulations on the cowpea aphid *Aphis craccivora* Koch. Arch Phytopathol Plant Protect 44:655–670

Schoonhoven AV (1978) The use of vegetable oils to protect stored beans from bruchid attack. J Econ Entomol 71:254–256

Sehrawat A, Phour M, Kumar R et al (2021) Bioremediation of pesticides: an eco-friendly approach for environment sustainability. In: Panpatte DG, Jhala YK (eds) Microbial rejuvenation of polluted environment. Microorganisms for sustainability, vol 25. Springer, Singapore

Sharaby A, El-Nujiban A (2015) Evaluation of some plant essential oils against the black cutworm *Agrotis ipsilon*. Global J Adv Res 2:701–711

Sharma HC, Sujana G, Rao DM (2009) Morphological and chemical components of resistance to pod borer, *Helicoverpa armigera* in wild relatives of pigeonpea. Arthropod Plant Interact 3: 151–161. https://doi.org/10.1007/s11829-009-9068-5

Sherer T, Richardson J, Testa C et al (2007) Mechanism of toxicity of pesticides acting at complex I: relevance to environmental etiologies of Parkinson's disease. J Neurochem 100: 1469–1479

Shivanandappa T, Rajashekar Y (2014) Mode of action of plant-derived natural insecticides. In: Singh D (ed) advances in plant biopesticides, 1st edn. Springer, New Delhi, pp 323–345

Singh P, Pandey AK (2018) Prospective of essential oils of the genus *Mentha* as biopesticides: a review. Front Plant Sci 9:1295

Singh D, Mehta SS, Neoliya NK et al (2003) New possible insect growth regulators from *Catharanthus roseus*. Curr Sci 84:1184–1186

Snoeren TA, Mumm R, Poelman EH et al (2010) The herbivore-induced plant volatile methyl salicylate negatively affects attraction of the parasitoid *Diadegma semiclausum*. J Chem Ecol 36:479–489

Solomon MG, Cross JV, Fitzgerald JD et al (2000) Biocontrol of pests of apples and pears in northern and central Europe-3, Predat. Biocontrol Sci Tech 2:91–128

Souto AL, Sylvestre M, Tölke ED et al (2021) Plant-derived pesticides as an alternative to pest management and sustainable agricultural production: prospects, applications and challenges. Molecules 26:4835

Srijita D (2015) Biopesticides: an eco-friendly approach for pest control. World J Pharm Pharm Sci 6:250–265

Stroh J, Wan MT, Isman MB et al (1998) Evaluation of the acute toxicity to juvenile Pacific coho salmon and rainbow trout of some plant essential oils, a formulated product, and the carrier. Bull Environ Contam Toxicol 60:923–930

Tabari MA, Rostami A, Khodashenas A et al (2020) Acaricidal activity, mode of action, and persistent efficacy of selected essential oils on the poultry red mite (*Dermanyssus gallinae*). Food Chem Toxicol 138:111207

Taghizadeh-Saroukolai A, Moharramipour S, Meshkatalsadat MH et al (2009) Repellent activity and persistence of essential oil extracted from *Prangos acaulis* to three storedproduct beetles. Am Eurasian J Sustain Agric 3:202–204

Tak JH, Jovel E, Isman MB (2016) Comparative and synergistic activity of *Rosmarinus officinalis* L. essential oil constituents against the larvae and an ovarian cell line of the cabbage looper, *Trichoplusia ni* (Lepidoptera: Noctuidae). Pest Manag Sci 72:474–480

Tak J, Jovel E, Isman MB (2017) Effects of rosemary, thyme and lemongrass oils and their major constituents on detoxifying enzyme activity and insecticidal activity in *Trichoplusia ni*. Pestic Biochem Phys 140:9–16

Tempfli B, Pénzes B, Fail J, Szabó Á (2015) The occurrence of tydeoid mites (Acari: Tydeoidea) in Hungarian vineyards. Syst Appl Acarol 20:937–954

Tomizawa M, Casida JE (2005) Neonicotinoid insecticide toxicology: mechanisms of selective action. Annu Rev Pharmacol Toxicol 45:247–268

Tong F, Coats JR (2010) Effects of monoterpenoid insecticides on [3H]-TBOB binding in house fly GABA receptor and 36Cl-uptake in American cockroach ventral nerve cord. Pestic Biochem Physiol 98:317–324

Towers G (2011) Interactions of light with phytochemicals in some natural and novel systems. Can J Bot 62:2900–2911

Tripathi AK, Prajapati V, Aggarwal KK et al (2000) Effect of volatile oil constituents of *Mentha* species against stored grain pests, *Callosobruchus maculatus* and *Tribolium castaneum*. J Med Arom Plant Sci 22:549–556

Tumlinson JH, Turlings TCJ, Lewis WJ (1992) The semiochemical complexes that mediate insect parasitoid foraging. Agric Zoo Rev 5:221–252

Turlings TC, Benrey B (1998) Effects of plant metabolites on the behavior and development of parasitic wasps. Ecoscience 5(3):321–333

Turlings TC, W¨ackers F (2004) Recruitment of predators and parasitoids by herbivore-injured plants. In: Adv Insect Chem Ecol, vol 2, pp 21–75

Turlings T, Tumilinson J, Lewis W (1990) Exploitation of herbivore-induced plant odors by host-seeking parasitic wasps. Science 250:1251–1253

United Nations, Department of Economic and Social Affairs, Population Division (2019) Probabilistic Population Projections Rev. 1 based on the World Population Prospects 2019 Rev 1: http://population.un.org/wpp/

Upadhyay N, Singh VK, Dwivedy AK et al (2019) Assessment of *Melissa officinalis* L. essential oil as an eco-friendly approach against biodeterioration of wheat flour caused by *Tribolium castaneum* Herbst. Environ Sci Pollut R 26:14036–14049

Usha Rani P, Jyothsna Y (2010) Biochemical and enzymatic changes in rice as a mechanism of defense. Acta Physiol Plant 32:695–701

van den Boom CEM, van Beek TA, Posthumus MA et al (2004) Qualitative and quantitative variation among volatile profiles induced by *Tetranychus urticae* feeding on plants from various families. J Chem Ecol 30:69–89

van Loon JJA, de Boer JG, Dicke M (2000) Parasitoid-plant mutualism: parasitoid attack of herbivore resistance increases plant reproduction. Entomol Exp Appl 97:219–227

Velázquez-Fernández JB, Martínez-Rizo AB, Ramírez-Sandoval M et al (2012) Biodegradation and bioremediation of organic pesticides. In: Soundararajan R (ed) Pesticides - recent trends in pesticide residue assay [Internet]. IntechOpen, London. https://www.intechopen.com/chapters/38062

Vendan SE, Manivannan S, Sunny AM et al (2017) Phytochemical residue profiles in rice grains fumigated with essential oils for the control of rice weevil. PLoS One 12(10):e0186020

Venzon M, Janssen A, Sabelis M (1999) Attraction of a generalist predator towards herbivore-infested plants. Entomol Exp Appl 93:305–314

Walia S, Saha S, Rana V (2014) Phytochemical pesticides. In: Singh D (ed) Advances in plant biopesticides. Springer, New Delhi, pp 295–322
War AR, Paulraj MG, War MY et al (2011a) Jasmonic acid-mediated induced resistance in groundnut (*Arachis hypogaea* L.) against *Helicoverpa armigera* (Hubner) (Lepidoptera: Noctuidae). J Plant Growth Regul 30:512–523. https://doi.org/10.1007/s00344-011-9213-0
War AR, Paulraj MG, War MY et al (2011b) Herbivore- and elicitor-induced resistance in groundnut to Asian armyworm, *Spodoptera litura* (Fab.) (Lepidoptera: Noctuidae). Plant Signal Behav 6:1769–1777
War AR, Paulraj MG, Ahmad T et al (2012) Mechanisms of plant defense against insect herbivores. Plant Signal Behav 7(10):1306–1320
War AR, Taggar GK, Hussain B et al (2018) Plant defence against herbivory and insect adaptations. AoB Plants 10:ply037. https://doi.org/10.1093/aobpla/ply037
Weinzierl RA (2000) Botanical insecticides, soaps and oils. In: Rechcigl JE, Rechcigl NA (eds) Biological and biotechnological control of insect pests. Lewis publishers, Boca Raton, pp 110–130
Whalon ME, Mota-Sanchez D, Hollingworth RM (2008) Analysis of global pesticide resistance in arthropods. Glob Pestic Resist Arthropods 5:31
Wu W, Li S, Yang M et al (2020) Citronellal perception and transmission by *Anopheles gambiae* s.s. (Diptera: Culicidae) females. Sci Reports-UK 10:18615
Yang X, Huang Q, Jiang T et al (2017) Degradation dynamics of Azadirachtin in cabbage and soil. J South China Agric Univ 38(4):37–40
Zehnder G, Warthen JD (1988) Feeding inhibition and mortality effects of neem-seed extract on the Colorado potato beetle (Coleoptera: Chrysomelidae). J Econ Entom 81(4):1040–1044

# Chapter 9
# Efficiency of Plant Metabolic Products on Insect Pests and Their Parasitoids

**Nabil El-Wakeil, Nawal Gaafar, and Saad Al-Kahtani**

## 9.1 Introduction

Over nearly 350 million years of together living, plants and insects have been locked in an evolutionary arms race. In response to insect herbivory, plants have developed a sophisticated immune system adept for detecting infestation of insect pests, foreign chemicals, and signals from plant injured tissues (Howe and Jander 2008; Verhage et al. 2010; Hare 2011). This defense manifests through an assorted arsenal of morphological structures, secondary metabolites, and proteins that significantly affect herbivore biology and benefit their life as confirmed by Usha Rani and Jyothsna (2010) and War et al. (2011a, b, 2020). Plant-herbivore interactions activate through two primary mechanisms: direct and indirect defense. Direct defenses comprise traits such as physical barriers or toxic complexes that directly influence an herbivore's host preference or reproductive achievement (Hanley et al. 2007; Züst and Agrawal 2016). Indirect defenses, in contrast, often include the emission of volatile organic compounds that attract many kinds of natural enemies of the herbivore, thereby enlisting a third trophic level for plant protection (Dudareva et al. 2006; Arimura et al. 2009).

Central to these defenses are plant secondary metabolites (PSMs). Beyond their direct toxicity, PSMs adjust defensive signaling pathways (Divekar et al. 2022; Erb et al. 2021), enable rapid threat detection and response, and mediate multi-trophic

N. El-Wakeil (✉) · S. Al-Kahtani
Department of Arid Land Agriculture, College of Agricultural & Food Sciences, King Faisal University, Al-Ahsa, Saudi Arabia
e-mail: nelwakeil@kfu.edu.sa

N. Gaafar
Pests & Plant Protection Department, Agricultural & Biological Sciences Institute, NRC, Cairo, Egypt

 
N. El-Wakeil, A. Abdelkhalek (eds.), *Microbiomes and Plant Secondary Metabolites for Insect Pest Management*, Progress in Biological Control 23,
https://doi.org/10.1007/978-3-032-24330-0_9

interactions that ultimately benefit the host plants. A deeper understanding of PSM biosynthesis is therefore critical for rising crops with improved herbivore tolerance (Afroz et al. 2021). Given their pivotal role, the efficiency of plant metabolic products (PMPs) against insect pests and their parasitoids establishes a major research focus in agricultural entomology, plant pathology, and Integrated Pest Management (IPM). This book chapter will explore this theme through two key directions:

1. The direct and indirect modes of action of PMPs against insect herbivores.
2. The influence of PMPs on the behavior and efficacy of parasitoids and other natural enemies within IPM frameworks.

## 9.2 Efficiency of Plant Metabolic Products on Insect Pests

### 9.2.1 *Direct Defenses*

#### 9.2.1.1 Plants Protected from Insect Infestations

Herbivore defense in plants contains structural barriers, chemical toxins, and the attraction of various natural enemy types. Direct and indirect defensive responses are critical for insect herbivore management (Hanley et al. 2007; Howe and Jander 2008; Karban 2011; War et al. 2012, 2020). Plant induced responses decrease stress and make plants phenotypically plastic, reducing the likelihood of insects adapting to induced chemicals. Insect attack changes plant defenses, creating unpredictability for herbivores and affecting fitness and behavior (Agrawal et al. 2002; Miranda et al. 2007; Agrawal 2011). Early induced responses benefit plants by reducing subsequent attacks and refining overall fitness (Miranda et al. 2007; War et al. 2011a, b). Plants use physical barriers, secondary metabolites, and synergistic effects to increase their defense system (Jamwal et al. 2018; War et al. 2020). Tomatoes contain alkaloids, phenolics, proteinase inhibitors, and oxidative enzymes to work together in nicotine plants; as mentioned by (Steppuhn and Baldwin 2007), who reported that there are compounds like trypsin and nicotine expression to protect against *Spodoptera exigua* (Santiago et al. 2018).

**Morphological Traits**

Herbivore resistance, or host plant resistance (HPR), is greatly influenced by the architecture of plants. Waxy cuticles, spines, setae, and trichomes are some of the physical barriers that plants use to ward off herbivores (Hanley et al. 2007; Agrawal et al. 2009). Suberization and lignification alter the thickness of cell walls, which in turn creates noticeable protrusions (Hanley et al. 2007; He et al. 2011; Kundu et al. 2018). Plants defend themselves from insect herbivores by presenting different of structural features such as spines, thorns, pubescence, hardened leaves, granular minerals, and divaricated branches (Chamarthi et al. 2010; He et al. 2011). According to Handley et al. (2005) and Hanley et al. (2007), herbivores cause less harm when tissues are less palatable and digestible due to sclerophylly, or thickened leaves. Spinescence, thorns, and prickles, and pubescence, trichomes on stems, leaves, and fruits, also protect against various insect species (Chamarthi et al. 2010); who stated that *Atherigona soccata* had been affected by leaf glossiness, plumule, and leaf sheath pigmentation.

**Trichomes**
Trichomes, which have both noxious and preventing properties, are important for plants' defense against insect herbivores. Insect behavior, feeding, and larval nutrition are all adversely affected by thick trichomes. Insects and other arthropods are unable to move easily across the plant's surface, which further restricts their contact to the epidermal of the plant leaves (Handley et al. 2005; Agrawal et al. 2009; Chamarthi et al. 2010). Trichomes are encouraged in many plants by insect injury, but only in newly formed leaves (Agrawal 1999). The mass of the existing leaf trichomes remains unchanged. In *Salix cinerea* L., Dalin and Björkman (2003) mentioned that leaf beetle adult (*Phratora vulgatissima* L.) initiated damage that increased the density of trichomes in the resulting young leaves. Similarly, it has been reported that *S. cinera* shows a rise in trichome numbers in response to coleopteran infestation (Björkman and Ahrne 2005). It has also been distinguished that *Lepidium virginicum* L. trichome density enlarged after insect damage. Following *Pieris rapae* (L.) feeding, black mustard's trichome density and glucosinolate levels increased (Agrawal 1999; Katz et al. 2015). Trichothome density rose in *Alnus incana* as a result of infestation of this beetle (Baur et al. 1991). Although 500–100% has been recognized, herbivory usually results in a 25–100% increase in trichome density. Within days or weeks, trichome density altered as a result of insect damage. Furthermore, herbivory changes the ratio of glandular to non-glandular trichomes. Trichothome mass has been found to positively associate with the abundance of biocontrol agents (Erb et al. 2021). In addition, scelonid egg parasitoids are provided with supplementary floral nectar (AFN) by trichome exudates in squash bugs, *Gryon pennsylvanicum* (Olson and Nechols 1995; Stevenson et al. 2017).

#### 9.2.1.2 Secondary Metabolites as Plant Defense Tools

PSMs are plants' main defense against a wide range of insect herbivores, hence they must be studied for biosynthesis and use as an eco-friendly and maintainable pest management substitutes. Recent analytical methods have discovered secondary metabolites' significance in plant guard system (Wink 2003; Jamwal et al. 2018). Four categories of metabolites exist: terpenoids, phenolics, nitrogen, and sulfur-containing chemicals (Table 9.1).

Without prompting growth or development, secondary metabolites reduce the palatability of plant tissues. They could be dormant or produced as reacting to attacks of insects or microbes (Howe and Jander 2008). Phytoalexins are the latter, and phytoanticipins the former. Myrosinases (endogenous β-thioglucoside glucohydrolases) hydrolyze glucosinolates during tissue rupture, releasing biocidal

**Table 9.1** Secondary plant metabolites

| Terpenes | | | Phenolics | | | | N containing compounds | | | S containing compounds | | | |
|---|---|---|---|---|---|---|---|---|---|---|---|---|---|
| Monoterpenes | Sesquiterpenes | Triterpenes | Coumarin | Furano-coumarin | Flavonoids | Lignin | Alkaloids | Cyanogenic glycosides | Non-protein amino acids | Glutathione | Glucosinolates | Phtyoalexins | Defensin |
| Citronellol | Caryophyllena | Azadirachtin | Hydroxycoumarins | Psoroline | Quercetin | Resveratrol | Nicotine | Dhurrin | Azetidine | | | | |

aglycone metabolites. During herbivory, β-glucosidase activates these phytoanticipins (Morant et al. 2008). Poaceae contains a large number of BXs, another phytoanticipin. According to Morant et al. (2008), when plants experience tissue injury, plastid-targeted β-glucosidases break down BX-glucosides, generating biocidal aglycone BXs. These composites play a vital role in the defensive mechanisms of plants against insect pests. Herbivores' performance and survival are compressed by phytoalexins such as isoflavonoids, terpenoids, alkaloids, and others (Walling 2000; Stevenson et al. 2017; Dey et al. 2020; Yang et al. 2021).

Secondary metabolites do twofold duties: they defense plants from danger and increase their fitness (Jamwal et al. 2018). Reports indicate that the main reason maize infested by corn earworm, *Helicoverpa zea* (Boddie) is because of the presence of the secondary metabolites maysin and chlorogenic acid, which are phenylpropanoid derivatives (Nuessly et al. 2007; War et al. 2020). It has been revealed that the compound 4, 4-dimethyl cyclooctene is what gives *Sorghum bicolor* its resistance to shoot fly *A. soccata* (Chamarthi et al. 2011; Shahabinejad et al. 2014; Alvarenga et al. 2017). Researchers have reflected worldwide the secondary metabolites as potential defense mediators, their signaling mechanisms on various insect pests (War et al. 2020). Investigating secondary metabolites might lead for identifying the new signaling molecules that have a role in herbivore and other stress resistance in host plants (Jamwal et al. 2018). These metabolites' biosynthesis genes and enzymes might be found.

**Phenols**

One of the most frequent secondary metabolites, plant phenols, are defensive chemicals that help HPR against herbivores, including insects and microorganisms. Phenols defend the host plants against herbivores, microbes, and competing plants. Phenols and oxidative enzyme activity change generally in response to insect pest attack as confirmed by Barakat et al. (2010), War et al. (2011a, b), and War et al. (2020). Plant protection against insects and diseases relies on phenolic heteropolymer lignin (Barakat et al. 2010). It blocks infections by physically or cumulative leaf hardness, reducing herbivore grazing and leaf nutritive content (Johnson et al. 2009). After infestation of insect pests or pathogens, lignin production and fast deposition prevent pathogen or herbivore growth (Johnson et al. 2009; Ramírez-Gómez et al. 2020). Pest- and pathogen-infected plants express more lignin-associated genes as reported by Barakat et al. (2010). Polyphenol oxidase (PPO) and peroxidase (POD) could, through phenol oxidation, protect plants from infestation by herbivorous insects. Quinones, generated by phenol oxidation, impair herbivore protein digestion and poison insects. By cyclically reducing reactive oxygen species (ROS) and reducing plant protein feeding, phenols manage insect growth and development; Phenols structures are shown in Fig. 9.1.

**Flavonoids**

In particular, flavonoids play an important role in plant-environment interactions. These shield plants from injurious insects, pathogens, and ultraviolet light (Treutter 2006; Afroz et al. 2021). Enzymes bind to flavonoids, making them cytotoxic. Plants are protected by flavonoids and isoflavonoids because they impact the behavior as

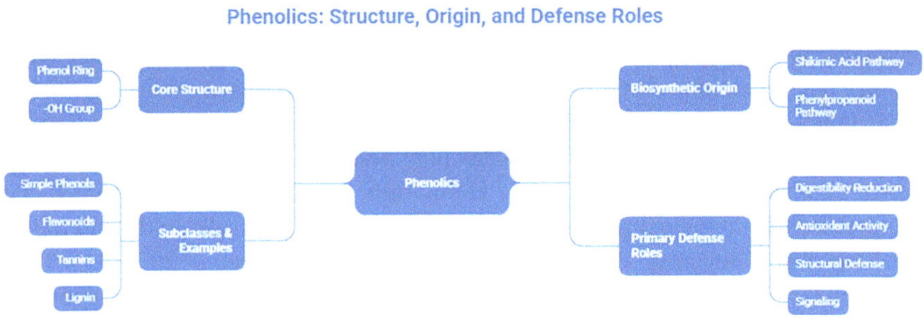


**Fig. 9.1** Phenol's structure and defense roles

well as development of insects. They lessen reactive oxygen species (ROS) and free radicals by chelating metals (Simmonds 2003). Anthocyanins, flavones, flavonols, flavanones, dihydroflavonols, chalcones, aurones, flavan, and proanthocyanidins are all types of flavonoids (Treutter 2006). Proanthocyanidins, flavan 3-ols, flavonones, flavans, and isoflavonoids have been discovered as deterrents for insect pest feeding. Simmonds et al. 1990) evaluated 11 plant-derived polyhydroxy alkaloids on 4 lepidopteran larvae for antifeedant effects. In a study of plant alkaloids, only pyrrolidine DMDP was operative against herbivores, while the piperidines fagomine, XZ-1, and alexine showed no activity. The researchers further confirmed that certain defensive alkaloids and common phagostimulatory sugars are structural analogs, yet they differentially affect insect neurons (Dey et al. 2020). These results have implications for Integrated Pest Management (IPM) approaches. In a related defensive appliance, the overexpression of a flavonoid transcription factor in Arabidopsis has been shown to confer resistance to *Spodoptera frugiperda* (Dowd and Johnson 2009).

Isoflavones, licoisoflavone A, luteone, licoisoflavone B, and wighteone prevent development of *Colletotrichum gloeosporiode* (Penz.) and *Cladosporium cladosporioides* (Fres.) (Lane et al. 1987). Antifeedants against *Helicoverpa armigera* with a concentration of 100 ppm include judaicin, judaicin-7-O-glucoside, 2-methoxyjudaicin, and maackiain, all of which initiate from wild chickpea. Judaicin and maackiain also deterred *S. littoralis* and *S. frugiperda* (Simmonds and Stevenson 2001; Bhonwong et al. 2009; Datta et al. 2020) Cyanopropenyl glycoside and alliarinoside suppress feeding in *Pieris oleracea* L., while isovitexin-6-″-D-β-glucopyranoside prevents late instar feeding (Renwick et al. 2001; Tian et al. 2021).

**Tannins**

Insects that feed on plants have their midguts damaged by tannins, which are astringent polyphenols that bind to proteins and decline absorption of nutrient stuff (Barbehenn and Constabel 2011). Through hydrogen or covalent interaction of protein -NH2 groups, they nonspecifically precipitate proteins, comprising digestive enzymes for herbivores. The bioavailability of herbivores is reduced because tannins bind metal ions. Because tannins reduce the digestibility of proteins, herbivores get less nutrients from plant parts. While *Quercus serrata* (Thunb.) and *Betula pendula* (Roth) do not have tannins, *Populus* species and *Pinus sylvestris* L. have tannins content. There is no effect of tannins on tannin content from factors such as oxidative stress or proteinase inhibitors (Tian et al. 2021).

Oligomeric or polymeric proanthocyanidins condensed tannins. They vary in structure and function. They keep *Lymantria dispar, Euproctis chrysorrhoea*, and *Operophtera brumata* from eating (Bernays 1981). *Quercus robur* leaves comprise rich tannins such (+)-catechin, (+)-gallocatechin, and vanillin. *Aphis Craccivora* (Koch) in groundnut is prevented by procyanindin polymers. Alaska paper birch tannins diminish *Rheumaptera hastata* (L.) pupal bulk and survival. Induction of *Populus tremuloides* Michx. tannin. Herbivores and wounds activate the flavonoid pathway. Injury encourages PtMYB134, a condensed tannins regulating gene, to produce tannins (Alvarenga et al. 2017). Light stress and UV exposure boost hybrid poplar tannin synthesis.

Tannins exemplify a quantitative, dosage-dependent chemical defense that improves plant efficiency primarily by imposing a metabolic burden on the insect pests. Their ability to bind proteins and decrease nutritional digestibility forces amplified consumption and prolonged feeding time, effectively using the herbivore's own physiology as a defensive implement (Barbehenn and Constabel 2011). This strategy is effective because it conserves the plant's nitrogen for growth while allocating carbon-rich compounds for defense. The inducible nature of many tannins, where their biosynthesis is upregulated following pest attack, further augments resource allocation by evading costly constitutive expression (Feeny 1976; Stamp 2003). While trade-offs with growth could occur, their broad-spectrum effectiveness and role as a digestibility reducer underscore their contribution to a durable and resource-efficient defensive tactic, effectively raising the ecological cost of herbivory.

#### 9.2.1.3 Proteins—Plant Defense

The interaction between insects and plants is vigorous to their survival. Healthy host plants attract insects for coupling, oviposition, and feeding of immature stages. Similar to other animals, insects need plant protein, and any imbalance in digestion and use greatly impacts insect physiology. Changes in gene expression affect protein quality and quantity, thereby altering signal transmission and oxidative defense under stress conditions such as insect attack. Stable plant proteins in the midgut cross the gut wall into the hemolymph in insects (Mellway et al. 2009; Afroz et al. 2021). Protein function depends on amino acid amount or sequence (Stevenson et al. 2017). Protease inhibitors (PIs) protect against insects by limiting proteolysis-susceptible protein degradation (Zhu-Salzman and Zeng 2015; Mrid et al. 2021). Studying protein structure and herbivore gut stabilizing post-translational alterations could expect plant resistance protein (PRP) toxicity and apparatus. Recent microarray and proteomic research indicates that many PRPs protect plants from insect herbivores. Arthropod-inducible proteins are controlled by numerous signaling pathways, including JA, SA, and ET, due to their varied eating patterns (Chen et al. 2005; Chen et al. 2009; Mrid et al. 2021).

**Lectins**

Carbohydrate-binding (glyco) proteins called lectins are everywhere and protect against many pests (Vandenborre et al. 2011; Mrid et al. 2021). Various plant lectins have been used as natural insecticides (Saha et al. 2006). Lectins' ability to survive in herbivore digestive systems makes them effective insecticides (Vandenborre et al. 2011). By attaching to digestive tract membrane glycosyl groups, they cause a variety of systemic side effects (Vandenborre et al. 2011). Lectins can have profound negative effects on insects by disrupting their digestive processes, damaging gut tissues, and altering their physiological states, eventually threatening their survival and generative success. This makes lectins a significant consideration in the study of plant-insect interactions and in the progress of pest control tactics (Saha et al. 2006).

Homopterans, lepidopterans, and coleopterans are all hypothetically killed by lectins (*Galanthus nivalis* L.). GNA, Phaseolus haemagglutinin, and wheat germ agglutinin are just a few samples of the carbohydrate-binding plant lectins that have been widely studied for their potential to deter insects (Dutta et al. 2005). Mannose-binding lectins engage with a cell membrane carbohydrate residue to efficiently kill sucking insects (Saha et al. 2006); as well an artificial diet in craccivora (Majumder et al. 2005).

Mannose-specific lectin GNA distinguishes gut glycoproteins and glycoconjugates on midgut epithelial cells (Powell et al. 2006). Rice fat bodies, ovarioles, and hemolymph comprise GNA, which controls *Nilaparvata lugens*. JA (Lannoo et al. 2007; De Freitas et al. 2019) and herbivores (Vandenborre et al. 2009) *S. littoralis*, *Manduca sexta* encourage tobacco leaves to express NICTABA lectin. Tobacco plants express HFR1, a mannose-binding jacalin-like lectin, and HFR2 and HFR3, two chimerolectin-like proteins, from Hessian fly larvae (Vandenborre et al. 2009). Many insect feeding patterns affect lectin expression, as in fall armyworm larvae (Giovanini et al. 2007; Subramanyam et al. 2008). The cherry-oat aphid *Rhopalosiphum padi* Kocchi activated HFR2 but not HFR3, while rice, barley, wheat, rye, and maize leaf tissues express multiple jasmonate-inducible lectins (Puthoff et al. 2005; Giovanini et al. 2007; War and Sharma 2014; De Freitas et al. 2019; Olson 2021). Transforming lectin genes into plants seems promising, yet some lectins may harm non-target animals, including mammals.

Lectins represent a qualitative, target-specific defense whose efficiency derives from high potency and precision targeting. By binding with high affinity to specific glycans in the herbivore gut, they disrupt critical functions like nutrient absorption and gut integrity at very low concentrations, minimizing the plant's metabolic investment per unit of defensive payoff (Vandenborre et al. 2011). This mode of action exemplifies an evolutionary arms race, where defense efficacy is depending on the presence of the specific receptor in the insect pest (Michiels et al. 2010). While this specificity could limit their spectrum of activity, it also decreases potential autotoxicity and non-target effects. Therefore, lectins are most effective as components of varied defensive assortments, where they provide a potent, "sniper"-- like strike against co-evolved or specialist herbivores, complementing the broader, "shotgun" approach of defenses like tannins.

#### 9.2.1.4 Proteinase Inhibitors

Plant defense proteins like proteinase inhibitors (PIs) are plentiful for host plants against insect pests. In seeds and tubers, PIs make approximately 1–10% of total proteins and block enzymes, protecting plants from insect infestation (Zhu-Salzman and Zeng 2015). Insect stomach digestive enzymes attach to PIs, decreasing protein digestion and causing amino acid deficiencies, sluggish development, and famine (Mrid et al. 2021). In response to stress, plants develop several PI classes. Serine PIs (SPIs) Kunitz proteinase inhibitors (KPIs) substantially upregulate defense genes in wounded or herbivore-fed plants (Miranda et al. 2007). Genome sequencing has shown many herbivore-induced proteinase inhibitors and defense components in plants, which could harmful for insect pests (Tian et al. 2021). Insect herbivory induces most plant KPIs, but the degree depends on insect-plant interaction. KPIs store PIs hereditarily to help plants fight many insect generations. Insects may create PI-insensitive proteases or inactivate PIs, avoiding plant defense (Steppuhn and Baldwin 2007).

**Enzymes**

Disruption of insect feeding is a significant HPR effect. A variety of peroxidases, including ascorbate peroxidases, polyphenol oxidases, and others, oxidize mono- or dihydroxyphenols to reactive o-quinones. These o-quinones can then polymerize or form covalent adducts with the nucleophilic groups found in proteins. Herbivory induces antioxidant enzymes in plants, according to recent studies (Chen et al. 2009; Gulsen et al. 2010; War et al. 2011b; Vandenborre et al. 2011).

**Peroxidases**

Host plant oxidation causes HPR to insects, generating ROS that antioxidant enzymes remove (Zhao et al. 2009; He et al. 2011). POD enzymes remove ROS and provide other defensive roles. PODs help plants respond quickly to insect pest attacks (Gulsen et al. 2010; War et al. 2011a, b; Mrid et al. 2021). There are many soluble, membrane-bound, and cell wall-bound isozymes of monomeric hemoproteins in plants. Tissue, stage of development, and environmental elements all influence their expression (Gulsen et al. 2010; He et al. 2011). Heng-Moss et al. (2004) and Sethi et al. (2009) found that PODs in plants control processes such as wound healing, auxin metabolism, somatic embryogenesis, lignification, and suberization. PODs' role in insect pest resistance has been inspected in different plant systems (Gulsen et al. 2010; War et al. 2011a, b). PODs and phenols produce phenoxy and other oxidative radicals that avoid insects from feeding and/or limit plant digestibility, which causes insect nutritional shortage and severely impacts their growth and development (Zhang et al. 2008; Chen et al. 2009). PODs also directly poison herbivore intestines; as well as insect-induced PODs have been purified and described from several plants as published by Stout et al. (2009), Gulsen et al. (2010), and He et al. (2011).

**Polyphenol Oxidases**

Phenolic polyphenols (PPOs) impact the plant nutrition, growth, and stress defense (Khare et al. 2020). PPO-derived quinones can harm plant nutrition by alkylating amino acids, generating intestinal oxidative stress, and releasing reactive oxygen species (ROS) (Teklić et al. 2021). PPOs are metallo-enzymes that oxidize monophenols and o-diphenols to quinones, highly reactive intermediate compounds that polymerize and react with nucleophilic side chains of amino acids and crosslink proteins, sinking protein availability and food nutritional quality (Zhang et al. 2008; Feng et al. 2019). Quinines poison plant herbivores more than phenols. PPOs develop plant tissue digestibility and palatability, while melanin production boosts insect and disease resistance (Zhao et al. 2009; Ramírez-Gómez et al. 2020). PPO activity induced by abiotic and biotic stressors and treatment with octadecanoid chemicals aids plant stress tolerance (Sethi et al. 2009; Khare et al. 2020; Teklić as al. 2021). Young tissues more susceptible to insect assault induce more. Spines, thorns, trichomes, hardened leaves, granular minerals, and divaricated branches defend plants from herbivory. High PPO levels help agricultural insect pests such as *S. litura, H. armsigera*, *Bemisia tabaci*, *Tetranychus cinnabarinus*, *Myzus persicae*, *Empoasca fabae*, *Aphis medicaginis*, *S. exigua*, and *Agelastica alni* resist (Tscharntke et al. 2001; Bhonwong et al. 2009).

**Lipoxygenases**

Antioxidant LOXs hydrolyze polyunsaturated fatty acids and generate fatty acid hydroperoxides to protect plants from stress (Bruinsma et al. 2009). Reactive substances like aldehydes, γ-ketols, epoxides, and ROS (e.g., hydroxyl, singlet oxygen, superoxide ion, peroxyl, acyl, and carbon-centered radicals) are produced during enzymatic and chemical lipid degradation, that affect protein amino acid assimilation. Plant defense relies on LOX oxidizing linolenic acid in the JA signaling pathway to create oxidative enzymes, protease inhibitors, and volatile organic compounds (VOC) that attract insect pest enemies (Mao et al. 2007; Bruinsma et al. 2009; Jones et al. 2019). It catalyzes polyunsaturated fatty acid oxidation.

Soybean herbivory stimulates LOX activity, as observed in tomato in response to aphids, *Macrosiphium euphorbiae* Thom., and *M. persicae* (Fidantsef et al. 1999), in *Nicotiana attenuata* following *Myzus nicotianae* Blackmann (Voelckel et al. 2004), and in wheat in response to grain aphid *Sitobion avenae* (F LOX3-mediated defense decreased larval growth, food consumption, and frass production in maize plants (Voelckel et al. 2004; Rayapuram and Baldwin 2007; Olson 2021). Mao et al. (2007) and Santiago et al. (2018) found that transgenic maize plants containing the wheat oxalate oxidase gene had higher levels of LOX transcripts and free phenolics by a factor of fourteen, which was positively linked to resistance to European corn borer *Ostrinia nubilalis*.

### 9.2.2 *Indirect Defenses*

Plant protection against herbivores depends on plant defenses (Dudareva et al. 2006; Mrid et al. 2021). Herbivore elicitors and mechanical damage can induce indirect defenses. Volatile synthesis and EFN secretion support plants interact with insect pest foes, lowering herbivore populations (Dudareva et al. 2006; Maffei 2010; Jones et al. 2019). These defenses have been studied genetically, biochemically, physiologically, and ecologically by many of researchers like Stevenson et al. (2017) and Turlings and Erb (2018).

#### 9.2.2.1 Herbivore-Induced Plant Volatiles

To prevent herbivores, plants release volatile and nonvolatile composites. Herbivore-induced plant volatiles (HIPVs) attract natural enemies of insect herbivore or discourage feeding and oviposition (Dudareva et al. 2006; Arimura et al. 2009; Jones et al. 2019). In reaction to herbivores, plants release lipophilic chemicals with increased vapor pressure from leaves, blooms, and fruits into the atmosphere and soil from the roots. HIPV production depends on plant hosts and insect species, growing stage, and ecological factors (Arimura et al. 2009; Maffei 2010). Plants release an optimal amount of volatile chemicals into the atmosphere, while herbivory produces a different blend (Arimura et al. 2009; Turlings and Erb 2018). Plants generate a volatile blend in reaction to insect assault that is distinctive to their insect-plant system, including biological control agents and neighbors Engelberth et al. 2004; Erb et al. 2021). HIPVs facilitate plant-arthropod, microbe, undamaged adjacent plant, and intraplant signaling connections (Arimura et al. 2009; Karban 2011). Different defense signaling pathways stimulate to produce volatile chemicals reliant on insect pest feeding behaviors (Walling 2000; Jones et al. 2019).

The hydroperoxide lyase (HPL) branch of the oxylipin pathway produces stress-inducible terpenes, GLVs, ethylene, methyl salicylate, and other VOCs (Dudareva et al. 2006; Maffei 2010; War et al. 2011c; Schweiger et al. 2014; War and Sharma 2014; Shahabinejad et al. 2014). Stress and defense are signaled by reactive electrophile GLVs. GLVs, such as C6-aldehydes like (Z)-3-hexenal and n-hexanal and their derivatives like (Z)-3-hexenol, hexen-1-yl acetate, and E-isomers, help plants protect themselves. GLV levels have been manipulated by elicitors or genetic alteration of plant HPL expression to inspect their role. Methyl salicylates and C16-homoterpene 4, 8, 12-trimethyl-1, 3(E), 7(E), 11-tridecatetraene [(E, E)-TMTT] attract predatory mites, helping plants defend. The most prevalent HIPV component is MeSA, found in insect-infested plants like lima bean (Dudareva et al. 2006; Matsui 2006; Arimura et al. 2009; Hare 2011; Schweiger et al. 2014). By MeSA created during infection, herbivores may identify it. Insect-infested plants contain MeBA, a chemically identical molecule to MeSA (Chen et al. 2003; Quintana-Rodriguez et al. 2018). MeSA and MeBA, among 30 volatiles released by rice *S. frugiperda* infection, attract different kinds of natural enemies (Turlings and Erb 2018). HIPV-designed natural enemies like the larval parasitoid *Cotesia marginiventris* (Cresson) may attract crop pests, which is environmentally harmful. *Leptinotarsa decemlineata* (Say), Colorado potato beetles, are lured to cis-3-hexenyl acetate, linalool, and MeSA, according to Dickens (2006) and Giongo et al. (2016).

Research suggests herbivores release chemicals such as Ester methyl salicylate (MeSA), myrcene, β-ocimene, TMTT, and α-farnesene hours after insect pest infestation (Dudareva et al. 2006; Schweiger et al. 2014). HIPVs deter herbivores or attract associated natural enemies, preventing plant harm (Maffei 2010). Terpenoids and lipid and shikimic acid metabolites are necessary to plant defense (Gill et al. 2010; Jamwal et al. 2018). Forest tent caterpillar attacks *S. littoralis* and hybrid poplar (*Populus trichocarpa* and *Oscularia deltaoides*) leaves. Differences in volatiles were found between (E)–β-ocimene and other mono-, sesqui-, and homoterpenes in maize plants exposed to (Z)-3-hexanol, which attracts natural enemies after caterpillar damage (Arimura et al. 2004; Ruther and Kleier 2005; Erb et al. 2021).

**Priming Volatile Emission Signals in Plants**
Engelberth et al. (2004) administered GLV compounds including (Z) -3-hexanal, (Z) -3-hexen-1-ol, and (Z) -3-hexenyl acetate to maize seedlings to help them adapt to wounding and beet armyworm regurgitation. Results indicated that JA and sesquiterpene accumulation compared to controls. In response to *M. sexta* infestation, Kessler et al. (2006) found no damage in *N. attenuata* primed with clipped sagebrush volatiles. Plant defense depends on priming to reduce biochemical investments until attack. However, volatile emissions from infested plants may attract non-target insects, increasing insect attack (Horiuchi et al. 2003; Engelberth et al. 2004; Hettenhausen et al. 2014; Kundu et al. 2018).

The predatory mite *Phytoseiulus persimilis* visits transgenic Arabidopsis overexpressing strawberry nerolidol synthase (Kappers et al. 2005; Katz et al. 2015). Overexpressing the corn TPS10 gene in transgenic maize plants attracted the parasitic wasp *Cotesia marginiventris* (El-Wakeil et al. 2003), which feeds on lepidopteran larvae, which produces volatiles that protect against root-feeding insects and act as antimicrobials, antiherbivores, and natural enemies (Dudareva et al. 2006). *Diuraphis noxia* (Mord.), a root-feeding insect, releases poisonous and repulsive 1,8-cineole. Megidi (Rasmann et al. 2005). To a greater extent than elongation, root phospholipid and sterol composition, root apical meristem DNA synthesis, and root-emitted volatiles such 1,8-cineole, root-emitted volatiles stunt the growth of *Brassica campestris* seedlings (Sharma 2009).

HIPVs epitomize an indirect, information-based defense that maximizes efficiency by recruiting carnivorous allies and priming systemic resistance. This strategy outsources the energetic cost of direct combat to predators and parasitoids, providing high defensive returns on a minimal investment in volatile signal production (Turlings and Ton 2006). Furthermore, HIPVs act as airborne signals that "prime" undamaged tissues and neighboring plants, enabling faster and stronger activation of direct defenses upon subsequent attack—a highly efficient form of pre-emptive protection (Frost et al. 2008). The efficiency of this tritrophic network is context-dependent, hinging on ecological community structure. Ultimately, HIPVs transform plant defense from an autonomous function into a recruiting and communicative network, showcasing a sophisticated ecological strategy that enhances fitness by manipulating the broader food web.

#### 9.2.2.2 Elicitors of Insect Oral Defense

Insect oral secretions (OS) and oviposition fluid components modify plant transcriptomes, proteomes, and metabolomes. The regurgitates and oral secretions of herbivores could improve the effects of insect-induced plant responses. Elictor defenses vary by nature and biological processes (Nishida et al. 2005; Pauwels et al. 2009; Aljbory and Chen 2018). *Piros brassicae* L. regurgitation may stimulate herbivore-induced plant volatiles. The β-glucosidase larvae release a volatile mixture from physically damaged cabbage leaves, attracting the parasitic wasp *Cotesia glomerata* (L.) (Mattiacci et al. 1994, 1995).

A FAC dominates insect oral secretions. Volicitin, N-(17-hydroxylinolenoyl)-L-glutamine, was the first FAC elicitor discovered in beet armyworm larvae. Volicitin application on *Zea mays* L. (Alborn et al. 1997). N-linolenoyl-glu from tobacco hornworm vomit stimulates larvae-feeding biocontrol agents (Alborn et al. 1997; Erb et al. 2021). *M. sexta* causes tobacco plant volatile emissions (Halitschke et al. 2001). Insect OS FACs activate MAPK, controlling plant growth and stress responses (Wu et al. 2007; Hettenhausen et al. 2014). In wild rice (*Oryza minuta* Presl.), sexta activates MAPKs, SIPK, WIPK, and bursts of JA, JA-Ile, SA, and ethylene on damaged leaves. Other lepidopteran species elicit FACs include N-acyl Gln/Glu (Yan et al. 2007) and 7-epi-jasmonic acid (Arimura et al. 2004). FACs also increase 7-epi-jasmonic acid, a phytohormone that activates herbivore-responsive genes in tobacco (Halitschke et al. 2001; Schweiger et al. 2014; Checker et al. 2018). Lepidopteran OS FACs induce distinct transcriptional and proteomic responses. In addition to FACs, insect oral secretions contain inceptins and caeliferins (Jan et al. 2021). Inceptins are disulfide-bonded peptides produced by damaging the γ-subunit of plastidic ATP synthase, while caeliferins are sulfated fatty acids detected in S′ oral secretions. The lipase activity of grasshopper oral secretions caused Arabidopsis to rapidly accumulate oxylipins as 13-hydroperoxy octadecatrienoic acid, 12-oxo-phytodienoic acid (OPDA), JA, and jasmonic acid-isoleucine. Grasshopper oral secretions have lipase activity, which leads Arabidopsis to quickly accumulate oxylipins such as OPDA, JA, jasmonic acid-isoleucine, and 13-hydroperoxy octadecatrienoic acid (Alborn et al. 2007; Schweiger et al. 2014; Sanches et al. 2017). The grasshopper oral secretions also raised cytosolic calcium, ethylene emission, and MAPK activation (Schaefer et al. 2011; Hettenhausen et al. 2014).

#### 9.2.2.3 Phytohormone-Induced Plant Resistance

Multiple signal transduction pathways in a phytohormone network protect plants from herbivores. Herbivore-damaged plants communicate intra- and inter-plant via several hormones (Verhage et al. 2010; Bonaventure 2018; Mrid et al. 2021). Signal-transduction pathways mediated by JA, SA, and ethylene drive many plant insect defense responses. Many of the reactions that plants have to pests are driven by signal-transduction pathways that are mediated by JA, SA, and ethylene. Injury or insect feeding activates defense genes. The attacker can use these hormones alone, synergistically, or antagonistically (Bonaventure 2018).

Jasmonic acid (JA), an essential phytohormone, strengthens plant defenses against many of insect herbivores through direct and indirect mechanisms (Schweiger et al. 2014; Checker et al. 2018). The octadecanoid pathway converts acid into jasmonic acid (JA), which accumulates in plant tissues after injury and herbivory. Herbivorous insects dioxygenate linoleic acid (18:1). Meanwhile, the enzyme OPR3 converts OPDA to JA in the peroxisome. Defensive reactions to jasmonates include antioxidant enzymes, PIs, VOCs, alkaloid synthesis, trichome development, and EFN release (Schweiger et al. 2014; Dey et al. 2020). Indigole glucosinolate content is regulated by JA, along with numerous herbivore defense genes (Rayapuram and Baldwin 2007; Hopkins et al. 2009; Katz et al. 2015). The OPDA signaling pathway influences both COI1-dependent and -independent transcription, which in turn changes the calcium and redox status of the intracellular environment (Ribot et al. 2008). Jasmonates, most likely the JA-amino acid conjugate jasmonoyl–isoleucine, interact with the COI1 unit of an E3 ubiquitin ligase complex known as SCFCOI1 (Skip/Cullin/F-box–COI1). This complex binds to JAZ (jasmonate ZIM-domain) proteins and destroys them, hence inhibiting the expression of JA-inducible genes (Sheard et al. 2010; Wei et al. 2021). In potato plants, JA affects CDPK transcript and activity (Ulloa et al. 2002; Schweiger et al. 2014). Signal transduction by CDPKs, serine/threonine kinases in plants (namely Arabidopsis) that belong to a large family, helps plants defend against biotic and abiotic stressors (Ludwig et al. 2004; Khare et al. 2020; Teklić et al. 2021). It is widely known that JA induces defense chemicals to directly and indirectly attack insect infestations (Ludwig et al. 2004). JA activates defense enzymes POD (Shivaji et al. 2010; War et al. 2011a), and PPO (Zhao et al. 2009; War et al. 2011b).

**Salicylic Acid**

Salicylic acid (SA), a benzoic acid derivative, affects plant defense and development via NPR1. SA pathway ROS generation may make tomato plants resistant to bugs like H. SA induces plant $H_2O_2$, which affects insect digestive systems and limits growth (Wouters et al. 2016). As with spider mite-infested Lima beans and tomatoes, it produces volatiles that attract biological control agents (de Boer et al. 2004; Schweiger et al. 2014).

**Ethylene**

Phytohormone ethylene supports plants combat insects and diseases. The ethylene signaling pathway is important in direct and indirect defense, but its significance in HIPV emission is unknown. ET signaling collaborates or opposes JA in express defense. ET and JA express tomato PI (Kunkel and Brooks 2002). *A. alni* infestation released ethylene and volatiles from *Alnus glutinosa* L. ET precursor, 1-aminocyclopropane-1-carboxylic acid boosts volatile emission from removed JA leaves. Volicitin, JA, or (Z) volatile emission is enhanced by ethylene.-3-hexen-ol in corn (Ruther and Kleier 2005; Turlings and Erb 2018).

#### 9.2.2.4 Defense of Plants by ROS

Transient ROS generation from biotic and abiotic factors helps plants battle stress (Maffei et al. 2007; Torres 2010; Khare et al. 2020). Superoxide (O-), hydrogen peroxide ($H_2O_2$), and hydroxyl radicals (HO-) activate MAPKs (Hettenhausen et al. 2014). Stress causes a sudden increase of ROS, dubbed an "oxidative burst." To reduce ROS and self-toxicity, plant cells scavenge them (Ludwig et al. 2005). Plants need steady, easily diffusible ROS like $H_2O_2$ to fight stress. Membrane-bound NADPH complex activation generates superoxide anion at the plasma membrane or apoplast during the oxidative burst. $H_2O_2$ triggers a chain of processes that express defense genes, protecting plants (Wu and Baldwin 2010). ROS activate defense genes and cross-link cell wall components (Van Breusegem and Dat 2006). The induction of $H_2O_2$ in oat, wheat, barley, and groundnut has been inspected in relation to a diversity of insect pests, suggesting it may initiate a chain of physiological and molecular events that produce further defenses and defend plants from additional infestation (Argandona et al. 2001; Torres 2010; Ye et al. 2019; Mrid et al. 2021).

#### 9.2.2.5 Gene Expression Underpins Plant Defense

A reaction to herbivory damage, hundreds or thousands of genes are up- or downregulated (Zheng and Dicke 2008; Chen et al. 2009). With whole-genome sequence data, ESTs, and microarrays, gene-expression profiles in response to insect infestation can be better understood (Kanchiswamy et al. 2010).

DNA microarrays scan thousands of genes simultaneously to reveal gene expression patterns and signaling responses. However, NGS technologies like RNA- and RAD-sequencing are projected to replace microarrays. Genexpression has changed thanks to eQTL mapping. Arabidopsis' inducible defenses for *P. rapae* and *Brassica oleracea* var. *capitata* L. were examined using eQTL mapping, which can handle dozens of features and *Brassica nigra* L. against *P. rapae* caterpillars and aphids (Huang et al. 2007; Olson 2021). Large-scale microarray investigation studied *Diuraphis noxia* (Mord.) feeding reactions (Broekgaarden et al. 2007; Nouri-Ganbalani et al. 2018). Gramine, *M. nicotiana*: *M. persicae* and *S. avenae* on

Arabidopsis, celery, sorghum, and *Apium graveolens* L. leaves. Planted wheat, tobacco, or cereal plants. (Zhu-Salzman et al. 2004; Voelckel et al. 2004; Zhao et al. 2009; Katz et al. 2015).

Herbivory changes gene expression, reallocating plant possessions to protection. Gene expression levels were compared to plant genotype transcriptional patterns; the aphid, expresses many genes (Jan et al. 2021). Persicae are more susceptible to *P. rapae* than caterpillars. Aphids regulate genes for cell wall changes, oxidative stress, calcium-dependent signaling, and glucosinolate production, while lepidopterans adjust genes for metabolism, detoxification, cell survival, and signal transduction (Katz et al. 2015). Different assailants react differently. *Bemisia tabaci* (Gen.) and persicae respond differently to herbivores, such as white cabbage cultivars' gene expression responses to *P. rapae* infestation (Voelckel and Baldwin 2004; Kempema et al. 2007). Combine genetic, genomic, deep sequencing, transcriptional profiling, and proteomics to understand molecular pathways (Jan et al. 2021).

#### 9.2.2.6 Transgenerational Herbivore Defense

Biogenic and abiotic stressors can cause mother and offspring plants to develop transgenerational immunity that protects offspring from insect pests and generates healthy seeds and seedlings (Agrawal 2001; Khare et al. 2020; Teklić et al. 2021). Few studies have examined transgenerational insect pest immunity. *P. rapae*-damaged or JA-treated wild radish plants (*Raphanus raphanistrum*) produce offspring with good insect resistance (Agrawal 2002). The offspring of stressed Arabidopsis plants are more resilient to stress due to amplified homologous recombination frequency and genome methylation (Boyko et al. 2010; Katz et al. 2015). Plant-insect interactions should also concentrate on the epigenetic regulation of plant defense systems and insect responses, according to mobile siRNA signals and inheritance of DNA methylation-based gene expression changes (Züst and Agrawal 2016; Afroz et al. 2021). More study is needed to influence maternal ecology for pest management tactics. Transgenerational resistance may explain herbivore resistance in plants.

#### 9.2.2.7 The Tritrophic Paradox: Attraction Versus Intoxication

A central and fascinating tension in tritrophic ecology is the paradox of attraction versus potential intoxication. Plants actively recruit parasitoids and predators via HIPVs, offering nectar and shelter, thereby investing in these insects as defensive allies. However, many of the same plants simultaneously produce defensive toxins (e.g., alkaloids, glucosinolates, cyanogenic glycosides) that herbivores may sequester for their own defense. When a parasitoid larva develops inside a host that has sequestered these plant-derived compounds, or when a predator consumes a toxic herbivore, the very defenders that the plant has recruited can be harmed (Ode 2006).

This paradox underscores that tritrophic interactions are not simplistic, harmonious relationships but rather complex webs of co-evolutionary trade-offs and aligned interests. The plant's defensive portfolio combining attractive signals with potent toxin, screates a selective landscape that shapes the evolution of both herbivores and their biocontrol agents. Ultimately, the "attraction-intoxication" tension highlights the sophisticated and sometimes risky nature of indirect plant defense, where efficiency is measured by the net outcome of dynamically interacting and evolving species.

## 9.3 Secondary Metabolites on Parasitoids

### *9.3.1 How Plant Metabolites Affect Parasitoid Wasps*

Feeding organisms could be significantly affected by secondary plant metabolites. Despite debate (Edwards and Wratten 1985; Bernays and Graham 1988; Schultz 1988; Baldwin 1994), many plant-produced secondary compounds have a distrustful role. These poisonous compounds are regularly produced by plants in response to infestations of insect herbivories and pathogens. Most main secondary metabolites could be produced by different plants species (Baldwin 1994). These plant compounds could influence herbivore development and their natural enemies. This book chapter examines the favorable and adverse plant impacts of plant compounds on natural enemies of herbivores, particularly parasitoid wasps.

The complicated biochemistry, intricate relationships, and important role in pest management make parasitoid wasps attractive ecological subjects. This wasp lays its eggs on or in many insect bodies, and its parasitic larvae eat and kill the host. Plants may benefit from parasitoids since parasitism kills the insect host. Before extensive research, it was assumed that plants had modified their chemical interaction to attract and benefit from this and other herbivore enemies. Several of these smart ecologists' forecasts have been confirmed recently by many studies (Price 1986; Vinson et al. 1987; Dicke and Sabelis 1992; Dicke et al. 1990b; Turlings et al. 1990).

While the insect host attack the host plant, plant metabolites begin to exertion, then the parasitoid wasps are presenting searching the different stage of insect herbivores. Common plant scents attract wasps to host habitats. Parasitoids can exploit herbivore-damaged plant volatiles effectively once they discovery habitations. These plant-specific volatiles are only released in substantial concentrations after herbivore infestation. Wasps use induced emissions as odorous signals to locate plants or plant sections afflicted with potential insect hosts. Plant poisons and digestibility reducers can also impede herbivore growth and extend their parasitoid vulnerability. Plant chemistry affects parasitoids following host discovery. Immature parasitoids on herbivorous hosts are similarly affected by plant metabolites. Plant compounds in the host's tissues can directly affect the emerging parasitoid wasps or indirectly affect the herbivore's potential as a host for immature parasitoids. Plant chemistry's impact on parasitoids according to others' studies. The parasitoid wasps fed on the nectar and pollen, which elongate the longevity of these parasitoids (Cortesero et al. 2000).

### 9.3.2 Parasitoids Use Diverse Sensory Cues to Locate Hosts

The largest parasitic insect group, Hymenoptera, feeds its offspring other arthropods. With over 50,000 species, life cycles are varied and each has its own biology (Godfray 1994). Some species are specialized and target only a few hosts, while others are generalists and could grow on numerous insect hosts. All stages of arthropods: eggs, larvae, pupae, and adults are attacked by parasitoids. There are egg, larval, pupal, adult parasitoids.

The reproductive success of parasitoids depends on gaining hosts for progeny. Female parasitoids can employ several sensory modalities, such as mechano-reception, vision, and chemo-reception (olfaction and taste) (Wäckers and Lewis 1994). The cues that a specific parasitoid relies on most depend on which cues are accessible and reliably associated with hosts (Turlings et al. 1993b). Chemicals used as prompts for host location are as assorted as the hosts that are host located, attack, and preyed. Knowledge of the biology of the host can provide good clues as to the stimuli used by its natural enemies (Vet et al. 1991; Vet and Dicke 1992).

As natural selection rewards individuals that are hard for natural enemies to determine, hosts would not show themselves needlessly (Tumlinson et al. 1992). In certain cases, hosts must interact with one another, a process that can be intercepted and manipulated by parasitoids and predators (Lewis 1982; Nordlund et al. 1983; Noldus and van Lenteren 1985). For instance, wasps that prey on the eggs of Lepidopteran insects utilize sex pheromones released by adult females to identify locations where eggs are likely to have been laid (Arakaki et al. 1996).

Parasitoids employ contact stimuli found in mouth secretions, excrement, or scales of insects to locate and recognize the appropriate hosts (Vinson and Piper 1986; Hu and Chen 1987; Lewis et al. 1991). Hosts offer minimal or no signals of their presence over extended distances, prompting wasps to depend on stimuli from alternate sources (Vet and Dicke 1992). A multitude of arthropods that succumb to parasitoids are herbivorous, and the flora they consume may significantly contribute to revealing their presence.

### 9.3.3 Parasitoids Use Herbivore-Induced Plant Volatiles to Locate Hosts

Parasitoids are known to utilize plant aromas to localize herbivorous hosts. Pine oil lures sexually mature female *Pimpla ruficollis*, a parasitoid that parasitizes larvae of the European pine shoot moth *Rhyaciona buoliana*, as noted by Thorpe and Caudle

(1938). Many studies have demonstrated that parasitoids are drawn to plant odors that their hosts may eat. Wasps seeking small, unobtrusive herbivores naturally find their hosts' food sources. This step-in host-searching is host habitat location (Doutt 1959; Vinson 1981, 1984; Turlings and Erb 2018). A host insect's habitat could be a field, clump or leaf of plants of a specific kind. In a world covered by green vegetation, wasps may use olfaction rather than eyesight to choose plants.

After entering potential hosts' habitats, biocontrol agents must use hints to detect hosts. The hosts' smells and other cues are cryptic. However, herbivore-attacked plants transmit reliable odorous signals (Röse et al. 1996). Herbivory has been known to cause plants to release huge amounts of sure volatiles for approximately a decade. Again, herbivore-induced volatiles vary by plant species (Turlings et al. 1993b). Even when attacked by the same species of herbivore, different herbivores produce diverse odors (Dicke 1994, 1995). Certain phases of the same herbivore can also produce different scents (Takabayashi et al. 1991a, b, 1995). Though, genotypic heterogeneity across plants of similar kinds tends to create more variable scents than herbivores (Turlings et al. 1990).

The earliest confirmation was mite-plant interactions, which stated by Dicke and Sabelis (1988). Terpenoids from spider mite-infested lima beans attract their predatory mites, according to this study. This has been observed in many plant-mite interactions; there are many plant-caterpillar-parasitoid systems with different crops such as maize, cabbage, or cotton plants. Maize plant research to validate herbivore-induced parasitoid attractant emissions has been assessed (Turlings and Benrey 1998; Erb et al. 2021).

### *9.3.4 Parasitoid Wasps Use Plant Volatiles to Locate Hosts*

Primarily, olfactometer and flight tunnel surveys showed that females of that parasitoid wasp use odor signals to find their host locations (Turlings et al. 1991a, b; McCall et al. 1993; Turlings and Erb 2018; Zhang et al. 2021). When caterpillars eat maize seedlings, they generate pleasant fragrances. Wasps remain drawn to injured seedlings after caterpillars and their byproducts are removed (Turlings et al. 1991a). Wasps prefer caterpillar-damaged plants more than the healthy ones (Wright et al. 2013).

Gas chromatographic examinations of maize seedling scents demonstrate that caterpillar-fed plants exhaled substantial levels of certain volatiles that healthy or mechanically damaged plants scarcely emit. Plants produce chemicals slowly after insect infestation. After caterpillar damage, terpenoids and indole are produced many hours later. Mechanically damaged plants behave similarly to caterpillar-infested plants when treated with oral excretions. For the response to occur, the caterpillar regurgitant must reach damaged plant matter, as healthy undamaged plants do not emit volatiles (Turlings et al. 1990). This treatment also shows that the plant's response is systemic, releasing terpenoids from both damaged and undamaged leaves (Turlings and Tumlinson 1992; El-Wakeil 2003). Oral secretions from several caterpillar species source maize plants to release the equivalent volatiles (Turlings et al. 1993a). Even grasshopper regurgitant is active, demonstrating that the plant response is broad and can be produced by many herbivores.

Caterpillars fed different diets produced active regurgitant, hence the cause of volatile emissions is not food. Even when forced to ingest filter paper, caterpillar regurgitant remains active (Turlings et al. 1993a). Recent research revealed *Spodoptera exigua* larvae regurgitant elicitor as N-(17-hydroxylinolenoyl)-L-glutamine as mentioned by Alborn et al. (1997). The plant guard cues jasmonic acid promotes the lipoxygenase pathway, which volicitin resembles. This discovery will help explain the interesting interactions between plants and insects at different trophic levels. *C. marginiventris* uses maize plant chemical cues to find its hosts, according to studies of Erb et al. (2021). Only after phytophagous bug mouth secretions react with plants do the signals appear. Tritrophic interaction examinations have proven the plant's active role in host and prey location; as well as, the plant volatile affects the biological control agents after herbivores are detected (El-Wakeil et al. 2012; Ode 2019).

## 9.4 What the Herbivore Diet Does to Parasitoids?

### *9.4.1 Insect Host Development and Parasitism Risk*

By affecting herbivore growth and development, host plant traits may indirectly affect parasitoid wasps. Herbivores eating different nutritionally rich plants grow at different degrees. Tannins in particular make proteins indigestible to insect pests. Duration of feeding can compensate for lower digestibility (Feeny 1976). Usually, parasitoid wasps can only attack their hosts at certain phases. If susceptible stages are extended, a host is more susceptible to parasitism. The slow growth-high mortality hypothesis (Clancy and Price 1987) states that natural enemies may kill plants faster as they grow (Moran and Hamilton 1980; Price et al. 1980; Grossmueller and Lederhouse 1985; Pappas et al. 2017). Benrey and Denno (1997) found conflicting evidence supporting this idea. Johnson and Gould (1992) discovered that *Campoletis sonorensis* boosted mortality on resistant tobacco plants during development of *Heliothis virescens* larvae.

Slow-growing larvae in slow-growing galls were less vulnerable to parasitoids in leaf-galling sawflies (*Pontania* spp.) on willow (Clancy and Price 1987). The slow growth-high mortality concept is more likely to apply to free-living insects that damage the same or closely similar plant species than it is to hidden herbivores or insects that consume varied host plants, according to a review of published research (Benrey and Denno 1997). Certain features of plants that directly avoid insect herbivores from accessing them might be more important than those that delay development. Certain host plants may be more appealing to parasitic wasps due to volatiles (Benrey et al. 1997), or feeding habits may produce structural refugia such as galls and leaf shelters that reduce insect pests' accessibility to natural enemies (Pappas et al. 2017).

### 9.4.2 Plant Components Affects Immature Parasitoids

For development, the parasitoid must locate all its nutrients in the host (Vinson 1975). The quality of an insect pest parasitoid host depends on its food intake (Vinson and Barbosa 1987). Thus, plant components could impact juvenile parasitoids' growth in several means.

### 9.4.3 The Host's Physical Defenses

Encapsulating parasitoids' eggs or larvae helps herbivores fight them. Hemocytes encapsulate foreign material in the host's hemolymph (Vinson and Iwantsch 1980). Plant chemistry affects herbivores' parasitoid egg encapsulation (Cheng 1970; Rhoades 1983; Benrey and Denno 1997). Stress caused by the host plant due to poor nutritional quality, starvation, or allelochemicals can affect the herbivore's vigor, which affects the encapsulation reaction. *Pieris rapae* has the ability to encapsulate *Cotesia glomerata* eggs, contingent upon the plant type that the host larvae consume (Potting et al. 1996; Benrey and Denno 1997). The encapsulation rates were greater on host plants where host larvae matured at the fastest rate. Subsequently, a parasitoid may benefit from a plant that has low nutritional value or poisons that damage the host, thereby reducing the insect's immune system.

### 9.4.4 Plant Chemical Defenses and Parasitoid Performance

Several research has discovered how plant chemistry affects parasitoids' survival (Kaur et al. 2021), development, growth, longevity, fecundity, and sex ratio (Godfray 1994). These consequences may be caused by the host's nutrition, harmful allelochemicals that injury the parasitoid, or their interplay. It can be hard to distinguish the effects of allelochemicals from those of nutrients due to their assorted ways of action. Four methods that plant allelochemicals may harm parasitoids are listed by Vinson and Barbosa (1987): (i) deplete host tissues by producing indigestible complexes; (ii) actively bind with the host's digestive enzymes; (iii) promote detoxification; and (iv) impede normal usage of nutrients via toxic effect on parasitoid organs or tissue. These effects are shown in a number of parasitoid fitness mechanisms.

The size of parasitoids is influenced by plant compounds, and this has an influence on fertility, longevity, and efficiency of searching ability, all of which are indicators of reproductive success (Visser 1994; Kazmer and Luck 1995). Parasitoids from hosts grown on several host plants vary widely in size. Pupal parasitoids *Brachymeria intermedia* from red oak leaves were larger than those from other host plants, according to Greenblatt and Barbosa (1981). When *P. rapae* larvae are grown on various plants, the body size of *C. glomerata* varies significantly. Larger parasitoids were produced by hosts cultivated on the best-resource plants.

### 9.4.5 Host Plant Chemistry Shapes Parasitoid Survival and Development

The nutritive properties of the host plant might potentially impact parasitoids' sex ratio by influencing sex allocation decisions or by differentially affecting sex survival (Turlings and Erb 2018; Zhang et al. 2021). Female wasps can control their offspring's sex by hatching males from unfertilized eggs and females from fertilized ones (Fox et al. 1990, 1996). The parasitoid *Aphytis melinus* produced more female progeny on *Aonidiella aurantti* raised on lemon than grapefruit, orange, or mandarin, according to studies by Hare and Luck (1991). The Diamondback moth (*Plutella xylostella*) parasitoid *Diadegma insulare* produced more females from hosts raised on more fertilized collards (Fox et al. 1996).

As expected, poisonous allelochemicals consumed by hosts decrease parasitoids' fitness. Well-adapted herbivores use plant compounds efficiently. These particular insects may store toxic allelochemicals to guard against natural adversaries (Rowell-Rahier & Pasteels 1990). Tomato and tobacco, two well-studied Solanaceae plants, provide sequesterable secondary plant compounds that affect parasitoid survival and development (Barbosa et al. 1986, 1991; Kester and Barbosa 1991; Kaur et al. 2021).

Tomatin, a tomato alkaloid, has no effect on larvae of *Helicoverpa zea*, the maize earworm, while its parasitoid, *Hyposoter exiguae*, is seriously damaged (Campbell and Duffey 1979, 1981). Herbivores consumed a greater quantity of tomatine, which resulted in a longer growing period for parasitoid larvae but resulted in a decrease in survival, adult size, and longevity. The parasitoid *Cotesia congregata* had a lower chance of surviving in its host when it was fed a diet that was high in nicotine as opposed to one that was low in nicotine plants, *Manduca sexta* (Barbosa et al. 1986; Thorpe and Barbosa 1986; Kaur et al. 2021).

Specialist and generalist parasitoids may be affected by plant allelochemicals in diverse ways (Barbosa et al. 1982; Barbosa 1988). Compared to the polyphagous *Hyposoter annulipes*, the specialized *C. congregata* is able to withstand and eradicate nicotine more effectively (Barbosa et al. 1986). Related studies show that while some allelochemicals have an impact on parasitoid performance, others do not (Barbosa et al. 1991; Barbosa and Saunders 1985). A study of rutin, hordenine, and nicotine revealed that each had a distinct effect on parasitoid performance. Adult mass, development time, and parasitoid survival were all adversely influenced by nicotine, whereas growth time was greatly impacted by rutin although neither of these variables were changed. Hordenine only had an impact on the mass of female parasitoids (Kaur et al. 2021). The harmful effects of these three allelochemicals on parasitoid wasps and their insect hosts are comparable.

Plant chemistry affects hosts and parasitoids differently. Lupine is better for unparasitized *Platyprepia virginalis* than hemlock. Hemlock-fed larvae are prospective to survive tachinid parasitoid *Thelairia bryanti* (English-Loeb et al. 1993). When *Pieris brassicae* was raised on several host plants, Karowe and Schoonhoven (1992) discovered poor correlation between its performance and that of its parasitoid *C. glomerata*. The greatest host plant for *P. rapae* also produced the best larval hosts for larval parasitoid *C. glomerata*, contrary to these results (Benrey 1993).

### 9.4.6 Plant-Parasitoid Mutualism

Some studies (Dicke et al. 1990b, Whitman and Eller 1990, Turlings et al. 1995) imply that plants convey signals to attract natural enemies (Pappas et al. 2017). This has directed to intriguing comments like "plants that are pleading for assistance" and "attraction of bodyguards" (Dicke and Sabelis 1988, 1989; Sabelis and de Jong 1988; Whitman 1988, 1994). Plant chemical emissions may be a reaction to natural selection favoring natural enemies (Faeth 1994; Godfray 1995; Turlings et al. 1995; Opitz and Müller 2009). If plant traits have evolved to boost parasitoids' efficacy, they must significantly contribute to plant fitness.

### 9.4.7 Does Parasitoid Help Plants?

Parasitoids terminate their hosts, yet the attacked plant may not benefit. Parasitoids, unlike predators, let hosts feed and grow for a long time. In maize trials, parasitoids like *C. marginiventris* reduced host feeding (Ashley 1983; Jalali et al. 1988). This is true for many parasitoid species (Beckage and Riddiford 1982; Beckage 1985), yet in many situations parasitized individuals feed more than unparasitized ones (Byers et al. 1993). For the plant, parasitism may cause more damage. In rare conditions, the host may survive a complete parasitism without fecundity loss (English-Loeb et al. 1993). Usually, the host dies, preventing herbivore offspring from attacking the same plant or its relatives. Indirect benefits could be quite insignificant and have no selective influence on the features of the plant. In situations where plants develop vegetatively, lowering herbivore populations will restrict future attacks on plant shoots that are closely connected to one another, and selection may be valuable to plants that attract parasitoid wasps more effectively. This also applies to plants with closely related individuals nearby and perennials that are attacked by multiple insect pests' generations.

A successful predator assault kills prey instantly, providing plants with more immediate protection than parasitoids. Dicke et al. (1990a, b) showed how spider mite infestation causes plants to release volatiles, which affect prey placement. The volatiles help predatory mites find their prey, spider mites. Spider mites can over-exploit and damage plants without predatory mites (Dicke and Sabelis 1988, 1989). In interactions where plants benefit from predators, selective pressures to develop features that promote their presence are stronger. Predator benefits may have driven plant features to attract natural enemies of insect herbivores (Sun et al. 2019). Some parasitoids may have taken advantage of plant-predator interactions and provided additional benefits to the plant.

It is difficult to prove that chemically mediated interactions, which increase parasitoids and predators, are the product of features evolved for this reason (Sun et al. 2019; Erb et al. 2021). Herbivore and parasitoid diets affect plant fitness; studies of interactions should evaluate the costs and advantages of generating the substances involved. Godfray (1995) found that evolutionary stability of plant signaling involves a cost, as uninfested plants would prefer the same method. Not only should energy costs be included (Gershenzon and Croteau 1993) for terpenoid synthesis cost estimates, but also additional consequences that the chemicals may have on adjacent organisms. For instance, the plants may attract herbivores, parasitoids, and predators (Sun et al. 2019).

### 9.4.8 How Do Induced Volatiles Affect Herbivores?

Many herbivores avoid induced volatile-emitting plants. Spider mites prefer healthy bean leaves to those that release volatiles (Dicke 1986). According to Pallini et al. (1997), *Teranycchus urticae* is drawn to cucumber plants afflicted with conspecifics. *T. urticae* avoids the stench of cucumber plants attacked by *Frankliniella occidentalis*, a herbivore that also preys spider mites. *Spodoptera* caterpillars damage less maize leaves with generated scents (Turlings and Tumlinson 1991). Cotton has an even stronger predilection, likely due to the simultaneous development of harmful defense chemicals (Alborn et al. 1996). Landolt (1993) examined how cotton and cabbage plant damage affected female *Trichoplusia ni*. *T. ni*-fed cotton plants attract foraging females, but they oviposit less on injured plants. In cabbage, larval injury reduced attractiveness and oviposition. Corn leaf aphids were deterred by induced maize volatiles (Bernasconi et al. 1998). Aphids damage maize plants in the field more easily than those that release induced volatiles, which could attract various predators (Dicke et al. 1993; Sun et al. 2019).

Herbivores may steer clear of plants that emit odors for a number of reasons. The volatiles may signal that the plant is (i) at risk of competition from other herbivores, (ii) that the plant has started to produce defense compounds, or (iii) that the plant is especially appealing to natural enemies (Opitz and Müller 2009; Pappas et al. 2017). Pallini et al. (1997) further suggest that herbivores may be drawn to afflicted plants because prior attacks may weaken their defenses. Much depends on herbivore biology and how they escape parasitoids and predators (Rafter et al. 2017). Herbivores can prefer infected plants; many bark beetles are initially drawn to conspecific-infested trees. Male pheromones and plant-derived volatiles and compounds from related diseases may contribute (Byers et al. 1985; Raffa et al. 1993). Different beetles agglomerate via plant volatiles; plants harmed by conspecific beetles release more volatiles, which attract scarabaeid and chrysomelid beetles (Harari et al. 1994; Loughrin et al. 1995; Bolter et al. 1997). Aggregation may increase defense chemical secondary metabolite sequestration. Adult herbivores can use sequestered plant scents to find mates (Birch 1984; Pasteels et al. 1988).

Aggregation reactions can be exploited to swiftly take advantage of a plant before it can assemble its chemical defenses, just like with bark beetles. Research on gregarious Lepidoptera revealed that larger groups of caterpillars grew more quickly (Denno and Benrey 1997). This suggests that aggregation may act as a food sink for herbivores. Herbivores react differently to generated scents. Most tritrophic studies show that smells repel herbivores (Ode 2019). This will benefit the plant and should be put to the positive side of herbivore-induced volatile emissions costs and benefits calculations (Agelopoulos and Keller 1994a, b, c; Turlings and Erb 2018).

### *9.4.9 Does Plant "Help" Immature Parasitoid Development?*

Plant metabolites may help immature parasitoids, but conflicting needs limit their potential. On one hand, the plant wants to help the parasitoid develop, but its chemical is partly designed to harm herbivores. These harmful effects are often passed on to the parasitoid. Herbivores may choose host plants with minimal parasitoid and predator mortality (Gross and Price 1988; Denno et al. 1990). Plant chemistry would be crucial. Willow-eating leaf beetles may lay their eggs on underperforming plants that contain substances that shield their young from biocontrol agents (Denno et al. 1990; Pappas et al. 2017). Plant chemistry can hurt herbivores but benefit parasitoids. Encapsulating parasitoid eggs can limit parasitoid development in some hosts. Toxic plants weaken the host's immune system, benefiting the parasitoid. Digestibility reducers and other compounds that are not poisonous to herbivores but can delay their development should not be underestimated. Slowing growth increases host vulnerability (Benrey and Denno 1997).

This plant impact benefits searching wasps but not parasitoid larvae on the host (Wiskerke and Vet 1994). These stages may be more susceptible to predators and hyperparasitoids because to extended development (Rafter et al. 2017). Plant metabolites affect immature parasitoids in diverse and variable ways. When herbivores are parasitized, plants thrive better, therefore natural selection should favor those that use chemical defenses that don't harm herbivores' natural enemies (Opitz and Müller 2009). Plant-produced poisons may not penetrate parasitoids or be broken down into less dangerous compounds before entering their tissues; because these immature stages of parasitoids didn't feed directly on plant host.

## 9.5 Role of Plant Secondary Metabolites among Herbivore-Natural Enemy Interactions

Secondary metabolites help plants engage with insect herbivores and their natural enemies (Mrid et al. 2021). Flavonoids, anthocyanins, and carotenoids give flowers color, while terpenoids, amines, and phenylpropanoids attract bees with their smells. Pollinators favor flowers for their sugar-rich nectar (Wink 2003).

As nectar is pollinators' main reward, defense metabolites in plant nectar are disagreeable. Modern researchers have revealed that PSMs help plants and pollinators (Stevenson et al. 2017). PSMs produced for herbivore protection are identical to those in floral nectar and pollen. This is because a single plant species' chemical defenses are regulated by various biochemical pathways and produces a specialized collection of secondary metabolites for diverse roles. The relationship between these metabolites for floral rewards and plant defense is unclear. Caffeine, aconitine, nicotine, thymol, linalool, lupanine, and grayanotoxins defend plants and attract pollinators (Stevenson et al. 2017). Research on caffeine and bees yielded startling findings. Citrus and Coffea floral nectar contain caffeine. Caffeine first deters honeybee feeding dose-dependently (Koch and Stevenson 2017). The nectar's caffeine levels were below bees' taste threshold. Caffeine naturally increased bees' memory and helped them discover nectar sources using floral fragrance, improving their olfactory learning and memory (Cardel and Koptur 2010; Wright et al. 2013; Richardson et al. 2015).

PSMs affect herbivore natural enemies' behavior, predation, parasitism, and physiology. After herbivore injury, predators and parasitoids use volatiles to find host insects (Turlings and Erb 2018). When herbivores selectively amass and sequester these metabolites for self-defense, they reduce natural enemies' effectiveness (Opitz and Müller 2009). Recent research reveals predators may develop cardenolide and glucosinolate resistance or tolerance (Rafter et al. 2017). Manipulating PSM synthesis or processing quantified natural enemy resistance. Silencing glucosinolate sulfatase to promote toxic isothiocyanate buildup in diamondback moth, *Plutella xylostella*, larvae allow *Chrysoperla carnea* to grow slowly without harming its biology (Sun et al. 2019). Predators and parasitoids that prey on adapted specialist herbivores can acquire PSM resistance. This resistance boosts fitness and predatory potential. PSMs can help or hinder biological control by finding hosts or reducing natural enemies (Pappas et al. 2017).

Long-term herbivore development may promote predation and parasitism (Benrey and Denno 1997). Prolongation of a sensitive growth stage can enhance parasitism, but this affects predation less if predators eat a variety of prey. Slow-growing herbivore hosts had more parasitism, supporting the slow growth-high mortality hypothesis (Cornelissen and Stiling 2006). Linalool deters *Nilaparvata lugens* and attracts its biocontrol agents (Xiao et al. 2012). The relevance of a single chemical in herbivore deterrence and natural enemy attraction is unknown.

Ribwort plantain, *Plantago lanceolate*, phytochemicals increase *Melitaea cinxia* growth and its sole endoparasitoid, *Hyposoter horticola*, on a high iridoid glycosides line. When grown on a high iridoid line, *Spodoptera exigua*, a generalist herbivore, and its single endoparasitoid, *Cotesia marginiventris*, had significantly lower pupal weights and adult masses (Harvey et al. 2005). *Spodoptera litura*, was exposed to phytochemicals from *Inula racemose*, which had several negative impacts on *Bracon hebetor*. Braconid parasitism, survival, and development decreased (Kaur et al. 2021). PSMs' negative impacts on natural enemies allow us to consider manipulating them for plant protection and understanding how they affect multitrophic interactions in nature.

## 9.6 Conclusion

The study of plant-herbivore-parasitoid interactions reveals a sophisticated defense system where plants are not passive victims but active participants in a chemical dialogue. Plants deploy a multi-layered arsenal, ranging from direct defenses like structural barriers and toxic secondary metabolites to indirect strategies such as emitting herbivore-induced plant volatiles (HIPVs) to recruit parasitoids. This review underscores that defense efficiency is not merely a function of toxin potency, but of strategic resource allocation, specificity, and ecological integration.

A central finding is the tritrophic paradox: plants simultaneously attract beneficial parasitoids while producing compounds that, when sequestered by the herbivore, may harm those same allies. This tension highlights that plant defense strategies are shaped by complex evolutionary trade-offs, where the net fitness benefit of reducing herbivory often outweighs the occasional cost to a third-party. The ability of specialized parasitoids to tolerate or adapt to host chemical defenses further illustrates the deep co-evolutionary arms race driving these interactions.

Ultimately, a plant's defensive phenotype whether constitutive or induced is a dynamic and plastic trait, finely tuned by environmental cues and herbivore attack patterns. Understanding this complexity is crucial for interpreting ecological networks and for translating fundamental knowledge into sustainable agricultural practice.

## 9.7 Future Prospects

Translating this ecological complexity into practical applications requires interdisciplinary research targeting several key frontiers:

1. *Molecular Decoding and Manipulation*

   – *Signal Perception and Transduction*: A primary challenge is the full identification of herbivore-specific elicitors and the characterization of their receptors and downstream signaling cascades in plants. Unlocking this "molecular dialogue" is essential.

- *Precision Metabolic Engineering*: Future efforts must move beyond simple overexpression of defense genes. The goal is the molecular modulation of specific metabolic pathways to enhance the production of key defensive compounds (e.g., specific alkaloids, flavonoids) or HIPV blends without incurring significant growth penalties. In vitro plant cell culture and advanced metabolic engineering offer promising biotechnological tools for this purpose.

2. *Eco-Evolutionary and Genomic Insights*.
   - *Ecological Genomics*: Employing population genomics, genome-wide association studies (GWAS), and selection mapping will be critical to understand the variable selection pressures on defense-related loci across different environments and herbivore communities.
   - *Resolving the Tritrophic Paradox*: The evolutionary dynamics of the plant-parasitoid mutualism require deeper study. To what extent has plant chemistry been optimized to *benefit from* parasitoids? Research must quantify the net fitness outcomes of these chemically mediated interactions under realistic ecological conditions.

**Acknowledgement** The authors are grateful to the Deanship of Scientific Research at King Faisal University in Al-Ahsa, Saudi Arabia, for providing all the necessary facilities under project number (KFU251143). This arrangement enabled them to provide all of the necessary literature, including books, manuscripts, and other literary materials.

**Use of AI Tools Declaration** The authors declare that they have not used Artificial Intelligence (AI) tools in the creation of the content of this article. However, AI tools have been utilized exclusively for grammar and spelling correction support. All the information and data in this paper have been ethically reviewed, edited, and studied. Furthermore, the complete work underwent independent peer review by specialists in the field.

## References

Afroz M, Rahman M, Amin R (2021) Insect plant interaction with reference to secondary metabolites: a review. Agric Rev 42:427–433. https://doi.org/10.18805/ag.R-200

Agelopoulos NA, Keller MA (1994a) Plant-natural enemy association in the tritrophic system *Cotesia rubecula-Pieris rapae*-Brassicaceae (Cruciferae). I: sources of infochemicals. J Chem Ecol 20:1725–1734

Agelopoulos NA, Keller MA (1994b) Plant-natural enemy association in the tritrophic system *Cotesia rubecula-Pieris rapae*-Brassicaceae (Cruciferae). II: preference of *C. rubecula* for landing and searching. J Chem Ecol 20:1735–1748

Agelopoulos NA, Keller MA (1994c) Plant-natural enemy association in the tritrophic system *Cotesia rubecula-Pieris rapae*-Brassicaceae (Cruciferae). III: collection and identification of plant and frass volatiles. J Chem Ecol 20:1955–1967

Agrawal AA (1999) Induced responses to herbivory in wild radish: effects on several herbivores and plant fitness. Ecology 80:1713–1723. https://doi.org/10.1890/0012-9658(1999)080[1713:IRTHIW]2.0.CO;2

Agrawal AA (2001) Transgenerational consequences of plant responses to herbivory: an adaptive maternal effect? Am Nat 157:555–569. PMID:18707262. https://doi.org/10.1086/319932
Agrawal AA (2002) Herbivory and maternal effects: mechanisms and consequences of transgenerational induced plant resistance. Ecology 83:3408–3415. https://doi.org/10.1890/0012-9658(2002)083[3408:HAMEMA]2.0.CO;2
Agrawal AA (2011) Current trends in the evolutionary ecology of plant defence. Funct Ecol 25: 420–432. https://doi.org/10.1111/j.1365-2435.2010.01796.x
Agrawal AA, Janssen A, Bruin J, Posthumus MA, Sabelis MW (2002) An ecological cost of plant defence: attractiveness of bitter cucumber plants to natural enemies of herbivores. Ecol Lett 5: 377–385. https://doi.org/10.1046/j.1461-0248.2002.00325.x
Agrawal AA, Fishbein M, Jetter R, Salminen JP, Goldstein JB, Freitag AE et al (2009) Phylogenetic ecology of leaf surface traits in the milkweeds (*Asclepias* spp.): chemistry, ecophysiology, and insect behavior. New Phytol 183:848–867. https://doi.org/10.1111/j.1469-8137.2009.02897.x
Alborn HT, Röse USR, McAuslane HJ (1996) Systemic induction of feeding deterrents in cotton plants by feeding of *Spodoptera spp.* larvae. J Chem Ecol 22:919–932
Alborn T, Turlings TCJ, Jones TH, Stenhagen G, Loughrin JH, Tumlinson JH (1997) An elicitor of plant volatiles from beet armyworm oral secretion. Science 276:945–949. https://doi.org/10.1126/science.276.5314.945
Alborn HT, Hansen TV, Jones TH, Bennett DC, Tumlinson JH, Schmelz EA et al (2007) Disulfooxy fatty acids from the American bird grasshopper *Schistocerca americana*, elicitors of plant volatiles. Proc Natl Acad Sci USA 104:12976–12981. PMID:17664416. https://doi.org/10.1073/pnas.0705947104
Aljbory Z, Chen MS (2018) Indirect plant defense against insect herbivores: a review. Insect Sci 25: 2–23. https://doi.org/10.1111/1744-7917.12436
Alvarenga R, Moraes JC, Auad AM, Coelho M, Nascimento AM (2017) Induction of resistance of corn plants to *Spodoptera frugiperda* by application of silicon and gibberellic acid. Bull Ѐntomol Res 107:527–533. https://doi.org/10.1017/S0007485316001176
Arakaki N, Wakamura s, Yasuda T (1996) Phoretic egg parasitoid, *Telenomus euproctidis*, uses sex pheromone of tussock moth *Euproctis taiwana* as a kairomone. J Chem Ecol 22:1079–1085
Argandona VH, Chaman M, Cardemil L, Munoz O, Zuniga GE, Corcuera LJ (2001) Ethylene production and peroxidase activity in aphid-infested barley. J Chem Ecol 27:53–68. PMID:11382067. https://doi.org/10.1023/A:1005615932694
Arimura G, Ozawa R, Kugimiya S, Takabayashi J, Bohlmann J (2004) Herbivore-induced defense response in a model legume. Two-spotted spider mites induce emission of (E)-beta-ocimene and transcript accumulation of (E)-beta-ocimene synthase in Lotus japonicus. Plant Physiol 135: 1976–1983. PMID:15310830. https://doi.org/10.1104/pp.104.042929
Arimura GI, Matsui K, Takabayashi J (2009) Chemical and molecular ecology of herbivore-induced plant volatiles: proximate factors and their ultimate functions. Plant Cell Physiol 50: 911–923. PMID:19246460. https://doi.org/10.1093/pcp/pcp030
Ashley TR (1983) Growth pattern alterations in fall armyworm, *Spodoptera frugiperda*, larvae after parasitization by *Apanteles marginiventris*, *Campoletis grioti, Chelonomus insularis*, and *Eiphosoma vitticole*. Florida Entomol 66:260–266
Baldwin IT (1994) Chemical changes rapidly induced by folivory. In: Bernays EA (ed) Insect-plant interactions, Vol. V. CRC Press, Boca Raton, pp 1–23
Barakat A, Bagniewska-Zadworna A, Frost CJ, Carlson JE (2010) Phylogeny and expression profiling of CAD and CAD-like genes in hybrid Populus (*P. deltoides* × *P. nigra*): evidence from herbivore damage for subfunctionalization and functional divergence. BMC Plant Biol 10: 100. PMID:20509918. https://doi.org/10.1186/1471-2229-10-100
Barbehenn RV, Constabel CP (2011) Tannins in plantherbivore interactions. Phytochemistry 72(13):1551–1565. PMID:21354580. https://doi.org/10.1016/j.phytochem.2011.01.040
Barbosa P (1988) Natural enemies and herbivore-plant interactions: influence of plant allelochemicals and host specificity. In: Barbosa P, Letourneau D (eds) Novel aspects of insect-plant interactions. Wiley, New York, pp 201–210

Barbosa P, Saunders JA (1985) Plant allelochemicals: linkages between herbivores and their natural enemies. Rec Adv Phytochem 19:107–137

Barbosa P, Saunders JA, Waldvogel M (1982) Plant mediated variation in herbivore suitability and parasitoid fitness. In: Proceedings of the 5th international symposium on plant-insect relationships. Wageningen, Pudoc, Wageningen, pp 63–71

Barbosa P, Saunders JA, Kemper J, Trumbule R, Olechno J, Martinat P (1986) Plant allelochemicals and insect parasitoids: effects of nicotine on *Cotesia congregata* and *Hyposoter annulipes*. J Chem Ecol 12:1319–1328

Barbosa P, Gross P, Kemper J (1991) Influence of plant allelochemicals on the tobacco hornworm and its parasitoid, *Cotesia congregata*. Ecology 72:1567–1575

Baur R, Binder S, Benz G (1991) Nonglandular leaf trichomes as short-term inducible defense of the grey alder, *Alnus incana* (L), against the chrysomelid beetle, *Agelastica alni* L. Oecologia 87:219–226. https://doi.org/10.1007/BF00325259

Beckage NE (1985) Endocrine interactions between endoparasitic insects and their hosts. Annu Rev Entomol 30:371–413

Beckage NE, Riddiford LM (1982) Effects of parasitism by *Apanteles congregatus* on the endocrine physiology of the tobacco hornworm *Manduca Sexta*. Gen Comp Endocrinol 47:308–322

Benrey B (1993) Host plant effects on the interaction of an insect herbivore and its larval parasitoid: the case of *Pieris rapae* and *Cotesia glomerata*. PhD Dissertation, Dept Entomol, Maryland Uni, College Park, Maryland

Benrey B, Denno RF (1997) The slow-growth–high-mortality hypothesis: a test using the cabbage butterfly. Ecology 78:987–999. https://doi.org/10.1890/0012-9658(1997)078[0987:TSGHMH]2.0.CO;2

Benrey B, Kaiser L, Denno RF (1997) The influence of plant species on attraction and host acceptance of *Cotesia glomerata*. J Insect Behav 10:619–630

Bernasconi ML, Turlings TCJ, Ambrosetti L, Bassetti P, Dorn S (1998) Herbivore-induced emissions of maize volatiles repel the corn-leaf aphid, *Rhopalosiphum maidis*. Entomol Exp Appl 187:133–142

Bernays EA (1981) Plant tannins and insect herbivores: an appraisal. Ecol Entomol 6:353–360. https://doi.org/10.1111/j.1365-2311.1981.tb00625.x

Bernays EA, Graham M (1988) On the evolution of host specificity in phytophagous arthropods. Ecol 69:886–892

Bhonwong A, Stout MJ, Attajarusit J, Tantasawat P (2009) Defensive role of tomato polyphenol oxidases against cotton bollworm (*Helicoverpa armigera*) and beet armyworm (*Spodoptera exigua*). J Chem Ecol 35:28–38. https://doi.org/10.1007/s10886-008-9571-7

Birch MC (1984) Aggregation in bark beetles. In: Bell WJ, Cardé RT (eds) Chemical ecology of insects. Chapman and Hall, London, pp 331–353

Björkman C, Ahrne K (2005) Influence of leaf trichome density on the efficiency of two polyphagous insect predators. Entomol Exp Appl 115:179–186. https://doi.org/10.1111/j.1570-7458.2005.00284.x

Bolter CJ, Dicke M, van Loon JJA, Visser JH, Posthumus MA (1997) Attraction of Colorado potato beetle to herbivore- damaged plants during herbivory and after its termination. J Chem Ecol 23:1003–1023

Bonaventure G (2018) Plants recognize herbivorous insects by complex signalling networks. Annu Plant Rev 47:1–35. https://doi.org/10.1002/9781118829783.ch1

Boyko A, Blevins T, Yao Y, Golubov A, Bilichak A, Ilnytskyy Y et al (2010) Transgenerational adaptation of Arabidopsis to stress requires DNA methylation and the function of dicer-like proteins. PLoS One 5:e9514. https://doi.org/10.1371/journal.pone.0009514

Broekgaarden C, Poelman EH, Steenhuis G, Voorrips RE, Dicke M, Vosman B (2007) Genotypic variation in genome-wide transcription profiles induced by insect feeding: *Brassica oleracea* – *Pieris rapae* interactions. BMC Genomics 8:239. PMID:17640338. https://doi.org/10.1186/1471-2164-8-239

Bruinsma M, Posthumus MA, Mumm R, Mueller MJ, van Loon JJA, Dicke M (2009) Jasmonic acid-induced volatiles of *Brassica oleracea* attract parasitoids: effects of time and dose, and comparison with induction by herbivores. J Exp Bot 60:2575–2587. PMID:19451186. https://doi.org/10.1093/jxb/erp101

Byers JA, Lanne LJ, Schlyter F, Bergström G (1985) Olfactory recognition of host-tree susceptibility by pine shoot beetles. Naturwissenschaften 72:324–326

Byers JR, Yu DS, Jones JW (1993) Parasitism of the army cutworm, *Euxoa auxiliaris* by *Copidosoma bakeri* and effect on crop damage. Can Entomol 125:329–335

Campbell BC, Duffey SS (1979) Tomatine and parasitic wasps: potential incompatibility of plant antibiosis with biological control. Sci 205:700–702

Campbell BC, Duffey SS (1981) Alleviation of a-tomatineinduced toxicity to the parasitoid *Hyposoter exiguae*, by phytosterols in the diet of the host, *Heliothis zea*. J Chem Ecol 7:927–946

Cardel YJ, Koptur S (2010) Effects of Florivory on the pollination of flowers: an experimental field study with a perennial plant. Bot Gaz 171:283–292. https://doi.org/10.1086/650154

Chamarthi SK, Sharma HC, Sahrawat KL, Narasu LM, Dhillon MK (2010) Physico-chemical mechanisms of resistance to shoot fly, *Atherigona soccata* in sorghum, *Sorghum bicolor*. J Appl Entomol 135:446–455. https://doi.org/10.1111/j.1439-0418.2010.01564.x

Chamarthi SK, Sharma HC, Vijay PM, Narasu LM (2011) Leaf surface chemistry of sorghum seedlings influencing expression of resistance to sorghum shoot fly, *Atherigona soccata*. J Plant Biochem Biotechnol 20:211–216. https://doi.org/10.1007/s13562-011-0048-3

Checker VG, Kushwaha HR, Kumari P, Yadav S (2018) Role of Phytohormones in plant defense: signaling and cross talk. In: Molecular aspects of plant-pathogen interaction. Springer, Singapore. https://link.springer.com/chapter/10.1007/978-981-10-7371-7_7

Chen F, Tholl D, D'Auria JC, Farooq A, Pichersky E, Gershenzon J (2003) Biosynthesis and emission of terpenoid volatiles from Arabidopsis flowers. Plant Cell 15:481–494. PMID:12566586. https://doi.org/10.1105/tpc.007989

Chen H, Wilkerson CG, Kuchar JA, Phinney BS, Howe GA (2005) Jasmonate-inducible plant enzymes degrade essential amino acids in the herbivore midgut. Proc Natl Acad Sci USA 102: 19237–19242. PMID:16357201. https://doi.org/10.1073/pnas.0509026102

Chen Y, Ni X, Buntin GD (2009) Physiological, nutritional, and biochemical bases of corn resistance to foliage-feeding fall armyworm. J Chem Ecol 35:297–306. PMID:19221843. https://doi.org/10.1007/s10886-009-9600-1

Cheng L (1970) Timing of attack of *Lypha dubia* on the winter moth, *Operophtera brumata* as a factor affecting parasite success. J Anmi Ecol 39:313–320

Clancy KM, Price PW (1987) Rapid herbivore growth enhances enemy attack: sublethal plant defenses remain a paradox. Ecology 68:736–738

Cornelissen T, Stiling P (2006) Does low nutritional quality act as a plant defence? An experimental test of the slow-growth, high-mortality hypothesis. Ecol Èntomol 31:32–40. https://doi.org/10.1111/j.0307-6946.2006.00752.x

Cortesero AM, Stapel JO, DeMoraes CM, Haney PB, Lewis WJ (2000) Understanding and manipulating plant attributes to enhance biological control. Biol Control 17:35–49

Dalin P, Björkman C (2003) Adult beetle grazing induces willow trichome defence against subsequent larval feeding. Oecologia 134:112–118. PMID:12647188. https://doi.org/10.1007/s00442-002-1093-3

Datta R, Kaur A, Saraf I, Singh IP, Chadha P, Kaur S (2020) Assessment of genotoxic and biochemical effects of purified compounds of *Alpinia galanga* on a polyphagous lepidopteran pest *Spodoptera litura*. Phytoparasitica 48:501–511. https://doi.org/10.1007/s12600-020-00813-8

de Boer JG, Posthumus MA, Dicke M (2004) Identification of volatiles that are used in discrimination between plants infested with prey or nonprey herbivores by a predatory mite. J Chem Ecol 30:2215–2230. PMID:15672666. https://doi.org/10.1023/B:JOEC.0000048784.79031.5e

De Freitas TFS, Stout MJ, Sant'Ana J (2019) Effects of exogenous methyl jasmonate and salicylic acid on rice resistance to *Oebalus pugnax*. Pest Manag Sci 75:744–752. https://doi.org/10.1002/ps.5174
Denno RF, Benrey B (1997) Aggregation facilitates larval growth in the neotropical butterfly *Chlosyne janais*. Ecol Entomol 22:133–141
Denno RF, Larsson S, Olmstead KL (1990) Role of enemyfree space and plant quality in host-plant selection by willow beetles. Ecology 71:124–137
Dey P, Kundu A, Kumar A, Gupta M, Lee BM, Bhakta T, Dash S, Kim HS (2020) Analysis of alkaloids (indole alkaloids, isoquinoline alkaloids, tropane alkaloids). In: Recent advances in natural products analysis. Elsevier, Amsterdam. https://doi.org/10.1016/B978-0-12-816455-6.00015-9
Dicke M (1986) Volatile spider-mite pheromone and host-plant kairomone, involved in spaced-out gregariousness in the spider mite *Tetranychus urticae*. Physiol Entomol 11:251–262
Dicke M (1994) Local and systemic production of volatile herbivore induced terpenoids: their role in plant-carnivore mutualism. J Plant Physiol 143:465–472
Dicke M (1995) Why do plants 'talk'? Chem 6:159–165
Dicke M, Sabelis MW (1988) How plants obtain predatory mites as bodyguards. Netherl J Zool 38: 148–165
Dicke M, Sabelis MW (1989) Does it pay plants to advertise for bodyguards? Towards a cost-benefit analysis of induced synomone production. In: Lambers H, Cambridge ML, Konings H, pons TL (eds) Causes and consequences of variation in growth rate and productivity of higher plants. SPB Academic Publication, The Hague, pp 341–358
Dicke M, Sabelis MW (1992) Costs and benefits of chemical information conveyance: proximate and ultimate factors. In: Roitberg BD, Isman MB (eds) Insect chemical ecology: an evolutionary approach. Chapman & Hall, New York, pp 122–155
Dicke M, Sabelis MW, Takabayashi J, Bruin J, Posthumus MA (1990a) Plant strategies of manipulating predator-prey interactions through allelochemicals: prospects for application in pest control. J Chem Ecol 16:3091–3118
Dicke M, van Beek TA, Posthumus MA, Ben Dom N, van Bokhoven H, de Groot AE (1990b) Isolation and identification of volatile kairomone that affects acarine predator-prey interactions: involvement of host plant in its production. J Chem Ecol 16:381–396
Dicke M, van Baarlen P, Wessels R, Dilkman H (1993) Herbivory induces systemic production of plant volatiles that attract predators of the herbivores: extraction of endogenous elicitor. J Chem Ecol 19:581–599
Dickens JC (2006) Plant volatiles moderate response to aggregation to pheromone in Colorado potato beetle. J Appl Entomol 130:26–31. https://doi.org/10.1111/j.1439-0418.2005.01014.x
Divekar PA, Narayana S, Divekar BA, Kumar R, Gadratagi BG, Ray A, Singh AK, Rani V, Singh V, Singh AK et al (2022) Plant secondary metabolites as defense tools against herbivores for sustainable crop protection. Internat J Molecr Sci 23:2690. https://doi.org/10.3390/ijms23052690
Doutt RL (1959) The biology of parasitic hymenoptera. Annu Rev Entomol 4:161–182
Dowd PF, Johnson ET (2009) Differential resistance of switchgrass Panicum virgatum L. lines to fall armyworms Spodoptera frugiperda (JE Smith). Genet Resour Crop Evol 56:1077–1089
Dudareva N, Negre F, Nagegowda DA, Orlova I (2006) Plant volatiles: recent advances and future perspectives. Crit Rev Plant Sci 25:417–440. https://doi.org/10.1080/07352680600899973
Dutta I, Saha P, Majumder P, Sarkar A, Chakraborti D, Banerjee S et al (2005) The efficacy of a novel insecticidal protein, *Allium sativum* leaf lectin (ASAL), against homopteran insects monitored in transgenic tobacco. Plant Biotechnol J 3:601–611. PMID:17147631. https://doi.org/10.1111/j.1467-7652.2005.00151.x
Edwards PJ, Wratten SD (1985) Induced plant defences against insect grazing: fact or artefact? Oikos 44:70–74

El-Wakeil N (2003) New aspects of biological control of *Helicoverpa armigera* in organic cotton production. PhD Disseration, Göttingen University, Germany (140pp); full text in http://webdoc.sub.gwdg.de/diss/2003/el-wakeil/index.html
El-Wakeil NE, Bernal J, Vidal S (2003) Effects of jasmonate applications on pest and natural enemy recruitment in BT and non BT-cotton fields. In: Proceeding of the world cotton research conference-3, from 9 to 13 March 2003. Cape Town, South Africa, pp 1239–1248
El-Wakeil NE, Volkmar C (2012) Effect of jasmonic acid application on economically insect pests and yield in spring wheat. Gesunde Pflanzen 64:107–116
Engelberth J, Alborn HT, Schmelz EA, Tumlinson JH (2004) Airborne signals prime plants against insect herbivore attack. Proc Natl Acad Sci USA 101:1781–1785. PMID:14749516. https://doi.org/10.1073/pnas.0308037100
English-Loeb GM, Brody AK, Karban RK (1993) Hostplant-mediated interactions between a generalist folivore and its tachinid parasitoid. J Anmi Ecol 62:465–471
Erb M, Züst T, Robert CAM (2021) Using plant chemistry to improve interactions between plants, herbivores and their natural enemies: challenges and opportunities. Curr Opin Biotechnol 70: 262–265. https://doi.org/10.1016/j.copbio.2021.05.011
Faeth SH (1994) Induced plant responses: effects on parasitoids and other natural enemies of phytophagous insects. In: Hawkins BA, Sheehan W (eds) Parasitoid community ecology. Oxford University Press, Oxford, pp 245–260
Feeny P (1976) Plant apparency and chemical defense. In: Wallace JW, Mansell RL (eds) Biochemical interaction between plants and insects. Plenum Press, New York, pp 1–40
Feng YX, Wang Y, You CX, Guo SS, Du YS, Du SS (2019) Bioactivities of patchoulol and phloroacetophenone from *Pogostemon cablin* essential oil against three insects. Int J Food Prop 22:1365–1374. https://doi.org/10.1080/10942912.2019.1648508
Fidantsef AL, Stout MJ, Thaler JS, Duffey SS, Bostock RM (1999) Signal interactions in pathogen and insect attack: expression of lipoxygenase, proteinase inhibitor II, and pathogenesis related protein P4 in the tomato, *Lycopersicon esculentum*. Physiol Mol Plant Pathol 54:97–114. https://doi.org/10.1006/pmpp.1998.0192
Fox LR, Letourneau DK, Eisenbach J, Van Nouhuys S (1990) Parasitism rates and sex ratios of a parasitoid wasp: effects of herbivore and plant quality. Oecologia 83:414–419
Fox LR, Kester KM, Eisenbach J (1996) Direct and indirect responses of parasitoids to plants: sex ratio, plant quality and herbivore diet breadth. Entomol Exp Appl 80:289–292
Frost CJ, Mescher MC, Dervinis C, Davis JM, Carlson JE, De Moraes CM (2008) Priming defense genes and metabolites in hybrid poplar by the green leaf volatile cis-3-hexenyl acetate. New Phytol 180(3):722–734
Gershenzon J, Croteau RB (1993) Terpenoid biosynthesis: the basic pathway and formation of monoterpenes, sesquiterpenes, and diterpenes. In: Moore TS Jr (ed) Lipid Metabolism in Plants. CRC Press, Boca Raton, pp 339–388
Gill RS, Gupta K, Taggar GK, Taggar MS (2010) Role of oxidative enzymes in plant defenses against herbivory. Acta Phytopathol Entomol Hung 45:277–290. https://doi.org/10.1556/APhyt.45.2010.2.4
Giongo AMM, Vendramim JD, Freitas SDL, Silva MFGF (2016) Toxicity of secondary metabolites from Meliaceae against *Spodoptera frugiperda*. Neotrop Entomol 45:725–733. https://doi.org/10.1007/s13744-016-0418-6
Giovanini MP, Saltzmann KD, Puthoff DP, Gonzalo M, Ohm HW, Williams CE (2007) A novel wheat gene encoding a putative chitin-binding lectin is associated with resistance against Hessian fly. Mol Plant Pathol 8:69–82. PMID:20507479. https://doi.org/10.1111/j.1364-3703.2006.00371.x
Godfray HCJ (1994) Parasitoids: behavior and evolutionary ecology. Princeton University Press, Princeton
Godfray HCJ (1995) Communication between the first and third trophic levels: an analysis using biological signalling theory. Oikos 72:367–374

Greenblatt JA, Barbosa P (1981) Effects of host's diet on two pupal parasitoids of the gypsy moth: *Brachymeria intermedia* and *Coccygomimus turionellae*. J Appl Ecol 18:1–10

Gross P, Price PW (1988) Plant influences on parasitism of two leafminers: a test of enemy-free space. Ecology 69:1506–1516

Grossmueller DW, Lederhouse RC (1985) Oviposition site selection: an aid to rapid growth and development in the tiger swallowtail butterfly *Papilio glaucus*. Oecologia 66:68–73

Gulsen O, Eickhoff T, Heng-Moss T, Shearman R, Baxendale F, Sarath G et al (2010) Characterization of peroxidase changes in resistant and susceptible warmseason turfgrasses challenged by Blissus occiduus. Arthropod Plant Interact 4:45–55. https://doi.org/10.1007/s11829-010-9086-3

Halitschke R, Schittko U, Pohnert G, Boland W, Baldwin IT (2001) Molecular interactions between the specialist herbivore *Manduca sexta* its natural host *Nicotiana attenuata*. III. Fatty acid-amino acid conjugates in herbivore oral secretions are necessary and sufficient for herbivore-specific plant responses. Plant Physiol 125:711–717. PMID:11161028. https://doi.org/10.1104/pp.125.2.711

Handley R, Ekbom B, Agren J (2005) Variation in trichome density and resistance against a specialist insect herbivore in natural populations of Arabidopsis thaliana. Ecol Entomol 30: 284–392. https://doi.org/10.1111/j.0307-6946.2005.00699.x

Hanley ME, Lamont BB, Fairbanks MM, Rafferty CM (2007) Plant structural traits and their role in antiherbivore defense. Perspec Plant Ecol Evol Syst 8:157–178. https://doi.org/10.1016/j.ppees.2007.01.001

Harari AR, Ben-Yakir D, Rosen D (1994) Mechanism of aggregation behavior in *Maladera matrida* Argaman (Coleoptera: Scarabeidae). J Chem Ecol 20:361–371

Hare JD (2011) Ecological role of volatiles produced by plants in response to damage by herbivorous insects. Annu Rev Entomol 56:161–180. PMID:21133760. https://doi.org/10.1146/annurev-ento-120709-144753

Hare JD, Luck RF (1991) Indirect effects of citrus cultivars on life history parameters of a parasitic wasp. Ecology 72:1576–1585

Harvey JA, van Nouhuys S, Biere A (2005) Effects of quantitative variation in allelochemicals in *Plantago lanceolata* on development of a generalist and a specialist herbivore and their endoparasitoids. J Chem Ecol 31:287–302. https://doi.org/10.1007/s10886-005-1341-1

He J, Chen F, Chen S, Lv G, Deng Y, Fang W et al (2011) Chrysanthemum leaf epidermal surface morphology and antioxidant and defense enzyme activity in response to aphid infestation. J Plant Physiol 168:687–693. PMID:21145126. https://doi.org/10.1016/j.jplph.2010.10.009

Heng-Moss TM, Sarath G, Baxendale F, Novak D, Bose S, Ni X et al (2004) Characterization of oxidative enzyme changes in buffalograsses challenged by Blissus occiduus. J Econ Entomol 97:1086–1095. PMID:15279295. https://doi.org/10.1603/0022-0493(2004)097[1086:COOECI]2.0.CO;2

Hettenhausen C, Schuman MC, Wu J (2014) MAPK signaling: a key element in plant defense response to insects. Insect Sci 22:157–164. https://doi.org/10.1111/1744-7917.12128

Hopkins RJ, van Dam NM, van Loon JJA (2009) Role of glucosinolates in insect-plant relationships and multitrophic interactions. Annu Rev Entomol 54:57–83. https://doi.org/10.1146/annurev.ento.54.110807.090623

Horiuchi JI, Arimura GI, Ozawa R, Shimoda T, Dicke M, Takabayashi J et al (2003) Lima bean leaves exposed to herbivore-induced conspecific plant volatiles attract herbivores in addition to carnivores. Appl Entomol Zoo 38:365–368. https://doi.org/10.1303/aez.2003.365

Howe GA, Jander G (2008) Plant immunity to insect herbivores. Annu Rev Plant Biol 59:41–66. PMID:18031220. https://doi.org/10.1146/annurev.arplant.59.032607.092825

Hu JS, Chen CM (1987) A study of the host-searching kairomone of *Apanteles cypris* Nixon. Acta Entomol Sin 30:31–40

Huang W, Zhikuan J, Qingfang H (2007) Effects of herbivore stress by *Aphis medicaginis* Koch on the malondialdehyde contents and activities of protective enzymes in different alfalfa varieties. Acta Ecol Sin 27:2177–2183. https://doi.org/10.1016/S1872-2032(07)60048-1

Jalali SK, Singh SP, Ballal CR (1988) Effect of parasitism by *Cotesia marginiventris* on consumption and utilization of artificial diet by larvae of *Spodoptera litura* (Lepidoptera: Noctuidae). Ind J Agric Sci 58:529–532

Jamwal K, Bhattacharya S, Puri S (2018) Plant growth regulator mediated consequences of secondary metabolites in medicinal plants. J Appl Res Med Aromat Plants 9:26–38. https://doi.org/10.1016/j.jarmap.2017.12.003

Jan R, Asaf S, Numan M, Kim K-M (2021) Plant secondary metabolite biosynthesis and transcriptional regulation in response to biotic and abiotic stress conditions. Agronomy 11:968. https://doi.org/10.3390/agronomy11050968

Johnson TM, Gould F (1992) Interaction of genetically engineered host plant resistance and natural enemies of *Heliothis virescens* in tobacco. Environ Entomol 21:586–597

Johnson MTJ, Smith SD, Rausher MD (2009) Plant sex and the evolution of plant defenses against herbivores. Proc Natl Acad Sci USA 106:18079–18084. PMID:19617572. https://doi.org/10.1073/pnas.0904695106

Jones AC, Seidl-Adams I, Engelberth J, Hunter CT, Alborn H, Tumlinson JH (2019) Herbivorous caterpillars can utilize three mechanisms to alter green leaf volatile emission. Environ Ѐntomol 48:419–425. https://doi.org/10.1093/ee/nvy191

Kanchiswamy CN, Takahashi H, Quadro S, Maffei ME, Bossi S, Bertea C et al (2010) Regulation of Arabidopsis defense responses against *Spodoptera littoralis* by CPK-mediated calcium signaling. BMC Plant Biol 10:97. PMID:20504319. https://doi.org/10.1186/1471-2229-10-97

Kappers IF, Aharoni A, van Herpen TWJM, Luckerhoff LL, Dicke M, Bouwmeester HJ (2005) Genetic engineering of terpenoid metabolism attracts bodyguards to Arabidopsis. Science 309:2070–2072. PMID:16179482. https://doi.org/10.1126/science.1116232

Karban R (2011) The ecology and evolution of induced resistance against herbivores. Funct Ecol 25:339–347. https://doi.org/10.1111/j.1365-2435.2010.01789.x

Karowe DN, Schoonhoven LM (1992) Interactions among three trophic levels: the influence of host plant on performance of *Pieris brassicae* and its parasitoid, *Cotesia glomerata*. Entomol Exp Appl 62:241–251

Katz E, Nisani S, Yadav BS, Woldemariam MG, Shai B, Obolski U, Ehrlich M, Shani E, Jander G, Chamovitz DA (2015) The glucosinolate breakdown product indole-3-carbinol acts as an auxin antagonist in roots of Arabidopsis thaliana. Plant J 82:547–555. https://doi.org/10.1111/tpj.12824

Kaur M, Saraf I, Kumar R, Singh IP, Kaur S (2021) Biological effects of secondary metabolites of *Inula racemosa* on the parasitoid *Bracon hebetor*. Ѐntomol Exp Appl 169:743–749. https://doi.org/10.1111/eea.13070

Kazmer DJ, Luck RF (1995) Field tests of the size-fitness hypothesis in the egg parasitoid *Trichogramma pretiosum*. Ecology 76:412–425

Kempema LA, Cui XP, Holzer FM, Walling LL (2007) Arabidopsis transcriptome changes in response to phloem-feeding silverleaf whitefly nymphs. Similarities and distinctions in responses to aphids. Plant Physiol 143:849–865. PMID:17189325. https://doi.org/10.1104/pp.106.090662

Kessler A, Halitschke R, Diezel C, Baldwin IT (2006) Priming of plant defense responses in nature by airborne signaling between *Artemisia tridentata* and *Nicotiana attenuata*. Oecologia 148:280–292. PMID:16463175. https://doi.org/10.1007/s00442-006-0365-8

Kester KM, Barbosa P (1991) Behavioral and ecological constraints imposed by plants on insect parasitoids: implications for biological control. Biol Cont 1:94–106

Khare S, Singh NB, Singh A, Hussain I, Niharika K, Yadav V, Bano C, Yadav RK, Amist N (2020) Plant secondary metabolites synthesis and their regulations under biotic and abiotic constraints. J Plant Biol 63:203–216. https://doi.org/10.1007/s12374-020-09245-7

Koch H, Stevenson PC (2017) Do linden trees kill bees? Reviewing the causes of bee deaths on silver linden (*Tilia tomentosa*). Biol Lett 13:20170484. https://doi.org/10.1098/rsbl.2017.0484

Kundu A, Mishra S, Vadassery J (2018) *Spodoptera litura*-mediated chemical defense is differentially modulated in older and younger systemic leaves of *Solanum lycopersicum*. Planta 248: 981–997. https://doi.org/10.1007/s00425-018-2953-3

Kunkel BN, Brooks DM (2002) Cross talk between signaling pathways in pathogen defense. Curr Opin Plant Biol 5:325–331. PMID:12179966. https://doi.org/10.1016/S1369-5266(02)00275-3

Landolt PJ (1993) Effects of host plant leaf damage on cabbage looper moth attraction and oviposition. Entomol Exp Appl 67:79–85

Lane GA, Sutherland ORW, Skipp RA (1987) Isoflavonoids as insect feeding deterrents and antifungal components from root of Lupinus angustifolius. J Chem Ecol 13:771–783. https://doi.org/10.1007/BF01020159

Lannoo N, Vandenborre G, Miersch O, Smagghe G, Wasternack C, Peumans WJ et al (2007) The jasmonate-induced expression of the *Nicotiana tabacum* leaf lectin. Plant Cell Physiol 48:1207–1218. PMID:17623741. https://doi.org/10.1093/pcp/pcm090

Lewis WJ (1982) Kairomones and their use for management of entomophagous insects. XIII. Kairomonal activity for *Trichogramma spp.* of abdominal tips, excretion, and a synthetic sex pheromone blend of *Heliothis zea* moths. J Chem Ecol 8:1323–1330

Lewis WJ, Tumlinson JH, Krasnoff S (1991) Chemically mediated associative learning: an important function of foraging behavior of *Microplitis croceipes* (Cresson). J Chem Ecol 17:1309–1325

Loughrin JH, Potter DA, Hamilton-Kemp TR (1995) Volatile compounds induced by herbivory act as aggregation kairomones for *Popilia japonica*. J Chem Ecol 21:1457–1467

Ludwig AA, Romeis T, Jones JD (2004) CDPK-mediated signalling pathways: specificity and cross-talk. J Exp Bot 55:181–188. PMID:14623901. https://doi.org/10.1093/jxb/erh008

Ludwig AA, Saitoh H, Felix G, Freymark G, Miersch O, Wasternack C et al (2005) Ethylene-mediated cross-talk between calcium-dependent protein kinase and MAPK signaling controls stress responses in plants. Proc Natl Acad Sci USA 102:10736–10741. PMID:16027369. https://doi.org/10.1073/pnas.0502954102

Maffei ME (2010) Site of synthesis, biochemistry and functional role of plant volatiles. S Afr J Bot 76:612–631. https://doi.org/10.1016/j.sajb.2010.03.003

Maffei ME, Mithoefer A, Boland W (2007) Insects feeding on plants: rapid signals and responses preceding the induction of phytochemical release. Phytochemistry 68:2946–2959. PMID:17825328. https://doi.org/10.1016/j.phytochem.2007.07.016

Majumder P, Mondal HA, Das S (2005) Insecticidal activity of *Arum maculatum* tuber lectin and its binding to the glycosylated insect gut receptors. J Agric Food Chem 53:6725–6729. PMID:16104791. https://doi.org/10.1021/jf051155z

Mao YB, Cai WJ, Wang JW, Hong GJ, Tao XY, Wang LJ et al (2007) Silencing a cotton bollworm P450 monooxygenase gene by plant-mediated RNAi impairs larval tolerance of gossypol. Nat Biotechnol 25:1307–1313. PMID:17982444. https://doi.org/10.1038/nbt1352

Matsui K (2006) Green leaf volatiles: hydroperoxide lyase pathway of oxylipin metabolism. Curr Opin Plant Biol 9:274–280. PMID:16595187. https://doi.org/10.1016/j.pbi.2006.03.002

Mattiacci L, Dicke M, Posthumus MA (1994) Induction of parasitoid attracting synomone in Brussel sprouts plants by feeding of *Pieris brassicae* larvae: role of mechanical damage and herbivore elicitor. J Chem Ecol 20:2229–2247

Mattiacci L, Dicke M, Posthumus MA (1995) Beta-glucosidase: an elicitor of herbivore-induced plant odor that attracts host-searching parasitic wasps. Proc Natl Acad Sci USA 92:2036–2040. PMID:11607516. https://doi.org/10.1073/pnas.92.6.2036

McCall PJ, Turlings TCJ, Lewis WJ, Tumlinson JH (1993) Role of plant volatiles in host location by the specialist parasitoid *Microplitis croceipes*. J Insect Behav 6:625–639

Mellway RD, Tran LT, Prouse MB, Campbell MM, Constabel CP (2009) The wound-, pathogen-, and ultraviolet B-responsive MYB134 gene encodes an R2R3 MYB transcription factor that regulates proanthocyanidin synthesis in poplar. Plant Physiol 150:924–941. PMID:19395405. https://doi.org/10.1104/pp.109.139071

Michiels K, Van Damme EJ, Smagghe G (2010) Plant–insect interactions: what can we learn from plant lectins? Arch Insect Biochem Physiol 73(4):193–212

Miranda M, Ralph SG, Mellway R, White R, Heath MC, Bohlmann J et al (2007) The transcriptional response of hybrid poplar (*Populus trichocarpa* × *P. deltoides*) to infection by *Melampsora medusae* leaf rust involves induction of flavonoid pathway genes leading to the accumulation of proanthocyanidins. Mol Plant-Microbe Interact 20:816–831. https://doi.org/10.1094/MPMI-20-7-0816

Moran N, Hamilton WD (1980) Low nutritive value as a defense against herbivores. J Theor Biol 86:247–254

Morant AV, Jørgensen K, Jørgensen C, Paquette SM, Sánchez-Pérez R, Møller BL et al (2008) Beta-glucosidases as detonators of plant chemical defense. Phytochemistry 69:1795–1813. PMID:18472115. https://doi.org/10.1016/j.phytochem.2008.03.006

Mrid RB, Benmrid B, Hafsa J, Boukcim H, Sobeh M, Yasri A (2021) Secondary metabolites as biostimulant and bioprotectant agents: a review. Sci Total Environ 777:146204. https://doi.org/10.1016/j.scitotenv.2021.146204

Nishida N, Tamotsu S, Nagata N, Saito C, Sakai A (2005) Allelopathic effects of volatile monoterpenoids produced by *Salvia leucophylla*: inhibition of cell proliferation and DNA synthesis in the root apical meristem of *Brassica campestris* seedlings. J Chem Ecol 31:1187–1203. PMID:16124241. https://doi.org/10.1007/s10886-005-4256-y

Noldus LPJJ, van Lenteren JC (1985) Kairomones for the egg parasite *Trichogramma evanescens*, Westword. I. Effect of volatile substances released by two of its hosts, *Pieris brassicae* L. and *Mamestra brassicae* L. J Chem Ecol 11:781–791

Nordlund DA, Lewis WJ, Gueldner RC (1983) Kairomones and their use for management of entomophagous insects XIV. Response of *Telenomus remus* to abdominal tips of *Spodoptera frugiperda*, (Z)-9-tetradecene-1-ol acetate, and (Z)-9-dodecene- 1-ol acetate. J Chem Ecol 9: 695–701

Nouri-Ganbalani G, Borzoui E, Shahnavazi M, Nouri A (2018) Induction of resistance against *Plutella xylostella* (L.) (Lep.: Plutellidae) by jasmonic acid and mealy cabbage aphid feeding in *Brassica napus* L. Front Physiol 9:859. https://doi.org/10.3389/fphys.2018.00859

Nuessly GS, Scully BT, Hentz MG, Beiriger R, Snook ME, Widstrom NW (2007) Resistance to *Spodoptera frugiperda* and *Euxesta stigmatias* in sweet corn derived from exogenous and endogenous genetic systems. J Econ Entomol 100:1887–1895. PMID:18232407. https://doi.org/10.1603/0022-0493(2007)100[1887:RTSFLN]2.0.CO;2

Ode PJ (2006) Plant chemistry and natural enemy fitness: effects on herbivore and natural enemy interactions. Annu Rev Entomol 51:163–185

Ode PJ (2019) Plant toxins and parasitoid trophic ecology. Cur Opin Insect Sci 32:118–123. https://doi.org/10.1016/j.cois.2019.01.007

Olson D (2021) The role of gibberellic acid in aphid-plant-arbuscular mycorrhizal fungus interactions. Available online: https://digscholarship.unco.edu/theses/164 (on 12.12.2021)

Olson DL, Nechols JR (1995) Effects of squash leaf trichome exudates and honey on adult feeding, survival, and fecundity of the squash bug (Coreidae) egg parasitoid Gryon pennsylvanicum (Hymenoptera: Scelionidae). Environ Entomol 24:454–458

Opitz SEW, Müller C (2009) Plant chemistry and insect sequestration. Chemoecology 19:117–154. https://doi.org/10.1007/s00049-009-0018-6

Pallini A, Janssen A, Sabelis MW (1997) Odour-mediated responses of phytophagous mites to conspecific and heterospecific competitors. Oekologia 110:179–185

Pappas ML, Broekgaarden C, Broufas GD, Kant M, Messelink GJ, Steppuhn A, Wäckers F, Van Dam NM (2017) Induced plant defences in biological control of arthropod pests: a double-edged sword. Pest Manag Sci 73:1780–1788. https://doi.org/10.1002/ps.4587

Pasteels JM, Rowell-Rahier M, Raupp MJ (1988) Plant derived defense in chrysomelid beetles. In: Barbosa P, Letourneau D (eds) Novel aspects of insect-plant interactions. Wiley, New York, pp 235–271

Pauwels L, Inzé D, Goossens A (2009) Jasmonate-inducible gene: what does it mean? Trends Plant Sci 14:87–91. PMID:19162528. https://doi.org/10.1016/j.tplants.2008.11.005

Potting RPJ, Vet LEM, Dicke M (1996) Host microhabitat location by the stemborer parasitoid *Cotesia flavipes*: the role of locally and systemically induced plant volatiles. J Chem Ecol 21: 525–539

Powell G, Tosh CR, Hardie J (2006) Host plant selection by aphids: behavioral, evolutionary, and applied perspectives. Annu Rev Entomol 51:309–330. PMID:16332214. https://doi.org/10.1146/annurev.ento.51.110104.151107

Price PW (1986) Ecological aspects of host plant resistance and biological control: interactions among three trophic levels. In: Boethel DJ, Eikenbary RD (eds) Interactions of plant resistance and parasitoids and predators of insects. Wiley, New York, pp 11–30

Price PW, Bouton CE, Gross P, McPheron BA, Thompson JN, Weis AE (1980) Interactions among three trophic levels: influence of plant on interactions between insect herbivores and natural enemies. Ann Rev Ecol Syst 11:41–65

Puthoff DP, Sardesai N, Subramanyam S, Nemacheck JA, Williams CE (2005) *Hfr-2*, a wheat cytolytic toxin-like gene, is up-regulated by virulent Hessian fly larval feeding. Mol Plant Biol 6:411–423. https://doi.org/10.1111/j.1364-3703.2005.00289.x

Quintana-Rodriguez E, Duran-Flores D, Heil M, Camacho X (2018) Damage-associated molecular patterns (DAMPs) as future plant vaccines that protect crops from pests. Sci Hortic 237:207–220. https://doi.org/10.1016/j.scienta.2018.03.026

Raffa KF, Phillips TW, Salmon SM (1993) Strategies and mechanisms of host colonization by bark beetles. In: Schowalter TD, Filip GM (eds) Beetle-pathogen interactions in conifer forests. Academic, London, pp 103–128

Rafter JL, Gonda-King L, Niesen D, Seeram NP, Rigsby CM, Preisser EL (2017) Impact of consuming 'toxic' monarch caterpillars on adult Chinese mantid mass gain and fecundity. Insects 8:23. https://doi.org/10.3390/insects8010023

Ramírez-Gómez XS, Jiménez-García SN, Campos VB, Campos MLG (2020) Plant metabolites in plant defense against pathogens- current threats and management trends. IntechOpen, London. https://doi.org/10.5772/intechopen.87958

Rasmann S, Köllner TG, Degenhardt J, Hiltpold I, Toepfer S, Kuhlmann U et al (2005) Recruitment of entomopathogenic nematodes by insect-damaged maize roots. Nature 434:732–737. PMID:15815622. https://doi.org/10.1038/nature03451

Rayapuram C, Baldwin IT (2007) Increased SA in NPR1-silenced plants antagonizes JA and JA-dependent direct and indirect defenses in herbivore-attacked Nicotiana attenuata in nature. Plant J 52:700–715. PMID:17850230. https://doi.org/10.1111/j.1365-313X.2007.03267.x

Renwick JAA, Zhang W, Haribal M, Attygalle AB, Lopez KD (2001) Dual chemical barriers protect a plant against different larval stages of an insect. J Chem Ecol 27:1575–1583. PMID:11521397. https://doi.org/10.1023/A:1010402107427

Rhoades DF (1983) Herbivore population dynamics and plant chemistry. In: Denno RF, McClure MS (eds) Variable plants and herbivores in natural and managed systems. Academic, New York, pp 155–204

Ribot C, Zimmerli C, Farmer EE, Reymond P, Poirier Y (2008) Induction of the Arabidopsis PHO1;H10 gene by 12-oxo-phytodienoic acid but not jasmonic acid via a CORONATINE INSENSITIVE1-dependent pathway. Plant Physiol 147:696–706. PMID:18434606. https://doi.org/10.1104/pp.108.119321

Richardson LL, Adler LS, Leonard AS, Andicoechea J, Regan KH, Anthony WE, Irwin RE (2015) Secondary metabolites in floral nectar reduce parasite infections in bumblebees. Proc R Soc B 282:20142471. https://doi.org/10.1098/rspb.2014.2471

Röse USR, Manukian A, Heath RR, Tumlinson JH (1996) Volatile semiochemicals released from undamaged cotton leaves. Plant Physiol 111:487–495

Rowell-Rahier M, Pasteels JM (1990) Phenolglycosides and interactions at three trophic levels: Salicaceae-herbivorespredators. In: Bernays EA (ed) Insect-plant interactions. CRC Press, Boca Raton, pp 76–94

Ruther J, Kleier S (2005) Plant-plant signaling: ethylene synergizes volatile emission in Zea mays induced by exposure to (Z)-3-hexen-1-ol. J Chem Ecol 31:2217–2222. PMID:16132223. https://doi.org/10.1007/s10886-005-6413-8

Sabelis MW, de Jong MCM (1988) Should all plants recruit bodyguards? Conditions for a polymorphic ESS of synomone production in plants. Oikos 53:247–252

Saha P, Majumder P, Dutta I, Ray T, Roy SC, Das S (2006) Transgenic rice expressing *Allium sativum* leaf lectin with enhanced resistance against sap-sucking insect pests. Planta 223:1329–1343. PMID:16404581. https://doi.org/10.1007/s00425-005-0182-z

Sanches P, Santos F, Peñaflor M, Bento J (2017) Direct and indirect resistance of sugarcane to *Diatraea saccharalis* induced by jasmonic acid. Bull Èntomol Res 107:828–838. https://doi.org/10.1017/S0007485317000372

Santiago R, López-Malvar A, Souto C, Barros-Ríos J (2018) Methods for determining cell wall-bound phenolics in maize stem tissues. J Agric Food Chem 66:1279–1284. https://doi.org/10.1021/acs.jafc.7b05752

Schaefer M, Fischer C, Meldau S, Seebald E, Oelmüller R, Baldwin IT (2011) Lipase activity in insect oral secretions mediates defense responses in Arabidopsis. Plant Physiol 156:1520–1534. PMID:21546453. https://doi.org/10.1104/pp.111.173567

Schultz JC (1988) Many factors influence the evolution of herbivore diets, but plant chemistry is central. Ecology 69:896–897

Schweiger R, Heise AM, Persicke M, Müller C (2014) Interactions between the jasmonic and salicylic acid pathway modulate the plant metabolome and affect herbivores of different feeding types. Plant Cell Environ 37:1574–1585. https://doi.org/10.1111/pce.12257

Sethi A, McAuslane HJ, Rathinasabapathi B, Nuessly GS, Nagata RT (2009) Enzyme induction as a possible mechanism for latex-mediated insect resistance in romaine lettuce. J Chem Ecol 35: 190–200. PMID:19184224. https://doi.org/10.1007/s10886-009-9596-6

Shahabinejad M, Shojaaddini M, Maserti B, Arvin SMJ, Seyedi SM (2014) Exogenous application of methyl jasmonate and salicylic acid increases antioxidant activity in the leaves of pistachio (*Pistacia vera* L. cv. Fandoughi) trees and reduces the performance of the phloem-feeding psyllid *Agonoscena pistaciae*. Arthropod Plant Interact 8:525–530. https://doi.org/10.1007/s11829-014-9334-z

Sharma HC (2009) Biotechnological approaches for pest management and ecological sustainability. CRC Press/Taylor and Francis, New York, p 526. https://doi.org/10.1201/9781420088489

Sheard LB, Tan X, Mao H, Withers J, Ben-Nissan G, Hinds TR et al (2010) Jasmonate perception by inositolphosphate- potentiated COI1-JAZ co-receptor. Nature 468:400–405. PMID:20927106. https://doi.org/10.1038/nature09430

Shivaji R, Camas A, Ankala A, Engelberth J, Tumlinson JH, Williams WP et al (2010) Plants on constant alert: elevated levels of jasmonic acid and jasmonate-induced transcripts in caterpillar-resistant maize. J Chem Ecol 36:179–191. PMID:20148356. https://doi.org/10.1007/s10886-010-9752-z

Simmonds MSJ (2003) Flavonoid-insect interactions: recent advances in our knowledge. Phytochemistry 64:21–30. PMID:12946403. https://doi.org/10.1016/S0031-9422(03)00293-0

Simmonds MSJ, Stevenson PC (2001) Effects of isoflavonoids from Cicer on larvae of *Heliocoverpa armigera*. J Chem Ecol 27:965–977. PMID:11471948. https://doi.org/10.1023/A:1010339104206

Simmonds MSJ, Blaney WM, Fellows LE (1990) Behavioural and electrophysiological study of antifeedant mechanisms associated with polyhydroxyalkaloids. J Chem Ecol 16:3167–3196. https://doi.org/10.1007/BF00979618

Stamp N (2003) Out of the quagmire of plant defense hypotheses. Quart Rev Biol 78(1):23–55

Steppuhn A, Baldwin IT (2007) Resistance management in a native plant: nicotine prevents herbivores from compensating for plant protease inhibitors. Ecol Lett 10:499–511. PMID:17498149. https://doi.org/10.1111/j.1461-0248.2007.01045.x

Stevenson PC, Nicolson S, Wright GA (2017) Plant secondary metabolites in nectar: impacts on pollinators and ecological functions. Funct Ecol 31:65–75. https://doi.org/10.1111/1365-2435.12761

Stout MJ, Riggio MR, Yang Y, MJ (2009) Direct induced resistance in *Oryza sativa* to *Spodoptera frugiperda*. Environ Entomol 38:1174–1181. PMID:19689897. https://doi.org/10.1603/022.038.0426

Subramanyam S, Smith DF, Clemens JC, Webb MA, Sardesai N, Williams CE (2008) Functional characterization of HFR1, a high-mannose N-glycan-specific wheat lectin induced by hessian fly larvae. Plant Physiol 147:1412–1426. PMID:18467454. https://doi.org/10.1104/pp.108.116145

Sun R, Jiang X, Reichelt M, Gershenzon J, Pandit SS, Vassão DG (2019) Tritrophic metabolism of plant chemical defenses and its effects on herbivore and predator performance. eLife 8:51029. https://doi.org/10.7554/eLife.51029

Takabayashi J, Dicke M, Posthumus MA (1991a) Induction of indirect defence against spider-mites in uninfested lima bean leaves. Phytochemistry 30:1459–1462

Takabayashi J, Dicke M, Posthumus MA (1991b) Variation in composition of predator-attracting allelochemicals emitted by herbivore-infested plants: relative influences of plant and herbivore. Chem 2:1–6

Takabayashi J, Takahashi S, Dicke M, Posthumus MA (1995) Developmental stage of herbivore *Pseudaletia separate* affects production of herbivore-induced synomone by corn plants. J Chem Ecol 3:273–287

Teklić T, Parađiković N, Špoljarević M, Zeljković S, Lončarić Z, Lisjak M (2021) Linking abiotic stress, plant metabolites, biostimulants and functional food. Ann Appl Biol 178:169–191. https://doi.org/10.1111/aab.12651

Thorpe KW, Barbosa P (1986) Effects of consumption of high and low nicotine tobacco by *Manduca sexta* on survival of gregarious endoparasitoid *Cotesia congregata*. J Chem Ecol 12:1329–1337

Thorpe WH, Caudle HB (1938) A study of the olfactory responses of insect parasites to the food plant of their host. Parasitology 15:523–528

Tian X, Li Y, Hao N, Su X, Du J, Hu J, Tian X (2021) The antifeedant, insecticidal and insect growth inhibitory activities of triterpenoid saponins from *clematis aethusifolia* Turcz against *Plutella xylostella* (L.). Pest Manag Sci 77:455–463. https://doi.org/10.1002/ps.6038

Torres MA (2010) ROS in biotic interactions. Physiol Plant 138:414–429. PMID:20002601. https://doi.org/10.1111/j.1399-3054.2009.01326.x

Treutter D (2006) Significance of flavonoids in plant resistance: a review. Environ Chem Lett 4:147–157. https://doi.org/10.1007/s10311-006-0068-8

Tscharntke T, Thiessen S, Dolch R, Boland W (2001) Herbivory, induced resistance, and interplant signal transfer in *Alnus glutinosa*. Biochem Syst Ecol 29:1025–1047. https://doi.org/10.1016/S0305-1978(01)00048-5

Tumlinson JH, Turlings TCJ, Lewis WJ (1992) The semiochemical complexes that mediate insect parasitoid foraging. Agric Zool Rev 5:221–252

Turlings TCJ, Benrey B (1998) Effects of plant metabolites on the behavior and development of parasitic wasps. Écoscience 5(3):321–333. https://doi.org/10.1080/11956860.1998.11682472

Turlings TCJ, Erb M (2018) Tritrophic interactions mediated by herbivore-induced plant volatiles: mechanisms, ecological relevance, and application potential. Annu Rev Èntomol 63:433–452. https://doi.org/10.1146/annurev-ento-020117-043507

Turlings TC, Ton J (2006) Exploiting scents of distress: the prospect of manipulating herbivore-induced plant odours to enhance the control of agricultural pests. Cur Opin Plant Biol 9(4):421–427

Turlings TCJ, Tumlinson JH (1991) Do parasitoids use herbivore-induced plant chemical defences to locate hosts? Florida Entomol 74:42–50

Turlings TCJ, Tumlinson JH (1992) Systemic chemical signaling by herbivore-injured corn. Proc Natl Acad Sci USA 89:8399–8402

Turlings TCJ, Tumlinson JH, Lewis WJ (1990) Exploitation of herbivore-induced plant odors by host-seeking wasps. Science 250:1251–1253

Turlings TCJ, Tumlinson JH, Eller FJ, Lewis WJ (1991a) Larval-damaged plants: source of volatile synomones that guide the parasitoid *Cotesia marginiventris* to the microhabitat of its hosts. Entomol Exp Appl 58:75–82

Turlings TCJ, Tumlinson JH, Heath RR, Proveaux AT, Doolittle RE (1991b) Isolation and identification of allelochemicals that attract the larval parasitoid *Cotesia marginiventris* to the micro-habitat of one of its hosts. J Chem Ecol 17:2235–2251

Turlings TCJ, McCall PJ, Alborn HT, Tumlinson JH (1993a) An elicitor in caterpillar oral secretions that induces corn seedlings to emit chemical signals attractive to parasitic wasps. J Chem Ecol 19:411–425

Turlings TCJ, Wäckers F, Vet LEM, Lewis WJ, Tumlinson JH (1993b) Learning of host-finding cues by hymenopterous parasitoids. In: Papaj DR, Lewis A (eds) Insect learning: ecological and evolutionary perspectives. Chapm & Hall, New York, pp 51–78

Turlings TCJ, Loughrin JH, Röse U, McCall PJ, Lewis WJ, Tumlinson JH (1995) How caterpillar-damaged plants protect themselves by attracting parasitic wasps. Proc Natl Acad Sci USA 92: 4169

Ulloa RM, Raíces M, MacIntosh GC, Maldonado S, Téllez-Iñón MT (2002) Jasmonic acid affects plant morphology and calcium-dependent protein kinase expression and activity in *Solanum tuberosum*. Physiol Plant 115:417–427. PMID:12081535. https://doi.org/10.1034/j.1399-3054.2002.1150312.x

Usha Rani P, Jyothsna Y (2010) Biochemical and enzymatic changes in rice as a mechanism of defense. Acta Physiol Plant 32:695–701. https://doi.org/10.1007/s11738-009-0449-2

Van Breusegem F, Dat JF (2006) Reactive oxygen species in plant cell death. Plant Physiol 141: 384–390. PMID:16760492. https://doi.org/10.1104/pp.106.078295

Vandenborre G, Miersch O, Hause B, Smagghe G, Wasternack C, Van Damme EJM (2009) *Spodoptera littoralis* induced lectin expression in tobacco. Plant Cell Physiol 50:1142–1155. PMID:19416954. https://doi.org/10.1093/pcp/pcp065

Vandenborre G, Smagghe G, Van Damme EJM (2011) Plant lectins as defense proteins against phytophagous insects. Phytochemistry 72:1538–1550. PMID:21429537. https://doi.org/10.1016/j.phytochem.2011.02.024

Verhage A, van Wees SCM, Pieterse CMJ (2010) Plant immunity: it's the hormones talking, but what do they say? Plant Physiol 154:536–540. PMID:20921180. https://doi.org/10.1104/pp.110.161570

Vet LEM, Dicke M (1992) Ecology of infochemical use by natural enemies in a tritrophic context. Annu Rev Entomol 37:141–172

Vet LEM, Wäckers FL, Dicke M (1991) How to hunt for hiding hosts: the reliability-detectability problem in foraging parasitoids. Netherl J Zool 41:202–213

Vinson SB (1975) Biochemical coevolution between parasitoids and their hosts. In: Price PW (ed) Evolutionary strategies of parasitic insects and mites. Plenum Press, New York, pp 14–48

Vinson SB (1981) Habitat location. In: Nordlund DA, Jones RL, Lewis WJ (eds) Semiochemicals: their role in pest control. Wiley, New York, pp 51–77

Vinson SB (1984) Parasitoid-host relationship. In: Bell WJ, Carde RT (eds) Chemical ecology of insects. Sinauer Associates, Sunderland, pp 111–124

Vinson SB, Barbosa P (1987) Interrelationships of nutritional ecology of parasitoids. In: Slansky F, Rodriguez JG (eds) Nutritional ecology of insects, mites, and spiders and related invertebrates. Wiley, New York, pp 673–695

Vinson SB, Iwantsch GF (1980) Host suitability for insect parasitoids. Annu Rev Entomol 25:397–419

Vinson SB, Piper GL (1986) Source and characterization of host recognition kairomones of *Tetrastichus hagenowii*, a parasitoid of cockroach eggs. Physiol Entomol 11:459–468

Vinson SB, Elzen GW, Williams HJ (1987) The influence of volatile plant allelochemicals on the third trophic level (parasitoids) and their herbivorous hosts. In: Labeyerie V, Fabres G, Lachaise D (eds) Insects-plants. Junk Publishers, Dordrecht, pp 109–114

Visser ME (1994) The importance of being large: the relationship between size and fitness in females of the parasitoid *Aphaereta minuta*. J Animal Ecol 63:963–978

Voelckel C, Baldwin IT (2004) Herbivore-induced plant vaccination. Part II. Array-studies reveal the transience of herbivore-specific transcriptional imprints and a distinct imprint from stress combinations. Plant J 38:650–663. PMID:15125771. https://doi.org/10.1111/j.1365-313X.2004.02077.x

Voelckel C, Weisser WW, Baldwin IT (2004) An analysis of plant-aphid interactions by different microarray hybridization strategies. Mol Ecol 13:3187–3195. PMID:15367131. https://doi.org/10.1111/j.1365-294X.2004.02297.x

Wäckers F, Lewis WJ (1994) Olfactory and visual learning and their combined influence on host site location by the parasitoid *Microplitis croceipes*. Biol Control 4:105–112

Walling LL (2000) The myriad plant responses to herbivores. J Plant Growth Regul 19:195–216. PMID:11038228

War AR, Sharma HC (2014) Effect of jasmonic acid and salicylic acid induced resistance in groundnut on *Helicoverpa armigera*. Physiol Èntomol 39:136–142. https://doi.org/10.1111/phen.12057

War AR, Paulraj MG, War MY, Ignacimuthu S (2011a) Jasmonic acid- mediated induced resistance in groundnut *Arachis hypogaea* against *Helicoverpa armigera*. J Plant Growth Regul 30:512–523. https://doi.org/10.1007/s00344-011-9213-0

War AR, Paulraj MG, War MY, Ignacimuthu S (2011b) Herbivore- and elicitor-induced resistance in groundnut to Asian armyworm, *Spodoptera litura*. Plant Signal Behav 6:1769–1777. PMID:22042128. https://doi.org/10.4161/psb.6.11.17323

War AR, Paulraj MG, War MY, Ignacimuthu S (2011c) Role of salicylic acid in induction of plant defense system in chickpea (*Cicer arietinum* L.). Plant Signal Behav 6:1787–1792. PMID:22057329. https://doi.org/10.4161/psb.6.11.17685

War A, Paulraj MG, Ahmad T, Buhroo A, Hussain B, Ignacimuthu S, Sharma HC (2012) Mechanisms of plant defense against insect herbivores. Plant Signal Behav 7:1306–1320. https://doi.org/10.4161/psb.21663

War AR, Buhroo AA, Hussain B, Ahmad T, Nair RM, Sharma HC (2020) Plant defense and insect adaptation with reference to secondary metabolites. In: Co-evolution of secondary metabolites, reference series in phytochemistry. Springer, Cham. https://doi.org/10.1007/978-3-319-96397-6_60

Wei X, Vrieling K, Kim HK, Mulder PP, Klinkhamer PG (2021) Application of methyl jasmonate and salicylic acid lead to contrasting effects on the plant's metabolome and herbivory. Plant Sci 303:110784. https://doi.org/10.1016/j.plantsci.2020.110784

Whitman DW (1988) Allelochemical interactions among plants, herbivores, and their predators. In: Barbosa P, Letourneau D (eds) Novel aspects of insect-plant interactions. Wiley, New York, pp 11–63

Whitman DW (1994) Plant bodyguards: mutualistic interactions between plants of the third trophic level. In: Ananthakrishan TN (ed) Functional dynamics of phytophagous insects. Oxford and IBH Publishing Co., New Delhi, pp 207–248

Whitman DW, Eller FJ (1990) Parasitic wasps orient to green leaf volatiles. Chem 1:69–75

Wink M (2003) Evolution of secondary metabolites from an ecological and molecular phylogenetic perspective. Phytochemistry 64:3–19. https://doi.org/10.1016/S0031-9422(03)00300-5

Wiskerke JSC, Vet LEM (1994) Foraging for solitarily and gregariously feeding caterpillars: a comparison of two related parasitoid species. J Insect Behav 5:585–603

Wouters FC, Blanchette B, Gershenzon J, Vassão DG (2016) Plant defense and herbivore counter-defense: Benzoxazinoids and insect herbivores. Phytochem Rev 15:1127–1151. https://doi.org/10.1007/s11101-016-9481-1

Wright GA, Baker DD, Palmer MJ, Stabler D, Mustard JA, Power EF, Borland AM, Stevenson PC (2013) Caffeine in floral nectar enhances a pollinator's memory of reward. Science 339:1202–1204. https://doi.org/10.1126/science.1228806

Wu J, Baldwin IT (2010) New insights into plant responses to the attack from insect herbivores. Annu Rev Genet 44:1–24. https://doi.org/10.1146/annurev-genet-102209-163500

Wu JQ, Hettenhausen C, Meldau S, Baldwin IT (2007) Herbivory rapidly activates MAPK signaling in attacked and unattacked leaf regions but not between leaves of *Nicotiana attenuata*. Plant Cell 19:1096–1122. PMID:17400894. https://doi.org/10.1105/tpc.106.049353

Xiao Y, Wang Q, Erb M, Turlings TCJ, Ge L, Hu L, Li J, Han X, Zhang T, Lu J et al (2012) Specific herbivore-induced volatiles defend plants and determine insect community composition in the field. Ecol Lett 15:1130–1139. https://doi.org/10.1111/j.1461-0248.2012.01835.x

Yan Y, Stolz S, Chételat A, Reymond P, Pagni M, Dubugnon L et al (2007) A downstream mediator in the growth repression limb of the jasmonate pathway. Plant Cell 19:2470–2483. PMID:17675405. https://doi.org/10.1105/tpc.107.050708

Yang JN, Wei JN, Kang L (2021) Feeding of pea leafminer larvae simultaneously activates jasmonic and salicylic acid pathways in plants to release a terpenoid for indirect defense. Insect Sci 28:811–824. https://doi.org/10.1111/1744-7917.12820

Ye M, Glauser G, Lou Y, Erb M, Hu L (2019) Molecular dissection of early defense signaling underlying volatile-mediated defense regulation and herbivore resistance in rice. Plant Cell 31: 687–698. https://doi.org/10.1105/tpc.18.00569

Zhang SZ, Hau BZ, Zhang F (2008) Induction of the activities of antioxidative enzymes and the levels of malondialdehyde in cucumber seedlings as a consequence of *Bemisia tabaci* (Hemiptera: Aleyrodidae) infestation. Arthropod Plant Interact 2:209–213. https://doi.org/10.1007/s11829-008-9044-5

Zhang X, Li L, Kesner L, Robert CAM (2021) Chemical cues shaping host-seeking by entomopathogenic nematodes. Curr Opin Insect Sci 44:72–81. https://doi.org/10.1016/j.cois.2021.03.011

Zhao LY, Chen JL, Cheng DF, Sun JR, Liu Y, Tian Z (2009) Biochemical and molecular characterizations of *Sitobion avenae*– induced wheat defense responses. Crop Prot 28:435–442. https://doi.org/10.1016/j.cropro.2009.01.005

Zheng SJ, Dicke M (2008) Ecological genomics of plant insect interactions: from gene to community. Plant Physiol 146:812–817. PMID:18316634. https://doi.org/10.1104/pp.107.111542

Zhu-Salzman K, Zeng R (2015) Insect response to plant defensive protease inhibitors. Annu Rev Èntomol 60:233–252. https://doi.org/10.1146/annurev-ento-010814-020816

Zhu-Salzman K, Salzman RA, Ahn JE, Koiwa H (2004) Transcriptional regulation of sorghum defense determinants against a phloem-feeding aphid. Plant Physiol 134:420–431. PMID:14701914. https://doi.org/10.1104/pp.103.028324

Züst T, Agrawal AA (2016) Mechanisms and evolution of plant resistance to aphids. Nat Plants 2: 15206. https://doi.org/10.1038/nplants.2015.206

# Chapter 10
# Role of Metabolic Substances in Plant Defense Against Herbivore or Predatory Insect and Mite Species

**Islam Mohammad Zidan, Marwa El-Desouky Barghout, Hanaa Embaby Sadek, Marwa Abd El-Aziz Ibrahim, Soumya Ghosh, Mohamed Lamlom, and Faten Mamdouh Momen**

## 10.1 Introduction

Plant community has evolved a variety of diverse chemical defenses covering nearly all kinds of secondary metabolites (SMs) that provide a significant line of defense against insect and mite pests (Smith 2005; Wink 2008; Fürstenberg-Hägg et al. 2013). Plant secondary metabolites (PSMs) are categorized according to their chemical structure (e.g., rings, sugar content), composition (nitrogen content or absence), solubility in different solvents, or synthesis pathway (e.g., phenylpropanoid, tannin production) (Bourgaud et al. 2001). Three main groups can be identified by a simple classification: terpenes (derived from mevalonic acid, which is almost entirely composed of carbon and hydrogen), phenolics (derived from simple sugars, which contain hydrogen, oxygen, and benzene rings) (Polumackanycz et al. 2020), and nitrogen-containing compounds (very diverse, such as alkaloids, which may also contain sulfur) (Hussein and El-Anssary 2019).

I. M. Zidan (✉) · M. E.-D. Barghout · H. E. Sadek · M. Lamlom · F. M. Momen
Agricultural and Biological Research Institute, Pests and Plant Protection Department, National Research Centre (NRC), Giza, Egypt
e-mail: im.zidan@nrc.sci.eg; me.barghout@nrc.sci.eg; he.sadek@nrc.sci.eg

M. A. E.-A. Ibrahim
Zoology and Agricultural Nematology Department, Faculty of Agriculture, Cairo University, Giza, Egypt
e-mail: marwa_abdelaziz_m@cu.edu.eg

S. Ghosh
Natural and Medical Sciences Research Center, University of Nizwa, Nizwa, Sultanate of Oman

Department of Genetics, University of the Free State, Bloemfontein, South Africa
e-mail: s.ghosh@unizwa.edu.om



N. El-Wakeil, A. Abdelkhalek (eds.), *Microbiomes and Plant Secondary Metabolites for Insect Pest Management*, Progress in Biological Control 23,
https://doi.org/10.1007/978-3-032-24330-0_10

These mixtures have a variable concentration and composition, rather than a constant composition. Mixtures therefore vary throughout organs, within populations, and during developmental phases (Guerriero et al. 2018). This variant might be recommended as a crucial tactic to prevent herbivores from adapting and becoming resistant to the chemical defense.

The plant defense response involving PSMs is adaptable. Some are induced after pest attack and both together affect pest's performance (Kant et al. 2015; Yang et al. 2020). Some are constitutive and primarily rely on carbon-based metabolites, such as terpenes and polyphenols, which are concentrated in special structures or compartments such as resin canals and significantly increase the dry matter content of plants (Sampedro et al. 2011; Santiago et al. 2018). These substances fall under the categories of chemicals resistant to antixenosis and antibiosis (Hondelmann et al. 2020; Gebretsadik et al. 2022).

## 10.2 Types of Secondary Plant Metabolites

### *10.2.1 Terpenes*

#### 10.2.1.1 Chemical Properties of Terpenes

The main class of PSMs is called terpenes. Certain terpenes, such as gibberellins and brassinosteroids, are involved in primary tasks, such as plant growth and development, even though the majority of terpenes are crucial in plant defense.

Terpenes are made up of over 25,000 different molecules Doughari (2015) have different volatile, semi-volatile, and volatile chemical structures (Figs. 10.1 and 10.2), which provides a classification and overview of the chemical structures of volatile terpenes as mention by Ninkuu et al. (2021). Terpenes have a variety of uses, such as preventing oviposition, preventing feeding, and direct poisoning. Because they can withstand terpenoids, specialized herbivores use them as stimulants for feeding as well as an attractant to help them find their host plants (Goldberg 2003). Insects that pollinate plants can be drawn to terpenes Doughari (2015).

### *10.2.2 Phenolic Compounds*

#### 10.2.2.1 Chemical Proprieties of Phenols

Plant phenols represent a diverse category of plant secondary metabolites (PSMs), encompassing nearly 10,000 different compounds. Phenolics are the most widespread secondary metabolites, characterized by a hydroxyl group attached to a phenyl group on an aromatic ring (Fig. 10.3).

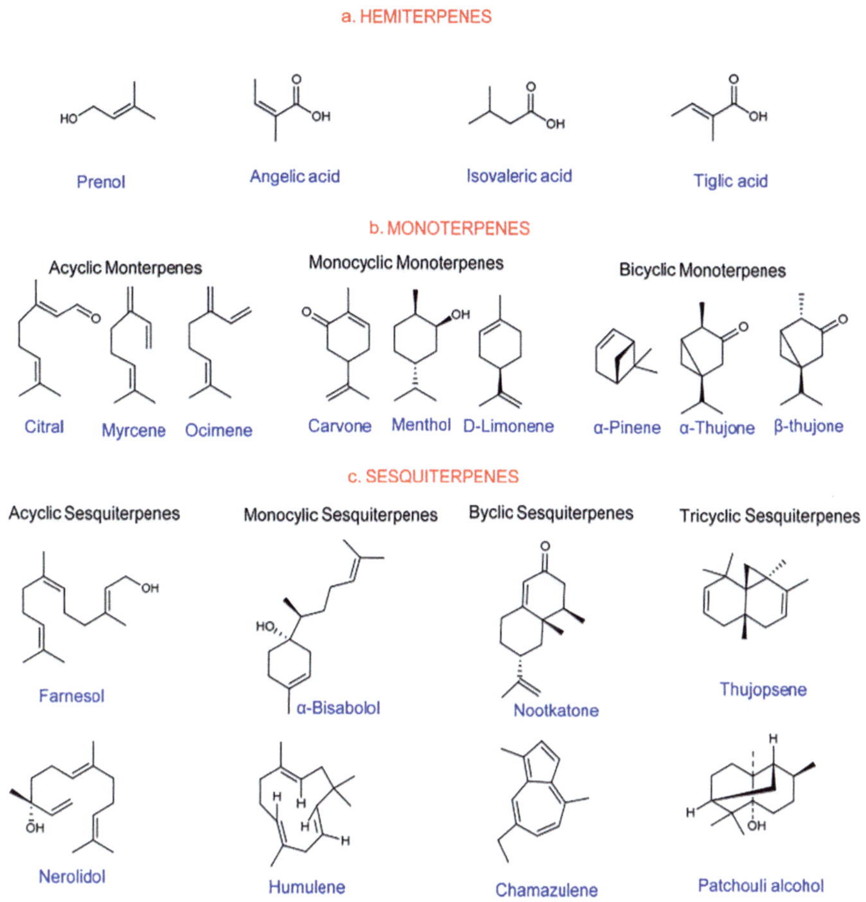


**Fig. 10.1** Classification and chemical structure of volatile terpenes. (Photo credits and source: Ninkuu et al. 2021, https://doi.org/10.3390/ijms22115710)

Simple phenols, such as catechols and their derivatives, flavonoids, catechol melanins, stilbenes, condensed tannins, and lignins are a few examples. These metabolites actively work to draw pollinators and shield plants from herbivores.

## *10.2.3 Sulfur-Containing PSMs*

### 10.2.3.1 Chemical Proprieties of Sulfur-Containing PSMs

Sulfur-containing plant secondary metabolites (PSMs) are primarily glucosides found in the Brassicaceae and Capparales plant families (Fig. 10.4). A total of 120 molecular structures for glycosides, which are derived from amino acids, have

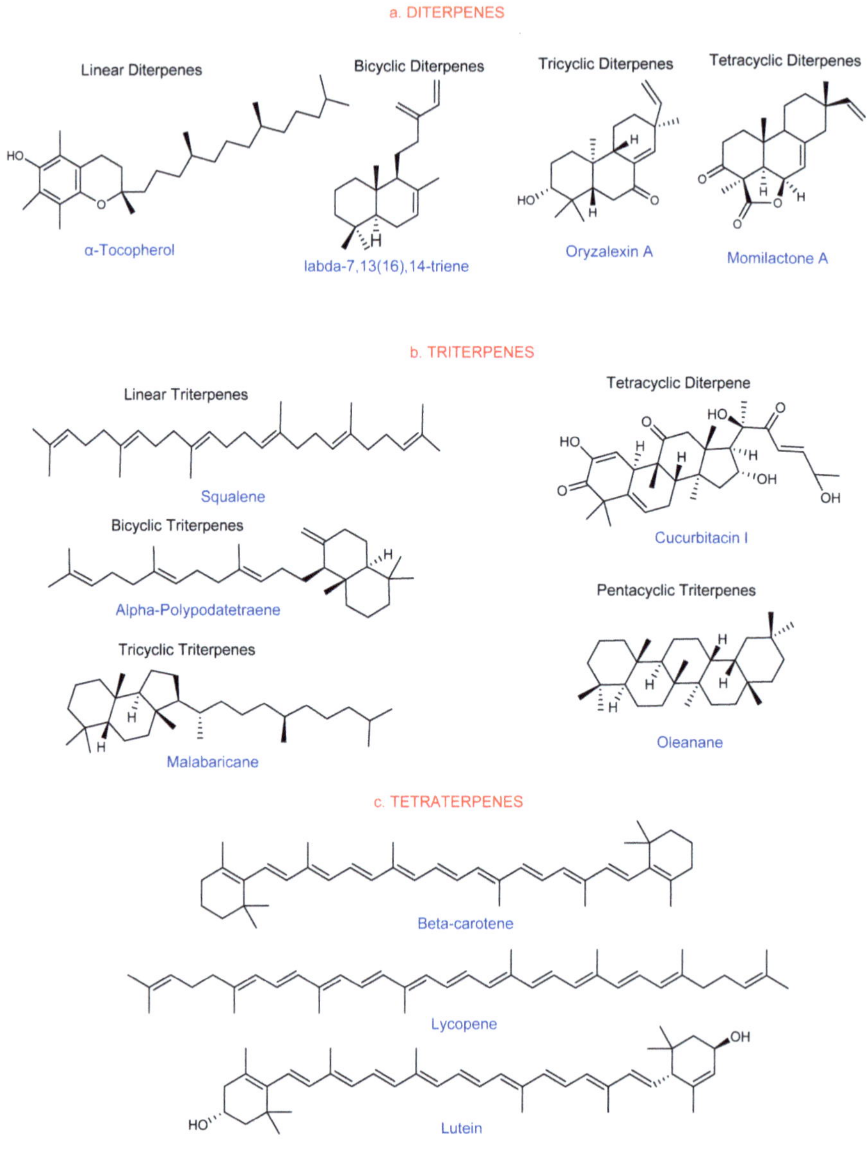


**Fig. 10.2** Classification and chemical structure of semi- and non-volatile terpenes. (Photo credits and source: Ninkuu et al. 2021, https://doi.org/10.3390/ijms22115710)

been described (Golubkina et al. 2022). The kind of glucoside is usually determined by the amino acid precursor from the side chain. Younger leaves and plant reproductive parts have been found to have higher glucosinolate concentrations. Methionine-derived metabolites (aliphatic glucosinolates), tryptophan-derived glucosinolates (indole glucosinolates), tyrosine or phenylalanine-derived glucosinolates (aromatic glucosinolates), and glucosinolates derived from various

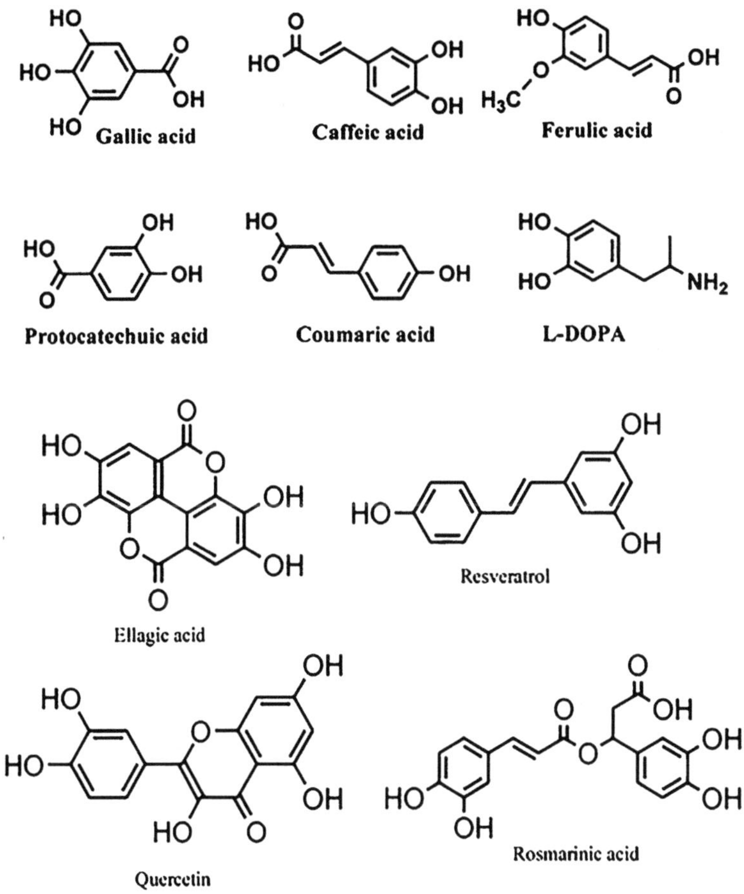


**Fig. 10.3** Common phenolic compounds in plant secondary metabolites. (Photo credits and source: https://doi.org/10.3390/molecules21101374)

amino acids or a single known amino acid are the four different groups of glucosides Golubkina et al. 2022. Myrosinases (thioglucosidase) cover glycosinolates, which are normally found in cell vacuoles (Kortbeek et al. 2019).

## *10.2.4 Nitrogen-Containing PSMs*

### 10.2.4.1 Chemical Properties of Nitrogen-Containing PSMs

Nitrogen-containing plant secondary metabolites (PSMs) primarily consist of alkaloids. Approximately 10,000 alkaloids' derivatives have been documented in the plant kingdom thus far (Elisovetskaya and Nastas 2013). Based on their mechanism

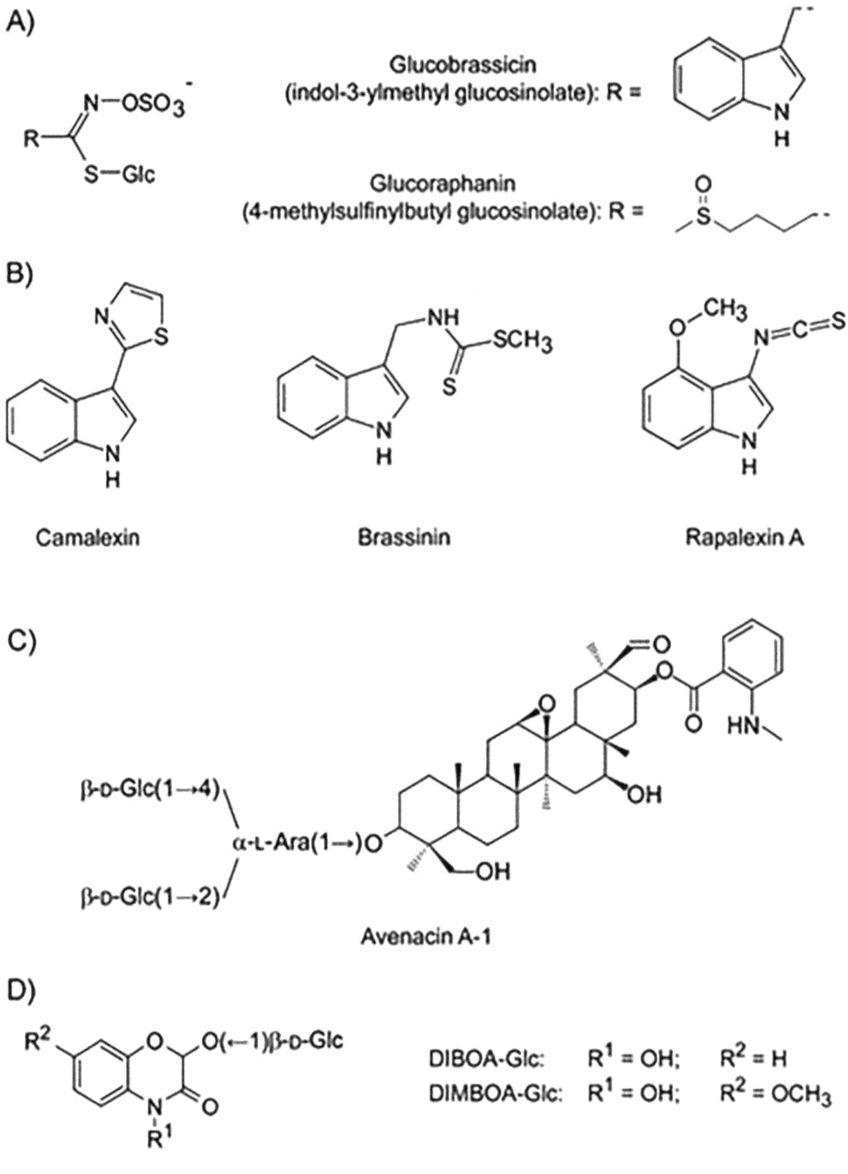


**Fig. 10.4** Plant secondary metabolite sulfuric compounds in Brassicaceae, (A) glucosinolates; (B) Brassicaceae phytoalexins; (C) avenacins; and (D) benzoxazinone glucosides: DIBOA (2,4-dihydroxy-1,4-benzoxazin-3-one) and DIMBOA (2,4-dihydroxy-7-methoxy-1,4-benzoxazin-3-one). (Photo credits and source: https://doi.org/10.1002/cbic.201200086)

of biosynthesis, alkaloids are classified into three types (Muhammad and Kashere 2020): (a) real alkaloids, which include nicotine, morphine, quinine, and atropine; (b) pseudo-alkaloids, which include capsaicin, solanidine, and caffeine; and

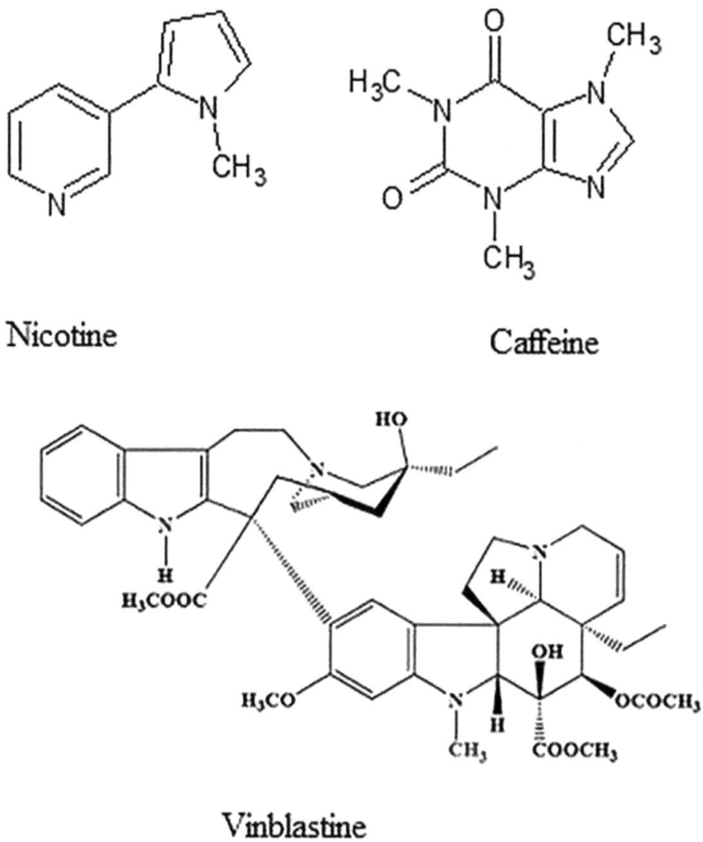


**Fig. 10.5** Plant secondary metabolite alkaloids. (Photo credits and source: https://doi.org/10.5772/intechopen.76139)

(c) proto-alkaloids, which include yohimbine, mescaline, and hordenine. The structures of vinblastine, caffeine, and nicotine are listed in Fig. 10.5.

Amino acids are the source of both proto- and real alkaloids; whereas pseudo-alkaloids are not derived from amino acids. A heterocyclic ring containing nitrogen is shared by true alkaloids derived from amino acids. They have strong biological activity and are quite reactive. When they conjugate with acids, they generate water-soluble salts, many of which are crystalline. With the exception of nicotine, which is a dark liquid, almost all real alkaloids are bitter and solid (Khan et al. 2015). The biosynthesis and deployment of these diverse PSM classes are not constitutive in all cases; rather, they are often precisely regulated by internal signaling networks, most notably plant hormone pathways.

## 10.3 Plant Hormones (Phytohormones phs)

Having outlined the major classes of defensive PSMs, we now turn to the primary regulatory systems that govern their production: plant hormones. These signaling molecules act as master switches, integrating environmental cues—such as herbivore attack—to orchestrate the targeted induction of PSMs detailed in the previous section. Plant hormones are chemical substances found in very small amounts within plants. They are derived from gases (ethylene), carotenoids (abscisic acid), terpenes (gibberellins), adenine (cytokinins), and indole (auxins). Every portion of the plant produces and distributes these hormones. The hormones can have complementary or opposing effects and work alone or in concert. Both hormones and external factors are crucial in processes such as vernalization, phototropism, seed germination, and dormancy (Widemann et al. 2021). Chemical messengers called plant hormones control a number of a plant's growth and development activities, including its defense mechanism against pests.

Plant defense mechanisms are influenced by five primary plant hormones, as follows:

### 10.3.1 *Salicylic Acid (SA)*

Salicylic acid (SA) ($C_7H_6O_3$) (Fig. 10.6) plays a crucial role in plant defense against various pathogens, including bacteria, fungi, and viruses. It is formed when a pathogen infects a host. This sets off a cascade reaction that produces antimicrobial chemicals like phytoalexins and activates defense genes. For instance, *Pseudomonas syringae* infection, which causes fire blight in pear trees, results in the production of SA. Antimicrobial molecules called phytoalexins are produced in response to the SA. Additionally, salicylic acid encourages the build-up of callose, a polymer that can impede the passage of diseases through plants.

### 10.3.2 *Jasmonic Acid (JA)*

Jasmonic acid (JA) (Fig. 10.7) is another hormone involved in plant defense. Jasmonic acid is synthesized in response to insect herbivores and other stressors, leading to the production of volatile compounds that attract natural enemies of the

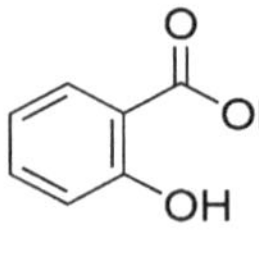


**Fig. 10.6** Salicylic acid chemical structure. (Photo credits: CC BY-SA 3.0 free source)

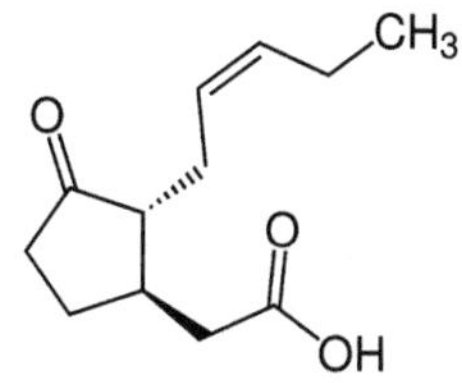


**Fig. 10.7** Chemical structure of jasmonic acid. (Photo credits: CC BY-SA 4.0 free source)

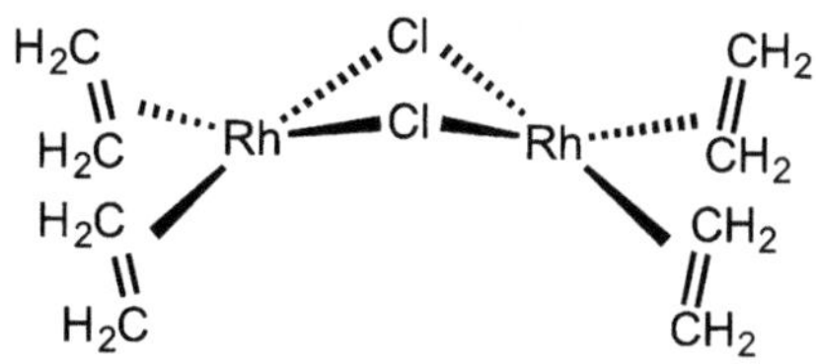


**Fig. 10.8** Chemical structure of ethylene. (Photo credits: CC BY-SA 4.0 free source)

herbivores, such as predators and parasites. For instance, jasmonic acid is produced when the tobacco budworm (*Helicoverpa virescens*) feeds on plants. Additionally, jasmonic acid promotes the production of antimicrobial compounds and activates defense-related genes (Ghorbel et al. 2021).

### *10.3.3 Ethylene (ET)*

Ethylene (ET) (Fig. 10.8), a gaseous hormone, plays a role in various plant responses, including defense against pests. Stressors such as insect herbivores cause the production of ethylene, which in turn triggers the creation of antimicrobial chemicals and activates genes involved in defense. For instance, the Colorado potato beetle, *Leptinotarsa decemlineata*, feeds on plants and produces ET in response. The growth of thicker cell walls caused by ethylene can increase a plant's resistance to insect Khatabi and Schäfer 2012.

### *10.3.4 Abscisic Acid (ABA)*

Abscisic acid is a hormone that plays a key role in plant responses, particularly in stress tolerance and the regulation of seed dormancy. This also contributes to plant defense by encouraging the production of callose and other protective compounds. For example, ABA is synthesized in response to infection by the nematode *Meloidogyne incognita*, which causes root knot disease in tomato plants. Abscisic acid promotes the production of callose, which can block the movement of nematodes through the plant (Bharath et al. 2021).

### *10.3.5 Gibberellins (GAs)*

Gibberellins play a role in various facets of germination as well as the natural process of breaking dormancy. The starch reserves provide energy to the seedling during the early phases of germination, before the photosynthetic machinery develops to a sufficient degree. Four forms of GAs are found, gibberellin A1 (GA1), *ent*-gibberellane, *ent*-kaurene, and gibberellic acid (GA3) (Fig. 10.9).

## 10.4 PSMs' Impacts on Herbivore (Insects and Mites) Physiology and Behavior

PSMs directly impact the growth, development, and survival of the pest in antibiosis; in contrast, chemical factors change the behavior of the insect in antixenosis defense, preventing it from feeding and laying eggs (War et al. 2018). Plant volatiles are crucial for indirect plant defense against insect and mite pests because they can either prevent pests from finding hosts or draw in their natural enemies (Belmain et al. 2013). The impacts of these PSMs could be adverse as well behavioral. These direct effects can damage cells, deform the herbivore's midgut, or act as channel blockers and enzyme inhibitors; whereas behavioral effects involve deterring or repelling herbivores, attract natural enemies, or disrupt the herbivore's ability to find hosts and lay eggs (Wink 2010, 2018; Divekar et al. 2022).

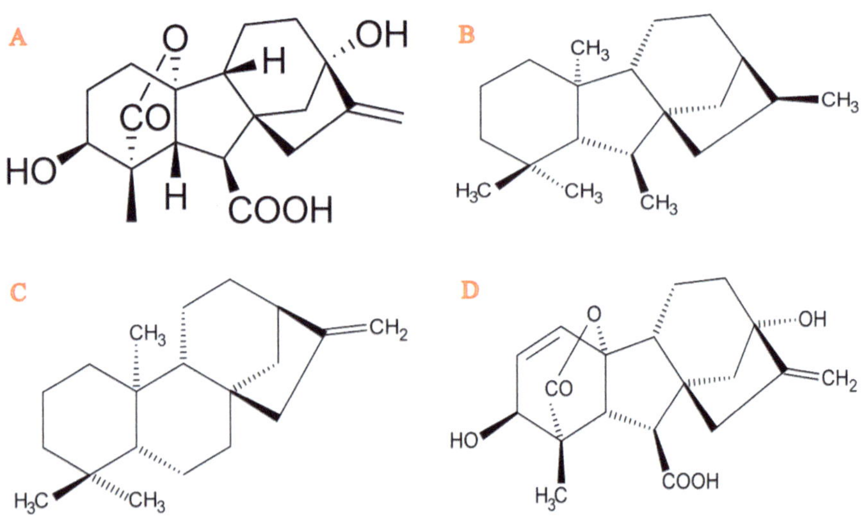


**Fig. 10.9** Chemical structure of gibberellins: (**a**) gibberellin A1, (**b**) *ent*-gibberellane, (**c**) *ent*-kaurene, and (**d**) gibberellic acid. (Photo credits: CC BY-SA 4.0 free source)

There are various PSMs that are fatal to the insects' herbivores. Alkaloids are detrimental to pests since they cause disruption of their cellular enzyme activities, nervous system, protein synthesis, and DNA replication (Cheng and Cheng 2015; Divekar et al. 2022). For example, terpenes can cause toxicity and deter feeding, while some phenolics interfere with digestion, slow growth, inhibit enzyme activity and cell division, or have an unpleasant taste (Wink 2015; Priyanka et al. 2021). The interactions between plant secondary metabolites (PSMs) and insect or mite pests are complex and can vary based on the specific compound, the pest species, the host plant, and environmental conditions.

The nitrogen-containing substances frequently build up in seeds due to their capacity to fix nitrogen from the atmosphere. Nitrogen can be found in secondary metabolites such as amines, cyanogenic glucosides, alkaloids, and non-protein amino acids (Al-Ani et al. 2022). Fabaceae plants have the capacity to create more nitrogen-containing secondary metabolites (Wink 2013).

Regarding the effects of alkaloids, there are numerous reports detailing the actions of secondary metabolites at the cellular or subcellular level; Glycoalkaloids are a family of steroidal chemicals found in potatoes and other plants in the Solanaceae family (Friedman 2006). These compounds have the ability to disrupt cellular membranes or decrease the function of acetylcholinesterase. Chowański et al. (2016) covered the bioactivity of economic and ecological species of the Family Solanaceae including alkaloid metabolites and crude extracts that possess insecticidal and acaricidal effect.

Also, Caffeine is one of the most recognized alkaloids of the methylxanthine family, due to its repellent activity, disrupting the herbivore metabolism and making it more susceptible to other toxins; A previous study has shown that caffeine has reduced the number of fruits attacked by *Hypothenemus hampei* (Ferrari) (Coleoptera: Curculionidae: Scolytinae) and *Monalonion velezangeli* Carvalho and Costa (Hemiptera: Miridae) by 70%. Notably, no phytotoxic effects were observed in coffee plants (Góngora et al. 2023).

In addition, nicotine, a major alkaloid found in *Nicotiana* spp., a family of tobacco plants, acts as a neurotoxin, disrupting the nervous system of the herbivore and ultimately leading to paralysis and death (Kim and Isman 2017). *Nicotinea* spp. is effective against a variety of insect pests, including aphids, beetles, and caterpillars. Additionally, *Tetranychus urticae* Koch (Acari: Tetranychidae) was not exposed to nicotine's harmful effects according to Akyazi et al. (2018). Among the alkaloids that modify neuronal signal transduction include erythrina alkaloids, tubocurarine, ergot alkaloids, muscarine, agroclavine, and theophylline (Lyubenova et al. 2023).

Pietrosiuk et al. (2003) assessed the effects of Pyrrolizidine Alkaloids derived from *Lithospermum canescens* on certain biological features of *T. urticae*; the treatments led to nymph mortality, a reduction in female reproductive capacity, and a loss in lifespan. Compounds that contain sulfur are another highly aromatic category that is employed to protect plants. These secondary metabolites are separated into two classes based on their synthesis. The hydrolysis of the myrosinase enzyme results in the production of glucosinolate in the first group. Members of the family Crucifereae display this pathway (Ahmed et al. 2017). The glucosinolate–myrosinase pathway is present in cabbage, nasturtium, and broccoli. The hydrolysis of the alliinase enzyme results in the production of alliin, which is commonly found in the genus Allium and belongs to the second group of secondary metabolites that also includes sulfur (Durán-Lara et al. 2020). According to Nwachukwu and Asawalam's (2014) research, *Allium sativum* exhibited a harmful effect in *Sitophilus zeamais*. According to Fatima et al. (2015), *Tetranychus evansi* in tomato crops can be suppressed by garlic extract. When applied to the plants, garlic extract was found to be effective against aphids, whiteflies, weevils, and mites, as well as larvae of Lepidoptera (Durán-Lara et al. 2020; Golubkina et al. 2022). According to Attia et al. (2011), extract from garlic leaves reduced female fertility and resulted in high mortality in T. urticae. It was discovered that *T. urticae* is fatally affected by the essential oils of *Tanacetum vulgare* L. and *Artemisia absinthium* L. (Chiasson et al. 2001). Elisovetskaya and Nastas (2013) have established the high insecticidal activity of *Veratrum lobelianum* Bernh. extract against the imago and larvae of *Leptinotarsa decemlineata*. It has been shown that *Galleria mellonella* larvae are repulsed by the extract.

A small but diversified family of sulfur-containing amino acid derivatives, glucosesinolates are produced from glucose itself. According to Hopkins et al. (2009), these chemicals are mostly restricted to members of the Brassicaceae family. Certain molecules function as allelochemicals, shielding plants against herbivorous organisms (Lobo et al. 2018). Indole glucosinolate prevents *T. urticae* from ovipositing on Arabidopsis (*Arabidopsis thaliana*), hence providing protection (Widemann et al. 2021).

Plant phenolic compounds encompass approximately 10,000 metabolites (Goldberg 2003). According to Lobo et al. (2018), key subgroups of phenols include simple phenols (phenolic acids and coumarins) and polyphenols (flavonoids and nonflavonoids). Phenolic compounds directly affect herbivores by acting as toxins that disrupt their growth and development, and by serving as feeding deterrents. Within plant cells, most phenolic compounds are bound to sugars to reduce their endogenous toxicity. These compounds can function as antioxidants, structural polymers (like lignin), attractants (such as flavonoids and carotenoids), UV protectants (flavonoids), signal molecules (salicylic acid and flavonoids), and defense response chemicals (like tannins and phytoalexins).

Terpenes represent the most diverse group of secondary metabolites. In plants, they play crucial roles in physiological and ecological processes, including allelopathy, insecticidal and acaricidal activity, attracting predators and pollinators, and the production of plant hormones (Tiwari and Rana 2015). Terpenoids, particularly monoterpenoids and sesquiterpenoids, are the primary volatile components of essential oils (EOs) (Doughari 2015). Terpenes can be toxic and act as feeding deterrents. Additionally, some phenolic compounds hinder digestion, slow growth, inhibit enzyme activity and cell division, or simply taste unpleasant. Alkaloids can disrupt ion channels, inhibit enzymes, or interfere with neurotransmission, leading to symptoms such as loss of coordination, vomiting, and even death.

Neem, *Azadirachta indica* A. Juss, (Meliaceae), is a notable source of biopesticides, standing out among the 2400 botanicals used for this purpose globally (Adhikari et al. 2020). Neem seed kernels contain significant amounts of limonoids, with azadirachtin being the most potent, demonstrating pesticidal effects (Hashmat et al. 2012; Khan et al. 2015; Acharya et al. 2017).

Neem's systemic properties make it highly effective in plant protection, providing defense for up to ten weeks against harmful pests in various crops such as rice, maize, wheat, barley, sugarcane, tomatoes, cotton, and brinjal. For example, in rice, pest infestations by yellow stem borer, green leaf hopper, gall midge, thrips, and leaf folder were reduced by 73.13%, 75.12%, 69.93%, 79.73%, and 85.57%, respectively, depending on treatment (Mohapatra et al. 2018). Many Lepidoptera species, including *Plutella xylostella* (Yponomeutidae), *Spodoptera litura*, *Sesamia calamistis*, *Heliothis armigera* (Noctuidae), *Ostrinia nubilais* (Crambidae), and *Eldana saccarina* (Pyralidae), are highly sensitive to azadirachtin, exhibiting strong anti-feeding behaviors (Muhammad and Kashere 2020).

Zhang et al. (2020) noted that jasmonic acid (JA) plays a significant role in plant–herbivorous arthropod interactions. To investigate how pepper (*Capsicum annuum* L.) responds to herbivory, they studied the leaf transcriptomes and metabolomes of two genotypes with different susceptibility to the two-spotted spider mite (TSSM), *T. urticae*. Mite infestations triggered both JA and salicylic acid (SA) signaling, but with distinct transcriptome profiles. Compared to JA, mites induced fewer differentially expressed genes in metabolic pathways, except for those involved in the phenylpropanoid and lipid metabolic processes. Zhang et al. hypothesized that the absence of SA-mediated suppression of JA-induced defenses might lead to increased resistance to spider mites in pepper.

Chemically, essential oils (EOs) are complex mixtures of alkaloids, flavonoids, isoflavones, monoterpenes, phenolic acids, carotenoids, and aldehydes, unique to each plant and extraction method. As of 2019, there were 2025 articles on insecticidal/insect repellent bioactivity and 140 articles on acaricidal bioactivity (Fierascu et al. 2019) (Table 10.1).

## 10.5 Tritrophic Interactions

The tritrophic interactions between host plants, herbivores, and predatory or parasitoid species have been metaphorically described as "a cry for help" by Dicke and Sabelis (1988b) who explored the question, "How do plants recruit predators as bodyguards?" to better understand these interactions (Dicke and Baldwin 2010).

Plants emit volatile compounds in response to herbivorous arthropods feeding on them. These herbivore-induced plant volatiles (HIPVs) consist of volatile organic compounds (VOCs), a unique class of metabolites naturally released by plants for communication and signaling purposes. As a result, HIPV being demonstrated to have an active physiological reaction, and is thought to have the ability to draw in herbivorous predators as natural opponents (Kutty and Mishra 2023). Blends of

**Table 10.1** Plant secondary metabolites (not commercialized) with defensive roles against herbivores

| PSMs type/group | Chemical components | Plant family | Herbivores (insect/mite species) | Physiological/behavioral effects | Reference |
|---|---|---|---|---|---|
| Alkaloids | Glycoalkaloids<br>Tropane alkaloids and ergot alkaloids<br>Nicotine, caffeine, quinine, quinine. Solanine, tomatine, and cucurbitacins | Solanaceae<br>Erythroxylaceae<br>Asteraceae<br>Convolvulaceae, Brassicaceae, Proteaceae,<br>Liliaceae<br>Euphorbiaceae<br>Rhizophoraceae<br>Ranunculaceae | *Leptinotarsa decemlineata, Manduca sexta*<br>*Diaprepes abbreviatus, Myzus persicae, Aphis gossypii*, *Frankliniella occidentalis*, *Trialeurodes vaporariorum, Sitophilus* spp | Feeding deterrents, growth inhibitors, interfere with neurotransmission, block enzyme action | Saunders et al. (1992), Hollingsworth et al. (2002), Adeyemi (2010), Rattan (2010), Ibanez et al. (2012), Pavela (2016), and Benelli et al. (2016) |
| Phenolic acids | Caffeic, chlorogenic | All plants | | Cause oxidative damage, browning in fruits and wine | |
| Coumarins | Umbelliferone | Carrots, parsnip | | Cross-link DNA, block cell division | |
| Terpens | Monoterpenes (carvone, 1,8-cineole, cuminaldehyde, linalool, eugenol and menthol, Pyrethrins), Sesquiterpenes (Hemigossypol)<br>Diterpens (gossypol)<br>Triterpens (Cucurbitacin-C, limonene) tetraterpenoid (carotene) | Asteraceae, Lamiaceae<br>Cucurbitaceae, Rutaceae, Meliaceae | *Sitophilus oryzae (L.)* (Coleoptera: Curculionidae)<br>*Myzus persicae, Aphis gossypii* glover (Hemiptera: Aphididae) *Frankliniella occidentalis,*<br>*Trialeurodesvaporariorum,*<br>*Glyphodes pyloalis* Walker (Lepidoptera: Pyralidae)<br>*Tetranychus urticae* Koch (Acari: Tetranychidae) | Repellent herbivore by masking the plant's odor causing irritation and damage the tissues, disrupt the nervous system, toxic | Benelli and Mugnaini (2016), Kim and Isman (2017), Kortbeek et al. (2019), Liu and Ren (2020a, b), Derbalah et al. (2021), Zhang and Chen (2021a, b), and Souza et al. (2022) |

| | | | | | |
|---|---|---|---|---|---|
| Lignans | Podophyllin urushiol | Mayapple poison ivy | | Cathartic, vomiting, allergic dermatitis | |
| Flavonoids | Anthocyanin, catechin | Almost all plants | | Flower, leaf color; inhibit enzymes, anti- and pro-oxidants, estrogenic | |
| Tannins | Gallotannin, condensed tannin | Oak, hemlock trees, birdsfoot trefoil, legumes | | Bind to proteins, enzymes, block digestion, antioxidants | |
| Lignin | Lignin | All land plants | | Structure, toughness, fiber | |

HIPV exhibit both qualitative and quantitative variation depending on the species of plants and herbivores that interact. Therefore, one would wonder if this enables natural enemies to distinguish between volatiles from herbivore species-infested plants that are either appropriate or inappropriate for the natural enemy to eat (Xiao et al. 2012).

Dicke et al. (1999) noted that predatory mites from the Phytoseiidae family encounter herbivore-induced plant volatiles (HIPVs) under various conditions, and their responses are context-dependent. For example, when a phytoseiid has experienced prolonged starvation, it is more likely to respond to cues from a less-preferred prey species compared to when it is well-fed. Factors such as predator and prey dispersal also influence these responses (Sabelis and Dicke 1985).

Additionally, Dicke et al. (1990) identified specific chemicals that attract the predatory mite *Phytoseiulus persimilis* to *Tetranychus urticae*-infested Lima bean plants (*Phaseolus lunatus*). These compounds include terpenoids such as (E)-β-ocimene, 4,8-dimethyl-1,3 (E), 7-nonatriene, linalool, and the phenolic methyl salicylate. These volatiles were produced by the bean plants in response to herbivory, but not by spider mites or mechanically damaged plants. Additionally, *P. persimilis* has shown attraction response to HIPV emitted from different *T. urticae*-infested plant systems; e.g., cucumber, *Cucumis sativus* L. (Takabayashi et al. 1994), tomato, *Lycopersicon esculentum* L., pear, *Pyrus communis* L., and rose, *Rosa X hybrida* (Dicke and Sabelis 1988b), gerbera, *Gerbera jamesonii* Bolus ex Hooker f. (Krips et al. 1999), and citrus, *Citrus* spp. (Cruz-Miralles et al. 2021).

Similar evidence of the tritrophic plant-mite systems, that conducted predatory species attraction to *T. urticae*-infested plants by HIPVs have been found in case of *Amblyseius swirskii* Athias-Henriot to Lima bean, *P. lunatus* (Dicke and Sabelis 1988a), *A. andersoni* Chant in both lima beans, *P. lunatus* (Dicke and Groeneveld 1986), and beans, *Phaseolus vulgaris* L. (Koveos et al. 1995), and the predatory species, *N. californicus* and *Euseius stipulatus* (Athias-Henriot) in citrus (Cruz-Miralles et al. 2021).

In addition to the attraction of phytoseiid mites to various herbivore-infested plants, studies have shown similar behavior in other predatory species. For example, *Amblyseius andersoni* is attracted to the western flower thrips (WFT), *Frankliniella occidentalis* infesting Lima bean plants (Dicke and Groeneveld 1986). Predatory mites like *Neoseiulus californicus* and *Amblyseius limonicus* respond to Cassava green mite (*Mononychellus progresivus*) (Janssen et al. 1990), while *A. andersoni*, *Typhlodromus pyri*, and *Euseius finlandicus* are attracted to apple trees infested with European red mites (*Panonychus ulmi*) and Apple rust mites (*Aculus schlechtendali*) (Takabayashi et al. 1991).

During the soybean growing season, many predatory seven-spotted lady beetles (*Coccinella septempunctata*) were observed on soybean plants heavily infested by soybean aphids (*Aphis glycines*). Zhu and Park (2005) investigated whether the beetle's attraction was due to specific volatiles emitted by aphid-infested soybean plants. Their gas chromatography-mass spectrometry analysis revealed distinct differences between volatiles from aphid-infested plants and undamaged ones. Methyl salicylate was significantly higher in infested plants at both V1 and V2 growth stages. However, (D)-limonene and (E,E)-α-farnesene release patterns varied by growth stage.

Electrophysiological tests confirmed that methyl salicylate elicited strong responses from *C. septempunctata*. In field tests, methyl salicylate-baited traps were highly attractive to adult *C. septempunctata*, while 2-phenylethanol attracted lacewings (*Chrysoperla carnea*) and syrphid flies. The multi-colored Asian lady beetle, *Harmonia axyridis* showed no preference for these compounds. These findings suggest that *C. septempunctata* uses methyl salicylate as an olfactory cue to locate prey. Additionally, field tests showed that traps baited with benzaldehyde attracted significantly higher numbers of alate soybean aphids.

**Tritrophic Interactions: The Defensive Cascade**
The true sophistication of PSM-based defense is fully realized at the tritrophic level, where plants recruit the natural enemies of their herbivores. As outlined in Sect. 10.3, herbivore attack triggers specific hormonal signatures, most notably the jasmonate pathway (Koo and Howe 2009). This signaling cascade does not only induce direct toxins; critically, it also upregulates the synthesis and emission of volatile organic compounds (VOCs), a major class of PSMs (Dudareva et al. 2006). These VOCs, including terpenoids, green leaf volatiles, and phenolic derivatives, form an airborne chemical "cry for help" that is detectable by predatory and parasitic insects.

For example, aphid infestation on broad bean plants induces the emission of methyl salicylate. This volatile is highly attractive to the seven-spot ladybird beetle (*Coccinella septempunctata*), a key aphid predator (El-Wakeil et al. 2003; Zhu and Park 2005). This demonstrates how an herbivore-induced volatile, methyl salicylate—a phenolic compound related to the hormone salicylic acid (SA)—can serve as a key foraging cue for specific predators, thereby completing a defense cascade initiated by plant hormones. Similarly, maize plants damaged by lepidopteran larvae release a blend of terpenoids that attract parasitic wasps, which then oviposit in the herbivore hosts (El-Wakeil et al. 2003; Turlings and Erb 2018). Through these mechanisms, plants actively shape their ecological environment, turning herbivore pressure into a signal that triggers top-down control, enhancing their fitness. The precise composition of the VOC blend can convey specific information about the herbivore's identity and severity of attack, allowing natural enemies to make efficient foraging decisions (Dicke and Baldwin 2010).

## 10.6 Conclusions

Besides their role in defense, PSMs have other purposes such as attracting pollinators and dispersing seeds. To attract pollinators, plants produce compounds with distinctive colors or odors, which are recognized by pollinators but are toxic and deters sap-flower feeders. Some PSMs are antioxidants, nitrogen storage compounds, and protect the plant from UV.

Plant secondary metabolites (PSMs) influence various biological targets, such as proteins, biomembranes, nucleic acids, and enzymes. They disrupt neuronal signal transduction, causing significant physiological and behavioral changes. Many PSMs can alter the bioactivity and three-dimensional structure of proteins, including specific inhibitors and phenolic compounds. Lipophilic PSMs affect the fluidity and permeability of biomembranes, while saponins can lyse membranes by forming complexes with membrane cholesterol. Additionally, antimicrobial peptides interfere with the biomembranes of both microbes and eukaryotic cells.

Plant secondary metabolites (PSMs) interact with nucleic acids and the enzymes responsible for their metabolism, including DNA intercalating and alkylating compounds. Intercalators stabilize DNA, inhibiting the function of helicases and RNA polymerases, which can result in mutagenicity, genotoxicity, and cytotoxicity. Alkylating agents bind directly to nucleotide bases, also causing mutations and genotoxic effects. Overall, the pharmacological and toxicological properties of PSMs are complex and varied, as different compounds target a wide range of biological processes.

**Declaration of AI Assistance** The authors used the artificial intelligence tools DeepSeek-V3, and Gemini-Google in a limited and assistive capacity to improve the linguistic phrasing and clarity of the following sections of the manuscript: the introduction, and the key elements of the chapter. Hereby, we declare that AI assistance was made only for enhancing sentence's structure and grammatical errors. All outputs were reviewed, edited, and their accuracy was carefully verified by the authors themselves, who bear full responsibility for the final content of the work. The chapter was subsequently peer-reviewed by specialist reviewers.

## References


Acharya P, Mir SA, Nayak B (2017) Competence of biopesticide and neem in agriculture. Int J Environ Agric Biotechnol 2(6):2958–2964. https://doi.org/10.22161/ijeab/2.6.23

Adeyemi MMH (2010) The potential of secondary metabolites in plant material as deterents against insect pests: a review. Afr J Pure Appl Chem 4:243–246

Adhikari K, Bhandari S, Niraula D, Shrestha J (2020) Use of neem (Azadirachta indica A. Juss) as a biopesticide in agriculture: a review. J Agric Appl Biol 1(2):100–117. https://doi.org/10.11594/jaab.01.02.08

Ahmed E, Arshad M, Khan MZ, Amjad MS, Sadaf HM, Riaz I, Sabir S, Ahmad N (2017) Secondary metabolites and their multidimensional prospective in plant life. Pharmacogn Phytochem 6:205–214. Available at https://www.phytojournal.com/archives/view-pdf/1152/6-2-2

Akyazi A, Soysal M, Altunc YE, Lisle A, Hassan E, Akyol D (2018) Acaricidal and sublethal effectsof tobacco leaf and garlic bulb extract and soft soap on *Tetranychus urticae*. Syst Appl Acarol 23:2054–2069

Al-Ani E, Al-Khazraji H, Al-Shaibani M (2022) The use of alkaloids as botanical insecticide. Res J Microbiol Biotechnol 2(2):33–41

Attia S, Grissa KL, Mailleux AC, Lognay G, Heuskin S, Mayoufi S, Hance T (2011) Effective concentrations of garlic distillate (*Allium sativum*) for the control of *Tetranychus urticae* (Tetranychidae). J Appl Entomol 136:302–312

Belmain SR, Haggar J, Holt J, Stevenson PC (2013) Managing legume pests in sub-Saharan Africa: challenges and prospects for improving food security and nutrition through agro- ecological intensification. Natural Resources Institute, University of Greenwich, Chatham Maritime. 34p

Benelli G, Mugnaini A (2016) The role of terpenoids in plant-insect interactions. Phytochemistry 124:1–14

Benelli G, Canale A, Toniolo C, Higuchi A, Murugan K, Pavela R, Nicoletti M (2016) Neem (Azadirachta indica): towards the ideal insecticide? Nat Prod Res 31:369–386

Bharath P, Gahir S, Raghavendra AS (2021) Abscisic acid-induced stomatal closure: an important component of plant defense against abiotic and biotic stress. Front Plant Sci 4(12):615114. https://doi.org/10.3389/fpls.2021.615114

Bourgaud F, Gravot A, Milesi S, Gontier E (2001) Production of plant secondary metabolites: a historical perspective. Plant Sci 161:839–851. https://doi.org/10.1016/S0168-9452(01)00490-3

Cheng F, Cheng Z (2015) Research progress on the use of plant allelopathy in agriculture and the physiological and ecological mechanisms of allelopathy. Front Plant Sci 6:1020. https://doi.org/10.3389/fpls.2015.01020

Chiasson H, Bélanger A, Bostanian N et al (2001) Acaricidal properties of Artemisia absinthium and Tanacetum vulgare (Asteraceae) essential oils obtained by three methods of extraction. J Econ Entomol 94(1):167–171. https://doi.org/10.1603/0022-0493-94.1.167

Chowański S, Adamski Z, Marciniak P, Rosiński G, Büyükgüzel E, Büyükgüzel K, Falabella P, Scrano L, Ventrella E, Lelario F, Bufo S (2016) A review of bioinsecticidal activity of Solanaceae Alkaloids. Toxins 8(3):60. https://doi.org/10.3390/toxins8030060

Cruz-Miralles J, Cabedo-López M, Guzzo M, Pérez-Hedo M, Flors V, Jaques JA (2021) Plant defense responses triggered by phytoseiid predatory mites (Mesostigmata: Phytoseiidae) are species-specific, depend on plant genotype and may not be related to direct plant feeding. BioControl 66:381–394. https://doi.org/10.1007/s10526-021-10077-8

Derbalah A, Keratum A, Darweesh M, Elebiary, Mostafa, Hegazy (2021) New trends for controlling *Sitophilus oryzae* concerning adult mortality, offspring production, mode of action, and grain quality. https://doi.org/10.1007/s00003-021-01339-9

Dicke M, Baldwin IT (2010) The evolutionary context for herbivore-induced plant volatiles: beyond the 'cry for help'. Trends Plant Sci 15(3):167–175. https://doi.org/10.1016/j.tplants.2009.12.002

Dicke M, Groeneveld A (1986) Hierarchical structure in kairomone preference of the predatory mite *Amblyseius potentillae*: dietary component indispensable for diapause induction affects. Ecol Entomol 11:131–138. https://doi.org/10.1111/j.1365-2311.1986.tb00287.x

Dicke M, Sabelis MW (1988a) Infochemical terminology: Should it be based on costbenefit analysis rather than origin of compounds? Funct Ecol 2:131–139

Dicke M, Sabelis M (1988b) How plants obtain predatory mites as bodyguards. Neth J Zool 38: 148–165. https://doi.org/10.1163/156854288X00111

Dicke M, Sabelis MW, Takabayashi J, Bruin J, Posthumus MA (1990) Plant strategies of manipulating predator-prey interactions through allelochemicals: prospects for application in pest control. J Chem Ecol 16(11):3091–3118. https://doi.org/10.1007/BF00979614

Dicke M, Gols R, Ludeking D, Posthumus MA (1999) Jasmonic acid and herbivory differentially induce carnivore-attracting plant volatiles in lima bean plants. J Chem Ecol 25:1907–1922. https://doi.org/10.1023/A:1020942102181

Divekar PA, Narayana S, Divekar BA, Kumar R, Gadratagi BG, Ray A, Singh AK, Rani V, Singh V, Singh AK (2022) Plant secondary metabolites as defense tools against herbivores for sustainable crop protection. Int J Mol Sci 23:2690. https://doi.org/10.3390/ijms23052690

Doughari JH (2015) An overview of plant immunity. J Plant Pathol Microbiol 6(11):322. https://doi.org/10.4172/2157-7471.1000322

Dudareva N, Negre F, Nagegowda DA, Orlova I (2006) Plant volatiles: recent advances and future perspectives. Critic Rev Plant Sci 25(5):417–440

Durán-Lara EF, Valderrama A, Marican A (2020) Natural organ-ic compounds for application in organic farming. Agriculture 10(2):41. Article 2. https://doi.org/10.3390/agri-culture10020041

Elisovetskaya D, Nastas T (2013) Biological activity of the extract of *Veratrum lobelianum* bernh. against harmful species of insects and mites and its impact on entomophages. Oltenia J Stud Natural Sci 29(1):185–192

El-Wakeil NE, Bernal J, Vidal S (2003) Effects of jasmonate applications on pest and natural enemy recruitment in BT and non BT-cotton fields. In: Proceeding of the world cotton research conference-3, from 9 to 13 March 2003. Cape Town, South Africa, pp 1239–1248

Fatima K, Lovejoy T, Wisdom K (2015) Efficacy of garlic (*Allium sativum*) and red chilli pepper (*Capsicum annum*) extracts in the control of red spider mite (Tetranychus urticae) in tomatoes (*Lycopersicon esculentum*). Am J Appl Sci 10(2):124–131. https://www.appliedsciencesjournal.com/pdfs/volume-10-issue-2/6.pdf

Fierascu R, Fierascu I, Dinu-Pirvu C, Fierascu I, Alina P (2019) The application of essential oils as a next-generation of pesticides: recent developments and future perspectives. Zeitschrift fur Naturforschung. C, J Biosci 75. https://doi.org/10.1515/znc-2019-0160

Friedman M (2006) Potato glycoalkaloids and metabolites: roles in the plant and in the diet. J Agric Food Chem 54:8655–8681

Fürstenberg-Hägg J, Zagrobelny M, Bak S (2013) Plant defense against insect herbivores. Int J Mol Sci 14:10242–10297. https://doi.org/10.3390/ijms140510242

Gebretsadik KG, Zhang Y, Chen J (2022) Screening and evaluation for Antixenosis resistance in wheat accessions and varieties to grain aphid, *Sitobion miscanthi* (Takahashi) (Hemiptera: Aphididae). Plants (Basel). 11(8):1094. https://doi.org/10.3390/plants11081094

Ghorbel M, Brini F, Sharma A et al (2021) Role of jasmonic acid in plants: the molecular point of view. Plant Cell Rep 40:1471–1494. https://doi.org/10.1007/s00299-021-02687-4

Goldberg G (2003) Plants: diet and health. The report of a British Nutrition Foundation Task Force. Blackwell, Oxford, p 347

Golubkina N, Zayachkovsky V, Sheshnitsan S et al (2022) Prospects of the application of garlic extracts and sele-nium and silicon compounds for plant protection against herbivorous pests: a review. Agriculture 12(1):64. https://doi.org/10.3390/agriculture12010064

Góngora CE, Tapias J, Jaramillo J, Medina R, González S, Restrepo T, Casanova H, Benavides P (2023) A novel caffeine Oleate formulation as an insecticide to control Coffee Berry Borer, Hypothenemus hampei, and other coffee pests. Agronomy 13:1554. https://doi.org/10.3390/agronomy13061554

Guerriero G, Berni R, Muñoz-Sanchez J, Apone F, Abdel-Salam E, Qahtan A, Faisal M (2018) Production of plant secondary metabolites: examples, tips and suggestions for biotechnologists. Genes 9(6):309. https://doi.org/10.3390/genes9060309

Hashmat I, Azad H, Ahmed A (2012) Neem (*Azadirachta indica* A. Juss)-A nature's drugstore: an overview. Int Res J Biolog Sci 1(6):76–79

Hollingsworth RG, Armstrong JW, Campbell E (2002) Caffeine as a repellent for slugs and snails. Nature 417:915–916

Hondelmann P, Paul C, Schreiner M, Meyhöfer R (2020) Importance of antixenosis and antibiosis resistance to the cabbage Whitefly (*Aleyrodes proletella*) in Brussels sprout cultivars. Insects 11(1):56. https://doi.org/10.3390/insects11010056

Hopkins RJ, van Dam NM, van Loon JJA (2009) Role of glucosinolates in insect-plant relationships and multitrophic interactions. Annu Rev Entomol 54:57–83

Hussein RA, El-Anssary AA (2019) Plants secondary metabolites: the key drivers of the pharmacological actions of medicinal plants. Herbal Med. https://doi.org/10.5772/intechopen.76139

Ibanez S, Gallet C, Despres L (2012) Plant insecticidal toxins in ecological networks. Toxins 4: 228–243

Janssen A, Hofker CD, Braun AR, Mesa N, Sabelis MW, Bellotti AC (1990) Preselecting predatory mites for biological control: the use of an olfactometer. Abstr Bull Entomol Res 80(2):177–181. https://doi.org/10.1017/S0007485300013390

Kant MR, Jonckheere W, Knegt B, Lemos F, Liu J, Schimmel BCJ, Villarroel CA, Ataide LMS, Dermauw W, Glas JJ, Egas M, Janssen A, Van Leeuwen T, Schuurink RC, Sabelis MW, Alba JM (2015) Mechanisms and ecological consequences of plant defence induction and suppression in herbivore communities. Ann Bot 115(7):1015–1051. https://doi.org/10.1093/aob/mcv054

Khan J, Anjum SI, Khan I, Rehman FU, Khan A (2015) Larvicidal & development retarding effects of hexane crude extract of *Otostegia limbata* on 3rd instar larvae of *Drosophila melanogaster* meign (Diptera: Drosophilidae). J Entomol Zool Stud 3(1):06–09

Khatabi B, Schäfer P (2012) Ethylene in mutualistic symbioses. Plant Signal Behav 7:1634–1638. https://doi.org/10.4161/psb.22471

Kim JH, Isman MB (2017) Menthol: a versatile natural insecticide. Pest Manag Sci 73(1):4–16

Koo AJ, Howe GA (2009) The wound hormone jasmonate. Phytochemistry 70(13–14):1571–1580

Kortbeek RWJ, van der Gragt M, Bleeker PM (2019) Endogenous plant metabolites against insects. Eur J Plant Pathol 154:67–90

Koveos DS, Kouloussis NA, Broufas GD (1995) Olfactory responses of the predatory mite *Amblyseius andersoni* chant (Acari, Phytoseiidae) to bean plants infested by the spider mite *Tetranychus urticae* Koch (Acari, Tetranychidae). J Appl Entomol 119:615–619. https://doi.org/10.1111/j.1439-0418.1995.tb01345.x

Krips OE, Willems PEL, Gols R, Posthumus MA, Dicke M (1999) The response of *Phytoseiulus persimilis* to spider mite-induced volatiles from Gerbera: influence of starvation and experience. J Chem Ecol 25(12):2623–2641. https://doi.org/10.1023/A:1020887104771

Kutty NN, Mishra M (2023) Dynamic distress calls: volatile info chemicals induce and regulate defense responses during herbivory. Front Plant Sci 14:1135000. https://doi.org/10.3389/fpls.2023.1135000

Liu J, Ren F (2020a) Caffeine as an insecticidal agent: from mode of action to application. Front Plant Sci 11:1004

Liu J, Ren F (2020b) Terpenes as insecticidal agents: from mode of action to application. Front Plant Sci 11:1251

Lobo MG, Hounsome N, Hounsome B (2018) Biochemistry of vegetables. In: Handbook of vegetables and vegetable processing, pp 47–82. https://doi.org/10.1002/9781119098935.ch3

Lyubenova A, Georgieva L, Antonova V (2023) Utilization of plant secondary metabolites for plant protection. Biotechnol Biotechnol Equip 37:1. https://doi.org/10.1080/13102818.2023.2297533

Mohapatra SD, Raghu S, Prasanthi G, Baite MS, Prabhukarthikeyan SR, Yadav MK, Banerjee A, Aravindan S, Patil N, Chatterjee S (2018) Bio-ecology of rice insect pests and diseases: paving the way to climate-smart rice protection technologies. ICAR-National Rice Research Institute: Cuttack Odisha, India

Muhammad A, Kashere MA (2020) Neem, *Azadirachta indica* L. (A. Juss): an eco-friendly botanical insecticide for managing farmers' insects pest problems – a review. FUDMA J Sci 4(4):484–491. https://doi.org/10.33003/fjs-2020-0404-506

Ninkuu V, Zhang L, Yan J, Fu Z, Yang T, Zeng H (2021) Biochemistry of terpenes and recent advances in plant protection. Int J Mol Sci 22(11):5710. https://doi.org/10.3390/ijms22115710

Nwachukwu ID, Asawalam EF (2014) Evaluation of freshly pre-pared juice from garlic (Allium sativum L.) as a biopesti-cide against the maize weevil, *Sitophilus zeamais* (motsch.) (coleoptera: curculionidae). J Plant Protect Res 54(2):132–138. https://doi.org/10.2478/jppr-2014-0021

Pavela R (2016) History, presence and perspective of using plant extracts as commercial botanical insecticides and farm products for protection against insects—a review. Plant Prot Sci 52:229–241

Pietrosiuk A, Furmanowa M, Kropczyńska D, Kawka B, Wiedenfeld H (2003) Life history parameters of the two-spotted spider mite *Tetranychus urticae* feeding on bean leaves treated with pyrrolizidine alkaloids. J Appl Toxicol 23:187–190

Polumackanycz M, Kaszuba M, Konopacka A, Marzec-Wróblewska U, Wesolowski M, Waleron K, Buciński A, Viapiana A (2020) Phenolic composition and biological properties of wild and commercial dog rose fruits and leaves. Molecules 25:5272. https://doi.org/10.3390/molecules25225272

Priyanka SL, Navya RN, Chandrasekaran M, Hasan W (2021) Role of secondary plant metabolites against insects. In: Hi-tech crop production and pest management. Biotech Books, New Delhi, 17 Pages
Rattan RS (2010) Mechanism of action of insecticidal secondary metabolites of plant origin. Crop Prot 29:913–920. [CrossRef]
Sabelis MW, Dicke M (1985) Long-range dispersal and searching behavior. In Helle W, Sabelis MW (eds) Spider mites, their biology, natural enemies and control. World Crop Pests (pp 141–160). Elsevier, Amsterdam
Sampedro L, Moreira X, Zas R (2011) Costs of constitutive and herbivore-induced chemical defences in pine trees emerge only under low nutrient availability. J Ecol 99:818–827
Santiago R, López-Malvar A, Souto C, Barros-Ríos J (2018) Methods for determining cell wall-bound Phenolics in maize stem tissues. J Agric Food Chem 66:1279–1284
Saunders JA, O'Neill NR, Romeo JT (1992) Alkaloid chemistry and feeding specificity of insect herbivores. In: Alkaloids: chemical and biological perspectives. Springer Science and Business Media LLC, Berlin/Heidelberg, pp 151–196. https://link.springer.com/chapter/10.1007/978-1-4612-2908-7_2 (25 December 2025)
Smith CM (2005) Plant resistance to arthropods. Springer, Dordrecht, pp 19–100
Souza LP, Zuim V, Stinguel P, Pinheiro PF, Zago HB (2022) Toxicity of essential oil of *Mentha piperita* (Lamiaceae) and its Monoterpenoid menthol against *Tetranychus urticae* Kogan 1836 (Acari: Tetranychidae). An Acad Bras Cienc 94(suppl 4):e20200427. https://doi.org/10.1590/0001-3765202220200427
Takabayashi J, Dicke M, Posthumus MA (1991) Variation in composition of predator-attracting allelochemicals emitted by herbivore-infested plants: relative influence of plant and herbivore. Chemoecology 2:1–6
Takabayashi J, Dicke M, Posthumus MA (1994) Volatile herbivore-induced terpenoids in plant-mite interactions: variation caused by biotic and abiotic factors. J Chem Ecol 20:1329–1354. https://doi.org/10.1007/BF02059811
Tiwari R, Rana CS (2015) Plant secondary metabolites: a review. Int J Eng Res Gen Sci 3(5):661–670
Turlings TC, Erb M (2018) Tritrophic interactions mediated by herbivore-induced plant volatiles: mechanisms, ecological relevance, and application potential. Ann Review Entomol 63:433–452
War AR, Taggar GK, Hussain B, Taggar MS, Nair RM, Sharma HC (2018) Plant defense against herbivory and insect adaptations. Plants Basel 11(8):1094
Widemann E, Bruinsma K, Walshe-Roussel B, Rioja C, Arbona V, Saha RK, Letwin D, Zhurov V, Gómez-Cadenas A, Bernards MA, Grbić M, Grbić V (2021) Multiple indole glucosinolates and myrosinases defend Arabidopsis against Tetranychus urticae herbivory. Plant Physiol 187(1):116–132. https://doi.org/10.1093/plphys/kiab247. PMID: 34618148; PMCID: PMC8418412
Wink M (2008) Plant secondary metabolism: diversity, function and its evolution. Nat Product Communic 3. https://doi.org/10.1177/1934578X0800300801
Wink M (2010) Introduction: biochemistry, physiology and ecological functions of secondary metabolites. Ann Plant Rev 40:1–19
Wink M (2013) Evolution of secondary metabolites in legumes (fabaceae). S Afr J Bot 89:164–175. https://doi.org/10.1016/j.sajb.2013.06.006
Wink M (2015) Modes of action of herbal medicines and plant secondary metabolites. Medicine 2:251–286. https://doi.org/10.3390/medicines2030251
Wink M (2018) Plant secondary metabolites modulate insect behaviour steps towards addiction? Front Physiol 9:364. https://doi.org/10.3389/fphys.2018.00364
Xiao Y, Wang Q, Erb M, Turlings TC, Ge L, Hu L, Li J, Han X, Zhang T, Lu J, Zhang G, Lou Y (2012) Specific herbivore-induced volatiles defend plants and determine insect community composition in the field. Ecol Lett 15(10):1130–1139. https://doi.org/10.1111/j.1461-0248.2012.01835.x

Yang F, Zhang Q, Yao Q, Chen G, Tong H, Zhang J, Li C, Su Q, Zhang Y (2020) Direct and indirect plant defenses induced by (Z)-3-hexenol in tomato against whitefly attack. J Pest Sci 93: 1243–1254. https://doi.org/10.1007/s10340-020-01234-6

Zhang Y, Bouwmeester HJ, Kappers IF (2020) Combined transcriptome and metabolome analysis identifies defence responses in spider mite-infested pepper (Capsicum annuum). J Exp Bot 71(1):330–343. https://doi.org/10.1093/jxb/erz422

Zhang L, Chen Z (2021a) Terpenes as insecticidal agents: a review of recent advances. J Agric Food Chem 69(16):5350–5362

Zhang L, Chen Z (2021b) Potential of plant alkaloids for insect pest control. Pest Manag Sci 75(1): 5–19

Zhu J, Park KC (2005) Methyl salicylate, a soybean aphid-induced plant volatile attractive to the predator *Coccinella septempunctata*. J Chem Ecol 31(8):1733–1746. https://doi.org/10.1007/s10886-005-5923-8

# Chapter 11
# Efficient of Plant Metabolic Products on Insects and the Associated Entomopathogenic Organisms

**Mamdouh I. Nassar and Hassan A. Dawah**

## 11.1 Introduction

The living organisms must deal with environmental and biotic problems throughout their lives. The strongest species have the most genetic ability to defend or fight in natural habitats, plants are constantly subjected to a variety of biological and abiotic stresses, including herbivore or pathogen, drought, salt, ultraviolet irradiation, high temperatures, and depletion (Freeman and Beattie 2008). According to estimates, phytophagous insects damage many of global agricultural production each year. Plants produced many of secondary metabolites as a resistant mechanism against phytophagous insects (Teklić et al. 2021). There are many types of phytogenous insects, some of them are parasitic or burrowing in plants. The type of crop damage an insect pest is liable to inflict is usually determined by these two features.

Whole plant tissue from the shoot, root, fruit, and flowers is torn off and swallowed by the chewing insects' mouthparts. The ability of the roots to absorb water and nutrients, the shoot tissues' capacity to retain carbon during photosynthetic reactions, and the plant's capacity to create seeds are just a few of the essential functions that this feeding activity can substantially impair mechanically. Many chewing insects feed on the surface of the plant and often spread to neighboring plants or different parts of the same plant. Leaves miners stem borers, and root borers are among the chewing insect species that have one or more life stages inside the tissues of plants. Depending on the species, larvae or nymph may burrow into the plant tissues, but adult females deposit their eggs deep within. Particularly when a

M. I. Nassar (✉)
Cairo University, Faculty of Science, Giza, Egypt
e-mail: mamdouh@sci.cu.edu.eg

H. A. Dawah
National Museum of Wales, Cardiff, UK

 
N. El-Wakeil, A. Abdelkhalek (eds.), *Microbiomes and Plant Secondary Metabolites for Insect Pest Management*, Progress in Biological Control 23,
https://doi.org/10.1007/978-3-032-24330-0_11

plant is heavily plagued with insects, these lesions and the removal of plant cell contents can cause significant direct damage. Furthermore, piercing sucking insects are significant carriers of germs and viruses that cause plant diseases. Lepidoptera, whose caterpillar larvae are almost uniformly and ancestrally phytophagous, and Coleoptera beetles, who's larval and phytophagy adult are undergone several evolutionary changes. Thousands of phytophagous species have found in four insect orders: the Hemiptera, which feed only on fluids by piercing and sucking; the Diptera, with a variety of feeding styles; and two groups with the chewing feeding habit, the sawflies and wood wasps within the Hymenoptera and the main phytophagous order of Orthoptera, e.g., Locust and Grasshoppers.

Plants have evolved several adaptive strategies through physiological changes in response to herbivore attacks. Plant secondary metabolites are synthesized to provide defensive functions and regulate defense signaling pathways to safe guard plants against herbivores. Herbivore injury initiates complex plant reactions which ultimately lead to synthesis and accumulation of PSMs. Phytohormones and other signaling molecules interact to control the formation of these metabolites. Hormonal defense signaling pathways can be directly induced by plant volatile metabolites, which are produced under herbivore attack. In a habitat and pest situation that is changing swiftly, secondary metabolites help plants detect herbivore attacks and react promptly. According to a number of studies, insect herbivores have little chance of adapting to or developing resistance to secondary metabolites. These compounds directly poison insect pests and activate plant antixenosis mechanisms against herbivores insects. Multitrophic interactions between host plants, herbivores, natural enemies, and pollinators are regulated by plant secondary metabolites. The metabolic engineering of biosynthetic pathway intermediates to increase plant resistance to herbivores will be made by molecular understanding of the regulatory genes and proteins involved in the manufacture of secondary metabolites. Compared of many plant species that have been reported, contain lots of PSMs have been isolated and characterized (Pimentel and Peshin 2014). Secondary metabolites build up more when exposed to herbivores and function as signaling molecules to enhance the expression of defensive response genes. Plant secondary metabolites guarantee and signaling plants, survival protectant and competitiveness under stressful situations (Khare et al. 2020). These SMs includes, alkaloids, terpenes, amines, glucosinolates, cyanogenic glucosides, quinones, phenolics, peptides, and polyacetylenes (Divekar et al. 2022). PSMs are essential for adaptability and defense against herbivores, which produced by a number of intermediates in plant defense. Plants' regular growth and development are actively regulated by primary metabolites. Despite having distinct functions, these plant metabolites are connected because primary metabolites serve as building blocks for the production of secondary metabolites (Schweiger et al. 2014). Recent genetic and chemical investigations have demonstrated the multifunctional nature of PSMs, which act as potent regulators of plant growth, defense and primary metabolism. Induced plant defenses are driven by the phytohormones jasmonic acid (JA) and salicylic acid (SA), and the associated pathways interact in a complicated fashion at the transcript and protein level. Adverse effects on the survival of chewing insects, *Heliothis virescens* and sucking insects, *Myzus persicae* were reported after JA and SA application (Katz et al. 2015). Secondary metabolites may also have hormone-like properties by binding to specific receptor proteins (Mason and Singer 2014). There is a synergistic effect of PSMs operating together to tackle herbivore damage. A combination of these secondary metabolites is likely to prevent or delay the development of resistance by insect herbivores (War et al. 2012). Although the role of secondary metabolites in plant defense is well established, in addition some metabolites are used to attract insect pollinators and parasitoids. Secondary metabolites are seen as not only a cost effective and ecologically friendly means to sustain agriculture, but they also compete with agrochemicals in terms of plant growth and protection.

Multiple strategies for improved crop resistance to insect pests, especially strategies relating to plant secondary metabolism, immunity and microbiome science, are becoming available. Recent advances in metabolic engineering of plant secondary chemistry offer the promise of specific toxicity or deterrence to insect pests; improved understanding of plant immunity against insects provides routes to optimize plant defenses against insects. The microbiomes of insect pests can be exploited, either as a target or as a vehicle for delivery of insecticidal agents. Implementation of these advances will be facilitated by ongoing advances in plant breeding and genetic technologies (Erb et al. 2021).

The main purpose of this article is to explore three broad routes for the discovery of novel sources of resistance to insects: to customize plant secondary metabolites (PSMs) for specific toxicity to insect pests with relation to entomopathogenic organisms, to harness plant immunity against insect pests, and to target the microbiome of insect pests. Some aspects of these topics are supported by well-developed research disciplines, and others are more speculative to provide context for assessing the opportunities and limitations of these novel approaches for sustainably maintaining plant health.

## 11.2 Application of Plant Secondary Metabolites as Defense Tools Against Herbivore Insects for Long-Term Crop Protection

In order to use plant secondary metabolites (PSMs) as an environmentally friendly and sustainable pest management option, it is vital to study their biosynthesis and implementation. PSMs are a key defensive mechanism used by plants to protect themselves against a variety of plant-eating insects. The significance of plant secondary metabolites in defense has been clarified thanks to recent analytical instruments and procedures (Stevenson et al. 2017). Terpenoids, alkaloids, phenolics, cyanogenic glucosides, and acylsugars are the five classes into which these metabolites are divided (Table 11.1).

The most often demonstrated pesticidal metabolites are terpenoids generated from isoprene, which are followed by alkaloids and phenolic compounds.

**Table 11.1** Types of plant secondary metabolites (PSMs)

| PSMs | Chemical Feature |
|---|---|
| Terpenoids | Monoterpenes, e.g., (citronellole)<br>Sesquiterpenes, e.g., (β caryophyllene)<br>Triterpenes, e.g., (azadirachtin) |
| Alkaloids | Morphine, caffeine, nicotine, and cocaine |
| Phenolica | Flavonoids, e.g., quercetin<br>Polyphenols, Lignin, e.g., resveratrol anthocyanine |
| Cyanogenic<br>Glucosides | Cyanogenic glycosides, e.g., dhurrin<br>Non-protein amino acids, e.g., azetidine<br>Glutathione<br>Glucosinolates<br>Isothiocyanates<br>Leptin |
| Acylsugars | Nicotiana<br>Aliphatic acylchains |

Importantly, defense against insects can also be dependent on the manufacture of defense proteins or polymers, such as cuticular waxes, which are metabolites of acylsugars and cyanogenic glycosides (Mrid et al. 2021). Over 200,000 distinct structures have been calculated to represent the divergence of specialized metabolites (Kuzina et al. 2009). Their precursor pathways, which differ from primary metabolism and include hormones, fatty acids, and amino acids, are comparatively conserved among plant species. Nevertheless, the diversity of the endproducts is mostly due to isomeric configuration, skeleton rearrangements, and modifications of the final product, which can be flexible (Douglas 2015). In order to safeguard post-harvest goods, the plant-derived metabolites are defined by their effects on insects, which may not be as applicable in the field.

### 11.2.1 Terpenoids

Terpenoids, often referred to as isoprenoids, are the well-characterized group of chemical substances in the plant kingdom. They serve as the foundation for a wide range of fundamental and specialized metabolic compounds. In addition to structural cell components like colors like carotenoids and relatively tiny and volatile compounds like mono and sesquiterpenes, these structures also contain hormones like cytokinins, abscisic acid, and gibberellins.

The identical isoprene components, such as isopentenyl diphosphate (IPP) and its isomer dimethylallyl diphosphate (DMAPP), are used to make all terpenoids. It is believed that about 26,000 distinct terpenoids are created in the plant kingdom overall (Aharoni et al. 2005). Plant species may be able to swiftly respond to current herbivorous challenges thanks to their high biosynthetic flexibility. Regarding insects, terpenoids are likewise among the most often reported bioactive substances.

By acting as repellents, volatile allelochemicals might stop or lessen plant-insect interaction and, consequently, the spread of viral diseases. For example, the silverleaf whitefly Bemisia tabaci, a begomovirus vector, is repelled by the sesquiterpene 7-epizingiberene from Solanum habrochaites (Bleeker et al. 2011b).

When the biosynthetic pathway was introduced into a tomato cultivar, 7-epizingiberene was produced, which reduced the plant's appeal to whiteflies and appeared to be harmful to spider mites, Tetranychus urticae (Balkema-Boomstra et al. 2003). (E)-β-caryophyllene is another well-characterized repellent sesquiterpene. Increased (E)-β-caryophyllene emission from the flowers and a reduction in the citrus psyllid Diaphorina citri, a vector of the pathogen that causes Citrus Greening Disease, are the results of overexpression of the β-caryophyllene synthase in Arabidopsis plants (Hall et al. 2013). A recent research found that the fruit's peel has a number of citrus accessions that produce the repellent (E)-β-caryophyllene, which might be used in breeding efforts (Zhang et al. 2017).

Tree resin contains concentrated combinations of volatile and non-volatile terpenes. As an example, the Norway spruce (Picea abies) accumulates terpenoids in its resin when attacked by the bark beetle, a natural adversary of the tree, Ips typographus (Schiebe et al. 2012; Erbilgin et al. 2006). The beetle does less harm to trees in a natural population that collect large levels of eucalyptol and D-limonene in their resin, indicating that these substances have poisonous and repulsive qualities. The interaction between the Sitka spruce, P. sitchensis, which stores the diterpene dehydroabietic acid for protection, and the white pine weevil, Pissodes strobe, seems to be influenced by these resin-based terpenoids (Robert et al. 2010). Nicotiana attenuate, a wild tobacco, synthesizes terpene derivatives known as diterpene 17-hydroxygeranyllinalool glucosides. Diterpene glucosides were reduced when a GGPP synthase that generates the diterpene precursor molecule geranylgeranyl pyrophosphate was silenced; primary metabolites were unchanged, but M. sexta larvae development increased tenfold on silenced plants (Heiling et al. 2010). These transgenic plants also fared far worse in field tests when attacked by other herbivores, including grasshopper and mirid insects.

Moreover, they are believed to inhibit insect steroid receptors (Alquezar et al. 2017). Triterpenes that influence insects include cucurbitacins, which are common specialized metabolites in cucurbit species. An oxygenated triterpene called cucurbitacin-C works well against spider mites in cucurbits, but it also seems to promote the growth of Spodoptera exigua larvae (Balkema-Boomstra et al. 2003). Water-soluble compounds called glycosylated triterpenes, or saponins, are believed to play a part in plant defense against diseases and herbivores (Jeschke et al. 2016). The saponin-aglycones hederagenin cellobioside and oleanolic acid cellobioside were discovered to be responsible for the observed resistance to the flea beetle Phyllotreta nemorum (Kuzina et al. 2009; Shinoda et al. 2002) due to natural variation in saponin production in a population of winter cress, Barbarea vulgaris. In several different studies, saponins were associated with resistance to the diamondback moth, Plutella xylostella, and they were found to be effective feeding deterrents against Brassicacea specialists.

### 11.2.2 Alkaloids

All plant species generate alkaloids, and many different structures have been found. These include cocaine, nicotine, caffeine, and morphine. Alkaloids are non-volatile chemical compounds that have a nitrogen base organized in a cyclic ring structure (Ashihara et al. 2008). They usually originate from nitrogen-containing compounds of the plant's basic metabolism, such as purins, putrescine, argenine, and phenylalanine. The most well-known alkaloid, nicotine, has been shown to be essential for protecting tobacco plants from a variety of native herbivorous insects, such as Spodoptera exigua and Diabrotica undecimpunctata. Feeding damage from a variety of insects increased threefold when the nicotine production pathway in N. attenuate decreased (Ashihara et al. 2008). Some other insects, like Helicoverpa zea and the caterpillar Manduca sexta, which not only tolerate nicotine but actively use the metabolite to deter their natural adversary, flourish on tobacco. The poisonous alkaloids found in the leaves of nightshade plants, such potatoes and Solanum tuberosum, are also well-known. The tubereating larvae of the Guatemalan potato moth, Tecia solanivora, are poisonous to the glycol-alkaloids α-solanine and α-chaconine (Karlsson et al. 2013).

α-solanine and α-chaconine, which are found in a number of wild potato accessions, are acetylated into leptines, a metabolite that has a significant impact on the growth of the Colorado potato beetle, Leptinotarsa decemlineata (Smith et al. 2009). Additionally, leptin production was effectively transmitted to potato hybrids, resulting in the successful introduction of insect resistance (Rangarajan et al. 2000). On the other hand, the Colorado potato beetle grew more when a critical step in the potato's alkaloid production pathway was blocked. The main alkaloids found in tomatoes are called tomatines, which are spyrosolane-type glycoalkaloids made up of dehydrotomatine and α-tomatine. There is no proof that tomatines have an insecticidal effect in nature, despite the fact that several studies on their insecticidal properties have been published. All of the findings are based on experiments using artificial diets (Barbour and Kennedy 1991).

### 11.2.3 Phenolics

Phenolics are molecules that have hydroxylated benzene rings. L-phenylalanine is converted to cinnamic acid by the vital enzyme L-phenylalanine ammonia-lyase (PAL). Flavonoids, anthocyanins, benzoic acids, and polyphenols like lignins are among the thousands of phenolics that have been identified as a result of this conversion, which takes place when the L-phenylpropanoid route enters (Wu et al. 2015). Flavonoids are polymers that include sugar conjugates, tannins, and aglycones. Even while the herbivore hormone jasmonic acid (JA) increases the formation of anthocyanins, flavonols, and anthocyanins themselves don't seem to have any direct insecticidal effects.

In addition, they are precursors of tannins, which are poisonous and insect-repelling substances that produce reactive oxygen species when they are oxygenated in the insect's stomach (Barbehenn and Constabel 2011). The observed effect on aphid fecundity was not associated with the increased concentration of anthocyanins at the sorghum feeding location (Costa-Arbulú et al. 2001).

Additionally, anthocyanin and polyphenolic buildup after the expression of a particular MYB transcription factor (AtMYB75/PAP1) seems to influence herbivore performance (Malone et al. 2009). Further investigation, however, showed that the precursor flow was rewired to include quercetin and kaempferol, which produced large amounts of anthocyanins but hindered the synthesis of kaempferol-3,7-dirhamnoside.

Since phenols are toxic to herbivorous insects like Lepidoptera, the metabolic pathway was in charge of plant defense. The phenolic compounds rutin, vanillic acid, and synaptic acid were formed in pepper Capsicum annuum as a result of feeding by the larvae of the leafworm Spodoptera litura. This changed both the behavior of the adult moth and the growth of the larvae. Nonetheless, prophenoloxidases in insect stomachs seem to be widely used for the efficient detoxification of various phenolics (Movva and Pathipati 2017).

## *11.2.4 Acylsugars*

It was discovered in the 1960s that Nicotiana gossei, a wild tobacco species, generated a toxic exudate that included substances that would later be referred to as acylsugars. Acylsugars are sugar molecules, often sucrose or glucose, that include one or more esterified C2-C12 aliphatic acylchains in the ring structure (Thurston and Webster 1962).

Acylsugars, which are produced in the Solanaceae family and can accumulate from the leaf dry weight up to 20% when released to the leaf surface, have a significant negative effect on a variety of insects. Exudates of acylsugars from wild potatoes. It has been demonstrated that S. berthaultii affects the sap-feeding green peach aphid. Myzus persicae (Neal et al. 1990) and shield N. attenuata from Manduca sexta, a gnawing insect. However, acrylsugar, which is produced by the tomato species S. galapagense, S. cheesmaniae, S. pimpinellifolium, and S. pennellii, affects a wide variety of insects, including many thrips species, B. tabaci, Tuta absoluta, and T. urticae. Even though acrylsugar's synthesis and chemical structure are relatively straightforward, a single extract has a vast array of different AS structures (Rakha et al. 2017). A plant's acylsugar composition can be complicated and varied due to variations in the sugar backbone, acylchain length, number, and location. The abundance of acrylsugar present in many plant species today is the consequence of genetic divergence among biosynthetic genes, which also altered enzymatic activity, enzyme efficiency, and substrate affinity (Smeda et al. 2018).

The mortality rate of B. tabaci and aphid Myzus persicae is influenced by the backbone, number, and length of acylchains, according to bioassays using synthetic and fractionated acrylsugar. Fractionated acylsugar extracts from wild tomatoes revealed a difference in mortality between acylglucoses with C5 acylchains and fractions with a mixture of C5/C10/C12 acylchains (Mithen 2001).

### 11.2.5 Cyanogenic Glucosides

The breakdown of amino acids produces cyanogenic glucosides, which are stable and relatively non-toxic sugar-linked cyanide groups that are stored in reserve within the vacuole. During cell disintegration, or herbivory, they combine with β-glycosidases, which cleave off the sugar moiety to produce instable cyanohydrins. Hazardous hydrogen cyanides (HCN) are produced during hydrolysis, which can happen spontaneously or as a result of α-hydroxynitrilase activity (Gleadow and Møller 2014).

It is thought that some plant species actively synthesize cyanogenic glucoside as part of their defense mechanism, even though all plants make some HCN as a consequence of producing ethylene. Similarly, Brassica plants, sometimes referred to as the mustard oil bomb, use myrosinases to hydrolyze glucosinolates to produce isothiocyanates (ICN) (Gleadow and Møller 2014). Instead of acting as poisons, cyanogenic glucosides are typically employed in defensive mechanisms as deterrents to insect eating. As phloem feeders, insects that are able to stop tissue damage avoid releasing chemicals called cyanogenic glucosides. This might be partly due to the fact that certain insects can detoxify cyanogenic glucoside and either store HCN as a protection against predators or utilize it as a carbon source (Pentzold et al. 2015). Numerous insect herbivores, both specialist and generalist, seem to be able to tolerate cyanogenic plants (Nair and Padmavathy 2014). Plutella xylostella, a crucifer specialist, employs cyanogenic glucosides as host signals for oviposition and is capable of detoxifying glucosinolates. Cyanogenic glucosides, however, could be useful in crop breeding. For instance, the development of the pest herbivore Pieris rapae was suppressed by greater ICN levels in Brassica olerace, or cabbage (Sun et al. 2009; Agrawal and Kurashige 2003). Additionally, high amounts of dhurrin, a cyanogenic monoglycoside, led flea beetle, Phyllotreta nemorum larval survival to decline to 2% in comparison to the nontransgenic control when Sorghum cyanogenic glucoside biosynthesis genes were expressed ectopically in Arabidopsis plants. The poplar plants react to herbivory by their natural enemy, the gypsy moth larva Lymantria dispar, by inducing the development of cyanogenic glucoside-related P450s. In order to discourage the gypsy moth caterpillar, the authors showed that this leads to the production of 2-methylbutyronitrile and 3-methylbutyronitrile (Pentzold et al. 2015; Jeschke et al. 2016).

## 11.3 Metabolites Involved in Secretion and Regulation Secretory Structures

The body of a plant can include a variety of structures, such as glandular trichomes (GTs), peltate glands, secretory chambers, and glandular ducts, depending on the species (Balcke et al. 2017). It is important to remember that herbivory might affect the formation of secretory structures and the synthesis of metabolites. On the surface of plants, glandular trichomes (GTs) are hairy structures that vary in size and shape. One or more glandular head cells on top of a stalk that attaches to the epidermis are what define GTs. Peltate glands are similar to GTs in that they have glandular cells that emerge from the epidermis as stalkless oil bodies (Balcke et al. 2017). Some studies show that gland cells use sunlight as a decreasing power source while actively metabolizing the sucrose received from the leaf. Although the secretion pathway is unclear, the finished product is then kept in intracellular cavities or subcuticular compartments. GTs and peltates, on the other hand, behave as boil bombs for insects that come into direct contact with them and subsequently come into contact with a high concentration of sticky and toxic exudates. Underlying the epidermal cells are extracellular spaces called secretory cavities, which are surrounded by a network of metabolically active secretory cells (Tissier et al. 2017). The cavity can arise by cell delamination, lysigenous cavities, schizogenous cavities, programmed cell death, or a mixture of these, known as schizolysigenous cavities. The contents of the essential oils found in these cavities are mostly volatile terpenoids, non-volatile terpene derivatives, and phenolics, as previously stated.

Defense substances are transported and secreted to the external environment via glandular ducts, which resemble veins and run through the plant body. Laticifers are these ducts, and secretory ducts are well-known for their function in the production and preservation of rubber (Glas et al. 2012). They grow firmly across the extracellular space between adjacent cells and are very elongated.

By bringing in materials from nearby cells, they create latex, an emulsion that resembles milk. The chemical makeup of latex reflects a variety of metabolic types; sesquiterpenes, triterpenes, phenolics, and alkaloids are commonly present (Sessa et al. 2000). The laticifers shatter and release their poisonous or deterring latex when they are damaged by herbivory. Secretory ducts are extracellular channels that are also created by schizogeny or lysigeny. The epithelial cells load resin into the lumen of these ducts.

Despite being common among vascular land plants, conifers are the focus of most studies, probably because of their high resin content. In response to an insect or pathogen attack, ducts form in a variety of vegetative tissues and enlarge. Resin serves several defensive purposes and is often a viscous combination of phenolics, monoterpenoids, sesquiterpenoids, and diterpene acids (Martin et al. 2002).

The insect is trapped by the sticky resin and the resin solidifies following the evaporation of the sesquiterpenoids, it plugs and closes the wound caused by herbivore assault. Third, as previously said, the resin could include poisonous or repulsive substances that hinder insect growth, eating, and reproduction.

## 11.4 Plant Metabolite Regulation by Transcription

Continuous synthesis of certain metabolites can therefore have a detrimental effect on plant growth and development, particularly when environmental resources are few, such as when photosynthetic capacity is limited. Consequently, it is essential to control the flow via metabolic pathways. Specialized metabolite production is intricately and dynamically regulated, reacting both spatially and temporally to environmental and hormonal signaling changes, such as herbivory (Gang et al. 2001).

In reaction against herbivores, the plant hormone jasmonic acid (JA) increases the production of nicotine (Quintana-Rodriguez et al. 2018). Plant development and various stressors are influenced by the jasmonic acid (JA) signaling branch, which is mostly regulated by myelocytomatosis (MYC) transcription factors. MYC and thylene responsive factors (ERFs), which are crucial regulators of a variety of plant developmental processes and stress responses, govern downstream reactions to jasmonic acid (JA), a crucial signal in herbivore defense (Pré et al. 2008).

When a plant is attacked repeatedly, its defense mechanisms may be primed by prior exposure to herbivorous pests. Furthermore, both transcriptional and post-transcriptional levels can be involved in the control of gene expression. Transcription factors (TFs) and small RNA regulation are the two main actors in these processes. The promoters of biosynthetic genes can be transactivated by transcription factors (TFs), which are regulatory proteins that affect metabolic flux. They frequently only have this action in the secretory structures. It has been discovered that a number of TF families contribute to the buildup of specialized metabolites, either by activating gene expression in response to internal or external stimuli (Spyropoulou et al. 2014). Many of these TFs are members of the WRKY gene family, which is involved in how the body reacts to abiotic stress. It has been demonstrated that SlWRKY7-RNAi in the tomato, Solanum lycopersicum line, is involved in the activation of the monoterpene synthase genes, and that N. attenuata WRKY3-RNAi and WRKY7-RNAi influence the formation of diterpene glycosides in response to herbivory (Neal et al. 1990).

Nematode infection causes WRKY23-RNAi, which is involved in flavonol production, to be expressed in Arabidopsis plants, whereas WRKY33-RNAi binds the promoter of the gene PAD4 that produces camalexin. MYBs are another kind of TF linked to specialized metabolism, and families of basic helix-loop-helix (bHLHs) have been found to govern secondary metabolism in plants. Members of this family are in charge of the phenylpropanoid pathway, which produces flavonoids and anthocyanins (Onkokesung et al. 2014).

For instance, AtMYB75 contributes to caterpillar resistance and regulates the manufacture of flavonoids. R2R3-MYBs frequently form a complex with bHLH TFs and a WD40 domain to exert their influence (Cavallini et al. 2015). In addition to controlling the synthesis of anthocyanins in Nicotiana benthamiana, this complex is reported to have an impact on the build-up of phenolamides and nornicotine conjugates, which are substances involved in herbicide protection (Outchkourov et al. 2014).

## 11.5 Toward the Development of Endogenous Metabolites Against Insects

A strategy that transfers high concentrations of synthetic pesticides to plants that develop their own defensive chemicals may be based on the fact that several kinds of metabolites produced by plants can be efficient in protecting against insects. However, there are several obstacles to overcome when breeding agricultural plants for metabolite-based insect resistance. The crop germplasm should have biosynthetic enzymes, their regulators, biosynthesis and storage structures, and be effectively transferred to a resistant variety in order to achieve biologically metabolite levels.

Introgression of many quantitative trait loci (QTLs) might be the result of this. Insect resistance is also present in crop germplasm; linkage drag, or the introduction of undesirable characteristics for which the underlying genes are situated in the same genomic area as the metabolic QTLs, can significantly complicate interspecific crosses with distant relatives.

Increased acylsugar levels in tomato lines' glandular trichomes due to the introgression of many unlinked wild tomato QTLs are the best instances of breeding for self-defense chemicals. Numerous metabolic QTLs, each with unique inheritance features and a positive or negative effect, underlie the quantitative synthesis of metabolites in that material (Alseekh et al. 2015).

Enhancing herbivore resistance in crops may be accomplished more quickly, effectively, and thoroughly through the use of genetic engineering. Numerous genetic technologies have been introduced, such as the use of RNA interference to target insect-specific sequences, the production of insecticidal Bacillus thuringiensis (Bt) in transgenic crops, and contemporary genome editing tools (like CRISPR-Cas, whose GMO status is still up for debate in the European Commission, EU) (Visweshwar et al. 2015). However, because of legal issues and public debate around the use of genetically modified organisms (GMOs) in agriculture, there aren't many examples of field applications, especially in the EU. The development of specific metabolites by GMOs is exemplified by tomato fruits' monoterpene biosynthesis, which increases resistance to maize earworm larvae. By increasing the quantities of (E)-β-farnesene and limonene in the headspace, overexpressing endogenous terpene synthase 46 (OsTPS46) in rice inhibits the cherryoat aphid. Discrepancies arise because the biological action of allelochemicals is not always evident. More than 30 years ago, it was first reported that the wild potato, S. berthaultii, emits (E)-β-farnesene, which repels the green peach aphid, M. persicae (Harvey et al. 2005). Aphids of M. persicae were prevented from reacting to the alarm pheromone volatile by transgenic Arabidopsis plants that overexpressed a (E)-β-farnesene synthase. Additionally, wheat plants modified for

(E)-β-farnesene synthesis that initially repelled several cereal aphid species under laboratory circumstances showed a loss of the phenotype in field trials, and later research was unable to demonstrate host-plant preference or the influence on insect performance (Elzinga et al. 2014). This example demonstrates the intricate way that insects react to volatiles and demonstrates that herbivore behavior is influenced by a balanced mix of several volatile cues, effective concentrations, and environmental visual cues rather than relying only on one signal.

One significant complicating element that affects the quantity and quality of specialized metabolite synthesis is the environment itself. Because plants cannot escape their surroundings, they must integrate physiology and metabolism in order to respond to rapidly changing biotic and abiotic factors like temperature, osmotic, nutrient, and water status, as well as light quality (photosynthetic carbon fixation and biomass accumulation) (Akula and Ravishankar 2011).

Cardenolide, a toxin included in milkweed latex, was found to be negatively correlated with plant growth rate. However, the compound's origin and the surrounding environment determine how to balance plant development and defense. While triterpenoid saponin levels are reliant on photosynthesis because of the need for nitrogen, the expense of manufacturing phenolics is based on carbon. When considering the possibility of turnover and re-allocation to primary activities, the allocation cost of specialized metabolites may be reduced (Massad et al. 2012).

According to recent studies, there is a regulatory growth defense network that can be uncoupled and, thus, theoretically, be influenced rather than an inherent agreement of resources. Additionally, it is a technique for creating compounds outside of the plant's edible parts, preventing taste or palatability interactions. This method could be suitable for fruit crops like cucumber, tomato, and pepper, but it might not always be suitable for raw leaf crops like lettuce. Additionally, before being adopted, the non-target effect of plant defense compounds on beneficial insects like pollinators and predators should be evaluated.

However, insects have developed tolerance to traditional pesticides and are able to thwart the plant's natural production of anti-insect metabolites. In addition to behavioral adaptation, co-adaptation to toxins, such as metabolite sequestration or product detoxification, is made possible by the adaptability of an insect population's genepool (Opitz and Müller 2009). After ten generations, whiteflies have demonstrated improved survival on nicotine-producing tobacco, as evidenced by the modification of genes related to muscular contraction and the metabolism of carbohydrates (Bleeker et al. 2011).

Additionally, they seem to alter their muscle strength and body volume to prevent becoming physically ensnared in glandular trichome exudates. It can take longer for adaptation to happen if the protection feature is engineered in an inducible insect attack. JA-responsive TFs that trigger particular metabolic pathways are known strategies to do this. In general, high resistance can be a serious problem as, regardless of the chemical source, severe selection pressure might cause an insect to develop resistance. Similar to industrial pesticides, plant-based antibiosis and antixenosis may provide selective pressure, accelerating the evolution of pest resistance (Carrasco et al. 2015).

On the other hand, excessive levels of resistance might not be desirable, and partial resistance could last for a long time. The degree to which a crop variety can tolerate herbivore damage while still yielding enough is known as its resistance mechanism. This is in contrast to sensitive cousins.

## 11.6 Assessing the Production of Secondary Metabolites in Plants

The complicated biosynthesis of plant secondary metabolites (PSMs) prevents these botanical substances from becoming resistant by acting competitively against certain stressors through a variety of mechanisms. The production of PSMs depends on certain routes and substrate enzymes. PSMs are produced using a number of precursors from primary metabolites, including acetyl-CoA, carbohydrates, fatty acids, and amino acids.

The tricarboxylic acid cycle and its routes produce these precursor chemicals. The biosynthesis pathways of primary metabolites are preserved in plants because they are found in all plant tissues. A limited fundamental metabolic framework is the outcome of maintaining these metabolic pathways. Numerous changes in fundamental structures have resulted from metabolic processes such as hydroxylation, glycosylation, methylation, oxidation, acylation, phosphorylation, and prenylation, as well as chemical changes brought about by enzyme customization.

PSM synthesis is a complex yet cooperative process that includes metabolite channeling and metabolon formation. In order to guarantee specialized metabolite manufacturing in a given cell and avoid metabolic interference, metabolites are channeled in various plant cells and cellular compartments (Selmar and Kleinwächter 2013).

Plant secondary metabolite genes are activated and transcribed by a chain of particular transcription factors, as evidenced by the organ, tissue, cell, and developmental stage-specific production of PSM. PSM biosynthesis, transport, and storage are all regulated by transcription factors. While certain alkaloids, such as coniine, caffeine, furanocoumarins, and a few terpenes, are formed in the chloroplast, the majority of biosynthesis routes take place in the cytoplasm. Sterols, sesquiterpenes, and dolichols, on the other hand, are generated in the cytosol or the endoplasmic reticulum (ER).

Isopentenyl diphosphate (IPP) and its isomer dimethyl diphosphate (DMAPP) are used to make terpenes. Terpenes are vital building blocks of sterols, pigments, and other phytohormones. They also act as herbivore deterrents, defense poisons, and allelochemicals (Checker et al. 2018).

Phosphoenolpyruvate and erythrose 4-phosphate combine to form shikimic acid, which is the starting point of the Shikimate pathway. The Shikimate pathway produces the amino acids phenylalanine, tyrosine, and tryptophan, which are building blocks for secondary metabolites that include nitrogen and phenolics. While tryptophan is a precursor of alkaloids, phytoalexins, and indole glucosinolates, phenylalanine is a precursor for phenolics, including flavonoids, lignin, condensed tannins, betalain pigments, and quinones.

The final result of the Shikimate pathway's several stages is chorismate. Enzymes like chorismate mutase, isochorismate synthase, anthranilate synthase, and aminodeoxychorismate synthase control chorismite, which is a precursor of tryptophan, tyrosine, phenylalanine, salicylate, phylloquinone, and folate in higher plants (Neilson et al. 2013). In plants, PSM storage can occasionally surpass 10% of the total dry weight. The vacuole is where water-soluble PSMs are kept. Instead of being trapped in vacuoles, lipophilic PSMs are discharged into trichomes, oil cells, raisin ducts, or cuticles (Opitz and Müller 2009).

Gene duplication and stress-specific adaptations allow for the control of several biosynthetic pathways and the individual enzymes involved. Numerous studies have shown that distinct plants evolved separately and are capable of producing structurally distinct metabolites or comparable chemicals. Consequently, a given PSM's distribution and presence within plants are neither universal nor clearly phylogenetically oriented (Akula and Ravishankar 2011). Research on the intricate cascades that go into the synthesis of PSMs is currently ongoing. Controlling the process of regulating plant defense will be possible with a comprehensive understanding of PSM biosynthesis.

## 11.7 The Function of Secondary Metabolites in Controlling Insect Defense in Plants

In general terms, plants that are bitten by herbivorous insects become more resilient while experiencing slower development. Through a network of phytohormonal signals, this priority is made possible. In response to an attack by the chewing herbivore Chilo suppressalis, the rice transcription factor WRKY70 protein is activated by herbivore-induced mitogen-activated protein kinase signaling. This protein plays a crucial role in preferring defense in growth by negatively regulating gibberellin (GA) levels and positively regulating jasmonic acid (JA) levels (Divekar et al. 2019).

WRKY70-dependent JA biosynthesis activates proteinase inhibitors and confers resistance to C. suppressalis. However, induction of WRKY70 makes rice plants more susceptible to Nilaparvata lugens, a rice brown planthopper (Wang et al. 2008). Research on rice lines lacking GA showed that WRKY70-dependent GA signaling is responsible for the susceptibility to N. lugens. Therefore, choosing defense above development compromises resistance, which has significant ramifications for the management of plant defense.

In order to preserve metabolic energy and avoid self-harm, plants regulate defensive activation. Feedback regulation is widely used to modify defense investment and hormonal networks, including both positive and negative feedback loops included in early defensive signals (Divekar et al. 2019). Including herbivorous insects directly in regulatory feedback might help plants monitor and regulate defense effectively since they can disrupt the production of defense chemicals at numerous levels, including the final stages of biosynthesis.

Secondary metabolites can also assist plants enhance the synergy across many defenses and compensate for defense system deficiencies, as they are defense activation outputs. Due to the fact that many secondary metabolites are stored in inactive forms or segregated, plants can detect damage to herbivore tissue. Plants that have been damaged perceive a range of endogenous chemicals as danger signals, which are referred to as damage-associated molecular patterns (DAMPs). As DAMPs, secondary metabolites would be used. Green-leaf volatiles are secondary metabolites that are also DAMPs (Akula and Ravishankar 2011).

Plant defenses can also be regulated by volatile compounds, such as terpenoids, green leaf volatiles, and aromatic chemicals, in addition to glucosinolates and benzoxazinoids. Most of these volatiles are generated in response to herbivore attacks, and they can directly induce or prime hormonal defense signaling pathways and resistance. Mutants in maize that are unable to produce volatile indole are unable to prepare their systemic tissues to rapidly release terpenes in reaction to herbivore attacks. Adding indole to the headspace of maize plants restores this priming phenotype (Ton et al. 2007).

In response to indole, rice and Oryza sativa plants activate early defensive signaling components, including the map kinase OsMPK3. The fact that transgenic plants devoid of the OsMPK3 gene are insensitive to indole suggests that indole works by triggering early defensive signaling (Strauss et al. 2002). It has now been shown that five different kinds of secondary metabolites—benzoxazinoids, glucosinolates, terpenes, aromatics, and leaf volatiles—may act as plant defense regulators to defend themselves against herbivores.

By triggering defense-related signaling pathways such mitogen activated protein kinase (MAPK) and hormone signaling, which causes the production of several defense-related genes and phytochemicals, they are able to identify and react to intruders. Induced resistance to herbivores is mediated through jasmonic acid (JA), salicylic acid (SA), and ethylene (ET)-mediated signaling (Kumar et al. 2015). To trigger a particular defense, plants need to be able to discriminate between physical harm and herbivorous insects.

The oral secretion fluid of insects has been found to contain specific active substances known as elicitors. They play a key role in the development of downstream signaling cascades related to defense and are recognized by plants. It has also been noted that oral secretions inhibit plant defense mechanisms. Elicitors are classified into six groups according to their chemical makeup and structure: benzyl cyanide, fatty acids, peptides, esters, fatty acid amino acid conjugate (FACs), enzymes, and fatty acids (Gilardoni et al. 2010).

Esters and benzyl cyanide are produced in oviposition fluids during egg laying, whereas oral secretions contain enzymes, fatty acids, FACs, and peptides. The effects and modalities of action of various elicitor types may vary. The first insect elicitor identified and documented was β-glucosidase, an enzyme elicitor from Pieris brassicae regurgitates (Agrawal and Kurashige 2003).

The plant defense is activated and a series of volatiles are released by β-glucosidase in cabbage, lima beans, and corn plants. The parasitoid Cotesia glomerata is drawn to these emitted volatiles. Therefore, P. brassicae's β-glucosidase promotes indirect plant defense. Along with chewing herbivores, sucking insects also exhibit this kind of elicitor. Nilaparvata lugens are primarily composed of β-glucosidase, which increases the levels of ethylene, hydrogen peroxide (H2O2), and jasmonic acid (JA) (Wang et al. 2008).

The parasite Anagrus nilaparvatae is drawn in as a result of the release of volatiles like dodecenal and tetradecane and several downstream signaling cascades. The inactive, glucosidically bound volatiles glucosinolates are separated from β-glucosidases. These two components work together to produce β-glucosidase activity, which causes insect feeding to release volatiles such thiocyanate, isothiocyanate, and nitriles.

The synthesis of defensive metabolites is triggered by a series of interconnected signaling pathways that are triggered by specific elicitors from insect oral secretions (OS) that enter plant tissue through wounds during insect feeding. All eukaryotic organisms have mitogen-activated protein kinases (MAPKs), which are involved in several cellular functions related to both stress reactions and regular growth.

Salicylic acid-induced protein kinase (SIPK) and wound-induced protein kinase (WIPK) are the two forms of MAPKs that are activated in plants by insect OS. The dynamics of phytohormones, ethylene, salicylic acid, and jasmonic acid generated by insect herbivores are actively regulated by MAPKs. Additionally, MAPKs are necessary for the transcriptional activation of genes linked to herbivore protection and the buildup of defensive chemicals.

Signaling mechanisms that control the subsequent reactions to insect herbivores, such as the transcriptional activation of genes encoding enzymes involved in the manufacture of defensive chemicals, sustain induced protection. The most important signaling pathway following insect assault is the activation of MAPKs, which alter phytohormone levels and reorganize the transcriptome and proteome in preparation for plant defense.

All herbivore-induced signaling chemicals that are recognized by the host crop and, as a result, trigger defensive responses are referred to as herbivore insects associated molecular patterns, or HAMPs. The most studied HAMPs for comprehending herbivore-plant interactions are those that contain elicitors from insects. When these elicitors penetrate plant tissue during insect feeding, they trigger insect reactions. N-(17-hydroxy linolenoyl)-L-glutamine (volicitin) was the first elicitor found in the oral secretions of the armyworm Spodoptera exigua (Barrett and Agrawal 2004).

## 11.8 Phytohormones and Proteins Regulate Plant Defense via Secondary Metabolites

Plants can identify hostile organisms and stop them before they do significant damage thanks to phytochemicals and protein-based defenses. In addition to their function in the growth, development, and physiological activities of the plant, phytohormones also trigger defense responses to herbivore assaults. One significant hormone thought to be in charge of initiating both direct and indirect defensive reactions against herbivores is jasmonic acid (JA). When plants are harmed by herbivores, the linoleic acid that is produced from the plants is transformed into active jasmonate through the octadecanoid pathway, which is regulated by a number of enzymes, including lipoxygenase (LOX), and ultimately triggers defense-related gene expression and resistance mechanisms (War et al. 2011). When chewing insects eat plant parts, LOX converts linoleic and linolenic acid to hydroperoxy derivatives, which are further converted by the plastids' allene oxide synthase (AOS) and allene oxide cyclase (AOC) into 12-oxo-phytodienoic acid (12-OPDA). Additionally, plants lacking LOX are vulnerable to insect pests, and it has been demonstrated that many types of crops may fend off significant pests by inducing lipoxygenase activity following insect infection. In order to limit the production of JA signals, Jasmonate Zim-domain (JAZ) proteins also bind to transcription factors, specifically MYC2. In contrast, JA levels will be lower when there is no stress and will rise quickly in response to any stressor, such as herbivory attack event injury (Dombrecht et al. 2007). Salicylic acid (SA), a crucial signaling molecule in plants, is produced from benzoic acid, indicating that plants synthesis SA via two different mechanisms. According to some accounts, the majority of SA is made from isochorismate, although plants use phenylalanine ammonialyase (PAL) to synthesis SA from cinnamonate. But in order to generate salicyclic acid (SA), both routes use the same product, chorismite, which is the byproduct of the Shikimate pathway. When insects pierce and sucking plants, the salicyclic acid pathway is triggered. Larval mining stimulates both the JA and SA defensive pathways, whereas puncturing a leaf-by-leaf minor adults promotes the JA route (Gilardoni et al. 2010).

Insects that pierce and feed on plants produce ethylene. There have been reports of plants accumulating jasmonic acid (JA) and emitting ethylene when they are fed herbivory (Pré et al. 2008). The ethylene pathway interacts with a network of plant hormones to trigger a defensive response rather than acting alone. The induction of plant defenses by insect feeding involves several communication cascades. For instance, Arabidopsis plants are more susceptible to the generalist herbivore Egyptian cotton worm (Spodoptera littoralis) when ethylene signaling is present. Phytohormones interact with one another through a complicated signaling network rather than acting individually. Both antagonistic and synergistic metabolites result from the interaction of the signaling pathways for salicylic acid (SA) and jasmonic acid (JA) (Gilardoni et al. 2010). In order to protect plants against herbivores, secondary metabolites are produced when phytohormones are applied endogenously or

**Table 11.2** Effects of PSM on the physiology and behavior of insect herbivores

| Adverse effect of PSM against insect | |
|---|---|
| Physiological effect behavioral effect | |
| Direct toxic effect leading to death | Avoidance; feeding deterrent and repellent effects on insect herbivores |
| Gut malformation and tissues disruption | Attract natural enemy (predators and parasites of herbivores) |
| Neorotoxic effects: as membrane channel blockers, and inhibit cholenstrase | Interrupt insect host finding behavior |
| Enzyme inhibitors—caffeic acid—trypsin inhibitors | Interruption of oviposition and colonization behavior |

externally (Table 11.2). Attacks by herbivores also trigger the production of other phytohormones, including auxin, gibberellins, cytokinins, and abscisic acid (ABA).

After herbivore assaults, it was discovered that plants' ABA levels increased. JA-driven defensive responses in plants against herbivores work in concert with the formation of abscisic acid (ABA) in plants (Erb et al. 2011). Following fatty acid-amino acid conjugate (FAC) elicitation, Nicotiana attenuata exhibits regulation of cytokinin-related genes, indicating a function for cytokinins in the hormonal regulatory network (Gilardoni et al. 2010). In plants, cytokinin priming raised JA, linolenic acid, and wound-inducible transcript levels while lowering weight in gypsy moth larvae, Lymantria dispar. JA-dependent secondary metabolites, anthocyanins, and phenolamides are formed in stems as a result of Manduca sexta's attack on Nicotiana attenuata, which raises indole-3-acetic acid (IAA) levels in plants (Dervinis et al. 2010). When combined with JA, indole-3-acetic acid (IAA) seems to play a part in plant defensive communication against herbivores. Gibberellins are naturally occurring phytohormones; the first gibberellin to be structurally characterized was gibberellic acid (GA). Significant insect pest resistance is provided by GA-downregulated poplar plants (Takahashi et al. 1955). By regulating the synthesis of plant secondary metabolites (PSMs), our understanding of phytohormones will enable us to boost plant resistance against herbivores.

## 11.9 Impact of Plant Secondary Metabolites on the Physiology and Behavior of Insect

Plants need to respond rapidly to herbivory and produce defense chemicals at the site of the attack. PSMs may repel or deter the feeding or cause direct toxic symptoms leading to inhibition of growth which ultimately results in the death of insects (Table 11.2). Recent molecular tools enable us to detect the target sites of these toxins at cellular or molecular level. Defense response of these PSMs compounds is flexible which modulated by herbivore damage. From these metabolites are constitutive while others are generated after herbivore attacks. Induced defense metabolites are produced only after insect attack. Constitutive defenses rely primarily on carbon-based metabolites, such as terpenes and polyphenols, which contribute significantly to plant dry matter content and are concentrated in specific structures or compartments, such as resin canals in the xylem of coniferous trees and cereal cell walls (Balcke et al. 2017).

PSMs are bioactive substances that hinder insects' ability to digest food and either repel or intoxicate them. As growth inhibitors, feeding deterrents, and neurotransmitter targets that alter neural signal transduction, alkaloids are often disliked by insect pests. Alkaloids hinder changes in concentrations as well as neurotransmitter expression. As a result, insects experience negative physiological and behavioral changes that directly affect them as toxins and repellents.

Nicotine, caffeine, erythrina alkaloids, tubocurarine, ergot alkaloids, muscarine, agroclavine, and theophylline are alkaloids that alter neuronal signal transmission. By inhibiting phosphodiesterase activity against herbivores, the alkaloid caffeine demonstrated insecticidal effects, including paralysis and intoxication, and has the potential to be a biopesticide. Tobacco's nicotine alkaloid was unintentionally employed to control insect pests (Steppuhn and Baldwin 2007).

Asteraceae members' pyrethrum, require seed's azadirachtin, and hot pepper extracts' capsaicin are examples of secondary metabolites that have been employed as insecticides. These metabolites have a variety of effects on the insect, such as blocking nervous system-related receptors and channels, inhibiting cellular respiration, and upsetting the insect's hormonal balance.

A few PSMs have protein-inhibiting properties. Ricine, albrine, lycorine, and emetine, for instance, block the ribosomes' ability to synthesize proteins; few other substances target the structure and function of proteins. Certain PSMs interfere with the process of cell division or target the cytoskeleton of cells. Certain inhibitory substances, including rotenone, sanguinarine, and colchicine, prevent mitotic cell division. By creating many hydrogen and ionic connections with the target organism's proteins, phenolic chemicals alter the three-dimensional structure of proteins.

On resistant maize germplasm, the biological characteristics of the maize stalk borer, Chilo partellus, were impacted, suggesting that the presence of secondary metabolites produced in the plants as a result of larval feeding was the source of the antibiosis (Divekar et al. 2019). Other PSMs are extremely reactive and include unstable functional groups that selectively target the amino, hydroxyl, or sulfhydryl groups of protein amino acid residues, changing the structural and functional properties of the proteins in the process. PSMs, which include steroids, mustard oils, phenylpropanoids, mono, sesqui, di, and triterpenes, are lipophilic. Globular proteins' three-dimensional structure is rearranged by lipophilic terpenes.

These lipophilic PSMs change the fluidity and permeability of biomembranes in addition to attacking the membranes that envelop live cells and those found inside intracellular compartments. DNA intercalation is the result of some PSMs. Nucleotide pairings, such as GC-pairs, are intercalated by PSMs with aromatic, hydrophobic, and planar functional groups. By preventing the actions of helicases and RNA and preventing the transitional phase of DNA replication, this intercalation stabilizes DNA replication.

Additionally, cell death results from frameshift mutations and deletions caused by intercalating PSMs. Protoberberine, benzophenanthridine alkaloids, including quinine, emetine, furanoquinoline alkaloids, furanocoumarins, anthraquinones, and betcarboline alkaloids are among the metabolites that intercalate DNA (Wink 2007). Through direct binding to nucleotide bases and the creation of covalent bonds, DNA intercalating agents induce mutations and genotoxicity.

Herbivore-induced plant volatiles (HIPVs) are a class of small-molecule volatile chemicals that are produced by plants. Terpenoids, benzenoids, phenylpropanoids, volatile fatty acid derivatives, and volatile amino acid derivatives are examples of HIPVs. Many plant pests' natural enemies can discover infected plants and seek victims by using HIPVs as signals.

Because they draw parasitic or predatory insects that harm herbivores, herbivore-induced plant volatiles (HIPVs) constitute an indirect defensive mechanism. Because indole can adversely affect the food consumption and survival of African cotton leafworms, HIPVs can also directly affect insect pests. Whiteflies (Bemisia tabaci) are repelled by the terpenes zingiberene, curcumene, p-cymene, terpinene, and phellandrene found in wild tomato species (Bleeker et al. 2011). It has been reported that volatile terpenes, like 7-epizingiberene (sesquiterpene) from Solanum habrochaites, repel the silver leaf whitefly Bemisia tabaci. This reduces plant-insect contact and, ultimately, the spread of begomovirus-caused viral diseases, allowing for their effective management. The plant became less appealing to whiteflies and poisonous to spider mites (Tetranychus urticae) after 7-epi-zingiberene biosynthesis pathways were introduced into tomato cultivars (Bleeker et al. 2011).

## 11.10 Functional of PSMs in Mediating the Multitrophic Interactions

Interactions between insect herbivores and their natural enemies are mediated by plant secondary metabolites (PSMs). The color or scent of flowers attracts insect pollinators; flavonoids, anthocyanins, or carotenoids provide color, while terpenoids, amines, and phenylpropanoids release distinct scents that bees can smell.

Pollinators are rewarded with sugar-rich nectar from flowers, which they prefer over other flowering materials. Since nectar serves as pollinators' main reward, it is undesirable for defensive metabolites to be present in plant nectar. PSMs can help pollinators and plants, according to recent research (Stevenson et al. 2017). PSMs produced to protect plants from herbivores are similar to those in pollen and nectar from flowers.

This is because a single plant species' chemical defenses are controlled by distinct biochemical pathways, and the plant generates a particular set of secondary metabolites to carry out different functions. It is not well understood how metabolites interact to provide floral rewards and protect plants. Phytochemicals such as caffeine, aconitine, nicotine, thymol, linalool, lupanine, and grayanotoxins aid in plant defense and pollinator attraction (Steppuhn and Baldwin 2007). Unexpected results came from studies looking at how coffee affected bees. The flower nectar of Coffea and citrus species contains caffeine. In a dose-dependent manner, caffeine prevents honeybees from eating. The nectar's caffeine content, however, was below the bees' threshold for taste recognition. Caffeine enhanced honeybees' ability to use floral scent to find a nectar source, ultimately enhancing their memory and olfactory learning. PSMs present in pollen and nectar from flowers can aid bees in fending off external diseases and halting their spread (Ashihara et al. 2008).

They can lower the load of pathogen infections in foraging bees and successfully combat bee stomach parasites. The intestinal parasite Crithidia bombi in eastern bumblebees (Bombus spp.) is effectively reduced by a number of phytochemicals present in nectar, such as iridoid glycosides, terpenoids, and alkaloids (Akutse et al. 2014). Floral characteristics including size, pollen concentration, and nectar volume have been shown to be influenced by insect herbivory, which encompasses root herbivory, folivory, and florivory. PSMs alter the physiology, behavior, parasitism, and predation of herbivores' natural enemies.

Herbivore predators and parasitoids use volatiles released following herbivore damage to find host insects. When these metabolites are conveyed by herbivores, who preferentially store and sequester them for self-defense, they negatively impact the efficacy of natural enemies. According to recent data, some predators may acquire tolerance or resistance to cardenolides and glucosinolates. The synthesis or processing of PSMs was altered in order to quantify resistance in natural enemies.

The predatory lacewing, Chrysoperla carnea, may develop slowly without any negative impacts on its biology when glucosinolate sulfatase is inhibited to promote the buildup of poisonous isothiocyanate in the larvae of the diamondback moth, Plutella xylostella (Sun et al. 2009). When predators and parasitoids feed on specialized herbivores, they may become resistant to PSMs. Alongside this resistance, there is an increase in predatory potential. By acting as host locating cues or inhibiting natural enemy performance, PSMs can either strengthen or weaken biological control. It is expected that herbivore development will increase their exposure to parasites and predators. Extending a vulnerable growth stage can boost parasitism since parasitoids target a host's developmental stage, but it also has some effects on predation if predators eat a wide range of food. An empirical test of the slow growth-high mortality hypothesis revealed that herbivore hosts with slower development were more parasitic. The rice brown planthopper, Nilaparvata lugens, is repelled by linalool, which also draws in its natural enemies. But it's still unclear how important a particular molecule is for repelling herbivores and attracting natural enemies.

Phytochemicals from the ribwort plantain, Plantago lanceolate, cause the specialized herbivore Melitaea cinxia and its lone endoparasitoid, Hyposoter horticola, to develop more quickly when cultivated on a line with high iridoid glycosides. On the other hand, the generalist herbivore Spodoptera exigua's pupal weight and the adult mass of its lone endoparasitoid, Cotesia marginiventris, were significantly decreased when cultivated on a high iridoid line (Wang et al. 2008).

Several detrimental effects on the biological parameters of Spodoptera litura, the host insect of Bracon hebetor, were noted during exposure to phytochemicals derived from Inula racemose. It has been demonstrated that the Braconid parasitoid has a longer life cycle, less survival, and less parasitism (Akutse et al. 2014). We may think about modifying PSMs for better plant protection and how PSMs impact multitrophic interactions in natural environments because of their detrimental effects on natural enemies.

## 11.11 Endophytes Sources of the Bioactive Structures

### *11.11.1 Bioactive Compounds of Botanical Endophytes*

Bioactive chemicals that are effective against a variety of bacteria have been found in plants. Microorganisms that are linked with plants, as opposed to the plants themselves, have demonstrated in recent years that they may provide materials and products with significant medical promise. Endophytes, which are typically bacteria or fungi, are a type of endosymbiotic microbe that reside inside the cells of plants. Since all or a portion of these organisms' life cycles occur inside their hosts, they do not show any visible symptoms of sickness. They have complex connections with their hosts that may involve mutualism and parasitism, and they are extensively spread in the natural environment (Hanan et al. 2020). It is well recognized that endophytes improve nutrient uptake and host growth. They could increase a plant's resilience to insects and its capacity to withstand different kinds of stress. Phytohormones and other bioactive substances of biotechnological relevance for pharmaceutical medicines and enzymes are produced by them. Endophytes may colonize plant stems, roots, petioles, and leaf segments, as well as weed inflorescences, fruit, buds, and hollow hyaline cells. Researchers have discovered endophytes in one or more forms in every plant they have studied. A plant species' endophyte population varies and is influenced by a number of factors, including the host species and stage, the density of the inoculum, and the environmental conditions. While plant sources are being thoroughly investigated for the discovery of new compounds for various therapeutic purposes, endophytic microorganisms play a significant role in this search with potential applications in the medical field and as drug-obligate endophytes. This has become the discovery rate of bioactive active novel chemicals for the past few decades (Joseph and Priya 2011). The genesis of endophytic organisms in various species has been the subject of several investigations in recent years. The rhizosphere or seed-born microbial communities are first selected by researchers as the primary endophyte suppliers. Typically, the genomic organization of endophytes provides information on their interactions with other plants as well as proof of their survival and transmission strategy (Strobel and Daisy 2003). By establishing a correlation between the size of the bacterial genome and its lifestyle, investigators have determined the genome sizes and origins of endophytes. The fact that endophytes reside inside plant species, where the environment is more stable than in the soil where the plant develops, may be a good characterization for them. Some endophytes, on the other hand, only show up in plants for a portion of their life cycle. As consequently, the endophytic community is composed of organisms from different origins; organisms with bigger genomes are more likely to survive in changeable settings, such soils, whereas organisms with smaller genomes are more likely to be vertically transmitted and remain in stable conditions (Joseph and Priya 2011).

### *11.11.2 Endophytes Fungal Entomopathogens*

Endophytes are widely spread members of the kingdom of plants that interact with a variety of species and provide an extra line of protection against several insects and pests. Endophytic fungi are plant-related microorganisms that live inside plant tissues like stems, leaves, or roots without posing a threat to the plant. They need to be categorized as either mutually helpful root endophytes or plant related endophytes (Gan et al. 2017). The term "symbiotic" refers to the relationship that exists between endophytic fungi and higher vascular plants. The former indirectly protect against herbivores in return for aiding in the restoration of nutrients to their hosts (Gan et al. 2017). By infecting lepidopteran larvae, sucking pest complexes, and other insects, the fungi endophytes in Table 11.3 can function as bioagents that cause disease in insects. Despite their ability to infect particular hosts, they are known to cause no harm to other species, even beneficial ones. It has been discovered that Beauveria bassiana is an endophyte in tomatoes, opium poppies, and bananas (Quesada-Moraga et al. 2006). Coffee and cocoa seedlings can be successfully inoculated with B. bassiana by applying a spore solution to the radicle as soon as it germinates.

**Table 11.3** Entomopathogenic fungi (EPF)

| Entomopathogenic endophytes fungi | Example |
|---|---|
| Ascomycota | *Metarhizium (M. anisopliae, M. robertsii, M. brunneum, M. lepidiotae, M and M. guizhouense).* |
| | *Beauveria (B. bassiana* and *B. brongniartii).* |
| | Hirsutella (*H. thompsonii, H. nodulosa, H. aphidis*). |
| | Lecanicillium (*L. lecani*-formerly *Verticillium lecanii, L. longisporum*). |
| Entomophtoromycota | Entomophtoromycota<br>*Furia, Conidiobolus, Entomophaga*, or *Erynia*. |

### 11.11.3 Isolation and Identification of Endophytes from Different Sources

Many endophytes that occur within plants are commonly identified as symbionts; they have a unique association with the plants and have long been disregarded and thought of as pathogens that cause contamination. Fluorescence in situ hybridization confocal laser scanning microscopy investigations and other cultivation-independent tests have recently verified the presence of endophytes. Any microbial or plant growth medium, such as agar, potato dextrose, and media containing carbon or nitrogen, may be readily isolated from endophytes. Isolating endophytes from surface-sterilized host plant tissue is the most used method for endophyte detection and counting (Nair and Padmavathy 2014). The genotype of the plant is the main factor that may affect entophyte colonization within a plant or microbial species. Endophytes are also impacted by the plant's development stage, physiological state, tissue type, soil conditions, surface sterility, selective medium, and culture conditions. Targeting a specific type of endophytic bioactivity with the highest potential for bioprospecting can also be aided by ecological understanding of the function endophytes play in nature (Strobel and Daisy 2003).

### 11.11.4 Therapeutic Compounds of Endophytes Against Pathogenic Microorganisms

Endophytes are a tremendous source of medicines for the treatment of diseases and have potential applications in food, medicine, cosmetics, and agriculture since they have been demonstrated to produce a diversity of bioactive metabolites in a single plant or microbe (Joseph and Priya 2011). Alkaloids, benzopyranones, chinones, flavonoids, phenolic acids, quinones, steroids, saponins, tannins, terpenoids, tetralones, xanthones, and many more functional classes were used to categorize these secondary metabolites. A number of variables, including the time of year the sample was collected, the local climate, and the geographic location, influence the extraction of metabolites from endophytes. However, extraction from plants and other natural sources is now more practical, effective, and easy thanks to a groundbreaking synthetic approach that has been devised (Hata and Sone 2008). The evolution of the host microorganisms, which may have incorporated genetic information from higher plants, has been directly linked to the production of bioactive substances by endophytes. This has allowed the microorganisms to better adapt to perform certain functions, such as protection from grazing animals, insects, and various pathogens. Table 11.4 lists a few of the secondary bioactive substances that are frequently discovered in endophytes.

**Table 11.4** Bioactive compounds from endophytes against pathogenic microorganisms

| Source of endophytes | Bioactive compounds from endophytes | Pathogen | Mode of transmission of the pathogen |
|---|---|---|---|
| *Boesenbergia rotunda Strepto-myces coelicolor* | Munumbicins | *Escherichia coli* | Ground meats, raw or under pasteurized milk |
| *Chloridium* sp. *Allamanda cathartica* | JavanicinMunumbicinsPhomopsilactone | *Pseudomonas* sp. | Contaminated water or surgi-cal instruments |
| *Diaporthe helianthi* | Fabatin, tyrosol | *Enterococcus hirae* | Nosocomial infection through hospi-talized patients |
| *Ganoderma boninense* | Rapamycin, cyclododecane, petalostemumol | *Bacillus subtilis* | Rice, pastas, raw milk, and meat products |
| *Hypericum perforatum, Diaporthe helianthi* | Hypericin, emodin, tyrosol | *Salmonella* sp. | Meat, eggs, and untreated tree nuts |
| *Phomopsis* sp. *Cinnamomum mollissimum* | Munumbicins, Saadamycin | *Aspergillus niger* | Maize, cereals, groundnuts, and tree nuts |
| *Thottea grandiflora* | Streptomyces | *Bacillus cereus* | Uncooked meat and raw milk |
| *Xylaria* sp. *Ginkgo bilobaFusarium proliferatum* | Sordaricin 7 amino-4-methylcoumarin, Beauvericin | *Yersinia enterocolitica* | Swine meat and meat products, milk and dairy products |

### *11.11.5 Mechanism of Endophytes Entomopathogenic Fungi (EEPF)*

Entomopathogenic fungi (EEPF) infection process is initiated when the conidia, penetrates the host plant and grows into hypha, forming a germ tube. Both in plants and insects, the infection continues has entered the host tissue. The EEPF and the plant must first overcome a few environmental obstacles for the plant to develop and thrive (Quesada Moraga 2020).

Nonetheless, the genera Beauveria and Metarhizium show similar mechanisms of entry and establishment within their plant and insect hosts, as well as comparable genetic involvement that may be the result of gene duplication, according to reports of general models for EEPF-insect interactions (Greenfield et al. 2016). After escaping the insect's defenses, the EEPF enters plant tissue. After searching for suitable plant hosts and identifying them using similar molecular patterns, invasive insects asexually adhere their spores to the surfaces of their hosts, penetrate the host, and proliferate within the plant tissue (Klieber and Reineke 2016). Moreover, some surface proteins started to degrade insect cuticular proteins following their identification. After adhering effectively, the conidia develop into hyphae, while the hemocel inside the insect undergoes differentiation into blastospores. Beauvericin and destruxins are insecticidal substances that kill insects. Additionally, to reduce microbial competition and guarantee proper resource allocation, EEPFs generate antimicrobial chemicals within the corpse. However, during EEPF colonization, a variety of plant strigolactones (SLs) and EEPF lipochitooligosaccharides (LCOs) function as signaling metabolites in plants. Following the first line of defense, plants transmit specific effector chemicals that trigger an immune system response (ETI) against the pathogen (Shrivastava et al. 2015). MAMP-triggered immunity (MTI) and/or pattern-triggered immunity (PTI) are activated when microbes or pathogen associated molecular patterns (MAMPs or PAMPs) recognize specific effector molecules secreted by plants that trigger effector-triggered immunity (ETI) against the potential pathogen (Liu et al. 2016). Another notable instance of pattern-triggered immunity (PTI) is the EEPF chitosan molecules, which have been shown in experiments to promote host defense. The carbohydrate-binding module LysM, found in many extracellular proteins and receptors, may recognize polysaccharides with N-acetylglucosamine residues, including EEPF chitosan. It has been shown that chitin and chitin polymers, together with their modified form, chitosan, induce host defense responses in a variety of horticultural crops. In the presence of fungal endophytes, chitin oligomers generated by the fungal cell wall are recognized by plant chitin-specific receptors (PR-3), which speeds up the development of subsequent defensive responses (Sánchez-Vallet et al. 2015). In the meanwhile, the fungus's endophytic development has no effect on the plant's immune system. According to a research, a raffinose transporter and an extracellular invertase that hydrolyzes sucrose are necessary for M. robertsii to establish itself in plants. Root exudates include raffinose and sucrose, which are essential for the growth of the M. robertsii rhizosphere and root competence. Plants might benefit from EEPFs' multimodal defense against phytophagous insects. This protection might come from creating lethal chemicals or volatile organic compounds (VOCs), which draw in natural enemies (Ye et al. 2019).

### *11.11.6 Multitrophic Interactions of Entomopathogenic Fungi*

Plant-associated bacteria enhance interactions between plants, herbivores, insects, and natural enemies. Plant-associated microbes facilitate interactions between plants, herbivores, and natural enemies. A host's tolerance to herbivores and natural enemies can be altered by endophytes. By altering the multitrophic relationships between insect species and plants, as well as those between insect pests and their natural enemies, parasitoids, and carnivores, systemic endophytic colonization offers plants multimodal protection. Endophytes affect and alter the metabolic processes of plants since they live and interact with them. There are instances where the endophyte and beneficial plant use the same way to make novel chemicals. To produce novel metabolites, both parties can alter the backbones of existing metabolites (Shukla et al. 2014).

### 11.11.7 Trophic Level of Entomopathogenic Fungus and Plants

The intimate relationship between endophytes and plants has garnered a lot of interest since endophytic fungi have a variety of beneficial impacts on their hosts. Fungus endophytes encourage increased plant development, photosynthesis, resistance, nutrient absorption capacity, and hormone levels (Zhang et al. 2021). According to recent studies, EEPFs improve plant height, weight, and growth rate while also having additive effects on host plant growth and development. Endophytic Metarhizium species supply the host plant with nitrogen from the insects, which the host plant then converts to carbon (Greenfield et al. 2016). Furthermore, the fresh weight of the roots, shoots, and height of the broad bean Vicia faba were observed to increase under foliar treatment with EPF, B. bassiana, B. brongniartii, and M. brunneum (González-Mas et al. 2019). Additionally, after being bathed in a solution containing fungal conidial suspension, the sweet pepper, Capsicum annum, grew more aggressively (Movva and Pathipati 2017). Numerous abiotic and biotic factors, such as the inoculation method, plant and microbial genotypes, climate, and the dynamics of interaction within the soil and plant biomes, have been discovered to affect the degree of endophytism (Busby et al. 2017). Additionally, it has been shown that EEPFs emit poisons that inhibit the development and multiplication of pathogenic organisms and other competitors, hence enabling plants to tolerate stressors like heat, drought, and salt (Pineda et al. 2010).

### 11.11.8 Trophic Level of Entomopathogenic Fungus and Insect

Endophytic entomopathogenic fungi have been shown to be able to suppress insect pests that consume plants colonized by microbes. By preventing insect growth acting as a feeding deterrent, delaying insect growth, and limiting insect survival and oviposition, endophytic fungi lessen the harm caused by insect herbivores. A decrease in Helicoverpa zea incidence in tomato plants treated with *B. bassiana*.

When B. bassiana was applied to tomato leaves, the larvae of Tuta absoluta (Meyrick) experienced 50% mortality and decreased lifespan. Furthermore, the damage of tomato plants' H. armigera was decreased in response to B. bassiana colonization (Wei et al. 2020). Most of this research relates the decrease of insect pest damage to the build-up of mycotoxin in plant tissues. Also, two EEPFs, *B. bassiana* (Bb1TS11) and *M. robertsii* (Mr4TS04), create volatile chemicals that have the ability to repel *C. sordidus*, the banana rhizome weevil (Lozano-Soria et al. 2020). Seven volatile organic compounds (VOCs) styrene, benzothiazole, camphor, borneol, 1,3-dimethoxy-benzene, 1-octen-3-ol, and 3-cyclohepten-1-one were found to repellent activity against, according to studies using gas chromatography mass spectrometry solid phase microextraction (GC/MS-SPME). Additionally, it has been demonstrated that *B.bassiana* possesses insect repelling qualities against *Rhynchophorus ferrugineus*, the palm weevil. When B. bassiana inoculates tomato seedlings produce the bioactive compounds chitin and glucans, the whitefly, Bemisia tabaci, does not like these plants. After Metarhizium brunneum colonization in various tissues of potato plants and melon leaves, destruxin was found to be effective against whiteflies. Recent research indicates that biting and sucking insects that eat exophytically on plants are killed by EPF's transient endophytic colonization of plant tissues (Bleeker et al. 2011).

### *11.11.9 Level of Trophic EEPF and Naturally Enemies*

The third trophic level must be included in order to fully comprehend herbivores and EEPF-colonized host plants. The altered nutrient allelochemistry of the EEPF-associated host plants delayed larval development, which in turn increased the qualitative and quantitative foraging efficiency of parasites and predators, in accordance with the slow growth high mortality hypothesis (SG-HM). In comparison to host plants that are not infected, the parasitoid that feeds on M. brunneum-infected plants has a 33% higher parasitization rate and a reduced ability for reproduction (Miranda-Fuentes et al. 2020). Additionally, the parasitoid Spodoptera littoralis (Lepidoptera: Noctuidae) and its natural adversary Hyposoter didymator interact multitrophically. Additionally, Chrysoperla carnea's consumption of Aphis gossypii prey in response to endophytic colonization by B. bassiana did not differ significantly from that of Aphidius colemani's mummification rate (González-Mas et al. 2019). In the meanwhile, when parasitoids and predators work together with EEPFs, the harm caused by insect herbivores and plants is reduced. EEPF B. bassiana, M. brunneum, and Aphidius colemani Viereck are used to suppress Myzus persicae Sulzer, the green peach aphid. Previous research indicates that EEPFs negatively affect natural enemies' development, reproduction, and adult survival (González-Mas et al. 2019). After being infected by plants, insects eventually pass on mycotoxins to their parasites.

### 11.11.10 Entomopathogenic Fungi Influence Insect-Natural Enemy Interactions and Metabolites

Several volatile organic compounds (VOCs) and secondary metabolites generated by plants colonized by EEPF have important impacts in addition to directly hindering tritrophic interactions. Lepidopteran insects, including Spodoptera littoralis, are susceptible to the phagodeterrent and larvicidal effects of the fungus Pimarane diterpene sphaeropsidin A (SphA) (Miranda-Fuentes et al. 2020). Depending on its capacity to produce SphA in vivo, SphA was found to exhibit both oral and contact toxicity against the chewing lepidopteran S. littoralis. An in vitro research found that the partially purified protein from the EEPF, Lecanicillium lecanii (Zimmerman), linked to solanaceous crops like tomato, is sublethal to the green peach aphid Myzus persicae (Sulzer). Furthermore, RT-qPCR expression analysis revealed that significant genes associated with the jasmonic acid (JA) and salicylic acid (SA) pathways were upregulated, which significantly harmed M. persicae fertility and survival. The SA-associated genes PR1, BGL2, AOS, PAL, LOX, and AOC were found to be significantly enhanced by the protein derived from L. lecanii. This suggests that the protein should be purified and characterized as a novel biomolecule against pest insects that feed on phloem, like aphids, and this indicates the enhancement of systemic resistance in plants (Hanan et al. 2020). It has been demonstrated that Trichoderma harzianum Tr6 increases plant immunity, is a strong inducer of induced systemic resistance (ISR), and negatively impacts the host selection and egg production of Trialeurodes vaporariorum. Aphid growth rate, offspring performance, and fertility have also been demonstrated to be impacted by the application of EEPFs (Trichoderma asperellum, Gibberella moniliformis, B. bassiana, M. anisioplaie, and Hypocrea lixi) on Vicia faba by seed treatment. When compared to the control treatment, endophyte seed treatment improved seedling survival. Applying endophyte topically has been shown to decrease aphid fabae and aphid pisum nymphs (Aldaghi et al. 2021). After being treated separately with the arbuscular mycorrhizal fungus Rhizophagus intraradices and the EEPF B. bassiana, tomato plants infected with beet armyworm, Spodoptera exigua, exhibited increased levels of monoterpenes and sesquiterpenes. This implies that the host plants have a more resilient defensive response mediated by terpenoid mediators. Both wild and farmed common beans colonized by Trichoderma isolates showed different host choice and had lower emergence with the E20 and T34 strains of Acanthoscelides obtectus.

## 11.12 Strategies to Improve Crop Resistance Against Insects

Insect pests are thought to be responsible for a significant portion of the global loss in agricultural productivity, both directly and indirectly through the spread of plant diseases. Since more food must be produced worldwide to feed the world's expanding population, reducing the impact of insect agricultural pests is of utmost importance. Improvements in agricultural output would require greater automation and biotechnology innovation, according to a growing agreement that this goal should be achieved through sustainable higher productivity without environmental damage (Grotewold 2008). In comparison to their wild cousins, the majority of agricultural plants, particularly modified cultivars that have benefited from large fertilizer inputs, have a higher nutritional content and less chemical and physical defenses (King 2017). Increasing crop resistance to insect pests without sacrificing crop quality and production is therefore a problem. The solutions to this problem are being made possible by recent developments in plant biotechnology and our knowledge of the molecular underpinnings of insect-plant interactions. Molecular approaches to crop resistance to insect pests are becoming commonplace. In particular, a gene or group of genes having a product or products that inactivate or disable the products of the target insect genes is known as plant resistance, whereas an insect characteristic is characterized in terms of one or more gene targets. The development of molecular techniques to study the makeup and function of microbial communities, including unculturable forms, has resulted from the genomic revolution. This has led to the realization that interactions with resident microorganisms, collectively referred to as the microbiome, are essential for the long-term health and fitness of both plants and insects (King 2017). The ability of microbiome members to affect insect-plant interactions and contribute to crop resistance tactics against insect pests is becoming more and clearer.

### *11.12.1 The Variability of Secondary Metabolism in Plants*

Terpenes, alkaloids, phenylpropanoids, aromatic flavonoids, stilbenoids, cyanogenic glycosides, and glucosinolates are among the more than 200,000 compounds in biosynthetic classes that are thought to constitute secondary metabolites of plants. By drawing parasitic wasps or other natural enemies of phytophagous insects, many of these substances are poisonous to insects or serve as indirect defenses. Nonetheless, certain insects use plant secondary chemicals as signals needed for oviposition or eating, while some insects can detoxify or tolerate specific plant secondary metabolites (Akhtar et al. 2003). When designing and managing crop resistance based on secondary metabolite chemistry, it is crucial to take into account the abundance of evidence that many phytophagous insects have coevolved with the secondary metabolite profile of their host plants. Compared to similar wild plant species, several crops have significantly lower levels of secondary metabolite production and animal toxicity. For instance, two genes (zFPP and ShZIS) that code for the enzymes that produce the sesquiterpene 7-epizingiberene were lost when the tomato, Lycopersicon esculentum, was domesticated. Transferring these genes from a wild tomato species to the cultivated tomato that expresses them in trichomes improved the plant's resistance to a variety of insects. In particular, the growth of the tobacco hornworm Manduca sexta (Lepidoptera) was postponed, L. decemlineata (Coleoptera) was discouraged from eating, and the whitefly Bemisia tabaci (Hemiptera) was prevented from ovipositing on the trichome-rich adaxial leaf surface (Garrido-Jurado et al. 2017). Due to domestication, certain plants have kept their toxicity. For instance, cyanogenic glycosides are found in cassava tubers, alkaloids are present in many legume seeds, and glycoalkaloids are abundant in potato tubers. Given that poisons provide defense against predators, such as insects, but are rendered inactive by cooking and other processing before being consumed by humans, it has been suggested that selection for toxicity had a role in the domestication of particular crops (Rangarajan et al. 2000).

### *11.12.2 Crop Metabolism Engineering for Insect Resistance*

Recent developments in molecular biology and plant genetics provide the chance to increase crop resistance to insect pests by addressing the genetically deficient and variable secondary metabolite profile. The objective is to more accurately control the amount of toxins present in crops in order to provide consumers with a longer-lasting protection against insect pests and a lower risk of acute or chronic harm. However, compared to the engineering of protein toxins like Bacillus thuringiensis (Bt), the relationship between an engineered gene activity and the secondary metabolite profile of the plant is much more complicated because the gene products are the enzymes that synthesize the bioactive compounds rather than the bioactive compounds themselves. There are two issues. First, a large number of secondary metabolism-related enzymes have low substrate specificity, which means that the distribution of metabolic flux throughout the entire metabolic network and the availability of several alternative substrates can both affect the composition of reaction products. Another way, secondary metabolite profiles can vary, and manipulation results are frequently unpredictable. To boost the production of squalene, for instance, genes for farnesyl diphosphate synthase and squalene synthase were inserted into tobacco. This led to a number of unanticipated metabolic alterations, including much higher levels of the alkaloid nicotine. The emergence of unexpected metabolic anomalies is not limited to developed plants. Because metabolic pathways are tightly regulated with flux controls dispersed across multiple reactions, it can be challenging to engineer plant metabolism by introducing one or more genes intended to increase flux through particular reactions or to introduce novel reactions. These limitations can be lessened by a variety of tactics. For instance, the selection of a response to accomplish the intended alteration in secondary metabolism can be guided by metabolic modeling. In order to find potential pathways for managing secondary metabolite profiles, it is also becoming feasible to either overexpress or silence certain transcription factors and miRNAs that control the spatiotemporal expression of secondary metabolism genes. Furthermore, the required metabolic alterations can be achieved by focusing expression to the plastid, where the metabolic network is often more tolerant to flux variation than in the cytoplasm (Grotewold 2008).

### *11.12.3 Plant Metabolites Enhance the Performance of Crops*

The use of metabolic engineering to increase crop resilience expands on the finding that some crops including the model plant Arabidopsis can enhance plant defenses against insects through genetic alteration of secondary metabolite profiles. Linalool and its glycosylated and hydroxylated derivatives were found in 40–60 times higher concentrations in the leaves of Arabidopsis plants that were transformed to express a terpene synthase (more precisely, a dual S-linalool (3S)-E-nerolidol synthase, FaNES1, isolated from strawberry Fragaria ananassa) in the plastid (Aharoni et al. 2003). The aphid Myzus persicae, the only insect studied, was significantly repelled by this metabolite profile. Comparing the altered plants to the wild-type and several transformed controls, the flea beetle Phyllotreta nemorum, a significant crucifer crop pest, showed significantly lower adult feeding and extremely high larval mortality. Similarly, the Arabidopsis plant has been modified to produce the cyanogenic glycoside dhurrin. Indirect plant defenses have also been engineered by manipulation of secondary metabolites. The sesquiterpene (E)-β-farnesene (Eβf), which serves as the aphid alarm pheromone, provides this method. When Eβf is released from an aphid, it triggers rapid dispersal of nearby aphids and also attracts aphid predators and parasitic wasps. Numerous plants also produce Eβf (Frago et al. 2017), and it has been shown that the potato species Solanum berthaultii's trichomes secrete this substance, which helps to protect against aphids (Neal et al. 1990). Aphids were repelled and Diaeretiella rapae, a parasitoid of crucifer-feeding aphids, was drawn to Arabidopsis plants that were transformed with the peppermint, Mentha piperita, Eβf synthase gene (Crock et al. 1997). Indirect plant defenses have also been engineered by manipulation of secondary metabolites. The sesquiterpene (E)-β-farnesene (Eβf), which serves as the aphid alarm pheromone, provides this method. When Eβf is released from an aphid, it triggers rapid dispersal of nearby aphids and also attracts aphid predators and parasitic wasps. Numerous plants also produce Eβf (Frago et al. 2017), and it has been shown that the potato species Solanum berthaultii's trichomes secrete this substance, which helps to protect against aphids (Neal et al. 1990). Aphids were repelled and Diaeretiella rapae, a parasitoid of crucifer-feeding aphids, was drawn to Arabidopsis plants that were transformed with the peppermint, Mentha piperita, Eβf synthase gene (Crock et al. 1997). Diverse outcomes from protecting Eβf-producing wheat against aphid pests demonstrated how difficult it is to control plant secondary chemistry in agriculture. The spatiotemporal pattern and concentration of the product affect the effect of secondary metabolites on the target insect, and several interactions can determine the overall impact of PSMs on crop output. For instance, Eβf cannot be used as an aphid deterrent in maize because it is a strong oviposition stimulant for the European corn borer Ostrinia nubilalis (Binder et al. 1995). Furthermore, modifying the terpene chemistry of plants to shield them against numerous phytophagous insects might interfere with pollinators' vital chemical signals.

## 11.13 Manipulation of Insect Microbiomes and Crop Resistance

### *11.13.1 Microbiomes in Interactions Between Plants and Insects*

There is now strong evidence that the activity of the microbial communities that reside on the surfaces and inside the tissues of plants and animals is essential for their nutrition, immune response, and general health. Given evidence that plant microbial communities, including one closed system, contain intracellular microbial symbionts in specialized somatic cells (bacteriocytes) with obligatory transmission to the ovary and, typically, insertion into the cytoplasm of the unfertilized egg, interactions with microorganisms are particularly significant in plant-insect interactions. The insect host cannot obtain a substitute microbiota from the environment or other insects if the maternally derived microbial cells are eliminated.

Open ecosystems contain gut microbes that are obtained by food, survive, and frequently multiply to create sustaining populations that are expelled in feces for variable periods of time. Fungal endophytes and mycorrhizal fungi can affect herbivory, and the microbiome of some insects shapes their plant range. Insects are frequently colonized by fecal oral cycling or from the environment, and their gut microbiota is typically diminished or destroyed when they molt (Shrivastava et al. 2015).

### *11.13.2 Inhibition of the Microbial Partners that Insects Need*

The main purpose of plants with microbicidal activity against insect-associated microorganisms is to fight bacterial pathogens spread by hemipteran insects, particularly Candidatus Liberibacter species spread by psyllids that feed on phloes and Xylella fastidiosa, the agent of Pierce's disease spread by xylem-feeding sharpshooters. Since they support the immune systems of all plants and animals, antimicrobial peptides (AMPs) are the preferred class of microbicides (Thao et al. 2000). Since AMPs are genetically encoded, it is possible to quickly construct them to improve their effectiveness against the target germ and integrate many activities into a single fusion protein. Additionally, they are much more destructive to microbes than to animal or plant cells. Grape has been converted in two different studies employing constructions that code for the insect AMP cecropin B, either with elastase or a portion of a second AMP, melittin. The removal of disease symptoms and Xylella titers in the plant's xylem vessels demonstrated that both designs offered effective protection against Pierce's disease under glasshouse conditions (Quintana-Rodriguez et al. 2018). Nevertheless, in field tests, the construct was unable to protect against Xylella, most likely as a result of decreased AMP expression in field plants exposed to abiotic stress—something that glasshouse plants do not experience the effectiveness of several AMPs in Ca regulation. Additionally, L. asiaticus has been studied (Quintana-Rodriguez et al. 2018), and two citrus-modified spinach defensins (SoD2 and SoD7) are now being tested in the field for resistance to huanglongbing disease.

### *11.13.3 Microbial Transport System for the New Pesticides*

Through open microbial interactions, microorganisms have a significant potential for transferring pesticides to insects. The gut microbiome of the great majority of insects is acquired orally and contains microorganisms linked to plants for phytophagous insects. It's very intriguing to use bacteria as RNAi delivery vehicles. After being consumed and colonizing the insect's stomach, a microbe that has been modified to produce dsRNA specific to an insect pest's crucial gene provides a consistent flow of dsRNA molecules. Microorganisms isolated from insect pests have been the subject of recent research because they allow colonization without triggering the insect defense mechanism. The strategy's effectiveness is demonstrated by two research. An expression plasmid encoding the eGFP gene or dsRNA targeting the insect α-tubulin gene was used to alter BFo2α, a gut bacterium linked to Pantoea ananatis, which was isolated from the western flower thrip Frankliniella occidentalis (Broderick et al. 2006). Given to flies, the BFo2α bacterium transformed with eGFP showed competition with the native microbiota by colonizing the proximal hindgut and remaining there for many days. When flower thrips were fed the ds-α-tubulin gene, their expression decreased by 60% over two days, and their mortality rose significantly over four days. The second study examines the spotted-wing fruit fly, Drosophila suzukii, an invasive pest of small fruits that consumes bacteria and yeasts linked to fruits. D. suzukii caused high larval mortality and poor adult fecundity by feeding on Saccharomyces cerevisiae yeasts that expressed the ds-γ-tubulin gene construct (Morimoto et al. 2017). Microbial RNAi is more effective than plant RNAi against insect pests under certain circumstances. The main advantage of microbial delivery is that, in contrast to the laborious and usually complex methods needed for continuous crop transformation, genetic alteration of microorganisms—especially bacteria—is cheap and simple. This difference is particularly noticeable for crops that have a market need for several cultivars or variations, as well as generalist insects like F. occidentalis and D. suzukii that need protection against a wide variety of plant species. One of the purported advantages of microbial RNAi delivery—microbial-encoded controls can reside in the insect permanently and spread to other individuals by fecal oral transmission—would, by definition, be hampered by environmental measures (Facey et al. 2015). Paratransgenesis, a comparable technique that uses genetically engineered bacteria to prevent disease transmission and has been used to impair vector competence in hematophagous insects, shares same difficulties. Resolving these fundamental problems in controlling the microbiomes of insect pests would create new opportunities for agriculture and human health.

### *11.13.4 Microbial Transmission Impact on Insect Features*

Numerous microbial communities are linked to insects, and these connections are frequently essential to the survival and development of insects. Numerous enzymes that provide specific tasks to insect hosts are delivered by symbiotic bacteria in the stomach, hemolymph, and specialized cells (Ben-Yosef et al. 2014). A range of crop pests, including plant sap-sucking Hemipteran insects like aphids, whiteflies, and psyllids (Thao et al. 2000), as well as human disease vectors and urban pests like tsetse flies in the genus Glossina and the common bed bug (Cimex lectularius), depend on endosymbionts for vital nutrients, especially amino acids and B vitamins. Certain symbionts may recycle nitrogen for termites and cockroaches or break down complex polysaccharides (Berlanga et al. 2016). As demonstrated in cotton leafworms and beewolf digger wasps, symbionts' generation of antimicrobials helps the immune system combat invasive diseases. In addition to providing immunological and nutritional support, symbionts have the ability to influence how insects and their natural adversaries interact ecologically. For instance, Hamiltonella defensa, the secondary symbiont of aphids, disrupted wasp embryogenesis by its bacteriophage-encoded poisons, increasing the likelihood of host survival after parasitoid wasp assaults (Crossley et al. 2017). Additionally, it has been shown that H. defensa attenuates volatile emission in aphid-infested plants, hence reducing the recruitment of parasitic wasps. Similarly, it has been demonstrated that the Colorado potato beetle and whiteflies work together to manipulate plant physiology in a way that facilitates insect colonization. The change in body color of facultative symbionts may indicate an aphid's vulnerability to parasitism or predation (Crossley et al. 2017). Specifically, Rickettisella infection caused the pea aphid Acyrthosiphon pisum to become green by increasing the production of blue-green polycyclic quinone pigments. The aphid's relative risks of predation and parasitism are thought to be impacted by this symbiont-dependent color variation because parasitoids preferentially attack the green morphs, while predators like ladybird beetles prefer to dine on the red morphs. It has been demonstrated that the presence of symbiotic midgut bacteria is necessary for Bacillus thuringiensis (Bt) to have insecticidal action in gypsy moth (Lymantria dispar) larvae. According to microbiome investigations conducted on field-collected samples, different microbiome compositions were linked to Bt resistance in Helicoverpa armigera, the cotton bollworm (Visweshwar et al. 2015). The diamondback moth exhibited varying degrees of vulnerability to chlorpyrifos. Following antibiotic treatment, Plutella xylostella was recolonized with several gut-associated bacteria (Xia et al. 2018). Mosquitoes that were fed with various gut commensal bacteria had decreased pesticide mortality, indicating the role of gut bacteria in insecticide resistance (Riehle and Jacobs-Lorena 2005). Microbes have been demonstrated to impact long-term dispersion, oviposition, mating, host hunting, and kin identification in insects. Additionally, research indicates that microbiomes may affect host behavior by producing metabolites that affect host neuro-endocrine circuits—a process known as the "gut-brain axis". The gut-brain axis has been extensively studied, with much of that study concentrating on mammalian systems. Additionally, research indicates that the host's olfactory-guided foraging preferences for meals with varying microbial content are influenced by the microbiome of Drosophila melanogaster (Qiao et al. 2019).

In order to investigate previously unidentified microbiomes in a wider range of insects, developments in high throughput sequencing and functional genomics must reach significant milestones in insect microbiome research. As a result, the variety and composition of insect microbiomes are better understood in relation to host and environmental variables. Microbial communities linked to mosquitoes and Drosophilid and Tephritid fruit flies (Hamden et al. 2013) are a couple of examples. Research conducted on the potent lab model Drosophila melanogaster revealed that its microbiome, which is made up of microbial species, influences host feeding choice, speeds up larval development, and influences reproduction—all of which are critical indicators of invasiveness. The development of the microbiome in ripening fruits is essential for its near relative, Drosophila suzukii, often known as spotted wing drosophila (SWD), a major pest of tiny fruits (Bing et al. 2018). Microbial symbiosis was initially identified in the olive fruit fly Bactrocera oleae, one of the most damaging agricultural pests in the Tephritidae family. In contrast to other species of Bactrocera that feed on fruit, B. oleae is linked to an obligatory bacterial symbiont (Candidatus Erwinia dacicola) that is kept in the midgut caeca of larvae. Although the symbiont is not yet grown, research has demonstrated that it aids in the growth and reproduction of flies by supplying necessary amino acids and breaking down urea from a variety of sources, including bird droppings, so that adult flies may get nitrogen (Ras et al. 2017). By inhibiting oleuropein, a plant defense molecule, it also aids in the larvae's development in unripe olives. It's interesting to note that domesticated B. oleae raised on artificial medium including antibiotics have been discovered to lack the symbiont, indicating the influence of upbringing on symbiont selection. Although their nutritional function is unknown, Pantoea sp. and Burkholderia sp. are other bacterial species linked to the intestines of B. oleae (Mitter et al. 2013). The apple maggot fly, Rhagoletis pomonella, and the medfly, Ceratitis capitata, have also been seen to exhibit microbiome-dependent larval development. In medflies, diazotrophs that express the nitrogen reductase gene (nifH) in the stomach are involved in microbial food supply. The genera Citrobacter, Klebsiella, Pectobacteria, Enterobacter, and Pantoea are among the bacteria that are linked to medflies. When given as probiotics, the community has been demonstrated to promote fly growth, reproduction, and lifespan (Puri et al. 2016), as well as boost male copulatory success. Along with the prevalent Enterobacteriaceae, Pseudomonas species are part of a small but steady ecosystem linked to the stomach of medflies. Variations in the microbial population at various medfly instar stages were noted using a metabarcoding technique (Puri et al. 2016). As proposed in Tetraponera ants, Burkholderia may be involved in nitrogen fixation and was shown to be prominent in early instars and adults (Borm et al. 2002). Leuconostoc, Weissella, Acetobacter, Gluconobacter, and an unidentified bacterium from the family Xanthomonadaceae were more prevalent at the pupal stage, while bacteria from the genera Sphingomonas, Pseudomonas, and an unidentified bacterium from the family Methylobacteraceae were more prevalent in the late instars of medflies. Additionally, medflies fed on several host plants showed a varied microbial population. Acinetobacter and Gluconobacter, for instance, were shown to be present in medfly larvae that were fed Ficus carica (common fig) fruits, but Acetobacter and Leuconostoc were found to be more prevalent when given Prunus persica (peaches) (Hanan et al. 2020). It has been proposed that Gluconostoc and Acinetobacter contribute to the detoxification of phenolic glycosides (Mason and Singer 2014). Likewise, a wide variety of Enterobacteria, including Pantoea, Klebsiella, and Enterobacter, are linked to the guts of apple maggot flies. During oviposition, the bacteria are deposited into the fruit, supplying vital proteins and nutrients needed for the growth of the larva. Symbiotic bacteria, which are also Tephritids of the sub-families Dacinae and Trypetinae, aid in the growth of larvae by metabolizing carbohydrates, raising the amount of organic nitrogen, and producing vitamins.

### *11.13.5 Antimicrobial Peptides (AMPs) and the Insect Management*

A wide range of resources that might be targeted for pest control are represented by the contribution of microbiomes to several features associated to insect invasiveness. Using biochemicals to eradicate or interfere with insect symbiosis is a straightforward tactic. Antibiotics like tetracycline and penicillin, for instance, have been demonstrated to inhibit the growth of young ticks, reduce the symbiont load of adult ticks, and make tsetse flies infertile by harming the obligate mutualist Wigglesworthia. Although antimicrobial peptides (AMPs) are more frequently utilized to inhibit plant or human infections that are vectored by insects, they have also been investigated for manipulating insect symbionts (Quintana-Rodriguez et al. 2018). Innate immunity in insects depends heavily on AMPs, which include a variety of amphiphilic and cationic oligopeptides. They offer defense against a range of microorganisms, such as viruses, fungus, and bacteria. While many AMPs, like moricin and gloverin, are exclusive to specific insect orders, like the Lepidoptera, others, including defensins, cecropins, attacins, and proline-rich peptides, are found in all insect orders. Because of their positively charged surface, which allows them to bind negatively charged microbial surfaces through charge-charge interactions and disrupt the integrity of bacterial cell walls, AMPs are thought to have antibacterial activity (Quintana-Rodriguez et al. 2018). In order to provide resistance against bacterial, fungal, and certain eukaryotic parasites, AMPs have been engineered into plants and some insects. In transgenic tobacco, for instance, it has been demonstrated that insect defensins (gallerimycin from Galleria mellonella) and cecropin (sarcotoxin-IA from Sarcophaga peregrina) give resistance to pathogenic fungi. Likewise, it has been demonstrated that transgenic production of cecropins confers resistance to bacterial and fungal diseases in tomato and rice plants (Sharma et al. 2000). Additionally, cecropins exhibit efficacy against protozoan parasites as Trypanosoma and Plasmodium. It has been demonstrated that transgenic expression of cecropin in Anopheles gambiae, a human parasite vector, lowers the quantity of Plasmodium berghei oocysts. Furthermore, it has been demonstrated that transgenic co-expression of cecropin-A and defensin-A in Aedes aegypti cooperatively prevents the spread of the Plasmodium parasite (Kokoza et al. 2010). The potential for non-target effects, which might disturb the native microbiome of non-target insects, is a drawback of utilizing antibiotics or AMPs. Low bioavailability, instability, and the development of antibiotic resistance are further drawbacks of AMP use. However, fusion with antibody fragments that are specific to particular antigens, protein engineering, and synthetic biology approaches (e.g., amino acid replacement, chemical changes) can all produce AMPs with improved specificities. Research on using nanotechnology to enhance AMP distribution to many life systems is also ongoing (Biswaro et al. 2018). Another effective method for modifying insect characteristics is to introduce a foreign microbe to the insects or swap out a symbiont with another bacterium. Aphids and stink bugs have shown experimental substitution of particular cultivated and uncultured insect symbionts. Poor M. punctatissima egg hatching on the plants was the consequence of switching the Ishikawaella symbionts between the stink insect Megacopta punctatissima, a common pest of soybean and other legumes, and a closely related non-pest species, Megacopta cribraria. It has also been demonstrated that the pea aphid's ability to withstand heat changes when the major symbiont Buchnera is experimentally replaced with a different genotype using microinjection (Varenhorst et al. 2015).

### *11.13.6 Exploring the Microbiome for Innovation of Biopesticides*

Crop security is threatened by the rise of pesticide resistance as many insect pests lack substitute methods of management. The challenges of turning a lead molecule into a product that can meet strict safety and environmental regulations have led to an increase in the cost and complexity of developing new synthetic pesticides. The creation of innovative pesticides is necessary to address resistance concerns and enhance sustainable agriculture. Interest in biopesticides has increased as a result of the hunt for novel insecticides with minimal negative effects on the environment. A-biochemical pesticides are naturally occurring chemicals that can change the physiology or behavior of pests; b-microbial pesticides use toxic or pathogenic microbes as the active ingredient; and c-plant incorporated protectants (PIPs) are transgenic plants that produce pesticidal compounds (Marrone 2013). Biopesticides can be divided into one of three categories.In contrast, microbial pesticides are applied directly to the field and take the form of live bacteria or spores. One of the most widely used microbial pesticides is entomopathogenic fungi, or EPFs. Fungal spores have a longer shelf life than other biological control agents, and many EPFs can be mass-cultured in vitro. These features make them perfect candidates for biological control agents. Soil-dwelling Hypocreales, including Metarhizium anisopliae, Beauveria bassiana, and Entomophaga maimaiga, comprise the majority of commercially significant species. Insect biological control is greatly aided by entomopathogenic nematodes (EPNs) belonging to the Heterorhabditidae and Steinernematidae families, as well as their bacterial symbionts, Xenorhabdus and Photorhabdus.

## 11.14 Pant Immunity Against Pathogens

### *11.14.1 Insect Elicitors*

The ability of plants to recognize and react to molecular patterns linked to external illnesses and cellular disturbances brought on by chemical or mechanical harm is known as plant immunity. It is suggested that plant cells utilize methods like to those employed to identify microbial diseases to identify insect compounds as alien. The basal immune system is thought to be activated by conserved insect elicitors in particular.

The two prevailing theories for how the plant immune system interacts with microbial pathogens are PTI and ETI. PTI is a conserved basic defensive reaction that occurs when transmembrane receptors (pattern recognition receptors) on the cell surface identify the conserved molecular patterns (elicitors) of pathogens. In particular, the basal immune system is believed to be activated by conserved insect elicitors. Pathogen-triggered immunity (PTI) and effector-triggered immunity (ETI) are the prevailing paradigms for how the plant immune system interacts with microbial pathogens. PTI, a conserved basal defense response, is triggered when transmembrane receptors (pattern recognition receptors) at the cell surface identify conserved molecular patterns of pathogens (elicitors).

Although it was developed for interactions between plants and pathogens, the molecular structure of plant immunity may also be applied to interactions with many of the natural enemies of plants, such as insects. It is believed that the plant possesses two complementary cell-autonomous systems: effector-triggered immunity (ETI), which is triggered against natural enemy molecules (usually referred to as insect effectors) that suppress PTI, and basal defense, also known as pathogen-triggered immunity (PTI), which is triggered against conserved molecular features of elicitor molecules released by the natural enemy.

Certain interactions between the plant immune system and their natural adversaries do not adhere strictly to the dichotomy of PTI/elicitor versus ETI/effector. Some elicitors display limited conservation between insect pests and bacterial pathogens (Cook et al. 2015), and the mechanisms of effector-triggered immunity (ETI) and pathogen-triggered immunity (PTI) are often not completely exclusive. It is anticipated that responses similar to the conserved pathogen-triggered immunity (PTI) and insect effectors will evolve alongside plant effector-triggered immunity (ETI), leading to the prevalence of dominant alleles of individual genes that provide resistance against the insect pest. Insects unintentionally make themselves known to plants through the components of their egesta (frass, honeydew) and oral secretions (gut regurgitant, saliva), as well as the oviposition fluids that female insects produce when they deposit their eggs into plant tissue. Fatty acid–amino acid conjugates (usually an unsaturated C18-fatty acid connected to glutamine or glutamate, e.g., volicitin), caeliferins (sulfooxy fatty acids), and other proteins (e.g., glucose oxidase) are examples of these potential insect elicitors of plant immunological responses. These elicitors often trigger the jasmonic acid (JA) signaling pathway, which results in indirect defenses like volatiles that draw certain natural enemies of the phytophagous insect and direct defenses like polyphenol oxidase activity (Schwartzberg and Tumlinson 2014). It is anticipated that single molecules derived from insects can be used to manage the immune defenses of multiple crops against multiple economically significant pests, according to the paradigm of plant immunity that emerged from research on plant pathogens. For instance, the tomato fruitworm's saliva contains the elicitor glucose oxidase. Caterpillar frass can either stimulate or inhibit plant defense responses that inhibit caterpillar growth, depending on the insect species, crop, and insect location on the plant. Helicoverpa zea also causes JA-mediated plant defenses in tomatoes while inhibiting the production of the defensive compound nicotine in tobacco. Despite the fact that there are cases of pathogen elicitor evolutionary diversification (Hardy et al. 2017), this variability implies that insect elicitor evolutionary diversification may be less restricted than pathogen elicitor evolutionary diversification.

The relationship between insect elicitors and plant immunity has been extensively studied in tandem with studies on the variety and identity of insect elicitors. Interestingly, contrary to what the plant-pathogen paradigm predicts, plant responses to some insect elicitors seem to benefit the insect rather than protect the plant. In particular, the phytohormone salicylic acid (SA) coordinates defenses against biotrophic infections through the activation of certain insect elicitors.

Because antagonistic cross talk between the JA-dependent system and the SA-dependent pathway suppresses JA-mediated protection against the insect, this plant response is often beneficial to the insect. These insect pests are essentially taking advantage of a flaw in the way the intricate signaling networks that govern plant immune responses are regulated. The SA-dependent pathway has several insect elicitors that have been identified.

For instance, two maize chitinases (Endochitinase A and Pr4) that are ingested by autumn armyworm caterpillars that feed on plant tissue and then released back onto the plant surface in caterpillar frass cause SA-dependent defense in maize. Insect performance is enhanced by the resulting SA-mediated inhibition of JA-mediated defense (Smith et al. 2009). Similarly, SA and likely other chemicals found in the honeydew secreted by the pea aphid Acyrthosiphon pisum promote SA-mediated defense in the fava bean Vicia faba, inhibiting JA signaling to the aphid's advantage.

However, insects also create microbial elicitors. Bacteria found in L. decemlineata's gut regurgitant cause tomato plants to undergo SA signaling, which lowers JA-regulated plant defenses including polyphenol oxidase and a cysteine protease inhibitor. The bacterial flagellin protein, a well-known bacterial elicitor of plant immunity, triggers this reaction.

The lower development rates of beetle larvae on plants where SA signaling is genetically inhibited or when antibiotic treatment suppresses the bacteria show how important this protein is for insect performance (Smith et al. 2009).

When combined, these basic investigations into the relationship between the immune system of plants and elicitors derived from insects show that certain phytophagous insect pests have the ability to alter the innate immune responses of plants. This manipulation is indicative of coevolutionary interactions between insects and plant immunity, raising the possibility of natural variation in plant immune responses to specific insect pests.

### *11.14.2 Insect Effectors*

Numerous pieces of evidence suggest that phytophagous insects produce specific effectors that diminish the basic plant defenses triggered by different elicitors in a way akin to pathogen effector molecules (Bos et al. 2010). Plant resistance genes (R genes) encode intracellular nucleotide-binding leucine-rich repeat receptors (NLRs), which serve as plant receptors for a wide range of pathogen effectors. Resistance genes offer good breeding opportunities for increased crop resistance because dominant alleles of R genes can confer resistance to a large number of genotypes of specific diseases.

The ETI paradigm is thought to be most applicable to insects that directly interact with the contents of living plant cells since NLRs, nucleotide-binding leucine-rich repeat receptors produced by resistance genes, (Table 11.5) against pathogens are intracellular. In line with this reasoning, insects like aphids and whiteflies that have mouthparts that puncture sieve elements and take phloem sap provide some of the strongest genetic evidence for gene coevolutionary interactions involving R genes (Bos et al. 2010).

Moreover, cecidomyiid midges produce nutrient-rich fluids by penetrating plant cells close to the meristem (Stuart et al. 2012). Like R genes against pathogens, plant

**Table 11.5** The putative plant R genes contributing resistance to insect infestations

| Crop | Insect pest | Candidate *R* gene | Evidence |
|---|---|---|---|
| Cultivated rice (*O. sativa*) | Rice gall midge (*Orseolia oryzae*) | Eleven rice genes (*Gm1–Gm11*) | Mapped to chromosomal regions that include *NLR* genes |
| Soybean (*Glycine max*) | Soybean aphid (*Aphis glycines*) | *Rag* (*Resistance to Aphis glycines*) genes | At least two of the five *Rag* genes map to chromosomal regions coding *NLR* (*nucleotide-binding leucine-rich repeat receptor*) genes |
| Cultivated wheat (*Triticum aestivum*) | Hessian fly (*Mayetiola destructor*) | Various | Mapped to clusters of NLR-coding genes |
| *Medicago truncatula* | Blue alfalfa aphid (*Acyrthosiphon kondoi*) | *AKR* (*Acyrthosiphon kondoi resistance*) gene | Gene fully characterized at the molecular level |
| Tomato (*Lycopersicon esculentum*) | Potato aphid (*Macrosiphum euphorbiae*), tomato psyllid (*Bactericerca cockerelli*), whitefly (*Bemisia tabaci*) | *Mi-1.2* gene | Gene fully characterized at the molecular level |

R genes against phloem-feeding hemipterans and cecidomyiid midges. When detected, these genes produce intracellular NLR proteins. The fact that R gene-mediated resistance encourages the evolutionary diversity of insect effector proteins, which breaks the resistance, is a major drawback for crop protection. The Hessian fly on wheat (Bos et al. 2010), Nilaparvata lugens on rice, and Aphis glycines on soybean have all shown evidence of this coevolutionary arms race.

Cecidomyiid midges infiltrate plant cells close to the meristem, causing them to secrete fluids that are rich in nutrients. The fact that R gene-mediated resistance encourages the evolutionary diversity of insect effector proteins, which breaks the resistance, is a major drawback for crop protection. The Hessian fly and aphids have yielded the most advancements in crop R gene research. The R gene, which provides resistance against Hessian flies with the vH13 gene, is genetically dominant in interactions with wheat that carries the H13 gene. The genome of the Hessian fly contains many potential effector genes vH13 (Aggarwal et al. 2014).

As the gene for paradigm had anticipated. On wheat with the H13 resistance gene, Hessian fly larvae with the vH13 gene specifically die at a very high rate, while larvae with null mutations in vH13 or treated with dsRNA that inhibits vH13 production are able to live and grow on the H13 wheat. However, as vH13 codes for a tiny taxon-specific protein that has little similarity to other publically accessible sequences, the molecular mechanism of this interaction remains unclear. Numerous studies have demonstrated that certain plant-produced salivary proteins significantly impact the fertility or lifespan of aphids.

Nevertheless, little is known about the physiological mechanisms via which R genes control insect resistance, and they seem to differ. For instance, melon's resistance to Aphis gossypii caused by virus aphid transmission (Vat) entails a hypersensitive reaction that includes localized cell death close to the feeding site and peroxidase activity that includes a micro-oxidative burst and callose deposition. These reactions happen quickly, manifesting 10–20 minutes after aphid infestation (Boissot et al. 2016).

While JA signaling has been linked to the R. gene-mediated resistance of Medicago truncatula to Acyrthosiphon kondoi, the resistance in tomatoes caused by Mi1.2 is independent of the hypersensitive response and includes SA and MAPK signaling cascades.

However, because the concentration and localization of the effector protein may vary between plants that bear salivating insects and plants that express the transgene, resulting in distinct plant immune responses, candidate effectors of insects expressed in plants R gen can produce artifactual results.

For instance, M. persicae's decreased performance after RNA interference against Mp55 confirms that this aphid performs better on plants that express Mp55 (Stuart et al. 2012).

## References


Aggarwal R, Subramanyam S, Zhao C, Chen MS, Harris MO, Stuart JJ (2014) Avirulence effector discovery in a plant galling and plant parasitic arthropod, the Hessian fly (*Mayetiola destructor*). PLoS One 9:e100958

Agrawal AA, Kurashige NS (2003) A role for isothiocyanates in plant resistance against the specialist herbivore Pieris rapae. J Chem Ecol 29:1403–1415. https://doi.org/10.1023/A:1024265420375

Aharoni A, Giri AP, Deuerlein S, Griepink F, de Kogel WJ et al (2003) Terpenoid metabolism in wild-type and transgenic *Arabidopsis* plants. Plant Cell 15:2866–2884

Aharoni A, Jongsma MA, Bouwmeester HJ (2005) Volatile science? Metabolic engineering of terpenoids in plants. Trends Plant Sci 10(12):594–602

Akhtar Y, Rankin CH, Isman MB (2003) Decreased response to feeding deterrents following prolonged exposure in the larvae of a generalist herbivore, *Trichoplusia ni* (Lepidoptera: Noctuidae). J Insect Behav 16(6):811–831

Akula R, Ravishankar GA (2011) Influence of abiotic stress signals on secondary metabolites in plants. Plant Signal Behav 6(11):1720–1731

Akutse KS, Fiaboe KK, Van Den Berg J, Ekesi S, Maniania NK (2014) Effects of endophyte colonization of Vicia faba (Fabaceae) plants on the life–history of leafminer parasitoids Phaedrotoma scabriventris (Hymenoptera: Braconidae) and Diglyphus isaea (Hymenoptera: Eulophidae). PLoS One 9:e109965. https://doi.org/10.1371/journal.pone.0109965

Aldaghi M, Allahyari H, Hosseininaveh V, Behboudi K (2021) Effect of *Trichoderma harzianum* Tr6 in inducing resistance in tomato against *Trialeurodes vaporariorum* (Hem: Aleyodidae). Plant Prot Sci J Agric 44(3):107–117. https://doi.org/10.22055/PPR.2021.17128

Alquezar B, Volpe HXL, Magnani RF, de Miranda MP, Santos MA, Wulff NA et al (2017) Beta-caryophyllene emitted from a transgenic Arabidopsis or chemical dispenser repels Diaphorina citri, vector ofCandidatus Liberibacters. Sci Rep 7(1):5639

Alseekh S, Tohge T, Wendenberg R, Scossa F, Omranian N, Li J, Kleessen S, Giavalisco P, Pleban T, MuellerRoeber B, Zamir D, Nikoloski Z, Fernie AR (2015) Identification and mode of inheritance of quantitative trait loci for secondary metabolite abundance in tomato. Plant Cell 27(3):485–512

Ashihara H, Sano H, Crozier A (2008) Caffeine and related purine alkaloids: biosynthesis, catabolism, function and genetic engineering. Phytochemistry 69(4):841–856

Balcke GU, Bennewitz S, Bergau N, Athmer B, Henning A, Majovsky P et al (2017) Multi-omics of tomato glandular Trichomes reveals distinct features of central carbon metabolism supporting high productivity of specialized metabolites. Plant Cell 29(5):960–983

Balkema-Boomstra A, Zijlstra S, Verstappen F, Inggamer H, Mercke P, Jongsma M et al (2003) Role of cucurbitacin C in resistance to spider mite (*Tetranychus urticae*) in cucumber (Cucumis sativus L.). J Chem Ecol 29(1):225–235

Barbehenn RV, Constabel CP (2011) Tannins in plant–herbivore interactions. Phytochemistry 72(13):1551–1565

Barbour JD, Kennedy GG (1991) Role of steroidal Glycoalkaloid alpha-Tomatine in host-plant resistance of tomato to Colorado potato beetle. J Chem Ecol 17(5):989–1005

Barrett RDH, Agrawal AA (2004) Interactive effects of genotype, environment, and ontogeny on resistance of cucumber (*Cucumis sativus*) to the generalist herbivore, Spodoptera exigua. J Chem Ecol 30(1):37–51

Ben-Yosef M, Pasternak Z, Jurkevitch E, Yuval B (2014) Symbiotic bacteria enable olive flies (Bactrocera oleae) to exploit intractable sources of nitrogen. J Evol Biol 27:2695–2705. https://doi.org/10.1111/jeb.12527

Berlanga M, Llorens C, Comas J, Guerrero R (2016) Gut bacterial community of the Xylophagous cockroaches Cryptocercus punctulatus and Parasphaeria boleiriana. PLoS One 11:e0152400. https://doi.org/10.1371/journal.pone.0152400

Binder BF, Robbins JC, Wilson RL (1995) Chemically mediated ovipositional behaviors of the European corn borer, *Ostrinia nubilalis* (Lepidoptera: Pyralidae). J Chem Ecol 21:1315–1327

Bing X, Gerlach J, Loeb G, Buchon N (2018) Nutrient-dependent impact of microbes on drosophila suzukii development. MBio 9:e02199–e02117. https://doi.org/10.1128/mBio.02199-17

Biswaro LS, da Costa Sousa MG, Rezende TMB, Dias SC, Franco O et al (2018) Antimicrobial peptides and nanotechnology, recent advances and challenges. Front Microbiol 9:855. https://doi.org/10.3389/fmicb.2018.00855

Bleeker PM, Diergaarde PJ, Ament K, Schütz S, Johne B, Dijkink J et al (2011) Tomato-produced 7-epizingiberene and R-curcumene act as repellents to whiteflies. Phytochemistry 72(1):68–73

Bleeker PM, Spyropoulou EA, Diergaarde PJ, Volpin H, De Both MT, Zerbe P et al (2011b) RNA-seq discovery, functional characterization, and comparison of sesquiterpene synthases from *Solanum lycopersicum* and *Solanum habrochaites* trichomes. Plant Mol Biol 77(4–5):323

Boissot N, Thomas S, Chovelon V, Lecoq H (2016) NBS-LRR-mediated resistance triggered by aphids: viruses do not adapt; aphids adapt via different mechanisms. BMC Plant Biol 16:25

Borm S, van Buschinger A, Boomsma JJ, Billen J (2002) Tetraponera ants have gut symbionts related to nitrogen–fixing root–nodule bacteria. Proc R Soc Lond B Biol Sci 269:2023–2027. https://doi.org/10.1098/rspb.2002.2101

Bos JI, Prince D, Pitino M, Maffei ME, Win J, Hogenhout SA (2010) A functional genomics approach identifies candidate effectors from the aphid species *Myzus persicae* (green peach aphid). PLoS Genet 6:e1001216

Broderick NA, Raffa KF, Handelsman J (2006) Midgut bacteria required for bacillus thuringiensis insecticidal activity. Proc Natl Acad Sci USA 103:15196–15199. https://doi.org/10.1073/pnas.0604865103

Busby PE, Soman C, Wagner MR, Friesen ML, Kremer J et al (2017) Research priorities for harnessing plant microbiomes in sustainable agriculture. PLoS Biol 15:e2001793

Carrasco D, Larsson MC, Anderson P (2015) Insect host plant selection in complex environments. Curr Opin Insect Sci 8:1–7

Cavallini E, Matus JT, Finezzo L, Zenoni S, Loyola R, Guzzo F et al (2015) The phenylpropanoid pathway is controlled at different branches by a set of R2R3-MYB C2 repressors in grapevine. Plant Physiol 167(4):1448–1470

Checker VG, Kushwaha HR, Kumari P, Yadav S (2018) Role of phytohormones in plant defense: signaling and cross talk. In: Molecular aspects of plant-pathogen interaction. Springer, Singapore

Cook DE, Mesarich CH, Thomma BPHJ (2015) Understanding plant immunity as a surveillance system to detect invasion. Annu Rev Phytopathol 53:541–563

Costa-Arbulú C, Gianoli E, Gonzáles WL, Niemeyer HM (2001) Feeding by the aphid *Sipha flava* produces a reddish spot on leaves of *Sorghum halepense*: an induced defense. J Chem Ecol 27(2):273–283

Crock J, Wildung M, Croteau R (1997) Isolation and bacterial expression of a sesquiterpene synthase cDNA clone from peppermint (*Mentha* × *piperita*, L.) that produces the aphid alarm pheromone (*E*)-β- farnesene. PNAS 94:12833–12838

Crossley MS, Chen YH, Groves RL, Schoville SD (2017) Landscape genomics of Colorado potato beetle provides evidence of polygenic adaptation to insecticides. Mol Ecol 26(22):6284–6300

Dervinis C, Frost CJ, Lawrence SD, Novak NG, Davis JM (2010) Cytokinin primes plant responses to wounding and reduces insect performance. J Plant Growth Regul 29:289–296

Divekar P, Kumar P, Suby SB (2019) Screening of maize germplasm through antiobiosis mechanism of resistance against *Chilo partellus* (Swinhoe). J Entomol Zool Stud 7:1111–1114

Divekar PA, Narayana S, Divekar BA, Kumar R, Ray BGGA, Singh AKSVRV, Singh AK, Kumar A, Singh RP, Meena RS, Behera TK (2022) Plant secondary metabolites as defense tools against herbivores for sustainable crop protection. Int J Mol Sci 23(5):2690

Dombrecht B, Xue GP, Sprague SJ, Kirkegaard JA, Ross JJ, Reid JB et al (2007) MYC2 differentially modulates diverse jasmonate-dependent functions in *Arabidopsis*. Plant Cell 19(7):2225–2245

Douglas AE (2015) Multiorganismal insects: diversity and function of resident microorganisms. Annu Rev Entomol 60:17–34

Elzinga DA, De Vos M, Jander G (2014) Suppression of plant defenses by a *Myzus persicae* (green peach aphid) salivary effector protein. Mol Plant-Microbe Interact 27:747–756

Erb M, Köllner TG, Degenhardt J, Zwahlen C, Hibbard BE, Turlings TCJ (2011) The role of abscisic acid and water stress in root herbivore-induced leaf resistance. New Phytol 189:308–320

Erb M, Züst T, Robert CAM (2021) Using plant chemistry to improve interactions between plants, herbivores and their natural enemies: challenges and opportunities. Curr Opin Biotechnol 70: 262–265. https://doi.org/10.1016/j.copbio.2021.05.011

Erbilgin N, Krokene P, Christiansen E, Zeneli G, Gershenzon J (2006) Exogenous application of methyl jasmonate elicits defenses in Norway spruce (*Picea abies*) and reduces host colonization by the bark beetle *Ips typographus*. Oecologia 148(3):426–436

Facey PD, Meric G, Hitchings MD, Pachebat JA, Hegarty MJ et al (2015) Draft genomes, phylogenetic reconstruction, and comparative genomics of two novel cohabiting bacterial symbionts isolated from *Frankliniella occidentalis*. Genome Biol Evol 7:2188–2202

Frago E, Mala M, Weldegergis BT, Yang C, McLean A, Godfray HCJ et al (2017) Symbionts protect aphids from parasitic wasps by attenuating herbivore-induced plant volatiles. Nat Commun 8:1860. https://doi.org/10.1038/s41467-017-01935-0

Freeman B, Beattie G (2008) Freeman an overview of plant defenses against pathogens and herbivores. Plant Health Instr. https://doi.org/10.1094/PHI-I-2008-0226-01

Gan H, Churchill ACL, Wickings K (2017) Invisible but consequential: root endophytic fungi have variable effects on belowground plant-insect interactions. Ecosphere 8:e01710

Gang DR, Wang J, Dudareva N, Nam KH, Simon JE, Lewinsohn E et al (2001) An investigation of the storage and biosynthesis of phenylpropenes in sweet basil. Plant Physiol 125(2):539–555

Garrido-Jurado I, Resquıń-Romero G, Amarilla SP, Rıós-Moreno A, Carrasco L, Quesada-Moraga E (2017) Transient endophytic colonization of melon plants by entomopathogenic fungi after foliar application for the control of *Bemisia tabaci* gennadius (Hemiptera: Aleyrodidae). J Pest Sci 90(1):319–330. https://doi.org/10.1007/s10340-016-0767-2

Gilardoni PA, Schuck S, Jüngling R, Rotter B, Baldwin IT, Bonaventure G (2010) SuperSAGE analysis of the *Nicotiana attenuate* transcriptome after fatty acid-amino acid elicitation (FAC): identification of early mediators of insect responses. BMC Plant Biol 10:66

Glas JJ, Schimmel BC, Alba JM, Escobar-Bravo R, Schuurink RC, Kant MR (2012) Plant glandular trichomes as targets for breeding or engineering of resistance to herbivores. Int J Mol Sci 13(12): 17077–17103

Gleadow RM, Møller BL (2014) Cyanogenic glycosides: synthesis, physiology, and phenotypic plasticity. Annu Rev Plant Biol 65:155–185

González-Mas N, Cuenca-Medina M, Gutiérrez-Sánchez F, Quesada-Moraga E (2019) Bottom-up effects of endophytic *B. bassiana* on multitrophic interactions between the cotton aphid, *Aphis gossypii*, and its natural enemies in melon. J Pest Sci 92(3):1271–1281. https://doi.org/10.1007/s10340-019-01098-5

Greenfield M, Gómez-Jiménez MI, Ortiz V, Vega FE, Kramer M, Parsa S (2016) *Beauveria bassiana* and *Metarhizium anisopliae* endophytically colonize cassava roots following soil drench inoculation. Biol Control 95:40–48. https://doi.org/10.1016/j.biocontrol.2016.01.002

Grotewold E (2008) Transcription factors for predictive plant metabolic engineering: are we there yet? Curr Opin Biotechnol 19(2):138–144

Hall DG, Richardson ML, Ammar ED, Halbert SE (2013) Asian citrus psyllid, *Diaphorina citri*, vector of citrus huanglongbing disease. Entomol Exp Appl 146(2):207–223

Hamden H, Guerfali MM, Fadhl S, Saidi M, Chevrier C (2013) Fitness improvement of mass-reared sterile males of ceratitis capitata (vienna 8 strain) (Diptera: Tephritidae) after gut enrichment with probiotics. J Econ Entomol 106:641–647. https://doi.org/10.1603/EC12362

Hanan A, Basit A, Nazir T, Majeed MZ, Qiu D (2020) Anti-insect activity of a partially purified protein derived from the entomopathogenic fungus *Lecanicillium lecanii* (Zimmermann) and its putative role in a tomato defense mechanism against green peach aphid. J Invertebr Pathol 170: 107282. https://doi.org/10.1016/j.jip.2019.107282

Hardy NB, Peterson DA, Ross L, Rosenheim JA (2017) Does a plant-eating insect's diet govern the evolution of insecticide resistance? Comparative tests of the pre-adaptation hypothesis. Evol Appl 11:739

Harvey JA, van Nouhuys S, Biere A (2005) Effects of quantitative variation in allelochemicals in *Plantago lanceolata* on development of a generalist and a specialist herbivore and their endoparasitoids. J Chem Ecol 31:287–302

Hata K, Sone K (2008) Isolation of endophytes from leaves of Neolitsea sericea in broadleaf and conifer stands. Mycoscience 49:229–232. https://doi.org/10.1007/S10267-008-0411-Y

Heiling S, Schuman MC, Schoettner M, Mukerjee P, Berger B, Schneider B et al (2010) Jasmonate and ppHsystemin regulate key malonylation steps in the biosynthesis of 17-hydroxygeranyllinalool diterpene glycosides, an abundant and effective direct defense against herbivores in Nicotiana attenuata. Plant Cell 22(1):273–292

Jeschke V, Gershenzon J, Vassao DG (2016) A mode of action of glucosinolate-derived isothiocyanates: detoxification depletes glutathione and cysteine levels with ramifications on protein metabolism in *Spodoptera littoralis*. Insect Biochem Mol Biol 71:37–48

Joseph B, Priya RM (2011) Bioactive compounds from endophytes and their potential in pharmaceutical effect: a review. Am J Biochem Mol Bio 1:291–309. https://doi.org/10.3923/ajbmb.2011.291.309

Karlsson MF, Birgersson G, Witzgall P, Lekfeldt JD, Nimal Punyasiri PA, Bengtsson M (2013) Guatemalan potato moth *Tecia solanivora* distinguish odour profiles from qualitatively different potatoes *Solanum tuberosum* L. Phytochemistry 85:72–81

Katz E, Nisani S, Yadav BS, Woldemariam MG, Shai B, Obolski U, Ehrlich M, Shani E, Jander G, Chamovitz DA (2015) The glucosinolate breakdown product indole-3-carbinol acts as an auxin antagonist in roots of Arabidopsis thaliana. Plant J 82:547–555. https://doi.org/10.1111/tpj.12824

Khare S, Singh NB, Singh A, Hussain I, Niharika K, Yadav V, Bano C, Yadav RK, Amist N (2020) Plant secondary metabolites synthesis and their regulations under biotic and abiotic constraints. J Plant Biol 63:203–216. https://doi.org/10.1007/s12374-020-09245-7

King A (2017) Technology: the future of agriculture. Nature 544:S21–S23

Klieber J, Reineke A (2016) The entomopathogen *Beauveria bassiana* has epiphytic and endophytic activity against the tomato leaf miner *Tuta absoluta*. J Appl Entomol 140(8):580–589. https://doi.org/10.1111/jen.12287

Kokoza V, Ahmed A, Woon Shin S, Okafor N, Zou Z, Raikhel AS (2010) Blocking of Plasmodium transmission by cooperative action of cecropin A and defensin A in transgenic Aedes aegypti mosquitoes. Proc Natl Acad Sci USA 107:8111–8116. https://doi.org/10.1073/pnas.1003056107

Kumar D, Haq I, Chapagai D, Tripathi D, Donald D, Hossain M, Devaiah S (2015) Hormone signaling: current perspectives on the roles of salicylic acid and its derivatives in plants. In: The formation, structure and activity of phytochemicals. Springer, Berlin/Heidelberg, Germany, pp 115–136

Kuzina V, Ekstrøm CT, Andersen SB, Nielsen JK, Olsen CE, Bak S (2009) Identification of defense compounds in *Barbarea vulgaris* against the herbivore *Phyllotreta nemorum* by an ecometabolomic approach. Plant Physiol 151(4):1977–1990

Liu L, Sonbol F-M, Huot B, Gu Y, Withers J, Mwimba M, Yao J, He SY, Dong X (2016) Salicylic acid receptors activate jasmonic acid signalling through a non-canonical pathway to promote effector-triggered immunity. Nat Commun 7:13099

Lozano-Soria A, Picciotti U, Lopez-Moya F, Lopez-Cepero J, Porcelli F, Lopez-Llorca LV (2020) Volatile organic compounds from entomopathogenic and nematophagous fungi, repel banana black weevil (*Cosmopolites sordidus*). Insects 11(8):509. https://doi.org/10.3390/insects11080509

Malone LA, Barraclough EI, Lin-Wang K, Stevenson DE, Allan AC (2009) Effects of red-leaved transgenic tobacco expressing a MYB transcription factor on two herbivorous insects, *Spodoptera litura* and *Helicoverpa armigera*. Entomol Exp Appl 133(2):117–127

Marrone PG (2013) Market opportunities for biopesticides. In: Proceedings of the American Chemical Society, 246th National Meeting and Exposition, Indianapolis, IN, USA, vol 84, p 104

Martin D, Tholl D, Gershenzon J, Bohlmann J (2002) Methyl jasmonate induces traumatic resin ducts, terpenoid resin biosynthesis, and terpenoid accumulation in developing xylem of Norway spruce stems. Plant Physiol 129(3):1003–1018

Mason PA, Singer MS (2014) Defensive mixology: combining acquired chemicals towards defence. Funct Ecol 29:441–450. https://doi.org/10.1111/1365-2435.12380

Massad TJ, Dyer LA, Vega G (2012) Costs of defense and a test of the carbon-nutrient balance and growth-differentiation balance hypotheses for two co-occurring classes of plant defense. PLoS One 7(10):e47554

Miranda-Fuentes P, Quesada-Moraga E, Aldebis HK, Yousef-Naef M (2020) Compatibility between the endoparasitoid *Hyposoter didymator* and the entomopathogenic fungus *Metarhizium brunneum*: a laboratory simulation for the simultaneous use to control spodoptera littoralis. Pest Manag Sci 76(3):1060–1070. https://doi.org/10.1002/ps.5616

Mithen R (2001) Glucosinolates–biochemistry, genetics and biological activity. Plant Growth Regul 34(1):91–103

Mitter B, Petric A, Shin MW, Chain PSG, Hauberg- Lotte L, Reinhold- Hurek B et al (2013) Comparative genome analysis of *Burkholderia phytofirmans* PsJN reveals a wide spectrum of endophytic lifestyles based on interaction strategies with host plants. Front Plant Sci 4:120. https://doi.org/10.3389/fpls.2013.00120

Morimoto J, Simpson SJ, Ponton F (2017) Direct and trans-generational effects of male and female gut microbiota in drosophila melanogaster. Biol Lett 13:20160966. https://doi.org/10.1098/rsbl.2016.0966

Movva V, Pathipati UR (2017) Feeding-induced phenol production in Capsicum annuum L. Influences Spodoptera litura F. larval growth and physiology. Arch Insect Biochem Physiol 95:e21387. https://doi.org/10.1002/arch.21387

Mrid RB, Benmrid B, Hafsa J, Boukcim H, Sobeh M, Yasri A (2021) Secondary metabolites as biostimulant and bioprotectant agents: a review. Sci Total Environ 777:146204. https://doi.org/10.1016/j.scitotenv.2021.146204

Nair DN, Padmavathy S (2014) Impact of endophytic microorganisms on plants, environment and humans. Sci World J 2014:250693. https://doi.org/10.1155/2014/250693

Neal JJ, Tingey WM, Steffens JC (1990) Sucrose esters of carboxylic acids in glandular trichomes of *Solanum berthaultii* deter settling and probing by green peach aphid. J Chem Ecol 16(2): 487–497

Neilson EH, Goodger JQ, Woodrow IE, Møller BL (2013) Plant chemical defense: at what cost? Trends Plant Sci 18(5):250–258

Onkokesung N, Reichelt M, van Doorn A, Schuurink RC, van Loon JJ, Dicke M (2014) Modulation of flavonoid metabolites in *Arabidopsis* thaliana through overexpression of the MYB75 transcription factor: role of kaempferol-3, 7-dirhamnoside in resistance to the specialist insect herbivore *Pieris brassicae*. J Exp Bot 65(8):2203–2217

Opitz SEW, Müller C (2009) Plant chemistry and insect sequestration. Chemoecology 19:117–154

Outchkourov NS, Carollo CA, Gomez-Roldan V, de Vos RC, Bosch D, Hall RD et al (2014) Control of anthocyanin and non-flavonoid compounds by anthocyanin-regulating MYB and bHLH transcription factors in *Nicotiana benthamiana* leaves. Front Plant Sci 5:519

Pentzold S, Zagrobelny M, Bjarnholt N, Kroymann J, Vogel H, Olsen CE et al (2015) Metabolism, excretion and avoidance of cyanogenic glucosides in insects with different feeding specialisations. Insect Biochem Mol Biol 66:119–128

Pimentel D, Peshin R (2014) Integrated Pest management, vol 3. Springer Science and Business Media LLC, Berlin/Heidelberg, Germany. Available online: https://link.springer.com/book/10.1007/978-1-4615-7269-5. Accessed on 12 December 2021

Pineda A, Zheng SJ, van Loon JJ, Pieterse CM, Dicke M (2010) Helping plants to deal with insects: the role of beneficial soil-borne microbes. Trends Plant Sci 15:507–514

Pré M, Atallah M, Champion A, De Vos M, Pieterse CM, Memelink J (2008) The AP2/ERF domain transcription factor ORA59 integrates jasmonic acid and ethylene signals in plant defense. Plant Physiol 147(3):1347–1357

Puri A, Padda KP, Chanway CP (2016) Evidence of nitrogen fixation and growth promotion in canola (Brassica napus L.) by an endophytic diazotroph Paenibacillus polymyxa P2b-2R. Biol Fertil Soils 52:119–125. https://doi.org/10.1007/s00374-015-1051-y

Qiao H, Keesey IW, Hansson BS, Knaden M (2019) Gut microbiota affects development and olfactory behavior in drosophila melanogaster. J Exp Biol 222:jeb192500. https://doi.org/10.1242/jeb.192500

Quesada Moraga E (2020) Entomopathogenic fungi as endophytes: their broader contribution to IPM and crop production. Biocontrol Sci Tech 30(9):864–877. https://doi.org/10.1080/09583157.2020.1771279

Quesada-Moraga E, Landa B, Muñoz-Ledesma J, Jiménez-Diáz R, Santiago-Alvarez C (2006) Endophytic colonisation of opium poppy, Papaver somniferum, by an entomopathogenic Beauveria bassiana strain. Mycopathologia 161:323–329. https://doi.org/10.1007/s11046-006-0014-0

Quintana-Rodriguez E, Duran-Flores D, Heil M, Camacho X (2018) Damage-associated molecular patterns (DAMPs) as future plant vaccines that protect crops from pests. Sci Hortic 237:207–220

Rakha M, Bouba N, Ramasamy S, Regnard J-L, Hanson P (2017) Evaluation of wild tomato accessions (*Solanum* spp.) for resistance to two-spotted spider mite (*Tetranychus urticae* Koch) based on trichome type and acylsugar content. Genet Resour Crop Evol 64(5):1011–1022

Rangarajan A, Miller AR, Veilleux RE (2000) Leptine glycoalkaloids reduce feeding by Colorado potato beetle in diploid *Solanum* sp. hybrids. J Am Soc Hortic Sci 125(6):689–693

Ras E, Beukeboom LW, Cáceres C, Bourtzis K (2017) Review of the role of gut microbiota in mass rearing of the olive fruit fly, bactrocera oleae, and its parasitoids. Entomol Exp Appl 164:237–256. https://doi.org/10.1111/eea.12609

Riehle MA, Jacobs-Lorena M (2005) Using bacteria to express and display anti-parasite molecules in mosquitoes: current and future strategies. Insect Biochem Mol Biol 35:699–707

Robert JA, Madilao LL, White R, Yanchuk A, King J, Bohlmann J (2010) Terpenoid metabolite profiling in Sitka spruce identifies association of dehydroabietic acid,(+)-3- carene, and terpinolene with resistance against white pine weevil. Botany 88(9):810–820

Sánchez-Vallet A, Mesters JR, Thomma BP (2015) The battle for chitin recognition in plant-microbe interactions. FEMS Microbiol Rev 39(2):171–183. https://doi.org/10.1093/femsre/fuu003

Schiebe C, Hammerbacher A, Birgersson G et al (2012) Inducibility of chemical defenses in Norway spruce bark is correlated with unsuccessful mass attacks by the spruce bark beetle. Oecologia 170:183–198. https://doi.org/10.1007/s00442-012-2298-8

Schwartzberg EG, Tumlinson JH (2014) Aphid honeydew alters plant defence responses. Funct Ecol *28:386*–394

Schweiger R, Heise A-M, Persicke M, Müller C (2014) Interactions between the jasmonic and salicylic acid pathway modulate the plant metabolome and affect herbivores of different feeding types. Plant Cell Environ 37:1574–1585. https://doi.org/10.1111/pce.12257

Selmar D, Kleinwächter M (2013) Stress enhances the synthesis of secondary plant products: the impact of stress-related over-reduction on the accumulation of natural products. Plant Cell Physiol 54(6):817–826

Sessa RA, Bennett MH, Lewis MJ, Mansfield JW, Beale MH (2000) Metabolite profiling of Sesquiterpene lactones from *Lactuca* Species. Major latex components are novel oxa- late and sulfate conjugates of lactucin and its derivatives. J Biol Chem 275(35):26877–26884

Sharma A, Sharma R, Imamura M, Yamakawa M, Machii H (2000) Transgenic expression of cecropin B, an antibacterial peptide from Bombyx mori, confers enhanced resistance to bacterial leaf blight in rice. FEBS Lett 484:7–11. https://doi.org/10.1016/S0014-5793(00)02106-2

Shinoda T, Nagao T, Nakayama M, Serizawa H, Koshioka M, Okabe H et al (2002) Identification of a triterpenoid saponin from a crucifer, *Barbarea vulgaris*, as a feeding deterrent to the diamondback moth, Plutella xylostella. J Chem Ecol 28(3):587–599

Shrivastava G, Ownley BH, Augé RM, Toler H, Dee M, Vu A et al (2015) Colonization by arbuscular mycorrhizal and endophytic fungi enhanced terpene production in tomato plants and their defense against a herbivorous insect. Symbiosis 65(2):65–74

Shukla ST, Habbu PV, Kulkarni VH, Jagadish KS, Pandey AR, Sutariya VN (2014) Endophytic microbes: a novel source for biologically/pharmacologically active secondary metabolites. Asian J Pharmacol Toxicol 2:1–16

Smeda JR, Schilmiller AL, Anderson T, Ben-Mahmoud S, Ullman DE, Chappell TM et al (2018) Combination of Acylglucose QTL reveals additive and epistatic genetic interactions and impacts insect oviposition and virus infection. Mol Breed 38(1):3

Smith JL, De Moraes CM, Mescher MC (2009) Jasmonate-and salicylate-mediated plant defense responses to insect herbivores, pathogens and parasitic plants. Pest Manag Sci 65:497–503

Spyropoulou EA, Haring MA, Schuurink RC (2014) RNA sequencing on *Solanum lycopersicum* trichomes identifies transcription factors that activate terpene synthase promoters. BMC Genomics 15(1):402

Steppuhn A, Baldwin IT (2007) Resistance management in a native plant: nicotine prevents herbivores from compensating for plant protease inhibitors. Ecol Lett 10(6):499–511

Stevenson PC, Nicolson S, Wright GA (2017) Plant secondary metabolites in nectar: impacts on pollinators and ecological functions. Funct Ecol 31:65–75. https://doi.org/10.1111/1365-2435.12761

Strauss SY, Rudgers JA, Lau JA, Irwin RE (2002) Direct and ecological costs of resistance to herbivory. Trends Ecol Evol 17(6):278–285

Strobel GA, Daisy B (2003) Bioprospecting for microbial endophytes and their natural products. Microbiol Mol Biol Rev 67:491–502. https://doi.org/10.1128/MMBR.67.4.491-502.2003

Stuart JJ, Chen MS, Shukle R, Harris MO (2012) Gall midges (Hessian flies) as plant pathogens. Annu Rev Phytopathol 50:339–357

Sun JY, Sønderby IE, Halkier BA, Jander G, de Vos M (2009) Non-volatile intact indole glucosinolates are host recognition cues for ovipositing *Plutella xylostella*. J Chem Ecol 35(12):1427–1436

Takahashi N, Kitamura H, Kawarada A, Seta Y, Takai M, Tamura S, Sumiki Y (1955) Biochemical studies on "bakanae" fungus. Part XXXIV. Isolation of gibberellins and their properties. Bull Agr Chem Soc 19:267–277

Teklić T, Parađiković N, Špoljarević M, Zeljković S, Lončarić Z, Lisjak M (2021) Linking abiotic stress, plant metabolites, biostimulants and functional food. Ann Appl Biol 178:169–191. https://doi.org/10.1111/aab.12651

Thao ML, Moran NA, Abbot P, Brennan EB, Burckhardt DH, Baumann P (2000) Cospeciation of psyllids and their primary prokaryotic endosymbionts. Appl Environ Microbiol 66:2898–2905. https://doi.org/10.1128/AEM.66.7.2898-2905.2000

Thurston R, Webster JA (1962) Toxicity of *Nicotian a gossei* domin to *Myzus persicae* (sulzer). Entomol Exp Appl 5(3):233–238

Tissier A, Morgan JA, Dudareva N (2017) Plant volatiles: going 'in'but not 'out'of trichome cavities. Trends Plant Sci 22:930

Ton J, D'alessandro M, Jourdie V, Jakab G, Karlen D, Held M et al (2007) Priming by airborne signals boosts direct and indirect resistance in maize. Plant J 49(1):16–26

Varenhorst AJ, McCarville MT, O'Neal ME (2015) Reduced fitness of virulent *Aphis glycines* (Hemiptera: Aphididae) biotypes may influence the longevity of resistance genes in soybean. PLoS One 10:e0138252

Visweshwar R, Sharma HC, Akbar SMD, Sreeramulu K (2015) Elimination of gut microbes with antibiotics confers resistance to bacillus thuringiensis toxin proteins in helicoverpa armigera (Hubner). Appl Biochem Biotechnol 177:1621–1637. https://doi.org/10.1007/s12010-015-1841-6

Wang X, Zhou G, Xiang C, Du M, Cheng J, Liu S, Lou Y (2008) β-Glucosidase treatment and infestation by the rice brown planthopper *Nilaparvata lugens* elicit similar signaling pathways in rice plants. Chin Sci Bull 53:53–57

War AR, Paulraj MG, War MY, Ignacimuthu S (2011) Jasmonic acid-mediated induced resistance in groundnut (*Arachis hypogaea* L.) against *Helicoverpa armigera* (Hubner) (Lepidoptera: Noctuidae). J Plant Growth Regul 30:512–523

War AR, Paulraj MG, Ahmad T, Buhroo AA, Hussain B, Ignacimuthu S, Sharma HC (2012) Mechanisms of plant defense against insect herbivores. Plant Signal Behav 7:1306–1320. https://doi.org/10.4161/psb.21663

Wei QY, Li YY, Xu C, Wu YX, Zhang YR, Liu H (2020) Endophytic colonization by *Beauveria bassiana* increases the resistance of tomatoes against *Bemisia tabaci*. Arthropod Plant Interact 14(3):289–300. https://doi.org/10.1007/s11829-020-09746-9

Wink M (2007) Chapter 1 Molecular modes of action of cytotoxic alkaloids: from DNA intercalation, spindle poisoning, Topoiso- merase inhibition to apoptosis and multiple drug resistance. In: Alkaloids Chem Biol, vol 64, pp 1–47

Wu K, Zhang J, Zhang Q, Zhu S, Shao Q, Clark KD et al (2015) Plant phenolics are detoxified by prophenoloxidase in the insect gut. Sci Rep 5:16823

Xia X, Sun B, Gurr GM, Vasseur L, Xue M, You M (2018) Gut microbiota mediate insecticide resistance in the diamondback moth, Plutella xylostella (L.). Front Microbiol 9:25. https://doi.org/10.3389/fmicb.2018.00025

Ye M, Glauser G, Lou Y, Erb M, Hu L (2019) Molecular dissection of early defense signaling underlying volatile-mediated defense regulation and herbivore resistance in rice. Plant Cell 31: 687–698

Zhang H, Xie Y, Liu C, Chen S, Hu S, Xie Z et al (2017) Comprehensive comparative analysis of volatile compounds in citrus fruits of different species. Food Chem 230:316–326

Zhang X, Li L, Kesner L, Robert CAM (2021) Chemical cues shaping host-seeking by entomopathogenic nematodes. Curr Opin Insect Sci 44:72–81

# Chapter 12
# Efficacy of Plant Metabolic Products to Suppress the Plant-Parasitic Nematodes

**Mahfouz M. M. Abd-Elgawad**

## 12.1 Introduction

Despite the assurance of existing myriad plant-parasitic nematode (PPN) species, the identification of new species is tremendously continuing via accurate and rapid biochemical and molecular tools to accurately complement morphological characteristics and morphometrics (Shao et al. 2023). This modern trend is further facilitated by setting deep learning patterns for their automatic identification based on both their characteristics and artificial intelligence in agriculture (Shabrina et al. 2023). From the standpoint of pest management, variations exist among these species in their economic importance, especially the damage they cause to important crops, their global spread, ecology, biology, and plant hosts (Abd-Elgawad 2022a). Therefore, nematologists and stakeholders must keep updated of every discipline related to these nematode species and choose the best inexpensive and safe techniques to control PPNs especially economically important ones.

Due to their abundance and distinctions, PPNs may be placed in four different groups (Fig. 12.1). Nonetheless, the most economic losses are brought about by a handful of sedentary PPN genera of the first group in addition to a few semi-endoparasites in another group (Abd-Elgawad 2022b). More specifically, the root-knot nematodes related to *Meloidogyne* spp. and the cyst nematodes contained in the species of the two genera *Heterodera* and *Globodera* rank high as sedentary PPN genera. Yet, the importance of the semi-endoparasitic group of nematodes should not be underestimated. They include globally serious species such as *Rotylenchulus reniformis* and *Tylenchulus semipenetrans*. They resemble the sedentary nematode group in sexual dimorphism since the body of nematode females always swells by

M. M. M. Abd-Elgawad (✉)
Plant Pathology Department, Agricultural and Biological Research Institute, National Research Centre, Giza, Egypt



N. El-Wakeil, A. Abdelkhalek (eds.), *Microbiomes and Plant Secondary Metabolites for Insect Pest Management*, Progress in Biological Control 23,
https://doi.org/10.1007/978-3-032-24330-0_12

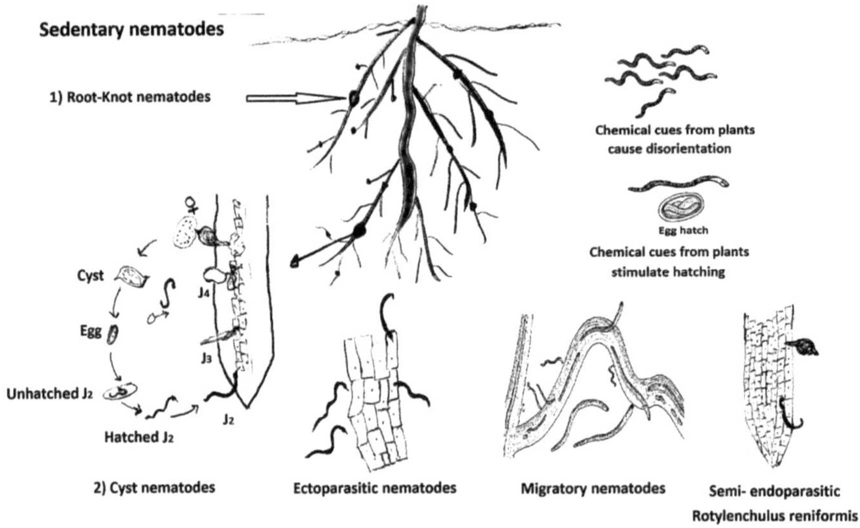


**Fig. 12.1** Categories of soil plant-parasitic nematodes depicting interactions with various plant hosts. Root metabolites can act as nematode attractants, repellents, inhibitors or killers (Abd-Elgawad 2022b)

feeding on root nutrients but the males keep the filiform shape. For instance, Duncan (2009) stressed that trees of citrus are usually injured by numerous PPNs, all but only one of them are of limited global distribution. That is because *T. semipenetrans* is the most economically remarkable species by virtue of its worldwide existence in almost every citrus producing area. Because *T. semipenetrans* and the likes of semi-endoparasites utilize only the nematode head to pierce the root, the nematode posterior part lies outside the root or in the soil; i.e., the female's body swells by feeding but outside the roots. Thus, members of this group are set at a fixed site on the root as well. Species of *Pratylenchus* and *Ditylenchus* genera represent examples of migratory endoparasites of a third group. Such species can break into and move within plant roots to feed. Contrary to the previous two groups, they can also quit the root to come in another one. The fourth group contains ectoparasites, e.g., *Helicotylenchus* spp. and other spiral nematodes. They do not get into the roots but feed on the nutrients found at root top and outermost cells via inserting their long/robust feeding stylets. These ectoparasites always move in soil looking for plant roots to insert the stylets only and absorb nutrients. Subsequent of feeding, second-stage juveniles (J2) develop to J3, J4, and finally the adult but without sexual dimorphism for two of the four groups, that is, ectoparasites and migratory endo-parasites. Usually, oviposition and the amount of eggs depend on the nematode species as well as biotic and abiotic factors in intimate contact with the feeding and development of nematode females. However, the eggs always hatch a subsequent generation of J2s. Ultimately, PPNs react with their plant hosts in various courses starting with the transient ectoparasite group, through the migratory endoparasitic and semi-endoparasitic groups, and ending with the sedentary endoparasitic group which involves an intimate sharing with their plant hosts.

## 12.2 Losses Brought About by the Nematodes and Methods of their Traditional Management

As nematode pests can be grouped into different categories, there are also challenges related to the plant species/cultivars rank concerning the losses that are inflicted on them. Overall, PPNs can cause substantial damage in qualitative and quantitative crop yields worldwide. Assessing crop damages brought about by PPNs for the twenty life-sustaining crops averaged 12.6% of global crop yield (Abd-Elgawad and Askary 2015). Such losses equaled $215.77 billion of annual yield. Other twenty crops with high rates for food/export have 14.45% annual loss of production amounting to $142.47 billion. Thus, total of the forty crops sustain a mean of 13.5% waste amounting to $358.24 billion annually (Abd-Elgawad and Askary 2015). Openly, these figures will probably increase on global adding of other PPN-infected plant species/cultivars to the records of losses.

Presently, these nematode pests are classically managed using various methods within crop production schemes. These methods comprise one or more of the following techniques: nematode-resistant plants and crop sequence, chemical nematicides, use of certified transplants, soil amendments, bionematicides, flooding, fallowing, solarization, and tillage. Utilizing nematode-resistant plants and crop sequence/rotation are reliable and benign techniques for controlling PPNs, were it not for the shortage in nematode-resistant/immune plant species and cultivars required in the crop rotation. Since most nematode pests wear out their lives within the soil of plant rhizosphere or within the roots, chemical nematicidal delivery to such surroundings of the nematodes is generally not easy (Chitwood 2003), yet, many of these chemicals are classically and frequently used for reliable PPN control (Abd-Elgawad et al. 2016).

Regrettably, the environmental and human health risks posed by many chemical nematicides, coupled with the emergence of resistance-breaking PPN strains due to their overuse, have necessitated a search for safer, more sustainable alternatives. While using certified seeds and transplants is a sound practice, they should not be planted directly into nematode-infested soils—a common and often unavoidable challenge. Furthermore, although soil amendments can promote plant growth in poor soils, they also risk inadvertently increasing PPN population densities or disrupting biocontrol agents (BCAs). Therefore, soil amendments must be applied with careful and informed management (Timper 2014).

Researchers have been addressing relevant additions such as plant extracts and botanical matrices with some promising results, but registration dealings and routine time-consuming have slowed down their commercial usage (Ntalli et al. 2020a). Fundamental conditions for using these materials are their effectiveness, safety, and advantageous economics (Dutta et al. 2019). Bionematicides are foremost friendly alternatives, but they are mostly slowly acting, less reliable, and more unstable than the synthetic nematicides (Abd-Elgawad and Askary 2020). Fallowing, solarization, and flooding are sometimes utilized in nematode control but are not mostly economic for nematode management plans. Tillage is advantageous not only against PPNs but also for many other plant pests/pathogens and weeds. However, tillage may directly affect populations of PPN-antagonistic organisms severely and consequently boost nematode damage on infected plants (Timper 2014). Finally, the aforementioned PPN cultural, chemical, and biocontrol methods are not soundly assumed/accepted and require tangible revisions for their safety and/or effectiveness (Hada et al. 2021a; Ali et al. 2017; Sasanelli et al. 2021). Additionally, their demerits and infirmity are mostly discouraging in connection with the overall low accuracy/precision in sampling the subterranean nematodes and their life stages within plant parts as well as a broad nematode-host range and their diversified and uneven distribution (Hada et al. 2021a).

Therefore, numerous researchers (Hada et al. 2021b) have recently evidenced that it is problematic to recommend a convenient nematode-management strategy that is effective, economical, environmentally friendly, and non-dangerous to the non-targets. Rather, growers and stakeholders would turn to use additional emerging products as agricultural practices for controlling nematode pests. Plant metabolic products to manage PPNs rank high among such emerging products. They are especially timely for numerous crops when the above-mentioned techniques and resources are not available or have environmentally unfriendly trends. In order to optimize their interests, such metabolic products can be involved in integrated nematode management (INM) schemes to function as complimentary or even eminent combinations in PPN management approaches. Advanced strategies for applying these metabolites should develop them into commercial products that can act additively/synergistically with other agricultural inputs in integrated pest management (IPM) programs. Such strategies aspire for combining bio-nematicides with other pesticides for more and multiple effectiveness against plant pests and pathogens to be applied on a wider basis. This is especially significant because there are numerous of such nematicidal products that are currently or are likely to become broadly available soon for commercial usage.

## 12.3 Plant Metabolic Products that Impact Plant-Parasitic Nematodes

### *12.3.1 General Idea About the Related Metabolic Products and Their Tests*

Plant metabolic products which suppress PPNs are divided into phytoalexins and phytoanticipins (Desmedt et al. 2020). Phytoalexins are mostly accumulated upon perception of pests such as PPNs but phytoanticipins are defense materials that are constitutively present (VanEtten et al. 1994). Yet, both groups may be constitutively available but increase in their levels after PPN attack. They may be found in definite plant parts but generated only upon pest/pathogen invasion in other tissues. Moreover, although phytoalexins may be found as an inactive arsenal such as some glycosides, they are freed upon pest/pathogen conception (VanEtten et al. 1994).

To study the influence of plant metabolic products on PPNs, researchers often expose the target nematodes to tested metabolite(s) in an in vitro assay to know whether it is directly toxic. Usually such an assay includes incubating the target nematodes in the product-water solutions of biologically proper concentrations for several hours/days. If the nematodes become rigid/immobile, the metabolic product is grouped as nematistatic (paralyzing the nematodes). Further test should transfer the nematodes from the solutions to clean water to know whether PPNs are reversibly paralyzed (nematistatic efficacy) or dead (nematicidal efficacy). The possible change in the PPN-motility/activity may distinguish whether the product possesses nematistatic or nematicidal efficacy. Nonetheless, nematicidal and nematistatic effects are just points on a continuum. Low rates and/or brief exposures of specific metabolite might be nematistatic, but higher rates and/or longer exposures are nematicidal (Desmedt et al. 2020). Other plant metabolites which also suppress PPNs may not be released outside the plant parts but are encoded via plant resistant gene(s) or R gene(s). These genes may have qualitative expression where the nematode cannot multiply or quantitative expression where the nematode multiplication is reduced, significantly lower reproduction rate than is typical for this host (Abd-Elgawad 1991). On the other hand, R genes may operate pre- or post-penetration of the PPNs. Pre-penetration R genes disable entering the nematode to the host plant such as those genes that express repellent host exudates. Post-penetration R genes allow the nematodes to enter the plant but make them unable to survive, develop, or reproduce. This inability may be due to various R gene expressions such as the existence of toxic metabolites that suppress PPN development/reproduction or blocking nematodes' feeding to kill the nematodes (Abd-Elgawad 2022b). For sedentary nematodes, the plant resistance can be either early or late resistance. Early resistance happens during PPN migration or early feeding site formation. Late resistance takes place after the PPN has set a feeding site (Fuller et al. 2008). In response, PPN may react to root exudate signals with host-specific gene expressions (Abd-Elgawad 2022b).

### *12.3.2 Impact Types of These Metabolites on the Nematodes*

As most PPN genera and species are subterranean parasites (Abd-Elgawad 2020a), the impact of plant root metabolites on these nematodes is mainly targeted in this chapter. Nonetheless, secondary metabolites can also be emitted as volatile compounds from various plant parts. Examples of plant volatile organic compounds (VOCs) that possess nematicidal properties are presented (Table 12.1). These

**Table 12.1** Examples of plant volatile compounds that can negatively affect different plant-parasitic nematodes

| Plant species/family | Volatile compound | Nematode species | References |
|---|---|---|---|
| *Rosmarinus officinalis*/Lamiaceae | 1,8-cineole, camphor, α-pinene | *Meloidogyne incognita* | Stadler et al. (1994) |
| *Origanum vulgare*/ Lamiaceae | Carvacrol, thymol, terpinen-4-ol | *M. incognita* | Ntalli et al. (2010, 2020b) |
| *Tagetes lucida*/ Asteraceae | Volatile oils | *M. incognita* and species of *Pratylenchus*, *Criconemella*, *Helicotylenchus* | Omer et al. (2015) |
| *Thymus caespititius*/ Lamiaceae | Carvacrol | *Bursaphelenchus xylophilus* | Barbosa et al. (2010) |
| *Eucalyptus meliodora*/ Myrtaceae | Trans-anethole, benzaldehyde | *M. incognita* | Ntalli et al. (2011) |
| *Armoracia rusticana*/ Brassicaceae | Allylisothiocyanate | *M. incognita* | Aissani et al. (2013) |
| *Mentha spicata*/ Lamiaceae | Carvone, volatile oils | *M. incognita* and species of *Hoplolaimuss Criconemella*, *Rotylenchulus* | Caboni et al. (2013), Abd-Elgawad and Omer (1995) |
| *Citrus reticulate*/ Rutaceae | Limonene | *M. incognita* | Goyal et al. (2021) |
| *Cymbopogon citratus*/Poaceae | Geranial, neral, ββ-myrcene | *B. xylophilus* | Barbosa et al. (2010) |
| *Zanthoxylum alatum*/ Rutaceae linalool, limonene, methyl | Linalool, limonene, methyl trans-cinnamate, 1,8-cineole | *B. xylophilus* | Kim et al. (2011) |
| *Citrus sinensis*/ Rutaceae | Limonene | *M. incognita* | Avato et al. (2017) |
| Brassica juncea/ Brassicaceae | Isothiocyanate | *M. incognita* | Barros et al. (2014b) |
| *Cuminum cyminum*/ Apiaceae | γ-terpinene, *p*-cymene | *M. incognita* | Pardavella et al. (2020) |
| *Cinnamomum zeylanicum*/ Lauraceae | (E)-cinnamaldehyde, eugenol | *M. incognita* | D'Addabbo et al. (2020), Barros et al. (2021) |

compounds can boost the resistance of plants to PPNs via stimulating hormone-dependent signaling pathways (Khan et al. 2023). Such types of secondary metabolites or VOCs possess a variety of tasks in nature, involving the ability to attract, repel, or even kill microorganisms. Also, numerous VOCs can be found in water for prolonged periods to function as nematicides for longer time (Silva et al. 2019). Many studies have suggested them as nematicidal materials (Khan et al. 2023). The nematicidal effects of VOCs from *Cymbopogon nardus* and *Dysphania ambrosioides* on the root-knot nematode (RKN) were so high that they killed more than 40% of *M. incognita*-J2 (de Freitas Silva et al. 2020). Such volatiles freed from *Azadirachta indica*, *Cajanus cajan*, *Brassica juncea*, *Mucuna pruriens*, and *Canavalia ensiformis* demonstrated suppression for *M. incognita* populations (Barros et al. 2014a). Others (Mei et al. 2021) proved the nematicidal potential of VOCs against economically important PPNs, e.g., VOCs of *Duddingtonia flagrans* on RKNs.

Generally, many plant metabolites generated by above- or below-ground plant parts are components used in common plant processes. About 5–21% of products generated via photosynthesis are freed in the type of root exudates into the rhizosphere (Haichar et al. 2014). Their release to the rhizosphere soil may be active via root secretion or passive via transmission of metabolites caused by osmotic differences between root cells and the soil solution. Although numerous root metabolites are beneficial in terms of repelling or suppressing PPN population, others may function in the opposite direction via attracting the nematodes to plant roots or/and stimulating their egg-hatching (Abd-Elgawad 2022b). Even those compounds that attract PPNs are sometimes beneficial; e.g., furfural acetone has attract-and-kill effect on the attacking nematodes (Cheng et al. 2021). Examples of plant metabolites that can affect (attract/repel) PPN movement as well as possess egg-hatching inhibition/stimulation for economically important PPN species are given (Tables 12.1 and 12.2). If more information is needed, recent references (Abd-Elgawad 2022b; Khan et al. 2023; Silva et al. 2019; de Freitas Silva et al. 2020; Sikder and Vestergård 2020; Sharma et al. 2023; Li and Zhang 2023; Maleita et al. 2023) can be consulted for further studies on structure, nematicidal performance, and regulation as well as biosynthesis of plant metabolic products that affect PPNs. Admittedly, there is an ongoing flow of such information resulting from continuous relevant investigations.

Revisions on the suppressing effects of phytoalexins on PPNs were given by Chitwood (2002). *Pratylenchus scribneri*, but not *M. javanica*, motility was impeded by coumestrol at 5–25 μg/ml. Coumestrol is produced by lima beans in response to *P. scribneri* infection (Rich et al. 1977). Piling up the phytoalexin glyceollin in roots of soybean plant was linked to an incompatible response to RKNs. This glyceollin could strongly but reversibly inhibit *M. incognita* movement in vitro at 15 μg/ml (Kanagy and Kaya 1996; Kaplan et al. 1980). Medicarpin, as the main phytoalexin of alfalfa, occurs in higher concentrations in the resistant cultivars and impede *P. penetrans* motility (Baldridge et al. 1998). It was extracted as nematicidal principle in *Taverniera abyssinica* (Fabaceae) roots (Stadler et al. 1994). Also, the sesquiterpenoid solavetivone as phytoalexin is genetically responsible for resistance for *Globodera rostochiensis* in potato (Desjardins et al. 1997).

On the other hand, plant hormones have key roles in the resistance to PPNs. It is generally known that ethylene (ET), in addition to two other classical hormones, salicylic acid (SA) and jasmonate (JA), can produce and transmit signals to assist host plants against attacking pathogens (Grant and Jones 2009; Robert-Seilaniantz et al. 2011). Almost a common consensus is that SA functions against threats from biotrophic/hemi-biotrophic plant pathogens, but JA and ET defend the plants against

**Table 12.2** Various effects of plant metabolites on plant-parasitic nematodes

| Root metabolites | Assay concentrations | Test type/Plant species | nematode Species | Effect | References |
|---|---|---|---|---|---|
| Methyl salicylate, α-pinene, limonene, and tridecane | 20, 40, 80 ng/μl dissolved in hexane as as negative control | Olfactory assay with sand/*Capsicum annuum* | *Meloidogyne incognita* | Attractant | Kihika et al. (2017) |
| Zeatin | 4, 15.6, 62.3, 250, 1000 ng/ml dissolved in 2% DMSO; 2% DMSO as negative control and 100 ng/ml methyl salicylate as positive control | Sand assay/*Solanum lycopersicum* | *M. incognita* | Attractant | Kirwa et al. (2018) |
| Palmitic acid and linoleic acid | 0.5,1, 2, 4 mM dissolved in methanol as control | In vitro assay/*Solanum lycopersicum* | *M. incognita* | Repellent | Dong et al. (2014) |
| 1,2-dehydropyrrolizidine alkaloids | 1, 0.1 and 0.01 mg/ml, water as control | In vitro assay/*Crotalaria* sp. and *Senecio jacobaea* | *Heterodera schachtii*, *Pratylenchus penetrans*, *Phasmarhabditis Hermaphrodita*, *Rhabditis* sp. | Nematicidal and ovicidal | Thoden et al. (2009) |
| Medicagenic acid | 125, 250, 500, 1000 mg/ml | In vitro assay/*Medicago sativa* | *M. incognita* and *Xiphinema index* | Nematicidal | D'Addabbo et al. (2011) |
| Root exudate compounds | Exudates of tested plants, distilled water as control | In vitro assay/*Solanum sisymbriifolium* (cv. Sharp) | *M. arenaria*, *M. chitwoodi*, *M. hapla* and *M. hispanica*; J2s not able to penetrate the plant and highly resistant | Hatching inhibitor of J2s of the five Meloidogyne species tested | Dias et al. (2012) |
| 2-propenyl isothiocyanate | 0.002% | In vitro assay/*Brassica juncea* | Globodera pallida | Hatching inhibitor | Brolsma et al. (2014) |
| Root exudate compounds | 0.0001, 0.001, 0.01, 0.1, 1 mg/ml root leachate, water and buffer as control | In vitro assay/*Solanum tuberosum* | G. rostochiensis and G. pallida | Hatching stimulants and attractants | Devine and Jones (2003) |
| 4-hydroxybenzeneethanol | 0.3, 0.6, 1.2, 2.4, 4.8, 9.6, and 19.2 mM dissolved in ethanol, water as control | In vitro assay/*Allium fistulosum* | *M. incognita* | Hatching inhibitor | Li et al. (2018) |

| | | | | | |
|---|---|---|---|---|---|
| Allyl isothiocyanate as commercial product | 221 to 367 kg $ha^{-1}$ | Field test/*Solanum lycopersicum* | *Criconemella* spp. | 81–84% reduction in nematode population | Yu et al. (2019) |
| Furfural or furfural acetone | 10, 50, and 250 mg furfural acetone dissolved in water/pot | Pot test/*Melia azedarach* or *Paenibacillus polymyxa* | *M. incognita* | Attract and kill 82.8% of the juveniles | Ntalli et al. (2010), Cheng et al. (2021) |
| *Solanum tuberosum* | Monosaccharides | Soil with cysts was saturated with a solution of glucose and fructose (17 and 35 μg/ml, respectively | *Globodera pallida* | Hatching stimulants | (Bell et al. (2021)) |

necrotrophic pathogens, insect pests, and RKNs (Nahar et al. 2011). To operate, the mentioned hormones may function synergistically/antagonistically. Generally, ET and JA signaling pathways can act synergistically in dicotyledons, but SA and JA/ET mutually function for defense (Xie et al. 2022). Accumulation of these hormones and others commands the expression of pathogenesis-related (PR) genes in plant tissues. For example, while JA pathway has a significant role in the defense of rice plants against *M. graminicola*, pathways of the three hormones are alike in terms of their significant defense against *Hirschmanniella oryzae*, the migratory root nematode, and *Ditylenchus angustus*, the rice stem nematode (Nahar et al. 2012; Khanam et al. 2018) as well as *Aphelenchoides besseyi*, the white tip nematode (Xie et al. 2022). Active metabolites should be chosen also to enhance efficacy of bioformulated nematicides. This is especially needed as plant rhizosphere usually comprises uncountable numbers of microbial species, whose efficacy is only partially known at the best assumptions. Admittedly, while some plant metabolites offer food sources necessary for the growth of many microbial populations in the rhizosphere, other metabolites can have positive or negative effect on PPNs. Although many studies addressed the influence of root exudates on PPNs and the characterization of their nematicidal compounds, such an influence on the overall nematode community of the rhizosphere is less studied (Wang et al. 2023).

## 12.4 Common Groups of Plant Metabolites Known to Suppress Plant-Parasitic Nematodes

Diversities of secondary metabolic structure and composition are known for the basic and needful role of these metabolites as factors for the plant-survival strategy in a mostly stressed/hostile environment (Li and Zhang 2023). They can participate in processes regarding biotic/abiotic stress tolerance such as environmental extremes and resistance to plant pests/pathogens (Hartmann 2007). However, this chapter centers on specific compounds of these metabolites that demonstrated evidence of being engaged in plant protection/defense against PPNs. These latter of secondary metabolites are mostly found in plant tissues affected by PPN parasitism. They could be correlated to PPN resistance as these metabolites are either induced upon PPN infection or constitutively found in plant cells (Desmedt et al. 2020). Although numerous other nematicidal syntheses have been characterized in plant extracts, they do not show obvious role in the actual settings of plant-nematode interactions (Chitwood 2002). On the other, as the PPN-relevant plant metabolic products contain many thousands of such secondary metabolites (Viant et al. 2017), they could be grouped based on various criteria into different categories (Desmedt et al. 2020). Common groups of these metabolites mostly contain phenolic compounds, alkaloids, saponins, terpenoids, organosulfur compounds, benzoxazinoids, and/or glucosinolates.

Although the majority of phenolic compounds are generated via the phenylpropanoid pathway, less popular phenolic ones, like alkylresorcinols, are generated via polyketide metabolism (Baerson et al. 2010). These latter metabolites should also be examined in plant defense against the nematodes. Phenolic syntheses through phenylpropanoid pathway whether induced upon infection or constitutively have proved to play active role not only in PPN resistance (Hölscher et al. 2014), but also against many other pests and pathogens (Pratyusha 2022). Flavonoids constitute the biggest family of phenolic compounds in the plant secondary metabolites (Desmedt et al. 2020). Although flavonoids are generally known to repel PPNs (Chin et al. 2018), recent feedbacks suggest that they are likely species-dependent (Sharma et al. 2023). The flavonoids kaempferol, quercetin, and myricetin are repellent for the RKN *M. incognita* and the burrowing nematode *Radopholus similis* but not for the lesion nematode *Pratylenchus penetrans*. Also, daidzein, genistein, and luteolin could repel *R. similis* but were ineffective against *M. incognita* and *P. penetrans* (Wuyts et al. 2006). However, as these phenolic compounds have been recorded to correlate with PPN resistance in a broad variety of plant-nematode interactions (Desmedt et al. 2020), they may be utilized in emerging strategies for controlling certain nematode species. For example, soil amendments should be further harnessed via fermentation procedures to upgrade their role in terms of boosting their PPN antagonistic compounds like phenolics (Abd-Elgawad 2021a).

A lot of alkaloids are quite toxic and therefore can contribute in controlling plant pests/pathogens (Mithöfer and Boland 2012). *Arabidopsis thaliana*-alkaloid camalexin is an important phytoalexin that is involved in the mechanism of *A. thaliana* resistance against many pests and diseases. In a double mutant that is highly impaired in camalexin generation, an increase ($P \leq 0.05$) in the cyst nematode *Heterodera schachtii* reproduction was noticed relative to its wild genotype (Shah et al. 2015). Also, a similar *pad3* mutant demonstrated more ($P \leq 0.05$) susceptibility to RKN *Meloidogyne incognita* (Teixeira et al. 2016). The main alkaloid of tobacco (*Nicotiana tabacum*), i.e., nicotine, in the *M. incognita*-resistant cultivar was more ($P \leq 0.05$) than that in in the susceptible cultivar 4 days after RKN infection which has likely prompted the resistance (Davis and Rich 1987). Pyrrolizidine alkaloids mostly derived from the aromatic amino acids L-phenylalanine and L-tyrosine can also function against PPNs (Desmedt et al. 2020).

Saponins were reported to be tentatively linked to *Heterodera avenae* resistance. Fractionation of root tip extracts via HPLC-MS analysis for single-seed descent lines of oats with different *H. avenae*-resistance levels revealed three peaks whose abundance was positively linked with *H. avenae* resistance (Bahraminejad et al. 2008). Two resulting subfractions comprised materials partially identified as avenacins, a group of saponins found in oats. Because saponins were examined in a limited range of host plants that showed nematicidal effects of few saponin-rich plant extracts (Ntalli and Caboni 2012), it is premature to document the significance of saponins as plant components for nematode resistance.

Terpenoid aldehydes (TAs) of cotton are the most broadly examined group of terpenoids for plant-PPN interactions. They comprise gossypol and its derivatives. Mixed results were recorded about the relation between TA concentrations and cotton resistance to *M. incognita.* Therefore, Khoshkhoo et al. (1994) gave three possible interpretations for this observation: (i) TA piling up could be one of other factors responsible for resistance to *M. incognita* in cotton, (ii) TAs might not have main role in this resistance, or (iii) the ratio of various TAs rather than whole TA concentrations might define resistance. A raw TA mixture taken out by *Gossypium hirsitum*-root extraction with subsequent partial purification proved robust nematistatic efficacy against *M. incognita* juveniles with half maximal inhibitory concentration ($IC_{50}$) of 10–50 μg/ml and 125 μg/ml showed nematicidal performance (Veech 1979). Genetically engineered cotton lines carrying a gene engaged in salicylic acid (SA)-mediated immunity manifested boosted resistance to *Rotylenchulus reniformis* (the reniform nematode) in addition to other fungal pathogens (Parkhi et al. 2010). Various terpenes concentrations in exudates of pepper (*Capsicum annuum*)-roots were correlated with the varietal response (susceptibility/resistance) to *M. incognita* (Kihika et al. 2017). The authors found that these terpenes-containing exudates may contribute in PPN susceptibility by boosting or inhibiting host-finding ability of *M. incognita*-J2. In addition to such functions, terpenoids are also engaged as plant hormones in plant-PPN relations. For instance, the brassinosteroids are derived from a triterpenoid (Choe 2006) and different gibberellin forms are diterpenoids (Hedden and Thomas 2012). Such hormones have varying (mostly antagonistic) services for PPN-plant resistance (Bauters et al. 2018; Yimer et al. 2018).

Organosulfur compounds are components of all organisms in the form of definite essential amino acids of the tripeptide glutathione, and of enzymes, coenzymes, vitamins, and hormones; a few of these compounds have sound nematicidal activities. For instance, α-terthienyl is recognized for its insecticidal, nematicidal, antiviral, fungicidal, and cytotoxic effects. It is the major compound in charge of the nematicidal efficacy of marigolds (*Tagetes* spp.). The inhibiting effect on nematode populations in *Tagetes*-cultivated soil was attributed to polythienyl synthesis in *Tagetes* roots as well as their exudates especially *α*-terthienyl (Chitwood 2002). Controlling *Pratylenchus penetrans* (the lesion nematode) populations were strongly correlated with polythienyl content in 175 Asteraceae species based on their meta-analysis and 15 out of 16 α-terthienyl-producing Asteraceae species could significantly lower *P. penetrans* population levels (Gommers and Voorin'tholt 1976). Several modes of action for α-terthienyl against PPNs have been proposed. Hamaguchi et al. (2019) documented that α-terthienyl produces reactive oxygen species upon stimulation by light and/or peroxidase enzymes which cause oxidative stress within the nematode body. These authors found that α-terthienyl could also penetrate the nematode hypodermis. Asparagusic acid generated by asparagus (*Asparagus officinalis*) as another organosulfur compound in plant roots has nematicidal activities. At a 50 ppm, this acid possesses robust nematicidal performance toward several PPN species, e.g., it inhibited *Heterodera* egg hatching (Takasugi et al. 1975). Sikder et al. (2021) reported stronger effects of benzoxazinoids on nematode communities in the maize roots than on those in the rhizosphere. The benzoxazinoid concentrations in plant roots often correlated negatively with the comparative abundance of nematode sequence reads. Nonetheless, such correlations between benzoxazinoids and several plant-parasitic nematodes were sometimes positive. Furthermore, while the nematode (*Pratylenchus crenatus*) population was reduced in the *bx1* mutant line, that of *P. neglectus* was enriched. Hence, differential and selective effects of benzoxazinoids on soil nematodes relying on both the benzoxazinoid compound and the PPN species were suggested (Sikder et al. 2021). In another report (Frew et al. 2018), blocking generation of benzoxazinoids could lead to significant boost in susceptibility of wheat roots to *P. neglectus* which reflects the role of benzoxazinoids against PPNs.

Glucosinolates are metabolites originated from an amino acid and glucose that contain nitrogen and sulfur atoms. Brassica plants are famous for producing varying levels of glucosinolates. When these metabolites hydrolyze, they release unstable isothiocyanates with a main role to control insect pests (van-Dam et al. 2009) and plant pathogens. Favorable levels of suppressing PPN populations through biofumigation with brassica plant tissues have been globally documented (Dutta et al. 2019).

## 12.5 Modes of Action of These Root Metabolites Against PPNs

Obviously, there are two main approaches of plant metabolites to suppress PPNs. These are evident by either boosting plant resistance through setting plant physiology to resist the invading nematodes or (and) inhibiting PPN development/population growth via antinematode-active molecules of the metabolites (Abd-Elgawad 2022b). Because investigations about the exact roles of these molecules have only recently been addressed, information is rather scanty about the molecular mechanisms of root metabolites to suppress PPNs. However, metabolomics as an analytical profiling technique to assess and compare various metabolites found in biological samples can rapidly advance this discipline. Adjoining multivariate data analysis to high-throughput analytical chemistry can provide an advanced window on such metabolic mechanisms. A recent conceivable mechanism is a modulation of certain genes and protein kinases of the suppressed PPN populations (Sharma et al. 2023). Inhibition of egg-hatching by these metabolites is brought about via disabling the hatching factors or manifesting opposition to bind with receptors of these hatching factors (Ochola et al. 2020). Teillet et al. (2013) reported that *Arabidopsis thaliana*-root exudates could affect the negative expression of *M. incognita*-J2 genes for the root penetration. Sikder and Vestergård (2020) documented such a gene expression of various plant metabolites even before RKN contacting the plant roots. As much as 63 nominated genes were characterized and differentially signified within 1 h of applying the exudate, offering a proof of inducing shifts in *M. incognita*-gene expression by the root exudates. Duarte et al. (2015) found also that several parasitism genes in *M. hispanica*-J2 were differentially upregulated via root exudates of tomato plants.

Because PPN released-degrading enzymes enable the nematodes to enter/parasitize on their host tissues, the related molecules secreted by the nematodes are effective in susceptible hosts (Abd-Elgawad 2022b). These nematode effectors are represented by the molecules that shift host-cell physiology to boost invasion and development in susceptible plants. For instance, effectors of *Pratylenchus coffeae* can perceive the components of root exudates and respond via tailoring gene expression, such as those encoding $\beta$-1,4-endoglucanase (*Pc-eng-1*) and $\beta$-1,4-endoxylanase (*Pc-xyl*), for adequate host-parasitism (Sharma et al. 2023). This capacity of parasitism may interpret the broad host range of *P. coffeae* which may be present in other PPNs of similar host range. Also, *flp* gene expression encoding FMRFamide-like peptides is regulated by root exudates. These latter are engaged in the PPN locomotive, feeding, and development, showing a significant role in nematode chemotaxis. Also, *M. incognita* was attracted to crown daisy, *Chrysanthemum coronarium*, by lauric acid secreted by this plant roots at the lowest concentration (0.5–2.0 mM). On the contrary, the nematode was repelled at its higher concentration (4.0 mM), probably this mode of action is dependent on lauric acid's concentration (Dong et al. 2014). Two components of castor root exudates, i.e., palmitic acid and linoleic acid, can function as nematode inhibitory compounds. Expression of the genes *Miflp-18* and *Mi-mpk-1* that encodes mitogen-activated protein kinase was also suppressed in a concentration-dependent behavior by palmitic and linoleic acid, *Meloidogyne incognita* is repelled by them (Dong et al. 2018). Thus, *flp* gene silencing can assist PPN-invaded plants in suppressing nematode populations (Kimber et al. 2007).

Structure-activity relationships of these metabolites can also refer to their nematicidal performance (Li and Zhang 2023). For instance, a few aromatic compounds manifested elevated nematicidal performance against *G. rostochiensis* and *G. pallida* due to their carbonyl- and aromatic-ring-conjugated systems. Thus, t-chalcone and 4-phenyl-3-butyn-2-one had the best performance among the tested compounds. In contrast, replacing the carbonyl group with an amide functional group (cinnamamide) led to inactivation. The best active aromatics show a carbonyl-conjugated unit, with more nematicidal activity being shown as the conjugation is stretched to aromatic groups. Therefore, salicylaldehyde, phthalaldehyde, and cinnamaldehyde with carbonyl-conjugated aromatics possess $EC_{50}$ rates of 11, 11, and 12 mg $mL^{-1}$ for suppressing *M. incognita* populations, respectively (Caboni et al. 2014).

## 12.6 Exploitation of Plant Metabolic Products to Suppress the Nematodes

### *12.6.1 Classical Breeding Programs*

Programs of traditional plant breeding should work to benefit from the accumulating data of such beneficial plant metabolites. The related plans may address reducing their PPN-attractant components and/or increasing inhibitor components in newly produced genotypes/cultivars. For instance, mutant roots with ethylene signaling deficiency attracted more RKNs of three important species (i.e., *Meloidogyne javanica*, *M. hapla*, and *M. incognita*) than the wild type (Čepulyte et al. 2018). Hence, it is likely that products of ethylene-responsive pathways or ethylene itself act for repelling. Such an impact of ethylene is not as easy to perceive in cyst nematodes as in RKNs (Sikder and Vestergård 2020). *Heterodera glycines*, the soybean cyst nematode, populations were more ($P \leq 0.05$) attracted to soybean roots treated with ethylene synthesis inhibitor where abundant nematodes could also penetrate these roots. On the other hand, ethylene-treated roots were not only more attractive to *H. glycines* but also faster infected than non-treated plant roots (Kammerhofer et al. 2015; Hu et al. 2017). Therefore, further studies are needed to show whether ethylene is a broad-spectrum repellant especially for the PPN species in the genera *Meloidogyne*, *Heterodera*, and *Globodera*. Another avenue for plant breeder is to exploit and breed for the metabolite attractants if such materials can attract and kill the nematodes. Furfural acetone recorded a control rate of 82.8% against *M. incognita* via attracting and killing the nematodes (Cheng et al. 2021). Such evidence should lead to further studies of furfural acetone as a new PPN control agent.

### *12.6.2 Molecular Genetic Techniques*

Molecular investigations revealed that these metabolites modulate the expression of PPN genes for favorable (in susceptible plants) or unfavorable (in resistant/non-host plants) processes related to plant infection by the nematodes (Sikder and Vestergård 2020). According to the types and constituents of plant metabolites within and/or exuded from roots, PPN invasion, survival, development, and behavior are modulated (Khan et al. 2023; Timper et al. 2006; Dandurand and Knudsen 2016; Wang et al. 2018) to ease nematode infection/plant damage or prevent/reduce such an infection and injury. In contrast, nematode effectors in terms of PPN molecules secreted into the plant can manipulate different aspects of plant metabolism and immunity to make a susceptible host whenever possible (Eves-van den Akker et al. 2021). In other words, PPNs generate effector proteins to modulate the molecular/physiological pathways of the susceptible plants for their own interest. For instance, RKNs manufacture effector proteins in plant cells to alter host cell processes and ease nematode parasitism. When the effector proteins are engaged in parasitism, it is known as “parasitomes” or “secretomes.” While these effectors are included in both compatible and incompatible plant-nematode interactions, RKNs can modulate roots of susceptible plant via various effectors (Khan et al. 2023).

Interestingly, plant metabolites that are unfavorable for the infection may also induce effective colonization of PPN-suppressive microorganisms in the plant rhizosphere (Elhady et al. 2018; Topalovic and Heuer 2019; Topalovic et al. 2019). Therefore, these metabolites should be considered in the microbial-aided control to optimize their efficacy against PPNs via better grasping of the tripartite (plants, nematodes, and rhizosphere microbiomes) interactions (Sikder and Vestergård 2020). Because PPN typically locate their favorable host via root exudate signals, several metabolites may constitute long and short distance attractants of nematodes so that PPN can rightly reach root-occupied soil strata and individual roots of a host, respectively (Bird 2004; Perry 2005). Moreover, Reynolds et al. (2011) found that infective juveniles of both *Meloidogyne graminicola* and *M. incognita* take the shortest route to their susceptible host but the longest route to their poor hosts. They concluded that definite root metabolites function as attractants and repellants, respectively. In contrast, tiny lipophilic molecules as root exudates of tomato and rice plants had repellent effect on *M. incognita* and *M. graminicola*, respectively (Dutta et al. 2012). Admittedly, progress in crop transformation techniques, and developments in high-throughput sequencing technologies will enable better and more economical use of them in sustainable agricultural systems. Emerging and novel biotechnological devices especially those related to molecular/genetic methods that ease/expedite the detailed studies of gene expression and setting of nematode-giant cells are needed. Of course, such technologies should address not only the host metabolites and related defense tools but also the nematode effectors for better optimization of plant-nematode interactions. This may enable researchers to develop durable resistance and maintain agricultural productivity. To introduce current molecular devices, concise excerptions of such alternative methods are as follows:

#### 12.6.2.1 Small Interfering RNA

Host-produced small interfering RNA (siRNA) can successfully transfer Double-stranded RNA (dsRNA) into feeding PPNs to silence nematodes-vital genes, essential for nematodes to initiate feeding processes. The RNA interference (RNAi) machinery of host plants can process the dsRNAs to introduce siRNAs for ingestion by the PPNs. Khan et al. (2023) recently reviewed such significant advances where the PPN intakes dsRNA-derived siRNAs via its stylet during host infection. This ingestion leads to the inhibition of intended gene expression and subsequently blocking effective nematode parasitism. Examples of success in its application are given for RKN control in transgenic tobacco (Yadav et al. 2006), and soybean (Ibrahim et al. 2011).

#### 12.6.2.2 RNAi Technology

This technique has recently been applied as a promising method for PPN management. Because it targets PPN genes and comprises both plant and PPN RNAi machinery, host-mediated RNAi technique is considered the most adequate strategy for introducing nematode resistance into plants (Joshi et al. 2022). Briefly, RNAi aims at silencing nematode-effector genes that facilitate PPN parasitism and successful development within the host plant. Rinsing PPN with dsRNA can silence its targeted gene. Parasitic success is negatively affected, while nematode effector genes that are required for parasitism are silenced. This technique was applied to the cyst nematodes, *H. glycines* and *G. pallida* (Urwin et al. 2002), and RKN *M. incognita* (Rosso et al. 2005).

#### 12.6.2.3 Quantitative Trait Loci

Because quantitative Trait Loci (QTLs) is a genetic position, fixed via marker loci, that is located near an operating gene affecting a quantitative trait, it has a specific merit. It can offer insights into whether phenotypic differences are resulting from a few loci with big effects, or by many loci, each with small effects. Also, QTL analysis was utilized to test the genetic perspectives that were manifested previously to impact the efficacy/durability of resistance (Khan et al. 2023). Many QTLs were found to offer resistance to soybean plants against *H. glycines*, yet; only rhg1 and Rhg4 are ordinarily utilized (Nissan et al. 2022). On the other hand, *M. incognita*-resistance QTL was detected in a distal part of pearl millet LG2 (Liu 2012). Khan et al. (2023) speculated that QTL introgression into boosted cultivars may not obstruct other important plant traits in terms of adaptation, yield quality, or other physiological attributes. Characterizing and mapping QTLs in charge of *M. graminicola*'s resistance/tolerance may offer benign and cost-effective management devices to farmers and stakeholders (Galeng-Lawilao et al. 2018).

#### 12.6.2.4 Genetic Engineering Strategies

Genetic engineering simply aims at inserting an isolated desired gene into the genome of the organism which needs to express this gene. Such an insertion to shape nematode resistant genotypes in advanced plant breeding, via genetic engineering strategy, is considered the best cost-effective and ecologically safe technique for PPN control. As plant resistance is only reliable against limited or relatively few RKN species/races, proper identification of the target species/races is crucial to the success of plant resistance (Khanal et al. 2018). Moreover, leveraging synergies between the genes expressing useful plant metabolites and other PPN control tactics should be exploited. This is especially important when molecular markers can help in integrated pest management. In this respect, genome editing methods especially those using the CRISPR/Cas9 system may facilitate the bases of such genetic engineering strategy (Abd-Elgawad 2022b; Khan et al. 2023). On the other hand, *Mi-1* homologs can bestow simultaneous resistance against several pests/pathogens, including the most widespread RKNs (*M. incognita*, *Meloidogyne javanica*, and *M. arenaria*), sweet potato whitefly (*Bemisia tabaci*), potato aphids (*Macrosiphum euphorbiae*), and oomycetes (*Phytophthora infestans*) in tomato plants (Abd-Elgawad 2022b). Also, the uses of several safe approaches for the integrated nematode management were recently reviewed for important crops such as pepper (Abd-Elgawad 2020b) potato (Abd-Elgawad 2020c), tomato (Abd-Elgawad 2020d), and others (Abd-Elgawad and Askary 2018; Abd-Elgawad 2021a; Li and Zhang 2023; Sharma et al. 2023). Hence, integrating such plant metabolites to operate synergistically or additively with other PPN control methods in emerging programs for pest management should be earnestly attempted.

### 12.6.3 Commercial Plant Metabolic Products to Suppress the Nematodes

Successful developments concerning secondary metabolites of several microorganisms have been achieved. Striking examples are abamectin and milbemectin. Abamectin is a commercial form of avermectin which comprises avermectins B1a and B1b. It is generated from the actinomycete *Streptomyces avermitilis*. New compounds were further derived from abamectin to avoid its defects. They include marketed derivatives such as emamectin, ivermectin, selamectin, eprinomectin, and doramectin in order to manifest more pest control range, insecticidal performance, and lower toxicity to humans and the environment (Abd-Elgawad 2021a). Also, milbemectin is an antibiotic obtained from the soil bacterium *Streptomyces hygroscopicus* subsp. *Aureolacrimosus*. It is developed as Milbeknock® which possesses acaricidal and nematicidal activities, e.g., against root-knot nematodes (Talavera et al. 2020).

Likewise, using purified PPN-inhibitory plant-derived compounds may prove to be an efficient nematode management strategy too (Dutta et al. 2019; Abd-Elgawad 2021a; Zanón et al. 2014). Factually, numerous plant species can be employed as antagonists of nematodes either via their intercropping or rotation with susceptible plant species or through extracting and applying their antagonistic compounds. The most important plants that can be utilized against significant PPN species comprise *Azadirahta indica*, *Tagetes* spp., *Allium* spp., *Crotalaria* spp., and *Brassica* spp. (Abd-Elgawad 2021a). For instance, applying a purified dimethyl disulfide (DMDS) as volatile organic compound created by the cellular degradation of specific amino acid precursors found within the cytoplasm of *Allium* spp. (e.g., garlic, onion, and leek) on roots of tomato plants could demonstrate negative effect on RKN-gall formation (Silva et al. 2018). A commercial biofumigant of DMDS has been developed and successfully applied against economically important PPN species

(Sharma et al. 2023). It could significantly decrease infestations of both *M. incognita* and *Heterodera* sp. in tobacco fields (Zanón et al. 2014). Such an effect was shown in two field experiments via its shank application at the rates of 200 and 400 L/ha in Talayuela (Spain) and in Villafontana (Italy). Moreover, the nematicidal impacts of DMDS were obvious against species of *Meloidogyne* and *Globodera* on plants of tomato and potato, respectively, (Coosemans 2004) and against *P. penetrans* and *M. hapla* that were negatively affecting strawberry plants (López-Aranda et al. 2009). Sharma et al. (2023) recently reviewed the favorable effects of DMDS against *H. carotae* and *M. incognita* on carrots (Curto et al. 2014) as well as *Meloidogyne* spp. on tomato (Myrta et al. 2018) and watermelon plants (Leocata et al. 2014).

On the other hand, other products of glucosinolates (GSLs), such as dehydroerucin, gluconapin, and glucotropeolin are secreted by members of the Brassicaceae family and were also found lethal to nematodes (Dutta et al. 2019). Obviously, biofumigation through the toxic volatile isothiocyanates (ITCs), the freed pesticidal compounds from brassicaceous plants, can suppress various populations of soil-borne pathogens such as PPNs. Therefore, favorable effects of Dazitol®, fabricated from mustard oil and oleoresin of pepper on *Meloidogyne javanica* infecting tomato plants were obtained (Hajji-Hedfi et al. 2018). Although Dazitol did not boost plant growth in the pot test, it lowered significantly *M. javanica* damage leading to a reduction in gall index on tomato roots (91%) and *M. javanica* population in soil (62%) relative to the untreated plants. Moreover, the greenhouse experiment demonstrated that Dazitol® significantly ($P < 0.05$) lowered *M. javanica* incidence on tomato too. The authors (Hajji-Hedfi et al. 2018) concluded that the best suppression of *M. javanica* populations was achieved by combining soil solarization with chemical or Dazitol nematicides via IPM scheme. Also, Cao et al. (2007) stressed that Dazitol contains as high as 4.37% ITC. Hence, it can be deemed as a good alternative to commercial fumigants and should be integrated in nematode management. Yet, the efficacy of solarization on suppressing PPNs (Hammam et al. 2022) should be practiced to improve such biofumigation impacts. In addition, accurate knowledge about GSL contents of various commercially available brassica genotypes is imperative prior to applying them in biofumigation programs (Dutta et al. 2019).

Recently, Li and Zhang (2023) discussed why there are only a few commercial metabolites. They indicated that some metabolites show high nematicidal activity in vitro against particular stages of PPN development, but they still manifest low performance or instability under actual field conditions. Hence, their application has been limited, and few metabolites have been fabricated as nematicidal products. Also, preparing nematicidal metabolites from plants is frequently costly. Although hermolides exhibit high nematicidal efficacy (Guo et al. 2012) its usage is limited due to its low production capacity. Improving the preparation of these hermolides and other natural compounds and enhancing the yields of active metabolites via bioengineering technology should be research priorities for advancing application of nematicidal metabolites (Li and zhang 2023). Eventually, emerging strategies should detect nematicidal metabolites that possess low-toxicity, highly-efficiency, ease of availability, and low-cost for PPN management.

## 12.7 Challenges Versus Opportunities for Utilizing Root Metabolites to Suppress PPNs

A good number of root metabolites that negatively affect the behavior, development, and more significantly survival of PPNs have been documented (e.g., Tables 12.1 and 12.2). Yet, the current recognition that root metabolites can offer satisfied PPN control under certain condition remains possibly unexploited by stakeholders and growers for reasons that have become more evident with increasing the related obstacles. One of such obstacles is materialized in the aforementioned capacity of root metabolites which is often examined on a single PPN species. Therefore, techniques and materials relevant to these metabolites have achieved mixed success. Hence, their exact function in regulating behavior, development, and survival of specific PPN should be further researched (Li and Zhang 2023; Sharma et al. 2023). Although their traditional inclusion in crop sequence, intercropping, and green manure treatments has been frequently met with success, upgrading such methods is a priority. As related metabolic research is in its early phases, further investigations to explore the leverage of root metabolites on a broader range of PPN taxa are imperative. These investigations on the impact of root metabolites should include their influence not only on harmful PPNS but also their side effects on beneficial nematodes; e.g., predatory and entomopathogenic nematodes (EPNs). Kanagy and Kaya (1996) found that $\alpha$-terthienyl at rates of 20 and 40 ppm had unfavorable impacts on non-target, beneficial nematodes. In other words, EPN numbers that infected insect pests were drastically reduced by synthetic $\alpha$-terthienyl at the aforementioned rates. Therefore, it is essential to prioritize metabolites that are safe or even advantageous to non-target nematodes. Moreover, screenings of plant species/genotypes with root metabolites that function as attractant/repellent for PPNs and possess supplemental merit(s) regarding intercropping attributes are direly needed (Abdel-Baset 2020; Sharma et al. 2023). It is interesting to identify root metabolic materials that suppress various species of PPN populations.

Infection levels of PPNs should be decreased via selecting and breeding of plant cultivars/genotypes that generate big quantities of repellents. It is interesting that such repellents have been detected in economically presumed host plants where RKNs are a main issue, e.g., tomato. In vegetables, plant resistance to PPNs may be obtained by grafting the above-ground scion with high yielding in quantity and/or quality onto an adequate rootstock with resistance to particular PPN species. Grafting offers a lucrative alternative to conventional and transgenic breeding for nematode resistance. In addition to grafting capability to alter plant metabolism and nutrient mobility, it can decrease the PPN damage via resistant gene-mediated plant responses, and alterations in plant vigor/physiological traits, as well as adjustment of microbial community in the plant rhizosphere (Phani et al. 2023). For instance, histopathological studies showed that *M. incognita*-resistant rootstock of melon *Cucumis metuliferus* could arrest RKN development in grafted plants. This was materialized in *M. incognita*-giant cells in terms of their smaller size and less volume with fewer nuclei relative to the susceptible *C. melo* cv. Paloma. Hence, Expósito et al. (2020) concluded that RKN-feeding cells in grafted plants could limit or adjust the root metabolic performance for *M. incognita* nourishment. Also, Sigüenza et al. (2005) found that a rootstock that releases PPN-repelling metabolites could show better PPN management.

Although a great deal of root metabolites-related research undergoes in vitro tests to obtain a conclusive evidence for the definite effects of the targeted compound, the physicochemical traits of soil can change/eliminate such effects (Sharma et al. 2023). Hol (2011) stressed that nutrient availability in soil can considerably affect the generation of root-defensive compounds. Additional edaphic factors like soil moisture and structure can modify the spread and hence impact the distribution of root volatile syntheses in the soil (Hiltpold and Turlings 2008). Obviously, these factors can impact the response of PPN to volatile cues. Therefore, in vivo/field experiments are essential to evaluate the impact of specific genotypes that are overexpressing/impaired in definite metabolic pathways under various abiotic conditions such as different soil types. Likewise, below/above-ground plant pests influence the production of numerous root metabolites (Karssemeijer et al. 2020) which in turn affect their impact on PPNs. Presence of *Solanum tuberosum* root exudates with *Myzus ersicae*, green peach aphid, led to not only a decrease in hatching of *G. pallida* but also an interference with cyst gene expression (Hoysted et al. 2018). Similarly, the inevitable microbial switch of released root metabolites may not adequately extrapolate the outputs attained by in vitro research for the sake of field settings (Zhou et al. 2020). Admittedly, the fauna and flora of the plant rhizosphere may decrease or disable the performance of nematicidal metabolites in soil.

Like other nematicides (Abd-Elgawad 2021a), three factors should govern the magnitude of using plant metabolic products against PPNs. These are safety to human health and the environment, reliable nematicidal impact, and supportive economics. In this respect, a lot of species in Brassicaceae family which comprise cruciferous plants such as mustard, cabbage, canola, broccoli, and rape can synthesize GSLs. Hydrolysis of GSLs as secondary metabolites of these plant roots leads to volatile/toxic ITCs which function as bionematicides. Their nematicidal traits in the plant rhizosphere may be related to indole GSLs and aromatic GSLs (Abd-Elgawad 2021a). Hence, cruciferous plants are rotated with nematode-susceptible plants as successful practice for PPN control. However, noxious impacts of ITCs as in mustard have led to vulnerability and suppression of useful organisms too (Ramirez et al. 2009; Dutta et al. 2019; Abd-Elgawad 2021a). Non-target soil organisms are negatively impacted as both ITCs released from the roots and natural ITCs have the same mode of action; both react in a non-specific/irreversible way with amino acids (Ntalli and Caboni 2012). Therefore, harnessing their benign usage in pest management plans should further be investigated. Conversely, azadirachtin syntheses are safe nematicides that can degrade rapidly in soil with low potential diffusion. Azadirachtin possesses comparative selectivity with non-toxicity to humans and is applied in IPM plans (Grubišić et al. 2018). Metabolic compounds of *Tagetes* spp. in soil require more studies to set their degradation periods under actual field conditions. Such studies should optimize sampling and extraction techniques for accurate PPN quantification (Abd-Elgawad 2021b). Also, these and similar studies should consider that the aforementioned safety is relative term and metabolic compounds should be quantified because they are still chemicals. Nonetheless, several plant metabolic products have been synthesized to ensure more safety for human application than the commonly known synthetic chemical nematicides.

Further opportunities are there as potentially new antagonistic plants with such metabolic products could be detected/identified (Sikora et al. 2018). Moreover, new approaches should be sought for their efficacious and multi-purpose usages. A striking example is the neem, *Azadirachta indica*, tree where its secondary metabolite azadirachtin can demonstrate significant effects in combating not only important PPN species but also other insect pests in soil (Grubišić et al. 2018). More projects are required to define the best conditions for biocontrol activities of such biopesticides, e.g., PPN species that are most vulnerable to specific metabolites. Also, advantageous plant species and tillage systems, suitable ambient temperature, planting date, and adequate growth stage as well as synergistic/additive effects of other production inputs, should be tested to be best tailored for favorable PPN control via these metabolites. For example, temperatures were too high for brassicas cover crops to suppress nematode populations in soil (Dutta et al. 2019). Moreover, studies of economic feasibility for using any of the plant metabolic products against PPN should be done. It should evaluate the grower's sight to combat PPN losses, the relative costs of the targeted nematicidal product relative to other ones, the cost of the commodity, and its commercial value. Agricultural extensions should demonstrate the direct and indirect merits of these metabolic products and encourage farmers to use them.

I have to unravel the complexities regarding the differential efficacy of these metabolites among PPN species. As aforementioned, for instance, while ethylene generally repels RKNs, it operates to attract some species of *Heterodera* and *Globodera*. Thus, in soil containing a mixture of root-knot and cyst nematodes, ethylene exudates can decrease the competition between both species by suppressing RKN species resulting in enhanced infestation by cyst nematode species (Jones et al. 2013; Kammerhofer et al. 2015; Hu et al. 2017; Čepulyte et al. 2018). Moreover, information is scanty about the response of ectoparasitic nematodes to root metabolites in general. Most work has been done on endoparasitic migratory and sedentary nematodes. However, these nematodes may compete with the less studied ectoparasitic nematodes in contrasting ecological settings. Sharma et al. (2023) speculated that strategies that aim to suppress these endoparasites can boost the ectoparasitic nematode populations and increase crop injury. Eventually, characterizing such ecological settings with various PPN communities will enable us to face the related challenges and recognize their solutions for efficient usage of root metabolites in the PPN management.

## 12.8 Future Prospects

Accelerated concern over the unsafe historical background and adverse effects of synthesized chemical nematicides on human health and environmental pollution has imposed searching for benign alternatives. Plant metabolic products especially root exudates are emerging as safe substitutional nematicides. Nonetheless, better understanding of the effects and outcomes of root exudates on various beneficial and harmful nematodes is critical for developing alternate management programs. This is especially important as many arable soils have polyspecific nematode communities. This study addressed the prospective role of a broad variety of root exudates and compounds in plant-nematode association. The chapter stressed that plant metabolic products possess bioactive materials that demonstrate a principal role in many nematode physiological processes. They can repel, attract, stimulate, or suppress nematode populations in the plant rhizosphere. These compounds can also affect egg-hatching or even survival of subterranean PPNs. Therefore, utilizing nematicides purified from plants and using nematicidal plants are being practiced. Yet, there is still a long way to go to extract and apply more of plant metabolic products for PPN control. Briefly, effectiveness, safety, and economics of the relevant nematicidal products should be assured. This does not negate that there are prospects for more safe and reliable PPN management utilizing root metabolites at reasonable prices. Numerous plant metabolites manifested efficacy gap between their impact on PPNs in the controlled and field conditions; less effective in the actual field than in controlled conditions. Such a gap may be imposed by various factors such as physicochemical attributes of field soil, microbial degradation, and excessive dilution of the high metabolic concentrations added to the soil. Therefore, more investigations should be carried out to bridge this gap via exploring PPN-antagonistic plant species and their related metabolites for durable agriculture systems. This will help in achieving environmentally benign practices to control harmful nematodes.

**Acknowledgements** This research was supported by In-House project No. 13050112 entitled "Pesticide alternatives against soil-borne pathogens and pests attacking economically significant export crops." The author acknowledges the support in part of this study by the US-Egypt Project cycle 17 (no. 172) entitled "Preparing and evaluating IPM tactics for increasing strawberry and citrus production." This article is derived from the Subject Data funded in part by NAS and USAID, and that any opinions, findings, conclusions, or recommendations expressed in it are those of the authors alone, and do not necessarily reflect the views of USAID or NAS. Special thanks to Dr. Ibrahim E. Shehata for re-arranging the reference section according to the book style.

**AI Assistance Declaration** The authors declare that no generative artificial intelligence tools were used in the writing of this chapter. The authors are solely responsible for writing, analysis, drafting, and final content. The work has undergone full review and editing by the authors and has been peer-reviewed by specialists in the field.

# References

Abdel-Baset SH (2020) Effect of intercropping onion with sugar beet on productivity of both crops and root-knot nematodes control under different onion plant densities and slow-release N fertilizer rates. J Plant Prod Sci 9:61–75. https://doi.org/10.21608/jpps.2020.157286

Abd-Elgawad MMM (1991) A new rating scale for screening plant genotypes against root-knot and reniform nematodes. Anz Schadlingskde Pflanzenschutz Umweltschutz/J Pest Sci 64:37–39

Abd-Elgawad MMM (2020a) Plant-parasitic nematodes and their biocontrol agents: current status and future vistas. In: Ansari RA, Rizvi R, Mahmood I (eds) Management of phytonematodes: recent advances and future challenges. Springer Nature Singapore Pte Ltd., Singapore, pp 171–204. https://www.springer.com/gp/book/9789811540868

Abd-Elgawad MMM (2020b) Biological control agents in the integrated nematode management of pepper in Egypt. Egypt J Biol Pest Cont 30:70. https://doi.org/10.1186/s41938-020-00273-6

Abd-Elgawad MMM (2020c) Biological control agents in the integrated nematode management of potato in Egypt. Egypt J Biol Pest Cont 30:121. https://doi.org/10.1186/s41938-020-00325-x

Abd-Elgawad MMM (2020d) Optimizing biological control agents for controlling nematodes of tomato in Egypt. Egypt J Biol Pest Cont 30:58. https://doi.org/10.1186/s41938-020-00252-x

Abd-Elgawad MMM (2021a) Optimizing safe approaches to manage plant-parasitic nematodes. Plants 10:1911. https://doi.org/10.3390/plants10091911

Abd-Elgawad MMM (2021b) Optimizing sampling and extraction methods for plant-parasitic and entomopathogenic nematodes. Plants 10:629. https://doi.org/10.3390/plants10040629

Abd-Elgawad MMM (2022a) Challenges in field application of biopesticides for successful crop pest management. In: Askary T (ed) Pest management: methods, applications and challenges. Nova Science Publishers, Inc., New York, pp 331–366. https://doi.org/10.52305/VCWF8745

Abd-Elgawad MMM (2022b) Understanding molecular plant-nematode interactions to develop alternative approaches for nematode control. Plants 11:2141. https://doi.org/10.3390/plants11162141

Abd-Elgawad MMM, Askary TH (2015) Impact of phytonematodes on agriculture economy. In: Askary TH, Martinelli PRP (eds) Biocontrol agents of phytonematodes. CAB International, Wallingford, pp 3–49

Abd-Elgawad MMM, Askary TH (2018) Fungal and bacterial nematicides in integrated nematode management strategies. Egypt J Biol Pest Cont 28:74. https://doi.org/10.1186/s41938-018-0080-x

Abd-Elgawad MMM, Askary TH (2020) Factors affecting success of biological agents used in controlling plant-parasitic nematodes. Egypt J Biol Pest Cont 30:17. https://doi.org/10.1186/s41938-020-00215-2

Abd-Elgawad MMM, Omer EA (1995) Effect of essential oils of some medicinal plants on phytonematodes. Anz Schadling Pflanzenschut Umweltschut 68(4):82–84

Abd-Elgawad MMM, Koura FFH, Montasser SA, Hammam MMA et al (2016) Distribution and losses of *Tylenchulus semipenetrans* in citrus orchards on reclaimed land in Egypt. Nematology 18:1141–1150. https://doi.org/10.1163/15685411-00003020

Aissani N, Tedeschi P, Maietti A, Brandolini V, Garau VL, Caboni P et al (2013) Nematicidal activity of allylisothiocyanate from horseradish (*Armoracia rusticana*) roots against *Meloidogyne incognita*. J Agric Food Chem 61:4723–4727

Ali MA, Azeem F, Abbas A, Joyia FA, Li H, Dababat AA et al (2017) Transgenic strategies for enhancement of nematode resistance in plants. Front Plant Sci 8:750. https://doi.org/10.3389/fpls.2017.00750

Avato P, Laquale S, Argentieri MP, Lamiri A, Radicci V, D'Addabbo T et al (2017) Nematicidal activity of essential oils from aromatic plants of Morocco. J Pest Sci 90:711–722

Baerson SR, Schröder J, Cook D, Rimando AM, Pan Z, Dayan FE et al (2010) Alkylresorcinol biosynthesis in plants: new insights from an ancient enzyme family? Plant Signal Behav 5: 1286–1289. https://doi.org/10.4161/psb.5.10.13062

Bahraminejad S, Asenstorfer RE, Williams KJ, Hayasaka Y, Zwer PK, Riley IT et al (2008) Metabolites correlated with cereal cyst nematode resistance in oats (*Avena sativa*) identified using single seed descent lines. Nematol Mediterr 36:145–152

Baldridge GD, O'Neill NR, Samac DA et al (1998) Alfalfa (*Medicago sativa* L.) resistance to the root-lesion nematode, *Pratylenchus penetrans*: defense-response gene mRNA and isoflavonoid phytoalexin levels in roots. Plant Mol Biol 38:999–1010

Barbosa P, Lima AS, Vieira P, Dias LS, Tinoco MT, Barroso JG, Fgueiredo AC, Mota M et al (2010) Nematicidal activity of essential oils and volatiles derived from Portuguese aromatic flora against the pinewood nematode, *Bursaphelenchus xylophilus*. J Nematol 42:8–16

Barros AF, Campos VP, da Silva JCP, Pedroso MP, Medeiros FHV, Pozza EA, Reale AL et al (2014a) Nematicidal activity of volatile organic compounds emitted by *Brassica juncea*, *Azadirachta indica*, *Canavalia ensiformis*, *Mucuna pruriens* and *Cajanus cajan* against *Meloidogyne incognita*. Appl Soil Ecol 80:34–43

Barros AF, Oliveira RDL, Lima IM, Coutinho RR, Ferreira AO, Costa A et al (2014b) Root-knot nematodes, a growing problem for Conilon coffee in Espírito Santo state, Brazil. Crop Prot 55: 74–79

Barros AF, Campos VP, de Paula LL, Pedroso LA, Silva FDJ, da Silva JCP, de Oliveira DF, Silva GH et al (2021) The role of Cinnamomum zeylanicum essential oil, (E)-cinnamaldehyde and (E)-cinnamaldehyde oxime in the control of Meloidogyne incognita. J Phytopathol 169:229–238

Bauters L, Hossain M, Nahar K, Gheysen G et al (2018) Gibberellin reduces the susceptibility of rice, *Oryza sativa*, to the migratory nematode *Hirschmanniella oryzae*. Nematology 20:703–709. https://doi.org/10.1163/15685411-00003198

Bell CA, Mobayed W, Lilley CJ, Urwin PE et al (2021) Monosaccharide constituents of potato root exudate influence hatching of the white potato cyst nematode. PhytoFrontiers 1:258–266. https://doi.org/10.1094/PHYTOFR-11-20-0034-R

Bird DMK (2004) Signaling between nematodes and plants. Curr Opin Plant Biol 7(4):372–376. https://doi.org/10.1016/j.pbi.2004.05.005

Brolsma KM, van der Salm RJ, Hoffland E, de Goede RGM et al (2014) Hatching of *Globodera pallida* is inhibited by 2-propenyl isothiocyanate in vitro but not by incorporation of *Brassica juncea* tissue in soil. Appl Soil Ecol 84:6–11. https://doi.org/10.1016/j.apsoil.2014.05.011

Caboni P, Saba M, Tocco G, Casu L, Murgia A, Maxia A, Menkissoglu-Spiroudi U, Ntalli N et al (2013) Nematicidal activity of mint aqueous extracts against the root-knot nematode *Meloidogyne incognita*. J Agric Food Chem 61:9784–9788. https://doi.org/10.1021/jf403684h

Caboni P, Tronci L, Liori B, Tocco G, Sasanelli N, Diana A et al (2014) Tulipaline A: structure-activity aspects as a nematicide and V-ATPase inhibitor. Pestic Biochem Physiol 112:33–39

Cao A, Zhang WJ, Liu JH et al (2007) Progress in the alternatives to methyl bromide in soil disinfestation. Plant Prot 33:15–18

Čepulyte R, Danquah WB, Bruening G, Williamson VM et al (2018) Potent attractant for root-knot nematodes in exudates from seedling root tips of two host species. Sci Rep 8(1):10847. https://doi.org/10.1038/s41598-018-29165-4

Cheng W, Chen Z, Zeng L et al (2021) Control of *Meloidogyne incognita* in three-dimensional model systems and pot experiments by the attract-and-kill effect of furfural acetone. Plant Dis 105:2169–2176. https://doi.org/10.1094/PDIS-07-20-1501-RE

Chin S, Behm CA, Mathesius U et al (2018) Functions of flavonoids in plant-nematode interactions. Plants 7:85. https://doi.org/10.3390/plants7040085

Chitwood DJ (2002) Phytochemical based strategies for nematode control. Annu Rev Phytopathol 40:221–249. https://doi.org/10.1146/annurev.phyto.40.032602.130045

Chitwood DJ (2003) Nematicides. In: Plimmer JR (ed) Encyclopedia of agrochemicals, vol 3. Wiley, New York, pp 1104–1115

Choe S (2006) Brassinosteroid biosynthesis and inactivation. Physiol Plan 126:539–548. https://doi.org/10.1111/j.1399-3054.2006.00681.x

Coosemans J (2004) Dimethyl disulphide (DMDS): a potential novel nematicide and soil disinfectant. In: VI international symposium on chemical and non-chemical soil and substrate disinfestation-SD2004, vol 698, pp 57–64. https://doi.org/10.17660/ActaHortic.2005.698.6

Curto G, Dongiovanni C, Sasanelli N, Santori A, Myrta A et al (2014) Efficacy of dimethyl disulfide (DMDS) in the control of the root-knot nematode *Meloidogyne incognita* and the cyst nematode *Heterodera carotae* on carrot in field condition in Italy. In: VIII international symposium on chemical and non-chemical soil and substrate disinfestation, vol 1044, pp 405–410. https://doi.org/10.17660/ActaHortic.2014.1044.55

D'Addabbo T, Carbonara T, Leonetti P, Radicci V, Tava A, Avato P et al (2011) Control of plant parasitic nematodes with active saponins and biomass from *Medicago sativa*. Phytochem Rev 10(4):503–519. https://doi.org/10.1007/s11101-010-9180-2

D'Addabbo T, Argentieri MP, Laquale S, Candido V, Avato P et al (2020) Relationship between chemical composition and nematicidal activity of different essential oils. Plants 9:1546

Dandurand LM, Knudsen GR (2016) Effect of the trap crop *Solanum sisymbriifolium* and two biocontrol fungi on reproduction of the potato cyst nematode, *Globodera pallida*. Ann Appl Biol 169:180–189. https://doi.org/10.1111/aab.12295

Davis EL, Rich JR (1987) Nicotine content of tobacco roots and toxicity to *Meloidogyne incognita*. J Nematol 19:23–29

de Freitas Silva M, Campos VP, Barros AF, da Silva JCP, Pedroso MP, de Jesus Silva F, Gomes VA, Justino JC et al (2020) Medicinal plant volatiles applied against the root-knot nematode *Meloidogyne incognita*. Crop Prot 130:105057. https://doi.org/10.1016/j.cropro.2019.105057

Desjardins AE, McCormick SP, Plaisted RL, Brodie BB et al (1997) Association between solavetivone production and resistance to *Globodera rostochiensis* in potato. J Agric Food Chem 45:2322–2326

Desmedt W, Mangelinckx S, Kyndt T, Vanholme B et al (2020) A phytochemical perspective on plant defense against nematodes. Front Plant Sci 11:602079. https://doi.org/10.3389/fpls.2020.602079

Devine KJ, Jones PW (2003) Investigations into the chemoattraction of the potato cyst nematodes *Globodera rostochiensis* and *G. pallida* towards fractionated potato root leachate. Nematology 5(5):65–75. https://doi.org/10.1163/156854102765216704

Dias MC, Conceição IL, Abrantes I, Cunha MJ (2012) *Solanum sisymbriifolium* – a new approach for the management of plant-parasitic nematodes. Eur J Plant Pathol 133(1):171–179. https://doi.org/10.1007/s10658-012-9945-0

Dong L, Li X, Huang C, Gao Y, Zhong L, Zheng Y, Zuo Y (2014) Lauric acid in crown daisy root exudate potently regulates root-knot nematode chemotaxis and disrupts *Mi-flp-18* expression to block infection. J Exp Bot 65:131–141. https://doi.org/10.1093/jxb/ert356

Dong L, Li X, Huang C, Lu Q, Li B, Yao Y, Liu T, Zuo Y et al (2018) Reduced *Meloidogyne incognita* infection of tomato in the presence of castor and the involvement of fatty acids. Sci Hortic 237:169–175. https://doi.org/10.1016/j.scienta.2018.03.066

Duarte A, Maleita C, Abrantes I, Curtis R et al (2015) Tomato root exudates induce transcriptional changes of *Meloidogyne hispanica* genes. Phytopathol Mediterr 54:104–108. https://doi.org/10.14601/Phytopathol_Mediterr-14595

Duncan LW (2009) Managing nematodes in citrus orchards. In: Ciancio A, Mukerji KG (eds) Integrated management of fruit crops and forest nematodes. Springer Science+Business Media B.V., Berlin, pp 135–173. https://doi.org/10.1007/978-1-4020-9858-1

Dutta TK, Powers SJ, Gaur HS, Birkett M, Curtis RH et al (2012) Effect of small lipophilic molecules in tomato and rice root exudates on the behaviour of *Meloidogyne incognita* and *M. graminicola*. Nematology 14:309–320. https://doi.org/10.1163/156854111X612306

Dutta TK, Khan MR, Phani V et al (2019) Plant-parasitic nematode management via biofumigation using brassica and non-brassica plants: current status and future prospects. Curr Plant Biol 17:17–32. https://doi.org/10.1016/j.cpb.2019.02.001

Elhady A, Adss S, Hallmann J, Heuer H et al (2018) Rhizosphere microbiomes modulated by pre-crops assisted plants in defense against plant-parasitic nematodes. Front Microbiol 9:1133. https://doi.org/10.3389/fmicb.2018.01133

Eves-van den Akker S, Stojilkovic B, Gheysen G et al (2021) Recent applications of biotechnological approaches to elucidate the biology of plant–nematode interactions. Curr Opin Biotechnol 70:122–130. https://doi.org/10.1016/j.copbio.2021.03.008

Expósito A, Pujolà M, Achaerandio I, Giné A, Escudero N, Fullana AM, Cunquer M, Loza-Alvarez P, Sorribas FJ et al (2020) Tomato and melon *Meloidogyne* resistant rootstocks improve crop yield but melon fruit quality is influenced by the cropping season. Front Plant Sci 11: 560024. https://doi.org/10.3389/fpls.2020.560024

Frew A, Powell JR, Glauser G, Bennett AE, Johnson SN et al (2018) Mycorrhizal fungi enhance nutrient uptake but disarm defences in plant roots, promoting plant-parasitic nematode populations. Soil Biol Biochem 126:123–132. https://doi.org/10.1016/j.soilbio.2018.08.019

Fuller VL, Lilley CJ, Urwin PE et al (2008) Nematode resistance. New Phytol 180:27–44. https://doi.org/10.1111/j.1469-8137.2008.02508.x

Galeng-Lawilao J, Kumar A, De Waele D et al (2018) QTL mapping for resistance to and tolerance for the rice root-knot nematode, *Meloidogyne graminicola*. BMC Genet 19:53

Gommers FJ, Voorin'tholt DJM (1976) Chemotaxonomy of compositae related to their host suitability for *Pratylenchus penetrans*. Neth J Plant Pathol 82:1–8. https://doi.org/10.1007/BF01977341

Goyal L, Kaushal S, Dhillon NK, Heena et al (2021) Nematicidal potential of *Citrus reticulata* peel essential oil, isolated major compound and its derivatives against *Meloidogyne incognita*. Arch Phytopathol Plant Protect 54:449–467

Grant MR, Jones JDG (2009) Hormone (dis)harmony moulds plant health and disease. Science 324: 750–752. https://doi.org/10.1126/science.1173771

Grubišić D, Uroić G, Ivošević A, Grdiša M et al (2018) Nematode control by the use of antagonistic plants. Agric Consp Sci 83:269–275

Guo JP, Zhu CY, Zhang CP, Chu YS, Wang YL, Zhang JX, Wu DK, Zhang KQ, Niu XM et al (2012) Thermolides, potent nematocidal pks-nrps hybrid metabolites from thermophilic fungus Talaromyces thermophiles. J Am Chem Soc 134(50):20306–20309

Hada A, Patil BL, Bajpai A, Kesiraju K, Dinesh-Kumar S, Paraselli B, Sreevathsa R, Rao U et al (2021a) Micro RNA-induced gene silencing strategy for the delivery of siRNAs targeting *Meloidogyne incognita* in a model plant *Nicotiana benthamiana*. Pest Manag Sci 77:3396–3405

Hada A, Singh D, Papolu PK, Banakar P, Raj A, Rao U et al (2021b) Host-mediated RNAi for simultaneous silencing of different functional groups of genes in *Meloidogyne incognita* using fusion cassettes in *Nicotiana tabacum*. Plant Cell Rep 40(8):2287–2302. https://doi.org/10.1007/s00299-021-02767-5

Haichar FZ, Santaella C, Heulin T, Achouak W et al (2014) Root exudates mediated interactions belowground. Soil Biol Biochem 77:69–80. https://doi.org/10.1016/j.soilbio.2014.06.017

Hajji-Hedfi L, Rebai E, Larayedh A, Regaieg H, Horrigue-Raouani N et al (2018) Biological control of *Meloidogyne javanica* on tomato with Dazitol® and soil solarization. Environ Sci Pollut Res Int 25(18):17278–17282. https://doi.org/10.1007/s11356-018-1962-x. Epub 2018 Apr 12. PMID: 29651727

Hamaguchi T, Sato K, Vicente CSL, Hasegawa K et al (2019) Nematicidal actions of the marigold exudate α-terthienyl: oxidative stress-inducing compound penetrates nematode hypodermis. Biol Open 8:bio038646. https://doi.org/10.1242/bio.038646

Hammam MMA, Abd-El-Khair H, El-Nagdi WMA, Abd-Elgawad MMM et al (2022) Can agricultural practices in strawberry fields induce plant-nematode interaction towards *Meloidogyne*-suppressive soils? Life 12:1572. https://doi.org/10.3390/life12101572

Hartmann T (2007) From waste products to ecochemicals: fifty years research of plant secondary metabolism. Phytochemistry 68:2831–2846. https://doi.org/10.1016/j.phytochem.2007.09.017

Hedden P, Thomas SG (2012) Gibberellin biosynthesis and its regulation. Biochem J 444:11–25. https://doi.org/10.1042/BJ20120245

Hiltpold I, Turlings TC (2008) Belowground chemical signaling in maize: when simplicity rhymes with efficiency. J Chem Ecol 34:628–635. https://doi.org/10.1007/s10886-008-9467-6

Hol WHG (2011) The effect of nutrients on pyrrolizidine alkaloids in Senecio plants and their interactions with herbivores and pathogens. Phytochem Rev 10:119–126. https://doi.org/10.1007/s11101-010-9188-7

Hölscher D, Dhakshinamoorthy S, Alexandrov T, Becker M, Bretschneider T, Buerkert A et al (2014) Phenalenone-type phytoalexins mediate resistance of banana plants (*Musa* spp.) to the burrowing nematode *Radopholus similis*. Proc Natl Acad Sci USA 111:105–110. https://doi.org/10.1073/pnas.1314168110

Hoysted GA, Bell CA, Lilley CJ, Urwin PE (2018) Aphid colonization affects potato root exudate composition and the hatching of a soil borne pathogen. Front Plant Sci 9:1278. https://doi.org/10.3389/fpls.2018.01278

Hu Y, You J, Li C, Williamson VM, Wang C (2017) Ethylene response pathway modulates attractiveness of plant roots to soybean cyst nematode *Heterodera glycines*. Sci Rep 7:41282. https://doi.org/10.1038/srep41282

Ibrahim HM, Alkharouf NW, Meyer SL, Aly MA, El-Din AEKYG, Hussein EH, Matthews BF (2011) Post-transcriptional gene silencing of root-knot nematode in transformed soybean roots. Exp Parasitol 127:90–99

Jones JT, Haegeman A, Danchin EGJ, Gaur HS, Helder J, Jones MGK, Kikuchi T, Manzanilla-López R, Palomares-Rius JE, Wesemael WML et al (2013) Top 10 plant-parasitic nematodes in molecular plant pathology. Mol Plant Pathol 14:946–961. https://doi.org/10.1111/mpp.12057

Joshi I, Kohli D, Pal A, Chaudhury A, Sirohi A, Jain PK et al (2022) Host delivered-RNAi of effector genes for imparting resistance against root-knot and cyst nematodes in plants. Physiol Mol Plant Pathol 118:101802

Kammerhofer N, Radakovic Z, Regis JMA, Dobrev P, Vankova R, Grundler FMW et al (2015) Role of stress-related hormones in plant defence during early infection of the cyst nematode *Heterodera schachtii* in *Arabidopsis*. New Phytol 207(3):778–789. https://doi.org/10.1111/nph.13395

Kanagy JMN, Kaya HK (1996) The possible role of marigold roots and ®-terthienyl in mediating host-finding by steinernematid nematodes. Nematologica 42:220–231

Kaplan DT, Keen NT, Thomason IJ et al (1980) Association of glyceollin with the incompatible response of soybean roots to *Meloidogyne incognita*. Physiol Plant Pathol 16:309–318

Karssemeijer PN, Reichelt M, Gershenzon J, van Loon J, Dicke M et al (2020) Foliar herbivory by caterpillars and aphids differentially affects phytohormonal signalling in roots and plant defence to a root herbivore. Plant Cell Environ 43:775–786. https://doi.org/10.1111/pce.13707

Khan A, Chen S, Fatima S, Ahamad L, Siddiqui MA et al (2023) Biotechnological tools to elucidate the mechanism of plant and nematode interactions. Plants 12:2387. https://doi.org/10.3390/plants12122387

Khanal C, McGawley EC, Overstreet C, Stetina SR et al (2018) The elusive search for reniform nematode resistance in cotton. Phytopathology 108:532–541

Khanam S, Bauters L, Singh RR, Verbeek R, Haeck A, Sultan SM et al (2018) Mechanisms of resistance in the rice cultivar Manikpukha to the rice stem nematode *Ditylenchus angustus*. Mol Plant Pathol 19:1391–1402. https://doi.org/10.1111/mpp.12622

Khoshkhoo N, Hedin PA, McCarty JC et al (1994) Terpenoid aldehydes in root-knot nematode susceptible and resistant cotton plants. J Agric Food Chem 42:204–208. https://doi.org/10.1021/jf00037a037

Kihika R, Murungi LK, Coyne D, Ng'ang'a M, Hassanali A, Teal PEA et al (2017) Parasitic nematode *Meloidogyne incognita* interactions with different *Capsicum annum* cultivars reveal the chemical constituents modulating root herbivory. Sci Rep 7:2903. https://doi.org/10.1038/s41598-017-02379-8

Kim J, Seo S-M, Park I-K et al (2011) Nematicidal activity of plant essential oils and components from *Gaultheria fragrantissima* and *Zanthoxylum alatum* against the pine wood nematode, *Bursaphelenchus xylophilus*. Nematology 13:87–93

Kimber MJ, McKinney S, McMaster S, Day TA, Fleming CC, Maule AG et al (2007) *flp* gene disruption in a parasitic nematode reveals motor dysfunction and unusual neuronal sensitivity to RNA interference. FASEB J 21:1233–1243. https://doi.org/10.1096/fj.06-7343com

Kirwa HK, Murungi LK, Beck JJ, Torto B et al (2018) Elicitation of differential responses in the root-knot nematode *Meloidogyne incognita* to tomato root exudate cytokinin, flavonoids, and alkaloids. J Agric Food Chem 66:11291–11300. https://doi.org/10.1021/acs.jafc.8b05101

Leocata S, Pirruccio G, Myrta A, Medico E, Greco N et al (2014) Dimethyl Disulfide (DMDS): a new soil fumigant to control root-knot nematodes, *Meloidogyne* spp., in protected crops in Sicily, Italy. Acta Hortic 1044:415–420. https://doi.org/10.17660/ActaHortic.2014.1044.57

Li G, Zhang K (2023) Natural nematicidal metabolites and advances in their biocontrol capacity on plant parasitic nematodes. Nat Prod Rep 40(3):646–675. https://doi.org/10.1039/d2np00074a

Li T, Wang H, Xia X, Cao S, Yao J, Zhang L et al (2018) Inhibitory effects of components from root exudates of Welsh onion against root knot nematodes. PLoS One 13(7):e0201471. https://doi.org/10.1371/journal.pone.0201471

Liu L (2012) Genetic mapping and quantitative trait locus (QTL) analysis of root-knot nematode resistance in pearl millet. PhD thesis, University of Georgia, Athens, GA, USA

López-Aranda JM, Miranda L, Soria C, Pérez-Jiménez RM, Zea T, Talavera M, Romero F, De Los Santos B, Vega JM, Páez JI et al (2009) Chemical alternatives to methyl bromide for strawberry in the area of Huelva (Spain): 2002–2007 results. Acta Horticult 2:957–960. https://doi.org/10.17660/ActaHortic.2009.842.212

Maleita C, Esteves I, Ciancio A, Oka Y et al (2023) Editorial: sustainable strategies for the management of phytoparasitic nematodes. Front Plant Sci 14:1148261. https://doi.org/10.3389/fpls.2023.1148261

Mei X, Wang X, Li G et al (2021) Pathogenicity and volatile nematicidal metabolites from *Duddingtonia flagrans* against *Meloidogyne incognita*. Microorganisms 9:2268. https://doi.org/10.3390/microorganisms9112268

Mithöfer A, Boland W (2012) Plant defense against herbivores: chemical aspects. Annu Rev Plant Biol 63:431–450. https://doi.org/10.1146/annurev-arplant-042110-103854

Myrta A, Santori A, Zanón M, Tsimboukis N, Vries RD, Tommaso ND et al (2018) Effectiveness of dimethyl disulfide (DMDS) for management of root-knot nematode in protected tomatoes in southern Europe. Acta Horticult 1207:123–128. https://doi.org/10.17660/actahortic.2018.1207.16

Nahar K, Kyndt T, De Vleesschauwer D, Höfte M, Gheysen G et al (2011) The jasmonate pathway is a key player in systemically induced defense against root knot nematodes in rice. Plant Physiol 157(1):305–316. https://doi.org/10.1104/pp.111.177576

Nahar K, Kyndt T, Nzogela YB, Gheysen G et al (2012) Abscisic acid interacts antagonistically with classical defence pathways in rice–migratory nematode interaction. New Phytol 196:901–913. https://doi.org/10.1111/j.1469-8137.2012.04310.x

Nissan N, Mimee B, Cober ER, Golshani A, Smith M, Samanfar B et al (2022) A broad review of soybean research on the ongoing race to overcome soybean cyst nematode. Biology 11:211

Ntalli NG, Caboni P (2012) Botanical nematicides: a review. J Agric Food Chem 60:9929–9940. https://doi.org/10.1021/jf303107j

Ntalli NG, Ferrari F, Giannakou I, Menkissoglu-Spiroudi U et al (2010) Phytochemistry and nematicidal activity of the essential oils from 8 Greek Lamiaceae aromatic plants and 13 terpene components. J Agric Food Chem 58:7856–7863

Ntalli NG, Ferrari F, Giannakou I, Menkissoglu-Spiroudi U et al (2011) Synergistic and antagonistic interactions of terpenes against *Meloidogyne incognita* and the nematicidal activity of essential oils from seven plants indigenous to Greece. Pest Manag Sci 67:341–351

Ntalli N, Adamski Z, Doula M, Monokrousos N et al (2020a) Nematicidal amendments and soil remediation. Plants 9:429

Ntalli NG, Ozalexandridou EX, Kasiotis KM, Samara M, Golfinopoulos SK et al (2020b) Nematicidal activity and phytochemistry of Greek Lamiaceae species. Agronomy 10:1119

Ochola J, Cortada L, Ng'ang'a M, Hassanali A, Coyne D, Torto B et al (2020) Mediation of potato-potato cyst nematode, *G. rostochiensis* interaction by specific root exudate compounds. Front Plant Sci 11:649. https://doi.org/10.3389/fpls.2020.00649

Omer EA, Hendawy SF, Nour El-Deen AM, Zaki FN, Abd-Elgawad MMM, Kandeel AM, Ibrahim AK, Ismail RF et al (2015) Some biological activities of *Tagetes lucida* plant cultivated in Egypt. Adv Environ Biol 9(2):82–88

Pardavella I, Daferera D, Tselios T, Skiada P, Giannakou I et al (2020) The use of essential oil and hydrosol extracted from *Cuminum cyminum* seeds for the control of *Meloidogyne incognita* and *Meloidogyne javanica*. Plants 10:46

Parkhi V, Kumar VLAM, Bell AA, Shah J, Rathore KS et al (2010) Resistance against various fungal pathogens and reniform nematode in transgenic cotton plants expressing Arabidopsis NPR1. Transgenic Res 19:959–975. https://doi.org/10.1007/s11248-010-9374-9

Perry RN (2005) An evaluation of types of attractants enabling plant-parasitic nematodes to locate plant roots. Russ J Nematol 13:83–88

Phani V, Gowda MT, Dutta TK et al (2023) Grafting vegetable crops to manage plant-parasitic nematodes: a review. J Pest Sci. https://doi.org/10.1007/s10340-023-01658-w

Pratyusha S (2022) Phenolic compounds in the plant development and defense: an overview. In: Hasanuzzaman M, Nahar K (eds) Plant stress physiology-perspectives in agriculture. IntechOpen, London. https://doi.org/10.5772/intechopen.102873

Ramirez RA, Henderson DR, Riga E, Lacey LA, Synder WE et al (2009) Harmful effects of mustard green manure on entomopathogenic nematodes. Biol Control 48:147–154

Reynolds AM, Dutta TK, Curtis RHC, Powers SJ, Gaur HS, Kerry BR et al (2011) Chemotaxis can take plant-parasitic nematodes to the source of a chemo-attractant via the shortest possible routes. J R Soc Interface 8(57):568–577. https://doi.org/10.1098/rsif.2010.0417

Rich JR, Keen NT, Thomason IJ et al (1977) Association of coumestans with the hypersensitivity of lima bean roots to *Pratylenchus scribneri*. Physio Plant Pathol 10:105–116

Robert-Seilaniantz A, Grant M, Jones JDG et al (2011) Hormone crosstalk in plant disease and defence: more than just jasmonate-salicylate antagonism. Annu Rev Phytopathol 49:317–343. https://doi.org/10.1146/annurev-phyto-073009-114447

Rosso MN, Dubrana MP, Cimbolini N, Jaubert S, Abad P et al (2005) Application of RNA interference to root-knot nematode genes encoding esophageal gland proteins. Mol Plant-Microbe Interact 18:615–620

Sasanelli N, Konrat A, Migunova V, Toderas I, Iurcu-Straistaru E, Rusu S, Bivol A, Andoni C, Veronico P et al (2021) Review on control methods against plant parasitic nematodes applied in southern member states (C Zone) of the European Union. Agriculture 11:602. https://doi.org/10.3390/agriculture11070602

Shabrina NH, Lika RA, Indarti S et al (2023) Deep learning models for automatic identification of plant-parasitic nematode. Artif Intell Agric 7:1–12. https://doi.org/10.1016/j.aiia.2022.12.002

Shah S, Anjam M, Mendy B, Anwer M, Habash SS, Lozano-Torres J et al (2015) Damage-associated responses of the host contribute to defence against cyst nematodes but not root-knot nematodes. J Exp Bot 68:5949–5960. https://doi.org/10.1093/jxb/erx374

Shao H, Zhang P, Peng D, Huang W, Kong L-a, Li C, Liu E, Peng H et al (2023) Current advances in the identification of plant nematode diseases: from lab assays to in-field diagnostics. Front Plant Sci 14:1106784. https://doi.org/10.3389/fpls.2023.1106784

Sharma N, Khanna K, Jasrotia S et al (2023) Metabolites and chemical agents in the plant roots: an overview of their use in plant-parasitic nematode management. Nematology 25:243–257. https://doi.org/10.1163/15685411-bja10220

Sigüenza C, Schochow M, Turini T, Ploeg A et al (2005) Use of Cucumis metuliferus as a rootstock for melon to manage Meloidogyne incognita. J Nematol 37:276–280

Sikder MM, Vestergård M (2020) Impacts of root metabolites on soil nematodes. Front Plant Sci 10:1792. https://doi.org/10.3389/fpls.2019.01792

Sikder MM, Vestergård M, Kyndt T, Fomsgaard IS, Kudjordjie EN, Nicolaisen M et al (2021) Benzoxazinoids selectively affect maize root-associated nematode taxa. J Exp Bot 72:3835–3845

Sikora RA, Coyne D, Hallmann J, Timper P et al (2018) Plant parasitic nematodes in subtropical and tropical agriculture, 3rd edn. CAB International, Cambridge, p 876

Silva JCP, Campos VP, Barros AF, Pedroso MP, Terra WC, Lopez LE, de Souza JT et al (2018) Plant volatiles reduce the viability of the root-knot nematode *Meloidogyne incognita* either directly or when retained in water. Plant Dis 102:2170–2179. https://doi.org/10.1094/PDIS-01-18-0143-RE

Silva JCP, Campos VP, Barros AF, Pedroso LA, Silva MF, Souza JT, Pedroso MP, Medeiros FHV et al (2019) Performance of volatiles emitted from different plant species against juveniles and eggs of *Meloidogyne incognita*. Crop Protect 116:196–203

Stadler M, Dagne E, Anke H et al (1994) Nematicidal activities of two phytoalexins from *Taverniera abyssinica*. Planta Med 60:550–552

Takasugi M, Yachida Y, Anetai M, Masamune T, Kegasawa K et al (1975) Identification of asparagusic acid as a nematicide occurring naturally in the roots of asparagus. Chem Lett 4:4–44. https://doi.org/10.1246/cl.1975.43

Talavera M, Vela-Delgado MD, Verdejo-Lucas S et al (2020) Nematicidal efficacy of milbemectin against root-knot nematodes. Plants 9(7):839. https://doi.org/10.3390/plants9070839

Teillet A, Dybal K, Kerry BR, Miller AJ, Curtis RH, Hedden P et al (2013) Transcriptional changes of the root-knot nematode *Meloidogyne incognita* in response to *Arabidopsis thaliana* root signals. PLoS One 8:e61259. https://doi.org/10.1371/journal.pone.0061259

Teixeira MA, Wei L, Kaloshian I et al (2016) Root-knot nematodes induce pattern-triggered immunity in *Arabidopsis thaliana* roots. New Phytol 211:276–287. https://doi.org/10.1111/nph.13893

Thoden TC, Boppré M, Hallmann J et al (2009) Effects of pyrrolizidine alkaloids on the performance of plant-parasitic and free-living nematodes. Pest Manag Sci 65(7):823–830. https://doi.org/10.1002/ps.1764

Timper P (2014) Conserving and enhancing biological control of nematodes. J Nematol 46:75–89

Timper P, Davis RF, Tillman PG et al (2006) Reproduction of *Meloidogyne incognita* on winter cover crops used in cotton production. J Nematol 38(1):83–89

Topalovic O, Heuer H (2019) Plant-nematode interactions assisted by microbes in the rhizosphere. Curr Issues Mol Biol 30:75–87. https://doi.org/10.21775/cimb.030.075

Topalovic O, Elhady A, Hallmann J, Richert-Poggeler KR, Heuer H et al (2019) Bacteria isolated from the cuticle of plant-parasitic nematodes attached to and antagonized the root-knot nematode *Meloidogyne hapla*. Sci Rep 9:11477. https://doi.org/10.1038/s41598-019-47942-7

Urwin PE, Lilley CJ, Atkinson HJ et al (2002) Ingestion of double-stranded RNA by preparasitic juvenile cyst nematodes leads to RNA interference. Mol Plant-Microbe Interact 15:747–752

van-Dam NM, Tytgat TOG, Kirkegaard JA et al (2009) Root and shoot glucosinolates: a comparison of their diversity, function and interactions in natural and managed ecosystems. Phytochem Rev 8:171–186. https://doi.org/10.1007/s11101-008-9101-9

VanEtten HD, Mansfield JW, Bailey JA, Farmer EE et al (1994) Two classes of plant antibiotics: phytoalexins versus "phytoanticipins". Plant Cell 6:1191–1192. https://doi.org/10.1105/tpc.6.9.1191

Veech JA (1979) Histochemical localization and nematoxicity of terpenoid aldehydes in cotton. J Nematol 11:240–246

Viant MR, Kurland IJ, Jones MR, Dunn WB et al (2017) How close are we to complete annotation of metabolomes? Curr Opin Chem Biol 36:64–69. https://doi.org/10.1016/j.cbpa.2017.01.001

Wang C, Masler EP, Rogers ST et al (2018) Responses of *Heterodera glyci*nes and *Meloidogyne incognita* infective juveniles to root tissues, root exudates, and root extracts from three plant species. Plant Dis 102(9):1733–1740. https://doi.org/10.1094/PDIS-09-17-1445-RE

Wang X, Wang C, Chen R, Wang W, Wang D, Tian X et al (2023) Plant genotype shapes the soil nematode community in the rhizosphere of tomatoes with different resistance to *Meloidognye incognita*. Plants 12:1528. https://doi.org/10.3390/plants12071528

Wuyts N, Swennen R, DeWaele D et al (2006) Effects of plant phenylpropanoid pathway products and selected terpenoids and alkaloids on the behaviour of the plant-parasitic nematodes *Radopholus similis*, *Pratylenchus penetrans* and *Meloidogyne incognita*. Nematology 8:89–101. https://doi.org/10.1163/156854106776179953

Xie J, Yang F, Xu X, Peng Y, Ji H et al (2022) Salicylic acid, jasmonate, and ethylene contribute to rice defense against white tip nematodes *Aphelenchoides besseyi*. Front Plant Sci 12:755802. https://doi.org/10.3389/fpls.2021.755802

Yadav BC, Veluthambi K, Subramaniam K et al (2006) Host-generated double stranded RNA induces RNAi in plant-parasitic nematodes and protects the host from infection. Mol Biochem Parasit 148:219–222

Yimer HZ, Nahar K, Kyndt T, Haeck A, Van Meulebroek L, Vanhaecke L et al (2018) Gibberellin antagonizes jasmonate-induced defense against *Meloidogyne graminicola* in rice. New Phytol 218:646–660. https://doi.org/10.1111/nph.15046

Yu J, Vallad GE, Boyd NS et al (2019) Evaluation of allyl isothiocyanate as a soil fumigant for tomato (*Lycopersicon esculentum* Mill.) production. Plant Dis 103:2764–2770. https://doi.org/10.1094/PDIS-11-18-2013-RE

Zanón MJ, Gutiérrez LA, Arbizzani A, Myrta A et al (2014) Control of tobacco nematodes with dimethyl disulfide (DMDS) in Spain and Italy. Acta Hortic 1044:375–380. https://doi.org/10.17660/ActaHortic.2014.1044.50

Zhou Y, Coventry DR, Gupta VV, Fuentes D, Merchant A, Kaiser BN, Li J, Wei Y, Liu H, Wang Y et al (2020) The preceding root system drives the composition and function of the rhizosphere microbiome. Genome Biol 21:1–19. https://doi.org/10.1186/s13059-020-01999-0

# Chapter 13
# Efficacy of Plant Metabolic Products to Suppress the Plant Pathogenic Fungi

**Tarek A. A. Moussa, Rehab M. Hafez, and Akram H. Mohamed**

## 13.1 Introduction

Plants naturally produce a wide range of metabolites with divergent chemical properties throughout their lives. Plants' metabolism is classified into two pathways: primary and secondary metabolism. The primary metabolite pathway synthesizes glucose, amino acids, fatty acids, and nucleotides, as well as macromolecules such as carbohydrates, proteins, lipids, nucleic acids, and hormones. They are found throughout the plant kingdom, and they are required for photosynthesis, nutrient assimilation, translocation, and respiration (Bernards 2010; Saltveit 2017).

Secondary metabolites, on the other hand, either from biosynthetic or evolutionary pathways, are typically by-products derived from primary metabolites that are not specifically related to growth and development. They represent the unique phytochemical profile of specific plant species, contributing to their adaptability for facing biotic and abiotic influences in their environment. These metabolites have a considerable effect on both belowground and aboveground defense mechanisms to neutralize any adverse consequences (Ashihara et al. 2010; Kaur et al. 2022). Also, they possessed taxonomically-restricted compositions and distributions that vary among plant species (González Mera et al. 2019). Additionally, their different impacts on both the plant and other living organisms gave them fascinating prospects as medicinal and/or antioxidant candidates (Bourgaud et al. 2001; Sanchez and Demain 2008).

T. A. A. Moussa (✉) · R. M. Hafez
Botany and Microbiology Department, Faculty of Science, Cairo University, Giza, Egypt
e-mail: tarekmoussa@cu.edu.eg

A. H. Mohamed
Department of Microbial Genetic Resources, National Gene Bank, Agricultural Research Center (ARC), Giza, Egypt

 
N. El-Wakeil, A. Abdelkhalek (eds.), *Microbiomes and Plant Secondary Metabolites for Insect Pest Management*, Progress in Biological Control 23,
https://doi.org/10.1007/978-3-032-24330-0_13

Plant secondary metabolites have generally been categorized according to various criteria: their chemical structure, solubility in water or organic solvents, composition, and biosynthesis pathways (Dhaniaputri et al. 2022).

They can be biosynthesized through one of the four pathways: the shikimic acid pathway, the malonic acid pathway, the mevalonic acid pathway, and the methylerythritol phosphate (MEP) pathway.

The shikimic acid pathway, called a metabolic tree with many branches is used by microbes, protozoans, and plants, but it is absent in animal cells (Starcevic et al. 2008). It provides an essential route for the synthesis of many aromatic metabolites, such as alkaloids, phenolics, and mycotoxins, in seven reaction steps (Santos-Sánchez et al. 2019). Its intermediate products and its branching pathway have intriguing qualities in a variety of industries, mainly the pharmaceutical industry (Wu et al. 2022).

Most plant phenolics are generated through the shikimic acid pathway. Although less relevant in higher plants, the malonic route is a substantial source of phenolics in bacteria and fungi (Saltveit 2017). The mevalonate pathway, known as the isoprenoid pathway, is a necessary route in eukaryotes, archaea, and certain eubacteria. It converts acetyl-CoA to isopentenyl 5-diphosphate, the key precursor of polyisoprenoid metabolites, sterol isoprenoids, and natural products (Miziorko 2011). It provides the cells with essential biomolecules for multicellular processes (Buhaescu and Izzedine 2007). The methylerythritol 4-phosphate (MEP) pathway was recently discovered in most bacteria, some eukaryotic parasites, and the plastids of plant cells. It is a source of isoprenoid precursors, isopentenyl diphosphate and dimethylallyl diphosphate, which lead to the formation of many isoprenoids with diverse biological processes (Banerjee and Sharkey 2014).

The glucose molecules generated during photosynthesis are involved in the metabolism of secondary metabolites (Ashihara et al. 2010; Saltveit 2017) through these four key routes in cell compartments (Saltveit 2017). The biosynthetic pathway of secondary metabolites proceeds by methylation, glycosylation, and hydroxylation processes, which give them a more complex structural makeup with more side chains (Ehrlich and Holdren 1971; Figueiredo et al. 2008).

Large concentrations of secondary metabolites are stored and compartmentalized in the vacuole, as some of them may be poisonous to the producing plant when present in greater quantities than required for a specific purpose, such as defense or signaling (Wink and Roberts 1998).

Secondary metabolites can also be divided into four major types based on their composition and chemical structures: (i) phenolic groups (composed of simple sugars and benzene rings); (ii) terpenes and steroids; (iii) nitrogen-containing compounds; and (iv) sulfur-containing compounds (Saxena et al. 2013; Twaij and Hasan 2022).

## 13.2 Plant Secondary Metabolites

### *13.2.1 Phenolic Groups (Composed of Simple Sugars and Benzene Rings)*

Phenolic chemicals are prevalent in the plant kingdom. Vascular plants are the primary source of a wide range of polyphenols, whereas lower plants, such as bryophytes, are the primary producers of flavonoids (Kumar et al. 2023). Several phenolic compounds have been identified as products of the phenylpropanoid pathway in plants. The phenolic compound consists of benzene rings with one or more hydroxyl substituents, which exhibit different structures and functions, ranging from single aromatic rings (caffeic acid and ferulic acid) to more complicated polymeric structures such as lignin/lignan, coumarins, quinones, tannins, and flavonoids (Twaij and Hasan 2022). Many phenolics, carotenoids, tocopherols, and even benzoxazinoids serve as defenses against herbivores, pathogens, and their generation of toxins in inhibiting their growth via reducing and scavenging the oxidative radicals and structural harm brought on by free radicals, thereby establishing a physical barrier against fungal invasions (Das and Roychoudhury 2014; Atanasova-Penichon et al. 2016; Altemimi et al. 2017; Saltveit 2017).

The most prevalent kinds of phenolic secondary compounds in plants are produced from phenylalanine through the formation of cinnamic acid by removing an ammonia molecule. This reaction is catalyzed by phenylalanine ammonia lyase (PAL). PAL also catalyzes a crucial regulatory step in the production of numerous phenolic compounds because it occurs at the junction of primary and secondary metabolism Fig. 13.1 (Khatri et al. 2023; Kumar et al. 2023). Environmental elements include low nutrition levels, light, and fungus infection, all boost the activity of PAL (Kumar et al. 2023).

Coumarins (1-benzopyran-2-one) are benzopyrone-type organic compounds found in a variety of plants. Coumarins are classified chemically into seven simple coumarin types, furanocoumarins, pyranocoumarins, pyrone-substituted coumarins, and isocoumarins, which are isomers of coumarin with the lactone ring reversed (Sharma et al. 2024). At least 1300 distinct coumarins have been discovered across a wide range of plant groups (tonka beans, woodruff, lavender oil, cassia, and sweet clover) and plant organs (seeds, fruits, flowers, roots, leaves, and stems). It can also be found in food plants like strawberries, cinnamon, peppermint, green tea, carrots, and celery (Tamokou et al. 2017; Batista et al. 2021).

Lignin is a monolignol-based polymer that forms on the surface of cell walls. Plant lignin contains three crucial monolignols: p-hydroxyphenyl, guaiacyl, and syringyl units (Boerjan et al. 2003). Lignin is vital for terrestrial plants because it provides structural support for upward development and allows for long-distance water movement; both of which are necessary for plants' evolutionary adaptation from aquatic to terrestrial environments. Lignin also provides physical and chemical protection to plants against disease invasion (Xie et al. 2018).

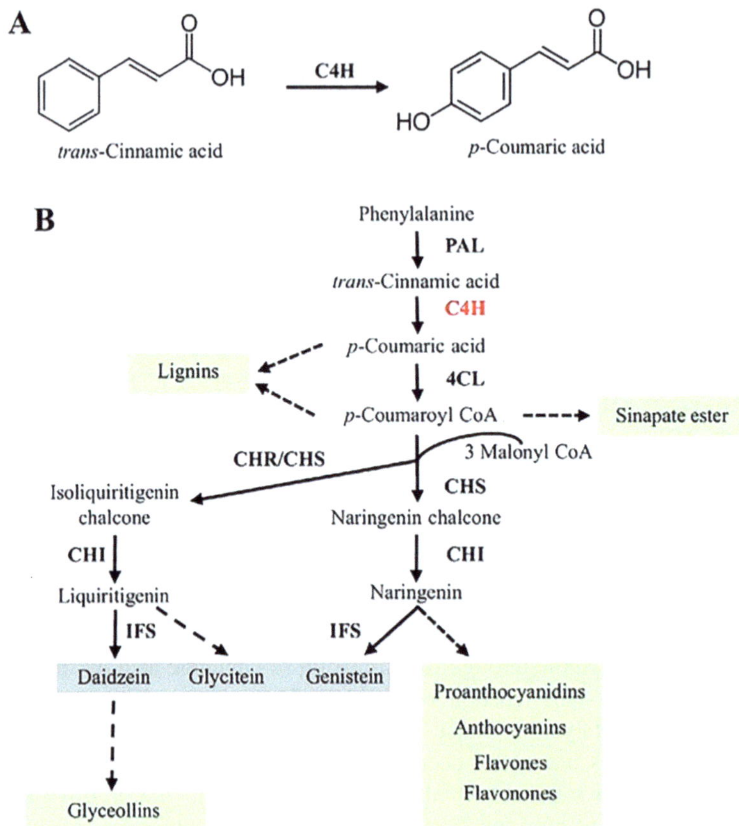


**Fig. 13.1** The phenylpropanoid biosynthetic pathway from phenylalanine to different phenolic secondary compounds is catalyzed first by phenylalanine ammonia lyase (PAL). (Adapted after Khatri et al. (2023))

Quinones are common natural biological pigments found in a variety of higher plants, lichens, microbes, and a few animals in various forms such as (1) isoprenoid quinones (in sea urchins), (2) anthraquinones (in insects), (3) plastoquinone, α-tocopherol quinones, and phylloquinone (in all photosynthetic tissues), (4) ubiquinone (in most plants and generally in animals), (5) benzoquinones, naphthoquinones, anthraquinones, and polycyclic quinones (in higher plants). These quinone-based antibiotics are gaining popularity due to their natural origins and are used in traditional African, Chinese, and Indian medicine to treat and prevent numerous diseases. Quinone-based antibiotics have diverse activity profiles based on their structures and exhibit specific biocidal and anti-biofilm capabilities (Martínez and Benito 2005; El-Najjar et al. 2011; Dahlem Junior et al. 2022).

More and more investigations have been conducted to identify plant extracts as dietary antioxidant supplements. Most of these natural antioxidants are found in fruits, vegetables, spices, grains, and herbs, including ginseng, curcuma, ginkgo, rosemary, green tea, grapes, ginger, and garlic. They include a diverse range of antioxidants that help protect cells from oxidative biotic stresses (Lü et al. 2010).

Flavonoids are polyphenolic compounds consisting of 15 carbon atoms and two aromatic rings. Flavonoids are the most varied class of phenolic chemicals, and they are found in all plant species (Mutha et al. 2021). They are mainly found in large concentrations in the leaf epidermis and fruit skins. Flavonoid biosynthesis is initiated primarily via the phenylpropanoid pathway. Flavonoids are divided into several subclasses, including flavonols, flavan-3-ols, isoflavones, flavones, anthocyanidins, and flavanones. Under normal conditions, flavonoids function as signaling molecules, protectants, and regulators of plant hormones such as auxin and cytokinin (Kurepa et al. 2018). However, under stress, they defend against oxidative damage (Shomali et al. 2022). Stress-induced dihydroxy B-ring-substituted flavonoids actively combat ROS (Agati et al. 2012).

Tannins are bioactive substances stored in plant cell vacuoles, protecting other cell components. Many plants naturally contain them in their roots, wood, bark, leaves, and fruit, particularly in the bark of oak, sumac, and myrobalan species (Khan et al. 2023).

### *13.2.2 Terpenes and Steroids*

Although terpenoids have poor natural yields, they are structurally varied and considered a large metabolite category with at least more than 35,000 distinct terpenes characterized thus far (Dewick 2002). They are synthesized from acetyl-coenzyme-A molecules or their glycolytic intermediates. They are divided into several groups according to the number of isoprene units in their carbon skeleton to mono-, di-, tri-, tetra-, sesterter-, sesqui-, and polyterpenes. There are at least two distinct processes for producing terpenes from primary metabolites: the mevalonic acid pathway and the methylerythritol phosphate (MEP) pathway, Fig. 13.2 (Abdallah and Quax 2017; Twaij and Hasan 2022).

### *13.2.3 Nitrogen-Containing Compounds*

Alkaloids contain nitrogen in their cyclic organic compounds which gives them complexity and varied structures. Alkaloids can be categorized by chemical structure, biochemical origin, or natural origin. Alkaloids can be biosynthesized through various processes, including the shikimate pathway; the ornithine, lysine, and nicotinic acid pathway; the histidine and purine pathway; and the terpenoid and polyketide pathway, Fig. 13.3 (Gutiérrez-Grijalva et al. 2020). Alkaloids make up nearly 50% of plant-derived medications. The plant families with the highest amounts of alkaloids are Papaveraceae, Berberidaceae, Leguminosae, Boraginaceae, Apocynaceae, Asclepiadaceae, Liliaceae, Gnetaceae, Ranunculaceae, Rubiaceae,

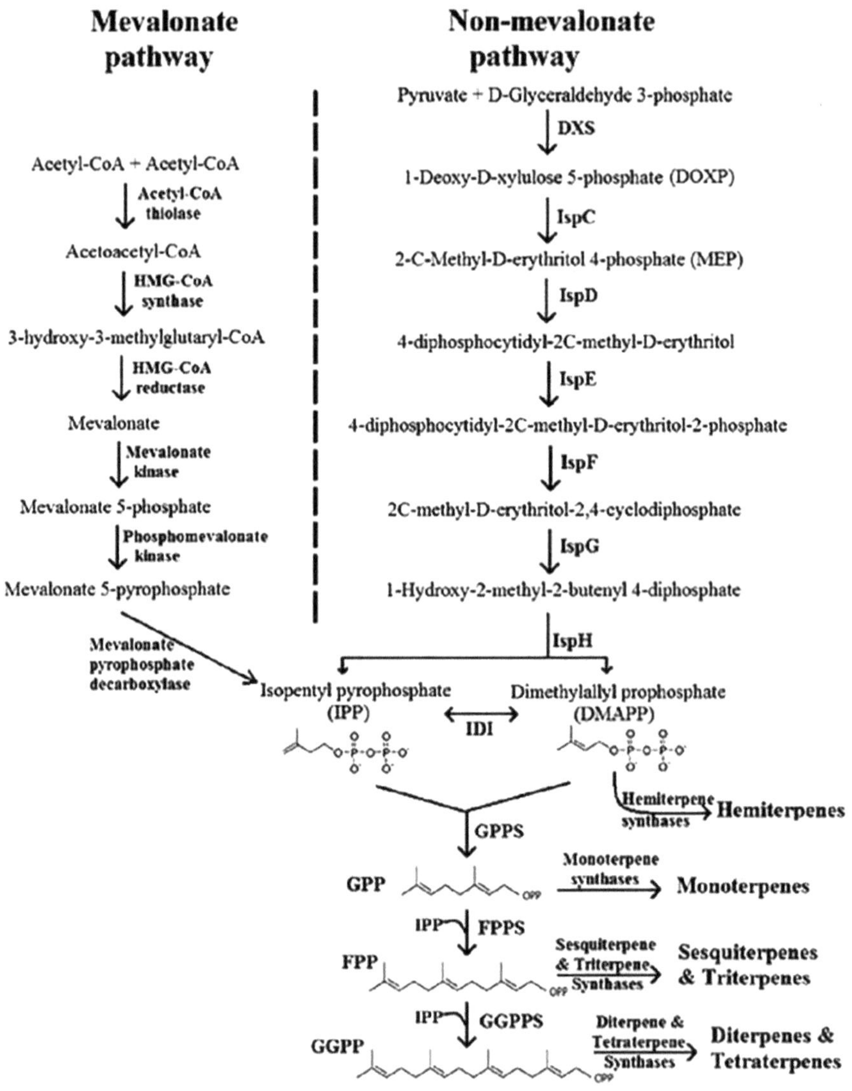


**Fig. 13.2** Biosynthetic pathways for various terpenoids. (Adapted from Abdallah and Quax (2017))

Solanaceae, and Rutaceae. Plant alkaloids have historically played an important part in folk medicine as purgatives, antitussives, sedatives, and remedies for a wide range of diseases. The most famous alkaloids employed in pharmacology include caffeine, nicotine, cocaine, morphine, codeine, brucine, ephedrine, and quinine (Oksman-Caldentey and Inzé 2004; Gutiérrez-Grijalva et al. 2020).

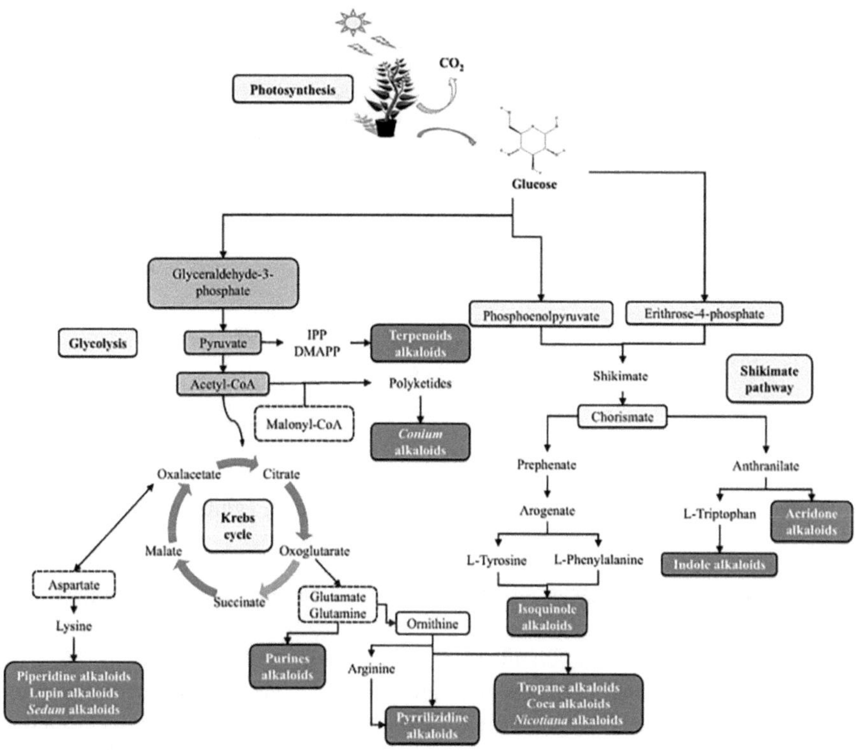


**Fig. 13.3** Biosynthesis pathway of different types of alkaloids. Abbreviations: *IPP* isopentenyl diphosphate, *DMAPP* dimethylallyl diphosphate. (Adapted from Gutiérrez-Grijalva et al. (2020))

Cocaine is a tropane-type pyrrolidine alkaloid, with the N-heterocycle formed from L-ornithine, an amino acid obtained from glutamate. Nicotine is a type of pyridine alkaloid made from aspartate or phenylalanine. Morphine is an isoquinoline alkaloid in which tyrosine can be used to synthesize isoquinoline alkaloids, which have an N-heterocycle that is isomeric to quinoline. Purine alkaloids, including caffeine and theobromine, are purine derivatives made from aspartate, glycine, and glutamine. Quinoline alkaloids, including quinine, can be produced from L-tryptophan (Ratnadewi 2017).

### *13.2.4 Sulfur-Containing Compounds*

Plants' potential to regulate stress tolerance through secondary metabolism has sparked renewed interest in S-containing bioactive chemicals. Among plant families that produce S-compounds, the Brassicaceae, which includes broccoli, cabbage, and

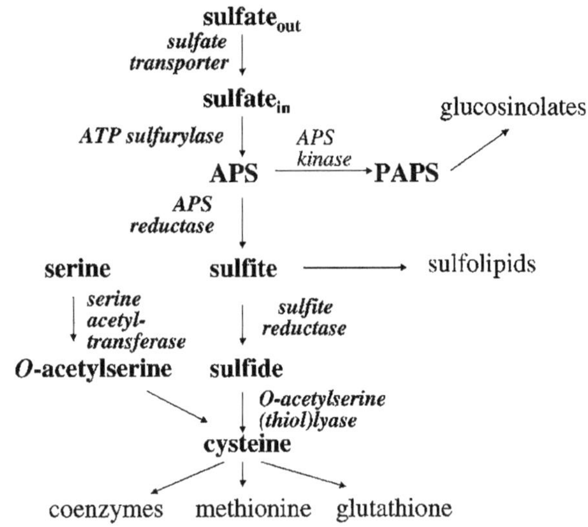


**Fig. 13.4** Biosynthesis of some sulfur-containing compounds from sulfate assimilation. (Adapted after Kopriva et al. (2009))

cauliflower, and the Liliaceae, which includes garlic and onion, are notable (Castro et al. 2023).

Major sulfur-containing chemicals implicated in plant defense mechanisms against phytopathogens include S-containing amino acids (mostly methionine and cysteine), glutathione, glucosinolates (GSLs), and their hydrolysis products such as isothiocyanates, as well as other S-containing substances such as phytoalexins, cysteine sulfoxides, antimicrobial peptides (as thionine and defensins), and thiosulfates, like allicin (Grubb and Abel 2006; Halkier and Gershenzon 2006). Sulfur-containing defense chemicals have multiple roles, including pathogen sensing and signal transduction pathways that are linked to diverse defense mechanisms regulated by plant hormones (salicylic acid, jasmonic acid, and ethylene) and reactive oxygen species (ROS) (Künstler et al. 2020).

Secondary metabolites are divided into two types based on their synthesis pathway. In the first category, the myrosinase enzyme hydrolyzes glucosinolates to produces glucosinolates. Members of the Cruciferae family exhibit this route. Cabbage, broccoli, and nasturtium all involve the glucosinolate–myrosinase pathway. In the second group of sulfur-containing secondary metabolites, the hydrolysis of the alliinase enzyme produces alliin, which is found in the *Allium* genus to avoid pathogenic attack Fig. 13.4 (Kopriva et al. 2009; Sayadi et al. 2020).

Glutathione is an essential tripeptide that contains the amino acids glutamate, cysteine, and glycine that are found in animal and plant kingdoms. It has a variety of roles, the most notable of which is redox-homeostatic buffering. Due to its antioxidant properties, it functions under stress to detoxify reactive oxygen species from the ascorbate-glutathione cycle by expressing defense genes (Zechmann and Müller 2010).

GSLs are stable, non-toxic molecules most commonly found in the Brassicaceae family, which has over 3000 species and many edible plants (Holst and Williamson

2004; Prieto et al. 2019). However, the distribution of GSLs in the plant is unequal, where the highest amounts observed were in young leaves, reproductive tissues, and roots (Agerbirk and Olsen 2012). Also, the composition of GSLs varies greatly even within the same family. GSLs are gaining popularity due to their unique biological properties, including antioxidant, anti-inflammatory, and antibacterial activities (Castro et al. 2023).

Thiosulfinates are another major phytoanticipin found in many *Allium* species. Among them is allicin, a dithiosulfinate, organosulfur molecule found in garlic that gives it its distinctive aroma and exhibits antifungal activities against *Candida albicans* (Borlinghaus et al. 2014; Zainal et al. 2021).

The flavanone intermediate naringenin serves as a precursor for isoflavonoids, which play a crucial role in plant development and defense. However, phytoalexins are low-molecular-weight isoflavones produced by plants in response to microbial challenges in their natural environment, providing them with resistance against such microbes. These compounds are poisonous and disrupt cellular structure or the metabolism of disease-causing organisms, yet they are frequently pathogen-specific in their negative effects (Teoh and Teoh 2016).

For example, orchids have produced more than 40 dihydrophenanthrene phytoalexins. The 9,10-dihydrophenanthrenes are the primary phytoalexins isolated from *Orchis militaris* (orchids) as a response to *Rhizoctonia repens* infection (Boller et al. 1957). Next, hircinol was isolated from *Loroglossum hircinum* infected with *R. versicolor*, followed by the discovery of its isomer, loroglossol (Hardegger et al. 1963). Phytoalexins, hydroxylated derivatives of coumarins, are formed in carrots in response to fungal infections and are thought to have antifungal properties (Hoult and Payá 1996).

Resveratrol, a stilbenoid, is a substance found on the skin of grapevines and helps prevent *Botrytis cinerea* from attacking the fruit (Favaron et al. 2009). Vitis also induces delta-viniferin following infection by *Plasmopara viticola* (Timperio et al. 2012). Papaya fruits contain phytoalexin called danielone, which protects them against the pathogen *Colletotrichum gloeosporioides* by its strong antifungal properties (Echeverri et al. 1997). *Avena sativa* produces avenanthramides, phytoalexins in reaction to *Puccinia coronata* (Mayama et al. 1995). Leaves of *Pinus densiflora* (pine) contain the natural stilbenoid pinosylvin, which has a higher potency than resveratrol as anti-*Saccharomyces cerevisiae* and anti-*Candida albicans* activity (Lee et al. 2005).

In contrast to phytoalexins which are synthesized during infection, stilbenoids, flavonoids, and others are pre-infectious compounds that are synthesized before infection as invasion-protection. Pinosylvin is a stilbenoid fungitoxin present in the heartwood of Pinaceae, protecting them from fungal infections (Lee et al. 2005). Sakuranetin is a flavanone (type of flavonoid) that prevents *Pyricularia oryzae* spores from infecting rice leaves (Kodama et al. 1992). The 3-deoxyanthocyanidin phytoalexins are produced to protect *Sorgum* from *Colletotrichum* infection (Shih et al. 2006).

Defensins are one of the most prevalent antimicrobial protein families (AMPs) in all eukaryotic groups; and they serve as an essential first line of defense for hosts against harmful microbes. A recent study found that defensins are present in the Fungi, Plantae, and Animalia kingdoms and that bacteria are their common ancestor. According to the Antimicrobial Peptide Database (APD1), about 2200 natural or artificial AMPs have been discovered so far, of which over 1900 have antibacterial activities and 800 have antifungal activities (Silva et al. 2014).

Garlic (*Allium sativum* L.) contains an oxygenated sulfur compound called allicin (S-(2-propenyl)-2-propene-1-sulfinothioate). It inhibits the growth of both types of bacteria and exhibits an antifungal effect against various fungi, including *Candida*, *Torulopsis*, *Trichophyton*, *Cryptococcus*, *Aspergillus*, *Trichosporon*, *Rhodotorula* species, and *Fusarium oxysporum* (El-Saber Batiha et al. 2020; Ogita et al. 2006; Marchese et al. 2016; Pârvu et al. 2019). Additionally, allicin and garlic oil exhibit strong antifungal effects against *Candida albicans*, *Ascosphaera apis*, and *A. niger* (Kuda et al. 2004). These compounds work by penetrating cell membranes as well as the membranes of organelles like the mitochondria, causing the destruction of organelles and cell death (Pai and Platt 1995). Allicin demonstrates potent anti-*C. neoformans* activity both in vitro and in vivo, primarily via impairing the permeability and associated functions of *Cryptococcus* cell membranes (Li et al. 2022).

Thionine is a small antimicrobial peptide present in infected plants as a defense mechanism, altering fungal hyphal cell walls or acting as a subcellular messenger in signal transduction. Purothionins (from *Triticum* sp. seeds), phoratoxins (from vegetative parts of *Viscum* sp. and leaves of *Phoradendron* sp.), and ligatoxins A (from *Phoradendron* sp.) (Plattner et al. 2015; Azmi and Hussain 2021) are some homologous examples of this peptide.

## 13.3 Mechanisms of the Plant Metabolic Products to Suppress Pathogenic Fungi

Continuous battles are usually between plants and pathogenic fungi, as higher plants must develop many weapons to withstand biotic and abiotic stresses via numerous biochemical, molecular, and morphological mechanisms (Nejat and Mantri 2017). Plants have different defense mechanisms, constitutive or induced that can be activated in response to pathogen and pest attacks. These mechanisms confer high levels of resistance against invaders (Pieterse et al. 2014). Plants can produce both primary and secondary metabolites: Primary metabolites mainly contain carbohydrates and proteins which are essential for plant growth and metabolism (Rojas et al. 2014), while secondary metabolites (terpenes, anthraquinones, phenols, alkaloids, etc.). These are produced in response to different environmental and pathogenic fungi (Suteu et al. 2020). Phytopathogenic fungi could be suppressed by different mechanisms (disruption of the cell membrane, suppression of fungal mitochondria, cell division inhibition, DNA/RNA synthesis suppression, inhibition of protein synthesis, and alteration of the functions of efflux pumps of fungal cells (Peter and Menon 2022; Kang et al. 2010; Kim et al. 2013; Walker and White 2017). So, it is important to discuss the types of plant metabolites and their roles in inhibiting phytopathogens. Plant metabolites could inhibit fungal invasion and infection through different mechanisms.

(a) *Cell Wall Formation Inhibition*

As the fungal cell wall mainly contains β-glucans, stopping their synthesis could affect the viability of the fungal cell (Peter and Menon 2022; Walker and White 2017).

(b) *Destruction of Fungal Cell Membrane*

Ergosterols are considered vital components in the cell membrane, so ergosterols inhibition causes the loss of the integrity of the fungal cell membrane, since many antifungal compounds depend mainly on ergosterol inhibition (Walker and White 2017).

(c) *Mitochondrial Dysfunction of Phytopathogen*

Mitochondria play a crucial role in different physiological processes of fungi. Its disruption could have occurred through electron transport. Inhibition results from the suppression of proton pumps in the respiratory chains, ultimately leading to the death of the fungal cell (Kim et al. 2013)

(d) *Cell Division Inhibition*

Preventing cell division is considered one of the fungal growth inhibitions, which could occur via depolymerization of microtubules, which finally leads to prevent the formation of the mitotic spindle (Walker and White 2017)

(e) *Suppression of DNA/RNA of Phytopathogen and Protein Deactivation*

Antifungal agents could enter the cell and cause interruption of RNA synthesis, leading to inhibition of DNA transcription, which finally results in suppression of protein synthesis (Freiesleben and Jäger 2014).

(f) *Disruption of Efflux Pumps in Fungal Cells*

Efflux pumps are known mechanisms by which the cells expel out the excess of toxic substances. Usually, this process is overexpressed, causing drug resistance in fungal cells; therefore, inhibition of efflux pumps leads to the inhibition of fungal growth (Kang et al. 2010).

### 13.3.1 Primary Metabolites as Inhibitors for Pathogenic Fungi

It is well known that plant could synthesize carbohydrates and proteins for essential processes of growth and developments, also there are different types of primary metabolites which perform their role in signaling transduction and aid in phytopathogen recognition processes during the fungal attack (Wang et al. 2012; Kachroo and Robin 2013).

#### 13.3.1.1 Pathogenesis-Related Proteins

One of the most important proteins is activated upon the fungal attack is peroxidase of the PR-9 family, which is responsible for increasing the lignification rate of the plant cell wall, making it more resistance for pathogenic attack (Passardi et al. 2004). In addition, PPR-15 and PR-16 generate hydrogen peroxide, which can stimulate plant defense and affect the fungal attack.

#### 13.3.1.2 Defensins

It is known as antimicrobial peptides (45–54 amino acids), which consists of βαββ pattern of highly conserved scaffold structure with cysteine motif that has the ability to bind with the negative charge of glycoprotein of fungi cell membrane (Silverstein et al. 2005). Defensins are located mainly in the stomatal and peripheral regions of plants at the entry points of phytopathogens, serving as the first line of defense against pathogenic fungi (Broekaert et al. 1995). Plant defensins are induced under abiotic stress (Lay and Anderson 2005). They are known also as proteins that induced upon the attack of phytopathogenic fungi, playing a vital role in the protecting vegetative tissues of plant. For example, tobacco and tomato plants could be protected against *Alternaria longipes* and *A. solani* by the expression of defensin (Terras et al. 1995).

#### 13.3.1.3 Phytoalexins Production

Phytoalexins are low-molecular-weight substances produced mainly by different plant families (Fabaceae, Solanaceae, Poaceae, Malvaceae, Vitaceae, and Brassicaceae), and are secreted in response to fungal pathogen invasion. Phytoalexins have antimicrobial activity, which allows them to suppress phytopathogen invasion (Jeandet 2015). Phytoalexins could be biosynthesized inside plant cells via different biosynthetic pathways (phenylpropanoic-polymalonic acid route, methylerythritol phosphate and geranyl-geranyl diphosphate pathways, and indole phytoalexin pathway) (Jeandet et al. 2014). More than 300 chemicals identified as phytoalexins have been found across more than 30 plant families. Phytoalexins are known as family-specific; for example, oryzalexins, zealexins, kauralexins, sakuranetin, and phenyl amides are most common to Poaceae, while brassinin is characteristic of Brassicaceae (Arruda et al. 2016).

Phytoalexins could be represented by different types of chemical groups; stilbenes are one of the most important phytoalexins correlated with the lignin pathway (Gabaston et al. 2020). The majority of leguminous phytoalexins are belonged to six classes of isoflavonoid; however, a few legume plants that produce furanoacetylenes (Jeandet 2015). Phytoalexins can exhibit antifungal activity against many pathogenic fungi, by suppressing germ tube elongation and altering the morphological and physiological characteristics of phytopathogic fungal cells. For example, rishitin, phytuberin, anhydro-β-rotunol, and solavetivone from the family Solanaceae showed good antifungal activity on *Phytophthora* sp. as it causes many changes in zoospores' motility and bursting cell membrane (Harris and Dennis 1977; Adrian et al. 1997; Adrian and Jeandet 2012).

#### 13.3.1.4 Lignin Formation

Lignification is considered one of the most important plant defense mechanisms as during the defense process lignin, a phenolic compound accumulated. Cell wall apposition which is also called papillae plays a vital role in plant defense against many pathogenic fungi. This process of defense makes plant cell wall less susceptible to degrading enzymes from pathogenic fungi, thereby preventing invasion of fungal mycelium to plant cell wall (Harris and Dennis 1977; Vance et al. 1980; Nicholson and Hammerschmidt 1992; Zeyen et al. 2002).

#### 13.3.1.5 Production of Melatonin

Melatonin is one of the most effective antimicrobial agents against fungal and bacterial phytopathogens, as low concentration of melatonin could enhance plant resistance (Tekbas et al. 2008). It functions through upregulating genes responsible for pathogen-related (PR) protein and salicylic acid (SA), while also regulating ethylene-associated genes. Besides, melatonin can decrease the severity of pathogen infection even after its entry (Arnao and Hernández-Ruiz 2019). Melatonin enhances genes signaling pathway of various salicylic acid, jasmonic acid, and ethylene (Zhu and Lee 2015). Interestingly, plants containing endophytic fungi or bacteria live mutually inside their tissues possess higher melatonin content than those without endophytic microorganisms (Jiao et al. 2016).

### *13.3.2 Polygalacturonase Inhibitor Proteins (PGIPs)*

Since the cell wall of plant is mainly composed of pectin (mainly consisted of galacturonic acid), plant pathogenic fungi should produce polygalacturonases (PGs) to facilitate their penetration to plant cell. Plant cells produce polygalacturonase inhibitor proteins (PGIPs) as defense response against fungal attack. Production levels of PGIPs by crop plants correlated with increased resistance to fungal pathogens (Howell and Davis 2005).

### 13.3.3 Volatile Organic Compounds (VOCs)

Plants are considered factories for different types of volatile organic compounds which mainly manufactured in epidermal cells and stored in especial secretory structures like glandular trichomes and secretory cavities. These secretions are lipophilic in nature which volatilized upon their contact to the air. It is represented by different phytochemical compounds like nitrogen-containing compounds, terpenes, aromatics, and fatty acid derivatives. VOCs are represented by volatile plant hormones like methyl jasmonate and methyl salicylate. These VOCs were produced in response to fungal attack as a method for defense. Plant volatiles are classified according to the chemical structures and biosynthetic pathways. Terpenes are considered as the largest and important group of volatiles which mainly stored in different plant organs, e.g., resin ducts, secretory cavities, and glandular trichomes (Bakker and Gerritsen 1990; Wang et al. 1999; Martin et al. 2003; Heskes et al. 2012). These terpenes are mainly secreted as constitutive or as response to external stimulus (Turlings and Erb 2018). The monoterpenes could be manufactured via methylerythritol phosphate pathway which mainly occurred in chloroplasts (Phillips et al. 2008).

Monoterpenes and sesquiterpenes of resistant grapevines genotypes are produced at higher levels than those produced by susceptible genotypes in response to downy mildew (Plasmopara viticola) (Algarra Alarcon et al. 2015). Besides, high levels of (−)-limonene terpene were obtained during the attack of Magnaporthe oryzae fungus on rice plant causing rice blast (Chen et al. 2018).

### 13.3.4 Indirect Action of Volatiles as Antifungal Agent

Methyl salicylate and methyl jasmonate are well-known volatile plant defense hormones that can induce resistance in different plant parts, regardless of the infected parts; it could induce resistance against pathogens (Hammerbacher et al. 2019). Systemic resistance could result from the induction of plant defense or it could activate plant against future infection to be ready for other infections (Conrath et al. 2015; Mauch-Mani et al. 2017). Volatile acetic, propionic, or butyric esters of GLVs are emitted by plants when plants are under pathogen attack. Wheat plant resistance could be activated through its exposure to (Z)-3-hexenyl acetate, which could induce resistance to Fusarium graminearum; in addition, this process activates transcription of jasmonate-responsive genes (Ameye et al. 2018). Also, GLVs are called as green vaccines which could have a role in protecting plants against future infection (Luna 2016).

**Table 13.1** Efficacy profile of major classes of plant antimicrobial metabolites

| Class of metabolite | Primary mechanism (s) | Example pathogens effective against | Comparative efficacy notes |
|---|---|---|---|
| Phenolics (e.g., thymol, eugenol) | Membrane disruption, enzyme inhibition | *Aspergillus* spp., *Pseudomonas aeruginosa, Botrytis cinerea* | Thymol shows fungicidal efficacy comparable to fluconazole against some *Aspergillus* strains in vitro. |
| Alkaloids (e.g., berberine, piperine) | Intercalation into DNA/RNA, metabolic interference | *Colletotrichum* spp., *Staphylococcus aureus, Plasmodium falciparum* | Berberine often requires higher concentrations than synthetic bacteriostats but shows synergistic benefits. |
| Terpenoids (e.g., azadirachtin in neem) | Growth regulation, antifeedant, endocrine disruption | *Puccinia triticina (wheat rust), Aedes aegypti larvae* | Neem oil extracts show protective efficacy rivaling chlorothalonil in field trials for certain rusts. |
| Sulfur Compounds (e.g., allicin from garlic) | Thiol reactivity, disabling key enzymes | *Candida albicans, Escherichia coli, Magnaporthe oryzae* | Allicin exhibits minimum inhibitory concentrations (MICs) against *C. albicans* similar to those of miconazole. |
| Flavonoids (e.g., quercetin, catechins) | Membrane integrity, virulence factor suppression | Influenza virus, *Vibrio cholerae, Plasmopara viticola (downy mildew)* | Synergistic effects observed; can enhance conventional antibiotic activity (e.g., with tetracycline). |

### 13.3.5 *From Mechanism to Practical Efficacy*

The detailed mechanisms of action explored in this chapter form the foundation for understanding the practical efficacy of plant-derived metabolites. To fully assess their potential, it is instructive to compare their effectiveness to conventional agents and to summarize their activity spectra. For example, research indicates that thymol (from thyme) can achieve mycological eradication rates similar to fluconazole in some models of aspergillosis (Ahmad et al. 2010), and specific flavonoid formulations have demonstrated protective efficacy against grape downy mildew on par with standard synthetic controls (Schnee et al. 2013). As summarized in Table 13.1, the efficacy of these compounds varies significantly by class, primary mechanism, and target pathogen, highlighting the need for a tailored application informed by comparative studies (Cowan 1999).

## 13.4 Conclusion

This chapter has demonstrated that plant-derived antimicrobial metabolites are potent, multifaceted agents capable of managing key agricultural pathogens, with efficacy benchmarks against conventional fungicides and bactericides firmly established. Their integration into crop protection strategies offers a sustainable pathway to reduce synthetic pesticide loads and delay resistance development. Looking ahead, the future of these biopesticides in agriculture hinges on translational innovation. Critical prospects include the formulation of field-stable and rainfast products, the development of synergistic combinations within integrated pest management (IPM) programs to enhance spectrum and reliability, and the breeding or engineering of crop varieties with elevated endogenous metabolite production for inherent resistance. Ultimately, harnessing the full potential of plant antimicrobials will require closing the gap between proven in vitro efficacy and consistent, cost-effective performance in the dynamic and challenging environment of the field.

**Declaration of AI Assistance** Authors did not use generative artificial intelligence (AI) and/or AI-assisted technologies in any of the writing processes.

## References

Abdallah II, Quax WJ (2017) A glimpse into the biosynthesis of terpenoids. In: International conference on natural resources and life sciences (NRLS-2016). KnE Life Sciences, pp 81–98

Adrian M, Jeandet P (2012) Effects of resveratrol on the ultrastructure of Botrytis cinerea conidia and biological significance in plant/pathogen interactions. Fitoterapia 83:1345–1350

Adrian M, Jeandet P, Veneau J, Weston LA, Bessis R (1997) Biological activity of resveratrol, a stilbenic compound from grapevines, against Botrytis cinerea, the causal agent for gray mold. J Chem Ecol 23:1689–1702

Agati G, Azzarello E, Pollastri S, Tattini M (2012) Flavonoids as antioxidants in plants: location and functional significance. Plant Sci 196:67–76

Agerbirk N, Olsen CE (2012) Glucosinolate structures in evolution. Phytochemistry 77:16–45

Ahmad A, Khan A, Manzoor N, Khan LA (2010) Evolution of ergosterol biosynthesis inhibitors as fungicidal against *Candida* and *Aspergillus* species. J Antimicrob Chemoth 65(5):1054–1061

Algarra Alarcon A, Lazazzara V, Cappellin L, Bianchedi PL, Schuhmacher R, Wohlfahrt G, Pertot I, Biasioli F, Perazzolli M (2015) Emission of volatile sesquiterpenes and monoterpenes in grapevine genotypes following Plasmopara viticola inoculation in vitro. J Mass Spectrom 50: 1013–1022

Altemimi A, Lakhssassi N, Baharlouei A, Watson DG, Lightfoot DA (2017) Phytochemicals: extraction, isolation, and identification of bioactive compounds from plant extracts. Plants 6:42

Ameye M, Allmann S, Verwaeren J, Smagghe G, Haesaert G, Schuurink RC, Audenaert K (2018) Green leaf volatile production by plants: a meta-analysis. New Phytol 220:666–683

Arnao MB, Hernández-Ruiz J (2019) Melatonin: a new plant hormone and/or a plant master regulator? Trends Plant Sci 24:38–48

Arruda RL, Paz ATS, Bara MTF, Côrtes MV d CB, de Filippi MCC, da Conceição EC (2016) An approach on phytoalexins: function, characterization and biosynthesis in plants of the family Poaceae. Ciência Rural 46:1206–1216

Ashihara H, Crozier A, Komamine A (2010) Plant metabolism and biotechnology. John Wiley & Sons

Atanasova-Penichon V, Barreau C, Richard-Forget F (2016) Antioxidant secondary metabolites in cereals: potential involvement in resistance to Fusarium and mycotoxin accumulation. Front Microbiol 7:187515

Azmi S, Hussain MK (2021) Analysis of structures, functions, and transgenicity of phytopeptides defensin and thionin: a review. Beni-Suef Univ J Basic Appl Sci 10:1–11

Bakker M, Gerritsen AF (1990) Ultrastructure and development of oil idioblasts in Annona muricata L. Ann Bot 66:673–686

Banerjee A, Sharkey TD (2014) Methylerythritol 4-phosphate (MEP) pathway metabolic regulation. Nat Prod Rep 31:1043–1055

Batista ÂG, da Silva-Maia JK, Maróstica MR Jr (2021) Generation and alterations of bioactive organosulfur and phenolic compounds. In: Chemical changes during processing and storage of foods. Elsevier, pp 537–577

Bernards MA (2010) Plant natural products: a primer. Can J Zool 88:601–614

Boerjan W, Ralph J, Baucher M (2003) Lignin biosynthesis. Annu Rev Plant Biol 54:519–546

Boller A, Corrodi H, Gäumann E, Hardegger E, Kern H, Winterhalter-Wild N (1957) Welkstoffe und Antibiotika. 20. Mitteilung. Über induzierte Abwehrstoffe bei Orchideen. I. Helv Chim Acta 40:1062–1066

Borlinghaus J, Albrecht F, Gruhlke MCH, Nwachukwu ID, Slusarenko AJ (2014) Allicin: chemistry and biological properties. Molecules 19:12591–12618

Bourgaud F, Gravot A, Milesi S, Gontier E (2001) Production of plant secondary metabolites: a historical perspective. Plant Sci 161:839–851

Broekaert WF, Terras FR, Cammue BP, Osborn RW (1995) Plant defensins: novel antimicrobial peptides as components of the host defense system. Plant Physiol 108:1353

Buhaescu I, Izzedine H (2007) Mevalonate pathway: a review of clinical and therapeutical implications. Clin Biochem 40:575–584

Castro V, Carpena M, Fraga-Corral M, Lopez-Soria A, Garcia-Perez P, Barral-Martinez M, Perez-Gregorio R, Cao H, Simal-Gandara J, Prieto MA (2023) Sulfur-containing compounds from plants. In: Natural secondary metabolites: from nature, through science, to industry. Springer, pp 363–402

Chen X, Chen H, Yuan JS, Köllner TG, Chen Y, Guo Y, Zhuang X, Chen X, Zhang Y, Fu J (2018) The rice terpene synthase gene Os TPS 19 functions as an (S)-limonene synthase in planta, and its overexpression leads to enhanced resistance to the blast fungus Magnaporthe oryzae. Plant Biotechnol J 16:1778–1787

Conrath U, Beckers GJM, Langenbach CJG, Jaskiewicz MR (2015) Priming for enhanced defense. Annu Rev Phytopathol 53:97–119

Cowan MM (1999) Plant products as antimicrobial agents. Clinic Microbiol Rev 12(4):564–582

Dahlem Junior MA, Nguema Edzang RW, Catto AL, Raimundo J-M (2022) Quinones as an efficient molecular scaffold in the antibacterial/antifungal or antitumoral arsenal. Int J Mol Sci 23:14108

Das K, Roychoudhury A (2014) Reactive oxygen species (ROS) and response of antioxidants as ROS-scavengers during environmental stress in plants. Front Environ Sci 2:53

Dewick PM (2002) Medicinal natural products: a biosynthetic approach. John Wiley & Sons

Dhaniaputri R, Suwono H, Amin M, Lukiati B (2022) Introduction to plant metabolism, secondary metabolites biosynthetic pathway, and in-silico molecular docking for determination of plant medicinal compounds: an overview. In: 7th international conference on biological science (ICBS 2021). Atlantis Press, pp 373–382

Echeverri F, Torres F, Quiñones W, Cardona G, Archbold R, Roldan J, Brito I, Luis JG, Lahlou E-H (1997) Danielone, a phytoalexin from papaya fruit. Phytochemistry 44:255–256

Ehrlich PR, Holdren JP (1971) Impact of population growth: complacency concerning this component of man's predicament is unjustified and counterproductive. Science (80- ) 171:1212–1217

El-Najjar N, Gali-Muhtasib H, Ketola RA, Vuorela P, Urtti A, Vuorela H (2011) The chemical and biological activities of quinones: overview and implications in analytical detection. Phytochem Rev 10:353–370

El-Saber Batiha G, Magdy Beshbishy A, Wasef LG, Elewa YH, Al-Sagan AA, El-Hack A, Mohamed E, Taha AE, Abd-Elhakim YM, Prasad Devkota H (2020) Chemical constituents and pharmacological activities of garlic (Allium sativum L.): a review. Nutrients 12:872

Favaron F, Lucchetta M, Odorizzi S, da Cunha ATP, Sella L (2009) The role of grape polyphenols on trans-resveratrol activity against Botrytis cinerea and of fungal laccase on the solubility of putative grape PR proteins. J Plant Pathol 91(3):579–588
Figueiredo AC, Barroso JG, Pedro LG, Scheffer JJC (2008) Factors affecting secondary metabolite production in plants: volatile components and essential oils. Flavour Fragr J 23:213–226
Freiesleben S, Jäger A (2014) Correlation between plant secondary metabolites and their antifungal mechanisms–a review. Med Aromat Plants 3:412–2167
Gabaston J, Valls Fonayet J, Franc C, Waffo-Teguo P, de Revel G, Hilbert G, Gomès E, Richard T, Mérillon J-M (2020) Characterization of stilbene composition in grape berries from wild vitis species in year-to-year harvest. J Agric Food Chem 68:13408–13417
González Mera IF, González Falconí DE, Morera Córdova V (2019) Secondary metabolites in plants: Main classes, phytochemical analysis and pharmacological activities. Bionatura 4:1000–1009
Grubb CD, Abel S (2006) Glucosinolate metabolism and its control. Trends Plant Sci 11:89–100
Gutiérrez-Grijalva EP, López-Martínez LX, Contreras-Angulo LA, Elizalde-Romero CA, Heredia JB (2020) Plant alkaloids: structures and bioactive properties. In: Plant-derived bioactive chemistry mode action. Springer Singapore, Singapore, pp 85–117
Halkier BA, Gershenzon J (2006) Biology and biochemistry of glucosinolates. Annu Rev Plant Biol 57:303–333
Hammerbacher A, Coutinho TA, Gershenzon J (2019) Roles of plant volatiles in defence against microbial pathogens and microbial exploitation of volatiles. Plant Cell Environ 42:2827–2843
Hardegger E, Schellenbaum M, Corrodi H (1963) Welkstoffe und Antibiotika. 27. Mitteilung. Über induzierte Abwehrstoffe bei Orchideen II. Helv Chim Acta 46:1171–1180
Harris JE, Dennis C (1977) The effect of post-infectional potato tuber metabolites and surfactants on zoospores of Oomycetes. Physiol Plant Pathol 11:163–169
Heskes AM, Lincoln CN, Goodger JQD, Woodrow IE, Smith TA (2012) Multiphoton fluorescence lifetime imaging shows spatial segregation of secondary metabolites in Eucalyptus secretory cavities. J Microsc 247:33–42
Holst B, Williamson G (2004) A critical review of the bioavailability of glucosinolates and related compounds. Nat Prod Rep 21:425–447
Hoult JRS, Payá M (1996) Pharmacological and biochemical actions of simple coumarins: natural products with therapeutic potential. Gen Pharmacol Vasc Syst 27:713–722
Howell JT, Davis MR (2005) Plant defense mechanisms against fungal pathogens: polygalacturonase inhibitor proteins. Can J Plant Pathol 27:5–15
Jeandet P (2015) Phytoalexins: current progress and future prospects. Molecules 20:2770–2774
Jeandet P, Hébrard C, Deville M-A, Cordelier S, Dorey S, Aziz A, Crouzet J (2014) Deciphering the role of phytoalexins in plant-microorganism interactions and human health. Molecules 19: 18033–18056
Jiao J, Ma Y, Chen S, Liu C, Song Y, Qin Y, Yuan C, Liu Y (2016) Melatonin-producing endophytic bacteria from grapevine roots promote the abiotic stress-induced production of endogenous melatonin in their hosts. Front Plant Sci 7:1387
Kachroo A, Robin GP (2013) Systemic signaling during plant defense. Curr Opin Plant Biol 16: 527–533
Kang K, Fong W-P, Tsang PW-K (2010) Novel antifungal activity of purpurin against Candida species in vitro. Med Mycol 48:904–911
Kaur S, Samota MK, Choudhary M, Choudhary M, Pandey AK, Sharma A, Thakur J (2022) How do plants defend themselves against pathogens-biochemical mechanisms and genetic interventions. Physiol Mol Biol plants an Int J Funct plant Biol 28:485–504
Khan S, Sahar A, Tariq T, Sameen A, Tariq F (2023) Essential oils in plants: plant physiology, the chemical composition of the oil, and natural variation of the oils (chemotaxonomy and environmental effects, etc.). In: Essential Oils. Elsevier, pp 1–36
Khatri P, Chen L, Rajcan I, Dhaubhadel S (2023) Functional characterization of Cinnamate 4-hydroxylase gene family in soybean (Glycine max). PLoS One 18:e0285698

Kim JH, Haff RP, Faria NCG, Martins M d L, Chan KL, Campbell BC (2013) Targeting the mitochondrial respiratory chain of Cryptococcus through antifungal chemosensitization: a model for control of non-fermentative pathogens. Molecules 18:8873–8894

Kodama O, Miyakawa J, Akatsuka T, Kiyosawa S (1992) Sakuranetin, a flavanone phytoalexin from ultraviolet-irradiated rice leaves. Phytochemistry 31:3807–3809

Kopriva S, Mugford SG, Matthewman C, Koprivova A (2009) Plant sulfate assimilation genes: redundancy versus specialization. Plant Cell Rep 28:1769–1780

Kuda T, Iwai A, Yano T (2004) Effect of red pepper Capsicum annuum var. conoides and garlic Allium sativum on plasma lipid levels and cecal microflora in mice fed beef tallow. Food Chem Toxicol 42:1695–1700

Kumar K, Debnath P, Singh S, Kumar N (2023) An overview of plant phenolics and their involvement in abiotic stress tolerance. Stress 3:570–585

Künstler A, Gullner G, Ádám AL, Kolozsváriné Nagy J, Király L (2020) The versatile roles of sulfur-containing biomolecules in plant defense—a road to disease resistance. Plants 9:1705

Kurepa J, Shull TE, Karunadasa SS, Smalle JA (2018) Modulation of auxin and cytokinin responses by early steps of the phenylpropanoid pathway. BMC Plant Biol 18:1–15

Lay FT, Anderson MA (2005) Defensins-components of the innate immune system in plants. Curr Protein Pept Sci 6:85–101

Lee SK, Lee HJ, Min HY, Park EJ, Lee KM, Ahn YH, Cho YJ, Pyee JH (2005) Antibacterial and antifungal activity of pinosylvin, a constituent of pine. Fitoterapia 76:258–260

Li Z, Li Z, Yang J, Lu C, Li Y, Luo Y, Cong F, Shi R, Wang Z, Chen H (2022) Allicin shows antifungal efficacy against Cryptococcus neoformans by blocking the fungal cell membrane. Front Microbiol 13:1012516

Lü J, Lin PH, Yao Q, Chen C (2010) Chemical and molecular mechanisms of antioxidants: experimental approaches and model systems. J Cell Mol Med 14:840–860

Luna E (2016) Using green vaccination to brighten the agronomic future. Outlooks Pest Manag 27: 136–140

Marchese A, Barbieri R, Sanches-Silva A, Daglia M, Nabavi SF, Jafari NJ, Izadi M, Ajami M, Nabavi SM (2016) Antifungal and antibacterial activities of allicin: a review. Trends Food Sci Technol 52:49–56

Martin DM, Gershenzon J, Bohlmann J (2003) Induction of volatile terpene biosynthesis and diurnal emission by methyl jasmonate in foliage of Norway spruce. Plant Physiol 132:1586–1599

Martínez MJA, Benito PB (2005) Biological activity of quinones. Stud Nat Prod Chem 30:303–366

Mauch-Mani B, Baccelli I, Luna E, Flors V (2017) Defense priming: an adaptive part of induced resistance. Annu Rev Plant Biol 68:485–512

Mayama S, Bordin APA, Morikawa T, Tanpo H, Kato H (1995) Association of avenalumin accumulation with co-segregation of victorin sensitivity and crown rust resistance in oat lines carrying the Pc-2 gene. Physiol Mol Plant Pathol 46:263–274

Miziorko HM (2011) Enzymes of the mevalonate pathway of isoprenoid biosynthesis. Arch Biochem Biophys 505:131–143

Mutha RE, Tatiya AU, Surana SJ (2021) Flavonoids as natural phenolic compounds and their role in therapeutics: an overview. Futur J Pharm Sci 7:1–13

Nejat N, Mantri N (2017) Plant immune system: crosstalk between responses to biotic and abiotic stresses the missing link in understanding plant defence. Curr Issues Mol Biol 23:1–16

Nicholson RL, Hammerschmidt R (1992) Phenolic compounds and their role in disease resistance. Annu Rev Phytopathol 30:369–389

Ogita A, Fujita K, Taniguchi M, Tanaka T (2006) Dependence of synergistic fungicidal activity of Cu2+ and allicin, an allyl sulfur compound from garlic, on selective accumulation of the ion in the plasma membrane fraction via allicin-mediated phospholipid peroxidation. Planta Med 72: 875–880

Oksman-Caldentey K-M, Inzé D (2004) Natural compounds derived from plants can be categorized into three different groups according to their final use in developing a drug. Trends Plant Sci 9: 433–440

Pai ST, Platt MW (1995) Antifungal effects of Allium sativum (garlic) extract against the Aspergillus species involved in otomycosis. Lett Appl Microbiol 20:14–18

Pârvu M, Moţ CA, Pârvu AE, Mircea C, Stoeber L, Roşca-Casian O, Ţigu AB (2019) Allium sativum extract chemical composition, antioxidant activity and antifungal effect against Meyerozyma guilliermondii and Rhodotorula mucilaginosa causing onychomycosis. Molecules 24:3958

Passardi F, Penel C, Dunand C (2004) Performing the paradoxical: how plant peroxidases modify the cell wall. Trends Plant Sci 9:534–540

Peter SP, Menon KS (2022) Phytochemical and Antifungal Evaluation of Capsicum frutescens. In: Recent Advances in Life Sciences. pp 37–47

Phillips MA, León P, Boronat A, Rodríguez-Concepción M (2008) The plastidial MEP pathway: unified nomenclature and resources. Trends Plant Sci 13:619–623

Pieterse CMJ, Zamioudis C, Berendsen RL, Weller DM, Van Wees SCM, Bakker PAHM (2014) Induced systemic resistance by beneficial microbes. Annu Rev Phytopathol 52:347–375

Plattner S, Gruber C, Stadlmann J, Widmann S, Gruber CW, Altmann F, Bohlmann H (2015) Isolation and characterization of a thionin proprotein-processing enzyme from barley. J Biol Chem 290:18056–18067

Prieto MA, López CJ, Simal-Gandara J (2019) Glucosinolates: molecular structure, breakdown, genetic, bioavailability, properties and healthy and adverse effects. Adv Food Nutr Res 90:305–350

Ratnadewi D (2017) Alkaloids in plant cell cultures. In: Alkaloids: alternatives in synthesis, modification and application. IntechOpen, pp 1–23

Rojas CM, Senthil-Kumar M, Tzin V, Mysore KS (2014) Regulation of primary plant metabolism during plant-pathogen interactions and its contribution to plant defense. Front Plant Sci 5:17

Saltveit ME (2017) Synthesis and metabolism of phenolic compounds. In: Fruit and vegetable phytochemicals: chemistry and human health, 2nd edn, pp 115–124

Sanchez S, Demain AL (2008) Metabolic regulation and overproduction of primary metabolites. Microb Biotechnol 1:283–319

Santos-Sánchez NF, Salas-Coronado R, Hernández-Carlos B, Villanueva-Cañongo C (2019) Shikimic acid pathway in biosynthesis of phenolic compounds. In: Plant Physiol Asp phenolic Compd, vol 1, pp 1–15

Saxena M, Saxena J, Nema R, Singh D, Gupta A (2013) Phytochemistry of medicinal plants. J Pharmacogn Phytochem 1:168–182

Sayadi V, Karimzadeh G, Rashidi Monfared S, Naghavi MR (2020) Identification and expression analysis of S-alk (en) yl-L-cysteine sulfoxide lyase isoform genes and determination of allicin contents in Allium species. PLoS One 15:e0228747

Schnee S, Queiroz EF, Voinesco F, Marcourt L, Dubuis PH, Wolfender JL, Gindro K (2013) *Vitis vinifera* canes, a source of stilbenoids against *Plasmopara viticola*. J Agricul & Food Chem 61(48):11709–11720

Sharma S, Perera KY, Jaiswal AK, Jaiswal S (2024) Natural antimicrobials from fruits and plant extract for food packaging and preservation. In: Food Packaging and Preservation. Elsevier, pp 133–152

Shih C-H, Chu IK, Yip WK, Lo C (2006) Differential expression of two flavonoid 3′-hydroxylase cDNAs involved in biosynthesis of anthocyanin pigments and 3-deoxyanthocyanidin phytoalexins in sorghum. Plant Cell Physiol 47:1412–1419

Shomali A, Das S, Arif N, Sarraf M, Zahra N, Yadav V, Aliniaeifard S, Chauhan DK, Hasanuzzaman M (2022) Diverse physiological roles of flavonoids in plant environmental stress responses and tolerance. Plants 11:3158

Silva PM, Gonçalves S, Santos NC (2014) Defensins: antifungal lessons from eukaryotes. Front Microbiol 5:65587

Silverstein KAT, Graham MA, Paape TD, VandenBosch KA (2005) Genome organization of more than 300 defensin-like genes in Arabidopsis. Plant Physiol 138:600–610

Starcevic A, Akthar S, Dunlap WC, Shick JM, Hranueli D, Cullum J, Long PF (2008) Enzymes of the shikimic acid pathway encoded in the genome of a basal metazoan, Nematostella vectensis, have microbial origins. Proc Natl Acad Sci 105:2533–2537

Suteu D, Rusu L, Zaharia C, Badeanu M, Daraban GM (2020) Challenge of utilization vegetal extracts as natural plant protection products. Appl Sci 10:8913

Tamokou JDD, Mbaveng AT, Kuete V (2017) Antimicrobial activities of African medicinal spices and vegetables. In: Medicinal spices and vegetables from Africa. Elsevier, pp 207–237

Tekbas OF, Ogur R, Korkmaz A, Kilic A, Reiter RJ (2008) Melatonin as an antibiotic: new insights into the actions of this ubiquitous molecule. J Pineal Res 44:222–226

Teoh ES, Teoh ES (2016) Secondary metabolites of plants. In: Medicinal Orchids of Asia. Springer International Publishing, Cham, pp 59–73

Terras FR, Eggermont K, Kovaleva V, Raikhel NV, Osborn RW, Kester A, Rees SB, Torrekens S, Van Leuven F, Vanderleyden J (1995) Small cysteine-rich antifungal proteins from radish: their role in host defense. Plant Cell 7:573–588

Timperio AM, d'Alessandro A, Fagioni M, Magro P, Zolla L (2012) Production of the phytoalexins trans-resveratrol and delta-viniferin in two economy-relevant grape cultivars upon infection with Botrytis cinerea in field conditions. Plant Physiol Biochem 50:65–71

Turlings TCJ, Erb M (2018) Tritrophic interactions mediated by herbivore-induced plant volatiles: mechanisms, ecological relevance, and application potential. Annu Rev Entomol 63:433–452

Twaij BM, Hasan MN (2022) Bioactive secondary metabolites from plant sources: types, synthesis, and their therapeutic uses. Int J Plant Biol 13:4–14

Vance CP, Kirk TK, Sherwood RT (1980) Lignification as a mechanism of disease resistance. Annu Rev Phytopathol 18:259–288

Walker GM, White NA (2017) Introduction to fungal physiology. In: Fungi Biol Appl, pp 1–35

Wang ET, Rogel MA, Santos AG, Martinez-Romero J, Cevallos MA, Martinez-Romero E (1999) *Rhizobium etli bv. mimosae*, a novel biovar isolated from *Mimosa affinis*. Int J Syst Bacteriol 49: 1479–1491

Wang W, Wang S-X, Guan H-S (2012) The antiviral activities and mechanisms of marine polysaccharides: an overview. Mar Drugs 10:2795–2816

Wink M, Roberts MF (1998) Alkaloids: biochemistry, ecology, and medicinal applications. Plenum Press

Wu S, Chen W, Lu S, Zhang H, Yin L (2022) Metabolic engineering of shikimic acid biosynthesis pathway for the production of shikimic acid and its branched products in microorganisms: advances and prospects. Molecules 27:4779

Xie M, Zhang J, Tschaplinski TJ, Tuskan GA, Chen J-G, Muchero W (2018) Regulation of lignin biosynthesis and its role in growth-defense tradeoffs. Front Plant Sci 9:407961

Zainal M, Zain NM, Amin IM, Ahmad VN (2021) The antimicrobial and antibiofilm properties of allicin against Candida albicans and Staphylococcus aureus–a therapeutic potential for denture stomatitis. Saudi Dent J 33:105–111

Zechmann B, Müller M (2010) Subcellular compartmentation of glutathione in dicotyledonous plants. Protoplasma 246:15–24

Zeyen RJ, Carver TLW, Lyngkjaer MF (2002) Epidermal cell papillae. In: Belanger RR, Bushnell WR (eds) The powdery mildew: a comprehensive treatise

Zhu Z, Lee B (2015) Friends or foes: new insights in jasmonate and ethylene co-actions. Plant Cell Physiol 56:414–420

# Chapter 14
# Efficacy of Plant Metabolic Products to Suppress the Plant Pathogenic Organisms

**Said Behiry, Dalia G. Aseel, Dina Elkobrosy, Ahmed Kenawy, Hesham A. El Enshasy, and Ahmed Abdelkhalek**

## 14.1 Introduction

Plant phytochemicals are broadly categorized based on their roles in metabolic processes, with traditional classification distinguishing between primary and secondary metabolites. However, contemporary research has refined this framework, dividing phytochemicals into three functional groups: (1) primary metabolites, (2) secondary metabolites, and (3) phytohormones (Hussein and El-Anssary 2019). Primary metabolites play a fundamental role in sustaining plant growth and development, serving as indispensable components of core metabolic pathways (Fernie and Pichersky 2015). In contrast, secondary metabolites are specialized compounds that mediate interactions with abiotic and biotic environmental factors, contributing to plant defense, stress tolerance, and ecological fitness (Lal et al. 2023). Phytohormones, the third category, act as regulatory molecules orchestrating various physiological processes, including growth, development, and stress responses. These hormones also influence the biosynthesis of both primary and secondary

S. Behiry
Agricultural Botany Department, Faculty of Agriculture (Saba Basha), Alexandria University, Alexandria, Egypt

D. G. Aseel · D. Elkobrosy · A. Abdelkhalek (✉)
Plant Protection and Biomolecular Diagnosis Department, Arid Lands Cultivation Research Institute, City of Scientific Research and Technological Applications, Alexandria, Egypt

A. Kenawy
Nucleic Acids Research Department, Genetic Engineering and Biotechnology Research Institute, City of Scientific Research and Technological Applications, Alexandria, Egypt

H. A. El Enshasy
Institute of Bioproduct Development (IBD), Universiti Teknologi Malaysia (UTM), Skudai, Johor Bahru, Malaysia

 
N. El-Wakeil, A. Abdelkhalek (eds.), *Microbiomes and Plant Secondary Metabolites for Insect Pest Management*, Progress in Biological Control 23,
https://doi.org/10.1007/978-3-032-24330-0_14

metabolites by interacting with receptor proteins to trigger specific signaling pathways (Upadhyay et al. 2022, 2023). This tripartite classification highlights the intricate interplay between phytochemicals in supporting plant survival and adaptation. Plants produce a variety of organic compounds, some of which do not directly participate in photosynthesis, growth, or development but play crucial indirect roles in plant survival. These include signaling molecules that help deter herbivores, act as antimicrobial agents, or support defense mechanisms. Secondary metabolites, synthesized through primary metabolic pathways, are notable for their toxicity and possess antifungal, antibiotic, and antiviral properties (Sahoo et al. 2024). These compounds significantly enhance the plant's defense by protecting against pathogens, herbivores, and other microorganisms. In recent years, plant secondary metabolites have garnered substantial attention from researchers, scientists, and industries due to their multifaceted contributions (Li et al. 2024). They influence plant characteristics such as odor, color, and taste, while also playing a role in human nutrition and having commercial applications in cosmetics, pharmaceuticals, dyes, polymers, herbicides, and insecticides (Verma et al. 2024). Biogenically, secondary metabolites are classified into four major groups: phenolics and polyphenolics, terpenes, nitrogen-containing compounds (alkaloids), and sulfur-containing compounds (Sharma et al. 2022). Secondary metabolites in plants are broadly categorized into three major groups: alkaloids, terpenoids, and phenolic compounds. These groups include diverse compounds with distinct biosynthetic origins and functional roles (Mishra 2024).

## 14.2 Classes of Plant Secondary Metabolites

Plants serve as renewable resources with diverse applications, encompassing the generation of raw materials like lignocellulosic biomass and the extraction of phytochemicals, particularly secondary metabolites (Guerriero et al. 2018). Plant secondary metabolites are a huge group of structurally different molecules, mainly terpenoids, phenolic compounds, and nitrogen-containing compounds (Fig. 14.1). They are crucial for plant development and its stress response (Sobhy et al. 2023; Deng et al. 2024).

### *14.2.1 Terpenes*

Terpenes represent one of the largest and most diverse classes of secondary metabolites, constructed from isoprene (C5) units (Siddiqui et al. 2024). These units combine to form various terpenoid subgroups, including monoterpenoids (two units), sesquiterpenoids (three units), diterpenoids (four units), and triterpenoids (six units). The functional diversity of terpenes is further expanded through processes like oxidation and methylation (Perveen and Al-Taweel 2018). Over 25,000

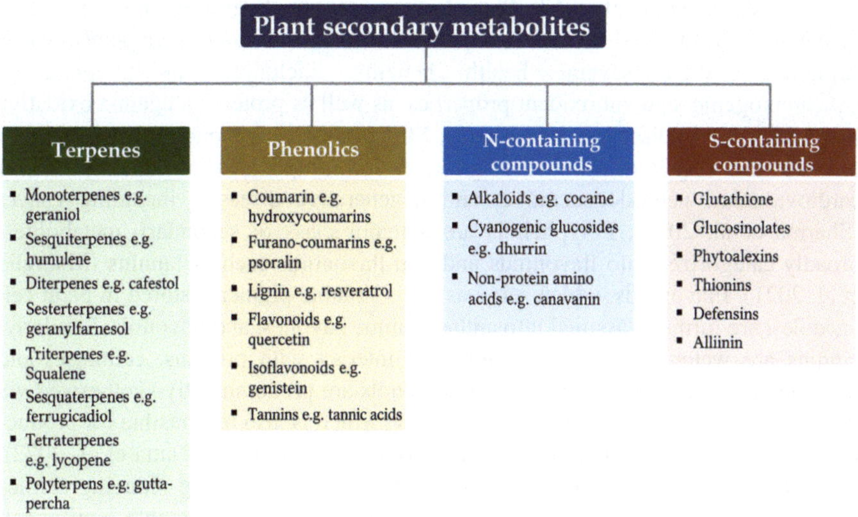


**Fig. 14.1** Categories of plant secondary metabolites (Twaij and Hasan 2022)

terpenoid types have been identified across different plant species (Khan et al. 2024). The biosynthesis of terpenes occurs via two primary pathways: the mevalonate (MVA) pathway and the methylerythritol phosphate (MEP) pathway, both of which generate isoprene (C5) precursors (Singh and Sharma 2015). Most terpenes are lipophilic, allowing them to integrate into biomembranes and influence membrane fluidity and permeability (Wink 2015). Terpenes play a variety of roles in plants, including defense against herbivores and pathogens, antimicrobial and insecticidal activities, growth regulation, and contributing to plant-specific odors (Gutiérrez-del-Río et al. 2018). Specific terpene subcategories include steroids, carotenoids, and gibberellic acid. Steroids, in particular, are vital for the physiology and biochemistry of living organisms. Synthetic steroids are extensively used in medicine for their applications in cancer treatment, contraception, anti-inflammatory therapies, cardiovascular care, and antibiotics (Naji et al. 2024). Additionally, some steroids are employed to promote muscle growth and fat loss with minimal side effects (Alfakje and Al-Mashhadane 2024).

### *14.2.2 Phenolics and Polyphenolics*

Phenolic compounds are multifaceted molecules that carry out a range of functions in plants. They readily undergo oxidation and function as antioxidants (Mohite et al. 2024). For instance, seeds accumulate phenols that restrict oxygen access to the embryo, thereby inhibiting germination (González Mera et al. 2019). Additionally, phenolic compounds can absorb up to 90% of UV radiation by accumulating on leaf surfaces (Verdaguer et al. 2017). Phenolics represent the largest group of secondary metabolites synthesized by fruits, vegetables, teas, cocoa, and other plants. These compounds offer several health benefits, including anti-inflammatory, anticarcinogenic, and antioxidant properties, as well as protection against oxidative stress and certain diseases (Vyas 2017). Structurally, polyphenols contain at least one aromatic ring with a hydroxyl group and are implicated in reducing risks of cardiovascular, neurodegenerative, and degenerative diseases, including cancer (Sharma et al. 2022). Polyphenols are a major class of secondary metabolites, broadly categorized into flavonoids and non-flavonoids such as tannins (Maniglia et al. 2021). Flavonoids, which serve as water-soluble pigments stored in plant cell vacuoles, are further classified into anthocyanins, flavones, and flavonols. Similarly, tannins are water-soluble compounds that interact with proteins, cellulose, and minerals to form stable complexes. Polyphenols are predominantly synthesized via the shikimic acid (phenylpropanoid) pathway, which is also responsible for producing isoflavones, coumarins, lignins, and aromatic amino acids (Datta et al. 2022). These compounds are known for their health benefits, including reducing cardiovascular disease risks (Pastor et al. 2019), and exhibit notable antifungal, antibacterial, and antiviral activities. The biosynthesis of phenolics is often upregulated under stress conditions, such as pathogen attacks (Kumar et al. 2014). Flavonoids are low-molecular-weight polyphenolic compounds widely distributed in fruits, vegetables, grains, and various plant tissues. These compounds are recognized for their potent biological activities, including anticancer, antiviral, and antimicrobial properties (Mandal and Domb 2024). Structurally, flavonoids consist of a flavin nucleus, two benzene rings, and a heterocyclic pyrene ring that connects the benzene rings, forming their characteristic framework (Dhamija and Mangla 2022).

### *14.2.3 Nitrogen-Containing Compounds*

Nitrogen-containing compounds represent a crucial group of plant secondary metabolites, encompassing alkaloids, cyanogenic glycosides, glucosinolates, and non-protein amino acids (Barreda et al. 2024). These metabolites are primarily synthesized from amino acids, reflecting their strong connection to nitrogen metabolism (Jamwal et al. 2018). Alkaloids are among the most prominent and chemically diverse classes of secondary metabolites, defined by the presence of at least one nitrogen atom, often accompanied by sulfur or oxygen atoms (Chauhan et al. 2017). Based on the position of the nitrogen atom, alkaloids are broadly categorized into heterocyclic and non-heterocyclic types (Eguchi et al. 2019). These compounds exhibit a wide array of biological activities, including anti-inflammatory, anti-psychotic, anti-plasmodic, and anticancer properties (Debnath et al. 2018). Notable examples include morphine (analgesic), berberine (antibiotic), and vinblastine (anticancer), as well as other significant alkaloids like codeine, nicotine, quinine, and strychnine (Lal et al. 2023). Saponins derive their name from the soapwort plant, historically used as a natural soap (from the Latin sapo meaning soap). These compounds play a protective role in plants by exhibiting antimicrobial activity and deterring insect attacks (Tatli Cankaya and Somuncuoglu 2021). Beyond their plant functions, saponins offer numerous health benefits, including hypoglycemic effects, cholesterol reduction, and antiviral, and anti-inflammatory properties (Sharma et al. 2023). As surface-active glycosides with foaming properties, they are commonly consumed through pulses, beans, lentils, chickpeas, and vegetables (Singh et al. 2017). Steroids are low-molecular-weight lipophilic compounds derived from cholesterol. They are produced by various terrestrial, marine, and synthetic sources and play essential roles in the biochemistry and physiology of living organisms (Naji et al. 2024). While steroids have significant medicinal applications, they are sometimes associated with adverse health effects. Medicinally, they serve as anesthetics, anticancer agents, anti-asthmatics, contraceptives, and cardiovascular drugs (Lopez et al. 2014; Badyal et al. 2020).

### *14.2.4 Sulfur-Containing Compounds*

Sulfur-containing compounds represent a relatively small but functionally significant class of plant secondary metabolites (Xu et al. 2025). Distinguished by their characteristic odor, these compounds serve critical roles as signaling molecules and defense agents against herbivores and pathogens (Majumdar et al. 2023). They are primarily divided into two major groups: glucosinolates, predominantly found in the *Brassicaceae* family, and alliins, characteristic of Allium species (Fritsch and Keusgen 2006). In response to herbivore damage, glucosinolates undergo enzymatic hydrolysis, yielding toxic derivatives such as nitriles, thiocyanates, and isothiocyanates. These breakdown products provide an efficient chemical defense against herbivory (Divekar et al. 2022; Lyubenova et al. 2023). Additionally, glucosinolates contribute to plant resistance against predators, parasites, and competitors by releasing volatile compounds that act as deterrents (Jamwal et al. 2018). These multifaceted functions highlight the ecological and biochemical importance of sulfur-containing metabolites in plant survival and adaptation.

## 14.3 Role of Plant Metabolites in Suppressing Plant Viral Infections

Plant metabolites play a critical role in defending plants from viral infections through diverse mechanisms, such as direct antiviral activity, enhancing plant immune responses, and influencing plant–virus interactions (Manjunatha et al. 2022). Secondary metabolites, in particular, are vital in this defense, as they are small organic compounds naturally synthesized by plants, often under biotic or abiotic stress (Fig. 14.2). These metabolites not only support defense signaling pathways but also counteract virus-induced disruptions in hormone biosynthesis, which affects growth, differentiation, and cellular communication (Mishra et al. 2020).

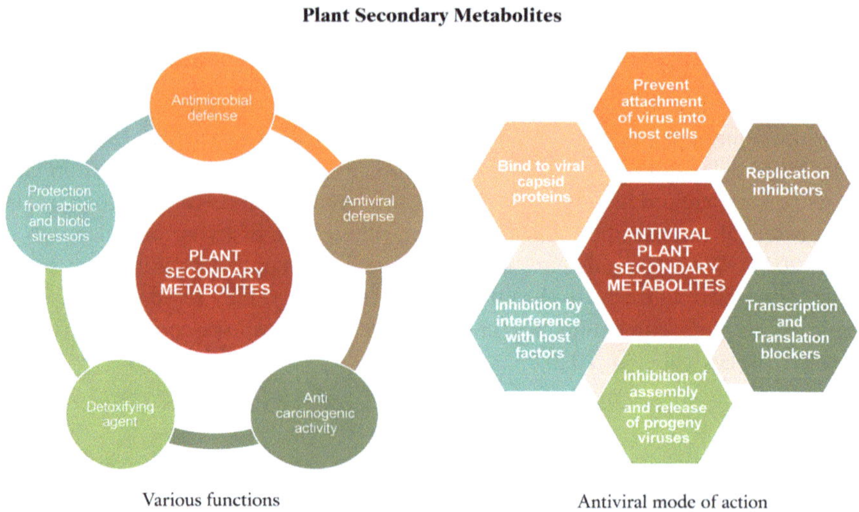


**Fig. 14.2** Diverse antibacterial properties of plant secondary metabolites (Moradi et al. 2018)

### *14.3.1 Direct Antiviral Activity*

Phenolic compounds, such as flavonoids, tannins, and phenolic acids, function as antiviral agents by interacting with viral proteins or genetic material, which disrupts the processes of viral replication and infection cycles (Zakaryan et al. 2017). Phenolic compounds demonstrate antiviral properties via various mechanisms, such as disrupting viral lipid bilayers, hindering viral attachment to host cells, and inhibiting viral protein activity, which interferes with replication, multiplication, and release (Montenegro-Landívar et al. 2021). The versatility of these mechanisms underscores the promise of phenolic compounds as potent antiviral agents (Kumar et al. 2023). Terpenoids are various plant chemicals that can be divided into six groups: hemiterpenes, monoterpenes, sesquiterpenes, diterpenes, sesterterpenes, triterpenes, and tetraterpenoids (Abdallah and Quax 2017). Found predominantly in the leaves and fruits of plants, terpenoids contribute to plant aroma, flavor, and pigmentation (Huang et al. 2021). These metabolites exhibit antiviral properties by targeting viral replication enzymes and destabilizing viral envelopes. Their role in plant-virus defense underscores their importance as bioactive compounds (Rodriguez-Concepcion et al. 2018).

Extensive research has demonstrated the potential of alkaloids, flavonoids, saponins, terpenoids, carotenoids, glucosides, and polyphenols from medicinal plants in inhibiting viral processes like adherence, penetration, replication, maturation, and packaging (Dhamija and Mangla 2022). Integrating these metabolites with current treatments could enhance antiviral efficacy and bioavailability (Kapoor et al. 2017). Certain secondary metabolites demonstrate direct antiviral properties, including flavonoids: Compounds such as quercetin, kaempferol, and apigenin demonstrate the ability to inhibit viral enzymes or interfere with viral replication processes. Saponins are compounds that interfere with viral membranes, thereby hindering systemic infection. Glucosinolates, present in plants of the Brassicaceae family, break down into compounds with antiviral properties. Phytoalexins: Metabolites produced under stress, like resveratrol, are generated in reaction to viral infections and demonstrate antiviral properties (Adeosun and Loots 2024). Alkaloids are nitrogen-containing organic compounds found in various plant tissues and are renowned for their therapeutic potential. This group includes well-known compounds like morphine, cocaine, and quinine (Abookleesh et al. 2022). Numerous alkaloids have also been evaluated for their antiviral efficacy (Dhamija and Mangla 2022). Alkaloids have been extensively studied for their broad-spectrum antiviral properties against both DNA and RNA viruses (Moradi et al. 2018). Alkaloids have antiviral effects by stopping DNA and RNA replication, RNA translation, protein synthesis, enzymatic activities that are necessary for viral survival, and the movement of ribonucleoprotein complexes (Peng et al. 2003). These mechanisms significantly impair the viral lifecycle, making alkaloids a key focus for antiviral research and applications. Plant metabolites, with their diverse chemical structures and multifaceted antiviral mechanisms, are pivotal in plant defense. Because these compounds directly inhibit viruses, change the immune system, and control hormones, they help us learn a lot about natural ways to fight viruses and possible uses in agriculture (Atampugbire et al. 2024).

### 14.3.2 Activation of Plant Defense Mechanisms

Secondary metabolites enable communication between hormonal pathways, including salicylic acid (SA), jasmonic acid (JA), and ethylene (ET), refining the plant's reaction to various viral infections according to the type of pathogen (Kaur et al. 2022). Beneficial microbes can stimulate plant metabolites, thereby activating induced systemic resistance. For example, bacteria or fungi associated with plants stimulate the production of signaling molecules such as SA and JA, thereby boosting the plant's antiviral resistance (Fig. 14.3). The diverse functions of primary and secondary metabolites highlight their significance in plant defense, providing valuable insights for potential applications in sustainable crop protection and antiviral strategies (Abdelkhalek et al. 2021; Al-Khayri et al. 2023; Rabie et al. 2024). Certain metabolites act as signaling molecules, initiating plant immune responses; for example, SA plays a vital role in systemic acquired resistance (SAR), enhancing defense mechanisms against various pathogens, including viruses. JA and ET are crucial in initiating defense responses, especially regarding necrotrophic pathogens and viruses. Secondary metabolites promote the generation of reactive oxygen species (ROS), resulting in direct harm to viral particles or enhancing cell walls to limit viral spread (Tripathi et al. 2019; Aseel et al. 2024).

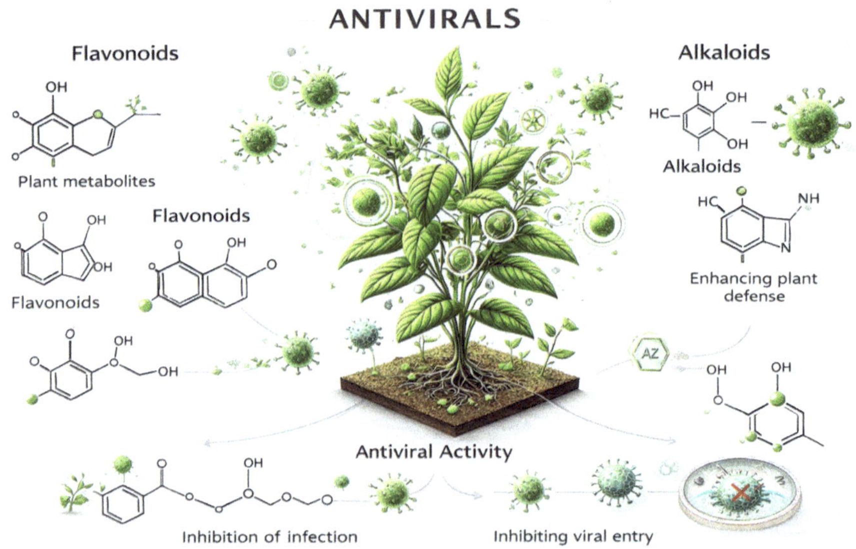


**Fig. 14.3** Plant metabolites in protecting plants against viruses, highlighting the interaction between metabolites and viruses

### *14.3.3 RNA Silencing and Antiviral Gene Regulation*

Certain secondary metabolites influence RNA silencing, a crucial mechanism for antiviral defense (Li and Wang 2019). It was indicated that small RNA molecules, including siRNAs, play a crucial role in the degradation of viral RNA (Suleiman et al. 2024). Furthermore, specific metabolites enhance the synthesis or stability of viral RNA, subsequently strengthening the plant's defenses. Phenolic compounds play a crucial role in lignin formation, contributing to the reinforcement of cell walls and providing protection against viral attacks (Saini et al. 2024). Furthermore, the occurrence of callose at plasmodesmata serves to obstruct the movement of viruses between cells (Chen et al. 2024a).

## 14.4 Role of Plant Metabolites as Antibacterial and Antifungal Agents

Plants, being stationary organisms that cannot alter their surrounding environment, depend on a wide range of metabolites to sustain their fitness, especially in stressful situations (Twaij and Hasan 2022). Among these, plant secondary metabolites play a pivotal role in defending against both biotic and abiotic stressors. These natural compounds exhibit a wide range of antimicrobial activities, which can be attributed to various chemical classes, including terpenes, phenolics, flavonoids, alkaloids, cyanogenic glucosides, quinones, amides, aldehydes, thiophenes, amino acids, saccharides, saponins, salicylates, glucosinolates, sulfur-containing compounds, and polyketides (Al-Askar et al. 2023; Aamer et al. 2024b; Behiry et al. 2024). Beyond their antimicrobial properties, these metabolites exhibit significant ecological functions, including acting as insect repellents, nematicides, insect growth regulators, anti-feedants, and allelopathic agents (Elsharkawy et al. 2022; Khamis et al. 2023). This versatility positions them as promising candidates for novel pest control agents or biopesticides (Elshafie et al. 2023). Plants strategically store these metabolites in cellular compartments, deploying them as needed to adapt to environmental challenges effectively. Moreover, the biochemical plasticity provided by secondary metabolites enables plants to respond rapidly to emerging stressors, showcasing their critical role in plant survival and adaptation to fluctuating environmental conditions (Weiszmann et al. 2023). These compounds not only highlight the complexity of plant defense mechanisms but also serve as potential sources for developing innovative antimicrobial and pest control solutions. The antimicrobial effect of these metabolites could be the sole effect of one compound and often enhanced in combination with other plant metabolites. For example, alkaloids or terpenoids could enhance phenolics' effect synergistically. This synergism can improve the overall effectiveness of the antimicrobial agents and reduce the likelihood of resistance development by the microbes toward these compounds. In addition, plants utilize secondary metabolites as a defense mechanism against biotic and abiotic stresses. These metabolites play crucial roles, such as deterring herbivores, inhibiting the growth of competing plants, protecting from ultraviolet radiation, attracting pollinators, aiding in seed dispersal, and fostering symbiotic relationships with root nodule bacteria, either directly or indirectly. Many secondary metabolites, derived from primary metabolites and their intermediates, are commercially valuable as dyes, flavors, pharmaceuticals, insecticides, and fragrances (Pagare et al. 2015). For instance, flavonoids extracted from *Zapoteca portoricensis*, a leguminous crop, exhibit antioxidant, anti-tumor, and anti-inflammatory properties. Similarly, the antifungal activity of dill seed oil against pathogens such as *Ustilago segetum var. tritici*, *Alternaria triticina*, and *Bipolaris sorokiniana* is attributed to key compounds like carvone and camphor (Kumar et al. 2023).

### *14.4.1 Mechanisms of Secondary Metabolites*

Plant phenolic metabolites represent a significant group of bioactive compounds recognized for their potent antimicrobial properties. Predominantly found in plant-based foods, herbs, and medicinal plants, these metabolites exhibit a wide range of biological activities. They are synthesized by plants as a defensive strategy to mitigate both biotic and abiotic stressors. The structural diversity of phenolic compounds gives rise to numerous subclasses, including phenolic compounds. These metabolites are primarily synthesized through the phenylpropanoid biosynthetic pathway, which is a crucial metabolic route in plants. Phenolic compounds display broad-spectrum antimicrobial activity, effectively targeting various plant pathogens, including Gram-positive and Gram-negative bacteria, as well as numerous fungal species. Their antimicrobial efficacy stems from diverse mechanisms of action, which may include many mechanisms (Twaij and Hasan 2022). These mechanisms include:

- Disruption of cellular membranes, where many phenolic compounds can interact with the lipid bilayer of cell membranes, causes leakage of essential cellular contents, leading to cell death. This membrane-disrupting activity has been observed in bacteria and fungi.
- Inhibition of enzymatic pathways, where phenolics can inhibit key enzymes essential for the survival and proliferation of microbes. Such as the enzymes involved in cell wall synthesis, protein synthesis, or metabolic processes in bacteria and fungi.
- Oxidative stress generation as phenolic compounds possess antioxidant properties, but in certain concentrations, they can shift from antioxidant activity to pro-oxidant activity. This can lead to the generation of reactive oxygen species (ROS), which damage microbial DNA, proteins, and lipids, contributing to microbial death in bacteria and fungi.
- Inhibition of biofilm formation, where biofilm formation is a critical factor in the virulence of many pathogens, especially in chronic infections. Plant phenolic compounds have been shown to interfere with biofilm formation by disrupting bacterial communication (quorum sensing) or by directly preventing the adhesion of microbes to surfaces while forming membrane layers.
- Metal Ion Chelation: Phenolics can chelate metal ions, which are crucial for bacterial enzyme activity and structural integrity.
- Reducing oxidative stress in plants indirectly helps in combating microbial infections. These mechanisms make phenolic compounds effective against a wide range of pathogens, including bacteria, fungi, viruses, and herbivores.

Despite their promising antimicrobial potential, the use of plant metabolites faces challenges such as low bioavailability, instability under environmental conditions (e.g., heat, light, or oxidation), and potential toxicity to plant cells in their free forms. Therefore, plants need to store these metabolites in glycoside forms. Flavonoids, predominantly found in glycosylated forms, are vital phenylpropanoid derivatives involved in plant protection, growth, and development. This class includes anthocyanins, flavonols, flavones, and isoflavones, which are implicated in UV protection, pollen fertility, ecological signaling, and hormone modulation (Samanta et al. 2011). Salicylate glycosides, abundant in Salicaceae family plants, serve as protective agents against herbivores and UV radiation, although the pathways and enzymes involved in their biosynthesis remain poorly understood (Boeckler et al. 2011). The induction of defense mechanisms in plants often involves recognition of elicitor molecules and subsequent signaling, leading to the production of phytoalexins such as isoflavonoids in legumes and sesquiterpenes in Solanaceae species. These bioactive compounds underscore the interplay between stress responses and the synthesis of secondary metabolites, highlighting their ecological and defensive significance (Isah 2019). This synthesis of secondary metabolites underpins the adaptive capacity of plants, contributing to their survival and ecological interactions in dynamic environments. Based on the previous information, plants can be considered a perfect laboratory to synthesize a wide range of secondary metabolites with unique structural motifs such as alkaloids, essential oils including terpenes, flavonoids, phenolics, phytosterols, and polyketides, as well as resins. These metabolites can provide antibacterial, antifungal, and other activities against plant pathogens (Aamer et al. 2024a; Khalil et al. 2024; Khamis et al. 2024).

## 14.5 Plant Metabolites Against Plant Parasitic Nematodes

There are various natural molecules act as biochemical immune responses beginning with the components of cell wall reinforcement extended to secondary metabolites, which have been detected to prevent the plant parasitic nematodes (PPNs) from plant invading toward affording nematode-specific resistance (Meresa et al. 2024). Plants have inclusive secondary metabolism specific for producing a large diversity of metabolites about 200,000 secondary metabolites (Viant et al. 2017). The production of plant metabolites as a protective system against nematode attacks is one of the defense systems termed anti-nematode phytochemicals (Desmedt et al. 2020). It is confirmed that secondary metabolites are not only depend on plant growth only, but also comprise combinations of different processes like tolerance of biotic stress, attraction of pollinators, and resistance to diseases, insects, and pests (Seigler 1998; Hartmann 2007). The defensive metabolites with a characterized vital impact on PPNs include phenolic compounds, tannins, organosulfur compounds, enzymatic antioxidants, alkaloids, and terpenoids (Desmedt et al. 2020). There are two kinds of secondary plant metabolite secretions that have been reported to prevent pathogens and pest attacks; some of these metabolites are employed as phytoalexins, and the other are phytoanticipins (Desmedt et al. 2020).

Constitutive "structural" defense barriers already existed regardless of pathogen infection involved phytoanticipins; this term refers to pre-formed mechanisms which can make a plant "non-host" by inhibiting the attacks of nematode (Desmedt et al. 2020). On the other hand, phytoalexins are activated as soon as pathogen perception is accomplished and PPN penetrations are achieved, which enhances the occurrence of inducible immune responses against intruders (De Lorenzo and Cervone 2022). The evictor host exudates generate resistance and act as a physical barrier where it inhibits the penetration of nematodes due to the absence of necessitated metabolites for host recognition; this term is "pre-penetration resistance (Lee et al. 2017). In addition, infective nematodes depend on the attractant metabolites secreted by plants to locate appropriate hosts relevant to natural conditions (Sikder and Vestergård 2020). The presence of toxic plant metabolites serves as a formidable barrier against parasitic nematodes, inducing a defense mechanism referred to as post-penetration resistance, or "metabolic resistance." This mechanism is triggered after nematodes penetrate the host plant, ultimately inhibiting their survival and reproduction (Desmedt et al. 2020). Both preformed structural defenses and inducible responses against parasitic nematodes are closely linked to the presence of secondary plant metabolites, which play a crucial role in anti-nematode activity (Chen et al. 2024b).

Certain plants possess resistance genes (R-genes) that encode receptors capable of recognizing specific parasitic nematode (PPN) effectors. These genes activate rapid and robust defensive responses characterized by a hypersensitive reaction (HR) upon PPN detection (Desmedt et al. 2020). Notable examples of anti-nematode R-genes include the *Mi* genes in tomatoes, which provide resistance against various root-knot nematode species, and the *Gro*, *Gpa*, and *Hero* genes in potatoes, which protect against potato cyst nematodes (Sato et al. 2019).

### 14.5.1 Phenolic Compounds

Phenolic compounds are key secondary metabolites that contribute significantly to plant defense against diseases and pests, often acting as phytoalexins associated with nematode resistance across diverse plant–nematode interactions (Hölscher et al. 2014). Chlorogenic acid, a prominent phenolic compound, plays a vital role in plant immune responses against nematodes. It is induced through the accumulation of caffeic acid and quinic acid at nematode infection sites in various plant species (Meher et al. 2015). In *Populus alba* × *Populus tremula*, chlorogenic acid biosynthesis is inhibited in susceptible clones during *Meloidogyne incognita* pathogenesis (Baldacci-Cresp et al. 2020). Similarly, caffeic acid accumulates in resistant tomato cultivars upon invasion by *M. incognita* and contributes to lignification, reinforcing structural defenses (Sato et al. 2019). Furthermore, caffeic acid and chlorogenic acid from *Artemisia annua* effectively inhibit *Xiphinema index* and *Globodera rostochiensis* even at low concentrations (D'Addabbo et al. 2013). Stilbenoids, a class of phytoanticipins derived from the phenylpropanoid pathway, have shown remarkable resistance to nematodes (Shah and Gupta 2023). For example, in pine trees, the accumulation of 3-O-methyldihydropinosylvin in bark and wood provides significant resistance to *Bursaphelenchus xylophilus*, the pine wood nematode (Hanawa et al. 2001). Similarly, stilbenoids confer resistance in susceptible and resistant rootstocks of grapevine (*Vitis vinifera*) against *M. incognita* (Wallis 2020). In bananas, resistant cultivars combat burrowing nematodes (*Radopholus similis*) by releasing phenylphenalenone phytoalexins, which interact with nematodes' lipid reserves at infection sites, leading to their death (Hölscher et al. 2014)).

### 14.5.2 Geraniol and Other Alkaloids

Geraniol, an alkyl alcohol, has demonstrated strong nematicidal properties, notably suppressing egg hatching in *M. javanica* and reducing the number of females infecting tomato roots (Nasiou and Giannakou 2018). The phenolic plant hormone SA also plays a critical role in plant resistance to *M. incognita* by acting as a signaling molecule that regulates various defense responses (Elkobrosy et al. 2022).

### 14.5.3 Flavonoids and Isoflavonoids

Flavonoids such as kaempferol effectively inhibit egg hatching of *R. similis*, while kaempferol, quercetin, and myricetin exhibit repellent activity against juvenile *M. incognita* (Wuyts et al. 2006; Parvaiz et al. 2024). Pterocarpans, including glyceollin I, derived from isoflavonoids, are released near the head region of *Heterodera glycines* in resistant soybean cultivars, triggering elicited responses post-penetration (Huang and Barker 1991). These secondary metabolites collectively represent an intricate defense strategy employed by plants to combat parasitic nematodes, highlighting their potential applications in agricultural nematode management. A high concentration of crude alkaloids recorded a strong activity against nematodes by killing their second-stage juveniles and eggs (Wen et al. 2013).

### 14.5.4 Tannins, Aldehydes, and Some Plant Hormones

Tannin compounds have been recorded a paramount role against several nematodes, where it inhibited the infection activity of *Globodera rostochiensis*; *Meloidogyne javanica; Meloidogyne incognita;* and *B. xylophilus* in field experiments and in vivo (Chen and Song 2021). A banana-resistant cultivar of *Radopholus similis* comprised high concentrations of tannins that were reported by analysis after 12 weeks of pathogenicity (Collingborn et al. 2000). Tannins showed a suppression role against the pinewood nematode via inhibition of some enzymes relevant to the nematode such as amylase, cellulase, and acetylcholine esterase (Chen and Song 2021). Tannins achieve good results against nematode recognition of the host plant. During the degradation of tannic acid in the soil, the tannic acid acts as an attracting chemical signal to root-knot nematode, which misleads the host plant to the wrong sites and causes them to cluster away from the crop by decrease their ability to search for the host plant just like that delaying and may preventing nematode attacks on the host plant (d'Errico et al. 2018; Chen and Song 2021).

Terpenoid aldehydes (TAs) play a crucial role in the defense of *Gossypium hirsutum* (cotton) against *M. incognita*. Their accumulation in root tissues, particularly in areas traversed by *M. incognita* juveniles and around developing feeding sites, contributes to resistance mechanisms (Veech 1979). Similarly, potato plants (*Solanum tuberosum*) produce high levels of terpenoid phytoalexins, such as the sesquiterpene solavetivone, to combat *Globodera rostochiensis* infestations (Desjardins et al. 1995). Terpenoids also participate in plant–nematode interactions indirectly by functioning as precursors for plant hormones. For example, abscisic acid is derived from tetraterpenoids (Nambara and Marion-Poll 2005), brassinosteroids originate from triterpenoids (Choe 2006), and gibberellins are classified as diterpenoids (Hedden and Thomas 2012). These terpenoid-based hormones often play diverse, and sometimes antagonistic, roles in plant resistance to parasitic nematodes (Song et al. 2018; Velez-Haro et al. 2024). Volatile terpenoid aldehydes, such as trans-2-hexenal, exhibit significant nematicidal effects across various nematode life stages. This compound has been shown to suppress nematode growth, egg hatching, movement, respiration, reproduction, and egg-laying (Cheng et al. 2017). Additionally, trans-2-hexenal disrupts nematode physiological processes by impairing digestive enzyme activity and nutrient metabolism while enhancing glutathione S-transferase activity, thereby interfering with their metabolism and vitality (Zhao et al. 2017). Furthermore, trans-2-hexenal has demonstrated suppressive effects against *M. incognita* and *Heterodera avenae*, highlighting its potential as a nematicidal agent (Miao JianQiang et al. 2012). These findings underscore the importance of terpenoids and their derivatives in plant defense strategies against nematodes, offering valuable insights into their potential applications in sustainable nematode management.

### *14.5.5 Organosulfur Compounds*

α-Terthienyl is one of the organosulfur compounds that exists in the roots of marigolds (*Tagetes* sp.) where it exhibits nematicidal activity toward plant-parasitic nematodes via secreting their exudates to attract nematodes and then inhibits its activity through penetration of nematode's hypodermis; this creates reactive oxygen species (ROS), inducing oxidative stress inside the nematode owing to activation by light or/and peroxidase enzymes (Hamaguchi et al. 2019). In addition, the roots of the asparagus plant (*Asparagus officinalis*) release one of the organosulfur compounds termed asparagusic acid that acts as a phytoanticipin where it suppresses egg hatching of *Heterodera* sp. and controls the attacks of plant-parasitic nematodes (Takasugi et al. 1975). Finally, the antimicrobial potential of plant secondary metabolites positions them as sustainable alternatives to synthetic pesticides and antibiotics. Their roles in pest management, crop protection, and pharmaceutical applications underscore their versatility and importance in addressing global challenges.

## 14.6 Conclusion: Translating Potential into Practice

Plant secondary metabolites—the sophisticated chemical arsenal of phenolics, terpenoids, alkaloids, and sulfur-containing compounds—are indispensable for plant survival, orchestrating defense against a formidable array of biotic and abiotic stresses. Their evolutionarily refined mechanisms, ranging from direct antimicrobial action (via membrane disruption or enzyme inhibition) to the elegant induction of systemic resistance in the host, provide a multi-target, ecologically integrated model for pest and disease suppression. This inherent multi-functionality, with its demonstrated antibacterial, antifungal, antiviral, and nematicidal activities, positions these compounds as cornerstone agents for the future of sustainable agriculture and environmental stewardship.

However, the journey from a potent in vitro compound to a reliable, commercial agrochemical is not trivial. The very properties that make these metabolites effective—their specific chemical structures and reactive nature—often render them susceptible to rapid degradation, photolysis, and limited systemic mobility within crops. These challenges of low environmental persistence, variable bioavailability, and inconsistent field efficacy underpin the current gap between laboratory promise and widespread agricultural adoption.

Closing this gap demands a concerted, interdisciplinary effort. The path forward hinges on advanced biotechnological and formulation sciences. Metabolic engineering and synthetic biology offer routes to enhance *in planta* production or create microbial biofactories for standardized compound supply. Meanwhile, innovative formulation technologies—such as nanoencapsulation, microemulsions, and carrier systems—are critical to protect these compounds, control their release, and ensure targeted delivery. Crucially, their integration must be validated through rigorous, multi-season field trials that benchmark performance against conventional standards.

Plant secondary metabolites are indispensable tools for plants, enabling them to thrive under biotic and abiotic stresses while offering remarkable applications in human industries. Their multifunctional properties, including antibacterial, antifungal, antiviral, and nematicidal activities, emphasize their importance in sustainable agriculture and environmental management. However, challenges such as low bioavailability and instability must be addressed through advanced biotechnological approaches. With continued research and innovation, plant secondary metabolites have the potential to revolutionize pest control, pharmaceuticals, and eco-friendly agricultural practices, thereby contributing significantly to global sustainability.

**Declaration of AI Assistance** The authors employed artificial intelligence tools, specifically ChatGPT and DeepSeek-V3, in a limited and supportive role to improve the clarity, linguistic quality, and organization of specific sections of this chapter, including the introduction, key conceptual components, and reference formatting. The authors thoroughly examined, revised, and confirmed all AI-generated suggestions, assuming complete responsibility for the accuracy, integrity, and scholarly quality of the final manuscript. The finalized chapter underwent standard peer review by qualified experts in the discipline.

## References

Aamer HA, Al-Askar AA, El-Tanbouly R, Abdelwahab EA, Kowalczewski PŁ, El-Messeiry S et al (2024a) Phytochemical analysis of *Bienertia sinuspersici* extract and its antioxidant and antimicrobial activities. Open Agric 9:20220399

Aamer HA, Elalem SF, Al-Askar AA, Sharaf OA, Gaber MA, Kowalczewski P et al (2024b) Antioxidant and antimicrobial activities of Salsola imbricata methanolic extract and its phytochemical characterization. Open Life Sci 19:20221011

Abdallah II, Quax WJ (2017) A glimpse into the biosynthesis of terpenoids. In: International conference on natural resources and life sciences (NRLS-2016). KnE Life Sciences, pp 81–98

Abdelkhalek A, Al-Askar AA, Alsubaie MM, Behiry SI (2021) First report of protective activity of *Paronychia argentea* extract against tobacco mosaic virus infection. Plants 10:2435

Abookleesh FL, Al-Anzi BS, Ullah A (2022) Potential antiviral action of alkaloids. Molecules 27: 903

Adeosun WB, Loots DT (2024) Medicinal plants against viral infections: a review of metabolomics evidence for the antiviral properties and potentials in plant sources. Viruses 16:218

Al-Askar AA, Bashir S, Mohamed AE, Sharaf OA, Nabil R, Su Y et al (2023) Antimicrobial efficacy and HPLC analysis of polyphenolic compounds in a whole-plant extract of *Eryngium campestre*. Separations 10:362

Alfakje T, Al-Mashhadane F (2024) Histopathological effects of anabolic androgenic steroids (Nandrolone Decanoate) on heart, liver and kidney of male local rabbits. Egypt J Vet Sci 55: 1907–1919

Al-Khayri JM, Rashmi R, Toppo V, Chole PB, Banadka A, Sudheer WN et al (2023) Plant secondary metabolites: the weapons for biotic stress management. Meta 13:716

Aseel DG, Ibrahim OM, Abdelkhalek A (2024) Biosynthesized silver nanoparticles mediated by Ammi visnaga extract enhanced systemic resistance and triggered multiple defense-related genes, including SbWRKY transcription factors, against tobacco mosaic virus infection. BMC Plant Biol 24:756

Atampugbire G, Adomako EEA, Quaye O (2024) Medicinal plants as effective antiviral agents and their potential benefits. Nat Prod Commun 19:1934578X241282923

Badyal S, Singh H, Yadav AK, Sharma S, Bhushan I (2020) Plant secondary metabolites and their uses. Plant Arch 20:3336–3340

Baldacci-Cresp F, Behr M, Kohler A, Badalato N, Morreel K, Goeminne G et al (2020) Molecular changes concomitant with vascular system development in mature galls induced by root-knot nematodes in the model tree host Populus tremula× P. alba. Int J Mol Sci 21:406

Barreda L, Brosse C, Boutet S, Perreau F, Rajjou L, Lepiniec L et al (2024) Specialized metabolite modifications in Brassicaceae seeds and plants: diversity, functions and related enzymes. Nat Prod Rep 41:834

Behiry S, Abdelwahab EA, Al-Askar AA, Kowalczewski P, Abdelkhalek A (2024) HPLC and GC–MS analyses of phytochemical compounds in Haloxylon salicornicum extract: antibacterial and antifungal activity assessment of phytopathogens. Open Chem 22:20240115

Boeckler GA, Gershenzon J, Unsicker SB (2011) Phenolic glycosides of the Salicaceae and their role as anti-herbivore defenses. Phytochemistry 72:1497–1509

Chauhan S, Kaur A, Vyas M, Khatik GL (2017) Comparison of antidiabetic and antioxidant activity of wild and cultivated variety of Rauwolfia serpentina. Asian J Pharm Clin Res 10:404–406

Chen J, Song B (2021) Natural nematicidal active compounds: recent research progress and outlook. J Integr Agric 20:2015–2031

Chen J, Xu X, Liu W, Feng Z, Chen Q, Zhou Y et al (2024a) Plasmodesmata function and callose deposition in plant disease defense. Plants 13:2242

Chen X, Li F, Wang D, Cai L (2024b) Insights into the plant response to nematode invasion and modulation of host defense by plant parasitic nematode. Front Microbiol 15:1482789

Cheng L, Xu S, Xu C, Lu H, Zhang Z, Zhang D et al (2017) Effects of trans-2-hexenal on reproduction, growth and behaviour and efficacy against the pinewood nematode, Bursaphelenchus xylophilus. Pest Manag Sci 73:888–895

Choe S (2006) Brassinosteroid biosynthesis and inactivation. Physiol Plant 126:539–548

Collingborn FMB, Gowen SR, Mueller-Harvey I (2000) Investigations into the biochemical basis for nematode resistance in roots of three musa cultivars in response to radopholus s imilis infection. J Agric Food Chem 48:5297–5301

D'Addabbo T, Carbonara T, Argentieri MP, Radicci V, Leonetti P, Villanova L et al (2013) Nematicidal potential of Artemisia annua and its main metabolites. Eur J Plant Pathol 137: 295–304

d'Errico G, Woo SL, Lombardi N, Manganiello G, Roversi PF (2018) Activity of chestnut tannins against the southern root-knot nematode Meloidogyne incognita. Redia 101:53–59

Datta R, Sharma A, Thakur A (2022) Secondary metabolites from plants: role in plant diseases and health care. In: Plant secondary metabolites: physico-chemical properties and therapeutic applications. Springer, pp 355–369

De Lorenzo G, Cervone F (2022) Plant immunity by damage-associated molecular patterns (DAMPs). Essays Biochem 66:459–469

Debnath B, Singh WS, Das M, Goswami S, Singh MK, Maiti D et al (2018) Role of plant alkaloids on human health: a review of biological activities. Mater Today Chem 9:56–72

Deng K, Li Z, Huang T, Huang J (2024) Noncoding RNAs in regulation of plant secondary metabolism. Plant Physiol Biochem 211:108718

Desjardins AE, McCormick SP, Corsini DL (1995) Diversity of sesquiterpenes in 46 potato cultivars and breeding selections. J Agric Food Chem 43:2267–2272

Desmedt W, Mangelinckx S, Kyndt T, Vanholme B (2020) A phytochemical perspective on plant defense against nematodes. Front Plant Sci 11:602079

Dhamija N, Mangla A (2022) Plant secondary metabolites in antiviral applications. In: Plant secondary metabolites: physico-chemical properties and therapeutic applications. Springer, pp 459–482

Divekar PA, Narayana S, Divekar BA, Kumar R, Gadratagi BG, Ray A et al (2022) Plant secondary metabolites as defense tools against herbivores for sustainable crop protection. Int J Mol Sci 23: 2690

Eguchi R, Ono N, Hirai Morita A, Katsuragi T, Nakamura S, Huang M et al (2019) Classification of alkaloids according to the starting substances of their biosynthetic pathways using graph convolutional neural networks. BMC Bioinformatics 20:1–13

Elkobrosy DH, Aseel DG, Hafez EE, El-Saedy MA, Al-Huqail AA, Ali HM et al (2022) Quantitative detection of induced systemic resistance genes of potato roots upon ethylene treatment and cyst nematode, Globodera rostochiensis, infection during plant–nematode interactions. Saudi J Biol Sci 29:3617–3625

Elshafie HS, Camele I, Mohamed AA (2023) A comprehensive review on the biological, agricultural and pharmaceutical properties of secondary metabolites based-plant origin. Int J Mol Sci 24:3266. https://doi.org/10.3390/ijms24043266

Elsharkawy MM, Al-Askar AA, Behiry SI, Abdelkhalek A, Saleem MH, Kamran M et al (2022) Resistance induction and nematicidal activity of certain monoterpenes against tomato root-knot caused by Meloidogyne incognita. Front Plant Sci 13:982414

Fernie AR, Pichersky E (2015) Focus issue on metabolism: metabolites, metabolites everywhere. Plant Physiol 169:1421–1423

Fritsch RM, Keusgen M (2006) Occurrence and taxonomic significance of cysteine sulphoxides in the genus Allium L. (Alliaceae). Phytochemistry 67:1127–1135

González Mera IF, González Falconí DE, Morera Córdova V (2019) Secondary metabolites in plants: main classes, phytochemical analysis and pharmacological activities. Bionatura 4:1000–1009

Guerriero G, Berni R, Muñoz-Sanchez JA, Apone F, Abdel-Salam EM, Qahtan AA et al (2018) Production of plant secondary metabolites: examples, tips and suggestions for biotechnologists. Genes (Basel) 9:309

Gutiérrez-del-Río I, Fernández J, Lombó F (2018) Plant nutraceuticals as antimicrobial agents in food preservation: Terpenoids, polyphenols and thiols. Int J Antimicrob Agents 52:309–315

Hamaguchi T, Sato K, Vicente CSL, Hasegawa K (2019) Nematicidal actions of the marigold exudate α-terthienyl: oxidative stress-inducing compound penetrates nematode hypodermis. Biol Open 8:bio038646

Hanawa F, Yamada T, Nakashima T (2001) Phytoalexins from Pinus strobus bark infected with pinewood nematode, Bursaphelenchus xylophilus. Phytochemistry 57:223–228

Hartmann T (2007) From waste products to ecochemicals: fifty years research of plant secondary metabolism. Phytochemistry 68:2831–2846

Hedden P, Thomas SG (2012) Gibberellin biosynthesis and its regulation. Biochem J 444:11–25

Hölscher D, Dhakshinamoorthy S, Alexandrov T, Becker M, Bretschneider T, Buerkert A et al (2014) Phenalenone-type phytoalexins mediate resistance of banana plants (Musa spp.) to the burrowing nematode Radopholus similis. Proc Natl Acad Sci 111:105–110

Huang J-S, Barker KR (1991) Glyceollin I in soybean-cyst nematode interactions: spatial and temporal distribution in roots of resistant and susceptible soybeans. Plant Physiol 96:1302–1307

Huang Y, Xie F-J, Cao X, Li M-Y (2021) Research progress in biosynthesis and regulation of plant terpenoids. Biotechnol Biotechnol Equip 35:1799–1808

Hussein RA, El-Anssary AA (2019) Plants secondary metabolites: the key drivers of the pharmacological actions of medicinal plants. Herb Med 1:11–30

Isah T (2019) Stress and defense responses in plant secondary metabolites production. Biol Res 52: 39

Jamwal K, Bhattacharya S, Puri S (2018) Plant growth regulator mediated consequences of secondary metabolites in medicinal plants. J Appl Res Med Aromat Plants 9:26–38

Kapoor R, Sharma B, Kanwar SS (2017) Antiviral phytochemicals: an overview. Biochem Physiol 6:7

Kaur S, Samota MK, Choudhary M, Choudhary M, Pandey AK, Sharma A et al (2022) How do plants defend themselves against pathogens-biochemical mechanisms and genetic interventions. Physiol Mol Biol Plants 28:485–504. https://doi.org/10.1007/s12298-022-01146-y

Khalil A, Abdelwahab EA, Sharaf OA, Al-Askar AA, Kowalczewski P, Abdelkhalek A et al (2024) Torilis arvensis ethanolic extract: phytochemical analysis, antifungal efficacy, and cytotoxicity properties. Open Chem. 22:20240113

Khamis WM, Heflish AA, El-Messeiry S, Behiry SI, Al-Askar AA, Su Y et al (2023) Swietenia mahagoni leaves extract: antifungal, insecticidal, and phytochemical analysis. Separations 10: 301. https://doi.org/10.3390/separations10050301

Khamis W, Al-Askar A, Kowalczewski P, Abdelkhalek A, Behiry S (2024) Phytochemical composition and insecticidal activity of Acokanthera oblongifolia (Hochst.) Benth & Hook. f. ex BD Jacks. extract on life span and biological aspects of Spodoptera littoralis (Biosd.). Open Agric 9:20220394

Khan F, Pandey P, Verma M, Upadhyay TK (2024) Terpenoid-mediated targeting of STAT3 signaling in cancer: an overview of preclinical studies. Biomolecules 14:200

Kumar L, Mahatma MK, Kalariya KA, Bishi SK, Mann A (2014) Plant phenolics: important bio-weapon against pathogens and insect herbivores. Pop Kheti 2:149–152

Kumar S, Korra T, Thakur R, Arutselvan R, Kashyap AS, Nehela Y et al (2023) Role of plant secondary metabolites in defence and transcriptional regulation in response to biotic stress. Plant Stress 8:100154

Lal N, Sahu N, Shirale AO, Gurav P, Rani K, Meena BP et al (2023) Plant secondary metabolites and their impact on human health. In: Nano-biofortification for human and environmental health. Springer, pp 295–321

Lee H-A, Lee H-Y, Seo E, Lee J, Kim S-B, Oh S et al (2017) Current understandings of plant nonhost resistance. Mol Plant-Microbe Interact 30:5–15

Li F, Wang A (2019) RNA-targeted antiviral immunity: more than just RNA silencing. Trends Microbiol 27:792–805

Li S, Jiang S, Jia W, Guo T, Wang F, Li J et al (2024) Natural antimicrobials from plants: recent advances and future prospects. Food Chem 432:137231

Lopez LM, Grimes DA, Schulz KF, Curtis KM, Chen M (2014) Steroidal contraceptives: effect on bone fractures in women. Cochrane Database Syst Rev 2014:CD006033

Lyubenova A, Georgieva L, Antonova V (2023) Utilization of plant secondary metabolites for plant protection. Biotechnol Biotechnol Equip 37:2297533

Majumdar S, Barman F, Paul A, Kundu R (2023) Role of sulfur in protection against major environmental stress in plants. In: Biology and biotechnology of environmental stress tolerance in plants. Apple Academic Press, pp 473–529

Mandal MK, Domb AJ (2024) Antimicrobial activities of natural bioactive polyphenols. Red 192: 193

Maniglia BC, Rebelatto EA, Andrade KS, Zielinski A, de Andrade CJ (2021) Polyphenols. Food Bioact Heal:1–39

Manjunatha L, Rajashekara H, Uppala LS, Ambika DS, Patil B, Shankarappa KS et al (2022) Mechanisms of microbial plant protection and control of plant viruses. Plants 11:3449

Meher HC, Gajbhiye VT, Singh G, Chawla G (2015) Altered metabolomic profile of selected metabolites and improved resistance of Cicer arietinum (L.) against Meloidogyne incognita (Kofoid & White) Chitwood following seed soaking with salicylic acid, benzothiadiazole or nicotinic acid. Acta Physiol Plant 37:1–12

Meresa BK, Matthys J, Kyndt T (2024) Biochemical defence of plants against parasitic nematodes. Plants 13:2813

Miao JianQiang MJ, Wang Meng WM, Li XiuHuan LX, Yang FaHui YF, Liu Feng LF (2012) Antifungal and nematicidal activities of five volatile compounds against soil-borne pathogenic fungi and nematodes. Acta Phytophyl Sin 39(6):561–566

Mishra J (2024) Naturally biosynthesized secondary plant products and its importance: an overview. Plant Arch 24:957–966

Mishra J, Srivastava R, Trivedi PK, Verma PC (2020) Effect of virus infection on the secondary metabolite production and phytohormone biosynthesis in plants. 3 Biotech 10:547

Mohite P, Puri A, Bharati D, Singh S (2024) Polyphenol-encapsulated nanoparticles for the treatment of chronic metabolic diseases. In: Role of flavonoids in chronic metabolic diseases: from bench to clinic, pp 375–416

Montenegro-Landívar MF, Tapia-Quirós P, Vecino X, Reig M, Valderrama C, Granados M et al (2021) Polyphenols and their potential role to fight viral diseases. Sci Total Environ 801:149719

Moradi MT, Karimi A, Lorigooini Z (2018) Alkaloids as the natural anti-influenza virus agents: a systematic review. Toxin Rev 37:11–18. https://doi.org/10.1080/15569543.2017.1323338

Naji EF, Abdulfatah HF, Hashim KS (2024) Plant secondary metabolites, their classification and biological roles: a review. J Univ Anbar Pure Sci 18

Nambara E, Marion-Poll A (2005) Abscisic acid biosynthesis and catabolism. Annu Rev Plant Biol 56:165–185

Nasiou E, Giannakou IO (2018) Effect of geraniol, a plant-based alcohol monoterpene oil, against Meloidogyne javanica. Eur J Plant Pathol 152:701–710

Pagare S, Bhatia M, Tripathi N, Pagare S, Bansal YK (2015) Secondary metabolites of plants and their role: overview. Curr Trends Biotechnol Pharm 9:293–304

Parvaiz S, Yousuf P, Lone R, Rather YA (2024) Plant phenolics in alleviating Root-Knot Disease in plants caused by Meloidogyne spp. In: Plant phenolics in biotic stress management. Springer, pp 417–439

Pastor RF, Mariani MLL, Villach Vaquer MG, Cascón Rodríguez PN, Giudice MPT, Pastor E, et al (2019) A novel approach to measure the total antioxidant power of wines through near infrared spectroscopy and its relevance in human nutrition. J Health Sci 7:209–214

Peng J, Hu J-F, Kazi AB, Li Z, Avery M, Peraud O et al (2003) Manadomanzamines A and B: a novel alkaloid ring system with potent activity against mycobacteria and HIV-1. J Am Chem Soc 125:13382–13386

Perveen S, Al-Taweel A (2018) Terpenes and terpenoids. BoD–Books on Demand

Rabie M, Aseel DG, Younes HA, Behiry SI, Abdelkhalek A (2024) Transcriptional responses and secondary metabolites variation of tomato plant in response to tobacco mosaic virus infestation. Sci Rep 14:19565

Rodriguez-Concepcion M, Avalos J, Bonet ML, Boronat A, Gomez-Gomez L, Hornero-Mendez D et al (2018) A global perspective on carotenoids: metabolism, biotechnology, and benefits for nutrition and health. Prog Lipid Res 70:62–93

Sahoo A, Dwivedi K, Almalki WH, Mandal AK, Alhamyani A, Afzal O et al (2024) Secondary metabolites in topical infectious diseases and nanomedicine applications. Nanomedicine 19:1

Saini N, Anmol A, Kumar S, Bakshi M, Dhiman Z (2024) Exploring phenolic compounds as natural stress alleviators in plants-a comprehensive review. Physiol Mol Plant Pathol 133: 102383

Samanta A, Das G, Das SK (2011) Roles of flavonoids in plants. Carbon N Y 100:12–35

Sato K, Kadota Y, Shirasu K (2019) Plant immune responses to parasitic nematodes. Front Plant Sci 10:1165

Seigler DS (1998) Plant secondary metabolism. Springer Science & Business Media

Shah AA, Gupta A (2023) Secondary metabolite basis of elicitor-and effector-triggered immunity in pathogen elicitation amid infections. In: Genetic manipulation of secondary metabolites in medicinal plant. Springer, pp 225–251

Sharma A, Sharma S, Kumar A, Kumar V, Sharma AK (2022) Plant secondary metabolites: an introduction of their chemistry and biological significance with physicochemical aspect. In: Plant secondary metabolites: physico-chemical properties and therapeutic applications. Springer, pp 1–45

Sharma K, Kaur R, Kumar S, Saini RK, Sharma S, Pawde SV et al (2023) Saponins: a concise review on food related aspects, applications and health implications. Food Chem Adv 2:100191

Siddiqui T, Sharma V, Khan MU, Gupta K (2024) Terpenoids in essential oils: chemistry, classification, and potential impact on human health and industry. Phytomed Plus 4:100549

Sikder MM, Vestergård M (2020) Impacts of root metabolites on soil nematodes. Front Plant Sci 10:1792

Singh B, Sharma RA (2015) Plant terpenes: defense responses, phylogenetic analysis, regulation and clinical applications. 3 Biotech 5:129–151

Singh B, Singh JP, Singh N, Kaur A (2017) Saponins in pulses and their health promoting activities: a review. Food Chem 233:540–549

Sobhy S, Al-Askar AA, Bakhiet EK, Elsharkawy MM, Arishi AA, Behiry SI et al (2023) Phytochemical characterization and antifungal efficacy of camphor (Cinnamomum camphora L.) extract against phytopathogenic fungi. Separations 10:189. https://doi.org/10.3390/separations10030189

Song L, Xu X, Wang F, Wang Y, Xia X, Shi K et al (2018) Brassinosteroids act as a positive regulator for resistance against root-knot nematode involving RESPIRATORY BURST OXIDASE HOMOLOG-dependent activation of MAPKs in tomato. Plant Cell Environ 41:1113–1125

Suleiman AA, Al-Chalabi R, Shaban SA (2024) Integrative role of small non-coding RNAs in viral immune response: a systematic review. Mol Biol Rep 51:107

Takasugi M, Yachida Y, Anetai M, Masamune T, Kegasawa K (1975) Identification of asparagusic acid as a nematicide occurring naturally in the roots of asparagus. Chem Lett 4:43–44

Tatli Cankaya II, Somuncuoglu EI (2021) Potential and prophylactic use of plants containing saponin-type compounds as antibiofilm agents against respiratory tract infections. Evidence-Based Complement Altern Med 2021:6814215

Tripathi D, Raikhy G, Kumar D (2019) Chemical elicitors of systemic acquired resistance—salicylic acid and its functional analogs. Curr Plant Biol 17:48–59

Twaij BM, Hasan MN (2022) Bioactive secondary metabolites from plant sources: types, synthesis, and their therapeutic uses. Int J Plant Biol 13:4–14. https://doi.org/10.3390/ijpb13010003

Upadhyay SK, Srivastava AK, Rajput VD, Chauhan PK, Bhojiya AA, Jain D et al (2022) Root exudates: mechanistic insight of plant growth promoting rhizobacteria for sustainable crop production. Front Microbiol 13:916488

Upadhyay SK, Rajput VD, Kumari A, Espinosa-Saiz D, Menendez E, Minkina T et al (2023) Plant growth-promoting rhizobacteria: a potential bio-asset for restoration of degraded soil and crop productivity with sustainable emerging techniques. Environ Geochem Health 45:9321–9344

Veech JA (1979) Histochemical localization and nematoxicity of terpenoid aldehydes in cotton. J Nematol 11:240

Velez-Haro JM, Velázquez-Márquez S, Velázquez-Márquez N (2024) Natural compounds for inhibiting oncogenic microbiota: a review of their therapeutic applications. In: Pathogens associated with the development of cancer in humans: OMICs, immunological, and pathophysiological studies, pp 369–410

Verdaguer D, Jansen MAK, Llorens L, Morales LO, Neugart S (2017) UV-A radiation effects on higher plants: exploring the known unknown. Plant Sci 255:72–81

Verma B, Choudhary J, Kumar R (2024) Chapter 12 Plant oleoresins as source of environment-friendly pesticides. In: Kumar R, de Oliveira MS, de Aguiar Andrade EH, Suyal DC, Soni R (eds) Pest Management. De Gruyter, Berlin, Boston, pp 237–260. https://doi.org/10.1515/9783111204819-012

Viant MR, Kurland IJ, Jones MR, Dunn WB (2017) How close are we to complete annotation of metabolomes? Curr Opin Chem Biol 36:64–69

Vyas M (2017) Nutritional profile of spinach and its antioxidant & antidiabetic evaluation. Int J Green Pharm 11

Wallis CM (2020) Grapevine (Vitis spp.) rootstock stilbenoid associations with host resistance to and induction by root knot nematodes, Meloidogyne incognita. BMC Res Notes 13:360

Weiszmann J, Walther D, Clauw P, Back G, Gunis J, Reichardt I et al (2023) Metabolome plasticity in 241 Arabidopsis thaliana accessions reveals evolutionary cold adaptation processes. Plant Physiol 193:980–1000. https://doi.org/10.1093/plphys/kiad298

Wen Y, Meyer SLF, Masler EP, Zhang F, Liao J, Wei X et al (2013) Nematotoxicity of drupacine and a cephalotaxus alkaloid preparation against the plant-parasitic nematodes Meloidogyne incognita and Bursaphelenchus xylophilus. Pest Manag Sci 69:1026–1033

Wink M (2015) Modes of action of herbal medicines and plant secondary metabolites. Medicine 2: 251–286

Wuyts N, Swennen R, De Waele D (2006) Effects of plant phenylpropanoid pathway products and selected terpenoids and alkaloids on the behaviour of the plant-parasitic nematodes Radopholus similis, Pratylenchus penetrans and Meloidogyne incognita. Nematology 8:89–101

Xu Z, Liu D, Zhu J, Zhao J, Shen S, Wang Y et al (2025) Catalysts for sulfur: understanding the intricacies of enzymes orchestrating plant sulfur anabolism. Planta 261:16

Zakaryan H, Arabyan E, Oo A, Zandi K (2017) Flavonoids: promising natural compounds against viral infections. Arch Virol 162:2539–2551

Zhao Y, Xu S, Lu H, Zhang D, Liu F, Lin J et al (2017) Effects of the plant volatile trans-2-hexenal on the dispersal ability, nutrient metabolism and enzymatic activities of Bursaphelenchus xylophilus. Pestic Biochem Physiol 143:147–153

# Chapter 15
# Antifungals Produced by Bacteria and Their Role in Biological Control

**Gaber A. Abo-Zaid, Shimaa Bashir, Toufic Elbeaino, and Ahmed Abdelkhalek**

## 15.1 Introduction

Many synthetic bactericides and fungicides are being used in large quantities to mitigate the reduction in crop productivity brought on by phytopathogens (Fenta and Mekonnen 2024). The application of such synthetic pesticides to address these infections results in adverse effects on human and environmental health (Ayilara et al. 2023). Extended exposure to these substances at low doses may result in immune suppression, hormonal imbalances, reproductive issues, and the development of cancerous diseases (Kharrazian 2021). Worries about the future of ecosystems have been heightened by the fact that this practice has made once fertile soil infertile, disrupted the soil's microflora composition, and caused chemical leaching into bodies of water (Miller et al. 2023). Biocontrol agents can manage phytopathogens, pests, weeds, and insects, providing an alternative to conventional pesticides while potentially reducing harm to the environment and human health (Pandit et al. 2022). A wide variety of microorganisms reside in soil, forming intricate relationships that enable them to thrive. The area surrounding plant roots is known as the rhizosphere, and it significantly affects the microbial relationships that exist in the soil (Wang et al. 2024a). The rhizosphere hosts a varied community of beneficial rhizobacteria, commonly referred to as plant growth-promoting rhizobacteria.

G. A. Abo-Zaid (✉)
Bioprocess Development Department, Genetic Engineering and Biotechnology Research Institute, City of Scientific Research and Technological Applications, Alexandria, Egypt

S. Bashir · A. Abdelkhalek
Plant Protection and Biomolecular Diagnosis Department, Arid Lands Cultivation Research Institute, City of Scientific Research and Technological Applications, Alexandria, Egypt

T. Elbeaino
Istituto Agronomico Mediterraneo di Bari (CIHEAM-IAMB), Bari, Italy



N. El-Wakeil, A. Abdelkhalek (eds.), *Microbiomes and Plant Secondary Metabolites for Insect Pest Management*, Progress in Biological Control 23,
https://doi.org/10.1007/978-3-032-24330-0_15

Numerous plant growth-promoting rhizobacteria (PGPR) have demonstrated their ability to combat bacterial and fungal diseases affecting plants in recent decades. Notable genera such as *Bacillus*, *Pseudomonas*, *Streptomyces*, *Rhizobium*, and *Serratia* have been identified as effective agents in inhibiting and eradicating plant pathogens (Sayyed et al. 2019; Al-Askar et al. 2024b). Alongside the extensive variety of prokaryote genes, PGPR significantly contributes to the improvement of crop production through the management of plant diseases. Using antagonistic PGPR for biocontrol of plant diseases presents a highly effective, economical, and environmentally friendly alternative to synthetic pesticides (Beyari 2024). The production of various secondary metabolites, including lipopeptides and hydrolytic enzymes, represents one of the numerous mechanisms through which PGPR operates (Vinay et al. 2016). Furthermore, PGPR can connect with particular root locations, release siderophores, actively compete for binding with target substrates, and induce systemic resistance (Mukherjee et al. 2018).

## 15.2 Major Antifungal Compounds Produced by PGPR

Antifungals produced by PGPR were classified into six main categories depending on their chemical nature and their biosynthetic pathways: the polyketides (PKs), the volatiles compounds, cell wall lytic enzymes, lipopeptides, siderophores, and phenazines.

### *15.2.1 Polyketides (PKSs)*

Numerous studies show that polyketides, a class of secondary metabolites, exhibit wide structural and functional diversity (Carroll and Moore 2018). Polyketides exhibit diverse bioactivities, including antibacterial, anti-cholesterol, antiviral, antifungal, anticancer, immunosuppressive, and anti-inflammatory effects. Examples of polyketides include tetracycline, amphotericin B, doxorubicin, rapamycin, lovastatin, and flavonoids (Chen et al. 2024b). The distinct synthetic approaches of type I, type II, and type III PKSs differentiate them from one another (Staunton and Weissman 2001; Robbins et al. 2016). Multifunctional peptides with covalently fused domains arranged linearly are known as type I PKSs. There are two subtypes of type I PKSs: iterative (iPKSs) and modular (mPKSs) (Keatinge-Clay 2012). Fungi are the most common hosts for iPKSs, and the domains are utilized again and time again to facilitate many rounds of elongation (Heimerl et al. 2018). However, like nonribosomal peptide synthases, mPKSs are mainly found in bacteria and carry out their functions in an assembly line (Shelest et al. 2015). Short spacer sections separate the several domains that constitute mPKSs, each of which performs a specific function. Carbon chain elongation and functional group regeneration are catalyzed by the different domains in a cooperative and non-iterative manner (Herbst et al. 2018). The monofunctional proteins form the multi-enzyme complexes known as type II PKSs., and these compounds are mostly produced by bacteria and can be described as aromatic polyketides (Hertweck et al. 2007). The third form of PKS differs from the first two in that it is primarily present in plants and is characterized by simple homodimers that anchor their chains to acyl-coenzyme A (CoA) instead of acyl carrier protein (ACP) (Shimizu et al. 2017). Numerous luminous pseudomonads produce 2,4-diacetyl phloroglucinol (DAPG or Phl), a phenolic polyketide molecule with antibacterial, antihelminthic, antifungal, and phytotoxic properties (Maurya et al. 2024). One of the main factors influencing the biocontrol activity of rhizobacteria that promote plant development is Phl. Studies indicate that Phl produced by specific strains of *P. fluorescens* can effectively manage the take-all disease of wheat, attributed to *Gaeumannomyces graminis* var. *tritici* (Kwak et al. 2009). One phenolic polyketide that stands out is pyoluteorin (Plt), which has a resorcinol ring structure. A bichlorinated pyrrole molecule is produced when the ring is coupled (Fernando et al. 2006). Phytopathogenic fungi-induced plant diseases were mitigated by various *Pseudomonas* sp. strains that produce Plt (Khoshru et al. 2024). Plt inhibited most oomycete infections, including *Pythium ultimum* (Wang et al. 2024b).

### *15.2.2 Volatile Compounds (VCs)*

PGPR synthesizes volatile inorganic compounds (VICs) mostly from by-products of primary metabolism. The compounds they encompass include hydrogenated, carbonated, nitrogen, or sulfur-containing substances such as $H_2$, $CO_2$, CO, HCN, NO, $NH_3$, $N_2$, and $H_2S$. A group of VIC nitrogenous compounds generated by *B. subtilis* displayed a broad spectrum of antifungal action (Al-Askar et al. 2024c). The breakdown of glycine generates hydrogen cyanide, which impedes aerobic bacteria by obstructing metal-dependent enzymes, including the crucial cytochrome c oxidase in the respiratory chain (Lachowicz et al. 2024). The volatile secondary metabolite hydrogen cyanide (HCN), produced by several rhizobacteria, affects a large array of microbial species. Because it blocks electron transport, HCN cuts off the cell's energy supply, which can kill a lot of cells. A diverse array of bacterial species, such as *Bacillus, Rhizobium*, *Pseudomonas*, *Alcaligenes*, and *Aeromonas* are known to produce various forms of HCN (El-Rahman et al. 2019; Ajinde et al. 2024). Volatile organic compounds (VOCs) consist of small molecules containing fewer than 20 carbon atoms. They can be differentiated based on their low molecular mass (100–500 Da), lipophilic characteristics, and low boiling point (Al-Askar et al. 2024a). Five distinct categories of VOCs exist: fatty acids and their derivatives; terpenoids; VOCs that contain sulfur; VOCs that contain nitrogen; and VOCs that include metalloids or halogenated compounds (Audrain et al. 2015). This database indicates that fatty acid derivatives, including ketones, alkanes, alcohols, alkenes, aldehydes, and acids, represent over 70% of the identified *Bacillus* VOCs, with sulfur- and nitrogen-containing chemicals following closely behind. The extensive range of volatile fatty acids and their derivatives constitutes as much as 87% of the identified antibacterial VOCs generated by *B. subtilis* (Caulier et al. 2019). The mycelial and sporological growth of *Aspergillus carbonarius* and *A. ochraceus* was demonstrated to be inhibited by VOCs (Farbo et al. 2018; Tilocca et al. 2019). Moreover, the VOC-produced *Burkholderia tropica* showed potential antifungal activity against some phytopathogenic fungi such as *Colletotrichum gloeosporioides*, *Fusarium oxysporum, F. culmorum*, and *Sclerotium rolfsii* (Tenorio-Salgado et al. 2013).

### 15.2.3 Cell Wall Lytic Enzymes

The PGPRs produce and release various lytic enzymes, including cellulases, glucanases, esterases, proteases, and chitinases. These enzymes represent a diverse collection capable of hydrolyzing and depolymerizing an extensive collection of polymeric compounds, such as cellulose, proteins, chitin, and hemicellulose (Kubicek et al. 2014). The existence of these enzymes could hinder the proliferation and functions of plant pathogens, either directly or indirectly. The synthesis of extracellular chitinase by *Paenibacillus* sp. is believed to show a decisive role in its antagonistic activity toward *F. oxysporum* f. sp. *cucumerinum*. It was discovered that the chitinase produced by *S. plymuthica* C48 prevented *Botrytis cinerea* from germination and germ tube extension effectively (Frankowski et al. 2001). *Streptomyces* sp. strain 385 and *Paenibacillus* sp. strain 300 produce the enzyme *β*-1,3-glucanase, leading to the lysis of the cell walls in *F. oxysporum* f. sp. *cucumerinum* (Singh et al. 1999). The formulations of chitinase-producing *S. cellulosae* Actino 48 demonstrated efficacy in mitigating root rot and pod rot in peanuts, which are caused by *S. rolfsii* and *R. solani* (Abo-Zaid et al. 2021, 2024). The activity of β-1,3-glucanases from biocontrol agents is associated with the degradation of cell walls in phytopathogenic fungi, as well as the inhibition of spore germination and mycelium development (Lin et al. 2007).

### 15.2.4 Lipopeptides (LPs)

Lipopeptides (LPs) are non-ribosomally synthesized peptides and serve as the primary antimicrobial agents produced by *Bacillus* and *Pseudomonas* species, exhibiting a wide range of antibacterial and antifungal properties (Ongena and Jacques 2008; Hamley 2015).

#### 15.2.4.1 Lipopeptides of *Bacillus* spp.

*Bacillus* spp. lipopeptides share a common structure with other amphiphilic molecules, which includes a lipid tail connected to a short cyclic oligopeptide. The multi-domain enzyme complexes known as non-ribosomal peptide synthetases (NRPSs) are responsible for facilitating the synthesis of lipopeptides (LPs) in *Bacillus*, which is conducted independently of mRNA. Iturins, surfactins, fengycins, and kurstakins are the four primary families of LPs produced by *Bacillus* species, as illustrated in Table 15.1 and Fig. 15.1. LPs are cyclic peptides that are amphiphilic and consist of either 7 α-amino acids (iturins, surfactins and kurstakins) or 10 α-amino acids (fengycins/plipastatin) (Maksimov et al. 2020). Each LP is associated with a distinct β-hydroxy fatty acid (surfactins, fengycins, and kurstakins) or β-amino fatty acid (iturins). Variations in amino acid composition or the length of hydrocarbon chains in fatty acids result in several isomeric forms (Jemil et al. 2024; Yu et al. 2024). Iturins, surfactins, and fengycins are synthesized by *B. subtilis*, whereas kurstakins

**Table 15.1** Categorization of lipopeptides from *Bacillus* species

| Family | Iturins | Surfactins | Fengycins | Kurstakins |
|---|---|---|---|---|
| Peptide length | Heptapeptide | Heptapeptide | Decapeptide | Heptapeptide |
| Chiral sequence | LDDLLDL | LLDLLDL | LDDDLDLLLL | Not described |
| Fatty acid type | *β*-amino FA | *β*-hydroxy FA | *β*-hydroxy FA | *β*-hydroxy FA or not |
| Fatty acid length | 14–17 carbons | 13–15 carbons | 16–19 carbons | 11–14 carbons |
| Cyclic type | Cyclic peptide | Cyclic lactone | Cyclic lactone | Cyclic lactone |

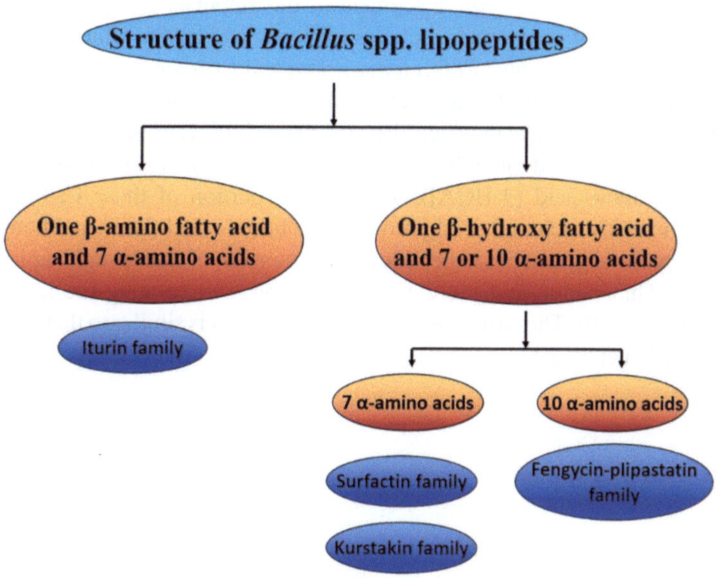


**Fig. 15.1** Structure of *Bacillus* spp. lipopeptides

are synthesized by *B. thuringiensis* (Saiyam et al. 2024). *B. subtilis* synthesizes three LP families, which include at least eight variations of fengycins, thirteen variations of surfactins, and fourteen variations of iturins (Caulier et al. 2019).

#### 15.2.4.2 Lipopeptides of *Pseudomonas* spp.

Fluorescent pseudomonads can synthesize many forms of lipopeptides (Nielsen et al. 2002). Lipopeptides (LPs) are crucial in various biological processes, including biofilm formation, swarming motility, seed and root colonization, and defense against predatory protozoa and competing microbes (Balleux et al. 2024). The production of LP is regulated by extensive multi-modular non-ribosomal peptide synthetases using a thiotemplate mechanism (Jan et al. 2023). A short oligopeptide, created in a lactone ring between two amino acids within the peptide chain, is associated with the fatty acid tail in the lipid composition generated by *Pseudomonas* spp. (Zhou et al. 2024). The lipopeptides (LPs) of *Pseudomonas* species are categorized into four primary groups based on the fatty acid tail length and the composition of it, the quantity and type of amino acids in the peptide moiety, and the configuration of these elements: amphisin, viscosin, syringomycin, and tolaasin. Cyclic lipopeptides (CLPs) from the viscosin group generally consist of nine amino acids linked at the N-terminus to 3-hydroxy decanoic acid (3-HDA) (Wang et al. 2024b). An example is the pectolytic strains of *P. fluorescens* that cause broccoli head rot; viscosin was characterized and discovered for these strains (Hildebrand et al. 1998). Massetolide A, produced by *P. fluorescens* SS101, was identified to be toxic to oomycete plant diseases (De Bruijn et al. 2007). Eleven amino acids constitute the peptide segment of the amphisin group associated with 3-HDA. This class of compounds includes amphisin and tensin, which were discovered to be antagonistic against *Rhizoctonia solani* and *Pythium ultimum*, respectively (Raaijmakers et al. 2006; Kenawy et al. 2019). In contrast to the viscosin and amphisin groups, the tolaasin group demonstrates significant variability in peptide chain composition and length (19–25 amino acids) as well as lipid tail (3-hydroxyoctanoic acid [3-HOA]). The peptide portion of these CLPs contains unique amino acids such as 2,3-dihydro-2-aminobutyric acid (Dhb) and homoserine (Hse). The cyclic segment of the peptide moiety consists of five to eight amino acids, with the all-Thr residue and the C-terminal amino acid creating the lactone ring (Kenawy et al. 2019). The structure of syringomycin LPs is similar to that of viscosin CLPs (Grgurina et al. 1994).

### *15.2.5 Siderophores*

Siderophores, iron-chelating compounds, are microbial metabolic products that have small molecular weight and a very specific affinity toward ferric iron ($Fe^{3+}$) (Terra et al. 2020). As of now, more than 500 distinct types of siderophores have been documented, with 300 of them having undergone structural characterization (Drożdżyński et al. 2024). The four different types of siderophores—carboxylate, hydroxamate, catecholates, and mixed type—display structural variety (Deb and Tatung 2024). Siderophores bind to $Fe^{3+}$ and assist in its transporting from the nearby environment inside the microbial cells (Barbeau et al. 2002). Iron is a substantial component of different proteins and pigments for both plants and phytopathogenic fungi (Greenshields et al. 2007; Kramer et al. 2019). By limiting available iron in the rhizosphere, siderophores can restrict the growth, development, abundance, and metabolic activity of phytopathogenic fungi and other deleterious microorganisms (Haggag and Abo-Sadera 2000; Pankhurst and Lynch 2004). Due to the vital roles of siderophores in disease suppression and phytopathogen prevention (iron competition), the selection of biocontrol agents can depend on their capability to produce large amounts of siderophores (Deb and Tatung 2024; Montoya-Martínez et al. 2024). The production of siderophores is widely recognized as a key mechanism in *P. fluorescent* spp. These organisms are capable of synthesizing various siderophores, including pyoverdines, pulcherrimin, and pseudobactins, which are crucial in inhibiting a range of pathogens (Pankhurst and Lynch 2004; Saravanakumar et al. 2008). It was reported that *Rahnella aquatilis*-producing siderophores inhibited *P. expansum* and *Botrytis cinerea* postharvest pathogens (Calvo et al. 2017). Formulations containing siderophore-generating *P. aeruginosa* and *P. fluorescens* reduced damping-off caused by *F. oxysporum* by 40% and 80%, respectively, while *R. solani* damping-off was reduced by 87.5% and 62.5%, respectively (Abo-Zaid et al. 2023). The capability of several species of *Streptomyces* to produce many siderophore molecules suggested that *Streptomyces* could be a rich source of siderophores which underlines their potential as antimicrobial agents toward various plant pathogens (Jones et al. 2019; Terra et al. 2020).

### *15.2.6 Phenazines*

Few bacterial genera, including *Burkholderia, Pseudomonas*, *Brevibacterium*, and *Streptomyces*, produce phenazines, which are nitrogen heterocyclic pigments and low molecular weight metabolites (Streletskii et al. 2024). Many compounds can be produced from the core structure of phenazines, which is a heterocyclic metabolite with the chemical formula $C_{12}H_8N_2$. This core structure consists of a pyrazine ring with two benzenes (Chaudhary and Khurana 2018; Yan et al. 2021). The physicochemical characteristics of phenazines, such as their redox properties, vibrant colors, and ability to change color based on pH and redox conditions, underscore their importance in biotechnology (Glasser 2017). There have been studies conducted on over 50 different phenazine chemicals found in nature. The simultaneous production of several phenazine derivatives by certain bacterial strains has been documented (Huang et al. 2024a). *P. fluorescens* 2–79 predominantly generates phenazine-1-carboxylic acid (PCA), in contrast to *P. aureofaciens* 30–84, which mainly produces PCA but also yields minor quantities of 2-hydroxyphenazine. One of the most important phenazine compounds biosynthesized by *P. aeruginosa* is 1-hydroxy-5-methyl phenazine (pyocyanin) (Wienberg 1970). Additionally, PCA, 1-hydroxyphenazine, and phenazine 1-carboxamide can be biosynthesized by *P. aeruginosa*. Various pseudomonads strains are responsible for the production of phenazines, which possess antibacterial and anticancer properties (Huang et al. 2024b). In the biocontrol of wheat disease, *G. graminis* var. *tritici*, PCA generated by *P. fluorescens* 2–79 and *P. aureofaciens* 30–84 plays a crucial role (Thomashow and Weller 1988; Yu et al. 2017). *P. chlororaphis* PCL1391 and *P. aeruginosa* MML2212 synthesize phenazine 1-carboxamide (PCN), which effectively suppresses the growth of rice pathogens, including *R. solani* Kühn, *Xanthomonas oryzae* pv. *oryzae*, and *F. oxysporum* f. sp. *radicis-lycopersici*, responsible for tomato foot and root rot (Chin-A-Woeng et al. 2000; Shanmugaiah et al. 2010). *P. aeruginosa* PNA1 synthesizes the antifungal compounds PCA acid and PCN, which play a vital role in inhibiting *P. myriotylum* (wild type) from inducing root rot in cocoyam. (Tambong and Höfte 2001). Pyocyanin and PCA, which are compounds generated by *P. aeruginosa*, have been found to have antagonistic effects on many plant pathogens, including *A. niger*, *F. oxysporum*, *S. rolfsii*, *R. solani*, and others (Rane et al. 2007; Abo-Zaid 2014). Phenazine compounds are essential for the suppression of plant pathogenic fungi by *P. chlororaphis* 30-84 (Yu et al. 2018).

## 15.3 Biosynthesis of PGPR Antifungals

### *15.3.1 Biosynthesis of Polyketides (PKSs)*

Polyketide synthases (PKSs) are very useful enzymes that are needed for the complex biosynthesis of polyketides. Fatty acid synthase and polyketide synthase exhibit a comparable mechanism. Enzymes like acyltransferase (AT) and ketosynthase (KS) are very important in many chemical reactions that happen during the process. AT helps substrates (like acetyl or malonyl) attach to the acyl carrier protein (ACP), and KS helps substrates linked to ACP condense. The subsequent steps involve the incorporation of ketoreductase (KR) to reduce the keto ester, dehydratase (DH) to facilitate the dehydration of the product, and enoylreductase (ER) to reduce the carbon–carbon double bond in the molecule, all following the condensation of the substrates (Fig. 15.2). In contrast to FAS, the processes facilitated by KR, DH, and ER, which are discretionary in PKSs, allow for the generation of diverse polyketide structures featuring keto groups, hydroxy groups, and/or double bonds at varying positions within the molecule (Staunton and Weissman 2001; Hopwood 2004; Shelest et al. 2015).

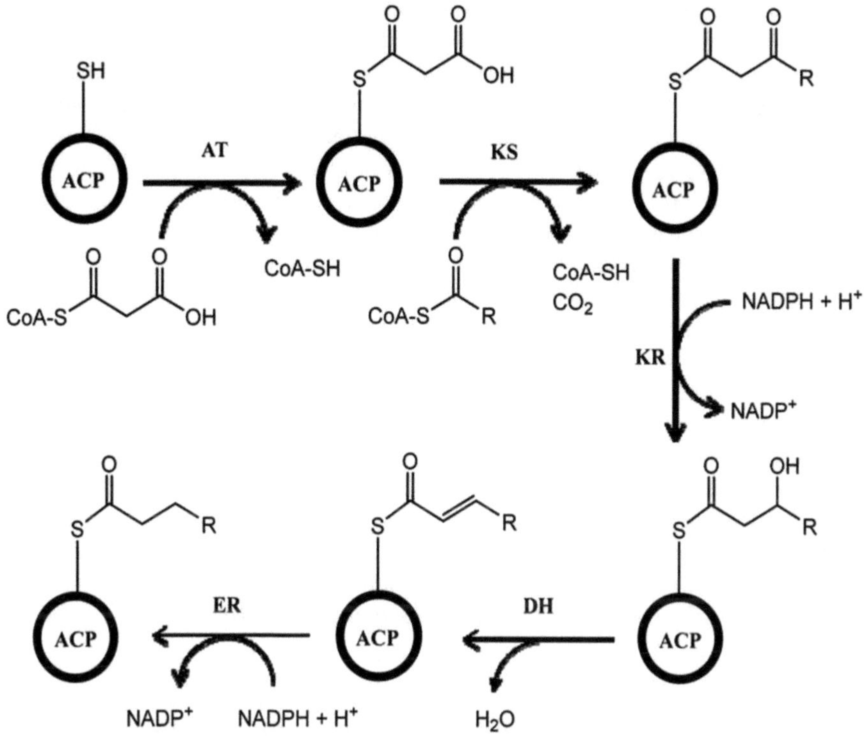


**Fig. 15.2** Process flow diagram for polyketide synthases (PKSs) (generated by Vance et al. 2016)

## *15.3.2 Biosynthesis of Volatile Organic Compounds (VOCs)*

The Embden–Meyerhof pathway, along with the Entner–Doudoroff pathway, as well as heterolactic and homolactic fermentation processes, primarily generate the precursors for volatile fatty acids and their derivatives through the metabolism of glucose (Peñuelas et al. 2014). The anaerobic digestion of pyruvate by *B. subtilis* results in the production of various ketone molecules, including acetoin (3-hydroxy-2-butanone) and 2,3-butanedione (Milanowski et al. 2019). Microbes can leverage different by-products from fatty acid biosynthesis or their β-oxidations to produce volatile organic compounds (VOCs) via decarboxylation or reduction of carboxyl groups (Caulier et al. 2019). They provide an array of fatty acid derivatives alongside hydrocarbons. Oxidative deamination of specific amino acids can yield alcohols, ketones, or volatile aldehydes. The degradation of L-phenylalanine is a notable illustration, whereas L-tyrosine often functions as the primary substrate for the synthesis of aromatic volatile compounds, including benzene and its carbohydrate derivatives. Ultimately, microorganisms possess the ability to produce benzenoid volatiles via the shikimate pathway, resulting in the creation of chorismate, which serves as a naturally occurring precursor for aromatic amino acids (Liu et al. 2023; Chen et al. 2024a). The generation of benzoin volatiles may also take place via the breakdown of aromatic amino acids or intermediates that originate from the shikimate pathway (Dickschat et al. 2005). Among the many secondary metabolites present in biological systems, terpenes and isoprenoids are among the most numerous (Tetali 2019). Two of the most essential precursors are allylic isomers: isopentenyl pyrophosphate and dimethylallyl pyrophosphate (Schulz and Dickschat 2007). The deoxy-xylulose phosphate pathway produces isopentenyl pyrophosphate and dimethylallyl pyrophosphate as by-products in addition to pyruvate and glyceraldehyde-3-phosphate, which are produced from the metabolism of glucose (Kant et al. 2024). It is also possible to produce terpenoids from isoprene molecules. The biosynthesis pathways of nitrogen-containing VOCs remain largely unexplored. Two primary pathways have been identified: One involves the non-enzymatic amination of acyloins, leading to pyrazines production; the other involves the utilization of a-aminoketone intermediates generated by AA catabolism (Zhu et al. 2010). There are mainly two pathways through which sulfur-containing VOCs generated by microbes originate: firstly, through the reduction of inorganic sulfate in methylated inorganic sulfide compounds; and secondly, through the catabolism of amino acids such as L-methionine or, less commonly, L-cysteine (Francioso et al. 2020). The synthesis of hydrogen sulfide and methanethiol yields certain VSCs as by-products known as secondary volatiles. Both of these substances serve as crucial building blocks for the synthesis of VSCs (Sourabié et al. 2012).

### *15.3.3 Biosynthesis of* Bacillus *spp. Lipopeptides (LPs)*

Lipopeptides are generally produced by the condensation of separately synthesized peptide and lipid components. Peptides are often produced by substantial enzymes known as non-ribosomal peptide synthetases (Mohan et al. 2024). A collection of genes that primarily encode antimicrobial peptides is referred to as an operon. The peptide chain results from the collaborative contributions of each gene and open reading frame (ORF) within the gene cluster. Various domains collaborate to facilitate each phase of biosynthesis. Each ORF within an operon constitutes a unique domain. In this context, there is a shift concerning the category of peptides that combat microbial entities (Sieber and Marahiel 2005). The four ORFs that constitute the 38–40 kb iturin operon are ituA, ituB, ituC, and ituD. The three main open reading frames of the iturin family—ituA, ituB, and ituC—contribute to the synthesis of each of the seven amino acids (Das et al. 2008). ItuD encompasses a domain of malonyl coenzyme during the final phases of fatty acid chain synthesis. This is a multi-component system that operates with β-keto acyl synthetase (KS-domain), acyl-CoA ligase (AL-domain), amino transferase (AMT domain), and the acyl carrier protein from ituA (ACP domain) (Velho et al. 2011). The surfactin operon includes ORFs: *srfA-A*, *srfA-B*, srfA-C, and a brief ORF *srfA-D* (Jacques 2011). A total of seven amino acids are produced by the first three ORFs. The third and sixth amino acid residues' adenylation domains include an epimerization region that changes L-Leu to D-Leu. The beginning condensation domain attaches the β-hydroxylated fatty acid chain to the first amino acid. The last domain, PCP, is responsible for secreting the produced lipopeptide by activating the thioesterase. The ORF *srfA-D* commences biosynthesis and encodes the second thioesterase domain.

The operon *fenA-E* and/or *ppsA-E* play a crucial role in the production of fengycin/plipastatin. In the initial three ORFs, *fenC*, *fenD*, and *fenE*, two amino acid residues may be involved: Gln associated with orn-*fenC*; Tyr linked with Thr-*fenD*; and Glu paired with Ala/Val-*fenE*. The fourth ORF (*fenA*) is designed to include three specific residues: Pro, Gln, and Tyr, whereas the final ORF (*fenB*) is responsible for the incorporation of the last amino acid residue, Ile. Like surfactin biosynthesis, the condensation domain facilitates the attachment of the β-hydroxylated fatty acid chain to the initial amino acid residue. The final ORF, which also secretes thioesterase, is responsible for facilitating the formation of an ester bond between the carboxylic group of Ile and the tyrosine that is located at position 3. Units 2, 4, 6, and 9 demonstrate epimerization, leading to the incorporation of amino acid residues in the D-form within the final product (Koumoutsi et al. 2004; Jacques 2011). The three ORFs constituting the kurstakin operon are integral to the synthesis of seven amino acids (Béchet et al. 2012). The biosynthesis of the lipopeptide's lipid component is carried out by a series of modules that allow the activation, coupling, and linking of the fatty acid to the peptide moiety. Five stages are essential for the biosynthesis of the lipopeptide. The acyl CoA-ligase (AL) domain initiated the first step, an ATP-dependent reaction, by coupling a coenzyme A to a fatty acid. Then, the acyl-carrier protein (ACP1), which is the cofactor for the first domain, receives the activated fatty acid. Likewise, the ACP2 domain is where amalonyl CoA is fixed. A β-ketoacyl thioester is formed when the malonyl and acyl thioesters are condensed, a process facilitated by the β-ketoacyl (KS) synthetase domain. Transamination converts the β-ketoacyl thioester into a β-amino fatty acid. The condensation domain is capable of transporting the β-amino fatty acid to the thiolation domain. This can be linked to an amino acid that is already attached to the initial module of the biosynthetic pathway of the peptidic component through the action of an aminotransferase (Duitman et al. 1999).

### *15.3.4 Biosynthesis of Siderophores*

The molecular characteristics dictate the selection of one of two pathways (NRPS and NIS synthetase) utilized in the production of the siderophore. A variety of enzymes, including monooxygenases, amino and acetyltransferases, decarboxylases, aldolases, and amino acid ligases synthesize siderophore building blocks through the NIS pathway. The final stage of synthesis necessitates the presence of enzymes classified as IucA and/or IucC. The straightforward condensation of citric acid or its derivative with an amine or alcohol group occurs through enzymatic catalysis in a single step (Carroll and Moore 2018). Usually, mechanisms independent of NRPS are used to assemble hydroxamate and α-hydroxy carboxylate siderophores (Miethke and Marahiel 2007; Al Shaer et al. 2020). In the NRPS pathway during synthesis, large multimodular enzyme complexes called NRPSs activate and assemble a wide variety of amino acids to create structurally varied peptides (Barry and Challis 2009). An NRPS module typically consists of three main domains: the A-domain, which is tasked with the selection, activation, and transfer of amino acids through thioesterification to the PCP domain; the C-domain, which facilitates the formation of amide or ester bonds between two thioesterified amino acids; and the PCP domain (Lin et al. 2009). Epimerization, methylation, and cyclization are three examples of substrate modifications that involve additional secondary domains that are part of the assembly line. Finally, the final peptide from the assembly line's terminal T domain is hydrolyzed by a thioesterase (TE-domain) (Duban et al. 2022). In specific contexts, the function of a polyketide synthase (PKS) domain is essential for the synthesis of aryl substrates or non-proteinogenic amino acids, a process that can be aided by non-integrated enzymes (Miller et al. 2002). Researchers have looked at how the enzymes that make siderophores are organized spatially (Gasser et al. 2020). Certain siderophores are thought to contain a siderosome, which is a complex of enzymes and proteins found inside the cell that is responsible for producing siderophores (Gasser et al. 2015). The connection of biosynthesis to secretion is an advantage of such compartmentalization that guarantees efficient siderophore biosynthesis. Biosynthetic gene clusters (BGCs) are groups of related genes that encode proteins involved in siderophore biosynthesis; iron plays a key role in regulating the transcription of these genes (Barry and Challis 2009; Gulick 2017). Reports indicate that the diphtheria toxin regulator (DtxR) and the ferric uptake regulator (Fur) serve as the primary regulators in the control of bacterial siderophores (Miethke and Marahiel 2007). The presence of an iron (II) regulatory binding site in both leads to the dimerization of Fur, which subsequently binds to specific DNA loci known as the Fur Box, thereby inhibiting the transcription of biosynthetic genes (Escolar et al. 1999).

### *15.3.5 Biosynthesis of Phenazines*

The seven genes phzABCDEFG play a crucial role in the synthesis of a 6.8 kb region in *P. fluorescens* 2-79, known as phenazine-1-carboxylic acid (PCA) (Mavrodi et al. 1998). The initial step in the biosynthesis of phenazine involves the synthesis of shikimic acid (Guo et al. 2017). The heterocyclic nucleus, resulting from the symmetrical condensation of two chorismic acid molecules, utilizes the amide nitrogen of glutamine as the primary nitrogen source. The initial step involves the conversion of chorismic acid into aminodeoxyisochorismate (ADIC), a reaction that is facilitated by amyladeoxyisochorismate synthase. Using ADIC lyase as a catalyst, the second step entails aromatization and the removal of pyruvate to produce 3-hydroxyanthranilic acid (Shende et al. 2024). The phenazine nucleus is formed through the condensation of two molecules of 3-hydroxyanthranilate, facilitated by the products of phzF and phzG. The conversion of phenazine-1,6-dicarboxylic acid to PCA appears to result from spontaneous non-enzymatic decarboxylation (Mavrodi et al. 1998). In addition to PCA, *P. aureofaciens* 30-84 and *P. chlororaphis* GP72 produce minor amounts of 2-hydroxyphenazine and 2-hydroxyphenazine, respectively. In strain 30-84, the transformation of PCA to 2-OH-PCA is facilitated by the phzO gene, responsible for encoding flavin-diffusible monooxygenase. In this process, a hydroxyl group is added to PCA in an orthogonal position to the carboxyl group (Kenawy et al. 2019). *P. aeruginosa* has two operons that are responsible for regulating the production of PCA. These operons are *phzA1B1C1D1E1F1G1* and *phzA2B2C2D2E2F2G2*. Additionally, three genes (*phzM*, *phzS*, and *phzH*) code for an enzyme complex responsible for converting PCA into 5-methyl-phenazine-1-carboxylic acid (5MPCA), 1-hydroxy-phenazine (1-OHPZ), PCN, and pyocyanin. The biosynthesis of PCA in *P. aeruginosa* is regulated by two operons (*phzA1B1C1D1E1F1G1* and *phzA2B2C2D2E2F2G2*). Additionally, three genes (*phzM*, *phzS*, and *phzH*) code for an enzyme complex that facilitates the conversion of PCA into 5-methyl-phenazine-1-carboxylic acid (5MPCA), 1-hydroxy-phenazine (1-OHPZ), PCN, and pyocyanin (Greenhagen et al. 2008; Jin et al. 2016).

## 15.4 Modes of Action of PGPR Antifungals

### *15.4.1 Modes of Action of Lytic Enzymes*

Fungal cell walls are primarily composed of structural polysaccharides like chitin and glucan, which provide critical rigidity and protection. Antifungal enzymes such as chitinases and glucanases act as precise molecular tools that degrade these essential components. Specifically, chitinases hydrolyze the β-1,4 glycosidic linkages within the chitin polymer, while glucanases target the β-1,3 and β-1,6 linkages in glucan (Adrangi and Faramarzi 2013; Gow et al. 2017). This enzymatic cleavage creates physical breaches in the cell wall, compromising its structural integrity. As a result, the cell can no longer withstand the internal osmotic pressure, leading to uncontrolled water influx, cell swelling, and eventual osmotic lysis, which is a direct cause of cell death.

The cell walls of fungi are generally composed of bilayered structures. The inner layer, present in the majority of fungi, consists mainly of branched (1-3) β-D-glucan, in addition to (1-6) β-D-glucan and a small amount of chitin. The outer layer is made up of glycoproteins, which are a complex of proteins and polysaccharides, as well as lipids and other components in lower quantities (De Beeck et al. 2021). Consequently, the fungal cell wall can be affected by various lytic enzymes synthesized by PGPR, such as *β*-1,3-glucanases, lipases, proteases, and chitinases (Santoyo et al. 2021). PGPR enzymes play a crucial role in the biocontrol of plant diseases by suppressing pathogenic fungi, hence limiting the spread of plant disease. These lytic enzymes prevent hyphal development and fungal penetration into plant tissues by cleaving the components of fungal cell wall such as cellulose, chitin, protein, and lipide. (Duhan et al. 2022). Several investigations reported that microbial strains such as *B. mojavensis*, *B. subtilis*, and *S. marcescens* act as inducing agents for the production of chitinase, protease, and glucanase, which are essential for the biocontrol of phytopathogens such as *R. cerealis*, *A. alternata*, and *F. oxysporum* (Brzezinska et al. 2020; Trejo-López et al. 2022; Yi et al. 2022). *Streptomyces* sp. NEAU-S7GS2 synthesizes glucanase to inhibit the growth of *Sclerotinia sclerotiorum*, a causing agent of sclerotinia stem rot disease (Liu et al. 2019). The cell wall of this fungus could also be degraded by *B. subtilis* BS-2301 and *P. fluorescens* PS1, which produce amylases, proteases chitinase, and $\beta$-1,3-glucanase (Aeron et al. 2011; Ayaz et al. 2024).

## *15.4.2 Modes of Action of* Bacillus *spp. Lipopeptides (LPs)*

The ability of lipopeptides produced by *Bacillus* spp. to interface with the cell membrane of their intended organisms and their amphiphilic nature define their mechanisms of action.

### 15.4.2.1 Mode of Action of Iturin

Lipopeptides' antagonistic activity is based on how they interact with cell membranes. This is associated with the Tyr residue at position 2 in the peptide ring (Fig. 15.3), which causes the membrane to solubilize and ion-conducting holes to develop at greater concentrations. Cell death and the release of cytoplasmic components and other electrolytes occur simultaneously upon pore formation (Volpon et al. 2000; Aranda et al. 2005). Several studies have demonstrated that iturins can

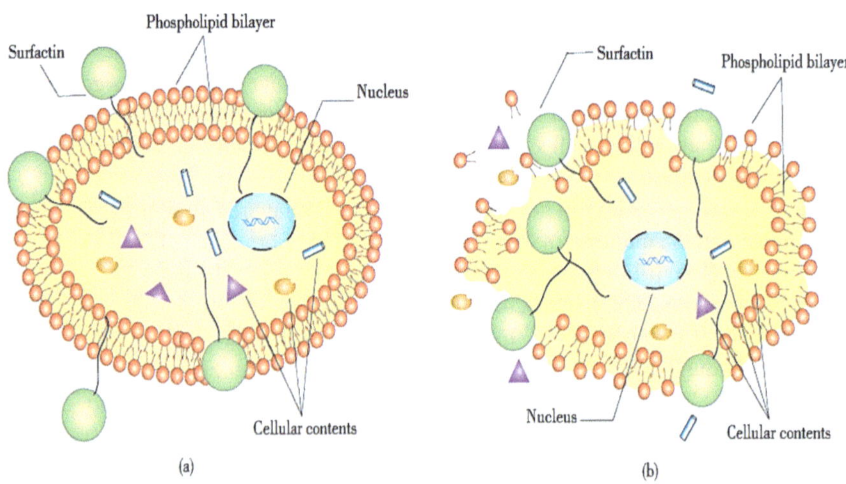


**Fig. 15.3** The proposed model of iturin mechanism (Balleza et al. 2019)

decrease the potential of the mitochondrial membrane, encourage the formation of reactive oxygen species, and induce the nucleus of the cell to condense (Olfa et al. 2015; Lei et al. 2019; Wang et al. 2020).

#### 15.4.2.2 Mode of Action of Surfactin

Extremely potent biosurfactants, surfactin-family lipopeptides readily interact with lipid bilayers, subsequently interfering with biological membrane structure in a dose-dependent manner; at high concentrations, these agents cause irreversible pore formation, followed by full lipid bilayer disruption and solubilization (Ongena and Jacques 2008; Mei et al. 2020). There has been a lot of research looking at how surfactin interacts with biological membranes. The initial step in the process of surfactin's incorporation into the cell membrane is the formation of hydrophobic contacts between the fatty acid and the hydrocarbon chains of the phospholipids that make up the cell membrane, as shown in Fig. 15.4. After this initial step, surfactin's cyclic peptide undergoes conformational modifications that aid in its incorporation into the cell membrane even more (Dexter and Middelberg 2008). The integration of surfactin dimers into the bilayer contributes to the phenomenon of membrane instability (Carrillo et al. 2003). This structural alteration, as the hypothesized mechanism of action states, dehydrates the polar heads of membrane phospholipids, which in turn disrupts the bilayer and allows cell content to flow out.

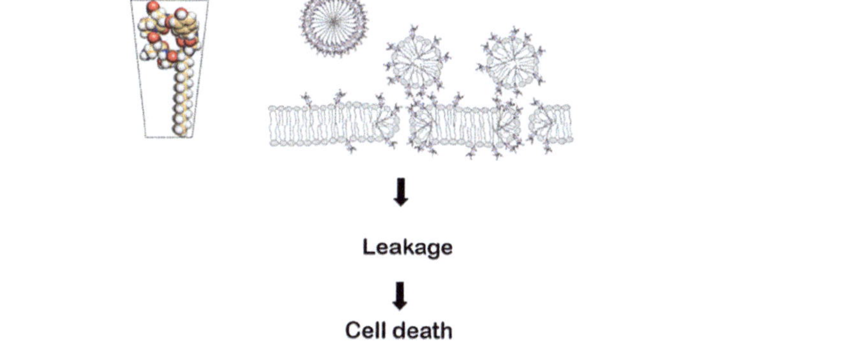


**Fig. 15.4** The suggested paradigm for the surfactin mechanism. (**a**) Schematic representation of insertion into the cell membrane and (**b**) schematic representation of disruption of the cell membrane (Mei et al. 2020)

#### 15.4.2.3 Mode of Action of Fengycin

The fengycin class of lipopeptides also alters the shape and permeability of the cell membrane through interactions with it (Deleu et al. 2008). Fungal pathogen hyphae were ultrastructurally destroyed by fengycin. Hyphae treated with fengycin showed cell walls that were gapped and/or detached from the cell membrane, as well as uncombined cytoplasm (Tao et al. 2011; Gong et al. 2015). The production of ROS by fengycin has the potential to cause cell death in living creatures by oxidizing lipids, proteins, DNA, and carbohydrates (Rodriguez and Redman 2005). The initial step in the proposed mechanism of action for fengycin involves the penetration of fengycin BS155 through the fungal hypha cell membrane, allowing it to access the intracellular region, where it inflicts damage on the plasma membrane (Zhang and Sun 2018). Second, fengycin BS155 upsets the MMP (mitochondrial membrane potential). Step 3 involves the buildup of ROS, a process that is intensified by the downregulation of ROS-scavenging enzymes like superoxide dismutase and catalase due to fengycin BS155. Step 4: Fengycin BS155 causes DNA condensation, which in turn increases the production of DNA repair proteins. As seen in step 5, the hyphal cells undergo apoptosis when fengycin BS155 triggers the cleavage of poly (ADP-ribose) polymerase (PARP). The convergence of organelle dysfunction, the buildup of ROS, and chromatin condensation result in cell death triggered by fengycin BS155. (Fig. 15.5).

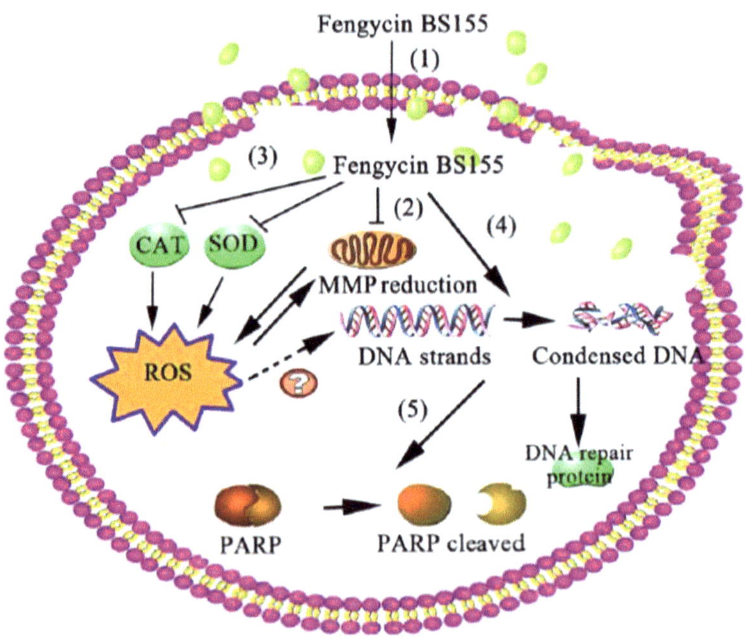


**Fig. 15.5** The suggested mechanism of fengycin BS155 action in the hyphal cells of the phytopathogen *Magnaporthe grisea* (Zhang and Sun 2018)

#### 15.4.2.4 Mode of Action of Kurstakin

Kurstakin is one of a novel family of lipopeptides as pore-forming compounds produced by *Bacillus thuringiensis* (Wang et al. 2024b). Kurstakins consist of lipoheptapeptides formed by seven amino acids linked to fatty acid chains at the C11 and C14 positions. The exact mechanism through which this family demonstrates antifungal properties against *Stachybotrys chartarum* remains unclear (Hathout et al. 2000). The kurstakin synthetase consists of three large multifunctional proteins that are encoded by the gene cluster krsABC (Béchet et al. 2012).

### 15.4.3 Mode of Action of Siderophores

PGPR engages in competition with pathogens for nutrients and physical spaces within the rhizosphere, serving as an indirect mechanism through which PGPR can influence specific phytopathogenic fungi (Kapoor et al. 2024). Since nutrients are essential for fungal population growth, PGPR's capacity to compete for these resources can limit the expansion of harmful fungi. When it comes to biocontrolling fungal infections, iron competition is a highly effective strategy (Wang et al. 2021). It was observed that the absence of iron ions leads to inhibited growth, reduced nucleic acid synthesis, and suppression of pathogen sporulation (Chowdhury et al. 2020). Thus, PGPR has an advantage in root colonization and pathogen exclusion due to its siderophore synthesis (Wang et al. 2021). *B. subtilis* MF497446 and *P. koreensis* MG209738 are known to produce siderophores and have been identified as antagonistic to *Cephalosporium maydis*, the pathogen that causes late wilt in maize (Ghazy and El-Nahrawy 2021). The siderophore-secreting *B. siamensis* Gxun-6 effectively halted the hyphal growth of *F. oxysporum* and protected banana seedlings against Fusarium wilt disease (Shen et al. 2022). It was observed that the incidence of damping-off disease was significantly diminished, and the mycelial growth of *F. oxysporum* and *R. solani* was impeded (Abo-Zaid et al. 2020).

### 15.4.4 Mode of Action of Phenazines

Phenazine pigments have shown some effect against fungus and other microorganisms; however, it is unclear how widespread this activity is. It is posited that pyocyanin has the capability to accept electrons, resulting in the formation of a steady anion radical that gladly integrates into the redox cycle. The formation of $O{\bullet}_2^-$ (superoxide radical) is increased by Mn-containing superoxide dismutase (MnSOD), a key enzyme, as shown in Fig. 15.6. The antimicrobial properties of pyocyanin may

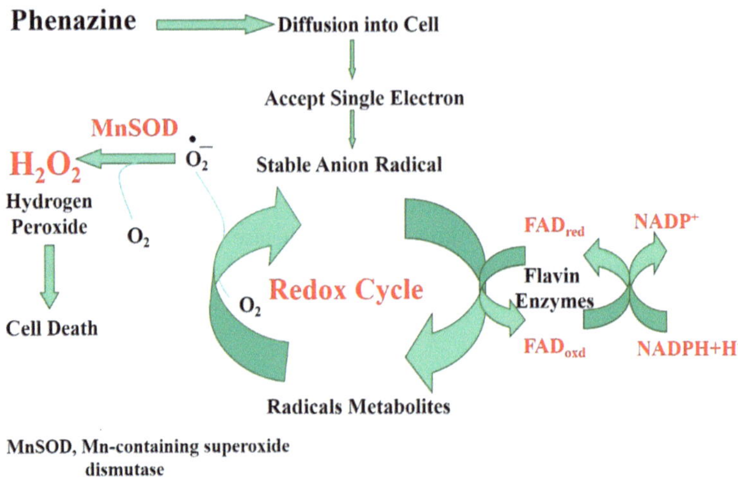


**Fig. 15.6** Mode of action of phenazine (Abo-Zaid 2014)

be endorsed by the poisonousness of $O{\bullet}_2^-$ and $H_2O_2$, which are generated in elevated concentrations in its occurrence (Mavrodi et al. 2001, 2006).

## 15.5 Role of PGPR Antifungals as Elicitors for Induced Systemic Resistance (ISR)

Jasmonic acid and ethylene play crucial roles in the regulation of induced systemic resistance, commonly referred to as ISR (Fig. 15.7). The ISR is initiated by beneficial bacteria or other molecules secreted by them, such as LPs, which can suppress pathogens and, in the end, plant diseases (Rakhonde et al. 2025). Secondary metabolites produced by beneficial microbes can enhance induced systemic resistance by functioning as immune elicitors and directly targeting harmful bacteria (Yassin et al. 2024). It was demonstrated that surfactins and fengycins lipopeptides serve as a new class of microbial-associated molecular patterns (MAMPs) that activate the defense signaling pathway (Ongena et al. 2007). These MAMPs protect bean and tomato plants through ISR. They also found that compounds of surfactin and fengycin may be more effectively used as ISR in tomato and bean plants when their genes were overexpressed in *B. subtilis* strain 168 (the poor producer). The attachment of surfactin molecules to the cell membrane of plants is the primary process that is responsible for the induction of ISR (Henry et al. 2011). The usage of semi-purified surfactin lipopeptide has been found to boost perennial ryegrass resistance to *M. oryzae*-caused grey leaf spot disease. This resistance is achieved by defense responses that are heightened by $H_2O_2$, which are in turn elicited by ISR (Rahman et al. 2015). Surfactin was found to induce ISR in response to *B. cinerea* in bean,

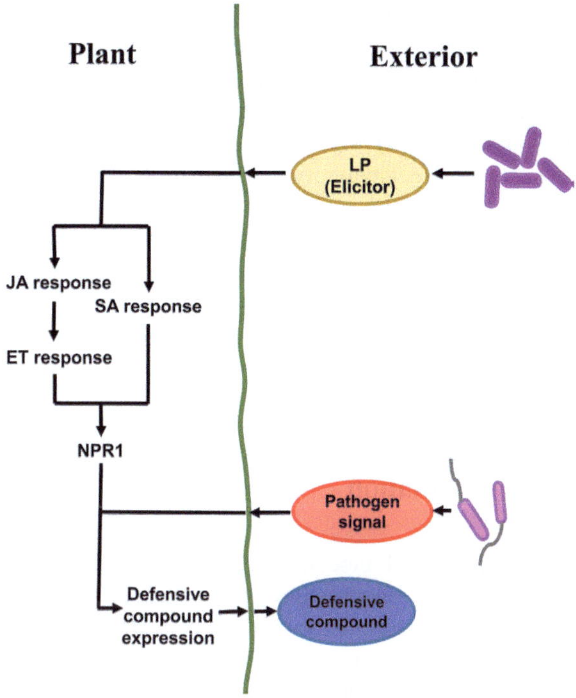


**Fig. 15.7** Induced systemic resistance in plants caused by LPs. These compounds activate a JA/ET response or an SA-dependent pathway, acting as an elicitor to set off the plant's systemic resistance. A defensive compound was produced by the plant as a result (Penha et al. 2020)

tomato, and tobacco, *Podosphaera fusca* in melon, and *S. rolfsii* in peanut (García-Gutiérrez et al. 2013; Nicolle 2023). ISR triggers biochemical changes that lead to plant cell wall reinforcements, antimicrobial phytoalexin production, and pathogenesis-related protein synthesis, including β-1,3-glucanases, peroxidases, and chitinases (García-Gutiérrez et al. 2013; Behiry et al. 2023). Surfactin mainly triggered a pathway regulated by SA, whereas mycosubtilin (from the iturin family) engaged both SA and JA signaling pathways (Akintayo 2024). It was observed that grapevine leaves developed a localized and persistent tolerance to the *B. cinerea* pathogen because of both LPs, surfactin, and iturine (Farace et al. 2015). The defense response of strawberries against *C. gloeosporioides* was largely constituted by iturin and surfactin, both of which played important roles. The levels of chitinase and *β*-1,3-glucanase that were expressed in strawberry leaves were dramatically impacted by the presence of LPs (Yamamoto et al. 2015). Phenazines produced by beneficial *Pseudomonas* exhibit antifungal and ISR-inducing activities (Chin-A-Woeng et al. 2003). Furthermore to their role in antibiosis, phenazines are crucial for the biocontrol of plant pathogens. Research has demonstrated that phenazines can cause ISR in various plant species, including tobacco, beans, and rice (Kang et al. 2007; Ma et al. 2016). Pyocyanin, a by-product of *P. aeruginosa* 7NSK2, has been previously elucidated as inducing systemic plant resistance to *B. cinerea*, a bacterium responsible for infections in tomatoes and grapevines (Audenaert et al. 2002). The induction of ISR in rice against *Magnaporthe oryzae* can be accomplished with PCN concentrations between 0.1 and 1 μmol/L, and this effect was negated in a CMR12a mutant that could not synthesize phenazine (Ma et al. 2016). Massetolide A, synthesized by *P. fluorescens* strain SS101, is crucial for root colonization and the initiation of a systemic resistance response in tomato plants (de Bruijn and Raaijmakers 2009). Microbial volatile compounds (MVCs) enhance photosynthetic rates, stimulate phytohormone signaling pathways, and strengthen the immune system, all of which play a significant role in promoting plant growth (Kong et al. 2018). The role of rhizobacterial siderophores in initiating ISR in host plants and enhancing soil suppressiveness is well recognized. Consequently, they hold the responsibility of biologically managing a diverse range of plant diseases (Deb and Tatung 2024; Kapoor et al. 2024; Rakhonde et al. 2025).

## 15.6 Role of PGPR Antifungals in Inducing Biofilm Formation

One way in which biocontrol agents work is by the production of biofilms. Microorganisms can form biofilms when their cells become embedded in a polymeric matrix that is present outside of the cells (Flemming and Wingender 2010). Several *Bacillus* species, including *B. subtilis*, *B. atrophaeus*, and *B. velezensis*, have been found to engage in biofilm formation as a means of biocontrol (Khezri et al. 2011; Karačić et al. 2024). The surfactin is a key component in biofilm formation and a biocontrol mechanism (Zhang et al. 2024). Bacillomycin D, which is a member of the iturin family, was discovered to be responsible for regulating the expression of genes that are linked with the development of biofilm in *B. amyloliquefaciens* (Jiao et al. 2021). The massetolide A is crucial for the swarming motility and biofilm development of *P. fluorescens* strain SS101, which is produced during the early exponential growth phase (De Bruijn et al. 2008). The biocontrol of various phytopathogenic fungi can be accomplished during the development of biofilms, with phenazines playing a crucial role in this process (Guo et al. 2024; Kumari et al. 2024).

## 15.7 Conclusion

Plant growth-promoting rhizobacteria (PGPR) represent a formidable and sustainable frontline defense against phytopathogenic fungi. Their efficacy stems from a multifunctional arsenal: the direct antagonism provided by a diverse suite of antifungal compounds (lipopeptides, phenazines, etc.), the indirect protection conferred by eliciting plant systemic resistance, and the ecological advantage secured through robust biofilm formation. Among the most promising strategies is the deployment of strains that combine multiple modes of action, such as *Bacillus* spp. producing biosurfactant lipopeptides (e.g., surfactin and iturin) alongside lytic enzymes, or *Pseudomonas* spp. synthesizing phenazines and iron-chelating siderophores. These combinations reduce the risk of pathogen resistance and enhance performance under variable environmental conditions.

However, significant gaps must be addressed to translate laboratory promise into reliable field applications. For many compounds, precise molecular modes of action remain unclear; for instance, the specific mechanism by which the lipopeptide kurstakin exerts its antifungal effect is not fully defined. Furthermore, a disconnect often exists between in vitro efficacy and consistent field performance, influenced by rhizosphere competition, soil chemistry, and climate variability.

## 15.8 Future Research and Application Challenges

Future research must address key application challenges to translate PGPR potential into reliable agricultural tools. Critical directions include:

*Fermentation and Formulation Science*: Scaling up production through optimized fermentation and developing advanced, carrier-based formulations (e.g., microencapsulation) to enhance field stability, prolong shelf life, and protect bacterial viability and metabolite efficacy.

*Synthetic Biology and Strain Enhancement*: Utilizing genetic engineering to create next-generation PGPR, such as by overexpressing key antifungal gene clusters (e.g., the phz operon), silencing competitive pathways, or introducing novel traits to improve rhizosphere fitness and biocontrol consistency.

*Mechanistic and Ecological Studies*: Resolving the precise mode of action for partially characterized compounds and decoding complex rhizosphere interactions to enable the rational design of synergistic microbial consortia.

**Declaration of AI Assistance** The authors declared that they utilized artificial intelligence tools, specifically ChatGPT and DeepSeek-V3, to support linguistic polishing and enhance readability in selected paragraphs of this chapter. The authors conducted a manual evaluation of all proposed edits to ensure the preservation of scientific accuracy, interpretation, and originality. The authors independently generated all substantive content, analyses, and conclusions. The authors independently peer-reviewed the final version.

## References

Abo-Zaid GA (2014) Scaling-up production of biocontrol agents from Pseudomonas spp. PhD disseration. Faculty of Agriculture at Alexandria University, Alexandria

Abo-Zaid GA, Soliman NA-M, Abdullah AS, El-Sharouny EE, Matar SM, Sabry SA-F (2020) Maximization of siderophores production from biocontrol agents, Pseudomonas aeruginosa f2 and Pseudomonas fluorescens JY3 using batch and exponential fed-batch fermentation. Processes 8:455

Abo-Zaid G, Abdelkhalek A, Matar S, Darwish M, Abdel-Gayed M (2021) Application of bio-friendly formulations of Chitinase-producing Streptomyces cellulosae Actino 48 for controlling Peanut soil-borne diseases caused by Sclerotium rolfsii. J Fungi 7:167

Abo-Zaid GA, Abdullah AS, Soliman NA-M, El-Sharouny EE, Al-Askar AA, Su Y et al (2023) Evaluation of bio-friendly formulations from Siderophore-producing fluorescent pseudomonas as biocontrol agents for the Management of Soil-Borne Fungi, Fusarium oxysporum and Rhizoctonia solani. Agriculture 13:1418

Abo-Zaid GA, Darwish MH, Ghozlan HA, Abdel-Gayed MA, Sabry SA (2024) Sustainable management of peanut damping-off and root rot diseases caused by Rhizoctonia solani using environmentally friendly bio-formulations prepared from batch fermentation broth of chitinase-producing Streptomyces cellulosae. BMC Plant Biol 24:760

Adrangi S, Faramarzi MA (2013) From bacteria to human: a journey into the world of chitinases. Biotechnol Adv 31(8):1786–1795

Aeron A, Dubey RC, Maheshwari DK, Pandey P, Bajpai VK, Kang SC (2011) Multifarious activity of bioformulated Pseudomonas fluorescens PS1 and biocontrol of Sclerotinia sclerotiorum in Indian rapeseed (Brassica campestris L.). Eur J Plant Pathol 131:81–93

Ajinde AO, Dayo-Olagbende OG, Akpor OB (2024) Direct and indirect mechanisms of growth promotion by plant growth promoting rhizobacteria. In: 2024 international conference on science, engineering and business for driving sustainable development goals (SEB4SDG) (IEEE), pp 1–16

Akintayo SO (2024) Bioprospecting for novel lipopeptide-producing strains for potential application in food and agriculture. Universität Hohenheim

Al Shaer D, Al Musaimi O, de la Torre BG, Albericio F (2020) Hydroxamate siderophores: natural occurrence, chemical synthesis, iron binding affinity and use as Trojan horses against pathogens. Eur J Med Chem 208:112791

Al-Askar A, Al-Otibi FO, Abo-Zaid GA, Abdelkhalek A (2024a) Diisooctyl phthalate, the major secondary metabolite of Bacillus subtilis, could be a potent antifungal agent against Rhizoctonia solani: GC-MS and in silico molecular docking investigations. Egypt J Chem 67:1137–1148

Al-Askar A, Al-Otibi FO, Abo-Zaid GA, Abdelkhalek A (2024b) Gas chromatography-mass spectrometry comparative analysis of secondary metabolites produced by some Bacillus spp. as biocontrol agents against phytopathogenic fungi. Egypt J Chem 67:997

Al-Askar A, Al-Otibi FO, Abo-Zaid GA, Abdelkhalek A (2024c) Pyrrolo [1, 2-a] pyrazine-1, 4-dione, hexahydro-3-(2-methylpropyl), as the primary secondary metabolite of Bacillus spp., could be an effective antifungal agent against the soil-borne fungus, Sclerotium bataticola. Egypt J Chem 67:1009

Aranda FJ, Teruel JA, Ortiz A (2005) Further aspects on the hemolytic activity of the antibiotic lipopeptide iturin A. Biochim Biophys Acta (BBA)-Biomembranes 1713:51–56

Audenaert K, Pattery T, Cornelis P, Höfte M (2002) Induction of systemic resistance to Botrytis cinerea in tomato by Pseudomonas aeruginosa 7NSK2: role of salicylic acid, pyochelin, and pyocyanin. Mol Plant-Microbe Interact 15:1147–1156

Audrain B, Farag MA, Ryu C-M, Ghigo J-M (2015) Role of bacterial volatile compounds in bacterial biology. FEMS Microbiol Rev 39:222–233

Ayaz M, Ali Q, Zhao W, Chi Y-K, Ali F, Rashid KA et al (2024) Exploring plant growth promoting traits and biocontrol potential of new isolated Bacillus subtilis BS-2301 strain in suppressing Sclerotinia sclerotiorum through various mechanisms. Front Plant Sci 15:1444328

Ayilara MS, Adeleke BS, Akinola SA, Fayose CA, Adeyemi UT, Gbadegesin LA et al (2023) Biopesticides as a promising alternative to synthetic pesticides: a case for microbial pesticides, phytopesticides, and nanobiopesticides. Front Microbiol 14:1040901

Balleux G, Höfte M, Arguelles-Arias A, Deleu M, Ongena M (2024) Bacillus lipopeptides as key players in rhizosphere chemical ecology. Trends Microbiol. https://doi.org/10.1016/j.tim.2024.08.001

Balleza D, Alessandrini A, Beltrán García MJ (2019) Role of lipid composition, physicochemical interactions, and membrane mechanics in the molecular actions of microbial cyclic lipopeptides. J Membr Biol 252:131–157
Barbeau K, Zhang G, Live DH, Butler A (2002) Petrobactin, a photoreactive siderophore produced by the oil-degrading marine bacterium Marinobacter hydrocarbonoclasticus. J Am Chem Soc 124:378–379
Barry SM, Challis GL (2009) Recent advances in siderophore biosynthesis. Curr Opin Chem Biol 13:205–215
Béchet M, Caradec T, Hussein W, Abderrahmani A, Chollet M, Leclère V et al (2012) Structure, biosynthesis, and properties of kurstakins, nonribosomal lipopeptides from Bacillus spp. Appl Microbiol Biotechnol 95:593–600
Behiry S, Soliman SA, Massoud MA, Abdelbary M, Kordy AM, Abdelkhalek A et al (2023) Trichoderma pubescens elicit induced systemic resistance in tomato challenged by Rhizoctonia solani. J Fungi 9. https://doi.org/10.3390/jof9020167
Beyari EA (2024) Alternatives to chemical pesticides: the role of microbial biocontrol agents in phytopathogen management: a comprehensive review. J Plant Pathol 107:291
Brzezinska MS, Kalwasińska A, Świątczak J, Żero K, Jankiewicz U (2020) Exploring the properties of chitinolytic Bacillus isolates for the pathogens biological control. Microb Pathog 148:104462
Calvo H, Marco P, Blanco D, Oria R, Venturini ME (2017) Potential of a new strain of Bacillus amyloliquefaciens BUZ-14 as a biocontrol agent of postharvest fruit diseases. Food Microbiol 63:101–110
Carrillo C, Teruel JA, Aranda FJ, Ortiz A (2003) Molecular mechanism of membrane permeabilization by the peptide antibiotic surfactin. Biochim Biophys Acta (BBA)-Biomembranes 1611:91–97
Carroll CS, Moore MM (2018) Ironing out siderophore biosynthesis: a review of non-ribosomal peptide synthetase (NRPS)-independent siderophore synthetases. Crit Rev Biochem Mol Biol 53:356–381
Caulier S, Nannan C, Gillis A, Licciardi F, Bragard C, Mahillon J (2019) Overview of the antimicrobial compounds produced by members of the Bacillus subtilis group. Front Microbiol 10:302
Chaudhary A, Khurana JM (2018) Synthetic routes for phenazines: an overview. Res Chem Intermed 44:1045–1083
Chen L, Liu R, Wu M, Ge Q, Yu H (2024a) A review on aroma-active compounds derived from branched-chain amino acid in fermented meat products: flavor contribution, formation pathways, and enhancement strategies. Trends Food Sci Technol 145:104371
Chen X, Li X, Zhang G, Wang C, Li C (2024b) Application of synthetic biology to the biosynthesis of polyketides. Synth Biol Eng 2:10012
Chin-A-Woeng TFC, Bloemberg GV, Mulders IHM, Dekkers LC, Lugtenberg BJJ (2000) Root colonization by phenazine-1-carboxamide-producing bacterium Pseudomonas chlororaphis PCL1391 is essential for biocontrol of tomato foot and root rot. Mol Plant-Microbe Interact 13:1340–1345
Chin-A-Woeng TFC, Bloemberg GV, Lugtenberg BJJ (2003) Phenazines and their role in biocontrol by Pseudomonas bacteria. New Phytol 157:503–523
Chowdhury FT, Zaman NR, Islam MR, Khan H (2020) Anti-fungal secondary metabolites and hydrolytic enzymes from rhizospheric bacteria in crop protection: a review. J Bangladesh Acad Sci 44:69–84
Das P, Mukherjee S, Sen R (2008) Genetic regulations of the biosynthesis of microbial surfactants: an overview. Biotechnol Genet Eng Rev 25:165–186
De Beeck MO, Persson P, Tunlid A (2021) Fungal extracellular polymeric substance matrices–highly specialized microenvironments that allow fungi to control soil organic matter decomposition reactions. Soil Biol Biochem 159:108304
de Bruijn I, Raaijmakers JM (2009) Regulation of cyclic lipopeptide biosynthesis in Pseudomonas fluorescens by the ClpP protease. J Bacteriol 191:1910–1923

De Bruijn I, De Kock MJD, Yang M, De Waard P, Van Beek TA, Raaijmakers JM (2007) Genome-based discovery, structure prediction and functional analysis of cyclic lipopeptide antibiotics in Pseudomonas species. Mol Microbiol 63:417–428
De Bruijn I, de Kock MJD, de Waard P, van Beek TA, Raaijmakers JM (2008) Massetolide A biosynthesis in Pseudomonas fluorescens. J Bacteriol. https://doi.org/10.1128/JB.01563-07
Deb CR, Tatung M (2024) Siderophore producing bacteria as biocontrol agent against phytopathogens for a better environment: a review. S Afr J Bot 165:153–162
Deleu M, Paquot M, Nylander T (2008) Effect of fengycin, a lipopeptide produced by Bacillus subtilis, on model biomembranes. Biophys J 94:2667–2679
Dexter AF, Middelberg APJ (2008) Peptides as functional surfactants. Ind Eng Chem Res 47:6391–6398
Dickschat JS, Bode HB, Wenzel SC, Müller R, Schulz S (2005) Biosynthesis and identification of volatiles released by the myxobacterium Stigmatella aurantiaca. Chembiochem 6:2023–2033
Drożdżyński P, Rutkowska N, Rodziewicz M, Marchut-Mikołajczyk O (2024) Bioactive compounds produced by endophytic bacteria and their plant hosts—an insight into the world of chosen herbaceous ruderal plants in Central Europe. Molecules 29:4456
Duban M, Cociancich S, Leclère V (2022) Nonribosomal peptide synthesis definitely working out of the rules. Microorganisms 10:577
Duhan L, Kumari D, Verma R, Pasrija R (2022) Fungal hydrolytic enzymes produced by plant growth-promoting rhizobacteria (PGPR). In: Secondary metabolites and volatiles of PGPR in plant-growth promotion. Springer, pp 313–333
Duitman EH, Hamoen LW, Rembold M, Venema G, Seitz H, Saenger W et al (1999) The mycosubtilin synthetase of Bacillus subtilis ATCC6633: a multifunctional hybrid between a peptide synthetase, an amino transferase, and a fatty acid synthase. Proc Natl Acad Sci 96: 13294–13299
El-Rahman A, Shaheen HA, El-Aziz A, Rabab M, Ibrahim DSS (2019) Influence of hydrogen cyanide-producing rhizobacteria in controlling the crown gall and root-knot nematode, Meloidogyne incognita. Egypt J Biol Pest Control 29:1–11
Escolar L, Pérez-Martín J, de Lorenzo V (1999) Opening the iron box: transcriptional metalloregulation by the Fur protein. J Bacteriol 181:6223–6229
Farace G, Fernandez O, Jacquens L, Coutte F, Krier F, Jacques P et al (2015) Cyclic lipopeptides from B acillus subtilis activate distinct patterns of defence responses in grapevine. Mol Plant Pathol 16:177–187
Farbo MG, Urgeghe PP, Fiori S, Marcello A, Oggiano S, Balmas V et al (2018) Effect of yeast volatile organic compounds on ochratoxin A-producing Aspergillus carbonarius and A. Ochraceus. Int J Food Microbiol 284:1–10
Fenta L, Mekonnen H (2024) Microbial biofungicides as a substitute for chemical fungicides in the control of Phytopathogens: current perspectives and research directions. Scientifica (Cairo) 2024:1
Fernando WGD, Nakkeeran S, Zhang Y (2006) Biosynthesis of antibiotics by PGPR and its relation in biocontrol of plant diseases. In: PGPR biocontrol biofertilization, pp 67–109
Flemming H-C, Wingender J (2010) The biofilm matrix. Nat Rev Microbiol 8:623–633
Francioso A, Baseggio Conrado A, Mosca L, Fontana M (2020) Chemistry and biochemistry of sulfur natural compounds: key intermediates of metabolism and redox biology. Oxidative Med Cell Longev 2020:8294158
Frankowski J, Lorito M, Scala F, Schmid R, Berg G, Bahl H (2001) Purification and properties of two chitinolytic enzymes of Serratia plymuthica HRO-C48. Arch Microbiol 176:421–426
García-Gutiérrez L, Zeriouh H, Romero D, Cubero J, de Vicente A, Pérez-García A (2013) The antagonistic strain B acillus subtilis UMAF 6639 also confers protection to melon plants against cucurbit powdery mildew by activation of jasmonate-and salicylic acid-dependent defence responses. Microb Biotechnol 6:264–274
Gasser V, Guillon L, Cunrath O, Schalk IJ (2015) Cellular organization of siderophore biosynthesis in Pseudomonas aeruginosa: evidence for siderosomes. J Inorg Biochem 148:27–34

Gasser V, Malrieu M, Forster A, Mély Y, Schalk IJ, Godet J (2020) In cellulo FRET-FLIM and single molecule tracking reveal the supra-molecular organization of the pyoverdine bio-synthetic enzymes in Pseudomonas aeruginosa. Q Rev Biophys 53:e1
Ghazy N, El-Nahrawy S (2021) Siderophore production by Bacillus subtilis MF497446 and Pseudomonas koreensis MG209738 and their efficacy in controlling Cephalosporium maydis in maize plant. Arch Microbiol 203:1195–1209
Glasser NR (2017) Physiological and biochemical mechanisms of Phenazine-mediated survival in Pseudomonas aeruginosa. California Institute of Technology
Gong A-D, Li H-P, Yuan Q-S, Song X-S, Yao W, He W-J et al (2015) Antagonistic mechanism of iturin A and plipastatin A from Bacillus amyloliquefaciens S76-3 from wheat spikes against Fusarium graminearum. PLoS One 10:e0116871
Gow NAR, Latge J-P, Munro CA (2017) The fungal cell wall: structure, biosynthesis, and function. Microbiol Spectrum 5(3). https://doi.org/10.1128/microbiolspec.funk-0035-2016
Greenhagen BT, Shi K, Robinson H, Gamage S, Bera AK, Ladner JE et al (2008) Crystal structure of the pyocyanin biosynthetic protein PhzS. Biochemistry 47:5281–5289
Greenshields DL, Liu G, Wei Y (2007) Roles of iron in plant defence and fungal virulence. Plant Signal Behav 2:300–302
Grgurina I, Barca A, Cervigni S, Gallo M, Scaloni A, Pucci P (1994) Relevance of chlorine-substituent for the antifungal activity of syringomycin and syringotoxin, metabolites of the phytopathogenic bacterium Pseudomonas syringae pv. syringae. Experientia 50:130–133
Gulick AM (2017) Nonribosomal peptide synthetase biosynthetic clusters of ESKAPE pathogens. Nat Prod Rep 34:981–1009
Guo S, Wang Y, Dai B, Wang W, Hu H, Huang X et al (2017) PhzA, the shunt switch of phenazine-1, 6-dicarboxylic acid biosynthesis in Pseudomonas chlororaphis HT66. Appl Microbiol Biotechnol 101:7165–7175
Guo S, Liu Y, Yin Y, Chen Y, Jia S, Wu T et al (2024) Unveiling the multifaceted potential of Pseudomonas khavaziana strain SR9: a promising biocontrol agent for wheat crown rot. Microbiol Spectrum 12:e00712–24
Haggag WM, Abo-Sadera SA (2000) Influence of iron sources and siderophore producing Pseudomonas on crown-rot disease incidence and seed contamination of peanut with pathogenic Aspergilli. Ann Agric Sci Moshtohor 38:1933–1948
Hamley IW (2015) Lipopeptides: from self-assembly to bioactivity. Chem Commun 51:8574–8583
Hathout Y, Ho Y-P, Ryzhov V, Demirev P, Fenselau C (2000) Kurstakins: a new class of Lipopeptides isolated from Bacillus t huringiensis. J Nat Prod 63:1492–1496
Heimerl N, Hommel E, Westermann M, Meichsner D, Lohr M, Hertweck C et al (2018) A giant type I polyketide synthase participates in zygospore maturation in Chlamydomonas reinhardtii. Plant J 95:268–281
Henry G, Deleu M, Jourdan E, Thonart P, Ongena M (2011) The bacterial lipopeptide surfactin targets the lipid fraction of the plant plasma membrane to trigger immune-related defence responses. Cell Microbiol 13:1824–1837
Herbst DA, Townsend CA, Maier T (2018) The architectures of iterative type I PKS and FAS. Nat Prod Rep 35:1046–1069
Hertweck C, Luzhetskyy A, Rebets Y, Bechthold A (2007) Type II polyketide synthases: gaining a deeper insight into enzymatic teamwork. Nat Prod Rep 24:162–190
Hildebrand PD, Braun PG, McRae KB, Lu X (1998) Role of the biosurfactant viscosin in broccoli head rot caused by a pectolytic strain of Pseudomonas fluorescens. Can J Plant Pathol 20:296–303
Hopwood DA (2004) Cracking the polyketide code. PLoS Biol 2:e35
Huang W, Wan Y, Su H, Zhang Z, Liu Y, Sadeeq M et al (2024a) Recent advances in Phenazine natural products: biosynthesis and metabolic engineering. J Agric Food Chem 72:21364–21379
Huang W, Wan Y, Zhang S, Wang C, Zhang Z, Su H et al (2024b) Recent advances in Phenazine natural products: chemical structures and biological activities. Molecules 29:4771

Jacques P (2011) Surfactin and other lipopeptides from Bacillus spp. In: Biosurfactants: from genes to applications, pp 57–91
Jan F, Arshad H, Ahad M, Jamal A, Smith DL (2023) In vitro assessment of Bacillus subtilis FJ3 affirms its biocontrol and plant growth promoting potential. Front Plant Sci 14:1205894
Jemil N, Besbes I, Gharbi Y, Triki MA, Cheffi M, Manresa A et al (2024) Bacillus methylotrophicus DCS1: production of different Lipopeptide families, in vitro antifungal activity and suppression of Fusarium wilt in tomato plants. Curr Microbiol 81:142
Jiao R, Cai Y, He P, Munir S, Li X, Wu Y et al (2021) Bacillus amyloliquefaciens YN201732 produces lipopeptides with promising biocontrol activity against fungal pathogen Erysiphe cichoracearum. Front Cell Infect Microbiol 11:598999
Jin X-J, Peng H-S, Hu H-B, Huang X-Q, Wang W, Zhang X-H (2016) iTRAQ-based quantitative proteomic analysis reveals potential factors associated with the enhancement of phenazine-1-carboxamide production in Pseudomonas chlororaphis P3. Sci Rep 6:27393
Jones SE, Pham CA, Zambri MP, McKillip J, Carlson EE, Elliot MA (2019) Streptomyces volatile compounds influence exploration and microbial community dynamics by altering iron availability. MBio 10:e00171-19
Kang B-R, Han S-H, Zdor RE, Anderson AJ, Spencer M, Yang K-Y et al (2007) Inhibition of seed germination and induction of systemic disease resistance by Pseudomonas chlororaphis O6 requires phenazine production regulated by the global regulator, GacS. J Microbiol Biotechnol 17:586–593
Kant G, Pandey A, Kumari S, Bux F, Srivastava S (2024) Enhancing bio-isoprene production in Escherichia coli through a combinatorial optimization approach. Process Biochem 144:210–219
Kapoor D, Sharma P, Sharma MMM, Yadav S, Husen A (2024) Exploring soil microbiota and their role in plant growth, stress tolerance, disease control and nutrient immobilizer. Biocatal Agric Biotechnol 178:103358
Karačić V, Miljaković D, Marinković J, Ignjatov M, Milošević D, Tamindžić G et al (2024) Bacillus species: excellent biocontrol agents against tomato diseases. Microorganisms 12:457
Keatinge-Clay AT (2012) The structures of type I polyketide synthases. Nat Prod Rep 29:1050–1073
Kenawy A, Dailin DJ, Abo-Zaid GA, Malek RA, Ambehabati KK, Zakaria KHN et al (2019) Biosynthesis of antibiotics by PGPR and their roles in biocontrol of plant diseases. In: Plant growth promoting rhizobacteria for sustainable stress management, vol 2, pp 1–35
Kharrazian D (2021) Exposure to environmental toxins and autoimmune conditions. Integr Med A Clin J 20:20
Khezri M, Ahmadzadeh M, Jouzani GS, Behboudi K, Ahangaran A, Mousivand M et al (2011) Characterization of some biofilm-forming Bacillus subtilis strains and evaluation of their biocontrol potential against Fusarium culmorum. J Plant Pathol 93:373–382
Khoshru B, Fallah Nosratabad A, Mahjenabadi VAJ, Knežević M, Hinojosa AC, Fadiji AE et al (2024) Multidimensional role of Pseudomonas: from biofertilizers to bioremediation and soil ecology to sustainable agriculture. J Plant Nutr 48:1–27
Kong HG, Shin TS, Kim TH, Ryu C-M (2018) Stereoisomers of the bacterial volatile compound 2, 3-butanediol differently elicit systemic defense responses of pepper against multiple viruses in the field. Front Plant Sci 9:90
Koumoutsi A, Chen X-H, Henne A, Liesegang H, Hitzeroth G, Franke P et al (2004) Structural and functional characterization of gene clusters directing nonribosomal synthesis of bioactive cyclic lipopeptides in Bacillus amyloliquefaciens strain FZB42. J Bacteriol 186:1084
Kramer J, Özkaya Ö, Kümmerli R (2019) Bacterial siderophores in community and host interactions. Nat Rev Microbiol 18:1–12
Kubicek CP, Starr TL, Glass NL (2014) Plant cell wall–degrading enzymes and their secretion in plant-pathogenic fungi. Annu Rev Phytopathol 52:427–451
Kumari S, Khanna V, Sharma N (2024) Characterization and biological evaluation of phenazine produced by antagonistic pseudomonads against Fusarium oxysporum f. sp. ciceris. Int J Pest Manag 70:1086–1099

Kwak Y-S, Bakker PAHM, Glandorf DCM, Rice JT, Paulitz TC, Weller DM (2009) Diversity, virulence, and 2, 4-diacetylphloroglucinol sensitivity of Gaeumannomyces graminis var. tritici isolates from Washington State. Phytopathology 99:472–479

Lachowicz JI, Alexander J, Aaseth JO (2024) Cyanide and cyanogenic compounds—toxicity, molecular targets, and therapeutic agents. Biomolecules 14:1420

Lei S, Zhao H, Pang B, Qu R, Lian Z, Jiang C et al (2019) Capability of iturin from Bacillus subtilis to inhibit Candida albicans in vitro and in vivo. Appl Microbiol Biotechnol 103:4377–4392

Lin C, Yang J, Sun H, Huang X, Wang R, Zhang K-Q (2007) Purification and characterization of a β-1, 3-glucanase from the novel mycoparasite Periconia byssoides. Biotechnol Lett 29:617–622

Lin S, Van Lanen SG, Shen B (2009) A free-standing condensation enzyme catalyzing ester bond formation in C-1027 biosynthesis. Proc Natl Acad Sci 106:4183–4188

Liu D, Yan R, Fu Y, Wang X, Zhang J, Xiang W (2019) Antifungal, plant growth-promoting, and genomic properties of an endophytic actinobacterium Streptomyces sp. NEAU-S7GS2. Front Microbiol 10:2077

Liu W-H, Chai L-J, Wang H-M, Lu Z-M, Zhang X-J, Xiao C et al (2023) Bacteria and filamentous fungi running a relay race in Daqu fermentation enable macromolecular degradation and flavor substance formation. Int J Food Microbiol 390:110118

Ma Z, Hua GKH, Ongena M, Höfte M (2016) Role of phenazines and cyclic lipopeptides produced by Pseudomonas sp. CMR12a in induced systemic resistance on rice and bean. Environ Microbiol Rep 8:896–904

Maksimov IV, Singh BP, Cherepanova EA, Burkhanova GF, Khairullin RM (2020) Prospects and applications of lipopeptide-producing bacteria for plant protection. Appl Biochem Microbiol 56: 15–28

Maurya S, Thakur R, Vighnesh R, Suresh S, Dang A, Raj D et al (2024) Eco-friendly management of plant pathogens through secondary metabolites released by fluorescent pseudomonads. J Pure Appl Microbiol 18:1471

Mavrodi DV, Ksenzenko VN, Bonsall RF, Cook RJ, Boronin AM, Thomashow LS (1998) A seven-gene locus for synthesis of phenazine-1-carboxylic acid by Pseudomonas fluorescens 2-79. J Bacteriol 180:2541–2548

Mavrodi DV, Bonsall RF, Delaney SM, Soule MJ, Phillips G, Thomashow LS (2001) Functional analysis of genes for biosynthesis of pyocyanin and phenazine-1-carboxamide from Pseudomonas aeruginosa PAO1. J Bacteriol 183:6454–6465

Mavrodi DV, Blankenfeldt W, Thomashow LS (2006) Phenazine compounds in fluorescent Pseudomonas spp. biosynthesis and regulation. Annu Rev Phytopathol 44:417–445

Mei Y, Yang Z, Yu F, Long X (2020) Recent progress on fermentation and antibacterial applications of surfactin. China Biotechnol 40:105–116

Miethke M, Marahiel MA (2007) Siderophore-based iron acquisition and pathogen control. Microbiol Mol Biol Rev 71:413–451

Milanowski M, Monedeiro F, Złoch M, Ratiu I-A, Pomastowski P, Ligor T et al (2019) Profiling of VOCs released from different salivary bacteria treated with non-lethal concentrations of silver nitrate. Anal Biochem 578:36–44

Miller DA, Luo L, Hillson N, Keating TA, Walsh CT (2002) Yersiniabactin synthetase: a four-protein assembly line producing the nonribosomal peptide/polyketide hybrid siderophore of Yersinia pestis. Chem Biol 9:333–344

Miller AL, Li S, Eichhorn CD, Zheng Y, Du L (2023) Identification and biosynthetic study of the siderophore lysochelin in the biocontrol agent Lysobacter enzymogenes. J Agric Food Chem 71: 7418–7426

Mohan MS, Salim SA, Pakhira P, Busi S (2024) Microbial production of polyketides and non-ribosomal peptides and their applications. In: Microbial products for health and nutrition. Springer, pp 365–390

Montoya-Martínez AC, Ruiz VV, Chávez-Luzanía RA, Villa-Rodríguez ED, de los Santos Villalobos S (2024) Biological control agents for mitigating plant diseases. In: New insights,

trends, and challenges in the development and applications of microbial inoculants in agriculture. Elsevier, pp 27–35
Mukherjee C, Beall CJ, Griffen AL, Leys EJ (2018) High-resolution ISR amplicon sequencing reveals personalized oral microbiome. Microbiome 6:1–15
Nicolle C (2023) Molecular mechanisms of the interaction of a rhizospheric Streptomyces sp. strain with plant roots. Université Paul Sabatier
Nielsen TH, Sørensen D, Tobiasen C, Andersen JB, Christophersen C, Givskov M et al (2002) Antibiotic and biosurfactant properties of cyclic lipopeptides produced by fluorescent Pseudomonas spp. from the sugar beet rhizosphere. Appl Environ Microbiol 68:3416–3423
Olfa T, Antonio DG, Sana A, Imen BS, Salem E, Mohamed Najib A et al (2015) Synergistic fungicidal activity of the lipopeptide bacillomycin D with amphotericin B against pathogenic Candida species. FEMS Yeast Res 15:fov022
Ongena M, Jacques P (2008) Bacillus lipopeptides: versatile weapons for plant disease biocontrol. Trends Microbiol 16:115–125
Ongena M, Jourdan E, Adam A, Paquot M, Brans A, Joris B et al (2007) Surfactin and fengycin lipopeptides of Bacillus subtilis as elicitors of induced systemic resistance in plants. Environ Microbiol 9:1084–1090
Pandit MA, Kumar J, Gulati S, Bhandari N, Mehta P, Katyal R et al (2022) Major biological control strategies for plant pathogens. Pathogens 11:273
Pankhurst CE, Lynch JM (2004) Biocontrol of soil-borne plant diseases. Encycl Soils Environ 4: 129–136. https://doi.org/10.1016/B0-12-348530-4/00137-5
Penha RO, Vandenberghe LPS, Faulds C, Soccol VT, Soccol CR (2020) Bacillus lipopeptides as powerful pest control agents for a more sustainable and healthy agriculture: recent studies and innovations. Planta 251:1–15
Peñuelas J, Asensio D, Tholl D, Wenke K, Rosenkranz M, Piechulla B et al (2014) Biogenic volatile emissions from the soil. Plant Cell Environ 37:1866–1891
Raaijmakers JM, De Bruijn I, De Kock MJD (2006) Cyclic lipopeptide production by plant-associated Pseudomonas spp.: diversity, activity, biosynthesis, and regulation. Mol Plant-Microbe Interact 19:699–710
Rahman A, Uddin W, Wenner NG (2015) Induced systemic resistance responses in perennial ryegrass against M agnaporthe oryzae elicited by semi-purified surfactin lipopeptides and live cells of B acillus amyloliquefaciens. Mol Plant Pathol 16:546–558
Rakhonde GY, Panda S, Ahale SR, Veratiya AK, Deshkar AM, Nayak L (2025) Role of microbes and organic amendments in stimulating ISR and SAR in plant disease management. In: Advances in organic farming. Apple Academic Press, pp 491–514
Rane MR, Sarode PD, Chaudhari BL, Chincholkar SB (2007) Detection, isolation and identification of phenazine-1-carboxylic acid produced by biocontrol strains of Pseudomonas aeruginosa. J Sci Ind Res 66(8):627–631
Robbins T, Liu Y-C, Cane DE, Khosla C (2016) Structure and mechanism of assembly line polyketide synthases. Curr Opin Struct Biol 41:10–18
Rodriguez R, Redman R (2005) Balancing the generation and elimination of reactive oxygen species. Proc Natl Acad Sci 102:3175–3176
Saiyam D, Dubey A, Malla MA, Kumar A (2024) Lipopeptides from bacillus: unveiling biotechnological prospects—sources, properties, and diverse applications. Braz J Microbiol 55:281–295
Santoyo G, Urtis-Flores CA, Loeza-Lara PD, Orozco-Mosqueda M d C, Glick BR (2021) Rhizosphere colonization determinants by plant growth-promoting rhizobacteria (PGPR). Biology (Basel) 10:475
Saravanakumar D, Ciavorella A, Spadaro D, Garibaldi A, Gullino ML (2008) Metschnikowia pulcherrima strain MACH1 outcompetes Botrytis cinerea, Alternaria alternata and Penicillium expansum in apples through iron depletion. Postharvest Biol Technol 49:121–128
Sayyed RZ, Seifi S, Patel PR, Shaikh SS, Jadhav HP, El Enshasy H (2019) Siderophore production in groundnut rhizosphere isolate, Achromobacter sp. RZS2 influenced by physicochemical factors and metal ions. Environ Sustain 2:117–124

Schulz S, Dickschat JS (2007) Bacterial volatiles: the smell of small organisms. Nat Prod Rep 24: 814–842

Shanmugaiah V, Mathivanan N, Varghese B (2010) Purification, crystal structure and antimicrobial activity of phenazine-1-carboxamide produced by a growth-promoting biocontrol bacterium, Pseudomonas aeruginosa MML2212. J Appl Microbiol 108:703–711

Shelest E, Heimerl N, Fichtner M, Sasso S (2015) Multimodular type I polyketide synthases in algae evolve by module duplications and displacement of AT domains in trans. BMC Genomics 16:1–15

Shen N, Li S, Li S, Zhang H, Jiang M (2022) The siderophore-producing bacterium, Bacillus siamensis Gxun-6, has an antifungal activity against Fusarium oxysporum and promotes the growth of banana. Egypt J Biol Pest Control 32:34

Shende VV, Bauman KD, Moore BS (2024) The shikimate pathway: gateway to metabolic diversity. Nat Prod Rep 41:604–648

Shimizu Y, Ogata H, Goto S (2017) Type III polyketide synthases: functional classification and phylogenomics. Chembiochem 18:50–65

Sieber SA, Marahiel MA (2005) Molecular mechanisms underlying nonribosomal peptide synthesis: approaches to new antibiotics. Chem Rev 105:715–738

Singh PP, Shin YC, Park CS, Chung YR (1999) Biological control of Fusarium wilt of cucumber by chitinolytic bacteria. Phytopathology 89:92–99

Sourabié AM, Spinnler H-E, Bourdat-Deschamps M, Tallon R, Landaud S, Bonnarme P (2012) S-methyl thioesters are produced from fatty acids and branched-chain amino acids by brevibacteria: focus on l-leucine catabolic pathway and identification of acyl–CoA intermediates. Appl Microbiol Biotechnol 93:1673–1683

Staunton J, Weissman KJ (2001) Polyketide biosynthesis: a millennium review. Nat Prod Rep 18: 380–416

Streletskii RA, Astaykina AA, Belov AA, Cheptsov VS, Vetrova AA (2024) Beneficial soil microorganisms and their role in sustainable agriculture. In: Sustainable agricultural practices. Elsevier, pp 293–333

Tambong JT, Höfte M (2001) Phenazines are involved in biocontrol of Pythium myriotylum on cocoyam by Pseudomonas aeruginosa PNA1. Eur J Plant Pathol 107:511–521

Tao Y, Bie X, Lv F, Zhao H, Lu Z (2011) Antifungal activity and mechanism of fengycin in the presence and absence of commercial surfactin against Rhizopus stolonifer. J Microbiol 49:146–150

Tenorio-Salgado S, Tinoco R, Vazquez-Duhalt R, Caballero-Mellado J, Perez-Rueda E (2013) Identification of volatile compounds produced by the bacterium Burkholderia tropica that inhibit the growth of fungal pathogens. Bioengineered 4:236–243

Terra L, Dyson P, Ratcliffe N, Castro HC, Vicente ACP (2020) Biotechnological potential of Streptomyces Siderophores as new antibiotics. Curr Med Chem. https://doi.org/10.2174/0929867327666200510235512

Tetali SD (2019) Terpenes and isoprenoids: a wealth of compounds for global use. Planta 249:1–8

Thomashow LS, Weller DM (1988) Role of a phenazine antibiotic from Pseudomonas fluorescens in biological control of Gaeumannomyces graminis var. tritici. J Bacteriol 170:3499–3508

Tilocca B, Balmas V, Hassan ZU, Jaoua S, Migheli Q (2019) A proteomic investigation of Aspergillus carbonarius exposed to yeast volatilome or to its major component 2-phenylethanol reveals major shifts in fungal metabolism. Int J Food Microbiol 306:108265

Trejo-López JA, Rangel-Vargas E, Gómez-Aldapa CA, Villagómez-Ibarra JR, Falfán-Cortes RN, Acevedo-Sandoval OA et al (2022) Isolation and molecular identification of Serratia strains producing chitinases, glucanases, cellulases, and prodigiosin and determination of their antifungal effect against Colletotrichum siamense and Alternaria alternata in vitro and on mango fruit. Int J Plant Biol 13:281–297

Vance S, Tkachenko O, Thomas B, Bassuni M, Hong H, Nietlispach D et al (2016) Sticky swinging arm dynamics: studies of an acyl carrier protein domain from the mycolactone polyketide synthase. Biochem J 473:1097–1110

Velho RV, Caldas DGG, Medina LFC, Tsai SM, Brandelli A (2011) Real-time PCR investigation on the expression of sboA and ituD genes in Bacillus spp. Lett Appl Microbiol 52:660–666
Vinay JU, Naik MK, Rangeshwaran R, Chennappa G, Shaikh SS, Sayyed RZ (2016) Detection of antimicrobial traits in fluorescent pseudomonads and molecular characterization of an antibiotic pyoluteorin. 3 Biotech 6:1–11
Volpon L, Besson F, Lancelin J-M (2000) NMR structure of antibiotics plipastatins A and B from Bacillus subtilis inhibitors of phospholipase A2. FEBS Lett 485:76–80
Wang Y, Zhang C, Liang J, Wu L, Gao W, Jiang J (2020) Iturin A extracted from Bacillus subtilis WL-2 affects Phytophthora infestans via cell structure disruption, oxidative stress, and energy supply dysfunction. Front Microbiol 11:536083
Wang H, Liu R, You MP, Barbetti MJ, Chen Y (2021) Pathogen biocontrol using plant growth-promoting bacteria (PGPR): role of bacterial diversity. Microorganisms 9:1–18. https://doi.org/10.3390/microorganisms9091988
Wang Q, Ding J, Zhang Z, Liang C, Lambers H, Zhu B et al (2024a) Rhizosphere as a hotspot for microbial necromass deposition into the soil carbon pool. J Ecol 113:168
Wang Y, Pei Y, Wang X, Dai X, Zhu M (2024b) Antimicrobial metabolites produced by the plant growth-promoting rhizobacteria (PGPR): Bacillus and Pseudomonas. Adv Agrochem 3:206
Wienberg ED (1970) Biosynthesis of secondary metabolites: roles of trace elements. Adv Microbiol Physiol 4:1–44
Yamamoto S, Shiraishi S, Suzuki S (2015) Are cyclic lipopeptides produced by Bacillus amyloliquefaciens S13-3 responsible for the plant defence response in strawberry against Colletotrichum gloeosporioides? Lett Appl Microbiol 60:379–386
Yan J, Liu W, Cai J, Wang Y, Li D, Hua H et al (2021) Advances in phenazines over the past decade: review of their pharmacological activities, mechanisms of action, biosynthetic pathways and synthetic strategies. Mar Drugs 19:610
Yassin Y, Aseel D, Khalil A, Abdel-Megeed A, Al-Askar A, Elbeaino T et al (2024) Foliar application of rhizobium leguminosarum bv. viciae strain 33504-Borg201 promotes Faba bean growth and enhances systemic resistance against bean yellow mosaic virus infection. Curr Microbiol 81:220. https://doi.org/10.1007/s00284-024-03733-6
Yi Y, Luan P, Wang K, Li G, Yin Y, Yang Y et al (2022) Antifungal activity and plant growth-promoting properties of Bacillus mojovensis B1302 against Rhizoctonia cerealis. Microorganisms 10:1682
Yu JM, Wang D, Pierson LS III, Pierson EA (2017) Disruption of MiaA provides insights into the regulation of phenazine biosynthesis under suboptimal growth conditions in Pseudomonas chlororaphis 30-84. Microbiology 163:94–108
Yu JM, Wang D, Pierson LS III, Pierson EA (2018) Effect of producing different phenazines on bacterial fitness and biological control in Pseudomonas chlororaphis 30-84. Plant Pathol J 34:44
Yu F, Shen Y, Pang Y, Fan H, Liu M, Liu X (2024) Effects of branched-chain amino acids on surfactin structure and antibacterial activity in Bacillus velezensis YA215. World J Microbiol Biotechnol 40:281
Zhang L, Sun C (2018) Fengycins, cyclic lipopeptides from marine Bacillus subtilis strains, kill the plant-pathogenic fungus Magnaporthe grisea by inducing reactive oxygen species production and chromatin condensation. Appl Environ Microbiol 84:e00445–e00418
Zhang R, Yu J, Yang L, Qiao J, Qi Z, Yu M et al (2024) Surfactins and Iturins produced by bacillus velezensis Jt84 strain synergistically contribute to the biocontrol of rice false smut. Agronomy 14:2204
Zhou L, Höfte M, Hennessy RC (2024) Does regulation hold the key to optimizing lipopeptide production in Pseudomonas for biotechnology? Front Bioeng Biotechnol 12:1363183
Zhu B-F, Xu Y, Fan W-L (2010) High-yield fermentative preparation of tetramethylpyrazine by Bacillus sp. using an endogenous precursor approach. J Ind Microbiol Biotechnol 37:179–186

# Chapter 16
# Biological Control of Viruses Using Actinomycetes Species

**Ahmed Abdelkhalek, Dalia G. Aseel, Lóránt Király, and Hassan Moawad**

## 16.1 Introduction

Plant diseases represent a significant danger to global food security and human health, leading to massive crop losses (Junaid and Gokce 2024). Crops of significant economic value, including wheat, rice, potatoes, soybeans, squash, and maize, are often impacted by severe diseases caused by various plant pathogens, such as bacteria, viruses, nematodes, and fungi (Savary et al. 2019). Plant viruses are a critical pathogen that can substantially affect agricultural yield (Abdelkhalek et al. 2018; İlbağı et al. 2024). The widespread occurrence of plant viruses in agriculture presents significant challenges that jeopardize the vitality and yield of crops. The presence of these diseases can result in various symptoms, such as stunted growth, diminished yield, and potential plant mortality, which may lead to financial challenges for farmers and affect global food security (Yassin et al. 2024). Also, the complicated connections between plant viruses and the insects that carry them show how important it is to use thorough pest control methods, especially since chemical methods are becoming harder to find because of environmental rules (Aseel et al. 2024a). Given the significant economic impact of these diseases, considerable effort has historically been directed toward other strategies like cross-protection and developing resistant plant varieties, alongside physical and chemical controls (Abdelkhalek and Hafez 2019; Aseel et al. 2024b). However, the need for more

A. Abdelkhalek (✉) · D. G. Aseel
Plant Protection and Biomolecular Diagnosis Department, Arid Lands Cultivation Research Institute, City of Scientific Research and Technological Applications, Alexandria, Egypt

L. Király
Centre for Agricultural Research, Plant Protection Institute, Budapest, Hungary

H. Moawad
Agriculture Microbiology Department, National Research Centre, Cairo, Egypt

 
N. El-Wakeil, A. Abdelkhalek (eds.), *Microbiomes and Plant Secondary Metabolites for Insect Pest Management*, Progress in Biological Control 23,
https://doi.org/10.1007/978-3-032-24330-0_16

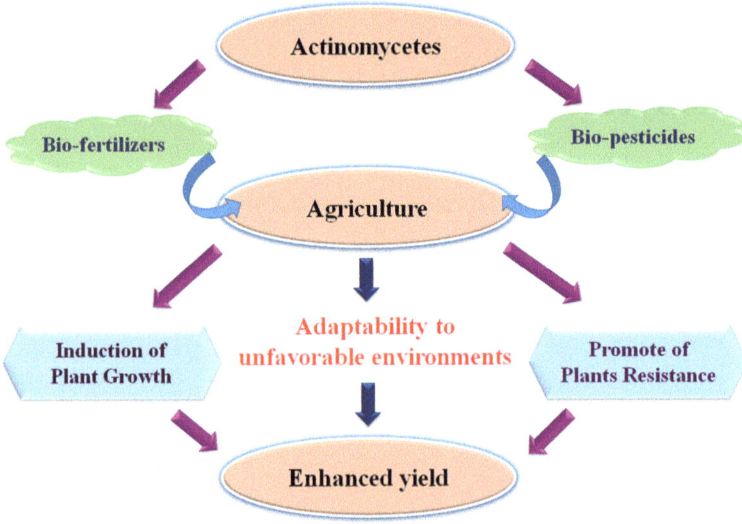


**Fig. 16.1** An illustration of how actinomycetes could potentially boost agricultural yield

sustainable solutions has driven the investigation of biological methods, among which actinomycetes have emerged as a particularly promising avenue for the holistic management of plant viruses.

Beneficial microorganisms have garnered significant interest for their effective and safe strategies in controlling viral plant diseases (Benaissa 2024). A significant proportion of biocontrol agents employed in the management of viral plant diseases are derived from the *Streptomyces* genus, encompassing around 780 species of actinobacteria and 30 subspecies (Al-Quwaie 2024). Actinobacteria are known to be effective biofertilizers and biocontrol agents for plant diseases (Abo-Zaid et al. 2020). They have shown promise as an alternative to chemical pesticides for controlling plant pathogen infections and increasing crop yields (Fig. 16.1). These species have the potential to combat these viruses effectively, resulting in increased crop yields and a reduced reliance on synthetic chemicals. This method is in harmony with the primary objectives of sustainable agriculture, highlighting the significance of investigating natural and ecological remedies to address plant viral infections. The complex relationships depicted reveal the diverse effects of viral infections in agricultural systems, emphasizing the pressing necessity for innovative and sustainable management strategies. Actinobacteria can mitigate plant diseases through several mechanisms. This encompasses the production of siderophores, hydrolytic enzymes such chitinases and glucanases, antibiotics, and hydrogen cyanide (Hajji-Hedfi et al. 2025; Sharma et al. 2025). Actinobacteria may enhance resistance to squash growth and postpone symptoms. Streptomyces spp. are significant in mitigating plant–pathogen interactions and have demonstrated efficacy in the biocontrol of fungal and bacterial diseases in plants (Wang et al. 2024). Foliar treatments on cucumbers with *Streptomyces albovinaceus* and *Streptomyces sparsogenes* achieved a 95% and 100% reduction in mosaic symptoms induced by the zucchini yellow mosaic virus, respectively (Askora 2005). *Streptomyces*, known for producing many bioactive metabolites, can function as biological control agents (Khan et al. 2023). Nonetheless, the antiviral effectiveness of several actinomycete species remains under examination. Understanding the complicated interactions between actinomycetes and plant systems allows researchers to develop novel biological control methods that lessen reliance on synthetic pesticides. These strategies not only promote sustainable agriculture but also protect environmental health, in line with global initiatives for environmentally friendly farming techniques. This combination of biology and agriculture highlights actinomycetes' promise in the broader context of plant disease management. The goal of this book chapter is to highlight the potential of different actinomycetes species to be applied as a biocontrol agents against plant viral infections.

## 16.2 Diversity and Classification of Actinobacteria

Actinobacteria are characterized by their high G+C DNA content and are classified as Gram-positive bacteria. They represent one of the most significant groups among the primary lineages recognized within the domain of bacteria (Călugăru and Țucă 2024). The genera within this phylum display a remarkable range of diversity regarding their morphology, physiology, and metabolic functions. There are 18 main lineages in the domain of bacteria. The phylum Actinobacteria makes up a large taxonomic unit. It is made up of six subclasses: *Rubrobacteria*, *Thermoleophilia*, *Coriobacteriia*, *Acidimicrobiia*, *Nitriliruptoria*, and *Actinobacteria* (Negi et al. 2024). The class Actinobacteria is subdivided into two orders: *Bifidobacteriales* and *Actinomycetales*. According to molecular and chemical makeup, the order *Actinomycetales* is split into 14 suborders. These are *Actinomycineae*, *Actinopolysporineae*, *Catenulisporineae*, *Corynebacterineae*, *Frankineae*, *Glycomycineae*, *Jiangellineae*, *Kineosporineae*, *Micrococcineae*, *Micromonosporineae*, *Propionibacterineae*, *Pseudonocardineae*, *Streptomycineae*, and *Streptosporangineae* (Mohammadipanah and Dehhaghi 2017). Most actinobacteria are free-living organisms that can be found in a wide range of terrestrial and aquatic (including marine) ecosystems. Nonetheless, the phylum has shown remarkable adaptability to various ecological environments: Actinomycetes can be found in soils, fresh and salt water, and even in the air. These organisms are more prevalent in soils than in other environments, especially in alkaline soils and those rich in organic matter, where they contribute significantly to the microbial community.

Actinobacteria are present not only on the soil surface but also at depths exceeding 2 meters below ground (Zahr et al. 2024). Some examples are *Streptomyces*, *Micromonospora*, *Rhodococcus*, and *Salinispora*. Some types, such as *Frankia* spp., coexist with plants, while others, like *Corynebacterium*, *Mycobacterium*, or *Nocardia* species, pose a threat to plants or animals. Other types live in the gut and help the body work (like *Bifidobacterium* spp.) (Barka et al. 2016). The members of this group that have been studied the most are soil-dwelling *Streptomyces* spp., which are known for producing a lot of antibiotics, and important human pathogens from the genus *Mycobacterium* (*M. tuberculosis* and *M. leprae*), which cause the most deaths from bacterial infections (Barka et al. 2016; Nobs 2024). Nonetheless, the genera *Streptomyces* and *Mycobacterium* represent just two of the numerous genera within this extensive phylum, which encompasses over 300 genera. Furthermore, there exist vast populations of inadequately researched actinobacteria that thrive in various habitats, including soil, water, deep-sea regions, and extreme environments like arctic ice, chemically contaminated areas, and radioactive sites. These microorganisms may also coexist with humans, animals, and plants, either in a symbiotic or antagonistic manner (Gao and Gupta 2012).

## 16.3 Actinobacteria as a Source of Bioactive Compounds

Actinobacteria have many important metabolic compounds for humans, such as antifungals, antibiotics, anticancer drugs, and other secondary metabolites that can be utilized in the medical sector or to enhance plant growth and disease resistance. (Sağlam et al. 2024). They also exhibit biocontrol capabilities against pests and function as plant growth promoters. The screening, isolation, and characterization of prospective actinomycete strains that produce potential secondary metabolites have been a significant focus of research for numerous groups globally for many years (Anavadiya et al. 2024). The *Streptomyces*, *Saccharopolyspora*, *Amycolatopsis*, *Micromonospora*, and *Actinoplanes* are the most prolific producers of commercially valuable biomolecules among *Actinomycetes* (Goel et al. 2024). Actinomycetes, particularly *Streptomyces* species, are recognized as industrially significant microorganisms due to their production of numerous secondary metabolites, many of which serve as useful antibiotics, antifungals, antiparasitics, herbicides, pesticides, anticancer or immunosuppressants, and essential industrial enzymes (Table 16.1).

Actinobacteria possess the ability to produce a diverse range of volatile compounds, antibiotics, and toxins. *Streptomyces* exhibit the capability to generate a diverse array of secondary metabolites, including those with antifungal and antimicrobial characteristics that actinomycetes can utilize to protect against plant diseases (Al-Quwaie 2024). Actinomycetes produce antibiotics such as tetracycline, spectinomycin, erythromycin, neomycin, and nystatin, which constitute nearly 80% of the total antibiotics available (Torres-Rodriguez et al. 2022). Furthermore, many compounds that share antibacterial properties with phytopathogenic fungi have been identified. These include amphotericin B, neopeptine, macrolides, natamycin, actinomycin D, and antimycin. The macrolides, a class of antibiotics produced by actinomycetes, prevent microorganisms from producing proteins (Vázquez-Laslop and Mankin 2018). By attaching itself to ergosterol in the fungal cell membrane, amphotericin B changes how cells can pass through and kill them (Hartmann et al. 2021). Furthermore, *Streptomyces* sp. strains that produce actinomycin D restrict

**Table 16.1** Various actinomycete species and their secondary metabolites have suitable uses in multiple fields

| Actinomycetes | Compound/bioactive metabolite | Activity/application |
|---|---|---|
| *Streptomyces griseus* | Streptomycin | Antibacterial |
| *Streptomyces aureofaciens* | Tetracycline | Antibacterial |
| *Saccharopolyspora erythraea* | Erythromycin | Antibacterial |
| *Cephalosporium acremonium* | Pyrroles | Antibacterial |
| *Streptomyces* sp. *YIM64018* | Peptide-cyclo | Antibacterial |
| *Micromonospora* sp. *GMKU326* | Spirotetronate Polyketides | Antibacterial |
| *Microbispora* sp. LGMB259 | β-Carboline | Antimicrobial and anticancer |
| *Streptomyces* sp. MA-12 | *di-O-prenylated flavone* | Antifungal |
| *Streptomyces griseus* | Candicidin | Antifungal and antiviral |
| *Streptomyces parvulus* | Actinomycin D | Antifungal and antiviral |
| *Streptomyces aureofaciens* | Arylcoumarins | Antifungal and antitumor |
| *Streptomyces peucetius* | Doxorubicin | Anticancer |
| *Streptomyces caespitosus* | Mitomycin | Anticancer |
| *Streptomyces* sp. *CA-129531* | Trichostatin derivatives | Tyrosinase inhibitors |
| *Streptomyces* sp. *M4* | Salvianolic acid B | Antifungal |
| *Streptomyces hygroscopicus TP-A0451* | δ-Lactone | Antitumor |
| *Streptomyces aureoverticillatus* | Aureoverticillactam | Antitumor |
| *Streptomyces hygroscopicus* | Rapamycin | Immunosuppressive |
| *Streptomyces avermitilis* | Phthoxazolin A | Antimicrobial |
| *Streptomyces antibioticus* | Quercetin 3-O-glucoside | Antimicrobial |

RNA synthesis and microbial proliferation (Xia et al. 2022). Antimycin acts as an inhibitor of the electron transport chain, specifically targeting the interaction between mitochondrial cytochromes b and c (Young et al. 2018). Fungal growth is suppressed when natamycin binds to the ergosterol present in the fungal cell membrane (Meena et al. 2021). Neopeptine enzymatically inhibits the synthesis of microbial cell walls (Zhang et al. 2020). A notable element of antibiotic synthesis is the relationship between actinomycetes and plants (Sharma et al. 2019). The plant exudates facilitate the growth of actinomycetes, while the antibiotic provides protection against phytopathogens. Reports indicate that actinomycetes spores may be found in the atmosphere. Furthermore, it has been proposed that airborne actinomycetes, including *Nocardia* spp., exhibit antibacterial properties (Seratnahaei et al. 2022). Scientists have found that certain bacteria living in the digestive tracts of insects can make a number of different bioactive compounds. This group comprises a variety of antibiotics such as alpiniamide A, streptanoate, alteramides A and B, coproporphyrin III, demethylenenocardamine, deferoxamine, surugamides A, B, C, D, and E, dihydropicromycin, nocardamine, valinomycin, tirandamycins A and B, as well as picromycin (Li et al. 2022; Al-Quwaie 2024).

Volatile organic compounds, commonly referred to as VOCs, are substances characterized by their low molecular weight, enabling them to evaporate readily under typical temperature and pressure conditions. This enables their distribution throughout both the soil and the atmosphere (Batty et al. 2024). Numerous *Streptomyces* species have been shown to produce it, demonstrating significant antimicrobial properties (Lu et al. 2024). The majority of VOCs exhibit lipid solubility, resulting in their low solubility in water. These VOCs traverse extensive distances within structurally diverse environments, including solid, liquid, or gaseous states (Carmona-Hernandez et al. 2019). Actinomycetes produce volatile organic compounds with various applications: They enhance plant growth, inhibit phytopathogenic fungi, exhibit nematocidal properties, and can trigger systemic resistance in plants (Abo-Zaid et al. 2021; Torres-Rodriguez et al. 2022). Dimethyl disulfide and 2-ethyl-5-methylpyrazine, which are produced by the *Streptomyces* species, prevent spore germination and mycelial growth (Gong et al. 2022). Some actinomycetes, like *Nocardiopsis* sp., make VOCs that stop fungi from growing mycelium. These VOCs include 1,2-dimethyldisulfane, S-methyl ethanethioate, 2-methyl propanoic acid, 3-methyl-butanoic acid, acetic acid, undecan-2-one, 2-isopropyl-5-methylcyclohexan-1-ol, and nonan-2-one. Ethers, alcohols, alkenes, furans, aldehydes, alkanes, and aromatic hydrocarbons were produced. Carvacrol, a compound identified as phenol, 2-methyl-5-(1-methylethyl), derived from *Streptomyces griseus*, exhibits cytotoxic properties against various phytopathogens (Danaei et al. 2014). The VOCs cyclohexanol, decanol, 2-ethyl-1-hexanol, nonanol, benzothiazole, and dimethyl trisulfide were found to inhibit the growth of *Sclerotinia sclerotiorum* mycelial cells and the germination of spores (Al-Quwaie 2024).

Actinobacteria generate diverse enzymes, including cellulases, amylases, and proteases, which find applications across industries such as food, textiles, and detergents (Nargotra et al. 2024). Furthermore, they play a role in the biodegradation of organic matter, aiding in environmental remediation efforts. Various bioactive compounds generated by Actinobacteria exhibit antimicrobial and antifungal characteristics, proving beneficial in agriculture for managing plant diseases (Silva et al. 2022). These lytic enzymes, which include chitinase, β-1,3-glucanase, and protease, break down the fungal cell wall, damage the integrity of the membrane, release parts inside the cell, and cause the cell to die (Cabrera et al. 2020). The microbial cell wall, composed of chitin, β-1,3-glucan, and protein, ensures the physical integrity of the cell. Some enzymes, called β-1,3 glucanases and chitinases, break down β-1,3 glucan β-D-glycosidic links and β-1,4 N-acetyl-β-D-glycosamide linkages, respectively (Torres-Rodriguez et al. 2022). This breaks down the cell walls of fungi. Proteases break down proteins, specifically mannoproteins, which make up the cell wall of phytopathogenic fungi. Siderophores enhance their transport into cells by binding to iron ions of low molecular weights (500–1000 Da) and establishing a connection with $Fe^{3+}$ ions (Maret 2024). They are released when there is not enough $Fe^{3+}$ available (Yan et al. 2025). Siderophores can be categorized into phenolate and catecholate types. Some siderophores display a mixed group composition, including hydroxamate and carboxylate (Deb and Tatung 2024). *Streptomyces* strains that make hydroxamate-type siderophores, specifically deferoxamine, have been shown to make siderophores. In addition, *Rhodococcus erythropolis* has been found to contain heterobactins, a type of mixed-type siderophores with catecholate and hydroximate properties (Retamal-Morales et al. 2021). In addition, albisporachelin is recognized as a hydroxamate-type siderophore synthesized by *Amycolatopsis albispora* (Wu et al. 2018). The amount of $Fe^{3+}$ that phytopathogenic fungi can get is limited when biocontrol agents produce enough siderophores. Because of this, microorganisms cannot grow or infect others as easily because the lack of iron in their environment stops important processes like respiration, transporting nutrients from photosynthesis, reducing nitrates, and getting rid of free radicals (Hernández Montiel et al. 2018).

## 16.4 Actinobacteria as Plant Growth-Promoting Agents

Actinobacteria can boost plant growth by producing phytohormones such as auxins, cytokinins, and gibberellins, which influence root development, cell division, and flowering (Abdellatif et al. 2024). These hormones can directly stimulate plant development by lengthening roots, boosting nutrient absorption, and accelerating the growth of shoots and leaves. *Streptomyces rochei*, *Streptomyces rimosus*, and *Streptomyces olivaceoviridis* are among the species in the *Streptomyces* genus that can produce exogenous auxins, such as indole-3-acetic acid (IAA), gibberellins, and cytokinins. Research indicates that *Streptomyces griseus* can be used to treat barley, oat, wheat, and carrot seeds to encourage their growth (Hungund et al. 2022). The culture filtrate of *Streptomyces* species plays a crucial role in enhancing shoot length, while the impact of *S. olivaceoviridis* on the increase in shoot mass in wheat plants is regarded as noteworthy. *S. rochei* and *Streptomyces cellulosae* have been recognized as significant sources of growth regulators that influence plant growth (Aldesuquy et al. 1998; Abo-Zaid et al. 2020). Certain actinobacteria can produce or change ethylene levels in plants, influencing root growth, stress responses, and fruit ripening (Kavya et al. 2024). Actinobacteria significantly contribute to the decomposition of organic matter in soil, facilitating nutrient recycling. Their activity enhances soil fertility by decomposing complex organic molecules, so rendering them more accessible to plants (Kumar et al. 2024). Some types of actinomycetes, like *Streptomyces*, *Micromonospora*, and *Micrococcus*, can dissolve phosphate, which makes it easier for plants to use and helps them grow faster. These actinomycetes are capable of inhibiting the growth of the organism responsible for damping off. Consequently, these provide dual benefits and function as plant growth-promoting rhizobacteria.

Plant growth-promoting rhizobacteria (PGPR) affects plant growth in two ways: (a) directly, by giving the plant a compound called PGPR that helps it take in nutrients; and (b) indirectly, by protecting against the harmful effects of microorganisms by making antibiotics and other secondary metabolites (Fig. 16.2). Actinomycetes can dissolve phosphorus (P) by making organic acids in the rhizosphere. This decreases the pH, transforming phosphorus from an insoluble form to a soluble

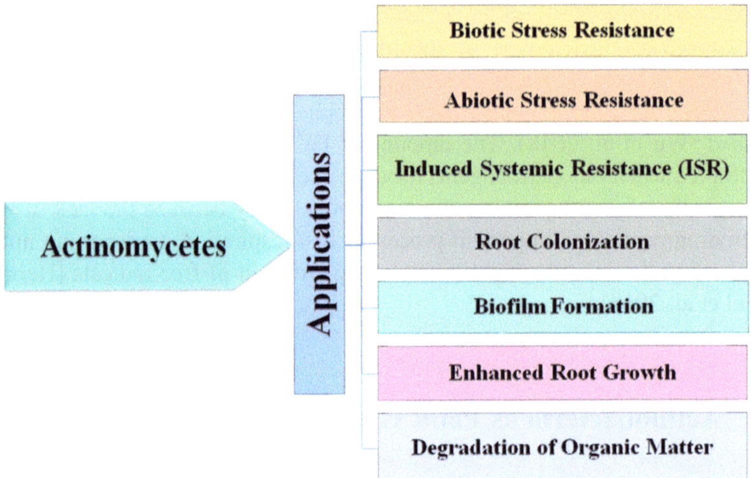


**Fig. 16.2** Actinomycetes with diverse applications as plant growth-promoting agents

one (Hungund et al. 2022; Zhu et al. 2024). Certain Actinobacteria, particularly those in the genus *Frankia*, establish symbiotic relationships with non-leguminous plants, converting atmospheric nitrogen into a bioavailable form for growth in plants (Bhattacharyya et al. 2024). Utilizing actinobacteria as biofertilizers presents numerous benefits, such as promoting sustainable agricultural practices, boosting plant growth and yield, enhancing soil health, and minimizing environmental pollution (Wei et al. 2024). Plants utilize actinomycetes to enhance their iron availability. Iron exists in soil primarily as ferric, which is its insoluble form represented by $Fe^{3+}$. Unlike ferric iron, plants take up iron as ferrous iron (soluble $Fe^{2+}$). Certain actinomycetes, including actinobacteria, can directly convert insoluble forms into soluble ones (Pan et al. 2024).

## 16.5 Actinomycetes as Biological Antiviral Control Agents

Using actinomycetes to control plant viruses biologically has many benefits for sustainable agriculture. Unlike chemical pesticides, actinomycetes are naturally occurring microorganisms that do not harm the environment or other living things (Trojak-Goluch 2024). Actinomycetes offer a sustainable source of bioactive substances that can be utilized for plant protection, hence decreasing reliance on synthetic chemicals (Bhatti et al. 2017). Given that actinomycetes generate a diverse array of bioactive compounds, the likelihood of plant pathogens developing resistance is diminished in comparison with traditional chemical treatments (Diab et al. 2024). Actinomycetes can be included in integrated pest management techniques in conjunction with other biological control agents, such as fungi or insects, and cultural practices to improve overall plant health (Abdellatif et al. 2024). Actinomycetes are soil-residing microorganisms that synthesize diverse bioactive compounds. They are distinguished for their antimicrobial capabilities and capacity to generate chemicals that impede the proliferation of many diseases, including viruses. *Actinomycetes* genera like *Streptomyces*, *Micromonospora*, *Nocardia*, and *Actinoplanes* are well known for making chemicals that may be able to fight viruses (Muazi Alenazi et al. 2023). These chemicals include antibiotics, enzymes, and secondary metabolites. Some actinomycetes can produce secondary compounds that prevent plant viruses from replicating. These substances may work by stopping the reproduction of the viral genome, inhibiting viral enzymes, or obstructing viral entry into plant cells (Trojak-Goluch 2024). Certain actinomycetes boost plant immunity by promoting the synthesis of defense-related enzymes, including peroxidases, chitinases, and glucanases (Ebrahimi-Zarandi et al. 2022). By strengthening cell walls, breaking down viral particles, or triggering defensive mechanisms, these enzymes aid plants in fending off viral infections. Actinomycetes are prolific biochemical factories, recognized for their capacity to generate a vast array of antimicrobial secondary metabolites. This capability directly supports their role in disease management. Notably, actinomycetes can inhibit the proliferation of viral vectors, such as insects, as well as fungal and bacterial pathogens. They achieve this by producing specific antimicrobial and insecticidal compounds that reduce vector populations and interfere with the transmission of viruses to host plants (Dhanasekaran and Thangaraj 2014). This positions them as a powerful tool within the biological control paradigm. Actinomycetes have the potential to engage with other microorganisms, including bacteria and fungi, to establish a biological barrier that can protect against plant viruses (Khan et al. 2023). Certain actinomycetes, for instance, can outcompete dangerous pathogens for resources and available space, thereby lowering the prevalence of viral infections. Marine actinomycetes hold promise as a source of various novel bioactive compounds, with marine organisms exhibiting distinct metabolic and physiological characteristics compared to terrestrial species (Lam 2006). Research has shown the antiviral properties of actinomycetes, such as *Streptomyces* sp. ZX01, *Streptomyces ahygroscopicus*, and *Streptomyces noursei* var. *xichangensis*, in combating tobacco mosaic virus infection (Manjunatha et al. 2022). The study investigated the efficacy of metabolites derived from 28 isolates of marine actinomycetes in combating the cucumber mosaic virus (Latake and Borkar 2017). *Streptomyces olivaceus* metabolite is particularly effective at treating CMV infections in the laboratory and open fields when used as seed and spray treatments separately. After applying a mixture of *Streptomyces* culture filtrates and virus sap to *Nicotiana glutinosa* and *Chenopodium quinoa*, the quantity of TMV and potato virus Y (PVY) necrotic lesions decreased (Mohamed and Galal 2005). Observations indicate that TMV can potentially be eradicated in *Nicotiana glutinosa* and *N. tabacum* through the combination of various elements, including fermented actinomycetes broth and TMV sap, over different time intervals (Zhang Xing et al. 2012). The antiviral compound ε-poly-L-lysine (ε-PL) from *Streptomyces ahygroscopicus*, with a molecular weight of 3454–4352 Da, was reported to be effective against TMV (Chen et al. 2019).

Actinomycetes, particularly those from the Streptomyces genus, may serve as a promising biological control agent for plant viral infections through soil treatment or foliar applications of their culture filtrates (Table 16.2). The application of *Streptomyces ovatisporus* LC597360 via foliar treatment on tomato plants demonstrated a biocontrol efficacy of 93.9% against tomato mosaic virus (ToMV) and resulted in a notable reduction in symptom severity, quantified at 17.6 ± 0.8%, when compared to infected control plants (Taha et al. 2021). It has been shown that applying *Streptomyces albovinaceus* and *Streptomyces sparsogenes* to the leaves of cucumbers lowers the number of mosaic symptoms caused by the zucchini yellow mosaic virus (ZYMV) to 95% and 100%, respectively (Askora 2005). Research has shown that diverse bioactive substances derived from different *Streptomyces* strains can mitigate TMV damage to *Datura metel* foliage (Ara et al. 2012). It was reported that the pre-inoculating tomato plants with Streptomyces pactum Act12 led to a 37.9% reduction in TYLCV (Li et al. 2019). The secondary metabolites compounds present in the culture filtrates of *Streptomyces* spp. demonstrated antiviral activity against the potato virus Y (PVY) and the TMV (Mohamed and Galal 2005). A cell-free suspension of *Streptomyces rochei* was utilized to assist *Datura metel* leaves in their recovery from TMV (Mansour et al. 1988). Specific *Streptomyces* strains inoculated in plants may lead to the development of systemic resistance against CMV infection (Shafie et al. 2016). A notable reduction in the severity of the banana bunchy top virus was observed in banana plants treated with *Streptomyces chinensis* ten days before viral inoculation (Hewedy et al. 2008). The application of culture filtrate from *Streptomyces albovianaceus* and *Streptomyces sparsogenes* to *Cucumis sativus* resulted in a significant reduction of zucchini yellow mosaic virus symptoms by 95–100% (Ghaly et al. 2005).

## 16.6 Diverse Mechanisms of Action Against Plant Viral Infections

Actinomycetes produce a diverse array of antimicrobial compounds with potential antiviral activity. However, the efficacy of specific metabolites depends on their mode of action. For instance, well-characterized antibiotics like actinomycin D and daunomycin primarily target DNA processes. Consequently, they offer limited direct utility against the majority of plant viruses, which are RNA-based. This underscores the importance of screening for compounds with direct RNA-targeting mechanisms or strong host-priming activity. Actinomycetes employ many techniques to suppress the virus. These techniques encompass the direct targeting of virus particles by antiviral agents, the activation of plant defense responses, and the production of signaling molecules (Fig. 16.3).

**Table 16.2** List of reported actinomycete species and their antiviral activity

| Actinomycetes bioagent | Plant virus | Host plant | References |
|---|---|---|---|
| *Streptomyces* sp. ZX01 | Tobacco mosaic virus | *Nicotiana glutinosa* | Zhang et al. (2015) |
| *Streptomyces noursei* var. *xichangensis* | Tobacco mosaic virus | *Nicotiana tobacum* | Han et al. (2014) |
| *Streptomyces olivaceus* | Cucumber mosaic virus | *Cucumis sativus* | Latake and Borkar (2017) |
| *Streptomyces erythraeus, Streptomyces naganishii, Streptomyces michigansis* | Tobacco mosaic virus | *Nicotiana glutinosa* | Mohamed and Galal (2005) |
| | Potato virus Y | *Chenopodium quinoa* | |
| *Actinomycete* strain ZX01 | Tobacco mosaic virus | *Nicotiana glutinosa, Nicotiana tabacum* | Xing et al. (2012) |
| *Streptomyces ahygroscopicus* | Tobacco mosaic virus | *Nicotiana glutinosa* | Chen et al. (2019) |
| *Streptomyces cellulosae* | Tobacco mosaic virus | *Solanum lycopersicum* | Abo-Zaid et al. (2020) |
| *Streptomyces albulus* SN40 | Tobacco mosaic virus | *Nicotiana glutinosa* | Shu et al. (2024) |
| | Potato virus Y | *Nicotiana benthamiana* | |
| *Streptomyces netropsis, Streptomyces ambofaciens, Streptomyces ambofaciens* | Potato virus Y | *Solanum tuberosum* | Nasr-Eldin et al. (2019) |
| *Streptomyces rochei* | Tobacco mosaic virus | *Datura metel* | Mansour et al. (1988) |
| *Streptomyces griseorebens, Streptomyces cavourensis* | Cucumber mosaic virus | *Cucumis sativus* | Shafie et al. (2016) |
| *Streptomyces chibaensis* | Banana bunchy top virus | Banana | Hewedy et al. (2008) |
| *Streptomyces albovianaceus, Streptomyces sparsogenes* | Zucchini yellow mosaic virus | *Cucumis sativus* | Ghaly et al. (2005) |
| *Streptomyces aureocirculatus, Streptomyces rochei, Streptomyces cacaoi* | Tobacco mosaic virus | *Nicotiana glutinosa* | Akl and El-Masry (2024) |
| *Streptomyces ovatisporus* | Tomato mosaic virus | *Solanum lycopersicum* | Taha et al. (2021) |
| *Streptomyces pactum* Act12 | Tomato yellow leaf curl virus | *Lycopersicon esculentum* | Li et al. (2019) |
| *Streptomyces fradiae* | Potato virus Y | *Solanum tuberosum* | Mohammed et al. (2024) |
| *Streptomyces* sp. | Tobacco mosaic virus | *Datura metel* | Ara et al. (2012) |

(continued)

**Table 16.2** (continued)

| Actinomycetes bioagent | Plant virus | Host plant | References |
|---|---|---|---|
| *Streptomyces carpaticus* | Cucumber mosaic virus | *Solanum lycopersicum* | Bataeva et al. (2024) |
| *Streptomyces chromofuscus* | Tomato yellow leaf curl virus | *Nicotiana benthamiana* | Chen et al. (2022) |
| *Streptomyces albovinaceus, Streptomyces sparsogenes* | Zucchini yellow mosaic virus | *Cucumis sativus* | Askora (2005) |
| *Streptomyces gibsonii* | Tobacco necrosis virus | *Phaseolus vulgaris* | Shoman et al. (2003) |

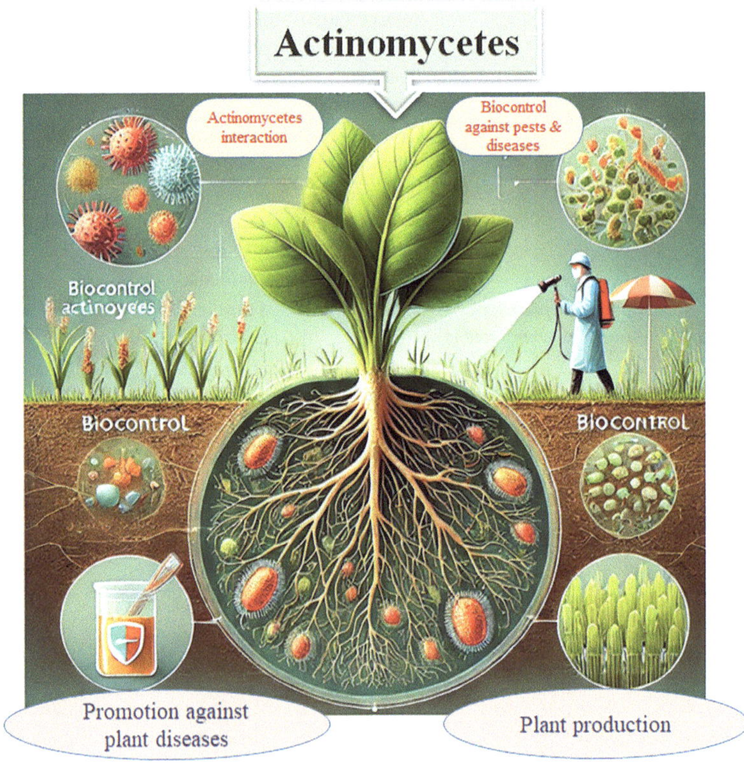


**Fig. 16.3** Actinomycetes with diverse applications in plant protection

### 16.6.1 Directly Targeting Viral Particles with Antiviral Agents

Antibiotics have long been considered to have a crucial impact on plant pathology, especially in plant virology (Manjunatha et al. 2022). Ningnanmycin, often known as NNM, is an antiviral agent that was initially isolated from *Streptomyces noursei* var. *xichangensis* (Chen et al. 2019). It is a novel cytosine nucleoside peptide antibiotic that effectively inhibits TMV infections and is extensively utilized in agricultural applications (Han et al. 2014). It has been observed that NNM interferes with the assembly of the TMV-CP through direct binding to multiple residues. These interactions break down the CP from discs into monomers, resulting in a nearly complete reduction in viral pathogenicity (Li et al. 2017). Although the underlying processes are unknown, NNM has been found to effectively suppress PVY infection in *Nicotiana benthamiana.* A recent study employing PVY movement-deficient mutants found that ningnanmycin inhibits PVY replication and binds to the PVY-encoded coat protein at a strength of 1.34 μmol/L. This interaction involves important amino acids $Glu^{168}$ and $Thr^{206}$ (Wu et al. 2024). In China, the cytosine peptidemycin, which was isolated from *Streptomyces ahygroscopicus*, has been commercialized, registered, and promoted as an anti-plant-viral agent due to its effective virucidal action (Wang et al. 2006). Furthermore, *Streptomyces* sp. ZX01 yielded a novel glycoprotein, GP-1, with an 80% anti-TMV rate at a dose of 1 mg/mL (Zhang et al. 2015). In addition, a bioactive substance called ε-poly-L-lysine (ε-PL) that is released by *Streptomyces ahygroscopicus* has shown strong TMV protection and healing abilities (Chen et al. 2019).

Anisomycin, one secondary metabolites of *Streptomyces* SN40, has recently been identified as an antiviral agent that successfully inhibits the infection of TMV and PVY in *Nicotiana glutinosa* and *Nicotiana benthamiana*, respectively (Shu et al. 2024). Molecular docking showed that anisomycin might be able to attach to the helicase domain of TMV replicase, the TMV coat protein, and the PVY helper component proteinase. Furthermore, studies have shown that a novel glycoprotein (GP-1) obtained from *Streptomyces kanasensis* exhibits a notable inhibitory effect on the transmission of TMV (Han et al. 2015). Actinomycin D, derived from various *Streptomyces* species, has been documented to suppress multiple plant viruses, including cowpea mosaic virus, cowpea chlorotic mottle virus, and tobacco mosaic virus (Lockhart and Semancik 1968; Turner and Dawson 1984). Comparable to actinomycin D, daunomycin exhibited minimal effects on virus-infected plants, impacting neither viral RNA synthesis nor the host plant (Manjunatha et al. 2022). This antibiotic was believed to have a direct impact on the virus particles. Research has shown that the antibiotics actinomycin D and mitomycin C influence nucleic acid metabolism and create complexes with DNA. In the initial phases of infection involving particular virus–host combinations, there is no hindrance to synthesizing plant pathogenic RNA viruses (Schwartz et al. 1965). Research indicates that the antibiotics mitomycin C and actinomycin D can bind to DNA and affect nucleic acid metabolism. Except for certain virus–host combinations during the initial phases of infection, they do not inhibit the production of the RNA virus responsible for causing disease in plants (Dawson and Schlegel 1976). Recent findings indicate that exopolysaccharides sourced from actinomycetes demonstrate significant promise as antiviral candidate compounds, aiding in the creation of new antiviral agents and offering insights for advanced antiviral management approaches (Wei et al. 2025).

Certain antimicrobial compounds from actinomycetes, such as actinomycin D and daunomycin, are potent inhibitors of DNA replication and transcription. While their primary mode of action targets DNA, which is not directly relevant to most plant viruses (which possess RNA genomes), they have been observed in some studies to exhibit indirect or secondary inhibitory effects on RNA virus replication. This suggests their utility may be limited or specific to particular viral contexts, highlighting the need to match the compound's mechanism with the pathogen's biology.

### 16.6.2 Activation of Plant Defense Responses

Actinomycetes have the potential to manage viral infections in plants by eliciting systemic resistance. Plants possess mechanisms to protect themselves from viral infections, primarily through systemic acquired resistance and induced systemic resistance (Abo-Zaid et al. 2020). Enzymes associated with defense are often considered the primary agents responsible for triggering systemic acquired resistance through beneficial microbes. *Streptomyces* enhances the activity of antioxidant enzymes in plants, including superoxidase, catalase, peroxidase, and chitinase, thereby improving their defense mechanisms against viral infections (Chen et al. 2022). Recently, actinomycetes have gained attention for their potential to effectively manage plant viral infections safely and environmentally sustainable (Rakhonde et al. 2025). There is evidence that *S. pactum Act12* prompted tomato plants to establish systemic resistance and increased peroxidase activity in response to the tomato yellow leaf curl virus. The observed resistance was characterized by elevated levels of salicylic and jasmonic acids, alongside the induction of PR-1 and increased activities of POD and PAL in the tomato leaves of the plants (Li et al. 2019). The application of *Streptomyces* spp. has resulted in the induction of peroxidase activity, serving as a defense enzyme in response to CMV (Shafie et al. 2016). Furthermore, it was noted that *Streptomyces* spp. promotes plant resistance to tobacco necrosis virus, characterized by an elevated level of chitinase enzymes as one of the pathogenesis-related proteins (Shoman et al. 2003). Ningnanmycin has been shown to play a significant role in enhancing resistance in TMV-infected plants by activating various plant defense mechanisms. This includes the induction of peroxidase, phenylalanine ammonia-lyase, and superoxide dismutase activities, as well as the activation of a group of acidic PR proteins. Additionally, the expression of NPR1 and Jaz3 contributes to the protective effects against TMV (Han et al. 2014). The foliar application of *Streptomyces cellulosae* isolate Actino 48 cell suspension clearly triggered the systemic resistance mechanism in tomatoes 48 hours before TMV infection. This resulted in a decrease in viral accumulation and symptoms within the infected plant tissues. Notable elevations in the activity of antioxidant enzymes, including peroxidase, along with the activation of transcription factors that enhance the expression levels of PAL, PR-1, and CHS, indicated the induction of systemic resistance (Abo-Zaid et al. 2020).

Various plant species from various families have defense-related enzymes, and it is proposed that activation of these enzymes is required for adaptation to biotic stresses. Furthermore, enzymes involved in defense in uninfected plant tissues are critical in preventing future viral infections and increasing the plant's ability to tolerate viral dissemination and replication (Taha et al. 2021). Reports indicate that the activity levels of peroxidase, polyphenol oxidase, and catalase were markedly elevated in the treated plants with *S. ovatisporus* LC597360 culture filtrate compared to the control, untreated plants. The findings clearly indicate that tomato plants subjected to treatment with *S. ovatisporus* demonstrated significantly elevated levels of antioxidant enzyme activity and protein content. This enhancement could play a crucial role in defending against infection by the tomato mosaic virus by promoting systemic resistance and encouraging plant development. It was noted that enzymes associated with defense mechanisms play a crucial role in the elicitation of systemic acquired resistance by *S. ovatisporus*, due to their effectiveness in addressing viral infections in plants while maintaining environmental safety (Taha et al. 2021). The *Streptomyces fradiae* QD3 filtrate appears to serve dual purposes as a protective and curative approach to PVY, demonstrating its ability to reduce disease symptoms and lessen virus infection both before and after PVY exposure. Species of *Streptomyces* are noted for their capacity to induce systemic resistance. The filtrate from *S. fradiae* appears to impede viral replication in the plant, as indicated by the notable reduction in PVY observed in both pre- and post-infection treatment groups (Mohammed et al. 2024). The analysis revealed that the majority of active compounds present in the crude actinobacteria filtrates originated from fatty acids and their derivatives, including Citroflex A, hexanoic acid, anhydride, hexadecanoic acid, ethyl ester, and diisooctyl phthalate. These compounds may serve as elicitor molecules that trigger the induction of systemic acquired resistance. The suggestion was made that utilizing the culture filtrate of *Streptomyces ovatisporus* LC597360 to address tomato mosaic virus could serve as an environmentally sustainable approach (Nasr-Eldin et al. 2019). The utilization of *Streptomyces* spp. exhibited notable antiviral properties against ZYMV infection, promoted plant growth, and triggered systemic resistance, resulting in the suppression of virus replication. This treatment significantly elevated levels of proline, phenols, and enzymes associated with defense mechanisms, including catalase, superoxide dismutase, and glutathione peroxidase, even in the presence of the ongoing infection (Ghanem et al. 2023). Furthermore, in the context of tomato yellow leaf curl virus infection, two strains of *S. chromofuscus* significantly enhanced the activities of defense-related enzymes, including peroxidase, catalase, and superoxidase (Chen et al. 2022).

## 16.7 Conclusions and Future Perspectives

Actinobacteria represent a crucial reservoir of bioactive compounds that hold considerable potential for applications in medicine, agriculture, and industry. The continuous investigation of these bacteria is producing new and significant compounds that could tackle urgent global health and environmental issues. Innovations in biotechnology and genomics are anticipated to significantly improve the identification and application of these important microbial products. In summary, actinobacteria offer various advantages for plant growth by utilizing mechanisms such as nutrient mobilization, hormone production, biocontrol, and enhancing stress resistance. Their applications in agriculture enhance crop health, minimize reliance on synthetic chemicals, and foster sustainable farming methods. Actinomycetes present significant promise as biological control agents for plant viruses, providing an environmentally friendly and sustainable option compared to chemical pesticides. The capacity to generate antiviral metabolites and bolster plant resistance positions them as a valuable asset in strategies for plant protection. Nevertheless, further investigation is essential to determine the most effective species, enhance their production, and address challenges associated with their application in various agricultural environments. With ongoing advancements in studies, actinomycetes may play a crucial role in integrated pest management strategies aimed at combating plant viral diseases.

**Declaration of AI Assistance** The authors declare that AI-based tools were used only for minor editorial tasks, such as sentence restructuring and formatting adjustments, without contributing to the intellectual or analytical content of this chapter. The authors carefully checked and edited all AI-assisted outputs. The authors bear sole responsibility for the scientific accuracy and scholarly quality of the chapter. Peer review was conducted in accordance with the standards of Springer Nature.

## References

Abdelkhalek A, Elmorsi A, Alshehaby O, Sanan-Mishra N, Hafez E (2018) Identification of genes differentially expressed in onion infected with Iris yellow spot virus. Phytopathol Mediterr 57. https://doi.org/10.14601/Phytopathol_Mediterr-2187

Abdelkhalek A, Hafez E (2019) Plant viral diseases in Egypt and their control. In: Cottage industry of biocontrol agents and their applications: practical aspects to deal biologically with pests and stresses facing strategic crops. Springer, pp 403–421. https://doi.org/10.1007/978-3-030-33161-0_13

Abdellatif AAM, Gebily DAS, Elmaghraby MMK, Sahu PK, Thakur B, Kaur S (2024) Transforming roles of Actinobacteria in sustainable agriculture: from soil health and plant productivity perspective. In: Metabolomics, proteomics and gene editing approaches in biofertilizer industry, vol II. Springer, Singapore, pp 299–338

Abo-Zaid GA, Matar SM, Abdelkhalek A (2020) Induction of plant resistance against tobacco mosaic virus using the biocontrol agent Streptomyces cellulosae isolate Actino 48. Agronomy 10:1620

Abo-Zaid G, Abdelkhalek A, Matar S, Darwish M, Abdel-Gayed M (2021) Application of bio-friendly formulations of chitinase-producing Streptomyces cellulosae Actino 48 for controlling peanut soil-borne diseases caused by Sclerotium rolfsii. J Fungi 7:167

Akl B, El-Masry SS (2024) Genetic characterization of three soil Streptomyces isolates exhibiting antiviral activity against TMV; insights from 16S rRNA gene sequencing and RAPD analysis. Egypt Acad J Biol Sci G Microbiol 16:33–53

Al-Quwaie DA (2024) The role of Streptomyces species in controlling plant diseases: a comprehensive review. Australas Plant Pathol 53:1–14

Aldesuquy HS, Mansour FA, Abo-Hamed SA (1998) Effect of the culture filtrates of Streptomyces on growth and productivity of wheat plants. Folia Microbiol (Praha) 43:465–470

Anavadiya B, Chouhan S, Saraf M, Goswami D (2024) Exploring endophytic actinomycetes: a rich reservoir of diverse antimicrobial compounds for combatting global antimicrobial resistance. The Microbe 4:100110

Ara I, Bukhari NA, Aref NM, Shinwari MMA, Bakir MA (2012) Antiviral activities of streptomycetes against tobacco mosaic virus (TMV) in Datura plant: evaluation of different organic compounds in their metabolites. Afr J Biotechnol 11:2130–2138

Aseel DG, Ibrahim OM, Abdelkhalek A (2024a) Biosynthesized silver nanoparticles mediated by Ammi visnaga extract enhanced systemic resistance and triggered multiple defense-related genes, including SbWRKY transcription factors, against tobacco mosaic virus infection. BMC Plant Biol 24:756

Aseel DG, Rabie M, El-Far A, Abdelkhalek A (2024b) Antiviral properties and molecular docking studies of eco-friendly biosynthesized copper oxide nanoparticles against alfalfa mosaic virus. BMC Plant Biol 24:1089

Askora AA (2005) Antiphytoviral studies from certain actinomycetal isolates. Available at: Ph.D. Thesis, Zagazig University, Zagazig, Egypt

Barka EA, Vatsa P, Sanchez L, Gaveau-Vaillant N, Jacquard C, Klenk H-P et al (2016) Taxonomy, physiology, and natural products of Actinobacteria. Microbiol Mol Biol Rev 80:1–43

Bataeva Y, Delegan Y, Bogun A, Shishkina L, Grigoryan L (2024) Whole genome analysis and assessment of the metabolic potential of Streptomyces carpaticus SCPM-OB-9993, a promising phytostimulant and antiviral agent. Biology (Basel) 13:388

Batty CA, Pearson VK, Olsson-Francis K, Morgan G (2024) Volatile organic compounds (VOCs) in terrestrial extreme environments: implications for life detection beyond Earth. Nat Prod Rep

Benaissa A (2024) Rhizosphere: role of bacteria to manage plant diseases and sustainable agriculture—a review. J Basic Microbiol 64:2300361

Bhattacharyya PN, Islam NF, Sarma B, Nath BC, Al-Ani LKT, Lesueur D (2024) Frankia-actinorhizal symbiosis: a non-chemical biological assemblage for enhanced plant growth, nodulation and reclamation of degraded soils. Symbiosis 92:1–26

Bhatti AA, Haq S, Bhat RA (2017) Actinomycetes benefaction role in soil and plant health. Microb Pathog 111:458–467

Cabrera R, García-López H, Aguirre-von-Wobeser E, Orozco-Avitia JA, Gutiérrez-Saldaña AH (2020) Amycolatopsis BX17: an actinobacterial strain isolated from soil of a traditional milpa agroecosystem with potential biocontrol against Fusarium graminearum. Biol Control 147: 104285

Călugăru EDC, Țucă O (2024) Insects as source of actinobacteria. Ann Univ Craiova Biol Hortic Food Prod Process Technol Environ Eng 29

Carmona-Hernandez S, Reyes-Pérez JJ, Chiquito-Contreras RG, Rincon-Enriquez G, Cerdan-Cabrera CR, Hernandez-Montiel LG (2019) Biocontrol of postharvest fruit fungal diseases by bacterial antagonists: a review. Agronomy 9:121

Chen D, Ali MNHA, Kamran M, Magsi MA, Mora-Poblete F, Maldonado C et al (2022) The Streptomyces chromofuscus Strain RFS-23 induces systemic resistance and activates plant defense responses against tomato yellow leaf curl virus infection. Agronomy 12:2419

Chen J, Liu H, Xia Z, Zhao X, Wu Y, An M (2019) Purification and structural analysis of the effective anti-TMV compound ε-poly-L-lysine produced by Streptomyces ahygroscopicus. Molecules 24:1156

Danaei M, Baghizadeh A, Pourseyedi S, Amini J, Yaghoobi MM (2014) Biological control of plant fungal diseases using volatile substances of Streptomyces griseus. Eur J Exp Biol 4:334–339

Dawson WO, Schlegel DE (1976) The sequence of inhibition of tobacco mosaic virus synthesis by actinomycin D, 2-thiouracil, and cycloheximide in a synchronous infection. Phytopathology 66: 177–181

Deb CR, Tatung M (2024) Siderophore producing bacteria as biocontrol agent against phytopathogens for a better environment: a review. South African J Bot 165:153–162

Dhanasekaran D, Thangaraj R (2014) Microbial secondary metabolites are an alternative approaches against insect vector to prevent zoonotic diseases. Asian Pacific J Trop Dis 4: 253–261

Diab MK, Mead HM, Ahmad Khedr MM, Abu-Elsaoud AM, El-Shatoury SA (2024) Actinomycetes are a natural resource for sustainable pest control and safeguarding agriculture. Arch Microbiol 206:268

Ebrahimi-Zarandi M, Saberi Riseh R, Tarkka MT (2022) Actinobacteria as effective biocontrol agents against plant pathogens, an overview on their role in eliciting plant defense. Microorganisms 10:1739

Gao B, Gupta RS (2012) Phylogenetic framework and molecular signatures for the main clades of the phylum Actinobacteria. Microbiol Mol Biol Rev 76:66–112

Ghaly MF, Awny AM, Galal AM, Askora A (2005) Characterization and action of antiphytoviral agent produced by certain Streptomyces species against zucchini yellow mosaic virus. Egypt J Biotechnol 19:209–223

Ghanem GAM, Mahmoud AMA, Kheder AA, Gebily DAS, Osamy A (2023) Antiviral activities of three Streptomyces spp. against Zucchini yellow mosaic virus (ZYMV) infecting squash (Cucurbita pepo L.) plants. Egypt J Biol Pest Control 33:113

Goel N, Choudhary P, Pattanaik L, Padhi SK, Chaudhary P (2024) Exploring the potential of microbial biomolecules in advancing human health and environmental sustainability. In: Industrial microbiology and biotechnology: an insight into current trends. Springer, pp 515–566

Gong Y, Liu J-Q, Xu M-J, Zhang C-M, Gao J, Li C-G et al (2022) Antifungal volatile organic compounds from Streptomyces setonii WY228 control black spot disease of sweet potato. Appl Environ Microbiol 88:e02317–21

Hajji-Hedfi L, Balbool BA, Rhouma A, Hassan A, Dali S, Perniola R et al (2025) Valorization of native rhizospheric vineyards' fungi to enhance sustainable biocontrol against Downy Mildew in Tunisia. Sydowia 77:161

Han L, Zhang G, Miao G, Zhang X, Feng J (2015) Streptomyces kanasensis sp. nov., an antiviral glycoprotein producing actinomycete isolated from forest soil around kanas lake of China. Curr Microbiol 71:627–631

Han Y, Luo Y, Qin S, Xi L, Wan B, Du L (2014) Induction of systemic resistance against tobacco mosaic virus by Ningnanmycin in tobacco. Pestic Biochem Physiol 111:14–18

Hartmann DO, Shimizu K, Rothkegel M, Petkovic M, Ferraz R, Petrovski Ž et al (2021) Tailoring amphotericin B as an ionic liquid: an upfront strategy to potentiate the biological activity of antifungal drugs. RSC Adv 11:14441–14452

Hernández Montiel LG, Rivas García T, Romero Bastidas M, Chiquito Contreras CJ, Ruiz Espinoza FH, Chiquito Contreras RG (2018) Antagonistic potential of bacteria and marine yeasts for the control of phytopathogenic fungi. Rev Mex ciencias agrícolas 9:4311–4321

Hewedy MA, Husseiny SM, Salama MI, NourEldein HA, Abdelsalam SM (2008) Reduction in Banana bunchy top virus infution rate by using Streptomyces chibaensi filtrates. Pak J Biotechnol 5:51–61

Hungund BS, Desai SS, Kamath KC, Tennalli GB (2022) Terrestrial ecology of actinobacteria. In: Actinobacteria: ecology, diversity, classification and extensive applications. Springer, pp 39–54

İlbağı H, Pocsai E, Çıtır A (2024) Economic significance of viruses in field crops. In: Viral diseases of field and horticultural crops. Elsevier, pp 27–32
Junaid MD, Gokce AF (2024) Global agricultural losses and their causes. Bull Biol Allied Sci Res 2024:66
Kavya T, Govindasamy V, Suman A, Abraham G (2024) Plant–Actinobacteria interactions for biotic and abiotic stress management in crops. In: Plant holobiome engineering for climate-smart agriculture. Springer, pp 441–463
Khan S, Srivastava S, Karnwal A, Malik T (2023) Streptomyces as a promising biological control agents for plant pathogens. Front Microbiol 14:1285543
Kumar P, Mehta H, Chandel V, Chib P, Passari AK (2024) Microbial communities in rhizosphere microbiome for sustainable soil health and productivity. In: Plant holobiome engineering for climate-smart agriculture. Springer, pp 367–386
Lam KS (2006) Discovery of novel metabolites from marine actinomycetes. Curr Opin Microbiol 9: 245–251
Latake SB, Borkar SG (2017) Characterization of marine actinomycete having antiviral activity against cucumber mosaic virus. Curr Sci 113:1442–1447
Li G, Zheng X, Zhu Y, Long Y, Xia X (2022) Bacillus symbiont drives alterations in intestinal microbiota and circulating metabolites of lepidopteran host. Environ Microbiol 24:4049–4064
Li X, Hao G, Wang Q, Chen Z, Ding Y, Yu L et al (2017) Ningnanmycin inhibits tobacco mosaic virus virulence by binding directly to its coat protein discs. Oncotarget 8:82446
Li Y, Guo Q, Li Y, Sun Y, Xue Q, Lai H (2019) Streptomyces pactum Act12 controls tomato yellow leaf curl virus disease and alters rhizosphere microbial communities. Biol Fertil Soils 55: 149–169
Lockhart BEL, Semancik JS (1968) Inhibition of the multiplication of a plant virus by actinomycin D. 36:504
Lu Y, Song W, Wang J, Cao Y, Han X, Xu C et al (2024) Biocontrol of Botrytis cinerea by Streptomyces noursei C27 and preliminary identification of antimicrobial metabolites. Biol Control 196:105561
Manjunatha L, Rajashekara H, Uppala LS, Ambika DS, Patil B, Shankarappa KS et al (2022) Mechanisms of microbial plant protection and control of plant viruses. Plants 11:3449
Mansour, F. A., Soweha, H. E., Shaaban-Desouki, S. A., and Mohamadin, A. H. (1988). Studies on the antiviral activity of some bacterial isolates belonging to streptomycetes
Maret W (2024) The extracellular metallometabolome: metallophores, metal ionophores, and other chelating agents as natural products. Nat Prod Commun 19:1934578X241271701
Meena M, Prajapati P, Ravichandran C, Sehrawat R (2021) Natamycin: a natural preservative for food applications—a review. Food Sci Biotechnol:1–16
Mohamed SH, Galal AM (2005) Identification and antiviral activities of some halotolerant Streptomycetes isolated from Qaroon Lake. Int J Agric Biol 7:747–753
Mohammadipanah F, Dehhaghi M (2017) Classification and taxonomy of Actinobacteria. In: Biol Biotechnol Actinobacteria, pp 51–77
Mohammed FA, Abu-Hussien SH, El Dougdoug NK, Koutb N, Korayem AS (2024) Streptomyces fradiae mitigates the impact of potato virus Y by inducing systemic resistance in two Egyptian potato (Solanum tuberosum L.) cultivars. Microb Ecol 87:131
Muazi Alenazi A, Anwar Y, EM Abo-Aba S, M Bataweel N (2023) A review on Actinomycetes distribution, isolation, and their medical applications. Nov Res Microbiol J 7:1918–1931
Nargotra P, Ortizo RGG, Wang J-X, Tsai M-L, Dong C-D, Sun P-P et al (2024) Enzymes in the bioconversion of food waste into valuable bioproducts: a circular economy perspective. Syst Microbiol Biomanufacturing:1–19
Nasr-Eldin M, Messiha N, Othman B, Megahed A, Elhalag K (2019) Induction of potato systemic resistance against the potato virus Y (PVY NTN), using crude filtrates of Streptomyces spp. under greenhouse conditions. Egypt J Biol Pest Control 29:62

Negi S, Rana D, Devi D, Pradhan S, Chitra (2024) Actinobacteria: potential natural synthesizers of bioactive compounds having multidimensional activities. In: Soil bacteria: biofertilization and soil health. Springer, pp 361–374
Nobs E (2024) Mycobacterial regulation of macrophage responses to infection: induction and functional role of type I interferon
Pan C, Yang H, Gao W, Wei Z, Song C, Mi J (2024) Optimization of organic solid waste composting process through iron-related additives: a systematic review. J Environ Manag 351:119952
Rakhonde GY, Panda S, Ahale SR, Veratiya AK, Deshkar AM, Nayak L (2025) Role of microbes and organic amendments in stimulating ISR and SAR in plant disease management. In: Advances in organic farming. Apple Academic Press, pp 491–514
Retamal-Morales G, Senges CHR, Stapf M, Olguín A, Modak B, Bandow JE et al (2021) Isolation and characterization of arsenic-binding siderophores from Rhodococcus erythropolis S43: role of heterobactin B and other heterobactin variants. Appl Microbiol Biotechnol 105:1731–1744
Sağlam NG, Rachid NA, Güngör ND (2024) Secondary metabolites and biological compounds of actinomycetes and their applications. In: Bacterial secondary metabolites. Elsevier, pp 123–145
Savary S, Willocquet L, Pethybridge SJ, Esker P, McRoberts N, Nelson A (2019) The global burden of pathogens and pests on major food crops. Nat Ecol Evol 3:430–439
Schwartz HS, Sodergren JE, Garofalo M, Sternberg SS (1965) Actinomycin D effects on nucleic acid and protein metabolism in intact and regenerating liver of rats. Cancer Res 25:307–317
Seratnahaei M, Eshraghi SS, Pakzad P, Ramazani AZ, Yaseri M (2022) Antimicrobial activities of the secondary metabolite extracted from a Nocardia strain. J Adv Biomed Sci
Shafie RM, Hamed AH, El-Sharkawy HHA (2016) Inducing systemic resistance against Cucumber Mosaic Cucumovirus using Streptomyces spp. Egypt J Phytopathol 44:127–142
Sharma A, Kaushal N, Dhiman K, Shukla S, Kumar S (2025) Soil microflora: as promisable approach for effective disease management. In: Plant diseases and their management. Apple Academic Press, pp 113–142
Sharma S, Fulke AB, Chaubey A (2019) Bioprospection of marine actinomycetes: recent advances, challenges and future perspectives. Acta Oceanol Sin 38:1–17
Shoman SA, Abd-Allah NA, El-Baz AF (2003) Induction of resistance to Tobacco necrosis virus in bean plants by certain microbial isolates. Egypt. J. Biol. 5
Shu J, Cao K, Fei C, Dai H, Li Y, Cao Y et al (2024) Antiviral mechanisms of Anisomycin produced by Streptomyces albulus SN40 on potato virus Y. J Agric Food Chem 72:3506–3519
Silva G d C, Kitano IT, Ribeiro IA d F, Lacava PT (2022) The potential use of actinomycetes as microbial inoculants and biopesticides in agriculture. Front Soil Sci 2:833181
Taha M, Ghaly M, Atwa H, Askoura M (2021) Evaluation of the effectiveness of soil Streptomyces isolates for induction of plant resistance against Tomato mosaic virus (ToMV). Curr Microbiol 78:3032–3043
Torres-Rodriguez JA, Reyes-Pérez JJ, Quiñones-Aguilar EE, Hernandez-Montiel LG (2022) Actinomycete potential as biocontrol agent of phytopathogenic fungi: mechanisms, source, and applications. Plants 11:3201
Trojak-Goluch A (2024) The use of bacteria, actinomycetes and fungi in the bioprotection of Solanaceous crops against tobacco mosaic virus (TMV). Agric. Basel 14
Turner MS, Dawson WO (1984) Specificity of the actinomycin-D-sensitive function of some RNA plant viruses. Intervirology 21:224–228
Vázquez-Laslop N, Mankin AS (2018) How macrolide antibiotics work. Trends Biochem Sci 43: 668–684
Wang M, Li H, Li J, Zhang W, Zhang J (2024) Streptomyces strains and their metabolites for biocontrol of phytopathogens in agriculture. J Agric Food Chem 72:2077–2088
Wang Y, Wu Y, Zhu C (2006) Studies on a new antiviral component of cytosinpeptidemycin. J Shenyang Agric Univ 37:44
Wei X-L, Mo S-Y, Ali-Saeed R, Zeng H (2025) Isolation, structure, and biological activities of exopolysaccharides from Actinomycetes: a review. Curr Microbiol 82:1–15

Wei X, Xie B, Wan C, Song R, Zhong W, Xin S et al (2024) Enhancing soil health and plant growth through microbial fertilizers: mechanisms, benefits, and sustainable agricultural practices. Agronomy 14:609

Wu Q, Deering RW, Zhang G, Wang B, Li X, Sun J et al (2018) Albisporachelin, a new hydroxamate type siderophore from the deep ocean sediment-derived actinomycete Amycolatopsis albispora WP1T. Mar Drugs 16:199

Wu Z, Huang M, Jiang J, Zhang C, Hao G, Chen M et al (2024) Ningnanmycin activates defense systems against potato virus Y in Nicotiana benthamiana. J Agric Food Chem 72:26633

Xia X, Liu J, Huang L, Zhang X, Deng Y, Li F et al (2022) Molecular details of actinomycin D-treated MRSA revealed via high-dimensional data. Mar Drugs 20:114

Xing Z, Lirong H, Dawei Z (2012) Anti TMV activity of the fermentation product of actinomycetes strain ZX01. J Northwest Agric For Univ

Yan H, Liu Y, Zhang H, Jin S, Han Z, Woo J et al (2025) Interaction of Ca2+ and Fe3+ in co-precipitation process induced by Virgibacillus dokdonensis and its application. J Environ Sci 147:131–152

Yassin Y, Aseel D, Khalil A, Abdel-Megeed A, Al-Askar A, Elbeaino T et al (2024) Foliar application of rhizobium leguminosarum bv. Viciae strain 33504-Borg201 promotes faba bean growth and enhances systemic resistance against bean yellow mosaic virus infection. Curr Microbiol 81:220. https://doi.org/10.1007/s00284-024-03733-6

Young DH, Wang NX, Meyer ST, Avila-Adame C (2018) Characterization of the mechanism of action of the fungicide fenpicoxamid and its metabolite UK-2A. Pest Manag Sci 74:489–498

Zahr R, Zahr S, El Hajj R, Khalil M (2024) Characterization of Actinobacteria strains in Lebanese soil with an emphasis on investigating their antibacterial activity. Braz J Microbiol 55:255–267

Zhang D, Lu Y, Chen H, Wu C, Zhang H, Chen L et al (2020) Antifungal peptides produced by actinomycetes and their biological activities against plant diseases. J Antibiot (Tokyo) 73:265–282

Zhang G, Han L, Zhang G, Zhang X, Feng J (2015) Purification and characterization of a novel glycoprotein from Streptomyces sp. ZX01. Int J Biol Macromol 78:195–201

Zhang Xing ZX, Han LiRong HL, Zhou DaWei ZD, Jing BingNian JB, Feng JunTao FJ (2012) Anti TMV activity of the fermentation product of actinomycetes strain ZX01

Zhu Y, Xing Y, Li Y, Jia J, Ying Y, Shi W (2024) The role of phosphate-solubilizing microbial interactions in phosphorus activation and utilization in plant–soil systems: a review. Plants 13: 2686

# Chapter 17
# Sustainability of Mediterranean Barley Production Under Water Stress

**Farid Hellal, Saied El-Sayed, Doaa M. Abou Basha, and Hanan H. Abdel Kader**

## Abbreviations

| | |
|---|---|
| Arg | Arginine |
| BGA | Blue-green algae |
| DI | Deficit irrigation |
| fed | feddan |
| GMP | Geometric mean productivity |
| HM | Harmonic means |
| MP | Mean productivity |
| ROS | Reactive oxygen species |
| SI | Sufficient irrigation |
| SSI | Stress susceptibility index |
| SSI | Stress tolerance index |
| TOL | Stress tolerance |
| WUE | Water use efficiency |
| YI | Yield index |
| YSI | Yield stability index |

F. Hellal (✉) · S. El-Sayed · D. M. Abou Basha · H. H. Abdel Kader
Department of Plant Nutrition, Institute of Agricultural and Biological Research, National Research Centre, Cairo, Egypt



N. El-Wakeil, A. Abdelkhalek (eds.), *Microbiomes and Plant Secondary Metabolites for Insect Pest Management*, Progress in Biological Control 23,
https://doi.org/10.1007/978-3-032-24330-0_17

## 17.1 Introduction

Rising sea levels and a lack of water are linked to several other problems in Egypt, such as pollution and population growth. Egypt is a crucial case study for understanding vulnerability to climate change. Beyond the Nile River, the nation has little internal renewable freshwater resources. Egypt may face a water crisis as the effects of climate change worsen. Egypt is on the verge of approaching "absolute water scarcity," having already surpassed the internationally recognized threshold for water scarcity. Projections indicate that by 2060, Egypt's total agricultural output could decline by 8–47%, accompanied by increased food costs. Egypt's lakes and rivers lose water due to evaporation as a result of rising temperatures and climate change. A drought is a type of climatic event that lowers agricultural yield by depriving crops of rain for an extended period, which lowers soil moisture content and reduces plant tissue water potential. According to Nezhadahmadi et al. (2013), 1.8 billion people will experience a complete water scarcity by 2025, and 65% of people would live in water-stressed environments. The major abiotic stresses such as drought, salinity, heat stress, and waterlogging are the main climatic factors for declining crop production in many developing countries. According to Passioura (2006), there are three factors that influence the grain production of crops grown under water-limited conditions: (i) the harvest index, which is the ratio of grain yield to total biomass; (ii) water use efficiency in creating biomass per unit of water utilized; and (iii) water usage as the quantity of water transpired by crops. The capacity of a plant to retain a high plant water status or cellular hydration despite a water shortage, such that plant activities are comparatively unexposed to tissue dehydration, is known as drought, which is a primary abiotic stress (Mitra 2001; Blum 2009). The ratio between the amounts of water required for the growth of the plant and the available water in its environment is known as water stress, whereby the useful reserve of water for the plant is the amount of water soil accessible by its root system. The genotype, the duration and intensity of the water shortage, and the developmental stage all affect how a plant reacts to water stress Nezhadahmadi et al. (2013). Barley ranked fourth in cereal crop after wheat, rice, and maize (FAO 2016). Among the cereal crops, barley is the least valued in Egypt, where it is only grown on marginal ground that experiences salinity and drought. It is mostly grown in northern coastal locations, with Northwest Coast and North Sinai seeing slightly greater average yearly precipitation of 135 mm and 135 mm, respectively (Noaman 2008). Barley includes three mechanisms for coping with drought stress (Tuberosa 2012): avoidance, escape, and tolerance. Varieties that originate from areas that experience terminal drought, or drought during the reproductive stage, finish their life cycle prior to the commencement of a significant water shortage, which is in line with a strategy for escaping drought. When there is less water available, plants that use a drought avoidance strategy keep their cells adequately hydrated (Kooyers 2015).

## 17.2 Grain Yield of Barley as Affected by Water Stress

One of the most detrimental environmental elements influencing agricultural productivity globally is water stress (Gontia-Mishra et al. 2016). Different barley genotypes experience growth inhibition results of drought-induced changes to plant morphology (induced senescence), physiological (stomata or osmotic adjustments and translocation of nutrients from older leaves to developing tissues and seeds), and biochemical processes (changes in the ratio of chlorophyll content, beta carotenoids, and reduced photosynthesis) (EL-Shawy et al. 2017; Temel et al. 2017). Barley ranks fifth among field crops in global cereal production after corn, wheat, rice and soybeans, and current soil management strategies are based on inorganic a chemical fertilizer, which poses a serious threat to human health and the environment. Different abiotic stressors have a major impact on how plants grow and develop. The most prevalent abiotic stressor, drought, is a significant factor that restricts agricultural output, particularly in arid and semiarid countries. It also poses a significant challenge to barley production (EL Sabagh et al. 2019). Due to reductions in photosynthesis, stomatal conductance, ATP synthesis, carbohydrate metabolism, and assimilate translocation, drought stress leads to decreased biomass production (Meeks et al. 2019). Additionally, during the reproductive stage, a limited water supply reduces the activity of enzymes involved in starch synthesis, downregulating the genes for invertase and sucrose synthase, which in turn affects pollen function. Insufficient pollen tube development resulting from the accumulation of diminishing starch in the pistils is the cause of reproductive failure (Hu et al. 2019; Li et al. 2020). Barley is planted on freshly reclaimed soils, in the northern coastal area of Egypt, as well as in the periphery portions of the Nile Valley and Delta, due to its greater drought tolerance than wheat. Hellal et al. (2019a) The purpose of the study was to compare the characteristics of Mediterranean barley types (Egyptian, Tunisian, Algerian, and Moroccan) cultivated under water stress conditions (40 percent of their water holding capacity) to those produced under normal conditions (75 percent). The most sensitive varieties were Giza 125 and Giza 126 (Egyptian), Tombari, Raihane (Tunisian), Ksar Megrine, Ras El-Mouch (Algerian), and Tamellalet and Adrar (Morocco). The results showed that the Egyptian varieties (Giza 131, Giza 127, Giza 2000, and Giza 130), Tunisian varieties (Kebili 3, Tozeur 2, and Sidi-Bou), Algerian varieties (Naïlia, Techedrett, and Temacine), and Moroccan varieties (Laanaceur, Massine, and Amira) gained the greatest values for most drought tolerance indices and produced greater grain yield under drought stress conditions. It can be suggested that the tolerant barley types be utilized as parents for gene mapping, genetic research, and enhancing drought tolerance in barley cultivars. Regarding to the grain yield (GY), Fig. 17.1 indicated that Ksar, Giza 131 (Egyptian), Kebili (Tunisian), Techedrett (Algerian), and Laanaceur (Morocco) gained the greatest values under normal and stress conditions, while the lowest values were attained at Giza123, Sabra, Ksar Megrine and Tamellalet (Morocco), Giza 131, Kebili-3, Naïlia, and Laanaceur in same countries mentioned above under both normal and stress conditions.

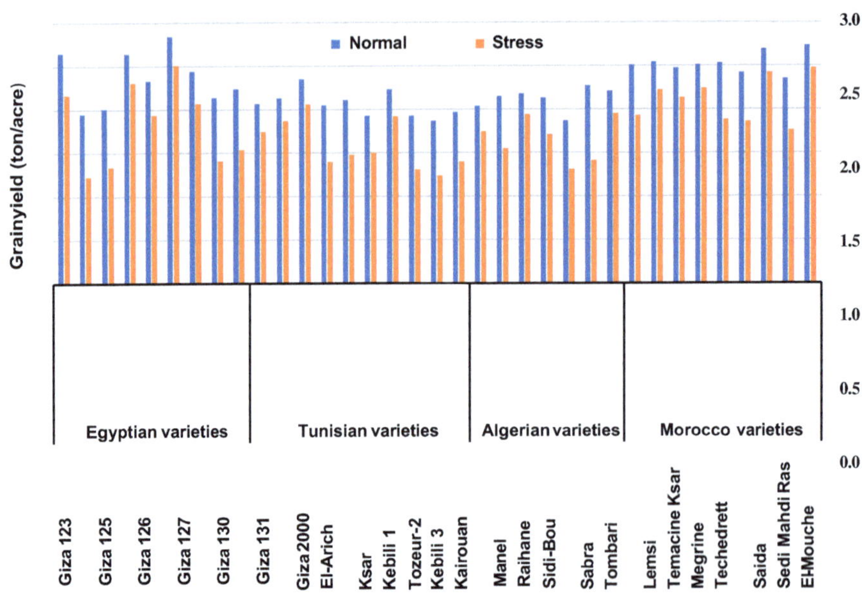


**Fig. 17.1** Grain yield of barley as affected by water stress

Giza 125 (Egyptian, Tombari (Tunisian), Sidi Mahdi (Algerian), and Tamellalet (Morocco) recorded the lowest values under normal and stress conditions. It is obvious that Laanaceur (Morocco) and Ksar (Egyptian) scored the highest grain values compared to other investigated varieties under both normal and stress conditions. Meanwhile, Giza 131 gained the opposite with its exception.

## 17.3 Water Use Efficiency of Barley as Affected by Water Stress

One of the most crucial metrics for choosing the best water management strategies is water usage efficiency, or WUE. Using this criterion, crop productivity under various applied treatments has been assessed per unit of water consumed by the crop. Furthermore, the obtained data demonstrated that WUE levels under stress were more prevalent than they were under normal circumstances. The mean values of WUE or water productivity for Morocco varieties were the greatest ones followed by Egyptian cultivars. The maximum and minimum values of the WUE were observed at Laanaceur (Morocco) with values 2.28, 3.43 kg $m^{-3}$ and Sidi Mahdi (Algerian) with values 1.56 and Giza 125 (Egyptian) 1.71 kg $m^{-3}$ for normal and stress treatments. Regarding to the differences within each group of the studied barley Mediterranean varieties could arrange as follows: 45–104 (Egyptian), 26–55 (Tunisian), 21–49 (Algerian), and 16–41% (Morocco) for normal and stress

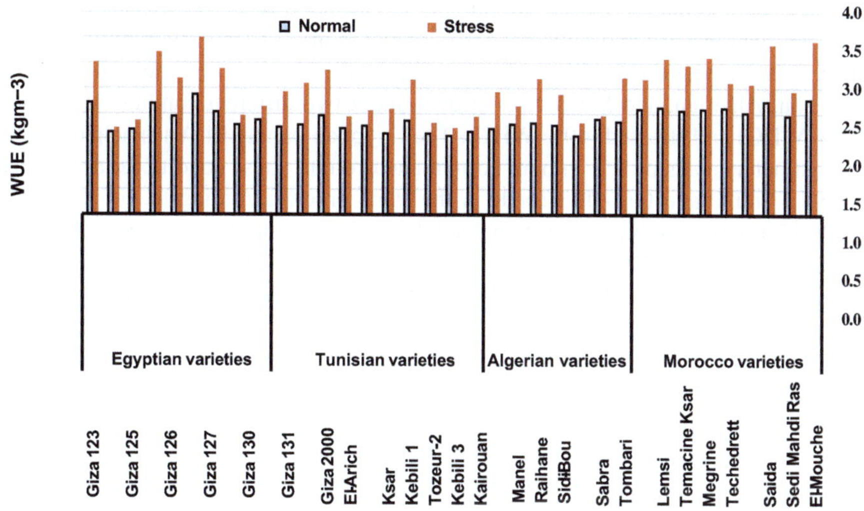


**Fig. 17.2** Water use efficiency of barley as affected by water stress

condition. Meanwhile, regarding the rate of change when comparing the greatest and the lowest WUE values within countries groups, Fig. 17.2 indicates that Giza 137 (Egyptian) was the superior cultivar where it scored the greatest value (2.37 kg $m^{-3}$). Also, data showed that Morocco varieties gained the greatest values as a mean value for groups and the opposite was true in case of Tunisian ones. Same trend was attained in case of stress treatments, where Giza 130 (3.49 $kgm^{-3}$, Egyptian) and Laanaceur (3.43 kg $m^{-3}$, Morocco) scored the greatest values (Hellal et al. 2019a).

## 17.4 Nutrient Status in Barley as Affected by Water Stress

### *17.4.1 Macronutrient Content of Barley Grains*

In Egypt, barley is produced in the north coastal regions under rainfed conditions and in the recently reclaimed fields under irrigation. It is also planted on salty soils with limited irrigation water. The amount and distribution of yearly rainfall in Egypt affects the overall area used for barley growing. Reduced nitrogen availability due to mineralization is one of the many causes of decreases in nutrient absorption during droughts (Sahnoune et al. 2004). Water scarcity is a global issue, especially in arid and semiarid regions where usage of water is rapidly increasing for both residential and agricultural purposes (El Mourid et al. 2010). Although this has not received much research, drought may potentially reduce nutrient intake by altering the kinetics of nitrogen uptake by roots (Christophe et al. 2011). Drought stress decreases the concentration of nitrogen (N) and phosphorus (P) in plant tissue, and several studies have shown that drought can decrease nutrient uptake from soil

(Sardans and Peñuelas 2012). Drought can also have a strong impact on plant nutrient relations (He and Dijkstra 2014). Hellal et al. (2020a). The purpose of the study was to assess the nutritional status of four different types of barley—Egyptian, Tunisian, Algerian, and Moroccan—grown under two different conditions: normal (75% WHC) and stressed (40% WHC). The Egyptian variety El-Arish achieved the highest percentages of N% in both normal and drought stress conditions, as well as Giza 127 for P and Giza 123 for K in the same order, according to the results in Table 17.1. Conversely, Giza 125 yielded the lowest levels of the preceding macronutrients. The Giza 2000 variety had the lowest known amounts of N, P, and Ca content. Regarding the Tunisian cultivars, Raihane achieved the lowest decrease percentages (9.38%, 3.54%, and 3.62%) for N, K, and Ca, respectively. The Tozeur-2 variety record for P content is (3.32%). On the other hand, Sabra (25.29%), Tombari (23.18%), Manel (15.88%), and Ca (16.37%) might be ranked in order of the percentage of theft decrease. Under normal and stress settings, the Algerian barley varieties N in Naïlia/Techedrett, P in Temacine/Ksar Megrine, K in Ras El-Mouche/Temacine, and Ca in Saida/Ksar Megrine obtained the highest and lowest results, respectively. Under normal circumstances, the lowest levels of Fe, Zn, and Mn were recorded for barley types Giza 125, Ksar for Zn, and Giza 127 for Mn. However, under stress, the highest and lowest values of Fe, Zn, and Mn were recorded for barley varieties Giza 2000, Giza 131, and Giza 127, respectively.

According to the results, the Tunisian varieties Kebili-3 and Tozeur-2, the Algerian varieties Techedrett and Naïlia, the Moroccan varieties Laanaceur and Amira, and the Egyptian varieties Giza 127, Giza131, and Giza 2000 registered the lowest reduction percentage in nutrient content, indicating their tolerance to sustain their productivity under conditions of water stress.

### *17.4.2 Micronutrient Content of Barley Grains*

Table 17.2 documents the micronutrient (Fe, Zn, and Mn) content of Egyptian barley cultivars under both normal and stress conditions. Hellal et al. (2020a) found that when stress conditions were applied, iron, zinc, and manganese increased (for iron, this was true; for zinc and manganese, this was not the case).

For the micronutrients Fe, Zn, and Mn, the highest concentrations were found in barley types Giza 2000 (339.0 ppm, 40.0 ppm) and Giza 131 (16.5 ppm) under normal conditions. Under typical conditions, the minimal values were reached for the barley types of Giza 125 for Fe, Ksar for Zn, and Giza 126 for Mn. But under stressful conditions, the barley types of Giza 2000, Giza 131, and Giza 127 had the highest and lowest values of Fe, Zn, and Mn, respectively. Both the Kairouan and Manel cultivars showed significant accumulation of Fe under normal and drought-stressed circumstances, respectively. It is also important to note that, except for Saida (normal) and Naïlia (stress), Techedrett had the highest Fe, Zn, and Mn levels in normal or drought conditions. Under normal conditions and under drought stress, respectively, Laanaceur and Firdaws barley types were strongly accumulated for Fe, but Oussama and Amira were not.

**Table 17.1** Macronutrient contents of barley grains as affected by water stress

| | N (%) | | P (%) | | K (%) | | Ca (%) | |
|---|---|---|---|---|---|---|---|---|
| Barley varieties | Normal | Stress | Normal | Stress | Normal | Stress | Normal | Stress |
| Giza123 | 1.958 | 1.742 | 0.742 | 0.629 | 0.516 | 0.485 | 0.107 | 0.100 |
| Giza125 | 1.338 | 0.995 | 0.615 | 0.513 | 0.600 | 0.453 | 0.128 | 0.117 |
| Giza126 | 1.370 | 1.111 | 0.662 | 0.551 | 0.556 | 0.470 | 0.124 | 0.119 |
| Giza127 | 1.655 | 1.327 | 0.672 | 0.525 | 0.590 | 0.434 | 0.216 | 0.176 |
| Giza130 | 1.396 | 1.238 | 0.532 | 0.427 | 0.562 | 0.463 | 0.137 | 0.129 |
| Giza131 | 1.407 | 1.172 | 0.557 | 0.457 | 0.660 | 0.513 | 0.139 | 0.123 |
| Giza2000 | 1.327 | 1.183 | 0.686 | 0.575 | 0.495 | 0.432 | 0.148 | 0.142 |
| El-Arich | 1.392 | 1.132 | 0.693 | 0.429 | 0.588 | 0.349 | 0.216 | 0.194 |
| Ksar | 1.753 | 1.438 | 0.768 | 0.626 | 0.426 | 0.404 | 0.202 | 0.186 |
| Kebili 1 | 1.742 | 1.552 | 1.061 | 0.966 | 0.490 | 0.450 | 0.229 | 0.208 |
| Tozeur-2 | 1.916 | 1.710 | 0.715 | 0.691 | 0.477 | 0.446 | 0.205 | 0.193 |
| Kebili 3 | 2.059 | 1.584 | 0.805 | 0.754 | 0.446 | 0.402 | 0.223 | 0.203 |
| Kairouan | 1.457 | 1.203 | 0.886 | 0.738 | 0.505 | 0.436 | 0.175 | 0.162 |
| Manel | 1.441 | 1.140 | 0.728 | 0.663 | 0.505 | 0.425 | 0.224 | 0.187 |
| Raihane | 1.520 | 1.378 | 0.753 | 0.694 | 0.503 | 0.485 | 0.219 | 0.211 |
| Sidi-Bou | 1.441 | 1.251 | 0.778 | 0.653 | 0.500 | 0.457 | 0.252 | 0.222 |
| Sabra | 1.378 | 1.029 | 0.731 | 0.588 | 0.458 | 0.386 | 0.201 | 0.168 |
| Tombari | 1.457 | 1.267 | 0.542 | 0.417 | 0.362 | 0.311 | 0.233 | 0.202 |
| Lemsi | 1.378 | 1.188 | 0.663 | 0.557 | 0.444 | 0.380 | 0.240 | 0.204 |
| Temacine | 1.629 | 1.428 | 0.612 | 0.537 | 0.430 | 0.393 | 0.340 | 0.274 |
| Ksar Megrine | 1.752 | 1.483 | 0.329 | 0.277 | 0.430 | 0.409 | 0.283 | 0.240 |
| Techedrett | 1.501 | 1.381 | 0.410 | 0.391 | 0.497 | 0.456 | 0.374 | 0.353 |
| Saida | 1.588 | 1.506 | 0.503 | 0.491 | 0.468 | 0.446 | 0.375 | 0.364 |
| Sidi Mahdi | 1.815 | 1.428 | 0.419 | 0.402 | 0.447 | 0.403 | 0.375 | 0.326 |
| Ras El-Mouche | 1.847 | 1.388 | 0.551 | 0.446 | 0.536 | 0.458 | 0.341 | 0.325 |
| Naïlia | 2.003 | 1.869 | 0.476 | 0.387 | 0.487 | 0.433 | 0.314 | 0.291 |
| Adrar | 1.378 | 1.148 | 0.645 | 0.432 | 0.447 | 0.379 | 0.222 | 0.199 |
| Oussama | 1.213 | 1.180 | 0.694 | 0.675 | 0.435 | 0.411 | 0.253 | 0.238 |
| Amalou | 1.180 | 1.085 | 0.607 | 0.568 | 0.431 | 0.391 | 0.215 | 0.195 |
| Massine | 1.059 | 0.980 | 0.625 | 0.574 | 0.377 | 0.347 | 0.227 | 0.206 |
| Taffa | 1.170 | 1.087 | 0.523 | 0.466 | 0.431 | 0.379 | 0.236 | 0.221 |
| Firdaws | 1.518 | 1.217 | 0.687 | 0.623 | 0.436 | 0.375 | 0.291 | 0.257 |
| Amira | 1.368 | 1.050 | 0.743 | 0.541 | 0.395 | 0.274 | 0.267 | 0.193 |
| Tamellalet | 1.281 | 0.901 | 0.706 | 0.472 | 0.393 | 0.234 | 0.236 | 0.100 |
| Laanaceur | 1.293 | 1.178 | 0.602 | 0.527 | 0.393 | 0.352 | 0.271 | 0.249 |
| LSD 0.05 | 0.017 | 0.013 | 0.020 | 0.019 | 0.021 | 0.014 | 0.012 | 0.010 |

**Table 17.2** Micronutrient content of barley grains as affected by water stress

| Barley varieties | Fe (ppm) | | Zn (ppm) | | Mn (ppm) | |
|---|---|---|---|---|---|---|
| | *Normal* | *Stress* | *Normal* | *Stress* | *Normal* | *Stress* |
| Giza 123 | 210.6 | 182.9 | 38.4 | 30.8 | 13.2 | 12.5 |
| Giza 125 | 89.0 | 85.0 | 27.6 | 26.0 | 11.4 | 8.5 |
| Giza 126 | 149.3 | 145.3 | 30.4 | 29.4 | 9.3 | 8.7 |
| Giza 127 | 171.0 | 118.6 | 28.9 | 21.4 | 10.5 | 7.4 |
| Giza 130 | 191.8 | 179.9 | 36.1 | 28.9 | 14.2 | 10.9 |
| Giza 131 | 240.2 | 195.7 | 30.1 | 29.0 | 16.3 | 12.1 |
| Giza 2000 | 335.1 | 229.4 | 50.1 | 34.9 | 14.8 | 12.2 |
| El-Arich | 245.2 | 190.8 | 39.4 | 28.0 | 11.8 | 8.5 |
| Ksar | 238.3 | 160.2 | 26.6 | 22.2 | 16.3 | 15.0 |
| Kebili 1 | 229.4 | 221.4 | 31.0 | 27.5 | 16.4 | 14.2 |
| Tozeur-2 | 170.0 | 159.2 | 31.4 | 29.5 | 18.0 | 11.2 |
| Kebili 3 | 160.2 | 120.6 | 25.2 | 20.9 | 17.7 | 15.1 |
| Kairouan | 286.7 | 200.7 | 29.7 | 20.9 | 15.5 | 12.5 |
| Manel | 264.9 | 231.3 | 27.4 | 26.2 | 16.3 | 13.3 |
| Raihane | 150.3 | 113.7 | 31.0 | 27.0 | 17.9 | 11.4 |
| Sidi-Bou | 158.2 | 146.3 | 32.8 | 29.3 | 10.4 | 9.1 |
| Sabra | 139.4 | 101.8 | 25.6 | 20.4 | 10.8 | 6.4 |
| Tombari | 160.2 | 114.7 | 45.1 | 32.0 | 16.5 | 11.5 |
| Lemsi | 195.7 | 97.9 | 26.9 | 21.5 | 10.6 | 6.6 |
| Temacine | 246.2 | 177.9 | 32.2 | 21.8 | 14.7 | 10.1 |
| Ksar Megrine | 158.2 | 114.7 | 24.0 | 22.9 | 17.9 | 11.5 |
| Techedrett | 287.7 | 225.4 | 36.7 | 26.7 | 22.2 | 18.4 |
| Saida | 254.1 | 174.0 | 38.9 | 28.4 | 18.8 | 16.0 |
| Sidi Mahdi | 129.5 | 90.0 | 22.9 | 21.7 | 14.9 | 11.4 |
| Ras El-Mouche | 135.4 | 118.6 | 24.5 | 22.5 | 8.7 | 7.6 |
| Naïlia | 266.9 | 194.8 | 38.4 | 26.4 | 26.9 | 17.1 |
| Adrar | 156.2 | 109.7 | 36.0 | 24.5 | 7.7 | 3.6 |
| Oussama | 144.3 | 110.7 | 26.4 | 23.2 | 14.1 | 10.2 |
| Amalou | 91.1 | 77.7 | 19.4 | 17.9 | 8.1 | 7.3 |
| Massine | 165.1 | 143.3 | 19.7 | 18.1 | 5.4 | 4.3 |
| Taffa | 165.1 | 111.7 | 22.3 | 20.7 | 9.1 | 4.4 |
| Firdaws | 188.8 | 114.7 | 29.2 | 19.2 | 15.3 | 11.5 |
| Amira | 133.5 | 162.1 | 24.8 | 13.7 | 12.1 | 9.2 |
| Tamellalet | 112.7 | 96.9 | 26.7 | 23.1 | 6.3 | 6.2 |
| Laanaceur | 242.2 | 215.5 | 36.3 | 32.1 | 17.9 | 12.3 |
| LSD 0.05 | 2.099 | 1.628 | 0.733 | 0.526 | 0.106 | 0.078 |

## 17.5 Drought Stress Tolerance Indices of Barley Cultivars

Grain yield of the varieties under both non-stress and stress conditions was measured for the purpose of calculating different sensitivity and tolerance indices. This allowed for the selection of stable genotypes based on their performance under stress and non-stress conditions, as well as the use of appropriate stress resistance indices for screening the various varieties under drought condition. As a result, low values of the SSI indicated resistance to stress circumstances and high levels of the indicators Yp, Ys, STI, MP, and GMP. According to Mevlüt and Sait (2011), genotypes with high STI often exhibit a significant yield differential under two distinct situations. In general, the GMP, MP, and STI parameters all showed identical rankings for the genotypes, indicating that these three factors are equivalent in their ability to select genotypes. Regarding stress susceptibility index (SSI), it calculates the rate of variation in yield for each genotype under stress and under non-stress circumstances in relation to the average variation for all genotypes. Low drought susceptibility (or high yield stability) is indicated by SSI values less than 1, while high drought susceptibility (or poor yield stability) is indicated by values more than one.

According to the drought tolerance indices in Table 17.3, (Hellal et al. 2019a, b) the maximum and minimum values were observed in Ras El-Mouche (Algeria) and Techedrett (Algeria), respectively. For **STI**, the highest and lowest values were found in Laanaceur (Morocco) and Giza 125 (Egypt). Comparing varieties by origin, the greatest and lowest values for SSI were: Egypt (Giza 125, Giza 131), Tunisia (Tombari-Tozeur-2, Kebili 3), Algeria (Ras El-Mouche, Techedrett), and Morocco (Taffa, Laanaceur). For STI, the respective varieties were: Egypt (Giza 131, Giza 125), Tunisia (Kebili 3, Tombari), Algeria (Naïlia, Sidi Mahdi), and Morocco (Laanaceur, Tamellalet).

When evaluating wheat genotypes' resistance to drought, the stress susceptibility index (SSI) was employed. It was discovered that the SSI for genotypes and their ranking pattern varied from year to year. Data of yield index (YI) and yield stability index (YSI) indicated that the greatest and lowest values within each group of the barley varieties were arranged as follows: Giza 131, Kebili 3 (Tunisian), and Laanaceur (Morocco) and Giza and Tombari, Ras El-Mouche (Algerian), and Tamellalet (Morocco), for YI and YSI, respectively. According to the drought indices tolerance index (TOL) and harmonic mean (HM), same previous trend was obtained with some exception, and the greatest and lowest values for the drought indices were observed in Giza 125 (Egypt) and Naïlia (Algeria) for TOL, and in Giza 131 (Egypt) and Tombari (Tunisia) for HM. Within specific origin groups, the greatest and lowest values for TOL were recorded for Giza 125/131 (Egypt), Kairouan/Tozeur 2 (Tunisia), Ras El-Mouche/Ksar Megrine (Algeria), and Taffa/Amira (Morocco). For HM, these values were observed in Giza 131/125 (Egypt), Kebili 3/Tombari (Tunisia), Naïlia/Sidi Mahdi (Algeria), and Laanaceur/Tamellalet (Morocco).

**Table 17.3** Drought stress tolerance indices of barley as affected by water stress

| Barley varieties | | SSI | STI | MP | GMP | YI | YSI | TOL | HM |
|---|---|---|---|---|---|---|---|---|---|
| Egyptian varieties | Giza 123 | 0.82 | 1.03 | 2.41 | 2.40 | 1.18 | 0.82 | 0.48 | 2.39 |
| | Giza 125 | 1.68 | 0.43 | 1.59 | 1.55 | 0.67 | 0.63 | 0.72 | 1.51 |
| | Giza 126 | 1.51 | 0.48 | 1.68 | 1.64 | 0.73 | 0.67 | 0.67 | 1.61 |
| | Giza 127 | 0.57 | 1.09 | 2.48 | 2.47 | 1.26 | 0.88 | 0.33 | 2.46 |
| | Giza 130 | 0.76 | 0.81 | 2.14 | 2.13 | 1.05 | 0.83 | 0.39 | 2.12 |
| | Giza 131 | 0.53 | 1.28 | 2.68 | 2.67 | 1.36 | 0.88 | 0.33 | 2.66 |
| | Giza 2000 | 0.69 | 0.91 | 2.26 | 2.25 | 1.13 | 0.85 | 0.37 | 2.24 |
| | El-Arich | 1.55 | 0.54 | 1.78 | 1.74 | 0.77 | 0.66 | 0.73 | 1.70 |
| | Ksar | 1.42 | 0.62 | 1.89 | 1.86 | 0.84 | 0.69 | 0.70 | 1.83 |
| Tunisian varieties | Kebili 1 | 0.68 | 0.85 | 1.91 | 1.90 | 1.10 | 0.85 | 0.32 | 1.90 |
| | Tozeur-2 | 0.54 | 0.94 | 2.00 | 2.00 | 1.18 | 0.88 | 0.26 | 1.99 |
| | Kebili 3 | 0.54 | 1.14 | 2.21 | 2.20 | 1.30 | 0.88 | 0.29 | 2.20 |
| | Kairouan | 1.39 | 0.68 | 1.73 | 1.69 | 0.88 | 0.68 | 0.65 | 1.66 |
| | Manel | 1.30 | 0.74 | 1.80 | 1.77 | 0.93 | 0.70 | 0.63 | 1.74 |
| | Raihane | 0.97 | 0.68 | 1.72 | 1.70 | 0.95 | 0.78 | 0.43 | 1.69 |
| | Sidi-Bou | 0.61 | 1.01 | 2.08 | 2.07 | 1.21 | 0.86 | 0.31 | 2.06 |
| | Sabra | 1.41 | 0.60 | 1.62 | 1.59 | 0.82 | 0.68 | 0.62 | 1.56 |
| | Tombari | 1.48 | 0.55 | 1.56 | 1.52 | 0.78 | 0.66 | 0.63 | 1.49 |
| | Lemsi | 1.27 | 0.65 | 1.69 | 1.66 | 0.88 | 0.71 | 0.57 | 1.64 |
| Algerian varieties | Temacine | 0.66 | 0.79 | 1.90 | 1.89 | 1.05 | 0.86 | 0.29 | 1.88 |
| | Ksar Megrine | 1.30 | 0.74 | 1.85 | 1.82 | 0.93 | 0.72 | 0.60 | 1.80 |
| | Techedrett | 0.51 | 0.94 | 2.06 | 2.06 | 1.17 | 0.89 | 0.24 | 2.05 |
| | Saida | 0.91 | 0.81 | 1.92 | 1.91 | 1.03 | 0.80 | 0.42 | 1.90 |
| | Sidi Mahdi | 1.37 | 0.55 | 1.59 | 1.57 | 0.79 | 0.70 | 0.56 | 1.54 |
| | Ras El-Mouch | 1.74 | 0.71 | 1.84 | 1.79 | 0.85 | 0.62 | 0.86 | 1.74 |
| | Naïlia | 0.54 | 0.96 | 2.08 | 2.08 | 1.17 | 0.88 | 0.26 | 2.07 |
| Morocco varieties | Adrar | 1.36 | 1.18 | 2.22 | 2.20 | 1.21 | 0.77 | 0.58 | 2.18 |
| | Oussama | 0.74 | 1.37 | 2.38 | 2.37 | 1.39 | 0.87 | 0.32 | 2.37 |
| | Amalou | 0.81 | 1.28 | 2.30 | 2.29 | 1.33 | 0.86 | 0.34 | 2.29 |
| | Massine | 0.63 | 1.36 | 2.38 | 2.37 | 1.40 | 0.89 | 0.27 | 2.37 |
| | Taffa | 1.51 | 1.15 | 2.21 | 2.18 | 1.18 | 0.74 | 0.65 | 2.16 |
| | Firdaws | 1.36 | 1.09 | 2.14 | 2.12 | 1.16 | 0.77 | 0.56 | 2.10 |
| | Amira | 0.59 | 1.58 | 2.55 | 2.55 | 1.51 | 0.90 | 0.27 | 2.55 |
| | Tamellalet | 1.48 | 1.00 | 2.06 | 2.03 | 1.10 | 0.75 | 0.59 | 2.01 |
| | Laanaceur | 0.56 | 1.64 | 2.60 | 2.60 | 1.54 | 0.90 | 0.26 | 2.59 |

## 17.6 Genetic Similarity and Phylogenetic Tree

Understanding the crop at the physiological and molecular biology levels provides the foundation for the advancement of breeding for drought stress resistance (Araus et al. 2008). Numerous morpho-physiological characteristics influence how a plant responds to stress; these characteristics interact, and each responds differently depending on the severity and length of the water deficit. Using the information produced by the expressing primers, the genetic similarity matrix (Table 17.4) was created using a straightforward matching coefficient (Hellal et al. 2019a, b). We were able to distinguish between each genotype and examine the genetic variation between the genotypes under investigation thanks to these primers. The similarity index showed stronger relationship and closer proximity for varieties from the same region. The greatest percentage of genetic similarity was (88%) detected between Giza 128 and Giza 127, and Raihane and Manel, followed by (86%). On the other hand, the lowest genetic similarity (20%) was obtained between Manel and Adrar. Based on phylogenic tree (Fig. 17.3) and banding pattern (Figs. 17.3 and 17.4), data showed that there are four main clusters that could summarized as follows: (i) The first cluster was divided into five clusters; the first cluster contained the genotypes (Algeria) Ksar Megrine and Naïlia, (ii) the second cluster contained Adrar and Amalou (Morocco), (iii) the third cluster was divided into subgroups containing genotypes from Tunisian Kairouan, Manel, and Raihane, while Sidi-Bou was highly diverged than other Tunisian genotypes, and finally (iv) the fourth cluster consists of Sidi Mahdi (Algeria) grouped with all genotypes from Egyptian (Giza 124 and Giza 125) which are six–row types, and other subculture consists of Giza 127 and Giza 128, which are two-row cultivars. Contrarily, the naked variety Giza 131 was grouped in diverged subcluster. The previous data suggested that this clustering is consistent with pedigree of these varieties.

## 17.7 Mitigating of the Adverse Effect of Water Stress on Barley

One of the most significant abiotic factors that might restrict agricultural productivity globally is water stress. Exploring genotypic selection in response to drought stress is now possible thanks to recent scientific developments, and it has the potential to increase output. One of the most detrimental conditions to plant development and production is drought stress as it causes morphological, physiological, and biochemical changes. It also reduces $CO_2$ absorption, increases leaf area and pigment content, expands stems, and disrupts water usage efficiency. To learn more about how plants tolerate dehydration, researchers have looked at the function that chemical component accumulation plays in drought-tolerant plants. Drought also increases the production of reactive oxygen species (ROS), which can be mitigated by increasing antioxidant enzyme levels (Boyer (2016).

**Table 17.4** Similarity among 15 barley varieties based on SRAP data using Jaccard methods

| | Giza 124 | Giza 125 | Giza 131 | Giza 128 | Giza 127 | Kebili 1 | Kairouan | Sidi-Bou | Manel | Rahiane | Sidi Mahdi | Ksar Megrine | Naïlia | Adrar |
|---|---|---|---|---|---|---|---|---|---|---|---|---|---|---|
| Giza 125 | 0.86 | | | | | | | | | | | | | |
| Giza 131 | 0.71 | 0.83 | | | | | | | | | | | | |
| Giza 128 | 0.75 | 0.86 | 0.71 | | | | | | | | | | | |
| Giza 127 | 0.67 | 0.75 | 0.63 | 0.88 | | | | | | | | | | |
| Kebili 1 | 0.56 | 0.44 | 0.50 | 0.40 | 0.50 | | | | | | | | | |
| Kairouan | 0.44 | 0.50 | 0.57 | 0.44 | 0.56 | 0.63 | | | | | | | | |
| S-di-Bou | 0.40 | 0.44 | 0.50 | 0.40 | 0.36 | 0.56 | 0.63 | | | | | | | |
| Manel | 0.56 | 0.44 | 0.50 | 0.40 | 0.50 | 0.75 | 0.56 | 0.56 | | | | | | |
| Rahiane | 0.50 | 0.40 | 0.44 | 0.36 | 0.45 | 0.67 | 0.75 | 0.67 | 0.88 | | | | | |
| Sidi Mahdi | 0.36 | 0.40 | 0.44 | 0.36 | 0.45 | 0.50 | 0.56 | 0.50 | 0.50 | 0.60 | | | | |
| Ksar Megrine | 0.29 | 0.21 | 0.23 | 0.29 | 0.36 | 0.38 | 0.31 | 0.29 | 0.38 | 0.46 | 0.46 | | | |
| Naïlia | 0.33 | 0.25 | 0.27 | 0.33 | 0.42 | 0.45 | 0.36 | 0.23 | 0.45 | 0.42 | 0.55 | 0.67 | | |
| Adrar | 0.31 | 0.33 | 0.25 | 0.31 | 0.29 | 0.31 | 0.23 | 0.31 | 0.21 | 0.20 | 0.38 | 0.31 | 0.46 | |
| Amalou | 0.38 | 0.31 | 0.23 | 0.29 | 0.27 | 0.38 | 0.21 | 0.29 | 0.29 | 0.27 | 0.27 | 0.47 | 0.43 | 0.75 |

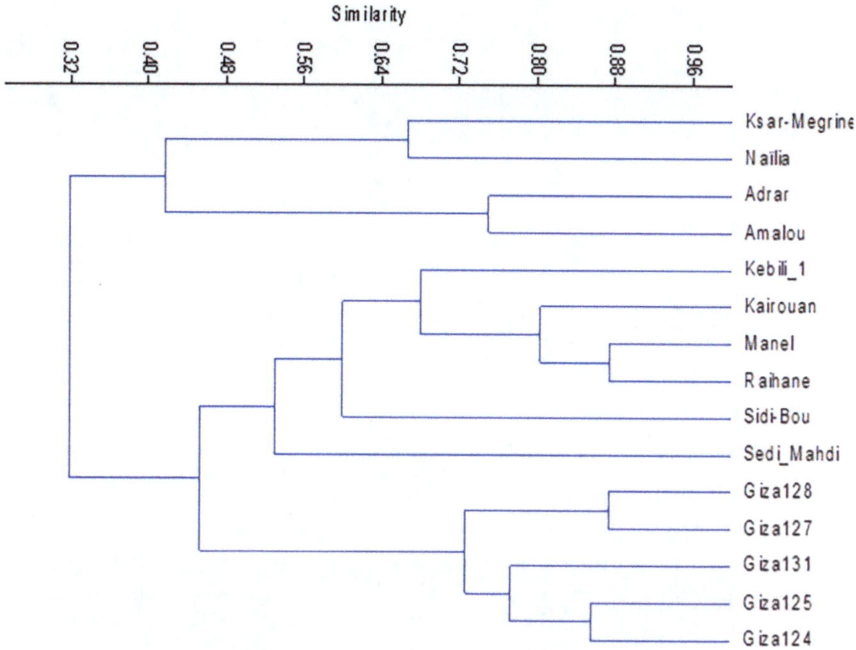


**Fig. 17.3** Neighbor joining (NJ) phylogenetic tree for barley based on SRAP makers using Jaccard method

## *17.7.1 Antitranspirants Application Under Water Stress*

Antitranspiration compounds or chemicals work by shrinking the number and size of stomata on plant leaves, hence lowering water loss from those leaves. Transpiration wastes all of the water that the plant absorbs. Three categories of anti-transpirations were identified: reflecting materials (which return part of the incident radiation that strikes the upper surface of the leaves), film-forming types (which cover the leaf surface with films impervious to water vapor), and stomatal closing types (which alter the metabolic processes in leaf tissues) (Prakash and Ramachandran 2000).

### 17.7.1.1 Antitranspirant Effects on Biological Yield of Barley

All the development and physiological parameters of the examined plants were positively impacted by foliar application of antitranspirants (ether kaolin at 6% and magnesium carbonate spray at 6%), which was also responsible for a significant decrease in transpiration rate. The use of antitranspirant significantly decreased water consumption and increased water yield during periods of irrigation water scarcity. Additionally, spraying Antitranspirants, which reduced the amount of water consumed, improved the majority of growth and yield metrics as well as

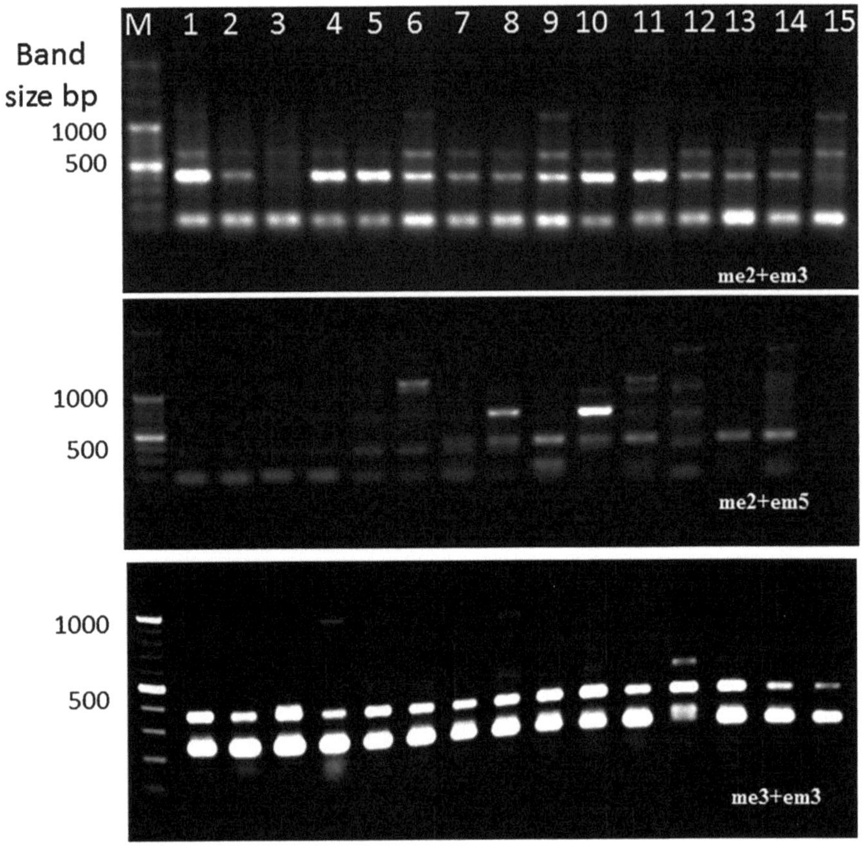


**Fig. 17.4** Agarose gel showing banding patterns of SRAP marker for barley genotypes from different origin 1–15. *M*: Molecular size standard 100 bp DNA ladder, *1*: Giza124, *2*: Giza125, *3*: Giza131, *4*: Giza128, *5*: Giza127, *6*: Kebili 1, *7*: Kairouan, *8*: Sidi-Bou, *9*: Manel, *10*: Raihane, *11*: Sidi Mahdi, *12*: Ksar-Megrine, *13*: Naïlia, *14*: Adrar, and *15*: Amalo

seed quality Rania and Maha (2018). Hellal et al. (2020a) conducted a field experiment at Nubaria Agricultural Research station, NRC, Egypt, to investigate how applying antitranspirants, such as salicylic acid (SA), potassium chloride ($K_2SO_4$), and magnesium oxide ($MgCO_3$), might help some vulnerable barley cultivars recover from the effects of drought. Results in Table 17.5 declared that the lowest values of biological yield, grain yield, harvest index, and water use efficiency were attained in control treatments, especially for Tamellat barley variety under both studied stress conditions. The greatest values were recorded for Giza 125 (SA) except for harvest index; Tombari ($MgCO_3$, SA) for biological yield and SA forgrain yield; and Ksar Megrine and Tamellat (SA) for WUE. Exceptions included the harvest index under control treatments and biological yield for Tamellat under K2SO4, across both normal and stress conditions. Same trend was observed if water stress is eliminated. However, it is clear to mention that the values under

**Table 17.5** Effect of antitranspirants on biological yield of barley under water stress

| | | Biological yield (ton/fed) | | Grain yield (ton/fed) | | Harvest Index | | WUE (kgm-3) | |
|---|---|---|---|---|---|---|---|---|---|
| Varieties | Treatment | Normal | Stress | Normal | Stress | Normal | Stress | Normal | Stress |
| Giza 125 | Control | 4.14 | 2.90 | 1.68 | 1.14 | 40.5 | 39.3 | 1.40 | 1.59 |
| | SA | 5.79 | 4.43 | 2.12 | 1.46 | 36.6 | 33.0 | 1.77 | 2.03 |
| | $K_2SO_4$ | 4.54 | 3.81 | 1.82 | 1.35 | 40.1 | 35.4 | 1.52 | 1.87 |
| | $MgCO_3$ | 4.81 | 3.34 | 1.98 | 1.42 | 41.2 | 42.6 | 1.65 | 1.97 |
| Tombari | Control | 3.71 | 2.99 | 1.39 | 1.24 | 37.6 | 41.5 | 1.16 | 1.72 |
| | SA | 5.09 | 3.67 | 1.86 | 1.34 | 36.6 | 36.5 | 1.55 | 1.86 |
| | $K_2SO_4$ | 4.76 | 3.51 | 1.68 | 1.30 | 35.2 | 37.0 | 1.40 | 1.81 |
| | $MgCO_3$ | 5.50 | 3.48 | 1.79 | 1.31 | 32.6 | 37.6 | 1.49 | 1.82 |
| Ksar Megrine | Control | 3.76 | 3.13 | 1.22 | 0.96 | 32.4 | 30.7 | 1.02 | 1.33 |
| | SA | 6.21 | 4.30 | 2.15 | 1.87 | 34.6 | 43.4 | 1.79 | 2.59 |
| | $K_2SO_4$ | 6.03 | 3.83 | 1.96 | 1.88 | 32.5 | 49.2 | 1.63 | 2.62 |
| | $MgCO_3$ | 5.90 | 4.33 | 1.71 | 1.26 | 29.0 | 29.1 | 1.43 | 1.75 |
| Tamellat | Control | 3.10 | 1.95 | 1.13 | 0.91 | 36.4 | 46.6 | 0.94 | 1.26 |
| | SA | 3.97 | 2.72 | 1.87 | 1.31 | 47.1 | 48.2 | 1.56 | 1.82 |
| | $K_2SO_4$ | 5.77 | 4.04 | 1.65 | 1.22 | 28.6 | 30.1 | 1.37 | 1.69 |
| | $MgCO_3$ | 4.70 | 3.68 | 1.56 | 1.17 | 33.2 | 31.7 | 1.30 | 1.62 |
| LSD 0.05 | | | | | | | | | |
| Varieties (V) | | 0.151 | 0.142 | 0.160 | 0.120 | | | | |
| Treatment (T) | | 0.132 | 0.122 | 0.120 | 0.110 | | | | |
| (V * T) | | 0.210 | 0.198 | 0.230 | 0.180 | | | | |

normal condition were higher than under water stress condition except WUE where the opposite was true. Regarding the effect of main factor (barley variety), Ksar Megrine gained the greatest values biological yield, grain yield (Giza 125 and Ksar Megrine), harvest index (Giza 125 and Tamellat), and WUE (Giza 125 and Ksar Megrine) under both stress condition, respectively (Table 17.5). Also, one can notice that the differences between the greatest values and lowest values were 25, 22, 23, and 22% under normal condition and 26, 30, 4, and 30% under water stress treatment.

The lowest values were attained at Tamellat (biological yield and grain yield), Ksar Megrine and Giza 125 (harvest index), and Tamellat (WUE) under normal and stress conditions, respectively. The plants utilize certain tactics to counteract the harm that comes from water stress. Applications of antitranspirant therapies might improve grain production and nutrient utilization while reducing the negative effects of water stress. Applying antitranspirants might help barley grow in semiarid and dry regions by lessening the negative impacts of water scarcity.

#### 17.7.1.2 Antitranspirant Effects on Macronutrient Contents

Hellal et al. (2020a) investigated the impact of an antitranspirant chemical on the N, P, and protein content of barley grains under normal irrigation and water stress situations. Results in Table 17.6 pointed out that the maximum and minimum values were attained under normal irrigation and water stress conditions, respectively, and control treatment for each examined barley varieties.

Additionally, data showed that under normal conditions, barley varieties Giza 125 + $K_2SO_4$ and Tombari + SA (salicylic acid) scored the highest values; under water stress conditions, on the other hand, Tombari + SA scored the highest values for N, P, and protein content. Results showed that under normal irrigation and water stress circumstances, the highest and lowest values were obtained for Giza 125 (N %), Tombari (P%), and Giza 125 (protein %), as well as Ksar Megrine-Tamellat, Tamellat-Giza 125, and Ksar Megrine-Tamellat, respectively. Comparison between the greatest with lowest values of N, P, and protein content yielded the following percentages: 27-8, 30-5, and 27-8% under normal irrigation and water stress conditions, respectively. Regarding the barley effect on the K and Ca content in barley grain under normal and water stress condition, data revealed that Ksar Megrine, Tombari, and Giza 125 gained the greatest values under both water stress conditions, respectively, whereas the reduction values were observed under control treatment. Also, the increment percentages when comparing the maximum with minimum values were 29, 11, 92, and 76% under normal irrigation condition and 24, 46, 50, and 17% under water stress conditions. The greatest values of the potassium and calcium content of barley grains were attained after spraying SA under all examined barley varieties and under examined water stress conditions. A similar trend was observed when the water stress condition was eliminated.

**Table 17.6** Effect of antitranspirants on N and P of barley grains under water stress

| Varieties | Treatment | Nitrogen (%) | | Phosphorus (%) | | Potassium (%) | | Calcium (%) | |
|---|---|---|---|---|---|---|---|---|---|
| | | *Normal* | *Stress* | *Normal* | *Stress* | *Normal* | *Stress* | *Normal* | *Stress* |
| Giza 125 | Control | 1.13 | 0.83 | 0.124 | 0.068 | 0.34 | 0.22 | 0.018 | 0.013 |
| | SA | 1.85 | 1.24 | 0.142 | 0.088 | 0.53 | 0.28 | 0.033 | 0.021 |
| | $K_2SO_4$ | 1.95 | 1.15 | 0.135 | 0.093 | 0.43 | 0.28 | 0.031 | 0.019 |
| | $MgCO_3$ | 1.32 | 1.05 | 0.137 | 0.085 | 0.49 | 0.25 | 0.030 | 0.024 |
| Tombari | Control | 1.01 | 0.76 | 0.106 | 0.071 | 0.26 | 0.25 | 0.023 | 0.018 |
| | SA | 1.70 | 1.46 | 0.196 | 0.105 | 0.42 | 0.36 | 0.048 | 0.036 |
| | $K_2SO_4$ | 1.32 | 0.98 | 0.140 | 0.092 | 0.39 | 0.28 | 0.042 | 0.030 |
| | $MgCO_3$ | 1.53 | 1.03 | 0.151 | 0.084 | 0.33 | 0.32 | 0.044 | 0.032 |
| Ksar Megrine | Control | 0.90 | 0.81 | 0.096 | 0.082 | 0.39 | 0.28 | 0.018 | 0.013 |
| | SA | 1.37 | 1.26 | 0.166 | 0.100 | 0.59 | 0.31 | 0.036 | 0.032 |
| | $K_2SO_4$ | 0.95 | 0.83 | 0.112 | 0.097 | 0.41 | 0.36 | 0.022 | 0.018 |
| | $MgCO_3$ | 1.70 | 1.14 | 0.119 | 0.089 | 0.43 | 0.33 | 0.025 | 0.018 |
| Tamellat | Control | 1.03 | 0.90 | 0.098 | 0.082 | 0.37 | 0.22 | 0.016 | 0.011 |
| | SA | 1.84 | 1.19 | 0.124 | 0.100 | 0.44 | 0.33 | 0.025 | 0.023 |
| | $K_2SO_4$ | 1.33 | 0.91 | 0.121 | 0.086 | 0.40 | 0.26 | 0.023 | 0.015 |
| | $MgCO_3$ | 1.53 | 0.97 | 0.112 | 0.097 | 0.39 | 0.30 | 0.018 | 0.013 |
| LSD (0.05) | (V) | 0.013 | 0.05 | 0.007 | 0.006 | 0.014 | 0.013 | 0.006 | 0.005 |
| | (T) | 0.011 | 0.03 | 0.004 | 0.003 | 0.012 | 0.011 | 0.003 | 0.002 |
| | (V * T) | 0.018 | 0.07 | ns | ns | 0.019 | 0.017 | 0.007 | 0.006 |

#### 17.7.1.3 Antitranspirant Effects on Micronutrient Contents

Hellal et al. (2020a) illustrated the effect of same antitranspirant compounds on the Fe, Mn, and Zn content in some examined barley varieties under normal and water stress conditions. Data in Table 17.7 revealed that stress condition had a negative effect on the studied micronutrients content.

The greatest values of the Fe, Mn, and Zn contents were observed under SA (salicylic acid) and $K_2SO_4$ in Tombari and Ksar Megrine, Tamellat and Ksar Megrine, and Ksar Megrine and Tamellat barley varieties under normal and water stress conditions, respectively. The lowest one was attained in control treatments for Tamellat + (Tombari) and Tombari under both normal and stress conditions. The Fe, Mn, and Zn content appeared the lightest values for the studied barley varieties at SA except Zn ($K_2SO_4$ + Ksar Megrine) and $MgCO_3$ + Tamellat, respectively. Also, data showed that the smallest values of the Fe, Mn, and Zn were attained at control, especially Tamellat, Tombari, and Tombari, respectively. Regarding the mean values of the Fe, Mn, and Zn of the examined barley varieties, data showed that Giza 125, Ksar Megrine, and Tamellat scored the greatest values, respectively, while the lowest values were obtained for barley varieties Tamellat, Tamellat, and Tombari in same sequences. The differences between the greatest and lowest micronutrients values in percentage were 26, 86, and 50% in same previous sequence. According to the effect of antitranspirant compound on the Fe, Mn, and Zn contents, resulted data

**Table 17.7** Antitranspirants on Fe, Mn, and Zn of barley grains under water stress

| | | Iron (ppm) | | Manganese (ppm) | | Zinc (ppm) | |
|---|---|---|---|---|---|---|---|
| Varieties | Treatment | *Normal* | *Stress* | *Normal* | *Stress* | *Normal* | *Stress* |
| Giza 125 | Control | 97.64 | 70.19 | 4.06 | 3.83 | 17.95 | 16.35 |
| | SA | 116.20 | 93.55 | 6.40 | 5.10 | 24.36 | 18.93 |
| | $K_2SO_4$ | 112.50 | 85.89 | 5.38 | 4.15 | 22.10 | 18.29 |
| | $MgCO_3$ | 104.40 | 92.23 | 5.30 | 4.31 | 19.10 | 18.05 |
| Tombari | Control | 82.42 | 57.64 | 2.82 | 2.48 | 15.76 | 12.03 |
| | SA | 109.04 | 89.20 | 4.08 | 3.11 | 21.95 | 19.51 |
| | $K_2SO_4$ | 92.32 | 60.47 | 3.05 | 2.89 | 17.49 | 16.06 |
| | $MgCO_3$ | 95.80 | 72.81 | 3.52 | 3.34 | 17.92 | 14.12 |
| Ksar Megrine | Control | 81.44 | 74.76 | 6.93 | 5.31 | 22.82 | 21.13 |
| | SA | 119.10 | 97.78 | 7.80 | 4.36 | 25.37 | 23.35 |
| | $K_2SO_4$ | 95.35 | 80.80 | 6.78 | 4.33 | 27.83 | 24.56 |
| | $MgCO_3$ | 92.26 | 87.58 | 6.21 | 4.69 | 23.31 | 22.91 |
| Tamellat | Control | 67.09 | 58.09 | 2.89 | 2.52 | 23.59 | 20.81 |
| | SA | 95.38 | 81.33 | 4.02 | 3.59 | 24.72 | 23.28 |
| | $K_2SO_4$ | 92.92 | 79.76 | 3.10 | 2.99 | 28.02 | 25.83 |
| | $MgCO_3$ | 75.97 | 62.64 | 2.96 | 2.83 | 28.33 | 27.28 |
| LSD (0.05) | Varieties (V) | 2.913 | 0.823 | 0.153 | 0.174 | 0.986 | 1.091 |
| | Treatment (T) | 1.862 | 0.612 | 0.147 | 0.123 | 1.123 | 1.324 |
| | (V * T) | 4.120 | 1.362 | 0.268 | 0.279 | 1.982 | 2.186 |

revealed that SA and control scored the greatest and lowest values, respectively, with increments of 36%, 25%, and 21%, respectively.

## 17.7.2 *Potassium Application Effects under Water Stress*

### 17.7.2.1 Potassium Application Effects on Chlorophyll Content

Numerous physiological activities, including photosynthesis, the transfer of photosynthesis into think organs, turgor maintenance, enzyme activation, and the reduction of certain ions in soils under stress from the environment, depend on potassium (Hussein et al. 2013). According to data in Table 17.8, under both normal irrigation conditions and water stress, the highest values of chlorophyll content were reported during the examined development stages when barley Giza 125 was sprayed with K citrate and barley Tombari with K silicate.

Conversely, the lowest were obtained with normal and water stress treatments under control (no potassium given). Based on available data, it was shown that the chlorophyll content was significantly reduced in barley variety Giza 125 and Tamellat control during stem elongation, booting, and ear emergency, respectively, when water stress treatments were applied. In contrast, the lowest decrease was seen

**Table 17.8** SPAD chlorophyll of barley varieties under water stress

| | | Elongation of stem | | Booting | | Ear emergency | |
|---|---|---|---|---|---|---|---|
| The varieties | Treatments | Normal | Stress | Normal | Stress | Normal | Stress |
| Giza 125 | Control without K | 43.1 | 28.2 | 47.5 | 39.6 | 45.1 | 38.2 |
| | Potassium citrate | 51.0 | 39.0 | 60.5 | 48.7 | 53.6 | 47.2 |
| | Potassium nitrate | 47.2 | 33.1 | 50.7 | 46.0 | 50.6 | 42.1 |
| | Potassium silicate | 44.7 | 31.6 | 50.1 | 42.3 | 48.4 | 41.7 |
| Tombari | Control without K | 45.1 | 35.2 | 37.6 | 30.7 | 44.9 | 34.9 |
| | Potassium citrate | 50.0 | 44.1 | 45.9 | 35.3 | 47.5 | 36.7 |
| | Potassium nitrate | 52.6 | 44.4 | 48.3 | 37.3 | 51.2 | 40.5 |
| | Potassium silicate | 64.5 | 51.6 | 51.7 | 39.4 | 52.9 | 42.6 |
| Ksar Megrine | Control without K | 44.8 | 37.7 | 37.8 | 30.4 | 38.0 | 32.1 |
| | Potassium citrate | 46.5 | 41.5 | 42.5 | 31.3 | 43.0 | 36.6 |
| | Potassium nitrate | 50.3 | 49.1 | 49.2 | 41.1 | 50.1 | 41.8 |
| | Potassium silicate | 49.2 | 43.9 | 44.1 | 37.1 | 48.9 | 37.1 |
| Tamellat | Control without K | 45.8 | 39.5 | 42.4 | 30.5 | 43.4 | 32.2 |
| | Potassium citrate | 55.1 | 48.1 | 50.4 | 38.0 | 50.5 | 44.4 |
| | Potassium nitrate | 54.5 | 46.6 | 47.4 | 36.1 | 48.5 | 36.5 |
| | Potassium silicate | 48.2 | 44.8 | 45.0 | 30.8 | 45.3 | 36.5 |
| LSD (0.05) | (Var.) | 0.129 | 0.741 | 0.026 | 1.058 | 0.013 | 1.157 |
| | (Treat.) | 0.033 | 0.059 | 1.285 | 0.554 | 0.138 | 0.079 |
| | Interaction | 2.037 | 1.572 | 2.303 | 3.272 | 3.154 | 2.679 |

at Giza 125 + K nitrate (9.2%), barley Ksar Megrine + K nitrate (2.2%), and Giza 125 + K citrate (12.1%).

Foliar application of potassium silicate provides several advantages, including increased photosynthetic efficiency, improved leaf erectness, and decreased grass lodging capacity (Ahmed et al. 2014). Moreover, it has advantages in a variety of agricultural uses, such as increasing growth and output, enhancing strength, reducing climatic stress, and providing resistance to mineral stress. Hellal et al. (2020b) demonstrated the impact on the chlorophyll content during the barley growth phases of foliar spraying potassium fertilizers (K citrate, K nitrate, and K silicate) and an untreated one (control).

#### 17.7.2.2 Potassium Application Effects on Relative Water Content

The significance of potassium as a cation (+) and/or citrate anion (−) in various physiological processes may be the cause of K citrate's beneficial effects on total soluble protein, phenolic compounds, and total soluble carbohydrates: Potassium is essential for the xylem's ability to carry nutrients and water (Malvi 2011; Capelo et al. 2012). Table 17.9 shows the relative water content (RWC) of barley types grown under water stress as influenced by the administration of potassium fertilizer during the booting development stage. In terms of RWC, the barley variety Ksar Megrine sprayed with K nitrate had the highest values (94.9%), whereas the barley variety Tombari, which was not potassium-treated, had the lowest values (72.3%). Conversely, the Giza 125 control treatment under water stress conditions had the lowest RWC value (62.0%) and the highest (85.8%) for Ksar Megrine sprayed with K citrate. Barley Giza 125 sprayed with K citrate showed the largest drop in RWC (24.5%), whereas Tombari treated with K silicate showed the lowest percentage (8.6%). Regardless of potassium treatment, Ksar Megrine and Barley Tombari had the lowest percentage decline in RWC when impacted by the water stress condition Hellal et al. (2020b).

#### 17.7.2.3 Potassium Application Effects on Proline Content

Energy for ATP production, which is necessary for several biochemical and physiological activities, is provided via the mitochondrial citric acid cycle (Taiz and Zeiger 2002). Proline accumulation involved in osmoregulation appeared to allow more water to be taken up from the environment, and the increase in proline concentration in cotton plants allowed the water-stressed plants to maintain low water potentials by reducing water potentials Gebaly et al. (2013). One of the most significant organic acids in plant cells' aspiratory routes is potassium citrate, which is the potassium salt of citric acid. During the booting and ear emergency periods, the proline content was positively impacted by the drought stress. According to Hellal et al. (2020b), Tamellat barley variety control (untreated) had the highest proline content values (3.98 mg/g fresh weight) at the ear emergence stage, while Tamellat

**Table 17.9** Effect of different potassium sources on relative water and proline content of barley varieties under water stress

| Varieties | Fertilizers | RWC% at booting stage | | Proline mg/g fresh weight at booting stage | | Proline mg/g fresh weight Ear emergence | |
|---|---|---|---|---|---|---|---|
| | | *SI* | *DI* | *SI* | *DI* | *SI* | *DI* |
| Giza 125 | Control without K | 78.2 | 61.3 | 1.57 | 3.93 | 1.97 | 4.15 |
| | Potassium citrate | 85.2 | 64.3 | 0.95 | 2.91 | 1.48 | 3.63 |
| | Potassium nitrate | 87.2 | 73.7 | 0.78 | 2.66 | 1.79 | 4.10 |
| | Potassium silicate | 90.2 | 78.2 | 0.63 | 2.43 | 0.79 | 3.27 |
| Tombari | Control without K | 71.5 | 63.5 | 1.27 | 3.90 | 2.01 | 4.57 |
| | Potassium citrate | 80.6 | 72.0 | 1.14 | 2.72 | 1.79 | 4.17 |
| | Potassium nitrate | 86.5 | 77.1 | 1.20 | 2.80 | 1.81 | 4.43 |
| | Potassium silicate | 83.9 | 76.7 | 0.65 | 2.49 | 0.63 | 4.15 |
| Ksar Megrine | Control without K | 85.6 | 73.4 | 1.68 | 3.20 | 2.64 | 4.70 |
| | Potassium citrate | 93.4 | 84.8 | 1.10 | 2.71 | 2.30 | 4.41 |
| | Potassium nitrate | 93.8 | 79.9 | 1.18 | 2.20 | 2.30 | 4.23 |
| | Potassium silicate | 88.0 | 77.5 | 0.80 | 2.10 | 1.37 | 3.15 |
| Tamellat | Control without K | 74.7 | 64.6 | 1.36 | 3.16 | 2.53 | 5.90 |
| | Potassium citrate | 84.7 | 73.9 | 0.43 | 3.83 | 2.34 | 4.81 |
| | Potassium nitrate | 82.5 | 65.7 | 0.52 | 2.25 | 1.90 | 3.20 |
| | Potassium silicate | 91.4 | 78.6 | 0.42 | 1.98 | 1.34 | 3.06 |
| LSD (0.05) | (Var.) | 1.5 | 2.6 | 0.29 | 0.28 | 0.61 | 0.70 |
| | (Treat.) | 1.0 | 1.6 | 0.27 | 0.29 | 0.52 | 0.78 |
| | Interaction | 2.2 | 3.8 | 0.49 | 0.51 | 1.04 | 1.34 |

*RWC* Relative water content, *SI* Sufficient irrigation, *DI* Deficit irrigation

variety underwent water stress conditions recorded the lowest proline content values after spraying K silicate. For Tamellat sprayed with K nitrate and Tombari sprayed with K silicate, the maximum and minimum percentage increments were noted, respectively. The results showed that the Giza 125 and Tamellat barley varieties under control (untreated) had the highest proline content values (3.98 mg/g fresh weight) at the booting stage and (5.97 mg/g fresh weight) at the ear emergence stage, respectively, while the Tamellat variety under water stress conditions had the lowest values after spraying K silicate (Table 17.9).

For Tombari sprayed with K silicate and Tamellat sprayed with K nitrate, the highest and minimum percentage increments were noted, respectively. The least quantity of proline was found when potassium citrate was sprayed, showing its significance in reducing plant water stress.

#### 17.7.2.4 Potassium Application Effects on Grain Yield

According to Ahmed et al. (2014), increases in grain output may be linked to potassium's beneficial effects on nutrient absorption and new growth, which in

**Table 17.10** Biological yield of barley affected by potassium source under water stress

| Varieties | Treatment | Biological yield (ton fed) | | Grain yield (ton/fed) | | Harvest index | | WUE (kgm-3) | |
|---|---|---|---|---|---|---|---|---|---|
| | | *SI* | *DI* | *SI* | *DI* | *SI* | *DI* | *SI* | *DI* |
| Giza 125 | Control | 3.64 | 3.24 | 1.81 | 1.06 | 49.13 | 32.43 | 1.50 | 1.47 |
| | K-citrate | 4.15 | 3.86 | 2.20 | 1.32 | 52.49 | 34.01 | 1.84 | 1.84 |
| | K-nitrate | 4.18 | 2.97 | 2.17 | 1.24 | 51.51 | 41.22 | 1.81 | 1.72 |
| | K-silicate | 4.01 | 3.40 | 2.14 | 1.27 | 52.59 | 36.78 | 1.78 | 1.76 |
| Tombari | Control | 3.52 | 2.61 | 1.32 | 1.15 | 37.37 | 43.30 | 1.11 | 1.58 |
| | K-citrate | 4.64 | 3.70 | 2.29 | 2.22 | 48.94 | 59.41 | 1.92 | 3.08 |
| | K-nitrate | 4.53 | 3.46 | 2.05 | 1.77 | 44.68 | 50.62 | 1.71 | 2.46 |
| | K-silicate | 4.00 | 3.42 | 2.16 | 1.73 | 53.19 | 50.02 | 1.80 | 2.40 |
| Ksar Megrine | Control | 3.44 | 2.49 | 1.36 | 1.30 | 39.15 | 51.60 | 1.14 | 1.81 |
| | K-citrate | 4.27 | 3.31 | 2.25 | 1.84 | 52.20 | 54.97 | 1.88 | 2.55 |
| | K-nitrate | 4.03 | 3.06 | 1.89 | 1.80 | 46.37 | 58.13 | 1.57 | 2.50 |
| | K-silicate | 3.91 | 3.27 | 2.11 | 1.70 | 53.19 | 51.41 | 1.76 | 2.36 |
| Tamellat | Control | 3.37 | 2.78 | 1.07 | 0.84 | 31.24 | 29.86 | 0.89 | 1.17 |
| | K-citrate | 4.23 | 3.62 | 2.17 | 1.39 | 50.62 | 38.06 | 1.81 | 1.94 |
| | K-nitrate | 3.92 | 2.92 | 1.54 | 1.03 | 38.75 | 34.70 | 1.29 | 1.42 |
| | K-silicate | 4.14 | 3.45 | 1.95 | 1.16 | 46.56 | 33.12 | 1.62 | 1.61 |
| LSD (0.05) | (V) | 0.022 | 0.104 | 0.671 | 0.062 | 0.21 | 0.20 | 0.093 | 0.094 |
| | (T) | 0.095 | 0.083 | 0.053 | 0.053 | 0.09 | 0.07 | 0.072 | 0.092 |
| | (V * T) | 1.96 | 1.44 | 0.97 | 0.88 | 3.18 | 3.05 | 0.12 | 0.13 |

*WUE* Water use efficiency, *SI* Sufficient irrigation, *DI* Deficit irrigation

turn led to favorable impacts on the number of opened bolls per plant and boll weight, resulting in increased cotton yield. Applying potassium citrate spray to mango trees was a highly successful way to increase fruit weight and quantity produced. (Ebeed and Abd El-Migeed 2005). Hellal et al. (2020b) showed that, when compared to the other varieties under investigation, the barley variety Giza 125 obtained the highest scores for grain yield, harvest index, and water usage efficiency (Table 17.10).

Under water stress treatment, the application of K silicate and K citrate produced the highest values of yield and yield components; in contrast, the case of the control showed the reverse trend. The investigated varieties of barley grain exhibited higher levels of nitrogen, phosphorus, and protein under normal conditions compared to those under water stress.

Notably, barley Giza 125 had the highest levels of nitrogen and protein under water stress conditions. After applying K citrate, the barley varieties Tombari and Ksar Megrine had the highest grain values for the majority of stress tolerance indicators evaluated, indicating an improvement in their drought tolerance; in contrast, when K silicate was treated, the barley types Giza 125 and Tamellat showed the reverse effect. With a few exceptions, potassium citrate foliar spray treatment produced the highest values of grain and biological yield as well as water usage efficiency under both normal irrigation and water stress conditions.

## 17.7.3 Nano-Silica Application Under Water Stress

### 17.7.3.1 Nano-Silica Effects on Chlorophyll Content

In plants cultivated in circumstances of water constraint, silicon can decrease the electrolyte leakage from rice leaves and thus increase photosynthetic activity. Higher water usage efficiency in wheat was observed by Gong et al. (2003) with the use of Si. According to Gao et al. (2006), Si impacts the mobility of stomata, which in turn affects the rate at which transpiration occurs via stomata. The highest increase percentage was obtained when nano-silica was increased in comparison with the control and 40% water stress, regardless of the kind of barley. Also, according to Zarafshar et al. (2015), large concentrations of Nano-$SiO_2$ applied to irrigated pear seedlings showed no harmful effects on plant biology. The results of a study conducted by Hellal et al. (2020c) on the impact of foliar application of nano-silica (0, 25, 50, and 100 ppm) on the SPAD chlorophyll of barley under varying water stress conditions are displayed in Fig. 17.5.

A significant reduction in the amount of chlorophyll in barley leaves was noted during periods of soil dryness. The highest SPAD chlorophyll levels under 80% WHC and 100 ppm nano-silica were recorded at Giza 133, followed by Giza 136 and Giza 129. The application of nano-silica to a substrate facilitated the induction of photosynthetic pigment production in barley leaf tissues or altered the processes involved in photosynthetic pigment biosynthesis to improve plant resilience to drought. A decrease in stress was indicated by an increase in chlorophyll in photosystems.

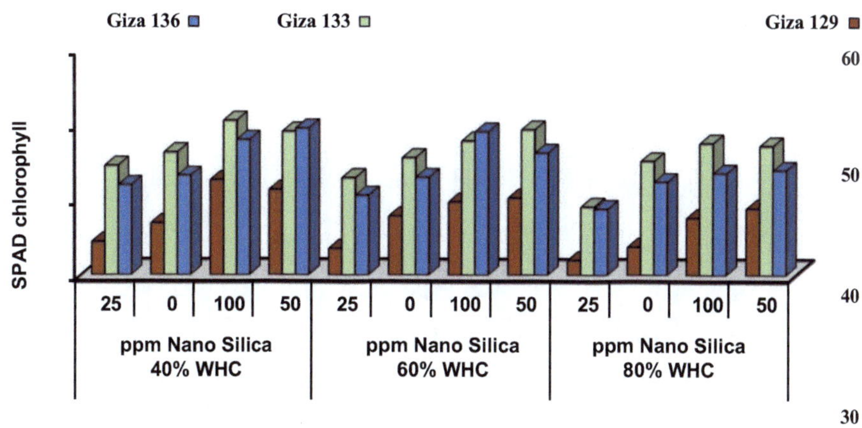


**Fig. 17.5** Chlorophyll content of barley as affected by nano-Si application under water stress

#### 17.7.3.2 Nano-Silica Effects on Proline Content

Proline buildup in plant cells aids in water storage and is a crucial physiological process for drought tolerance in plants. The proline concentration increases considerably in treatments with low soil moisture (Fig. 17.6). Under the application of 100 ppm nano-silica, the highest values for proline were found at Giza 129 > Giza 133 > Giza 136 in descending order. It is also obvious to note that for the types under investigation, the lowest proline levels were obtained at 40% WHC and zero nano-silica. The sequence might be Giza 133 < Giza 136 < Giza 129, in decreasing order. These findings are consistent with studies that have been published in the literature (Mauad et al. 2016; Shanan and El Sadek 2017), which have shown that higher plants' proline content increases in response to various drought stressors. This finding lends credence to the theory that proline buildup is a sign of damage caused by stress. According to Hellal et al. (2020c), foliar spraying of nano-silica reduced the proline concentration in the leaves of maize and wheat seedlings across all treatments. This suggested that the treatment of nano-silica to seedlings resulted in a reduced stress level and that the buildup of proline is unrelated to the protective effect of nano-silica.

#### 17.7.3.3 Nano-Silica Effects on Macronutrient Content

Good crop yields can result from cereal plants having an adequate supply of silicon (Si), as Si can increase cell strength and durability. In response to heavy nitrogen fertilization, a supply of Si causes leaves to become more upright, which increases photosynthesis. By reducing the permeability of water vapor from leaf epidermal cell walls, adequate use of silicon (Si) may lessen the probability of wilting plants during drought circumstances. Additionally, it influences phosphorus fixation, increasing its availability (Sumida 2002). A fundamental need for healthy growth, development, and survival under a variety of environmental stress situations is mineral nutrition.

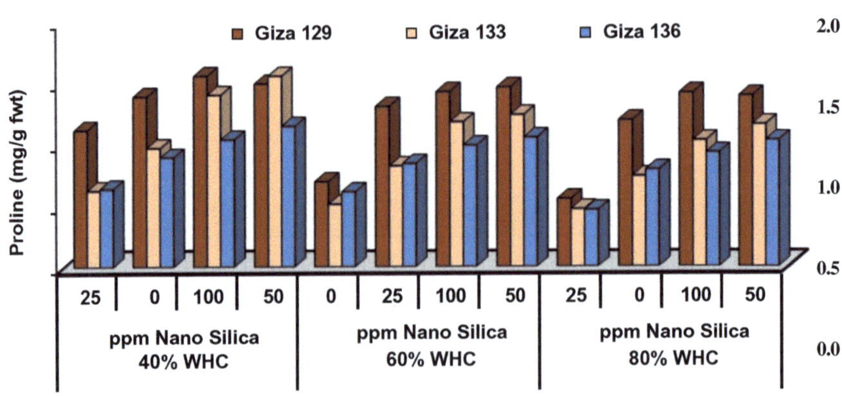


**Fig. 17.6** Proline content of barley as affected by nano-Si application under water stress

Research has demonstrated that a plant's mineral and nutritional status is crucial for mitigating the effects of abiotic stress (Nazar et al. 2015). Hellal et al. (2020c) provided examples of how water stress (80, 60, and 40% from water holding capacity) and nano-silica (0: 100 ppm) affected the plant nutrients (P, K, and Ca) content of a few chosen Egyptian barley types (Table 17.11). Most of the highest values were found at barley types of Giza 133 and Giza 129 at both 80% WHC and 100 ppm nano-silica, while the lowest values were found at 40% of WHC and 0% nano-silica, according to the data. The highest and lowest P, K, and Ca content values were noted at 80% WHC and 100 ppm nano-silica for P, K, and Ca, while the lowest values were noted at 40% WHC and zero nano-silica treatments, regardless of the types that were investigated. It goes without saying that increased water stress is linked to a decrease in the amount of nutrients that plants contain. Data on the second research component, nano-silica, showed that when examined plant nutrients in barley were increased, the concentration of the material increased to 100 ppm. Giza 129 under 40% stress and 100 ppm nano-silica had the highest concentration of calcium content, whereas Giza 129 under 80% water stress and zero nano-silica had the lowest concentrations.

#### 17.7.3.4 Nano-Silica Effects on Micronutrient Content

The target of applying nano-silica fertilizer to plants' leaves is to supply silica, a micronutrient element that plants require and can adequately absorb to ensure optimal development. It is possible to use nanotechnology to improve plants' capacity to absorb nutrients. Fertilizers that are nano-sized are more efficient and easier to absorb than those that are chemically based (Alejandro and Rubiales 2009). The effects of water stress and the application of nano-silica on micronutrient content (Fe, Zn, and Mn) were demonstrated by Hellal et al. (2020c). The results showed that under 80% water stress and 100 ppm nano-silica, the Giza 129 variety increased its Fe and Zn content to the highest levels. In the same sequences, Giza 136 experienced the highest value gain. In contrast, the lowest values for Fe and Zn under 40% water stress and 0 ppm nano-silica were found at Giza 129, while the lowest values for Mn Content were found at Giza 133 variety in the same sequence (Table 17.12).

#### 17.7.3.5 Nano-Silica Effects on Yield Parameters

The distinct physiochemical characteristics of nanoscale silicon particles have practical uses in a variety of industries, with the agriculture sector showing great promise. Si-NPs' special qualities enable them to withstand abiotic stress and/or agricultural harm brought on by climate change Tripathi et al. (2012) By using nano-fertilizers, one may increase nutrient efficiency, decrease contaminants that seep into soil and groundwater, eliminate the possible side effects of overdosing, and apply less often. According to Siddiqui and Al-Whaibi (2014), Si-NPs may have a direct

**Table 17.11** Nano-silica effects on P, K, and Ca content of barley under water stress

| Treatments | | P (%) | | | K (%) | | | Ca (%) | | |
|---|---|---|---|---|---|---|---|---|---|---|
| WHC | Nano-silica | G 129 | G 133 | G 136 | G 129 | G 133 | G 136 | G 129 | G 133 | G 136 |
| 80.0% | 0.0 ppm | 0.180 | 0.569 | 0.190 | 1.58 | 0.88 | 1.30 | 1.08 | 1.63 | 1.40 |
| | 25 ppm | 0.268 | 0.419 | 0.195 | 1.73 | 1.27 | 1.52 | 1.44 | 2.05 | 1.58 |
| | 50 ppm | 0.417 | 0.510 | 0.277 | 1.91 | 1.47 | 1.93 | 1.52 | 2.44 | 1.86 |
| | 100 ppm | 0.432 | 0.448 | 0.261 | 1.89 | 1.84 | 1.95 | 1.54 | 3.35 | 2.47 |
| 60.0% | 0.0 ppm | 0.598 | 0.875 | 0.455 | 1.47 | 0.84 | 1.21 | 1.58 | 1.50 | 1.37 |
| | 25 ppm | 0.658 | 0.757 | 0.587 | 1.69 | 1.16 | 1.50 | 1.77 | 1.90 | 1.57 |
| | 50 ppm | 0.750 | 0.763 | 0.621 | 1.96 | 1.30 | 1.78 | 1.90 | 2.35 | 1.84 |
| | 100 ppm | 0.711 | 0.745 | 0.648 | 1.94 | 1.42 | 1.91 | 1.93 | 2.64 | 2.06 |
| 40.0% | 0.0 ppm | 0.104 | 0.246 | 0.244 | 1.35 | 0.74 | 1.11 | 2.03 | 1.27 | 1.15 |
| | 25 ppm | 0.198 | 0.295 | 0.263 | 1.66 | 0.90 | 1.41 | 2.39 | 1.74 | 1.54 |
| | 50 ppm | 0.293 | 0.395 | 0.314 | 1.91 | 1.28 | 1.63 | 2.66 | 2.12 | 1.76 |
| | 100 ppm | 0.306 | 0.448 | 0.257 | 2.01 | 1.54 | 1.80 | 3.50 | 2.55 | 1.99 |
| LSD (0.05) | | (V) 0.040 | WHC 0.038 | NS 0.047 | (V) 0.066 | WHC 0.062 | NS 0.076 | (V) 0.027 | WHC 0.024 | NS 0.031 |
| | | V*WHC 0.070 | V*NS 0.081 | WHC*NS ns | V*WHC ns | V*NS 0.132 | WHC*NS ns | V*WHC ns | V*NS 0.054 | WHC*NS ns |
| | | V*WHC*NS 0.140 | | | V*WHC*NS ns | | | V*WHC*NS ns | | |

*V* Varieties, *WHC* Water holding capacity, *NS* Nano-silica, *ns* Non-significant

**Table 17.12** Nano-silica effects on Fe, Zn, and Mn content of barley under water stress

| Treatments | | Fe (ppm) | | | Zn (ppm) | | | Mn (ppm) | | |
|---|---|---|---|---|---|---|---|---|---|---|
| WHC | Nano-silica | G 129 | G 133 | G 136 | G 129 | G 133 | G 136 | G 129 | G 133 | G 136 |
| 80.0% | 0.0 ppm | 62.7 | 46.4 | 45.5 | 17.13 | 13.37 | 11.07 | 3.79 | 3.30 | 3.76 |
| | 25 ppm | 76.0 | 79.0 | 55.3 | 18.16 | 16.11 | 12.51 | 4.55 | 4.22 | 4.77 |
| | 50 ppm | 86.5 | 91.9 | 68.2 | 19.11 | 16.83 | 14.31 | 5.26 | 5.01 | 5.83 |
| | 100 ppm | 92.1 | 94.7 | 89.5 | 21.13 | 18.56 | 16.83 | 5.31 | 5.83 | 5.67 |
| 60.0% | 0.0 ppm | 58.7 | 39.1 | 43.1 | 16.63 | 13.17 | 10.68 | 3.55 | 3.13 | 3.46 |
| | 25 ppm | 67.7 | 66.4 | 51.6 | 17.75 | 15.79 | 11.66 | 4.33 | 4.05 | 4.62 |
| | 50 ppm | 87.6 | 88.6 | 65.0 | 18.71 | 16.48 | 13.62 | 4.96 | 4.79 | 5.55 |
| | 100 ppm | 85.7 | 86.3 | 86.2 | 19.62 | 17.73 | 16.05 | 5.08 | 5.58 | 5.32 |
| 40.0% | 0.0 ppm | 49.9 | 34.0 | 39.5 | 15.36 | 12.49 | 10.33 | 3.12 | 2.81 | 3.22 |
| | 25 ppm | 66.0 | 58.7 | 46.3 | 17.58 | 14.85 | 11.37 | 4.06 | 3.65 | 4.28 |
| | 50 ppm | 81.2 | 81.0 | 61.1 | 18.39 | 16.28 | 13.03 | 4.79 | 4.45 | 5.04 |
| | 100 ppm | 82.6 | 89.0 | 76.8 | 19.57 | 17.31 | 14.59 | 4.71 | 5.35 | 5.08 |
| LSD (0.05) | | (V) 0.33 | WHC 0.321 | NS 0.381 | (V) 0.233 | WHC 0.228 | NS 0.269 | (V) 0.233 | WHC 0.213 | NS 0.269 |
| | | V*WHC 0.552 | V*NS 0.572 | WHC*NS 0.64 | V*WHC 0.403 | V*NS 0.466 | WHC*NS ns | V*WHC 0.403 | V*NS 0.466 | WHC*NS ns |
| | | V*WHC*NS 0.66 | | | V*WHC*NS ns | | | V*WHC*NS ns | | |

*V* Varieties, *WHC* Water holding capacity, *NS* Nano-silica, *ns* Non-significant

effect on plants, changing their morphology and physiology in a variety of ways, such as giving the plants a structural color and enhancing plant growth and production. Furthermore, a recent study by Zarafshar et al. (2015) revealed that watering of pear seedlings with high concentrations of Nano-SiO2 had no harmful effects on plant biology.

Applications of nano-SiO2 solutions in semiarid regions have been shown to increase barley seed production and can be introduced as beneficial fertilizer for foliar application (Hamid et al. 2019). Due to their unique physicochemical properties, silicon nanoparticles enhance plant development and yield in adverse environmental circumstances. To investigate how the dosage and moisture levels of nano-silica affect drought resistance, Hellal et al. (2020d) carried out an experiment. The water regime was 40, 60, and 80% of the water holding capacity (WHC), and Nano-silica was administered at varying rates (0.0, 25, 50, and 100 ppm) (Table 17.13).

The results showed that the right concentration of 100 ppm of nano-silica applied topically under water stress might improve yield component and increase barley plants' tolerance to drought stress.

## *17.7.4 Ascorbic Acid Effects Under Water Stress*

### 17.7.4.1 Ascorbic Acid Effects on Plant Height

Reduced photosynthesis, transpiration, and other biochemical processes—which are closely linked to plant growth, development, and agricultural productivity—were caused by the water deficit Tiwari & Associates (Tiwari et al. 2010). According to Asada (2006), abiotic stress that caused oxidative stress in the plant cell caused a greater electron leakage toward O2 during photosynthetic and respiratory activities, which had an impact on the growth and development of the plant. According to Hellal et al. (2019a, b), barley Giza 130 (81.3 cm) showed an increase in plant height values, whereas Giza 124 (79.5 cm) showed the most gains following ascorbic acid spraying, which was reflected in the production of straw. Spraying ascorbic acid has a detrimental effect on the majority of the barley cultivars under examination, regardless of water stress. Straw yield might be reduced because of this. Giza 132 (10.4%), Giza 128 (6.3%), and the lowest ones, Giza 136 and Giza 131, are the Egyptian barley cultivars that are impacted by ascorbic acid spraying. In terms of water stress and how it affects plant height, the ranking of plant height in relation to water stress treatment can be arranged in descending order as follows: WHC at 75% > 60 > 40, with Giza 124 recording the highest values and Giza 126 recording the lowest, at 40% of WHC (Fig. 17.7).

**Table 17.13** Nano-silica effects on yield parameters of barley under water stress

| Treatment | | 1000 grain weight | | | Grain weight (g/pot) | | | Straw weight (g/pot) | | |
|---|---|---|---|---|---|---|---|---|---|---|
| WHC | Nano-silica | G 129 | G 133 | G 136 | G 129 | G 133 | G 136 | G 129 | G 133 | G 136 |
| 80.0% | 0.0 ppm | 25.6 | 35.6 | 37.5 | 23.92 | 16.51 | 17.60 | 52.49 | 71.18 | 60.40 |
| | 25 ppm | 28.8 | 38.5 | 39.8 | 24.32 | 17.79 | 21.35 | 62.28 | 86.60 | 75.63 |
| | 50 ppm | 29.8 | 40.7 | 41.3 | 26.00 | 18.98 | 24.02 | 66.24 | 93.52 | 86.70 |
| | 100 ppm | 32.2 | 42.1 | 42.2 | 28.97 | 20.56 | 27.78 | 69.40 | 92.73 | 84.82 |
| 60.0% | 0.0 ppm | 23.6 | 33.8 | 35.2 | 19.28 | 14.14 | 16.71 | 51.80 | 67.42 | 59.91 |
| | 25 ppm | 28.7 | 38.2 | 39.4 | 24.12 | 17.40 | 20.86 | 56.45 | 74.15 | 74.15 |
| | 50 ppm | 29.4 | 40.6 | 39.8 | 25.51 | 18.78 | 23.73 | 62.68 | 90.16 | 81.76 |
| | 100 ppm | 31.2 | 41.4 | 41.4 | 28.27 | 20.46 | 27.09 | 68.41 | 89.57 | 80.37 |
| 40.0% | 0.0 ppm | 21.9 | 33.2 | 34.9 | 18.09 | 13.15 | 15.82 | 43.79 | 62.38 | 56.55 |
| | 25 ppm | 28.0 | 37.5 | 37.7 | 23.92 | 16.71 | 18.09 | 54.18 | 71.57 | 67.42 |
| | 50 ppm | 29.0 | 39.0 | 39.0 | 25.01 | 18.19 | 21.65 | 62.58 | 88.08 | 77.61 |
| | 100 ppm | 29.7 | 39.8 | 39.3 | 25.41 | 20.66 | 20.17 | 66.73 | 87.29 | 80.87 |
| LSD (0.05) | | (V) 0.201 | WHC 0.237 | NS 0.274 | (V) 0.322 | WHC 0.384 | NS 0.444 | (V) 0.413 | WHC 0.487 | NS 0.562 |
| | | V*WHC 0.411 | V*NS 0.427 | WHC*NS 0.475 | V*WHC 0.666 | V*NS 0.746 | WHC*NS 0.769 | V*WHC 0.843 | V*NS 0.913 | WHC*NS 0.974 |
| | | V*WHC*NS 0.822 | | | V*WHC*NS 1.3312 | | | V*WHC*NS 1.687 | | |

*V* Varieties, *WHC* Water holding capacity, *NS* Nano-silica, *ns* Non-significant

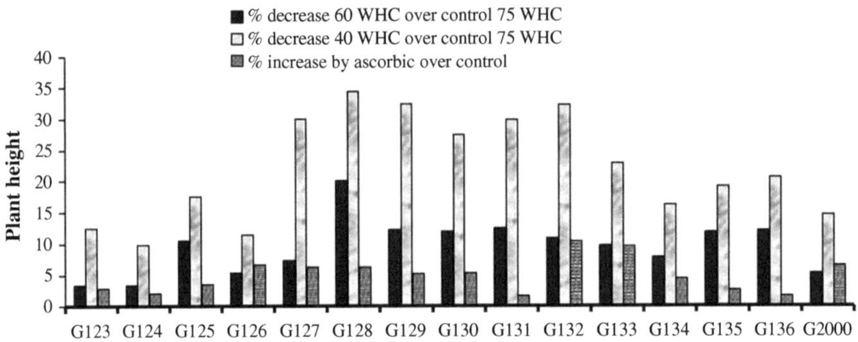


**Fig. 17.7** Water stress effects on plant height of barley cultivars

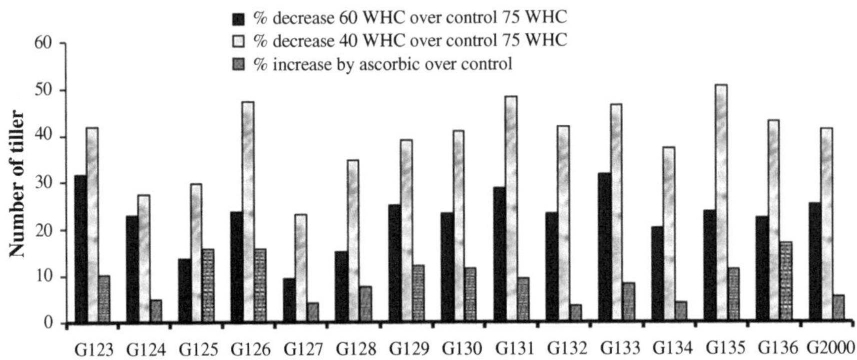


**Fig. 17.8** Number of tiller/pot of barley cultivars as affected by water stress

#### 17.7.4.2 Ascorbic Acid Effects on Tiller Number per Plant

According to Metwally et al. (2003), ascorbic acids have an impact on the metabolism of plant responses, which in turn accounts for adaptabilities that increase a plant's tolerance to environmental influences. The results demonstrated a substantial drop in the number of tillers per plant and per pot during harvest when there was water stress (drought). Water stress treatments had an impact on the tiller number/plant for most barley cultivars, even when the number of plants/pots was predetermined before the drought treatments were applied. Hellal et al. (2019a, b) observed that when soil humidity was reduced to 60% WHC, the proportion of tillers/plants decreased from 11% (Giza 127) to 37% (Giza 123). As soil humidity decreased to 40% of WHC which ranged from 24% (Giza 125) to 58% (Giza 135), this phenomenon increased. The Giza 127, Giza 125, and Giza 128 cultivars all demonstrated a high degree of tolerance for irrigation water deficits when compared to the other cultivars under study (Fig. 17.8). This resulted in a minimum decrease in the number of plants/pots under water stress of 60% WHC and 40% WHC when compared to the control treatment (75% of WHC). By applying ascorbic acid spray, the barley cultivars Giza 136, Giza 123, Giza 129, and Giza 135 demonstrated a noteworthy increase in the number of tillers per pot.

#### 17.7.4.3 Effect of Ascorbic Acid Application on Grain Yield

Plant responses would be impacted by ascorbic acid and undergo several alterations as a result. These alterations are occasionally explained as adaptations that increase a plant's tolerance or resistance to external stimuli (Metwally et al. 2003). By reducing the number of viable spikes and grains per plant, drought stress during the seed filling stage decreased grain output, as demonstrated by Samarah (2005). Grain yield was lowered by late drought stress, which also reduced grain weight and the quantity of grains per spick. Ascorbic acid contributes to both the process of photosynthesis and the body's defense against oxidative stress.

One of the most important cereal crops farmed in many developing nations is barley, which frequently experiences severe drought stress that has a negative impact on yield (Ceccarelli et al. 2007). Hellal et al. (2019a, b) investigated the effects of ascorbic acid and water stressors on a few yield components in an experiment involving many Egyptian barley cultivars (Table 17.14). Under open field circumstances, two amounts of ascorbic acid (zero and 10 ppm) were used as a subprimary therapy for the three main levels of water stress (75, 60, and 40% of water holding capacity, or WHC). When barley cultivars experience water stress, their grain production metrics significantly fall, regardless of whether they retain 60% or 40% of their water capacity in comparison with the control.

Under conditions of water stress, the application of 10 ppm ascorbic acid increased the yield parameter. Growing water stress has a significant negative impact on barley cultivars (Giza 125, Giza 126, Giza 128, Giza 129, Giza 133, and Giza 135). Grain yield in barley Giza 123, Giza 124, Giza 131, Giza 132, and Giza 2000 decreased when water stress increased to 40% of WHC. Ascorbic acid spraying increased the grain yield of the barley cultivars Giza 125, Giza 131, Giza 132, Giza 135, and Giza 2000.

### *17.7.5 Compost and Arginine Application Under Water Stress*

#### 17.7.5.1 Compost and Arginine Effects on Chlorophyll Content

Using compost appears like a practical approach to mitigate the negative effects of water scarcity by enhancing plant water status, SPAD value, chlorophyll fluorescence, plant growth traits, and irrigation efficiency, according to Taia et al. (2019). Data about the arginine impact showed that the least amount of chlorophyll was created by untreated arginine. Hellal et al. (2024) demonstrated the impact of arginine application and compost application rates (0.0, 2.0, 4.0, and 6.0-ton fed-1) on the pigment content (chlorophyll) reading of some barley varieties under

**Table 17.14** Grain yield per pot of barley cultivars as affected by ascorbic acid under water stress

| | Water holding capacity (a) | | | | | | LSD at 0.05 | | |
|---|---|---|---|---|---|---|---|---|---|
| | 75% | | 60% | | 40% | | | | |
| Barley cultivars | Control | Ascorbic acid | Control | Ascorbic acid | Control | Ascorbic acid | WHC (a) | Ascorbic acid (b) | (a*b) |
| Giza 123 | 41.2 | 45.5 | 37.0 | 39.1 | 30.4 | 31.9 | 0.37 | 0.23 | 0.55 |
| Giza 124 | 47.5 | 48.3 | 41.8 | 43.0 | 28.9 | 34.7 | 0.95 | 0.23 | 1.13 |
| Giza 125 | 47.9 | 48.3 | 33.6 | 46.7 | 21.8 | 29.9 | 0.85 | 1.05 | 1.84 |
| Giza 126 | 56.4 | 56.6 | 39.4 | 46.0 | 21.6 | 24.9 | 1.40 | 0.36 | 1.70 |
| Giza 127 | 55.8 | 59.4 | 39.8 | 46.3 | 30.0 | 30.3 | 0.67 | 0.38 | 1.00 |
| Giza 128 | 55.1 | 55.2 | 39.1 | 39.8 | 23.5 | 25.8 | 1.20 | 0.34 | 1.09 |
| Giza 129 | 43.0 | 46.0 | 34.8 | 37.6 | 21.9 | 22.4 | 0.95 | 0.13 | 1.03 |
| Giza 130 | 50.8 | 52.7 | 42.6 | 44.9 | 29.1 | 35.7 | 1.00 | 0.40 | 1.35 |
| Giza 131 | 35.5 | 38.4 | 31.1 | 35.5 | 25.1 | 25.5 | 0.31 | 0.21 | 0.47 |
| Giza 132 | 46.6 | 51.0 | 32.2 | 42.4 | 22.7 | 30.6 | 0.64 | 1.10 | 1.67 |
| Giza 133 | 50.4 | 53.9 | 34.7 | 48.9 | 25.4 | 27.2 | 0.61 | 0.92 | 1.48 |
| Giza 134 | 38.4 | 42.2 | 35.7 | 35.9 | 26.2 | 26.3 | 0.63 | 0.20 | 0.58 |
| Giza 135 | 41.2 | 45.5 | 37.0 | 39.1 | 30.4 | 31.9 | 0.37 | 0.23 | 0.55 |
| Giza 136 | 47.5 | 48.3 | 41.8 | 43.0 | 28.9 | 34.7 | 0.95 | 0.23 | 1.13 |
| Giza 2000 | 47.9 | 48.3 | 33.6 | 46.7 | 21.8 | 29.9 | 0.85 | 1.05 | 1.84 |

(Giza 125, Tombari, Ksar Megrine, and Tamellat) under sufficient irrigation (SI) and deficit irrigation (DI). The results in Table (15) showed that under SI and DI treatments, compost application rates, with or without arginine, considerably increase chlorophyll levels. The compost treated with arginine at 6.0 tons fed-1 yielded the highest and most significant quantities of chlorophyll. The barley types under study may be easily arranged in declining order as follows: The irrigation water treatments for Ksar Megrine, Tamellat, Giza 125, and Tombari are combined. Furthermore, the findings demonstrated that, under both irrigation regimes under study, arginine increased chlorophyll; however, the augmentation under DI was greater than that under SI.

#### 17.7.5.2 Compost and Arginine Effects on Proline of Barley

Proline and free amino acid synthesis is enhanced by arginine supplementation, regardless of the growth condition of wheat plants (Hussein et al. 2022). By improving physiological and metabolic features, arginine treatment had a major impact on reducing the impacts of drought stress on wheat plants.

Hellal et al. (2024) found that for all the examined barley types under both irrigation treatments, there was a significant drop in proline concentration as compost application rates increased and treated barley plants by arginine. Additionally, results showed that proline content during DI treatment had significant values higher than those under SI treatment. In all the barley varieties under study, the highest and most significant values of proline content were found at control. Tombari had the highest value (2.81, 2.57) under SI without arginine, while Giza 125 (3.85) and Ksar Megrine (3.53) had the highest values under DI without Arginine treatments, respectively (Table 17.15).

#### 17.7.5.3 Compost and Arginine Effects on Grain Yield

By supplying nutrients, organic matter directly affects crop development and production. It also indirectly affects these factors by altering the physical characteristics of the soil, which can enhance the root environment and promote plant growth. Applying compost helps maintain the highest level of water usage efficiency for plant development while also enhancing the physical and chemical characteristics of the soil (Bandyopadhyay et al. 2010). The growth of vegetation, stomatal activity, and osmotic potential have all been related to arginine. Additionally, it contributes to the production of proline, polyamines, and proteins (Nasibi et al. 2011).

Du Jardin (2015) stated that the most versatile amino acid, arginine is necessary for stress tolerance and is linked to the synthesis of signaling molecules. (Hamid et al. 2019). Hussein et al. (2022) concluded that arginine supplementation enhanced the synthesis of soluble sugars, proline, free amino acids, phenols, and flavonoids in wheat plants under both normal and stressed circumstances. Hellal et al. (2024) showed that applying compost (0.0, 2.0, 4.0, and 6.0-ton fed-1) and arginine (0.00

**Table 17.15** Chlorophyll and proline of barley as affected by arginine (Arg) and irrigation regimes

| Varieties | Compost | SPAD chlorophyll | | | | Proline (mg $g^{-1}$ fresh weight) | | | |
|---|---|---|---|---|---|---|---|---|---|
| | | Sufficient irrigation | | Deficit irrigation | | Sufficient irrigation | | Deficit irrigation | |
| | | Arg (−) | Arg (+) | Arg (−) | Arg (+) | Arg (−) | Arg (+) | Arg (−) | Arg (+) |
| Giza 125 | 0.0-ton $fed^{-1}$ | 41.50 | 42.20 | 31.29 | 36.00 | 2.45 | 2.21 | 3.85 | 2.99 |
| | 2.0-ton $fed^{-1}$ | 42.60 | 45.40 | 31.87 | 38.80 | 2.30 | 2.16 | 2.84 | 2.38 |
| | 4.0-ton $fed^{-1}$ | 46.10 | 50.10 | 34.45 | 42.80 | 1.55 | 1.46 | 2.60 | 2.21 |
| | 6.0-ton $fed^{-1}$ | 47.30 | 51.60 | 35.86 | 45.70 | 1.28 | 1.24 | 2.38 | 2.07 |
| Tombari | 0.0-ton $fed^{-1}$ | 40.50 | 40.61 | 30.21 | 34.46 | 2.81 | 2.57 | 3.44 | 2.85 |
| | 2.0-ton $fed^{-1}$ | 41.40 | 42.00 | 31.21 | 35.87 | 2.33 | 2.16 | 2.74 | 2.62 |
| | 4.0-ton $fed^{-1}$ | 41.10 | 43.40 | 31.70 | 37.53 | 2.06 | 1.95 | 2.66 | 2.57 |
| | 6.0-ton $fed^{-1}$ | 42.40 | 45.80 | 36.02 | 41.85 | 2.00 | 1.87 | 2.18 | 2.09 |
| Ksar Megrine | 0.0-ton $fed^{-1}$ | 38.20 | 39.64 | 26.89 | 31.56 | 2.65 | 2.12 | 3.53 | 2.78 |
| | 2.0-ton $fed^{-1}$ | 41.22 | 45.10 | 30.05 | 35.79 | 2.09 | 1.87 | 2.74 | 2.60 |
| | 4.0-ton $fed^{-1}$ | 45.76 | 51.90 | 32.62 | 38.69 | 1.75 | 1.53 | 2.65 | 2.52 |
| | 6.0-ton $fed^{-1}$ | 48.97 | 55.41 | 36.52 | 44.75 | 1.48 | 1.43 | 2.29 | 2.20 |
| Tamellat | 0.0-ton $fed^{-1}$ | 40.60 | 41.77 | 28.10 | 33.47 | 2.03 | 1.81 | 3.91 | 2.79 |
| | 2.0-ton $fed^{-1}$ | 43.10 | 44.20 | 32.29 | 39.40 | 1.67 | 1.57 | 2.33 | 1.97 |
| | 4.0-ton $fed^{-1}$ | 44.30 | 47.80 | 33.90 | 42.80 | 1.45 | 1.36 | 2.11 | 1.87 |
| | 6.0-ton $fed^{-1}$ | 50.70 | 54.80 | 36.77 | 44.67 | 1.11 | 1.06 | 1.54 | 1.41 |
| LSD (0.05) | Varieties | 0.36 | 0.44 | 0.26 | 0.53 | 0.10 | 0.06 | 0.14 | 0.19 |
| | Compost | 0.31 | 0.38 | 0.22 | 0.46 | 0.07 | 0.05 | 0.04 | 0.25 |
| | Interaction | 0.63 | 0.76 | 0.45 | 0.91 | 0.15 | 0.11 | 0.17 | 0.41 |

and 1000 ppm) under deficit irrigation might mitigate the detrimental effects of water stress on the yield of Mediterranean barley cultivars (Giza 125, Tombari, Ksar Megrine, and Tamellat) (Table 17.16). Tombari barley saw the highest values with adequate irrigation, while Tamellat saw the largest values under insufficient irrigation. Tombari had the highest production percentage at 24.6%, while Tamellat saw

**Table 17.16** Grain and biological yield of barley as affected by arginine and compost application

| Varieties | Compost | Grain yield (ton/fed) | | | | Biological yield (ton/fed) | | | |
|---|---|---|---|---|---|---|---|---|---|
| | | SI | | DI | | SI | | DI | |
| | | Arg (−) | Arg (+) | Arg (−) | Arg (+) | Arg (−) | Arg (+) | Arg (−) | Arg (+) |
| Giza 125 | 0 ton/fed | 1.44 | 1.49 | 1.27 | 1.33 | 3.37 | 3.50 | 2.22 | 2.27 |
| | 2 ton/fed | 1.59 | 1.68 | 1.30 | 1.39 | 4.04 | 4.26 | 2.95 | 3.08 |
| | 4 ton/fed | 1.84 | 1.96 | 1.32 | 1.42 | 4.61 | 5.00 | 3.31 | 3.72 |
| | 6 ton/fed | 1.93 | 2.10 | 1.33 | 1.51 | 4.98 | 5.58 | 4.05 | 4.61 |
| Tombari | 0 ton/fed | 1.63 | 1.67 | 1.29 | 1.33 | 2.99 | 3.19 | 2.69 | 2.84 |
| | 2 ton/fed | 1.74 | 1.84 | 1.30 | 1.37 | 3.78 | 4.04 | 3.24 | 3.46 |
| | 4 ton/fed | 1.82 | 1.94 | 1.31 | 1.45 | 3.92 | 4.18 | 3.45 | 3.79 |
| | 6 ton/fed | 1.94 | 2.07 | 1.39 | 1.56 | 4.73 | 5.23 | 3.54 | 3.98 |
| Ksar Megrine | 0 ton/fed | 1.49 | 1.58 | 1.28 | 1.31 | 2.84 | 2.98 | 3.12 | 3.20 |
| | 2 ton/fed | 1.54 | 1.65 | 1.25 | 1.33 | 3.53 | 3.78 | 3.38 | 3.53 |
| | 4 ton/fed | 1.69 | 1.82 | 1.30 | 1.42 | 4.13 | 4.37 | 3.61 | 3.88 |
| | 6 ton/fed | 1.75 | 1.91 | 1.36 | 1.51 | 4.47 | 4.77 | 4.33 | 4.76 |
| Tamellat | 0 ton/fed | 1.47 | 1.49 | 1.34 | 1.42 | 3.24 | 3.36 | 2.68 | 2.84 |
| | 2 ton/fed | 1.62 | 1.73 | 1.37 | 1.48 | 4.69 | 5.05 | 4.41 | 4.82 |
| | 4 ton/fed | 1.65 | 1.79 | 1.40 | 1.53 | 4.97 | 5.38 | 5.27 | 5.88 |
| | 6 ton/fed | 1.86 | 2.04 | 1.41 | 1.56 | 5.17 | 5.70 | 5.64 | 6.50 |
| LSD (0.05) | Varieties | 0.04 | 0.05 | 0.03 | 0.04 | 0.17 | 0.26 | 0.16 | 0.15 |
| | Compost | 0.06 | 0.07 | 0.04 | 0.05 | 0.27 | 0.36 | 0.17 | 0.30 |
| | Interaction | 0.09 | 0.10 | 0.07 | 0.08 | 0.41 | 0.56 | 0.30 | 0.41 |

the lowest at 15.4%. When there was enough irrigation water, the biological yield values generated were higher than when there was insufficient irrigation water. Additionally, the amount of decomposed material is linked to an increase in biological production, particularly following the application of 6-ton fed$^{-1}$ compost. The highest water consumption efficiency values were obtained when arginine was sprayed topically in both normal irrigation and water stress scenarios. It has been demonstrated that the application of compost and arginine combined enhances plant growth and production qualities in irrigated barley varieties.

## *17.7.6 Biofertilizers Application Under Water Stress*

### 17.7.6.1 Biofertilizer Effects on Chlorophyll Contents

One of the most significant metabolic processes in plants that is impacted by water stress is photosynthesis. According to Grzesik et al. (2017), the availability of a large amount of ambient nitrogen that is digested by BGA-Sp extract and delivered to plant tissues explains the increase in levels of various photosynthetic pigments in response to the extract. Degradation of proteins and membranes is caused by reduced

**Table 17.17** Chlorophyll content (SPAD reading) at growth stages of barley varieties

| Varieties (V) | Biofertilizer (Bio) | Elongation stage | | | | Heading stage | | | |
|---|---|---|---|---|---|---|---|---|---|
| | | SI | DI | Mean | Red. % | SI | DI | Mean | Red. % |
| Giza 125 | *Control* | 46.7 | 35.2 | 41.0 | 24.6 | 49.1 | 33.7 | 24.6 | 31.4 |
| | *Yeast* | 54.6 | 44.3 | 49.5 | 18.9 | 50.7 | 38.4 | 25.4 | 24.3 |
| | *BGA-Sp* | 58.2 | 48.3 | 53.3 | 17.0 | 53.1 | 41.3 | 26.6 | 22.2 |
| | *GPB12* | 57.6 | 45.5 | 51.6 | 21.0 | 56.7 | 41.7 | 28.4 | 26.5 |
| Tombari | *Control* | 44.9 | 35.4 | 40.2 | 21.2 | 47.6 | 33.2 | 23.8 | 30.3 |
| | *Yeast* | 48.4 | 39.1 | 43.8 | 19.2 | 52.9 | 40.5 | 26.5 | 23.4 |
| | *BGA-Sp* | 50.2 | 42.7 | 46.5 | 14.9 | 55.3 | 43.2 | 27.7 | 21.9 |
| | *GPB12* | 49.8 | 41.9 | 45.9 | 15.9 | 52.4 | 39.4 | 26.2 | 24.8 |
| Ksar Megrine | *Control* | 43.8 | 32.5 | 38.2 | 25.8 | 49.6 | 32.0 | 24.8 | 35.5 |
| | *Yeast* | 51.8 | 43.2 | 47.5 | 16.6 | 50.4 | 36.7 | 25.2 | 27.2 |
| | *BGA-Sp* | 50.2 | 40.2 | 45.2 | 19.9 | 55.3 | 42.3 | 27.7 | 23.5 |
| | *GPB12* | 50.7 | 39.6 | 45.2 | 21.9 | 52.7 | 36.4 | 26.4 | 30.9 |
| Tamellat | *Control* | 44.1 | 33.8 | 39.0 | 23.4 | 45.3 | 33.2 | 22.7 | 26.7 |
| | *Yeast* | 47.4 | 39.4 | 43.4 | 16.9 | 48.2 | 39.2 | 24.1 | 18.7 |
| | *BGA-Sp* | 51.9 | 44.3 | 48.1 | 14.6 | 52.8 | 44.3 | 26.4 | 16.1 |
| | *GPB12* | 50.3 | 40.5 | 45.4 | 19.5 | 51.6 | 40.3 | 25.8 | 21.9 |
| LSD (0.05) | Varieties(V) | 2.02 | 1.44 | | | 1.54 | 1.11 | | |
| | Biofertilizer (Bio) | 0.89 | 0.23 | | | 0.79 | 1.31 | | |
| | V*Bio | 2.65 | 1.52 | | | 2.12 | 2.51 | | |

*SI* 100% of water requirement, *DI* 50% of water requirement, *Red* reduction percentage

photosynthesis, which lowers plant growth under drought stress and increases the formation of reactive oxygen species (ROS) El-Bassiouny et al. (2018). According to Li et al. (2019), there was a notable increase in the chlorophyll content of the cucumber and wheat seedlings in the biofertilizer treatment group when compared to the equivalent control. According to Hellal et al. (2021), water stress resulted in a lower chlorophyll content, whereas adequate irrigation produced a greater level (Table 17.17). When BGA-Sp and GPB12 were applied under adequate irrigation and water stress during the investigated growth stages, the highest chlorophyll levels were obtained when compared to the control group.

Following the application of BGA-Sp during elongation under FI and WS, respectively, the Giza 125 and Tamellat barley varieties showed the highest chlorophyll levels (58.2, 48.3) and (51.9, 44.3), respectively. Following the application of BGA-Sp under FI and WS, respectively, the Tamellat barley varieties showed the highest chlorophyll levels (52.8, 44.3) at the heading stage. The SPAD chlorophyll measurement is negatively impacted by water stress, with estimated decrease percentages ranging from 16.6% (BGA-Sp) to 23.7% (control). Based on the mean growth stage data, biofertilizer application decreased the reduction rate to 16.6%, 17.9%, and 21.3% for BGA-Sp, yeast, and GPB12, respectively, compared to 20.9%, 23.4%, and 26.0% in the control group.

#### 17.7.6.2 Biofertilizer Effects on Relative Water Content

Plants with BGA-Sp applied topically or through soil improved in terms of their physiological state and drought resistance. To keep the stomata open and water flowing through the plant to the evaporating surfaces, higher RWC in plants treated with biofertilizer may be advantageous. Plant growth-promoting rhizobacteria (PGPR) are soil bacteria that, in the vicinity of the rhizosphere, produce and secrete a variety of regulatory compounds that either directly or indirectly aid in the promotion of plant growth and development (Ahemad and Kibret 2013). Since it indicates the relative quantity of water present on the plant tissues, relative water content (RWC) is a valuable indication of plant water balance and a trustworthy metric for evaluating plant water status (Table 17.18). According to Hellal et al. (2021), RWC values with adequate irrigation were greater than those under water stress treatment. The researched barley varieties were found to be significantly impacted by irrigation treatments; the Ksar Megrine variety showed the highest values (92.5, 80.2%) following the application of BGA-Sp treatment, while the barley Giza 125 control variety showed the lowest values of the WRC (81.4, 66.6%) under sufficient irrigation and water stress irrigation treatments at the elongation stage, respectively. Furthermore, it is evident that the Tombari barley

**Table 17.18** Relative water content (%) at growth stages of barley varieties

| Varieties (V) | Biofertilizer (Bio) | Elongation stage | | | | Heading stage | | | |
|---|---|---|---|---|---|---|---|---|---|
| | | SI | DI | Mean | Red. % | SI | DI | Mean | Red. % |
| Giza 125 | *Control* | 81.4 | 66.6 | 78.8 | 18.1 | 77.9 | 58.1 | 68.0 | 25.3 |
| | *Yeast* | 85.4 | 75.6 | 80.5 | 11.4 | 81.5 | 63.2 | 72.3 | 22.4 |
| | *BGA-Sp* | 88.5 | 81.6 | 77.5 | 7.8 | 82.3 | 66.1 | 74.2 | 19.6 |
| | *GPB12* | 87.6 | 76.3 | 84.6 | 12.9 | 85.1 | 65.3 | 75.2 | 23.3 |
| Tombari | *Control* | 83.6 | 64.3 | 74.0 | 23.1 | 77.0 | 56.3 | 66.6 | 26.8 |
| | *Yeast* | 86.2 | 76.5 | 80.3 | 11.3 | 81.8 | 65.8 | 73.8 | 19.6 |
| | *BGA-Sp* | 91.2 | 82.6 | 84.4 | 9.4 | 85.4 | 71.3 | 78.4 | 16.5 |
| | *GPB12* | 84.1 | 72.3 | 81.8 | 14.1 | 80.1 | 63.2 | 71.7 | 21.1 |
| Ksar Megrine | *Control* | 81.8 | 61.2 | 71.5 | 25.1 | 74.9 | 53.3 | 64.1 | 28.9 |
| | *Yeast* | 82.6 | 79.4 | 81.0 | 3.9 | 79.0 | 65.4 | 72.2 | 17.3 |
| | *BGA-Sp* | 92.5 | 80.2 | 84.4 | 13.3 | 80.8 | 67.4 | 74.1 | 16.6 |
| | *GPB12* | 88.6 | 82.6 | 87.6 | 6.8 | 83.9 | 64.6 | 74.3 | 23.1 |
| Tamellat | *Control* | 83.4 | 70.5 | 77.0 | 15.4 | 76.8 | 59.0 | 67.9 | 23.2 |
| | *Yeast* | 89.9 | 77.4 | 80.0 | 13.9 | 81.5 | 64.5 | 73.0 | 21.0 |
| | *BGA-Sp* | 90.7 | 81.7 | 86.2 | 10.0 | 86.5 | 69.3 | 77.9 | 19.9 |
| | *GPB12* | 82.6 | 80.2 | 85.0 | 2.9 | 84.9 | 66.4 | 75.6 | 21.8 |
| LSD (0.05) | Varieties (V) | 1.13 | 0.55 | | | 0.28 | 0.30 | | |
| | Biofertilizer (Bio) | 0.41 | 0.63 | | | 0.41 | 0.27 | | |
| | V*Bio | 1.40 | 1.08 | | | 0.63 | 0.52 | | |

variety's RWC values (85.4, 71.3%) were highest at the heading stage and lowest at the Ksar Megrine control treatment. Adequate irrigation positively impacts RWC; in contrast, stress conditions resulted in a greater percentage of RWC loss throughout the growth stage under study. The enhancement in RWC relative to the biofertilizer treatments was 18.1, 20.1, and 22.3% for BGA-Sp, yeast, and GPB12 comparing with control treatment (26.1%), respectively.

#### 17.7.6.3 Biofertilizer Effects on Proline Content of Barley

According to Hellal et al. (2021), water regime and biofertilizer both have a detrimental impact on proline content when compared to the control conditions of water stress throughout the heading and elongation growth phases and no applied biofertilizer (Table 17.19). The interaction between barley varieties and biofertilizer treatments further determined proline levels under both sufficient irrigation (SI) and water stress (DI) conditions. The data showed that the proline content was highest at SI and DI (2.30, 3.11 mg/g fresh weight) and lowest (1.44, 1.66 mg/g fresh weight) and (1.55, 1.82 mg/g fresh weight) for the BGA-Sp treatment of Tamellat at the elongation and heading growth stages, respectively. These values were registered for the control treatment of Tombari. For all barley cultivars under study, the water

**Table 17.19** Proline content (mg $g^{-1}$ fresh weight) at growth stages of barley varieties

| Varieties (V) | Biofertilizer (Bio) | Elongation stage | | | | Heading stage | | | |
|---|---|---|---|---|---|---|---|---|---|
| | | SI | DI | Mean | Red. % | SI | DI | Mean | Red. % |
| Giza 125 | *Control* | 2.12 | 2.88 | 2.37 | 35.85 | 2.61 | 3.47 | 3.18 | 32.95 |
| | *Yeast* | 1.95 | 2.42 | 2.14 | 24.10 | 2.33 | 2.88 | 2.65 | 20.60 |
| | *BGA-Sp* | 1.87 | 2.22 | 2.09 | 18.72 | 2.30 | 2.81 | 2.52 | 25.22 |
| | *GPB12* | 2.01 | 2.54 | 2.16 | 26.37 | 2.36 | 2.93 | 2.74 | 24.15 |
| Tombari | *Control* | 2.30 | 3.11 | 2.62 | 35.22 | 2.93 | 3.75 | 3.43 | 27.99 |
| | *Yeast* | 2.34 | 2.90 | 2.66 | 23.88 | 2.58 | 3.11 | 3.28 | 20.54 |
| | *BGA-Sp* | 1.99 | 2.43 | 2.27 | 21.93 | 2.55 | 3.03 | 2.77 | 18.82 |
| | *GPB12* | 2.07 | 2.71 | 2.32 | 31.17 | 2.97 | 3.65 | 2.87 | 22.90 |
| Ksar Megrine | *Control* | 2.15 | 2.89 | 2.30 | 34.42 | 2.44 | 3.46 | 3.18 | 41.80 |
| | *Yeast* | 1.98 | 2.40 | 2.02 | 21.21 | 2.06 | 2.74 | 2.57 | 33.01 |
| | *BGA-Sp* | 1.87 | 2.23 | 1.96 | 19.25 | 2.05 | 2.59 | 2.62 | 26.34 |
| | *GPB12* | 1.96 | 2.45 | 2.04 | 25.00 | 2.11 | 2.78 | 2.41 | 31.94 |
| Tamellat | *Control* | 1.95 | 2.67 | 1.99 | 36.92 | 2.03 | 2.78 | 2.73 | 36.95 |
| | *Yeast* | 1.49 | 1.88 | 1.54 | 26.17 | 1.58 | 1.97 | 1.93 | 24.68 |
| | *BGA-Sp* | 1.44 | 1.66 | 1.50 | 15.28 | 1.55 | 1.82 | 1.94 | 17.42 |
| | *GPB12* | 1.51 | 1.87 | 1.56 | 23.71 | 1.61 | 2.03 | 1.95 | 26.09 |
| LSD (0.05) | Varieties (V) | 0.22 | 0.16 | | | 0.09 | 0.04 | | |
| | Biofertilizer (Bio) | 0.14 | 0.07 | | | 0.13 | 0.07 | | |
| | V*Bio | 0.33 | 0.21 | | | 0.20 | 0.10 | | |

stress condition increases proline content in a manner similar to adequate irrigation treatment. According to Wang et al. (2012), plant enzyme activity is enhanced by biofertilizer strain inoculation, which reduces oxidative damage brought on by salt and drought. It was also found by Noorieh et al. (2013) that biofertilizers such as Pseudomonas sp. increase the biomass and development of canola plants by controlling the enzymes that cause oxidative stress and the vital nutrients that are lacking in water.

#### 17.7.6.4 Biofertilizer Effects on Barley Yield of Barley

According to Mozafar et al. (2000), when plants are under water stress, biofertilizers assist them overcome the adverse effects of the stress, increasing their growth parameters, chemical composition, and consequently yield and attributes. Sharma and Singh (2011) demonstrated how the combined application of chemical and biological fertilizers changed the flow of photosynthetic substances to stems and roots and altered the plant physiology by increasing amounts of these substances. This change also affected the plant's ability to absorb minerals from the soil. When applied under water stress, biological fertilizers can give corn plants all the nutrients they require for growth and metabolism without carrying the same dangers as chemical fertilizers (Koliai et al. 2012). The issue of feeding the world's growing population at a time when agriculture is under a lot of environmental stress can be resolved with the use of biofertilizers. Understanding the benefits of biofertilizers and incorporating their usage into contemporary farming methods is crucial (Deepak et al. 2014).

But there are still restrictions on how some biofertilizers may be used (Grady et al. 2016). One important limiting element is drought, and the use of biofertilizers in agriculture has been hampered by a shortage of water. This study's goal is to calculate the effects of various biofertilizer formulas on barley growth and production under various water irrigation techniques. Although biological fertilization is desperately required to protect the environment and lower agricultural production costs, water stress frequently impedes its effectiveness. Hellal et al. (2021) colleagues conducted an assessment to determine how biofertilizers (GPB12, Yeast (Saccharomyces cerevisiae {Sc}) and Blue-Green BGA-Sp{BGA-Sp}) affected the biochemical parameters, growth, and yield of North African barley varieties (Giza 125, Tombari, Ksar Megrine, and Tamellat) grown under both sufficient and deficit irrigation.

The results in Table 17.20 showed that for various barley types, the control group with no applied biofertilizer gained the lowest values of the grain yield under both irrigation regimes. Additionally, for every barley variety under study, the highest grain production values were seen following the administration of BGA-Sp under both irrigation regimes. The Giza 125 control treatment exhibited the highest percentage drop in grain production (37.2%) because of water stress, whereas the Tombari barley variety saw the lowest percentage reduction (19.2%) following the application of BGA-Sp treatment. The barley grain output at Tombari was

**Table 17.20** Biofertilizer application on grain and biological yield of barley varieties

| Varieties (V) | Biofertilizer (Bio) | Grain yield (ton fed)$^{-1}$ | | | | Biological yield (ton fed)$^{-1}$ | | | |
|---|---|---|---|---|---|---|---|---|---|
| | | SI | DI | Mean | Red. % | SI | DI | Mean | Red. % |
| Giza 125 | *Control* | 1.72 | 1.08 | 1.40 | 37.2 | 4.01 | 2.43 | 3.22 | 39.4 |
| | *Yeast* | 1.82 | 1.38 | 1.60 | 24.2 | 5.04 | 3.45 | 4.24 | 31.5 |
| | *BGA-Sp* | 1.83 | 1.47 | 1.65 | 19.7 | 6.68 | 5.33 | 6.01 | 20.2 |
| | *GPB12* | 1.79 | 1.33 | 1.56 | 25.7 | 4.73 | 3.21 | 3.97 | 32.2 |
| Tombari | *Control* | 1.91 | 1.19 | 1.55 | 37.7 | 4.43 | 2.69 | 3.56 | 39.3 |
| | *Yeast* | 2.03 | 1.45 | 1.74 | 28.6 | 5.47 | 4.02 | 4.74 | 26.5 |
| | *BGA-Sp* | 2.13 | 1.72 | 1.93 | 19.2 | 6.15 | 4.77 | 5.46 | 22.4 |
| | *GPB12* | 1.96 | 1.29 | 1.63 | 34.2 | 5.07 | 3.53 | 4.30 | 30.3 |
| Ksar Megrine | *Control* | 1.75 | 1.11 | 1.43 | 36.6 | 4.17 | 2.61 | 3.39 | 37.5 |
| | *Yeast* | 1.83 | 1.34 | 1.59 | 26.8 | 5.56 | 4.53 | 5.04 | 18.5 |
| | *BGA-Sp* | 1.97 | 1.62 | 1.80 | 17.8 | 5.82 | 4.80 | 5.31 | 17.5 |
| | *GPB12* | 1.82 | 1.26 | 1.54 | 30.8 | 5.54 | 4.44 | 4.99 | 19.9 |
| Tamellat | *Control* | 1.73 | 1.18 | 1.46 | 31.8 | 4.10 | 2.72 | 3.41 | 33.7 |
| | *Yeast* | 1.82 | 1.33 | 1.58 | 26.9 | 5.05 | 3.89 | 4.47 | 22.9 |
| | *BGA-Sp* | 1.92 | 1.46 | 1.69 | 24.0 | 5.50 | 4.42 | 4.96 | 19.7 |
| | *GPB12* | 1.78 | 1.27 | 1.53 | 28.7 | 4.41 | 3.11 | 3.76 | 29.6 |
| LSD (0.05) | Varieties (V) | 0.052 | 0.038 | | | 0.288 | 0.173 | | |
| | Biofertilizer (Bio) | 0.076 | 0.042 | | | 0.405 | 0.191 | | |
| | V*Bio | 0.113 | 0.074 | | | 0.633 | 0.332 | | |

significantly impacted by water stress conditions by 29.9% and at Giza 125, by 26.7%. We may infer that, in situations when water stress is present, the biological fertilization strategy is strongly advised to enhance barley output and yield qualities. Increased bacterial counts in biofertilizers are what allowed the atmosphere to fix nitrogen.

## 17.8 Conclusion

The Egyptian barley cultivars (Giza 123, Giza 127, Giza 130, Giza 131, and Giza 2000) were high drought tolerance as affected by water stress. These cultivars could be suggested to be grown in such arid and semiarid conditions. The cultivars from Algeria (Nailia, Techedrett, and Temacine), Morocco (Laanaceur, Massine, and Amira), and Tunisia (Kebili 3, Tozeur 2, and Sidi-Bou) had the highest values for drought tolerance indices and the best capacity to withstand drought stress. The results of this study are highly helpful for planning future barley breeding projects, and barley breeders, agronomists, and grain producers may be advised to employ the tolerant barley cultivars to increase grain output under various water regimes. The Egyptian barley cultivars (Giza 125, Giza 126, Giza 128, and Giza 129), Tombari,

Raihane (Tunis), Ksar Megrine, Ras El-Mouche (Algeria), and Tamellat and Adrar (Morocco) were severely affected by increasing water stress and classified as sensitively to drought stress. Also, their yield improved by spraying ascorbic acid. Under conditions of water stress, the application of the anti-transpiration chemicals, MgCO3, salicylic acid, and KSiO4 showed good results for barley cultivars.

In dry and semiarid regions, using antitranspirants might help barley output by lessening the negative consequences of water scarcity. Nano-silica improved barley growth, yield, and nutrient status in plants under water stress. Foliar application of K-citrate was the superior potassium fertilizers for alleviating the impact of water stress on barley cultivars as comparable to potassium silicate and potassium nitrate fertilizers. Application of compost, combined with arginine, improves plant growth and yield of sensitive barley cultivars under water stress condition. Compost application has important role such as enhancement of the physical and chemical properties that led soil maintain a higher water use efficiency for plant. However, the administration of arginine has been shown to lessen the harmful effects of biotic and abiotic stressors by producing proline, polyamine, and nitric oxide as by-products; therefore, it plays a significant role in boosting barley production during periods of water scarcity.

**Acknowledgments** The authors warmly thank the Agricultural Research in the Mediterranean Area 2 (ARIMNet 2) and Academy of scientific research and technology (ASRT) and National Research Centre (NRC), Egypt, who have funded the experiments included in this research work.

**Declaration of AI Assistance** The authors confirm that no artificial intelligence tools were used in writing this chapter. All authors contributed to the work in its entirety, read the final manuscript, and approved it for publication. The authors are solely responsible for its content. This chapter has undergone double-blind peer review by two independent international experts.

**Funding** This work was supported and funded by Egypt's National Research Center.

## References

Ahemad M, Kibret M (2013) Mechanisms and applications of plant growth promoting rhizobacteria: current perspective. J King Saud Univ Sci 74:174–194

Ahmed HSA, Zewail RMY, Hassan AA (2014) Effect of PIX and potassium citrate on growth, productivity fiber quality and yarn on Egyptian cotton. Bull Fac Agric Cairo Univ 65:420–430

Alejandro PDL, Rubiales D (2009) Pest Manag Sci 65:540–545

Araus JL, Slafer GA, Royo C, Serret MD (2008) Breeding for yield potential and stress adaptation in cereals. Crit Rev Plant Sci 27:377–412

Asada K (2006) Production and scavenging of reactive oxygen species in chloroplasts and their functions. Plant Physiol 141:391–396

Bandyopadhyay KK, Misra AK, Ghosh PK, Hati KM (2010) Effect of integrated use of farmyard manure and chemical fertilizers on soil physical properties and productivity of soybean. Soil Tillage Res 110:115–125

Blum A (2009) Effective use of water (EUW) and not water-use efficiency (WUE) is the target of crop yield improvement under drought stress. Field Crop Res 112:119–123

Boyer JS (2016) Plant productivity and environment. Science 218:443–448. https://doi.org/10.1126/science.218.4571.443. http://science.sciencemag.org/content/218/4571/443/tab-article-info

Capelo A, Santos C, Loureiro S, Pedrosa AM (2012) Phytotoxicity of lead on *Lactuca sativa*: effects on growth, mineral nutrition, photosynthetic activity and oxidant metabolism. Fresenius Environ Bull 21(2):450–459

Ceccarelli S, Grando S, Baum M (2007) Participatory plant breeding in water-limited environments. Exp Agric 43:1–25

Christophe S, Jean-Christophe A, Annabelle L, Alain O, Marion P, Anne-Sophie V (2011) Plant N fluxes and modulation by nitrogen, heat and water stresses: a review based on comparison of legumes and non-legume plants. In: Shanker A, Venkateswarlu B (eds) Abiotic stress in plants—mechanisms and adaptations. Intech, Rijeka, pp 79–118

Deepak B, Ansari MW, Sahoo RK, Tuteja N (2014) Biofertilizers function as key player in sustainable agriculture by improving soil fertility, plant tolerance and crop productivity. Microb Cell Factories 13:66

Du Jardin P (2015) Plant bio stimulants: definition, concept, main categories and regulation. Sci Hortic 2015(196):3–14

Ebeed S, Abd El-Migeed MM (2005) Effect of spraying sucrose and some nutrient elements on fagrikalan mango trees. J Appl Sci Res 1(5):341–346

El Mourid M, De-Pauw E, Tahar F (2010) Agriculture in the North Africa region and constraints to sustainable productivity. In: El Mourid M, Gomez-Macpherson H, Rawson HM (eds) Explore on-farm: the case of North Africa, Tunisia, pp 7–15

EL Sabagh A, Hossain A, Islam MS, Barutcular C, Hussain S, Hasanuzzaman M, Akram T, Mubeen M, Nasim W, Fahad S, Kumar N, Meena RS, Kızılgeçi F, Yıldırım M, Ratnasekera D, Saneoka H (2019) Drought and salinity stresses in barley: Consequences and mitigation strategies. AJCS 13(06):810–820

El-Bassiouny HMS, Abd El-Monem AA, Abdallah MMS, Soliman KM (2018) Role of arbuscular mycorrhiza, α-tocopherol and nicotinamide on the nitrogen containing compounds and adaptation of sunflower plant to water stress. Bio Sci Res 15(3):2068–2088

EL-Shawy EE, EL Sabagh A, Mansour M, Barutcular C (2017) A comparative study for drought tolerance and yield stability in different genotypes of barley (*Hordeum vulgare* L.). J Exp Biol Agric Sci 5(2):151.162

FAO (2016) Food and Agriculture Organization of the United Nations. Statistic divisions. Accessed 1 Feb 2015

Gao X, Zou C, Wang L, Zhang F (2006) Silicon decreases transpiration rate and conductance from stomata of maize plants. J Plant Nutr 29(9):1637–1647

Gebaly S, Ahmed F, Namich A (2013) Effect of spraying some organic, amino acids and potassium citrate on alleviation of drought stress in cotton plant. J Plant Prod 4:1369–1381

Gong H, Chen K, Chen G, Wang S, Zhang C (2003) Effects of silicon on growth of wheat under drought. J Plant Nutr 26(5):1055–1063

Gontia-Mishra I, Sapre S, Sharma A, Tiwari S (2016) Amelioration of drought tolerance in wheat by the interaction of plant growth-promoting rhizobacteria. Plant Biol 18:992–1000. https://doi.org/10.1111/plb.12505

Grady EN, MacDonald J, Liu L, Richman A, Yuan ZC (2016) Current knowledge and perspectives of Paenibacillus: a review. Microbial cell factories 15(1):203. https://doi.org/10.1186/s12934-016-0603-7

Grzesik M, Romanowska-Duda Z, Kalaji HM (2017) Effectiveness of cyanobacteria and green BGA-Sp in enhancing the photosynthetic performance and growth of willow (*Salix viminalis* L.) plants under limited synthetic fertilizers application. Biotech 15(4):239–240

Hamid ZH, Amer M, Wahab A (2019) Effect of arginine on growth and yield of tomato plant (*Lycopersicon esculentum* L.) under drought stress. Plant Arch 19:4441–4444

He M, Dijkstra FA (2014) Drought effect on plant nitrogen and phosphorus: a meta-analysis. New Phytol 204:924–931

Hellal F, Abdel-Hady M, Khattab I, El-Sayed S, Abdelly C (2019a) Yield characterization of Mediterranean barley under drought stress condition. AIMS Agric Food 4(3):518–533

Hellal FA, El Sayed SA, Amer KA, EL Azab KM (2019b) effect of water stress and ascorbic acid on yield components of Egyptian Barley cultivars. Biosci Res 16(1):814–821

Hellal F, El-Sayed S, Gad A, Abdel Karim G, Abdelly C (2020a) Antitranspirants application for improving the biochemical changes of barley under water stress. Iraqi J Agric Sci 51(1): 287–298

Hellal FA, El-Sayed S, Abou Basha DM, Abdelly C (2020b) Mineral nutrient status of some Mediterranean barley varieties as affected by drought stress in Egypt. Iraqi J Agric Sci 51 (Special Issue):138–147

Hellal F, El-Sayed S, Abdel Hady M (2020c) Barley responses to potassium fertilization under water stress condition. Plant Arch 20(1):3140–3147

Hellal F, Amer AK, El-Sayed S, El-Azab K (2020d) Mitigating the negative effect of water stress on barley by Nano silica application. Plant Arch 20(1):3224–3231

Hellal F, El-Sayed S, Khattab AA (2021) Beneficial effects of novel biofertilizers formula on growth and yield component of barley grown under water stress condition. J Biochem Int 8(1): 33–44. ISSN: 2454-4760

Hellal F, El Sayed S, Abo Basha DMR, Abdel Kader HH (2024) Mitigation of water stress by compost and arginine application and its impacts on barley production. Bull Nat Res Cent 48:25. https://doi.org/10.1186/s42269-024-01178-2

Hu W, Liu Y, Loka DA, Zahoor R, Wang S, Zhou Z (2019) Drought limits pollen tube growth rate by altering carbohydrate metabolism in cotton (*Gossypium hirsutum* L.) pistils. Plant Sci 286: 108–117

Hussein MM, Mehanna HM, El-Lethy SM (2013) Water deficit and foliar fertilization and their effect on growth and photosynthetic pigments of Jatropha plants. World Appl Sci J 27(4): 454–461

Hussein H-AA, Alshammari SO, Kenawy SKM, Elkady FM, Badawy AA (2022) Grain-priming with L-arginine improves the growth performance of wheat (*Triticum aestivum* L.) plants under drought stress. Plants 11:1219

Koliai AA, Akbari G, Akbari G, Armandpisheh O, Tarighaleslami M (2012) Effect of biological and chemical phosphorus fertilizers on yield components of maize (*Zea mays* L.) in different water stress conditions. Ann Biol Res 3:4204–4208

Kooyers NJ (2015) The evolution of drought escape and avoidance in natural herbaceous populations. Plant Sci 234:155–162

Li Y, Shi H, Zhang H, Chen S (2019) Amelioration of drought effects in wheat and cucumber by the combined application of super absorbent polymer and potential biofertilizer. Peer J 7:e6073. https://doi.org/10.7717/peerj.6073

Li X, Smith R, Choat B, Tissue DT (2020) Drought resistance of cotton (*Gossypium hirsutum* L.) is promoted by early stomatal closure and leaf shedding. Funct Plant Biol 47(2):91–98

Malvi RU (2011) Interaction of micronutrients with major nutrients with special reference to potassium. J Agric Sci 24(1):106–109

Mauad M, Carlos ACC, Adriano SN, HélioG.F., P.P L. Giuseppina (2016) Effects of silicon and drought stress on biochemical characteristics of leaves of upland rice cultivars. Rev Ciênc Agron 47(3):532–539

Meeks CD, Snider JL, Babb-Hartman ME, Barnes TL (2019) Evaluating the mechanisms of photosynthetic inhibition under growth-limiting, early-season water deficit stress in cotton. Crop Sci 59(3):1144–1154

Metwally A, Finkemeier I, Georgi M, Dietz K (2003) Salicylic acid alleviates the cadmium toxicity in barley seedlings. Physiol Biochem Plant 132:272–281

Mevlüt A, Sait C (2011) Evaluation of drought tolerance indices for selection of Turkish oat (*Avena sativa* L.) landraces under various environmental conditions. Zemdirbyste Agric 98(2):157–166

Mitra J (2001) Genetics and genetic improvement of drought resistance in crop plants. Curr Sci 80: 758–763

Mozafar A, Anken T, Ruh R, Frossard E (2000) Tillage intensity, mycorrhizal and non-mycorrhizal fungi and nutrient concentrations in maize, wheat and canola. Agron J 92:1117–1124

Nasibi F, Yaghoobi MM, Kalantari KM (2011) Effect of exogenous arginine on alleviation of oxidative damage in tomato plant underwater stress. J Plant Interact 6:291–296

Nazar R, Umar S, Khan NA (2015) Exogenous salicylic acid improves photosynthesis and growth through increment in ascorbate-glutathione metabolism and S assimilation in mustard under salt stress. Plant Signal Behav 10:e1003751

Nezhadahmadi A, Prodhan ZH, Faruq G (2013) Drought tolerance in wheat. Sci World J 2013: 610721

Noaman MM (2008) Barley development in Egyptian. In: Proceedings of the 10th international barley genetics symposium Alexandria, Egyptian, pp 3–15

Noorieh B, Arzanesh MH, Mahlegha G, Maryam S (2013) The effect of plant growth promoting rhizobacteria on growth parameters, antioxidant enzymes and microelements of canola under salt stress. J Appl Environ Biol Sci 3:17–27

Passioura J (2006) Increasing crop productivity when water is scarce—from breeding to field management. Agric Water Manag 80:176–196

Prakash M, Ramachandran K (2000) Effects of chemical ameliorants in brinjal (*Solanum melongena* L.) under moisture stress conditions. J Agro Crop Sci 185:237–239

Rania FEM, Maha EB (2018) Effect of antitranspirants application on growth and productivity of sunflower under soil moisture stress. Nat Sci 16(2):92–106

Sahnoune M, Adda A, Soualem S, Harch MK, Merah O (2004) Early water-deficit effects on seminal roots morphology in barley. C R Biol 327(4):389–398

Samarah NH (2005) Effects of drought stress on growth and yield of barley. Agron Sustain Dev 25: 145–149

Sardans J, Peñuelas J (2012) The role of plants in the effects of global change on nutrient availability and nutrient stoichiometry in the plant-soil system. Plant Physiol 160:1741–1761

Shanan NT, El Sadek ZH (2017) Influence of silicon on tuberose plants under drought conditions. Middle East J Agric Res 06(02):348–360

Sharma VK, Singh RP (2011) Organic matrix based slow-release fertilizer enhances plant growth, nitrate assimilation and seed yield of Indian mustard (*Brassica juncea* L.). J Environ Biol 32: 619–624

Siddiqui MH, Al-Whaibi MH (2014) Role of Nano-SiO2 in germination of tomato (*Lycopersicum esculentum* seeds Mill.). Saudi. J Biol Sci 21(1):13–17

Sumida H (2002) Plant available silicon in paddy soil. In: National Agricultural Research Center for Tohoku Region Omagari second silicon in agriculture conference (Tsuruoka, Yamagata, Japan)

Taia AAE-M, Ahmed MAE-S, Shimaa AAE-M, Nasr MA (2019) A novel compost alleviates drought stress for sugar beet production grown in Cd-contaminated saline soil. Agric Water Manage 226:105831

Taiz L, Zeiger E (2002) Plant physiology, 3rd edn. Sinauer Associates Inc., Sunderland, p 224

Temel A, Janack B, Humbeck K (2017) Drought stress-related physiological changes and histone modifications in barley primary leaves at HSP17 gene. Agron 7(2):43

Tiwari JK, Munshi AD, Kumar R, Pandey RN, Arora A, Bhat JS, Sureja AK (2010) Effect of salt stress on cucumber: Na+/K+ ratio, osmolyte concentration, phenols and chlorophyll content. Acta Physiol Plant 32:103–114

Tripathi DK, Singh VP, Kumar D, Chauhan DK (2012) Impact of exogenous silicon addition on chromium uptake, growth, mineral elements, oxidative stress, antioxidant capacity, and leaf and root structures in rice seedlings exposed to hexavalent chromium. Acta Physiol Plant 34(1): 279–289

Tuberosa R (2012) Phenotyping for drought tolerance of crops in the genomics era. Front Physiol 3: 347

Wang CJ, Yang W, Wang C, Gu C, Niu DD, Liu HX, Wang YP, Guo JH (2012) Induction of drought tolerance in cucumber plants by a consortium of three plant growth-promoting rhizobacterium strains. PLoS ONE 7(12):e52565

Zarafshar M, Akbarinia M, Askari H, Hosseini SM, Rahaie M, Struve D (2015) Toxicity assessment of SiO2 nanoparticles to pear seedlings. Int J Nano Sci Nanotechnol 11(1):13–20

# Part III
# Conclusions

# Chapter 18
# Discussion and Future Directions: Integrating Microbiomes and Plant Metabolites for Sustainable Pest Management

**Nabil El-Wakeil and Ahmed Abdelkhalk**

General discussion of these book chapters in a book entitled *Microbiomes and Plant Secondary Metabolites for Agricultural Pest Management*. The essential purpose of this final chapter is to create the transformative visions showed across the preceding 18 chapters and to chart the serious future trajectory of the field. Through its detailed exploration of plant–microbe–insect interactions, this volume has systematically revealed the sophisticated roles of root and foliar microbiomes and the vast repertoire of plant secondary metabolites in shaping plant health, defense, and climate resilience. Moving from foundational concepts to advanced research and practical applications, the discussion has integrated microbial ecology, plant biochemistry, and agroecology. Here, we distill these key findings into a cohesive framework and propose a forward-looking agenda focused on translating fundamental knowledge into practical, scalable, and maintainable pest management policies. The ultimate goal is to reduce reliance on synthetic chemicals and enhance the resilience of agricultural systems.

---

N. El-Wakeil (✉)
Department of Arid Land Agriculture, College of Agricultural & Food Sciences, King Faisal University (KFU), Al-Ahsa, Saudi Arabia

Pests & Plant Protection Department, Agricultural & Biological Sciences Institute, NRC, Cairo, Egypt
e-mail: nelwakeil@kfu.edu.sa

A. Abdelkhalk
Plant Protection and Biomolecular Diagnosis Department, City of Scientific Research & Technological Applications, Alexandria, Egypt



N. El-Wakeil, A. Abdelkhalek (eds.), *Microbiomes and Plant Secondary Metabolites for Insect Pest Management*, Progress in Biological Control 23,
https://doi.org/10.1007/978-3-032-24330-0_18

## 18.1 The First Book Part: Phytomicrobiome and Insect Microbiome for Pest's Management

The phytomicrobiome (plant-associated microbial communities) and the insect microbiome (microorganisms inhabiting insect bodies) play critical roles in shaping plant–insect interactions and offer inventive strategies for supportable pest management. These microbiomes impact plant health, insect behavior, and ecosystem dynamics, presenting opportunities for biological pest control.

### *18.1.1 The Role of Tree–Insect–Microbiome Interactions in Herbivory and Applications in Pest Management*

Agricultural biodiversity (crop rotation, polycultures, and hedgerows) influences the microbial communities associated with insect pests. An assorted landscape endorses beneficial microbes that can improve insect immunity against pathogens or insect pests. For example, some symbiotic bacteria in insects produce antimicrobial compounds that protect against fungal infections. Understanding these interactions could help design agroecosystems that naturally terminate some pest outbreaks while reducing reliance on chemical pesticides.

**Practical Trends**

Biological Pest Control

- Farmers can design various cropping systems (intercropping and cover crops) to substitute beneficial insect microbiomes that could kill many of insect pest species.
- Example: Ladybugs with symbiotic *Rickettsia* bacteria have enhanced predatory efficiency against varied aphid genius.

Reduced Pesticide Use

- By promoting natural microbial defenses in insects, farmers can decline reliance on chemical insecticides.

Habitat Restoration

- Planting hedgerows and wildflower strips near fields could increase microbial diversity, improving insect health and pollination of many plant species.

### *18.1.2 Tritrophic Interactions: The Role of the Plant and Insect Microbiome in Shaping Pest–Parasitoid Dynamics*

The plant microbiome (rhizobacteria and endophytes) can indirectly affect insect pests by:

*Inducing systemic resistance* (e.g., triggering plant defenses that prevent herbivores).
*Altering plant chemistry* (e.g., changing volatile organic compounds that attract more parasitoid wasps).
*Directly antagonizing pests* (e.g., entomopathogenic fungi such as *Beauveria bassiana*).

This tritrophic interaction (plant–microbe–insect–parasitoid) can be harnessed for biocontrol approaches.

**Applications**

- *Microbial Biocontrol Agents*:
  - Applying entomopathogenic fungi like *Metarhizium* and *Beauveria* as biopesticides.
  - Using plant growth-promoting rhizobacteria (PGPR) to encourage systemic resistance in different crops.
- *Tritrophic Interactions for Pest Management*:
  - Example: *Tomato plants inoculated with Bacillus subtilis* emit volatiles that attract parasitoid wasps, reducing caterpillar damage and safe the environment from pesticide hazards.
- *Precision Agriculture*:
  - Sensor-based microbiome monitoring to expect pest outbreaks and apply targeted biocontrol agents.

### *18.1.3 Use of Plant Beneficial Microbes to Control Bayoud Disease in Date Palms*

Bayoud disease, caused by *Fusarium oxysporum* f. sp. *albedinis*, devastates date palms in North Africa. Potential microbiome-based solutions include:

*Antagonistic microbes* (e.g., *Pseudomonas* spp., *Bacillus* spp.) that suppress *Fusarium* via competition or antibiotic production.
*Endophytic fungi* that increase palm resistance against many insect pests.
*Soil microbiome engineering* to promote disease-suppressive soils.

Research focuses on isolating and applying these microbes as bioinoculants.

**Applied Strategies**

- *Biofungicides*:
  - *Fusarium oxysporum* could be suppressed by applying many of commercial formulations of *Pseudomonas fluorescens* or *Trichoderma* spp.
- *Soil Probiotics*:
  - Introducing microbial consortia to date palm rhizospheres to develop disease resistance.
- *Genetic and Microbiome Breeding*:
  - Selecting date palm cultivars that recruit beneficial microbes naturally.
- *Economic Impact*:
  - Saving North Africa's date industry (~$500M/year losses due to Bayoud disease).

### 18.1.4 Effect of Plant Microbiome on Plant Physiology to Overcome Climate Changes

Climate change (drought, salinity, extreme temperatures, and others) stresses crops, but the microbiome can enhance resilience by:

*Improving water uptake* (via the vital role of mycorrhizal fungi).
*Enhancing nutrient acquisition* (capability of phosphate-solubilizing bacteria).
*Producing stress-alleviating compounds* (e.g., ACC deaminase bacteria decrease ethylene stress).

Microbiome-assisted breeding or synthetic microbial communities (SynComs) could future-proof agriculture.

New Trends in Applications

- *Drought-Resistant Crops*: Inoculating seeds with mycorrhizal fungi to enhance water uptake (e.g., culturing maize varieties in arid regions).
- *Salt-Tolerant Microbiomes*: Using halotolerant bacteria (e.g., *Halomonas*) in saline soils.
- *Heat Stress Mitigation*: *Bacillus* and *Pseudomonas* strains produce heat shock proteins that protect plants from such infestation.
- *Carbon Sequestration*: Soil microbes improve organic matter retention, improving climate resilience.

### *18.1.5 The Interconnected Microbiomes of Plants and Insects: Implications for Yield Quality and Pest Management*

*Plant microbiome* affects yield by improving nutrient use efficiency and stress tolerance.
*Insect microbiome* influences behavior of insect pests (e.g., gut microbes degrade plant toxins, enabling herbivory).

Strategies include:

- Disrupting pest microbiomes (e.g., antibiotic baits).
- Enhancing crop microbiomes for superior defense (e.g., seed coatings with beneficial microbes).

Applied Tools

- *Pest-Specific Microbial Disruption*: Using RNAi-based microbial sprays to silence pest genes via their gut microbiome.
- *Enhanced Pollination*: Probiotic supplements for bees to develop hive health and pollination efficiency among various plant types.
- *Post-Harvest Quality*: Microbiome management to reduce fungal spoilage in stored grains (e.g., *Aspergillus* control).

### *18.1.6 The Plant Microbiome: An Engine for Sustainable Crop Growth and Yield*

Commercial microbial inoculants (e.g., *Rhizobia* and *Azospirillum*) are already used to increase yield of many crops. Advances in microbiome sequencing and machine learning allow:

*Predictive modeling* of microbial functions for specific crops or soil types.
*Tailored biofertilizers/biopesticides* for precision agriculture, especially in the last years.

Challenges include variability in field efficacy and regulatory hurdles.

Practical Trends

- *Commercial Microbial Inoculants*: Products like *Rhizobium* biofertilizers for legumes or *Azospirillum* for cereals.
- *AI-Driven Microbial Selection*: Machine learning models forecast the best microbial consortia for specific soil–crop combinations and their effects on yields.

- *Seed Coating Technologies*: Nano-encapsulated microbes for slow release and higher germination rates.
- *Economic Benefits*: Estimated 10–30% yield increase with optimized microbiome applications.

### *18.1.7 Effectiveness of Honeybee Microbiota on Their Activities and Productivity*

Honeybee gut microbes (*Lactobacillus*, *Bifidobacterium*) are critical for:

*Digestion* (pollen breakdown and detoxification).
*Immunity* (pathogen resistance, e.g., against *Nosema*).
*Colony health* (microbial imbalances related to colony collapse disorder).

Probiotic supplements and hive microbiome management are emerging as tools for sustainable apiculture in different regions over the World.

Applications

- *Probiotic Hive Supplements*: Beekeepers use *Lactobacillus* and *Bifidobacterium* mixes to enhancement immunity against *Nosema* parasites.
- *Pathogen Blocking*: Engineered bee gut microbes to neutralize deformed wing virus (DWV).
- *Honey Production Boost*: Healthy microbiomes develop nectar digestion, increasing honey yields in different hive regimes.
- *Colony Collapse Disorder (CCD) Mitigation*: Microbiome restoration as a policy to fighting CCD.

## 18.2 The Second Book Part: Secondary Metabolism of Plants for Managing the Agricultural Pests

Plants produce a vast array of metabolic compounds that play essential roles in defense, communication, and relations with other organisms, including insect pests, their natural enemies, and microbial pathogens. These compounds such as alkaloids, terpenoids, flavonoids, and phenolics serve as chemical barriers against insect herbivores, attractants for biological control agents (parasitoids or predators), and inhibitors of pathogenic microorganisms. Understanding these interactions is important for developing maintainable pest and disease management strategies in organic agriculture.

### *18.2.1 Relationship among Plants, Insect Pests, and Their Natural Enemies*

Plant metabolic products mediate complex tritrophic interactions involving plants, herbivorous insects, and their parasitoids or predators. Some compounds act as direct toxins or feeding deterrents against various insect pests, while others serve as indirect defenses by attracting many of natural enemy species. For example, herbivore-induced plant volatiles (HIPVs) can signal the presence of host to parasitoid wasps, enhancing biological control policies. The balance between direct and indirect defenses influences pest population dynamics and the efficiency of different species of natural enemies.

Practical Trends

Increased use of metabolomics to detect volatile organic compounds (VOCs) that could attract parasitoid wasps and predators of insect pests.
Developing semiochemical-based pest management strategies (e.g., push–pull systems).
Genetic engineering of crops to enhance emission of defensive volatiles.

### *18.2.2 Efficiency of Plant Metabolic Products on Insect Pests and Their Parasitoids*

While secondary metabolites like nicotine, azadirachtin, and pyrethrins exhibit strong insecticidal properties, their impact on parasitoid wasps and predators varies. Some composites may deter or harm beneficial insects, whereas others have minimal effects or even enhance their efficiency by weakening host insects. Understanding selectivity is critical for integrating botanical pesticides with biological control agents.

Practical Trends

Bioassays to assess plant extracts (e.g., alkaloids and terpenoids) for selective toxicity against pests while conserving parasitoids for saving the environment.
Botanical insecticides (e.g., neem and pyrethrum) integrated with different natural enemies.
Nanoencapsulation of plant metabolites for slow release and enhanced efficacy.

### 18.2.3 Role of Secondary Metabolites in Plant Defense and Their Effects on Insect Pests and Predatory Arthropods

Plants activate defense-related metabolic pathways in response to insect pests, producing compounds that deter insect feeding, inhibit digestion, or disrupt development of insect pests or pathogens. However, some pests have evolved resistance or detoxification mechanisms. Predators and parasitoids may also be affected, either positively (through weakened prey) or negatively (via toxicity). The ecological consequences be contingent on the specific compounds and their concentrations.

Practical Trends

Studying induced systemic resistance (ISR) via secondary metabolites (e.g., phenolics and glucosinolates) and their role of IPM approaches.
Breeding crops with enhanced defensive chemistry (e.g., high-tannin varieties).
Use of elicitors (e.g., jasmonic acid) to increase plant defense responses against insect pests.

### 18.2.4 Efficiency of Plant Metabolic Products on Insect Pests and Associated Entomopathogenic Organisms

Entomopathogenic fungi could be synergized with certain plant metabolites, bacteria, or nematodes by weakening insect immunity or enhancing microbial virulence. For instance, phenolics and terpenoids may suppress insect immune responses, increasing the efficacy of *Beauveria bassiana* or *Metarhizium* infections. However, some compounds might also inhibit beneficial microbes, necessitating careful selection in integrated pest management (IPM).

Practical Trends

Synergistic interactions between plant metabolites (e.g., flavonoids) and entomopathogenic fungi (e.g., *Beauveria bassiana*).
Combined use of botanical pesticides and microbial biocontrol agents, which it should be one of important tactics.
Metabolomic profiling to detect plant compounds that enhance fungal virulence.

### *18.2.5 Efficacy of Plant Metabolic Products Against Plant Pathogenic Nematodes*

Many plant-derived compounds, such as glucosinolates (from Brassicaceae) and saponins, exhibit nematicidal properties by disrupting nematode membranes or nervous systems. Their use in soil modifications or biofumigation offers an eco-friendly alternative managing tool to synthetic nematicides, though field efficacy depends on environmental stability and application methods.

Practical Trends

Screening of nematicidal compounds (e.g., saponins and terpenes) from medicinal plants.
Soil amendments with plant residues (e.g., marigold and neem cake) for nematode management.
CRISPR-edited crops producing anti-nematode metabolites.

### *18.2.6 Efficacy of Plant Metabolic Products Against Plant Pathogenic Fungi*

Phytoalexins, flavonoids, and essential oils (e.g., thymol and eugenol) possess antifungal properties, inhibiting spore germination and mycelial growth. These compounds can be harnessed for organic disease control, though their persistence and spectrum of activity may limit broad-spectrum management compared to synthetic fungicides.

Practical Trends

Antifungal phytochemicals (e.g., allicin from garlic and resveratrol from grapes) as eco-friendly fungicides.
Essential oils (e.g., thyme and clove) for post-harvest disease control.
Nanocarrier systems for improved delivery of antifungal metabolites.

### *18.2.7 Efficacy of Plant Metabolic Products Against Plant Pathogens*

Different bacteria, oomycetes, and viruses could be suppressed by applying the beyond fungi and nematodes as well as the plant metabolites. For example, allicin from garlic has broad antimicrobial effects, while lignin derivatives enhance physical barriers against infection. Research into synergistic formulations could improve their practical application in different regions under various weather conditions.

Practical Trends

Broad-spectrum antimicrobial activity of plant extracts (e.g., berberine and curcumin).
Seed coatings with plant-derived antimicrobials to decrease soilborne disease infections.
Metagenomic approaches to discover novel antimicrobial plant metabolites.

### *18.2.8 Antifungals Produced by Bacteria and Their Role in Biological Control*

Beneficial bacteria (e.g., *Pseudomonas* and *Bacillus*) produce antifungal metabolites (e.g., lipopeptides and polyketides) that could manage pathogens like *Fusarium* and *Rhizoctonia*. These bacteria are key components of biocontrol strategies, offering supportable alternative tools to chemical fungicides while promoting plant growth.

Practical Trends

Discovery of novel antifungal lipopeptides (e.g., bacillomycin and iturin) from Bacillus spp.
Biofungicides (e.g., *Pseudomonas fluorescens*) for managing *Fusarium* and *Rhizoctonia*.
Genetic engineering of bacteria to overexpress antifungal genes.

### *18.2.9 Biological Control of Viruses Using Actinomycetes Species*

Actinomycetes (e.g., *Streptomyces*) produce antiviral secondary metabolites and induce systemic resistance in plants. Their potential in managing viral diseases (e.g., tobacco mosaic virus) is promising but requires further exploration in field conditions.

Practical Trends

Antiviral secondary metabolites (e.g., actinomycin and avermectin derivatives) from *Streptomyces*.
Foliar sprays with actinomycete extracts to induce systemic acquired resistance (SAR).
CRISPR-based modification of actinomycetes for enhanced antiviral production.

### *18.2.10 Effect of Mineral Nutrition and Metabolic Products on Barley Yield Under Drought Conditions*

Drought stress alters plant metabolism, affecting secondary compound production and nutrient uptake. Optimizing mineral nutrition (e.g., silicon and potassium) could encourage and enhance drought tolerance by improving osmotic regulation and antioxidant activity. Metabolic adaptations, such as increased proline or phenolic accumulation, may also lessen yield losses.

Practical Trends

- Silicon and potassium nutrition enhancing drought tolerance via osmolyte accumulation (e.g., proline and glycine betaine).
- Foliar application of seaweed extracts and salicylic acid to improve stress resilience.
- Metabolic engineering of barley for increased synthesis of drought-responsive metabolites.

**Challenges and Considerations**

- *Specificity*: The effects of PMPs can be highly specific to certain insect species. What is toxic to one pest may be harmless or even beneficial or biocontrol agents.
- *Non-Target Effects*: There is a risk of PMPs affecting non-target organisms, including beneficial insects like pollinators and biological control agents of insect pests.
- *Resistance Development*: Prolonged exposure to PMPs can lead to the improvement of resistance in pest populations, reducing the long-term effectiveness of these complexes.
- *Environmental Impact*: The application of PMPs in agricultural settings must consider potential environmental impacts, including effects on soil health and non-target organisms (pollinators as well as natural enemies).
- *Integrated Pest Management (IPM)*
- Incorporating PMPs into IPM strategies can improve sustainable pest control. This can be achieved through:
- *Crop Rotation and Intercropping*: Using crops that produce effective PMPs in rotation or intercropping systems can naturally decrease populations of insect pests.
- *Breeding for Resistance*: Developing crop cultivars with enhanced production of defensive PMPs can provide long-term pest resistance.
- *Biopesticides*: Formulating PMPs into biopesticides can offer a more targeted and environmentally friendly alternative strategies to synthetic pesticides.

**Final Conclusion**

Harnessing the phytomicrobiome and insect microbiome offers a promising, eco-friendly tactic to pest management. By leveraging microbial interactions, agriculture can decrease reliance on chemical pesticides while enhancing and increasing crop resilience. Future research should focus on microbiome engineering, field efficiency trials, and integration with other biological control strategies for maintainable pest management. Plant metabolic products are essential in ecological interactions, offering sustainable solutions for management of insect pests and diseases. However, their efficiency depends on ecological context, application methods, and potential non-target effects. Integrating these compounds with biological control agents, resistant crop varieties, and optimized agronomic practices can enhance agricultural sustainability while decreasing reliance on synthetic inputs.

Finally, the track forward for sustainable pest management lies not in pursuing microbiome engineering or plant metabolite utilization in isolation, but in the strategic, synergistic integration of both. The research synthesized in this book discloses that these systems are not merely parallel pathways but are deeply interconnected components of the plant's holistic defense network. A plant primed with a beneficial microbiome exhibits improved metabolic responsiveness and resilience, while its inherent and encouraged secondary metabolites shape and support that very microbial community. This reciprocity generates a powerful, multilayered defense system where microbial biological control agents and plant biochemical defenses act in concert, offering greater effectiveness and durability than either approach alone. Embracing this integrative paradigm, where the authors manage the plant as a complex meta-organism, would be fundamental to developing the next generation of agricultural solutions that are productive, sustainable, and resilient in the face of environmental changes.

MIX
Papier aus verantwortungsvollen Quellen
Paper from responsible sources
FSC® C105338

If you have any concerns about our products,
you can contact us on
**ProductSafety@springernature.com**

In case Publisher is established outside the EU,
the EU authorized representative is:
**Springer Nature Customer Service Center GmbH**
**Europaplatz 3, 69115 Heidelberg, Germany**

Printed by Libri Plureos GmbH
in Hamburg, Germany